EPIGENETIC MECHANISMS OF GENE REGULATION

COLD SPRING HARBOR MONOGRAPH SERIES

The Lactose Operon
The Bacteriophage Lambda
The Molecular Biology of Tumour Viruses
Ribosomes
RNA Phages
RNA Polymerase
The Operon
The Single-Stranded DNA Phages
Transfer RNA:
- Structure, Properties, and Recognition
- Biological Aspects

Molecular Biology of Tumor Viruses, Second Edition:
- DNA Tumor Viruses
- RNA Tumor Viruses

The Molecular Biology of the Yeast *Saccharomyces:*
- Life Cycle and Inheritance
- Metabolism and Gene Expression

Mitochondrial Genes
Lambda II
Nucleases
Gene Function in Prokaryotes
Microbial Development
The Nematode *Caenorhabditis elegans*
Oncogenes and the Molecular Origins of Cancer
Stress Proteins in Biology and Medicine
DNA Topology and Its Biological Implications
The Molecular and Cellular Biology of the Yeast *Saccharomyces:*
- Genome Dynamics, Protein Synthesis, and Energetics
- Gene Expression

Transcriptional Regulation
Reverse Transcriptase
The RNA World
Nucleases, Second Edition
The Biology of Heat Shock Proteins and Molecular Chaperones
Arabidopsis
Cellular Receptors for Animal Viruses
Telomeres
Translational Control
DNA Replication in Eukaryotic Cells
Epigenetic Mechanisms of Gene Regulation

EPIGENETIC MECHANISMS OF GENE REGULATION

Edited by

Vincenzo E.A. Russo
Max-Planck-Institut für Molekulare Genetik

Robert A. Martienssen
Cold Spring Harbor Laboratory

Arthur D. Riggs
Beckman Research Institute of the City of Hope

COLD SPRING HARBOR LABORATORY PRESS
1996

EPIGENETIC MECHANISMS OF GENE REGULATION

Monograph 32

Printed in the United States of America
Book design by Emily Harste

ISBN 0-87969-490-4
ISSN 0270-1847
LC 96-85974

All Cold Spring Harbor Laboratory Press publications may be ordered directly from Cold Spring Harbor Laboratory Press, 10 Skyline Drive, Plainview, New York 11803-2500. Phone: 1-800-843-4388 in Continental U.S. and Canada. All other locations: (516) 349-1930. FAX: (516) 349-1946. E-mail: cshpress@cshl.org. For a complete catalog of all Cold Spring Harbor Laboratory Press publications, visit our World Wide Web Site http://www.cshl.org/

This book is dedicated to Barbara McClintock,
who lived to see epigenetics become a recognized discipline,
and also to numerous investigators who contributed to the field
before the importance of epigenetic mechanisms
was widely understood.

Contents

REPEATED GENES AND GENE SILENCING

NUCLEAR ORGANIZATION AND CHROMATIN STRUCTURE

Preface

The study of epigenetics is almost as old as the study of genetics, although epigenetic phenomena were often not recognized for what they were. Indeed, it has usually been through the genetic manipulation of epigenetic phenomena that some understanding has been achieved: Position-effect variegation in *Drosophila* and mutable loci in maize were first described more than 60 years ago, and these systems provided some of the earliest clues as to the nature of epigenetic mechanisms of gene regulation.

The study of plant, mammalian, insect, fungal, and microbial epigenetics has for the most part been carried out by different groups that go to different meetings, publish in different journals, and often have a different focus. The plan for this book was to collect for the first time under one cover reviews on the epigenetics of microorganisms as well as higher organisms. The chapters of this book have been organized into several divisions. Some chapters are intended to provide historical insights, and toward this purpose three early papers on epigenetic mechanisms are reprinted at the end of the volume. Consideration was given to making other main divisions along organismic lines, such as plants, animals, and microorganisms. However, since one intent of this book is to lessen the barriers of communication between workers studying various organisms, we decided that this purpose would be better served by attempting to categorize the chapters according to function and mechanisms. It will be seen that the divisions are in many cases rather arbitrary, but we hope the deliberate mixing of organisms will serve the spirit of unity and will lead to productive cross-fertilization. Finally, the reader should note that at the end of the book we have included a glossary and two short chapters covering the essentials of plant and mammalian development. These chapters are intended to help students and the nonspecialist reader tackle some of the terminology found in the more specialized chapters. They are included as appendices to allow easy reference without interrupting the main text.

Apology should, of course, be made for failing to include numerous examples of epigenetic phenomena that did not find their way into this book. Our intention was to provide as comprehensive a review as possible, given inevitable space limitations, of those systems understood in greatest molecular detail.

We are grateful to the staff of Cold Spring Harbor Laboratory Press, especially John Inglis, Nancy Ford, Patricia Barker, and Inez Sialiano for their cooperative and competent collaboration.

V.E.A. Russo
R.A. Martienssen
A.D. Riggs

EPIGENETIC MECHANISMS OF GENE REGULATION

Introduction

Arthur D. Riggs
Biology Department
Beckman Research Institute of the
City of Hope, Duarte, California 91010

Robert A. Martienssen
Cold Spring Harbor Laboratory
Cold Spring Harbor, New York 11724

Vincenzo E.A. Russo
Max-Planck-Institut für Molekulare Genetik
D14195 Berlin, Germany

What is epigenetics? This book emphasizes epigenetic control of gene expression and development, so a useful definition is: *The study of mitotically and/or meiotically heritable changes in gene function that cannot be explained by changes in DNA sequence.* In both plant and mammalian tissue culture, for example, it is known that a significant percentage of clonally derived variants are not true mutations because they can be efficiently reverted by agents that inhibit enzymatic DNA methylation (see Riggs and Jones 1983; Harris 1986; Meins). Stable epigenetic control of gene function also is an essential component of normal development of most, and perhaps all, complex organisms, especially mammals and higher plants, which have unending division of differentiated cells. The epigenetic changes that take place during differentiation are normally erased in the germ line, but there are examples in both plants and transgenic mice of epigenetic variants that are transmitted through meiosis (see McClintock 1965; Sapienza 1990; Chaillet 1994; and the paper by Brink et al. reprinted at the end of this volume). This book provides a survey through early 1996 of many of the epigenetic systems now being studied.

Most research since the advent of genetics—especially in the post-Watson and Crick era—has focused on the study of (and use of) true mutations. A search of the National Library of Medicine literature database revealed that in the late 1960s and early 1970s only about 3 papers per year used the term epigenetics. However, there has been an accelerating increase in papers concerned with epigenetics, and more than 130 papers per year were published in 1993–1995. Thus, it is no longer prac-

Authors cited without dates refer to papers in this volume.

tical for a scientist interested in this field to attempt to read all of the primary publications. The intent of this book is to provide under one cover reviews of epigenetic studies being done in a wide variety of organisms—microorganisms as well as higher plants and animals.

As with classic genetics, it was in plants that many key observations were made. Barbara McClintock's discovery of transposable "controlling" elements at Cold Spring Harbor Laboratory in the 1940s and 1950s led to some of the most detailed descriptions of epigenetic mechanisms of gene regulation in any eukaryote (McClintock 1965). For a long time, transposons were thought to be unique to plants, but transposable elements have now been widely recognized in many organisms, and McClintock was awarded the Nobel Prize in 1983 for her discoveries. R.A. Brink, also working with maize in the mid 1950s, described the curious phenomenon of paramutation. Paramutation is the heritable alteration of one allele following exposure to another allele in a heterozygote. With the advent of transgenes and transgene silencing, paramutation and related phenomena in plants have been revisited in great detail in recent years, and some of these studies are described in this book. To many readers, paramutation may be less familiar than transposable elements, and so a review of paramutation published in 1968 has been reproduced at the end of this volume to introduce early work on plant epigenetics.

The study of epigenetic phenomena in mammals seems to have developed independently of the plant genetics field and can be considered as beginning with the discovery and correct genetic interpretation of X-chromosome inactivation by Susumu Ohno (Ohno et al. 1959) and Mary Lyon (1961). Not only are the two alternate "on" or "off" activity states of the X chromosome somatically heritable, but it was learned in 1971 that epigenetic differences exist between the maternally and paternally derived X chromosomes: The paternal X chromosome is preferentially inactive in the extraembryonic tissues of the mouse and in all tissues of marsupials (Cooper et al. 1971). The term "chromosome imprinting" was first introduced by Crouse (1960), who found that certain insects mark chromosomes according to parental source (male or female) and use this "imprint mark" to control activity in the next generation. It is now known that in mammals epigenetic marking of parental source is not confined to X-linked genes but is also seen for many autosomal genes. Heritable differences in gene function dependent only on whether the gene comes from the male or female parent comprise parental or genomic imprinting, which is known to be essential for mouse and human development. Fifteen imprinted mouse autosomal genes already have been identified (see Ainscough and Surani; Walter et al.), and it is likely that many more will be found.

The study of epigenetic phenomena in microorganisms has recently seen a great increase in activity, although as early as 1949 some persistent changes in bacterial cultures were proposed by Delbruck to be based on dynamic metabolic circuits with positive feedback (see Overview of Epigenetic Mechanisms by Riggs and Porter). Yeast mating type and telomeric silencing clearly have epigenetically heritable features, and these are the focus of intense study (see Holmes et al.; Allshire). *S. cerevisiae* and *S. pombe* are eukaryotes, but epigenetic phenomena are also well known in bacteria (see Casadesús and Torreblanca).

A detailed understanding of epigenetic mechanisms at the molecular level has been slow to emerge but is the ultimate goal of most of the work described in this book. Stable circuits dependent on *trans*-acting factors and feedback loops can explain rather well some epigenetic phenomena, but others, including X-chromosome inactivation and imprinting, are thought to require additional *cis*-acting, chromatin-based mechanisms. Twenty years ago, and about twenty years after the discovery of epigenetic phenomena, a mechanism dependent on DNA modification was hypothesized independently by Holliday and Pugh and by Riggs (see reprints at the end of this volume). Patterns of methylated bases were proposed to be heritable, and properties of a "maintenance methylase" were correctly predicted. Methylation-dependent epigenetics is now accepted as being important for higher plants and especially important for mammals, where gene knockout studies have established that DNA methyltransferase is essential for normal development and the maintenance of genomic imprinting (Li and Jaenisch 1993; Beard et al. 1995). Similar experiments also indicate that DNA methylation is involved in significant aspects of plant development and epigenetic gene control (see Finnegan). Publications on enzymatic DNA methylation related to gene function have grown from just a handful in the 1970s to several hundred per year. Despite this research effort, many details of methylation epigenetics remain poorly understood. Moreover, most investigators, including the editors of this book, do not think that there is only one epigenetic mechanism. On the contrary, several methylation-independent epigenetic mechanisms of gene regulation have been identified, for example, in yeast, *Drosophila*, and plants. The role of nuclear organization and chromatin structure in these methylation-independent mechanisms is described in several chapters in this book. It seems likely that for complex organisms with large genomes, several mechanisms acting together are needed to enable adequately stable and fail-safe epigenetic programs.

REFERENCES

Beard, C., E. Li, and R. Jaenisch. 1995. Loss of methylation activates *Xist* in somatic but not in embryonic cells. *Genes Dev.* **9:** 2325–2334.

Chaillet, J.R. 1994. Genomic imprinting: Lessons from mouse transgenes. *Mutat. Res.* **307:** 441–449.

Cooper, D.W., J.L. VandeBerg, G.B. Sharman, and W.E. Poole. 1971. Phosphoglycerate kinase polymorphism in kangaroos provides further evidence for paternal X inactivation. *Nat. New Biol.* **230:** 155–157.

Crouse, H.V. 1960. The controlling element in sex chromosome behavior in *Sciara. Genetics* **45:** 1429–1443.

Harris, M. 1986. Simultaneous induction of epigenetic variants. *Somatic Cell Mol. Genet.* **12:** 567–573.

Li, E.B.C. and R. Jaenisch. 1993. Role of DNA methylation in genomic imprinting. *Nature* **366:** 362–365.

Lyon, M.F. 1961. Gene action in the X-chromosome of the mouse. *Nature* **190:** 372–373.

McClintock, B. 1965. The control of gene action in maize. *Brookhaven Symp. Biol.* **18:** 162–184.

Ohno, S., W.D. Kaplan, and R. Kinosita. 1959. Formation of the sex chromatin by a single X-chromosome in liver cells of *Rattus norvegicus. Exp. Cell Res.* **18:** 415–418.

Riggs, A.D. and P.A. Jones. 1983. Methylcytosine, gene regulation, and cancer. *Adv. Cancer Res.* **40:** 1–30.

Sapienza, C. 1990. Parental imprinting of genes. *Sci. Am.* **230:** 52–60.

DNA Methylation in Eukaryotes: 20 Years On

Robin Holliday
CSIRO Division of Biomolecular Engineering
Sydney Laboratory
NSW 2113, Sydney, Australia

The proposals published in 1975, and reprinted in this volume, were based on certain features of development that were not understood at the time and, in large part, are still not understood. One such feature is the stability of the differentiated state; that is, the fact that differentiated cells, whether they are postmitotic or commonly undergo mitosis, stably retain their phenotype. Any differentiated cell has a large number of active genes coding for "housekeeping" enzymes or other proteins and a subset of genes coding for specialized "luxury" functions. This also means that genes coding for other luxury functions, appropriate to another specialized cell type, are stably repressed. A second feature of development, obviously related to the first, is the formation of different cell types by the switching on of specific sets of genes and the inactivity of other sets. The 1975 paper in *Cytogenetics and Cell Genetics* concentrated on the switch that inactivates one X chromosome in female mammals. The 1975 paper in *Science* discussed several switching mechanisms, including stem-line situations, the inactivation of the X chromosome, transdetermination in imaginal discs of *Drosophila*, and other developmental contexts. A third feature of development we[1] considered to be important was the likely existence of developmental clocks; that is, the existence of specific controls on the number of cell divisions in particular situations, before some developmental switch occurred. In this connection, we referred to some of the concepts developed by Holtzer, in which differentiation events are tied to cell division (Holtzer et al. 1972).

There was absolutely no direct evidence for the proposals in the 1975 publications, and the only indirect evidence was the well-known fact that restriction enzymes in bacteria could distinguish between specific, short, modified and unmodified sequences of DNA. We argued that there might be short, specific "controlling sequences" in or adjacent to promoters of genes subject to switching in activity, which are either in a modified or

[1]Here and elsewhere, "we" refers specifically to Holliday and Pugh, authors of the *Science* paper.

nonmodified state. Either the former or the latter would be associated with transcription, because there was at the time no preference for either possibility. 5-Methyl cytosine (m^5C) is the major modified base in higher organisms, so we proposed that specific methylases recognized particular controlling sequences. We also mentioned the possibility of specific demethylating enzymes but favored the idea that methyl groups might be simply lost by dilution in the absence of the appropriate methylase, for example, during early cleavage divisions. Both publications emphasized that a methylated sequence could be stably maintained, provided the maintenance enzyme recognized the hemimethylated DNA after replication and did not act on a nonmethylated DNA. Our general model therefore depended on pairs of enzymes: One was a switch, setting or controlling enzyme that recognized and methylated short sequences, and the other was a maintenance enzyme. In a stem-line situation, where cell division can segregate a cell determined to a differentiation pathway, and another stem cell, we proposed only one strand of a controlling sequence is methylated. In other situations, both strands might be methylated, so all the progeny of a cell would have a switched gene.

CpG doublets are the normal sites for methylation in mammalian cells, but we did not emphasize their significance, because these two bases alone do not provide specificity. By analogy with many modification-restriction systems in bacteria, we thought that specific sites of methylation would be in symmetrical, short palindromic sequences. Such sequences could, of course, contain CpG doublets, but we only briefly referred to this possibility. Since it was important in our argument that sequences longer than CpG would be necessary for the regulation of gene expression by methylation, it also followed that only a minority of m^5C residues in total DNA would have a regulatory function. The remainder might have some other functions, perhaps related to chromatin structure or chromosome behavior, but this was not explicitly stated. In this connection, it is relevant that replication origins in growing mammalian cells have islands of cytosine methylation, which include non-CpG methylation (Tasheva and Roufa 1994). In another study, CpNpG methylation has been demonstrated in such cells (Clark et al. 1995).

A substantial part of our paper dealt with the possibility that modification of DNA bases could actually lead to base-pair transitions. The methylation of cytosine followed by deamination can convert a C-G to a T-A base pair. This had previously been proposed by Scarano (1971) as a developmental mechanism and has been shown to occur during the RIP process in *Neurospora* (Selker 1990 and this volume). Another modification, originally proposed by D.C. Hawthorne (pers. comm.), is the deamination of adenine to inosine, which pairs with cytosine. In principle, both these transitions, mediated by enzymes, are reversible, and in this context we discussed the experiments of Gurdon and associates with *Xenopus* (Gurdon and Laskey 1970 and references therein). They showed that nuclei from specialized cells could be reprogrammed when transplanted to anucleate zygotes. However, we wrote: "Nevertheless we find it hard to believe that reprogramming could actually involve base changes in the DNA, and for this reason we tend to prefer the version of our model that depends on methylation of particular bases." It remains a possibility that enzyme-mediated base sequence changes do occur in particular contexts; for example, in mutation in the variable regions of immunoglobulin genes (Holliday 1994). The commonest changes seen are, in fact, AT-GC and GC-AT transitions (Betz et al. 1993).

Our proposal that there may be many types of DNA methylases has not been substantiated by subsequent studies, and it is now generally believed that there is only one or a family of related DNA methyltransferases (see Bestor, this volume). It is presumed that specificity in methylation may be provided by other proteins which are sequence-specific, but with regard to the switching of gene activities and establishing a specific pattern of methylation (see below), almost nothing is known. We did point out that specific methylases might be very difficult to detect, if only one or a few copies of their substrate were present in the genome or if their activity was transient in development.

With regard to switches in dividing cells, we realized that diploidy would present a problem, since methylation switches could be + and – for one chromosome and – and + for the other. (This opposite segregation of switches would occur in 50% of cases.) We proposed that allelic exclusion might operate in important controlling sequences; that is, only one homolog would have a switch and the other would be inactive. This situation is obviously related to X-chromosome inactivation, which we briefly discussed (a further manuscript specifically related to methylation and X-chromosome inactivation is extant, but was never published). We did recognize, as did Riggs, that X-chromosome inactivation requires a slow initial switching event, followed by some rapid process that "fixes" the difference in modification between the active and inactive X chromosomes. More recently, it has been proposed that genomic imprinting might be related to the need to have switching mechanisms at particular loci in only one of two chromosomes in the developing embryo (Holliday 1990a).

We realized that switching in single cells produced cell lineages with particular properties in the developing organism, but that development is often not based on clonal expansion. We showed how specific events could occur in groups of cells and also recognized that switches or clock mechanisms could be triggered by hormonal or other external influences. We briefly discussed programmed cell death during development and mentioned the possibility developed much more fully in the same year by Sager and Kitchin (1975) that DNA restriction mechanisms might exist in eukaryotic organisms. It is interesting that DNA degradation is known to be an important component of programmed cell death by apoptosis, but this degradation is due to nonspecific intranucleosomal cleavage.

Our proposals in 1975 were entirely based on the belief that special mechanisms were required to explain specific changes in gene activity during development or in stem cells. We wrote: "A continual interaction between cytoplasmic enzymes and DNA sequences is an essential part of the model to be presented." No one could disagree with that today, if "proteins" is substituted for "enzymes." Our proposal was more than a question of interactions; it incorporated the major feature of *chemical* changes to DNA; that is, modification of cytosines in DNA or other base changes. Although we did not use the word epigenetic, in effect, we proposed mechanisms whereby epigenetic information, or an epigenetic code, could be superimposed on the normal DNA code.

Since 1975, hundreds of papers on DNA methylation in eukaryotes have been published, and these cover a wide variety of biological topics and contexts. There have been many reviews of much of this material, including Razin and Riggs (1980), Doerfler (1981; 1983), Ehrlich and Wang (1981), Razin et al. (1984), Taylor (1984), Adams and Burdon (1985), Volpe and Nicolini (1989), Holliday et al. (1990), and, most recently, Jost and Saluz (1993a). In the following sections, only salient advances are mentioned that are related to the two 1975 papers reprinted in this volume.

ACCUMULATING EVIDENCE AND LANDMARK DISCOVERIES

Methods for Studying DNA Methylation

The methods available for studying DNA methylation before 1975 were primitive; namely, chromatographic analysis, coupled in some cases with nearest-neighbor analysis. The latter showed that CpG doublets were a common site for cytosine methylation in mammalian DNA. After 1975, important new methods were introduced, particularly the use of the isoschizomer restriction enzymes *Hpa*II and *Msp*I. The former does not cut CCGG sequences in DNA where the inner C is methylated, whereas the latter does. Similarly, *Hha*I does not cut GCGC when the inner C is methylated. Provided a DNA probe is available, the methylation status of these sequences can be examined by Southern blots, using these or other methyl-sensitive enzymes. This made it possible to study a variety of genes, viruses, and so on, where probes were available. In a very large number of cases, it was found that methylation of these sites was associated with gene inactivity, and lack of methylation with activity. These studies did not always acknowledge the fact that methyl-sensitive restriction enzymes only screen a subset of cytosine residues. For *Hpa*II or *Hha*I, this is significantly less than 10% for non-CG island DNA, where CG doublets are depleted. A related method, developed after the advent of polymerase chain reaction (PCR), depends on the cutting of genomic DNA by a methyl-sensitive enzyme. If there is a single restriction enzyme site in the amplified region, a PCR product will be seen if that site is methylated, but there will be no product if it is not.

The screening of all cytosines in the region of interest depends on special techniques in genomic sequencing. Starting with the original procedure of Church and Gilbert (1984), these have been consistently improved (see Saluz and Jost 1993), and the latest method provides a positive signal on a sequencing gel for every cytosine that is methylated (Frommer et al. 1992). Genomic sequencing is essential for any full analysis of cytosine and m^5C in any DNA region of interest.

It should be noted that a methyl-sensitive restriction enzyme, *Taq*I, in conjunction with PCR, has recently been used to detect 6-methyl adenine in mouse genomic DNA (Kay et al. 1994). This shows that there is more than one modification in the DNA of higher organisms, and it remains to be seen whether this has a role in the control of gene expression.

Heritability of DNA Methylation

Early studies provided direct evidence for the inheritance of a given pattern of cytosine methylation (Wigler et al. 1981; Harland 1982). More recent studies based on genomic sequencing have also provided an estimate of the fidelity of transmission of methylated and nonmethylated sites in a mammalian cell line (Pfeifer et al. 1990). However, the bulk of the evidence is indirect. It comes from the fact that methylated DNA has often been shown to remain methylated after many cell divisions, and nonmethylated DNA also remains unchanged (for review, see Holliday 1993). There are many examples, but a few will suffice. The inactive X chromosome in somatic cells has methylated CpG islands in dividing cells. The fact that particular tissues have specific patterns of DNA methylation in the genes that have been studied shows that whatever set up this pattern is subsequently maintained with considerable fidelity (Konachek et al. 1990; Behn-Krappa et al. 1991). Methylated or unmethylated DNA introduced into cells remains in its original state. This has been shown, for example, for the γ-globin gene transfected and integrated into chromosomal DNA (Busslinger et al. 1983). There are exceptions to a strict pattern of inheritance, since it is known that CpG islands can become methylated and silenced in mammalian cell lines, by the poorly understood process of de novo methylation (Antequera et al. 1990). Once methylated, they stably inherit the silent gene. Such genes can be demethylated and reactivated by the demethylating agent 5-azacytidine (5AC) and these reactivants are also stable (see below and Holliday et al. this volume).

The biochemical basis of maintenance in vivo is not understood, because known DNA methyltransferases have some de novo methylation activity in standard enzyme assays (Bestor, this volume). Presumably, there are additional controls in normal cells that prevent random or semi-random de novo methylation. Almost nothing is known about the mechanisms that set up a pattern of methylation in a particular cell type, which is subsequently maintained. However, there is some information about specific demethylation of DNA which is summarized in the following section.

Specific Methylation and Demethylation

A very important advance by Busslinger et al. (1983) exploited a novel in vitro DNA methylation technique. Briefly, single-stranded phage M13 DNA containing the γ-globin gene was used as a template for *Escherichia coli* pol 1 and 5-methyl dCTP with the other three standard deoxynucleotide triphosphates. The method took advantage of the fact that hemimethylated DNA becomes fully methylated in mammalian cells, at least at CpG sites. With the use of appropriate templates, it was possible to methylate either the whole gene, the 5′ promoter region, or the coding region. The various double-stranded molecules were cotransfected with TK^+ DNA into TK^- L cells. Stable transformants were analyzed for globin gene expression and the methylation pattern of the integrated globin DNA. It was found that the fully methylated and the promoter-methylated genes were inactive, whereas nonmethylated globin DNA, or DNA with methylation in the structural gene, was active. These experiments are described here because they demonstrate a rather direct relationship between specific methylation of the promoter region and lack of transcription. The experiments also demonstrate the heritability of the DNA methylation pattern, since the input DNA was maintained in its initial methylated state many replication cycles after transfection. Detailed analysis has also revealed the relationship between the methylation of promoters of adenovirus genes and their inactivity (for review, see Doerfler 1993).

Further work on the γ-globin gene suggested that the inhibition of transcription by methylation of this promoter was a general rather than a specific effect (Murray and Grosveld 1987). That is, the extent of transcription related to the number of CpGs methylated in the region, rather than to specific CpG sites. This is an important issue that has not been resolved. The existence of proteins that bind to methylated DNA irrespective of sequence suggests that specificity of methylation may not be important (Boyes and Bird 1991, 1992; Hsieh 1994). On the other hand, some studies show that individual methylated sites block transcription (Tasseron-de Jong et al. 1989; Graessman et al. 1994). More important, the existence of transcription factors that have differential affinity for methylated and nonmethylated CpGs within the specific binding sequence (see Ehrlich and Ehrlich 1993) strongly indicates that there are critical targets for methylation control of gene expression. It can be anticipated that many more examples will come to light in the future. In this connection, it is important to realize that many of the classic studies of promoter sequences in high eukaryotes were based on the alteration of DNA sequence (by deletion, mutation, and so on) in plasmid DNA, which is nonmethylated. In all these studies, the possible role of DNA methylation in controlling transcription was totally ignored. Many of these studies should be repeated, with experiments that take account of the possible significant role of DNA methylation in the altered, or naturally occurring, promoter sequence.

Several important experiments have demonstrated specific demethylation in vitro. It was shown that the induction of the egg protein vitellogenin by the hormone estradiol in the liver of egg-laying hens is associated with the specific demethylation of CpG doublets in the promoter region and the estradiol response element (ERE) (Jost et al. 1990; Jost and Saluz 1993b). This demethylation is an active process, not dependent on DNA synthesis. Specific demethylation has also been demonstrated in experiments in which genes methylated in vitro are introduced into cells that normally synthesize the gene product (Yisraeli et al. 1986; Frank et al. 1990). Thus, a methylated immunoglobulin gene is demethylated and activated in B lymphocytes, but not in fibroblasts. Similarly, a methylated insulin gene is specifically demethylated and activated in insulinoma cells; and a methylated α-actin gene is specifically demethylated and activated in myoblasts but remains silent and methylated in fibroblasts.

It has also been shown that a gene for crystallin becomes demethylated at the time the protein is synthesized in the lens, and this is not dependent on DNA replication (Sullivan and Grainger 1987). In another important study, which does not appear to have been followed up, teratoma cells were treated with retinoic acid and butyrated cAMP. The collagen IV gene was demethylated and activated (Burbelo et al. 1990).

All these studies and others indicate that there are cell-specific factors capable of demethylating regulatory DNA sequences in the appropriate cell type (for review, see Jost, this volume). This is in contrast to global demethylation by 5AC or 5-azadeoxycytidine, briefly reviewed in the next section.

The Amazing Properties of 5AC

In 1975, we referred to the well-known effects of bromodeoxyuridine (BrdU) on cell differentiation. We proposed that the substitution of BrdU for thymine in DNA might prevent the gain or loss of DNA methylation and thereby disrupt normal developmental processes. It also seemed possible that a cytosine analog, such as fluorocytidine, would have important effects on methylation and gene expression. This was later shown to be true (Jones and Taylor 1990), but a more important analog is 5AC, which has been shown to have a remarkable effect on biological systems. It was shown to induce the differentiation of CH3 10T1/2 or 3T3 cells into myotubes, adipocytes, and chondrocytes (Taylor and Jones 1979; Jones 1984; Jones et al. 1990). 5AC, or 5-azadeoxycytidine, is incorporated into DNA and has been shown to bring about extensive demethylation. This is due to covalent binding of DNA methyltransferase to 5AC in DNA (Santi et al. 1984), and the subsequent reduction of the enzyme activity in the cell results in the dilution-out and loss of methylation at many sites in the genome. This in turn accounts for gene reactivation by 5AC. In the case of muscle differentiation, it is clear that the CpG island of the Myo D1 gene is methylated in 10T1/2 and 3T3 cells and is demethylated and activated by 5AC (Jones et al. 1990; Rideout et al. 1994). However, non-muscle cells in vivo do not have methylated CpG islands, although until genomic sequencing is done, it remains possible that a subset of CpGs are methylated in such cells.

It has been demonstrated in many other contexts that methylation of DNA is associated with gene inactivity and demethylation by 5AC with gene reactivation (for review, see Jones 1984, 1985; Holliday 1987). These contexts include the reactivation of genes on the inactive X chromosome, reactivation of silent genes in cultured mammalian cell lines, and reactivation of latent viruses integrated into chromosomes. More recently, it has been shown to reactivate silent genes in fungi (Goyon and Faugeron 1989; Pandit and Russo 1992). 5AC also induces vernalization in plants, probably by demethylation of DNA (Burn et al. 1993). 5AC is a carcinogen, can induce the formation of fetal hemoglobin in adults with abnormal hemoglobin, and at higher concentrations, can alter chromosome condensation. It is not known whether all these effects are due to changes in DNA methylation.

X-Chromosome Inactivation

It was known in 1975 that cytosine methylation occurred predominantly in CpG doublets and also that CpGs were depleted in the genome. Subsequently, it was shown that regions of DNA did not have depleted CpGs and, furthermore, that these CpGs were not usually methylated. These nonmethylated CpG "islands" are commonly at the 5′ end of housekeeping genes, but they can also be associated with specialized genes (see Gardiner Garden and Frommer 1987; Antequera and Bird 1993). Several housekeeping genes on the inactive X chromosome have been shown to have methylated CpG islands. Thus, on the active X chromosome, genes with nonmethylated islands are actively transcribed, but the same regions of DNA are methylated and inactive on the other X chromosome. Since X inactivation is very stable, this strongly implies that the methylation is transmitted through cell division, and experimental evidence that this is the case has been obtained. It has also been shown by DNA transfection experiments that genes on the active X chromosome can change the phenotype of recipient cells, whereas the inactive X does not do so (see Singer-Sam and Riggs 1993; Riggs and Porter, this volume [X Chromosome Inactivation and Epigenetic Mechanisms]).

There is continuing controversy about the role of methylation in switching off the inactive X chromosome (see Singer-Sam and Riggs 1993; Gartler and Goldman 1994; Avner, this volume). Lock et al. (1987) provided evidence that the methylation of the HPRT gene occurs later than the time of X-chromosome inactivation, but this is not the case for PGK (Singer-Sam et al. 1990). PGK is near the X-inactivation center, whereas HPRT is a considerable distance from it. Any methylation model for X inactivation demands an initial switch, followed by a spreading of methylation to the ends of the chromosome arms, and this might occur over several cell divisions. There is some evidence that the inactive X in the germ line and in the placenta is not associated with DNA methylation, but in these studies, only one or a few CpG sites were examined (Gartler and Goldman 1994). Until appropriate genomic sequencing is done, one cannot know whether a subset of methylated cytosines is sufficient for inactivation, and additional methylation in somatic cells imposes further stability on the inactive X.

Genomic Imprinting

Genomic imprinting is due to the imposition of different information on DNA in male and female gametes, which is transmitted to the zygote. This information is heritable but also reversible before the gametes of the subsequent generation are formed. That is, a maternal imprint is reversed in a subsequent male gamete, and vice versa. Imprinting was first discovered in insects and subsequently in mouse and humans. With regard to the mechanism, it was clear from the beginning that DNA methylation might be the source of the imprinted signal; namely, that one gamete might have specific methylated cytosines, whereas these same cytosines would be unmethylated in the other gamete. Specific proposals were outlined in the important early review by Sager and Kitchin (1975).

In molecular studies of imprinting in mammals, evidence for methylation differences between gametes has been obtained. Initially, this was shown in transgenes, which are artificial insertions of DNA into the chromosomes. Later, it was shown that normal imprinted genes are associated with methylation differences, either at the location of the gene or some distance from it (see Ainscough and Surani; Shemer and Razin; Walter et al.; all this volume). It is presumed that the heritability of DNA methylation provides an essential requirement for imprinting, but the actual relationship of methylation to the activity or inactivity of a particular imprinted gene is not yet clear. In addition, it is not known how an imprinting pattern of methylation is erased and/or reestablished. It is certainly possible that mechanisms other than DNA methylation are responsible for chromosomal differences between male and female gametes, but none has so far been identified.

Maize Controlling Elements

The first evidence that the activity of transposable elements in maize might be subject to methylation control came from studies with *Activator* (*Ac*). An inactive *Ac* element, which arose in rather unusual circumstances, was found to have methylated cytosines, which were nonmethylated in the active element (Schwartz and Dennis 1986). Several restriction enzymes were used to detect such methylation, and it was also evident that gross changes in DNA sequence were not responsible. Although possibly the inactive *Ac* might have changes in base sequence, further studies on this and other controlling elements made this unlikely (Finnegan et al. 1993). *Spm* has been studied in particular detail, and it is clear that epimutations changing the activity of the element are due to changes in DNA methylation (see Fedoroff, this volume).

The importance of these studies is that genetic analysis of these controlling elements is very sophisticated. Thus, it is possible to study transmission through cell division within single plants and also transmission from one generation to the next. This means that a change of phenotype which is heritable and due to a change in methylation status is exactly analogous to a change in phenotype due to a true mutation. Thus, there is much more than a "correlation" between methylation and a change in phenotype. This is discussed further below.

Mammalian Cells in Culture

Similar considerations apply to certain heritable changes in phenotype in mammalian cell lines. It was found that many recessive mutations could be induced by standard mutagens in the Chinese hamster ovary (CHO) pseudo diploid cell line. This led to the concept of "functional hemizygosity," namely, that many parts of the genome are present in one rather than two copies (Siminovitch 1978). It is fairly clear that one basis for this is the silencing of one of two autosomal genes by de novo DNA methylation (see Holliday 1987, 1991a; Holliday et al., this volume). The single active gene can be mutated to give a recessive phenotype (as shown by complementation in hybrids), and the silent methylated gene can be reactivated by 5AC. In other cases, neither or both copies of a given gene may be silenced.

The first clear evidence for gene silencing by DNA methylation came from studies of a metallothionein gene in mouse lymphoid cells (Compere and Palmiter 1981). Since then, there have been many studies in which the methylation status of restriction enzyme sequences has been related to gene expression and one important study in which genomic sequencing of the promoter region of the PGK gene demonstrated methylation of part of the CpG island of inactive genes (Pfeifer et al. 1990).

DNA can become methylated and genes can become silenced by the uptake of 5-methyl dCTP into DNA (Holliday and Ho 1991; Nyce 1991). Genes silenced by this procedure can be reactivated by 5AC. All these observations suggest that there is a rather direct relationship between DNA methylation and gene expression, which is discussed further below and also in Holliday et al. (this volume).

CORRELATIONS AND CAUSES

In the 1975 paper there was no prediction that DNA methylation would be associated with gene inactivity, or vice versa. In subsequent years it soon became apparent that such methylation was indeed associated with lack of gene expression, with occasional exceptions. In hundreds of publications where such an association has been found, it is commonly stated that DNA methylation is *correlated* with gene inactivity or lack of expression. This evades the question of whether DNA methylation is a cause of gene silencing, or whether a gene is inactivated by some other mechanism and subsequently methylated. In fact, it has often been stated or implied that DNA methylation is an indirect or secondary mechanism (see, e.g., Tilghman 1993). In many instances, where no other analysis is possible, it is very reasonable to refer to correlations rather than causes, and this is probably the case for the majority of studies on DNA methylation. However, in some situations where genetic analysis is possible, a causal relationship between methylation and gene silencing can be established. An extreme example is RIP (repeat-induced point mutation) in *Neurospora crassa*, in which duplicated sequences of DNA in a haploid genome become extensively methylated premeiotically. This cytosine methylation (which is not restricted to CG or CNG sites) is apparently a substrate for a deoxycytidine deaminase that converts m^5C to thymidine (Selker 1990). Thus, in this instance, methylation leads to extensive mutation, and inactivation of the gene. A better demonstration of DNA methylation itself leading to gene silencing is the related process of MIP (methylation-induced premeiotically) in *Ascobolus immersus*. Again, duplicated genetic sequences are methylated, but in this species, mutation by deamination does not occur. Instead, the methylated gene is silenced and remains methylated through subsequent nuclear divisions. Sequencing of the gene shows that no mutations are present (Rhounim et al. 1992). Gene activity is regained by spontaneous demethylation at low frequency, or a higher frequency after 5AC treatment. These results demonstrate a very direct relationship between DNA methylation and gene inactivity. In effect, the methylated form of the gene is a null allele, although it has the wild-type sequence. The processes of RIP and MIP in fungi have their counterpart in higher plants, often referred to as cosuppression. Again, duplicated segments can lead to methylation and inactivation of a gene or genes, but in this case, the process is not premeiotic (see Matzke and Matzke; Meyer; both this volume). Genetic analysis strongly indicates that the changes in gene activity are brought about by methylation of the gene. Other results in *Neurospora* indicate that multiple-copy transformants can become inactivated in vegetative cells by DNA methylation and that these silent genes can be reactivated (Pandit and Russo 1992; Russo et al., this volume).

The study of maize transposable elements, which was briefly discussed above, also shows that changes in phenotype can be brought about by changes in methylation, which are heritable. The sophisticated genetic analysis possible in these systems establishes the reality of epimutations, that is, altered phenotypes which are not due to changes in DNA sequence. It can be said that methylation changes the epigenotype, which is imposed on a normal genotype. Finally, the experiments with cultured mammalian cells can be most readily interpreted in terms of gene silencing by DNA methylation, and gene reactivation by demethylation. In many cases, molecular studies support this interpretation, without the need to invoke other unknown mechanisms for gene inactivation and reactivation.

It is apparent in all these examples that genes can exist in different epigenetic states which are stably inherited, and that DNA methylation is an essential feature of the change in state. Similarly, the development of genetics in many organisms in this century was based on heritable gene mutations, which produced phenotypes geneticists could successfully exploit in many ways. It should be noted that the mutations were not correlated with their phenotypes but could be said, with every justification, to *cause* the phenotype. Prior to the realization that DNA was the genetic material, there was no biochemical understanding for the physical structure of the gene or its function. Nevertheless, genetic experiments could be successfully executed without worrying about the underlying biochemistry. In the same way, changes in gene activity due to DNA methylation can be studied in appropriate experimental systems, even though it is not understood how the methylation has such profound effects on gene expression. As in the case of mutation, it is therefore logical to conclude that a change in methylation causes a change in phenotype.

Since these experimental systems exist, it is also reasonable to suppose that in other situations, which are less amenable to genetic analysis (e.g., X-chromosome inactivation, latent viruses, imprinting), DNA methylation is also a primary cause of gene inactivation, rather than merely being correlated with it.

DNA methylation control (Burn et al. 1993). More generally, there is much interest in the role of DNA methylation in plant development (see Finnegan et al. 1993; Finnegan and several other contributions to this volume).

8. The pattern of DNA methylation is rigorously conserved in normal human tissues and the pattern is different in different cell types (Kochanek et al. 1990; Behn-Krappa et al. 1991). This must mean that there are specific mechanisms to set up this pattern during development, although we do not know how they operate.
9. There are remarkable changes in DNA methylation during gametogenesis and early development (see Monk et al. 1987; Razin and Cedar 1993). Again, the reasons for these changes are not understood.

These nine examples of the relevance of DNA methylation to development are probably the tip of the iceberg, and it is only a matter of time before the role of DNA methylation in development is clarified.

The importance of diffusible signaling mechanisms in the development of higher organisms is not in doubt. What the methylation model provides is a possible basis for induced changes in epigenotype, which may be heritable. Thus, a morphogen, growth factor, or hormone could interact with surface receptors, and by signal transduction, induce a change in methylation of a particular sequence of DNA. This, in turn, changes gene expression and the cell phenotype. The new phenotype would be due to an intrinsic, heritable, change at the DNA level. In this manner, external influences can alter the properties of the genome, contrary to the "central dogma" of molecular biology (Crick 1958). Such possibilities have been more fully explored by Jablonka and Lamb (1989; Jablonka et al. 1992). The closest example we have of an external agent inducing specific changes in DNA methylation is the estradiol-induced change in the vitellogenin gene in hens (Jost et al. 1990; Jost and Saluz 1993).

CANCER, AGING, AND THE HUMAN GENOME

Cancer is a derangement of normal developmental controls. One of the important epigenetic features of complex organisms is the intrinsic stability of determined cells (e.g., those constituting a specific stem-cell compartment) or differentiated cells. Whatever the epigenetic controls may be, it is clear that the emergence of tumors depends on the loss of one or more of these controls. Recently, almost all attention has been focused on mutations in oncogenes and tumor suppressor genes, or chromosome abnormalities. All these can obviously induce important changes in phenotype. The hypothesis that DNA methylation may be important for gene regulation was naturally followed by the suggestion that DNA damage might induce changes in DNA methylation, which in turn might alter gene expression (Holliday 1979, 1987). This possibility might be better able to explain notable differences between rodent and human cells in their susceptibility to neoplastic transformation. Mutation frequencies and chromosomal abnormalities are similar in human and rat or mouse cells, yet the latter are much more readily transformed. The epigenetic reactivation of the inactive X chromosome also appears to occur much more frequently in mouse tissues than in human tissues (see Holliday 1995). These results suggest that epigenetic controls, at least in part based on DNA methylation, are stricter in large, long-lived mammals than in small, short-lived animals.

There have been several important reviews of DNA methylation and cancer (Riggs and Jones 1983; Jones 1986; Jones and Buckley 1990; Spruck et al. 1993; Laird and Jaenisch 1994), but despite this, there appears to have been limited effort in looking at DNA methylation changes, for example, in the p53 tumor suppressor gene. Hundreds of studies of mutation in this gene have been published, but there has been no concerted study of the methylation status of the promoter region in tumor cells. It is perfectly possible that the loss of expression of the p53 protein is due to methylation, particularly in those cases where no mutation has been detected. There is some evidence that DNA methylation of the Rb tumor suppressor gene can lead to reduced expression in tumor cells (Sakai et al. 1991). More definitive evidence has come from studies of the p16 tumor suppressor gene, where it has been found that the CpG island is frequently methylated and silenced in human cancers (Merlo et al. 1995).

The current focus on mutation as a cause of cancer neglects the well-known examples of reversibility of the tumor cell phenotypes. Teratocarcinomas can be produced by placing a normal mouse embryo under the kidney capsule of an adult. These tumor cells can be transplanted into embryos and can then form apparently normal tissues (Mintz and Illmensee 1975). Clearly, in this case epigenetic changes in phenotype have occurred which are reversible. There is no evidence that DNA methylation is involved in these switches of phenotype, but this may simply be due to an ignorance of the genes with changed activity.

What has become clear is that tumor cells have lost the strict control of DNA methylation. Regions of the DNA may become hypomethylated, and CpG islands can become hypermethylated by de novo methylase activity. Most important, tumor cells in one population are clearly heterogeneous in their methylation pattern (Achten et al. 1991; Kochanek et al. 1991). This situation may well introduce epigenetic variability in the population and can facilitate the selection of faster-growing cells, or cells with metastatic potential. A great deal more information is needed about this type of epigenetic variation, particularly the relationship between DNA methylation and cell phenotype.

Another area in which DNA methylation is likely to be important is in cell and organism aging. The reactivation of the inactive X occurs much more frequently in aging female mice, and it is therefore quite possible that other normal epigenetic controls are lost during aging (for review, see Holliday 1995). This does not provide an all-encompassing theory of aging but certainly suggests that epimutations or dysdifferentiation may be one important component of the aging process.

The level of DNA methylation declines as mammalian fibroblasts proceed through their limited life span in vitro, and the rate of loss correlates in three species with the life span of these cells (Wilson and Jones 1983). There is also less substantial evidence that DNA methylation declines in vivo during aging (Wilson et al. 1987). Cooney (1993) has argued that disturbance of normal DNA methylation may also result from abnormalities in methylation metabolism, and he has thereby attempted to extend the methylation hypothesis into a general theory of aging. A major problem exists in trying to identify stochastic changes in DNA methylation during aging. Although the changes may be relatively frequent, each cell may have a different subset of alterations, and to identify these by standard procedures is impossible. It would be necessary to use methods applicable to single cells, such as PCR of DNA treated with a methyl-sensitive restriction enzyme, or even better, the bisulfite genomic sequencing procedure.

One approach which is already possible is to look for the ectopic expression of genes in cells where the gene in question is usually silent. This can be done by FACS analysis using fluorescent antibodies to the gene product. Thus, demethylation of promoter regions during aging may give rise to ectopic expression of genes. Of course, demethylation of silent repressor genes could have the reverse effect, which would be hard to detect.

Finally, it is important to consider DNA methylation in the context of the sequencing of the human genome. There are not four bases in human DNA, but at least five, and very likely others. To learn the position of every cytosine, but not to know whether it is methylated or not, will clearly fail to answer many important questions about the human genome. In particular, it will not address the question of those epigenetic controls in human DNA that are due to base modification. The sequence of the genome is one level of information and can be said to comprise the full genotype of our species. The time has come to recognize the importance of the epigenotype, both in the germ-line cells and in somatic tissues. When the human genome is finally sequenced, there will remain the enormous challenge of deciphering the epigenetic code that is superimposed on the four-base genetic code. This will surely occupy the attention of many investigators in the next 20 years.

REFERENCES

Achten, S., A. Behn-Krappa, M. Jücker, J. Sprengel, I. Hölker, B. Schmitz, H. Tesch, V. Diehl, and W. Doerfler. 1991. Patterns of DNA methylation in selected human genes in different Hodgkin's lymphoma and leukemia cell lines and in normal human lymphocytes. *Cancer Res.* **51:** 3702–3709.

Adams, R.L.P. and R.H. Burdon. 1985. *Molecular biology of DNA methylation.* Springer Verlag, Berlin.

Antequera, F. and A. Bird. 1993. CpG islands. *Exper. Suppl.* **64:** 169–185.

Antequera, F., J. Boyes, and A. Bird. 1990. High levels of *de novo* methylation and altered chromatin structure at CpG islands in cell lines. *Cell* **62:** 503–514.

Behn-Krappa, A., I. Hölker, U. Sandaradura de Silva, and W. Doerfler. 1991. Patterns of DNA methylation are indistinguishable in different individuals over a wide range of human DNA sequences. *Genomics* **11:** 1–7.

Betz, A.G., M.S. Nenberger, and C. Milstein. 1993. Discriminating intrinsic and antigen-selected hotspots in immunoglobulin V genes. *Immunol. Today* **14:** 405–411.

Boyes, J. and A.P. Bird. 1991. DNA methylation inhibits transcription indirectly via a methyl CpG binding protein. *Cell* **64:** 1123–1134.

———. 1992. Repression of genes by DNA methylation depends on CpG density and promoter strength. *EMBO J.* **11:** 327–333.

Burbelo, P.D., S. Horikoshi, and Y. Yamada. 1990. DNA methylation and collagen IV gene expression in F9 teratocarcinoma cells. *J. Biol. Chem.* **265:** 4839–4843.

Burn, J.E., D.J. Baguall, J.-D. Metzger, E.S. Dennis, and W.J. Peacock. 1993. DNA methylation, vernalisation and initiation of flowering. *Proc. Natl. Acad. Sci.* **90:** 287–291.

Busslinger, M., J. Hurst, and R.A. Flavell. 1983. DNA methylation and the regulation of globin gene expression. *Cell* **34:** 197–206.

Church, G.M. and W. Gilbert. 1984. Genomic sequencing. *Proc. Natl. Acad. Sci.* **81:** 1991–1995.

Clark, S.J., J. Harrison, and M. Frommer. 1995. CpNpG methylation in mammalian cells. *Nature Genet.* **10:** 20–27.

Compere, S.J. and R.D. Palmiter. 1981. DNA methylation controls the inducibility of the mouse metallothionein I gene in lymphoid cells. *Cell* **25:** 233–240.

Cooney, C.A. 1993. Are somatic cells inherently deficient in methylation metabolism? A proposed mechanism for DNA methylation loss, senescence and aging. *Growth Dev. Ageing* **57:** 261–272.

Crick, F.H.C. 1958. On protein synthesis. *Symp. Soc. Exp. Biol.* **12:** pp. 138–163.

Doerfler, W. 1981. DNA methylation—A regulatory signal in eukaryotic gene expression. *J. Gen. Virol.* **57:** 1–20.

———. 1983. DNA methylation and gene activity. *Annu. Rev. Biochem.* **52:** 93–124.

———. 1993. Pattern of *de novo* methylation and promoter inhibition: Studies on the adenovirus and the human genome. In *DNA methylation: Molecular biology and biological significance* (ed. J.P. Jost and H.P. Saluz), pp. 262–299. Birkhauser, Basel.

Donoghue, M.J., B.L. Patton, J.R. Sanes, and J.P. Merlie. 1992. An axial gradient of transgene methylation in skeletal muscle genomic imprint of rostrocaudal position. *Development* **116:** 1101–1112.

Ehrlich, M. and K.C. Ehrlich. 1993. Effect of DNA methylation and the binding of vertebrate and plant proteins to DNA. In *DNA methylation: Molecular biology and biological significance* (ed. J.P. Jost and H.P. Saluz), pp. 145–168. Birkhauser, Basel.

Ehrlich, M. and R.Y.H. Wang. 1981. 5-methyl cytosine in eukaryotic DNA. *Science* **212:** 1350–1357.

Finnegan, E.S., R.I.S. Brettell, and E.S. Dennis. 1993. The role of DNA methylation in the regulation of plant expression. In *DNA methylation: Molecular biology and biological significance* (ed. J.P. Jost and H.P. Saluz), pp. 218–261. Birkhauser, Basel.

Frank, D., N. Lightenstein, Z. Paroush, Y. Bergman, M. Shani, A. Razin, and H. Cedar. 1990. Demethylation of genes in animal cells. In *DNA methylation and gene regulation* (ed. R. Holliday et al.), pp. 63–73. Royal Society, London.

Frommer, M., L.E. McDonald, D.S. Millar, C.M. Collis, F. Watt, G.W. Grigg, P.L. Molloy, and C.L. Paul. 1992. A genomic sequencing protocol which yields a positive display of 5-methylcytosine residues in individual DNA strands. *Proc. Natl. Acad. Sci.* **89:** 1827–1831.

Gardiner-Garden, M. and M. Frommer. 1987. CpG islands in vertebrate genomes. *J. Mol. Biol.* **196:** 261–282.

Gartler, S.M. and M.A. Goldman. 1994. Reactivation of inactive X-linked genes. *Dev. Genet.* **15:** 504–514.

Goyon, C. and G. Faugeron. 1989. Targeted transformation of *Ascobolus immersus* and *de novo* methylation of the resulting duplicated DNA sequences. *Mol. Cell Biol.* **9:** 2818–2827.

Graessman, A., G. Sandberg, E. Guhl, and N. Graessman. 1994. Methylation of single sites within the Herpes Simplex Virus *tk* coding region and the Simian virus 40T-antigen intron causes gene inactivation. *Mol. Cell Biol.* **14:** 2004–2010.

Gurdon, J.B. and R.A. Laskey. 1970. The transplantation of nuclei from cultured cells into anucleate frogs eggs. *J. Embryol. Exp. Morph.* **24:** 227–248.

Harland, R.M. 1982. Inheritance of DNA methylation in microinjected eggs of *Xenopus laevis*. *Proc. Natl. Acad. Sci.* **79:** 2323–2327.

Holliday, R. 1979. A new theory of carcinogenesis. *Br. J. Cancer* **40:** 513–522.

———. 1987. Inheritance of epigenetic defects. *Science* **238:** 163–170.

———. 1990a. Genomic imprinting and allelic exclusion. *Development suppl.,* pp. 125–129.

———. 1990b. Mechanisms for the control of gene activity during development. *Biol. Rev.* **65:** 431–471.

———. 1991a. Mutations and epimutations in mammalian cells. *Mutat. Res.* **250:** 345–363.

———. 1991b. Quantitative genetic variation and developmental clocks. *J. Theoret. Biol.* **151:** 351–358.

———. 1993. Epigenetic inheritance based on DNA methylation. In *DNA methylation: Molecular biology and biological significance* (ed. J.P. Jost and H.P. Saluz), pp. 452–468. Birkhauser, Basel.

———. 1994. Epigenetics: An overview. *Dev. Genet.* **15:** 453–457.

———. 1995. *Understanding ageing*. Cambridge University Press, England.

Holliday, R. and T. Ho. 1991. Gene silencing in mammalian cells by uptake of 5 methyl deoxycytidine 5′ triphosphate. *Somatic Cell Mol. Genet.* **17:** 537–542

Holliday, R., M. Monk, and J.E. Pugh, eds. 1990. *DNA methylation and gene regulation*. Royal Society, London.

Holtzer, H., H. Weintraub, R. Mayne, and B. Mochan. 1972. The cell cycle, cell lineages and cell differentiation. *Curr. Topics Dev. Biol.* **7:** 229–256.

Hsieh, C.-L. 1994. Dependence of transcriptional repression on CpG methylation density. *Mol. Cell Biol.* **14:** 5487–5494.

Jablonka, E. and M.J. Lamb. 1989. The inheritance of acquired epigenetic variations. *J. Theoret. Biol.* **139:** 69–84.

Jablonka, E., M. Lachman, and M.J. Lamb. 1992. Evidence, mechanisms and models for the inheritance of acquired characters. *J. Theoret. Biol.* **158:** 245–268.

Jones, P.A. 1984. Gene activation by 5-azacytidine. In *DNA methylation: Biochemistry and biological significance* (ed. A. Razin et al.), pp. 165–187. Springer Verlag, Berlin.

———. 1985. Altering DNA methylation with 5-azacytidine. *Cell* **40:** 485–486.

———. 1986. DNA methylation and cancer. *Cancer Res.* **46:** 461–466.

Jones, P.A. and J.D. Buckley. 1990. The role of DNA methylation in cancer. *Adv. Cancer. Res.* **54:** 1–23.

Jones, P.A. and S.M. Taylor. 1980. Cellular differentiation, cytidine analogs and DNA methylation. *Nucleic Acids Res.* **9:** 2933–2947.

Jones, P.A., M.J. Wolkowicz, W.M. Rideout, F.A. Gonzalez. C.M. Marziasz, G.A. Coetzee, and S.J. Tapscott. 1990. *De novo* methylation of the *Myo*D1 CpG island during the establishment of immortal cell lines. *Proc. Natl. Acad. Sci.* **87:** 6117–6121.

Jost, J.P. and H.P. Saluz, eds. 1993a. *DNA methylation: Molecular biology and biological significance*. Birkhauser, Basel.

———. 1993b. Steroid hormone-dependent changes in DNA methylation and its significance for the activation and silencing of specific genes. In *DNA methylation: Molecular biology and biological significance* (ed. J.P. Jost and H.P. Saluz), pp. 425–451. Birkhauser, Basel.

Jost, J.P., H.P. Saluz, I. McEwan, I.M. Feavers, M. Hughes, S. Reiber, H.M. Lang, and M. Vaccaro. 1990. Tissue specific expression of avian vitellogenin gene is correlated with DNA hypomethylation and *in vivo* specific protein-DNA interactions. In *DNA methylation and gene regulation* (ed. R. Holliday et al.), pp. 53–62. Royal Society, London.

Kay, P.H., E. Pereira, S.A. Marlow, G. Turbett, C.A. Mitchell, P.F. Jacobson, R. Holliday, and J.M. Papadimitriou. 1994. Evidence for adenine methylation within the mouse myogenic gene Myo-D1. *Gene* **151:** 89–95.

Kochanek, S., A. Radbruch, H. Tesch, D. Renz, and W. Doerfler. 1991. DNA methylation profiles in the human genes for tumor necrosis factors α and β in subpopulations of leukocytes and in leukemias. *Proc. Natl. Acad. Sci.* **88:** 5759–5763.

Kochanek, S., M. Toth, A. Dehmel, D. Renz, and W. Doerfler. 1990. Interindividual concordance of methylation profiles in human genes for tumor necrosis factors α and β. *Proc. Natl. Acad. Sci.* **87:** 8830–8834.

Laird, P.W. and R. Jaenisch. 1994. DNA methylation and cancer. *Hum. Mol. Genet.* **3:** 1487–1495.

Li, E., T.H. Bestor, and R. Jaenisch. 1992. Targeted mutation of the DNA methyl transferase gene results in embryonic lethality. *Cell* **69:** 915–926.

Lock, L.F., N. Tagaki, and G.R. Martin. 1987. Methylation of the Hprt gene on the inactive X occurs after chromosome inactivation. *Cell* **48:** 8–16.

Merlo, A., J.G. Herman, L. Mao, D.J. Lee, E. Gabrielson, P.C. Burger, S.B. Baylin, and D. Sidransky. 1995. 5′ CpG island methylation is associated with transcriptional silencing of the tumor suppressor *p16/CDKN2/MTS1* in human cancers. *Nat. Med.* **1:** 686–692.

Mintz, B. and K. Illmensee. 1975. Normal genetically mosaic mice produced from malignant teratoma cells. *Proc. Natl. Acad. Sci.* **72:** 3585–3589.

Monk, M., M. Boubelik, and S. Lehnert. 1987. Temporal and regional changes in DNA methylation in embryonic, extra-embryonic and germ cell lineages during mouse embryo development. *Development* **99:** 371–382.

Murray, E.J. and F. Grosveld. 1987. Site specific demethylation in the promoter of human δ globin does not alleviate methylation mediated suppression. *EMBO J.* **6:** 2329–2335.

Nyce, J. 1991. Gene silencing in mammalian cells by direct incorporation of electroporated 5-methyl-2′-deoxycytidine 5′ triphospate. *Somatic Cell Mol. Genet.* **17:** 543–550.

Pandit, N.N. and V.E.A. Russo. 1992. Reversible inactivation of a foreign gene, hph, during the asexual cycle in *Neurospora crassa* transformants. *Mol. Gen. Genet.* **234:** 412–422.

Pfeifer, G.P., S.D. Steigerwald, R.S. Hansen, S.M. Gartler, and A.D. Riggs. 1990. Polymerase chain reaction-aided genomic sequencing of an X chromosome-linked CpG island: Methylation patterns suggest clonal inheritance, CpG site autonomy, and an explanation of activity state stability. *Proc. Natl. Acad. Sci.* **87:** 8252–8256.

Razin, A. and H. Cedar. 1993. DNA methylation and embryogenesis. In *DNA methylation: Molecular biology and biological significance* (ed. J.P. Jost and H.P. Saluz), pp. 343-357. Birkhauser, Basel.

Razin, A. and A.D. Riggs. 1980. DNA methylation and gene function. *Science* **210:** 604–610.

Razin, A., H. Cedar, and A.D. Riggs, eds. 1984. *DNA methylation: Biochemistry and biological significance.* Springer Verlag, Berlin.

Rhounim, L., J.L. Rossignol, and G. Faugeron. 1992. Epimutation of repeated genes in *Ascobolus immersus. EMBO J.* **11:** 4451–4457.

Rideout, W.M., P. Eversole-Cire, C.H. Spruck, C.M. Hustad, G.A. Coetzee, F.A. Gonzalez, and P.A. Jones. 1994. Progressive increase in the methylation status and heterochromatinization of the MyoD CpG island during oncogenic transformation. *Mol. Cell Biol.* **14:** 6143–6152.

Riggs, A.D. and P.A. Jones. 1983. 5-methylcytosine, gene regulation and cancer. *Adv. Cancer Res.* **40:** 1–30.

Sager, R. and R. Kitchin. 1975. Selective silencing of eukaryotic DNA. *Science* **189:** 426–433.

Sakai, T., J. Toguchida, N. Ohtani, D.W. Yandell, J.M. Rapaport, and T.P. Dryja. 1991. Allele specific hypermethylation of the retinoblastoma tumor suppressor gene. *Am. J. Hum. Genet.* **48:** 880–888.

Saluz, H.P. and J.P. Jost. 1993. Major techniques for the study of DNA methylation. In *DNA methylation: Molecular biology and biological significance* (ed. J.P. Jost and H.P. Saluz), pp. 11–26. Birkhauser, Basel.

Santi, D.V., A. Norment, and C.E. Garrett. 1984. Covalent bond formation between a DNA-cytosine methyl transferase and DNA containing 5-aza-cytidine. *Proc. Natl. Acad. Sci.* **81:** 6993–6997.

Scarano, E. 1971. The control of gene function in cell differentiation and in embryogenesis. *Adv. Cytopharmacol.* **1:** 13–24.

Schwartz, D. and E. Dennis. 1986. Transposase activity of the *Ac* control element in maize is regulated by its degree of methylation. *Mol. Gen. Genet.* **205:** 476–482.

Selker, E.V. 1990. Premeiotic instability of repeated sequences in *Neurospora crassa*. *Annu. Rev. Genet.* **24:** 579–613.

Simonovitch, L. 1978. On the nature of heritable variation in cultured mammalian cells. *Cell* **7:** 1–11.

Singer-Sam, J. and A.D. Riggs. 1993. X chromosome inactivation and DNA methylation. In *DNA methylation: Molecular biology and biological significance* (ed. J.P. Jost and H.P. Saluz), pp.358–384. Birkhauser, Basel.

Singer-Sam, J., M. Grant, J.M. LeBon, K. Okuyama, V. Chapman, M. Monk, and A.D. Riggs. 1990. Use of a *Hpa*II-polymerase chain reaction assay to study DNA methylation in the *Pgk-1* CpG island of mouse embryos at the time of X-chromosome inactivation. *Mol. Cell. Biol.* **10:** 4987–4989.

Spruck, C.H., III, W.M. Rideout III, and P.A. Jones. 1993. DNA methylation and cancer. In *DNA methylation: Molecular biology and biological significance* (ed. J.P. Jost and H.P. Saluz), pp. 487–509. Birkhauser, Basel.

Sullivan, C.H. and R.M. Grainger. 1987. δ crystallin genes become hypomethylated in post-mitotic lens cells during chicken development. *Proc. Natl. Acad. Sci.* **84:** 329–333.

Tasheva, E.S. and D.J. Roufa. 1994. Densely methylated DNA islands in mammalian chromosome replication origins. *Mol. Gen. Genet.* **14:** 5636–5644.

Tasseron de Jong, J., J. Aker, H. Der Dulk, P. van de Putte, and M. Giphart-Gassler. 1989. Cytosine methylation in the *Eco*RI site of active and inactive herpes virus thymidine kinase promoters. *Biochim. Biophys. Acta* **1008:** 62–70.

Taylor, J.H. 1984. *DNA methylation and cellular differentiation*. Springer Verlag, Berlin.

Taylor, S. and P.A. Jones. 1979. Multiple new phenotypes induced in $10T^1/_2$ cells and 3T3 cells treated with 5-azacytidine. *Cell* **17:** 771–779.

Tilghman, S.M. 1993. DNA methylation: A phoenix rises. *Proc. Natl. Acad. Sci.* **90:** 8761–8762.

Volpe, P. and C. Nicolini. 1989. Forum on eukaryotic DNA methylation. *Cell Biophys.* **15:** 1–155.

Wigler, H., D. Levy, and M. Perucho. 1981. The somatic replication of DNA methylation. *Cell* **24:** 33–40.

Wilson, V.L., and P.A. Jones. 1983. DNA methylation decreases in ageing but not in immortal cells. *Science* **220:** 1055–1057.

Wilson, V.L., R.A. Smith, S. Ma, and R.G. Cutler. 1987. Genomic 5-methyl deoxycytidine decreases with age. *J. Biol. Chem.* **262:** 9948–9951.

Yisraeli, J., R. Adelstein, D. Melloul, U. Nudel, D. Yaffe, and H. Cedar. 1986. Muscle specific activation of a methylated chimeric actin gene. *Cell* **46:** 409–416.

Overview of Epigenetic Mechanisms

Arthur D. Riggs and Thomas N. Porter
Biology Department
Beckman Research Institute of the City of Hope
Duarte, California 91010

How is activity state information of a particular gene or chromosomal domain passed unchanged from parent cells to progeny cells when both the active and inactive states have the same DNA sequence? Cell memory—the faithful transmission of determined states to progeny cells—is essential for development of multicellular organisms, including plants, insects, and mammals. What epigenetic mechanisms account for cell memory? With regard to genomic imprinting, how are chromosomal domains marked as to parental source, and how is monoallelic expression in diploid cells accomplished? How can one X chromosome in mammalian cells be genetically silenced without affecting its homolog? Although these important questions have been addressed by numerous scientists for many years, we still are without complete answers. Several models to explain epigenetic phenomena have been proposed, however, and these are categorized in Table 1. These models are discussed below, and where possible, they are explained in the context of experimental examples.

Relatively simple feedback circuits can explain at least some clonally heritable epigenetic phenomena in which both alleles in diploid cells are treated similarly. This is also true for phenomena involving co-regulation of all gene copies in polytene *Drosophila* cells or multicopy transgenic animal or plant cells. However, meiotic heritability and phenomena showing differential treatment of identical alleles, such as X-chromosome inactivation and genomic imprinting, seem to need additional *cis*-limited mechanisms. It should be clearly understood that models listed in Table 1 are not mutually exclusive. On the contrary, it is very likely that more than one mechanism is used for epigenetic switches that must maintain tight genetic silence, be very stable, and show binary switching between "full on" and "full off" states.

DYNAMIC, STEADY-STATE SYSTEMS WITH FEEDBACK LOOPS

Feedback circuits that depend on interactions with diffusible intracellular and/or intercellular factors can explain many epigenetic effects; for ex-

Table 1 Epigenetic mechanisms

Stable feedback loops	
assumptions:	free diffusion of *trans*-acting factors
comments:	several examples exist, including the lac operon, the λ lysogenic/lytic switch, and *Drosophila* sex determination
weakness:	does not adequately explain monoallelic expression in diploid cells, as seen for X-chromosome inactivation and imprinted genes
DNA cytosine methylation	
assumptions:	a maintenance methylase system exists; methylation affects protein binding and chromatin function
comments:	established mechanism; essential for mammals; used by plants
weakness:	not used by *Drosophila*, *Caenorhabditis elegans,* or yeast
Nuclear compartmentalization	
assumptions:	spatial separation of alleles with alternate activity states into biochemically different compartments
comments:	conceptually attractive; there is substantial evidence for compartments within the nucleus
weaknesses:	how is the compartment replicated and maintained through mitosis? requires one active compartment (entity) per diploid set of autosomes
Temporal separation by differential replication timing	
assumptions:	early-replicating chromatin is different from late-replicating chromatin
comments:	conceptually attractive; inactive genes tend to replicate late in S phase
weakness:	little experimental evidence
Heritable chromatin structure	
assumption:	somatically heritable nucleoprotein complexes
comments:	suggested by a large body of evidence from yeast and *Drosophila*
weakness:	more molecular detail needed to explain replication and heritability

See text for discussion and references.

ample, the existence of metastable variants in bacterial (and mammalian/plant?) cultures. Delbruck recognized in the late 1940s that, analogous to electrical circuits, metabolic circuits with positive or negative feedback loops can exist in alternate metastable states, and that this might explain alternate, heritable phenotypes in unicellular organisms such as paramecia (see Jablonka and Lamb 1995). Three examples of alternate, metastable states dependent on feedback circuits are described below. Cell-heritable changes in plants and a biochemical switch model for gene silencing are discussed in more detail by Meins (this volume).

The Lac Operon

Novick and Weiner (1957) showed that *Escherichia coli* can maintain alternate, heritable states of induction of the lac operon. Figure 1 illustrates an explanation of this epigenetic phenomenon. If cells are grown in lactose, the enzymes of the lac operon are induced about 1000-fold. However, prior to induction (in the absence of lac permease), the cell membrane is only weakly permeable to lactose. Moreover, in order for lactose to function as an inducer of the lac operon, it must be converted by β-galactosidase into allolactose, which is the natural inducer of the lac operon (Jobe and Bourgeois 1972). Thus, there is positive feed-

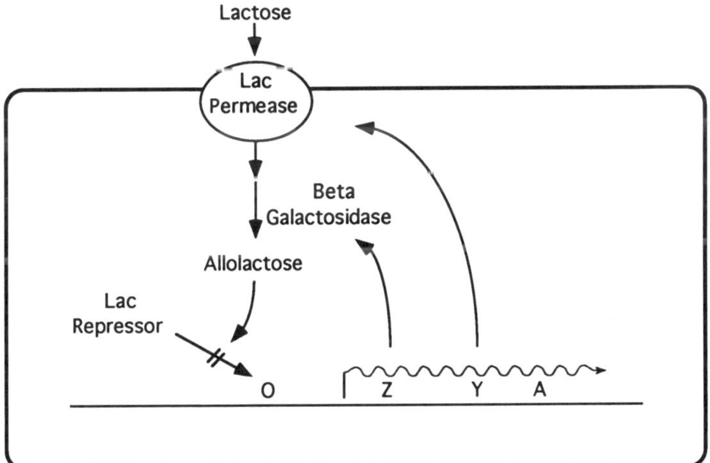

Figure 1 Positive feedback loops and lac operon induction. The lac operon encodes three proteins, two of which, β-galactosidase and lac permease, are required for the induction of the lac operon (see text for details). (O) lac operator; (Z) β-galactosidase; (Y) lac permease; (A) lac acetylase.

back at two points in the pathway to induction. An important question is: Why is the operon inducible at all? The operon is inducible only because the lac repressor does not completely prevent transcription, and there are always a few molecules of β-galactosidase produced even in the off state. Thus, relatively high concentrations of lactose in the growth medium induce the operon. Then, because the permease concentrates the lactose in the cell, only low maintenance levels of lactose are needed to keep the operon in the on state. If fully induced cells are mixed with uninduced cells and the culture is maintained at low lactose concentrations, the alternate, on or off phenotype will be faithfully transmitted from parent to progeny cells; i.e., there is clonal inheritance. The applicability of this type of model to mammalian and plant development should be obvious.

The λ Genetic Switch

The bacteriophage λ depends on an epigenetic mechanism to determine the course of its life cycle. After infection of a bacterial cell, a temperate bacteriophage such as λ may integrate into the host genome where it (the prophage) is passively replicated as part of the host genome, a process termed lysogeny. At a later time, the prophage may be induced to a lytic state. The lytic/lysogenic switch state is determined by an epigenetic circuit, which is sensitive to the internal environment of the host (Ptashne 1992). Figure 2 illustrates the basic components of the λ epigenetic switch. The cI gene codes for the λ repressor protein, which is necessary for the establishment and maintenance of the "off" or lysogenic state. The cro gene product is necessary for adoption of the "on" or lytic state. The cI repressor represses cro transcription, and the cro protein shuts off transcription of cI. Thus, the two proteins behave in an almost mirror-image fashion. An operator region, containing three binding sites, overlaps the two promoters for these genes. The two promoter regions, P_{RM} (the cI promoter) and P_R (the cro promoter), initiate transcription of their respective products in opposite directions. Upon infection of a cell, the cI repressor sometimes predominates and the lysogenic, genetically silent state is established. The prophage may exist in the dormant (off) state for many rounds of host cell division. However, in response to changes in the intracellular environment, for example, induction of the SOS response, cI protein may become inactivated. This leads to transcription of the cro gene, which shuts down transcription of cI and leads to the production of proteins involved in the lytic response (Fig. 2B). This simple circuit controls two mutually exclusive patterns of transcription that lead to two distinct epiphenotypes.

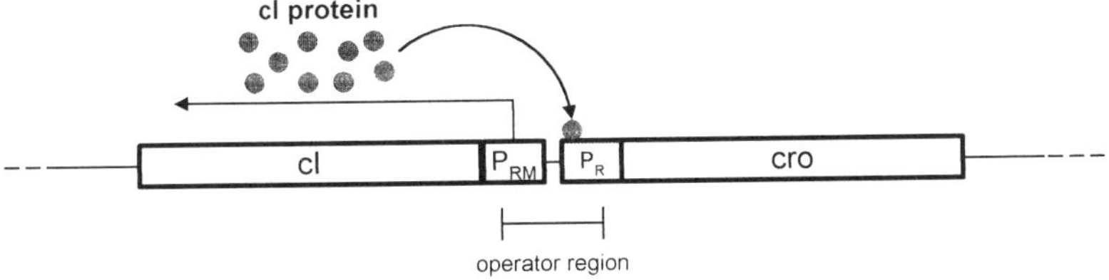

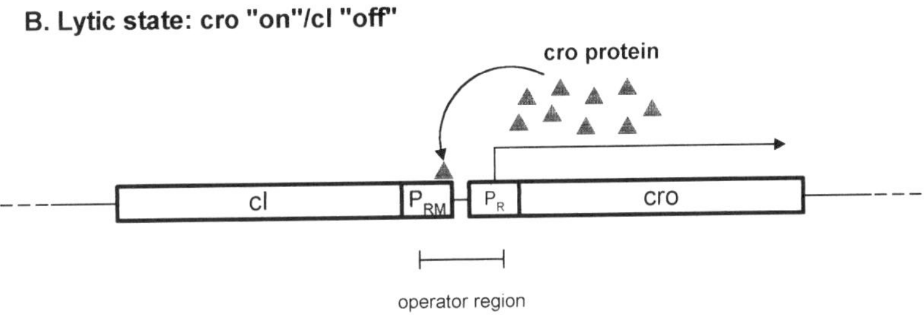

Figure 2 The λ lysogenic/lytic switch. The region of bacteriophage λ involved in determining the fate of the phage upon entering a host bacterium is shown. The left and right open rectangles represent the cI gene and cro gene, respectively. The shaded box adjacent to cI is P_{RM} (promoter for maintenance of repression. The shaded box adjacent to the cro gene (P_R) represents the cro promoter. The small solid circles represent molecules of cI protein (λ repressor) in *A* and the triangles represent cro protein in *B*. The operator region is indicated by the bar below the promoter region.

Drosophila Sex Determination

In *Drosophila*, sex state is a cell-autonomous, somatically heritable feature; all cells, including somatic cells, are either male or female. Soon after fertilization, the ratio of autosomes to X chromosomes is measured, and then, depending on this ratio, either the male or female state of a stable epigenetic switch is established (AA/XY or AA/XO embryos become male; AA/XX embryos become female). As illustrated in Figure 3, a dynamic feedback loop at the mRNA splicing level maintains male/female differences (Bell et al. 1991). Sxl protein is a master regulatory protein for sex determination, controlling the activity of several other proteins. Only in the presence of Sxl does the cell express the proteins and enzymes characteristic of a female. In the absence of Sxl, the default sex state is male. The key aspect of this system is that the Sxl protein is required in all somatic cells for correct splicing of the Sxl transcript. A cell with active Sxl protein passes enough active protein (or

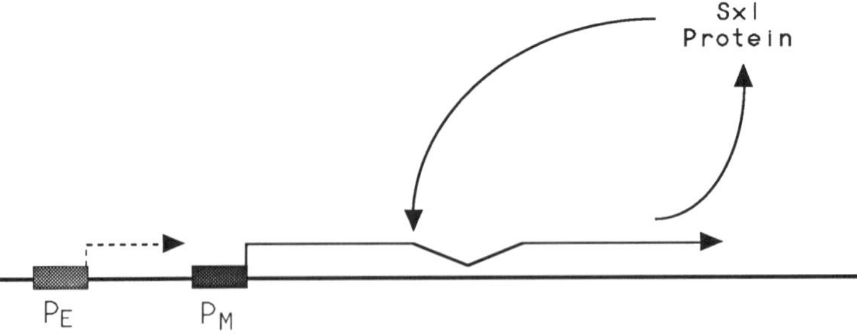

Figure 3 The epigenetic switch controlling *Drosophila* sex determination. A feedback loop at the mRNA splicing level maintains male/female differences in *Drosophila* somatic cells. Active Sxl protein is necessary for maintenance of the female state. The transcript that encodes the Sxl protein in adult somatic cells initiates only at the P_M promoter. Sxl protein is required for the proper splicing of this transcript. In the early embryo, transcription initiates at P_E. The early Sxl transcript splices correctly in the absence of Sxl protein, thus allowing establishment of the female state.

mRNA) to its progeny cells to maintain functional splicing and production of functional protein; therefore, the progeny of a cell in the female state remain in the female state. How is the "on" or female state established? In the early embryo a different promoter is used, the P_E promoter, and this early Sxl transcript splices correctly without Sxl protein. This sex determination system illustrates a potentially general mechanism for epigenetic maintenance of a particular phenotypic state. It is important to note that this example illustrates that epiphenotypes can be maintained by feedback control at levels other than transcription.

DNA METHYLATION

The epigenetic mechanism illustrated in Figure 4 depends on the presence of 5-methylcytosine at certain CpG sites and a DNA methyltransferase enzyme, often termed a maintenance methylase, to maintain preexisting methylation patterns. This model is supported by a large body of experimental evidence (for review, see Jost and Saluz 1993) and is generally accepted as being a component mechanism of some epigenetic systems in mammals and higher plants. The case is strongest for X-chromosome inactivation in mammalian cells (see other chapter by Riggs and Porter, this volume). Mammalian cells have a DNA (cytosine-5)-methyltransferase enzyme that postreplicationally converts certain cytosines to 5-methylcytosine. This enzyme preferentially methylates

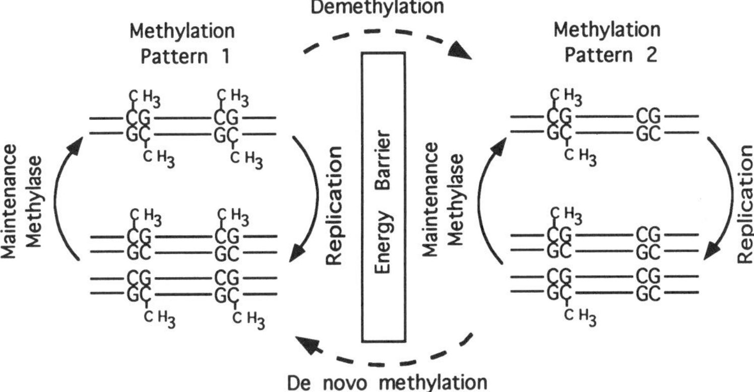

Figure 4 Cell memory and epigenetic inheritance by DNA methylation. According to this model, alternate chromatin states survive DNA replication because a maintenance DNA methyltransferase acts, to a first approximation, only on the hemimethylated sites that are formed by the replication of a symmetrically methylated site. As a result of this mechanism, the methylation pattern is self-templating. The dashed arrows indicate that the pattern of methylated cystosines can, however, be altered either by de novo methylation of unmethylated sites or by demethylation of methylated sites. The rectangle labeled energy barrier indicates that change is not rapid but does occur infrequently under normal circumstances. More rapid change can take place at specific stages of development (see text for details).

cytosine in CpG sites, and, in vitro, shows up to a 100-fold preference for hemimethylated sites, such as those generated by the replication of a CpG site symmetrically methylated in both strands (Fig. 4). All mammalian cells have cell-type-specific patterns of methylated and unmethylated CpG sites; on average, about 20% of CpG sites are unmethylated and 80% are methylated in both strands. Hemimethylated sites are rare and usually not detected in normal genomic DNA.

The maintenance methylase system tends to maintain a preexisting methylation pattern on progeny chromosomes, and methylation patterns have been experimentally shown to be clonally heritable. There are several hundred papers correlating a certain cell-type-specific methylation pattern, usually a hypermethylated promoter or other key site, with genetic silence. In vitro methylation often results in inhibition of in vitro transcription and also genetic silencing of the transgene upon integration after transfection. Importantly, inhibitors of the DNA methyltransferase system cause heritable changes in methylation patterns and concomitant heritable alteration in cellular phenotype. Detailed reviews of this evidence can be found in Jost and Saluz (1993) and in various chapters of this volume. The most important recent addition to our understanding of mammalian DNA methylation is that studies with transgenic mice deficient in methyltransferase have established that there is one, and perhaps only one, major DNA methyltransferase gene and that it is essential for normal development and maintenance of genetic silence of imprinted genes, including *Xist*, an X-linked gene (Li and Jaenisch 1993; Beard et al. 1995).

Here, we emphasize the dynamic aspects of the DNA methylation system (dotted arrows in Fig. 4). How are methylation patterns established and changed? Methylation patterns can be changed during development either (1) by de novo methylation of unmethylated sites or (2) by demethylation of previously methylated sites. The maintenance methylation system provides an energy barrier between two alternate states (Fig. 4). Energy is required to change one state to the other, a property that seems ideally suited for epigenetic switches used to guide development. Mechanisms for the specific removal of methylated cytosines have been shown to exist, and the detailed mechanism is currently under study (see Jost, this volume). The addition of methyl groups to unmethylated sites (de novo methylation) also occurs during mammalian development. Methylation patterns clearly are dynamic during development (see Shemer and Razin, this volume), with the emerging picture suggesting that de novo methylation predominates during late meiosis and/or gametogenesis, global demethylation predominates shortly after fertilization, and then global remethylation predominates at the time of implantation. During later cellular differentiation, one sees only gene-specific demethylation. It is generally assumed (but evidence is lacking) that stage-specific proteins and or specific chromatin/DNA structure determines the specificity of methylation patterns and the relative levels of maintenance methylation, de novo methylation, and demethylation.

In tissue culture, previously unmethylated sites occasionally become de novo methylated, including CpG islands which in vivo are unmethylated. Tumors, for example, often have hypermethylated CpG islands (for review, see Laird and Jaenisch 1996). Pfeifer et al. (1990) found that de novo methylation at the CpG island of PGK on the X_i can be as high as 5% per CpG site per generation. We also have recently found that de novo methylation of a *Hpa*II site in the differentially methylated region 3 kb upstream of *Igf2*, an imprinted autosomal gene, is also about 3% per generation (Z. Xiong and A.D. Riggs, unpubl.). Thus, it should be appreciated that although the methylation system is stable enough to aid inheritance, it is to a considerable extent dynamic, with stochastic as well as cell- and tissue-specific aspects. Equations describing the level of methylation at a specific site in a population of cells have been published (Pfeifer et al. 1990). The methylation level of any given CpG depends on (1) the previous methylation state, (2) the de novo activity of DNA methyltransferase, (3) the level of general and specific demethylation activity, and (4) local chromatin structure, which includes the influence of nucleosomes, specific and nonspecific protein factors, DNA structure, and higher-order chromatin structure. All of these parameters are likely to be specifically varied as part of a developmental program.

A postulate necessary for the epigenetic mechanism of Figure 4 is that cytosine methylation affects chromatin function. This postulate is now very well supported, as both in vitro and in vivo studies have shown that methylation inhibits transcription either directly or indirectly (Tate and Bird 1993). Adding to these observations are more than 20 reports of proteins whose DNA binding is affected by cytosine methylation (see Ehrlich and Ehrlich 1993). Most methylation-sensitive protein factors are sequence-specific and inhibited by methylation at key sites, but some proteins, such as MeCP1 and MeCP2, are not sequence-specific but do require methylated cytosines (Tate and Bird 1993). The influence of methylation on the binding of specific transcription factors was first studied using Sp1, and, although they were initially reported to be methylation-insensitive, more recent reports indicate that at least some members of the consensus binding sequence bind Sp1 more strongly if they are methylated (Jane et al. 1993; Joel et al. 1993).

NUCLEAR COMPARTMENTALIZATION

If active and inactive genes are spatially separated into separate nuclear compartments, then the maintenance of differential activity states can easily be explained by, for example, the inactive compartment having repressive proteins and/or not having necessary transcription factors. Nuclear compartmentalization was, in fact, the earliest model proposed (Comings 1966) to explain X-chromosome inactivation and continues to be an attractive explanation for this phenomenon. Recently, Schlossher et al. (1994) have discussed specific sequestration into separate nuclear territories as a possible mechanism for other epigenetic phenomena such as position-effect variegation (see Henikoff, this volume) and polycomb-mediated cell memory in *Drosophila* (see Pirrotta; Paro and Harte; both this volume). The structure of the nucleus is complex, and not homogeneous, with evidence for subnuclear domains (Ascoli and Maul 1991; Saunders et al. 1991; Cremer et al. 1993).

A model for the replication and inheritance of nuclear compartments has not been discussed, however, except in the context of the counting mechanism for X-chromosome inactivation (Gartler and Riggs 1983). In this system, apparently as a result of an X/autosome ratio-measuring mechanism, one active X is maintained per autosomal set. If one active compartment per diploid set re-forms after cell division, this provides a partial explanation. However, without some feature such as chromosome marking, this model provides no explanation as to why the previously active X always becomes associated with the active compartment. One possibility is that the structure specifying a nucleation center for the compartment never dissociates from the chromosome during replication and mitosis, but the central question then no longer concerns compartmentalization. Thus, although interesting, nuclear compartmentalization models are at present incomplete.

REPLICATION TIMING/TEMPORAL SEPARATION

Instead of spatial separation as discussed in the previous section, temporal separation may be used to aid maintenance of alternate epigenetic states (Riggs and Pfeifer 1992). In mammals and also yeast, active or potentially active genes tend to replicate in the first half of the S phase, whereas their inactive counterparts replicate in the second half of S phase (Holmquist 1987). If the intranuclear milieu changes with respect to available factors during the switch from early to late S, then active, early-replicating genes exist in a different environment during replication than do inactive, late-replicating genes, making it easy to imagine that functionally different chromatin structure could be established. Late replication has been proposed as the ancestral, most fundamental mechanism for X-chromosome inactivation, because late replication is the only readily identifiable feature of X inactivation seen in all mammals, including the egg-laying mammals. A similar model was earlier proposed to explain the switch from oocyte to somatic 5S gene transcription in *Xenopus* (Gottesfeld and Bloomer 1982; Wormington et al. 1982). Although this model is no longer thought to explain the differential activity of 5S genes (Bouvet et al. 1994), the temporal separation model is still an attractive model for X-chromosome inactivation. The key chromosomal elements for this model are likely to be replication origins.

In yeast, there is good evidence for a connection between replication timing and mating-type silencing. In contrast to most yeast genes, the genetically silent *HML* and *HMR* loci of the mating-type system replicate late, and both biochemical and genetic analyses have indicated that origin of replication (ORC) complexes are involved in the chromatin-based silencing. ORC-binding elements are contained in the silencer elements known to be necessary for the heritability of the silent state and probably also for its establishment. However, the system is not simple and not yet fully understood, since mutants have been obtained that affect silencing but not replication function (for review, see Pillus and Grunstein 1995; also see Holmes et al., this volume). It should be noted that replication timing models, although effective in explaining maintenance of differential chromatin states, do not explain meiotic transmission of this information, nor transmission of the information through metaphase (see other chapter by Riggs and Porter, this volume).

HERITABLE CHROMATIN STRUCTURE

Heritable chromatin structure, the last class of models listed in Table 1, has received considerable experimental attention and is the epigenetic model favored to explain somatic inheritance of yeast mating type, *Drosophila* cell memory, and position-effect variegation (PEV) (see Allshire; Holmes et al.; Pirrotta; Paro and Harte; Henikoff; all this volume). Chromatin architecture and its dynamics are emerging as key determinants in the developmental regulation of gene expression. Chromatin is a dynamic complex of DNA, RNA, histone, and nonhistone proteins embedded within the eukaryotic nucleus and nuclear matrix. The nuclear matrix is thought to provide the spatial arrangement and the structural framework needed for DNA replication, transcription, recombination, and nuclear transport. During mitosis, chromatin and the supporting nuclear matrix are efficiently disassembled, partitioned, and subsequently reassembled into daughter nuclei (He et al. 1995). How this is done is of crucial importance for an understanding of epigenetic mechanisms, but, aside from a relatively detailed understanding of nucleosome formation, structure, and modification, attempts to comprehend these mechanisms have been frustrated by the complexity of chromatin and its relationship with other nuclear structures.

Chromatin structure is reviewed by Kass and Wolffe (this volume). In brief outline, it is known that a large portion of the DNA in the eukaryotic nucleus is wrapped around a core histone octamer (two molecules each of H3, H4, H2A, and H2B) to form a nucleosomal particle. Under low-salt conditions, a chain of nucleosome beads 10 nm in width can be clearly seen in the electron microscope, but under more physiologically relevant conditions, the nucleosome-containing DNA fiber is seen as a more condensed 30-nm fiber. The 30-nm chromatin fiber consists of core nucleosomes, histone H1, other nonhistone proteins, and internucleosomal linker DNA packed either into a solenoid structure or, according to the most recent data, a less ordered, three-dimensional zigzag structure (Woodcock et al. 1993; Woodcock 1994). Whatever the exact configuration of the 30-nm chromatin fiber, which may vary according to activity state, it is thought to be attached periodically to the nuclear matrix/protein scaffold and folded into variable size loops (75 kb average size), the ends of which are marked by matrix attachment regions (MARs) for interphase chromatin, or scaffold associated regions (SARs) for metaphase chromosomes. The folded chromatin fiber is about 250 nm wide during interphase, and then during metaphase, it is further coiled to form the 700-nm chromatid of a metaphase chromosome.

Since histones, nucleosomes, and higher-order chromatin structure are involved in gene control, it is interesting that histones are subject to a variety of modifications, including phosphorylation, protein methylation, ubiquitinylation, and acetylation (for review, see Grunstein et al. 1995; Kass and Wolffe, this volume). In particular, the acetylation level (and pattern) of histone H4, and other core histones, is highly correlated with transcriptional activity or potential activity. An active domain is more highly acetylated than is its inactive counterpart, and it is likely that the acetylation of core histones leads to structural transitions as well as specific communication within related chromatin domains. Thus, alteration or modification of chromatin-related structural proteins may provide a dominant means of controlling the transcriptional activity of individual genes, domains, and entire chromosomes.

Basal transcription complexes are likely to be removed, at least transiently, from DNA by passage of a replication fork. Thus, epigenetic information in the form of altered chromatin states is temporarily erased by the physical passage of the replication fork. How does previous structural information survive replication? To accomplish the replication of a chromatin state, it seems likely that the existing structure must somehow guide the reestablishment of the same structure; that is to say, the chromatin structure must be self-templating. If only proteins and protein modifications are involved, how is this accomplished? Several mechanisms have been suggested (see Kass and Wolffe, this volume). For example, it is likely that at the replication fork, ordered nucleosomes are transiently removed and segregated randomly to daughter strands. However, formerly modified nucleosomes remain in the vicinity of their previous location, thus allowing the "seeding" of partially modified structures within a newly replicated domain. After replication, this generates a discontinuous pattern of histone acetylation over the patch of maturing, newly synthesized DNA, and this may provide a framework leading to the cooperative recruitment of domain-specific factors such as chromodomain proteins or histone modification enzymes, which complete the process of re-marking both daughter strands.

Most models proposed for heritable chromatin structure are deficient in details, especially with regard to explaining the *cis*-limited action and self-templating needed for a complete explanation of imprinting and XCI. However, one model (Wolffe 1994) is worthy of special emphasis. As illustrated in Figure 5, physically separated, *cis*-linked, multipartite structures could be involved in the restoration of chromatin-state information that is obliterated by the passage of a replication fork. This attractive model takes advantage of the multipartite nature of gene control elements; transcriptional regulation is determined not by a simple promoter, but rather by a *cis*-located network of elements including the promoter, nearby activator and repressors, and other, more distant sequence motifs including locus control regions, replication origins, and matrix attachment sites. A key aspect of this model is illustrated in Figure 5. Element B of this complex resides downstream from element A, with respect to replication fork movement. After the structure of element A is disrupted by passage of the replication fork, but before element B is disturbed, element B can pass activity-state information back to the newly replicated As (probably involving some form of looping or intradomain communication), thereby activating the recently duplicated upstream element. Then, when the replication fork disrupts downstream element B, element A can guide the proper re-formation of downstream element B. This model has the experimentally useful prediction that at least one complex must always be present on the DNA; if both structures are lost, then activity states must be reestablished de novo. According to this model, protein/DNA complexes that are involved in self-templating activity must be retained on metaphase chromosomes; thus, the study of such complexes on metaphase chromosomes has become a new approach to the study of epigenetic mechanisms (Hershkovitz and Riggs 1995).

CONCLUSIONS

Epigenetic regulation of gene expression results from differential alterations in chromatin structure rather than changes within the DNA sequence itself. In this overview, we have reviewed evidence that some epigenetic phenomena can result from relatively simple feedback circuits. The components of these circuits include *trans*-acting, diffusible intracellular components (proteins or RNA), whose interaction deter-

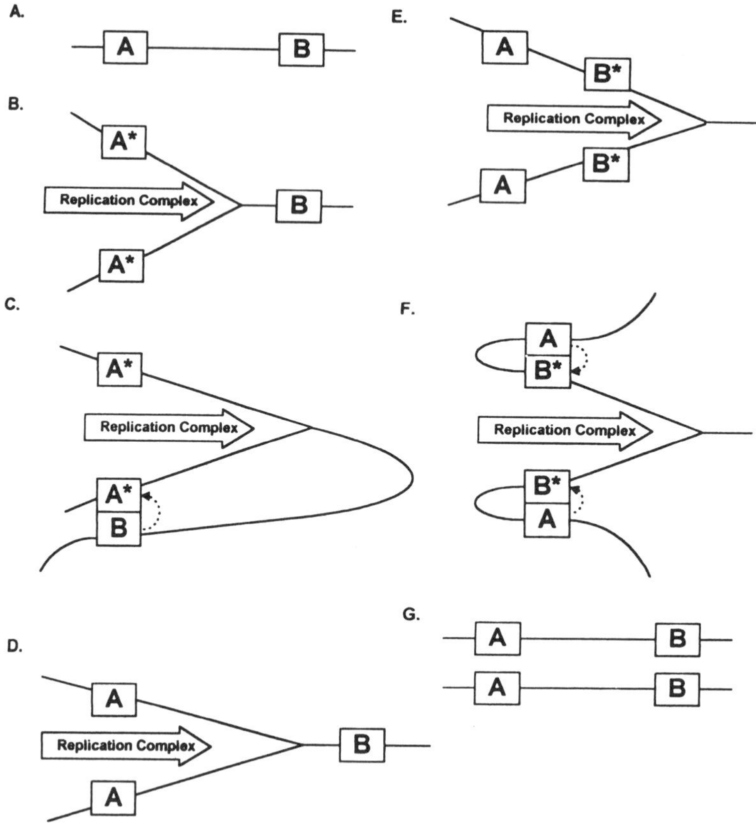

Figure 5 (*See facing page for legend.*)

mines whether the gene is somatically inherited as active or inactive. In addition to circuits of *trans*-acting factors, it seems clear that for mammals and plants, cytosine methylation supports the differential treatment of otherwise sequence-identical DNA. However, despite the solid evidence that cytosine methylation is one epigenetic system essential for mammalian development and the maintenance of genomic imprinting, there is no reason to think that it is sufficient for these duties. This is especially so because *Drosophila* depends on cell memory for development, and yeast shows chromatin-state heritability for mating type; yet neither of these organisms has any detectable 5-methylcytosine.

Epigenetic mechanisms for genomic imprinting and X-chromosome inactivation must also explain the chromosomal mark carried by gametes that allows gene function to be adjusted according to parental source. Epigenetic states of plant elements and transgenes also are sometimes

Figure 5 Inheritance of chromatin structure. This illustration is based on a model proposed to explain the inheritance of alternate chromatin states by interaction between distantly separated multipartite elements (Wolffe 1994; Kass and Wolffe, this volume). In this model, box A represents a gene-proximal regulatory element (e.g., a promoter) in a fully mature state (which can be either active or inactive). Box B is a downstream element (with respect to the direction of movement of the replication complex) that may comprise one or more enhancer elements, matrix attachment sites, replication origins, or locus control regions. The passage of the replication complex distributes the modification-state information of nucleosomes and associated non-histone proteins randomly between the parent and daughter strands. The boxes marked with an asterisk represent newly formed "immature" regions, which need not be active or inactive, just immature, until newly synthesized nucleosomes and associated proteins are integrated into the nucleoprotein complex. Panel *A* represents the state of chromatin organization prior to passage of the replication fork. In *B*, the fork has moved through the A region, disordering the chromatin organization within this region. Note that element B is unperturbed. In *C*, element B has looped back and associated with A*. The dotted line indicates that information in the form of structural alteration(s), or covalent modification such as acetylation, phosphorylation, or ubiquinylation is passed from the intact B element to A*, converting A* to the same A configuration it had before replication. It may be that the association of these two elements is sufficient to recruit ancillary factors that lead to the conversion of A* to A (*D*). In *E*, the replication complex has passed through the region containing the B element, disorganizing the information within this complex. *F* illustrates element A looping back to associate with B*. The dotted arrows indicate that information is passed from the now intact A to B*, transforming B* to B. This looping-back procedure should enable a chromatin region to be replicated with inheritance by the progeny chromatin fibers of the activity-state information of the parental chromatin fiber (*G*).

maintained through meiosis. DNA methylation is obviously one mark that can be passed by the gametes to the zygote. Replication and nuclear compartmentalization seem rather unlikely candidates for explaining the mechanism(s) involved in the meiotic inheritance of alternative chromatin states. Is heritable chromatin structure another mechanism for meiotic transfer? Conceivably it could be, but no experimental evidence for this has yet been obtained.

ACKNOWLEDGMENTS

This work was supported by a National Institutes of Health grant (GM-50575) to A.D.R.

REFERENCES

Ascoli, C.A. and G.G. Maul. 1991. Identification of a novel nuclear domain. *J. Cell Biol.* **112:** 785–795.

Beard, C., E. Li, and R. Jaenisch. 1995. Loss of methylation activates *Xist* in somatic but not in embryonic cells. *Genes Dev.* **9:** 2325–2334.

Bell, L.R., J.L. Horabin, P. Schedl, and T.W. Cline. 1991. Positive autoregulation of sex-lethal by alternative splicing maintains the female determined state in *Drosophila*. *Cell* **65:** 229–239.

Bouvet, P., S. Dimitrov, and A.P. Wolffe. 1994. Specific regulation of *Xenopus* chromosomal 5S gene transcription in vivo by histone H1. *Genes Dev.* **8:** 1147–1159.

Comings, D.E. 1966. The inactive X chromosome. *Lancet* **2:** 1137–1138.

Cremer, T., A. Kurz, R. Zirbel, S. Dietzel, B. Rinke, E. Schrock, M.R. Speicher, U. Mathieu, A. Jauch, P. Emmerich, H. Scherthan, T. Ried, C. Cremer, and P. Lichter. 1993. Role of chromosome territories in the functional compartmentalization of the cell nucleus. *Cold Spring Harbor Symp. Quant. Biol.* **58:** 777–792.

Ehrlich, M. and K.C. Ehrlich. 1993. Effect of DNA methylation on the binding of vertebrate and plant proteins to DNA. In *DNA methylation: Molecular biology and biological significance* (ed. J.P. Jost and H.P. Saluz), pp. 145–168. Birkhauser Verlag, Berlin.

Gartler, S.M. and A.D. Riggs. 1983. Mammalian X-chromosome inactivation. *Annu. Rev. Genet.* **17:** 155–190.

Gottesfeld, J. and L.S. Bloomer. 1982. Assembly of transcriptionally active 5S RNA gene chromatin in vitro. *Cell* **28:** 781–791.

Grunstein, M., A. Hecht, G. Fisher-Adams, J. Wan, R.K. Mann, S. Strahl-Bolsinger, T. Laroche, and S. Gasser. 1995. The regulation of euchromatin and heterochromatin by histones in yeast. *J. Cell Sci. Suppl.* **19:** 29–36.

He, D., C. Zeng, and B.R. Brinkley. 1995. Nuclear matrix proteins as structural and functional components of the mitotic apparatus. *Int. Rev. Cytol.* **162B:** 1–74.

Hershkovitz, M. and A.D. Riggs. 1995. Metaphase chromosome analysis by ligation-mediated PCR: Heritable chromatin structure and a comparison of active and inactive X chromosomes. *Proc. Natl. Acad. Sci.* **92:** 2379–2383.

Holmquist, G.P. 1987. Role of replication time in the control of tissue-specific gene expression. *Am. J. Hum. Genet.* **40:** 151–173.

Jablonka, E. and M.J. Lamb. 1995. *Epigenetic inheritance and evolution: The Lamarckian dimension*, pp. 81–82. Oxford University Press, England.

Jane, S.M., D.L. Gumucio, P.A. Ney, J.M. Cunningham, and A.W. Nienhuis. 1993. Methylation-enhanced binding of Sp1 to the stage selector element of the human gamma-globin gene promoter may regulate developmental specificity of expression. *Mol. Cell. Biol.* **13:** 3272–3281.

Jobe, A. and S. Bourgeois. 1972. Lac repressor-operator interaction. VI. The natural inducer of the lac operon. *J. Mol. Biol.* **69:** 397–408.

Joel, P., W. Shao, and K. Pratt. 1993. A nuclear protein with enhanced binding to methylated Sp1 sites in the AIDS virus promoter. *Nucleic Acids Res.* **21:** 5786–5793.

Jost, J.P. and H.P. Saluz, eds. 1993. *DNA methylation: Molecular biology and biological significance*. Birkhauser Verlag, Boston.

Laird, P.W. and R. Jaenisch. 1996. The role of DNA methylation in cancer genetics and epigenetics. *Annu. Rev. Genet.* **30:** (in press).

Li, E.B.C. and R. Jaenisch. 1993. Role of DNA methylation in genomic imprinting. *Nature* **366:** 362–365.

Novick, A. and M. Weiner. 1957. Enzyme induction as an all-or-none phenomenon. *Proc. Natl. Acad. Sci.* **43:** 553–566.

Pfeifer, G.P., S.D. Steigerwald, R.S. Hansen, S.M. Gartler, and A.D. Riggs. 1990. Polymerase chain reaction-aided genomic sequencing of an X chromosome-linked CpG island: Methylation patterns suggest clonal inheritance, CpG site autonomy, and an explanation of activity state stability. *Proc. Natl. Acad. Sci.* **87:** 8252–8256.

Pillus, L. and M. Grunstein. 1995. Chromatin structure and epigenetic regulation in yeast. In *Chromatin structure and gene expression* (ed. C.R. Elgin), pp. 123–146. Oxford University Press, New York.

Ptashne, M. 1992. *A genetic switch.* Blackwell Scientific, Cambridge.

Riggs, A.D. and G.P. Pfeifer. 1992. X chromosome inactivation and cell memory. *Trends Genet.* **8:** 169–174.

Saunders, W.S., C.A. Cooke, and W.C. Earnshaw. 1991. Compartmentalization within the nucleus: Discovery of a novel subnuclear region. *J. Cell Biol.* **115:** 919–931.

Schlossher, J., H. Eggert, R. Paro, and S. Cremer. 1994. Gene inactivation in *Drosophila* mediated by the polycomb gene product or by position-effect variegation does not involve major changes in the chromatin fibre. *Mol. Gen. Genet.* **243:** 453–462.

Tate, P.H. and A. Bird. 1993. Effects of DNA methylation on DNA-binding proteins and gene expression. *Curr. Opin. Genet. Dev.* **3:** 226–231.

Wolffe, A.P. 1994. Inheritance of chromatin states. *Dev. Genet.* **15:** 463–470.

Woodcock, C.L. 1994. Chromatin fibers observed in situ in frozen hydrated sections. *J. Cell Biol.* **125:** 11–19.

Woodcock, C.L., S.A. Grigoryev, R.A. Horowitz, and N. Whitaker. 1993. A chromatin folding model that incorporates linker variability generates fibers resembling the native structures. *Proc. Natl. Acad. Sci.* **90:** 9021–9025.

Wormington, J.L., I.C.A. Taylor, and D.D. Brown. 1982. Developmental regulation of *Xenopus* 5S RNA genes. *Cold Spring Harbor Symp. Quant. Biol.* **47:** 879–884.

Gene Silencing in Mammalian Cells

Robin Holliday, Thu Ho,
and Richard Paulin
CSIRO Division of Biomolecular Engineering
Sydney Laboratory
NSW 2113, Sydney, Australia

In the early genetic studies of mammalian cells in culture there was ongoing controversy about the sources of heritable variability (for review, see Holliday 1991). On the one hand, it was believed that mammalian cells could be handled essentially in the same way as microorganisms; that is, by the isolation and study of gene mutations. On the other hand, it was maintained that these cells were derived from tissues containing specialized genes, most of which were not transcribed, as well as standard housekeeping genes. Changes in the activity of specialized genes in cells in culture would be expected to have an epigenetic basis. This controversy was apparently ended by two definitive reviews (De Mars 1974; Siminovitch 1976), which established beyond doubt that classic mutations could be induced in cultured mammalian cells and that these had the inherited stability expected of such mutations. There was still a problem, because established cell lines had at least a diploid number of chromosomes, and often a higher number. It would be expected that autosomal recessive mutations would be expressed only rarely, and yet when particular mutants were sought they were often easily obtained. To resolve this problem, at least for the Chinese hamster ovary (CHO) cell line, Siminovitch suggested that these cells were "functionally hemizygous," that they were haploid for a considerable part of their genome, owing to chromosome rearrangements, which, in effect, produced monosomy either for whole chromosomes or for parts of chromosomes.

The situation changed when it was discovered that a silent metallothionein gene in mouse lymphoid cells could be reactivated by the demethylating agent 5-azacytidine (5AC) (Compere and Palmiter, 1981), and a stable thymidine kinase-deficient (TK^-) CHO strain could also be reactivated by 5AC (Harris 1982). CHO cells are proline auxotrophs, and Harris (1984a) showed that these were easily reactivated to prototropy by 5AC. In addition, asparagine and glutamine auxotrophs of the V79 or CHO cell lines were revertible by the same procedure (Harris 1984b, 1986). It was also clear from these experiments that the cells without gene activity were extremely stable and that the reactivated form of the gene was also stably inherited.

It became evident that classic mutants, induced by a standard mutagen such as ethyl methane sulfonate (EMS), were often revertible by 5AC. For example, a series of X-ray-sensitive (*xrs*) mutants were isolated after EMS treatment at a frequency of about 10^{-3} by Jeggo and Kemp (1983), and these were found to be reverted by 5AC (Jeggo and Holliday 1986). This made no sense in terms of functional hemizygosity based on chromosome rearrangement.

The explanation came from studies in which the DNA methylation status of the gene in question was examined by methyl-sensitive restrictive enzymes (with or without a non-methyl-sensitive isoschizomer). It was evident in several cases that when 5AC strongly reactivated a given null phenotype, a methylated silent copy of the gene was present (for review, see Holliday 1987). It thus became apparent that "functional hemizygosity" was probably due to one of a pair of autosomal genes becoming silenced by spontaneous de novo methylation. The other gene could be mutated by EMS to give a desired phenotype, but 5AC reactivated the silent gene to produce a nonmutant phenotype (Jeggo and Holliday 1986). On the basis of further studies with TK^- strains, it was suggested that *allelic exclusion* was a better term than functional hemizygosity to describe those cases where one of a pair of autosomal genes was silenced by DNA methylation and the other remained active (Holliday and Ho 1990). The term *epimutation* was also introduced to describe heritable changes in cells which are not due to changes in base sequence of a gene (Holliday 1984).

This interpretation of genetic and epigenetic events in cultured somatic cells was convincingly confirmed by molecular studies which showed that CpG islands in mammalian cell lines were often de novo methylated (Antequera et al. 1990). It is very probable that such cells sometimes have one of two autosomal genes silenced, sometimes both (as in the case of pro^- auxotrophy and cadmium sensitivity, due to silencing of metallothionein genes) and sometimes neither (as in the case of the adenine phosphoribosyl transferase [APRT] gene; see below). The extent of gene silencing presumably depends on both de novo DNA methylation and cell selection. Thus, CHO cells have silenced genes for proline synthesis because the amino acid is provided in standard media, and these cells grow slightly faster if the enzymes are not synthesized (Holliday and Ho 1990). Finally, it is clear that de novo methylation is one feature of the intrinsic genetic instability of mammalian cell lines, which are transformed cancer cells. There is no evidence that it occurs in primary diploid cells. In fact, it is not expected that it should occur either in vivo or in vitro, since normal differentiated somatic cells have strictly conserved patterns of DNA methylation (Kochanek et al. 1990; Behn-Krappa et al. 1991).

GENE SILENCING BY UPTAKE OF 5-METHYL DEOXYCYTIDINE TRIPHOSPHATE

A new technique was introduced by Nyce (1991) and Holliday and Ho (1991), in which cells are permeabilized by electroporation in the presence of 5medCTP. This experimental approach is the in vivo counterpart of the method used by Busslinger et al. (1983) to methylate DNA in vitro. 5medCTP is never synthesized by cells, for good reasons, but the replication machinery can incorporate it into DNA (Nyce 1991). Such incorporation is random and produces hemimethylated DNA. The maintenance methylase therefore methylates the parental strands at CpG sites, and probably CpNpG sites (see Clark et al. 1995). By exploiting selective procedures, it was shown that two autosomal genes, TK$^+$ and APRT$^+$, and one X-linked gene, HPRT$^+$, could be effectively silenced at a frequency of up to 100-fold above the spontaneous background. These isolates are very stable, but can in almost all cases be reactivated by 5AC. (Note that spontaneous loss of enzyme activity can be due either to mutation or to gene silencing.)

This new procedure makes it possible to manipulate gene activity by methylation: 5medCTP imposes methylation and 5AC removes it. This enables a fuller analysis of both genetic and epigenetic inheritance in mammalian cells, which we refer to as dual inheritance (a term introduced by Maynard-Smith [1990]). Such studies should be done in conjunction with the bisulfite genomic sequencing procedure, which distinguishes cytosine and 5-methyl cytosine in the region of interest (Frommer et al. 1992; Grigg and Clark 1994; Clark et al. 1995). We have chosen the APRT gene for this investigation, because it is possible to select cells lacking enzyme activity by plating in medium containing 2, 6-diamino purine (DAP) and cells with enzyme activity in medium containing adenine, aminopterin, and thymidine (AAT). In addition, the gene is relatively small and there have been extensive molecular studies of gene mutations (see Meuth 1990 and references therein). We have used the CHO K1 strain which has two APRT$^+$ genes and a derivative D422, which has only one active copy, because the other is deleted.

Early studies on mutation showed that cells partially resistant either to DAP or to 8-aza adenine (AA) had 50% of enzyme activity, and a further mutation conferred full resistance to DAP and AA with no enzyme activity (Chasin 1974; Jones and Sargent 1974). Starting with CHO K1 with 100% APRT enzyme activity, we have obtained derivatives by the 5medCTP gene-silencing technique that have no detectable enzyme activity and presumably have two silenced genes. Such a strain can be reactivated to APRT$^+$ by 5AC, regaining 50% enzyme activity, or less if reactivation is incomplete. By the use of 5medCTP, the mutagen EMS, and 5AC, we have been able to follow the epigenetic and genetic events shown in Figure 1. This analysis illustrates dual inheritance at the APRT locus.

In parallel with these studies, we have used the bisulfite genomic sequencing procedure to examine the part of the CpG island at the 5′ end of the gene, which includes the promoter region and transcription start site. It contains 17 CpG doublets, and in the CHO K1 "wild type" strain, all these are unmethylated, as well, of course, as all other cytosines. In APRT$^-$ derivatives obtained by the 5medCTP silencing procedure, variable numbers of these 17 CpG doublets are methylated. We have not yet identified specific methylated sites necessary for gene silencing, but neither can we exclude them. This analysis is currently in progress and will be published in detail elsewhere, together with the results summarized in Figure 1 (R. Paulin et al., in prep.).

Figure 1 Dual inheritance at the APRT locus in CHO K1 cells. The starting "wild type" cell line (*top*) has two active copies of the gene. Filled circles indicate mutations induced by standard mutagens, and filled rectangles the silencing of the gene by methylation of CpG doublets in the promoter region. The events in the pathways are as follows: (*A*) mutation produces a derivative with 50% enzyme activity. (*B*) A second mutation produces a strain fully resistant to DAP and AA, with no enzyme activity (for an analysis of events *A* and *B*, see Chasin 1974; Jones and Sargent 1974). (*C*) Electroporation in the presence of 5medCTP produced APRT$^-$ derivatives with no enzyme activity in a single step (see text and Holliday and Ho 1991). Note that both APRT genes are likely to be replicating at the same time in the cell cycle and both will be targets for silencing (but see step *E*). (*D*) treatment with 5AC reactivates one copy to give an APRT$^+$ derivative, typically with 50% enzyme activity. (*E*) This strain can be fully silenced by electroporation and 5medCTP treatment, at a much higher frequency than step *C*. (*F*) Treatment of cells with EMS produces an APRT$^-$ derivative with one silent and one mutant gene. (*G*) Treatment with 5AC again produces an APRT$^+$ derivative, comparable to that produced by step *A*. (*H*) Further mutagenic treatment (EMS) finally yields a stable APRT$^-$ strain, which is nonreactivatable by 5AC. Steps *B* and *H* are the same. The small arrows (*right*) indicate noncomplementing mutations at any position in the gene.

GENE SILENCING IN NORMAL HUMAN DIPLOID CELLS

So far, we have discussed genetic and epigenetic events in transformed cell lines. It is important to know whether the 5medCTP technique is also applicable to normal diploid cells. To investigate this, we have chosen to study the haploid human HPRT (hypoxanthine phosphoribosyl transferase) gene in male primary diploid fibroblasts, strain MRC-5. Human fibroblasts, whether male or female, are known to have a very low spontaneous frequency of $HPRT^-$ mutations (De Mars 1974). We treated

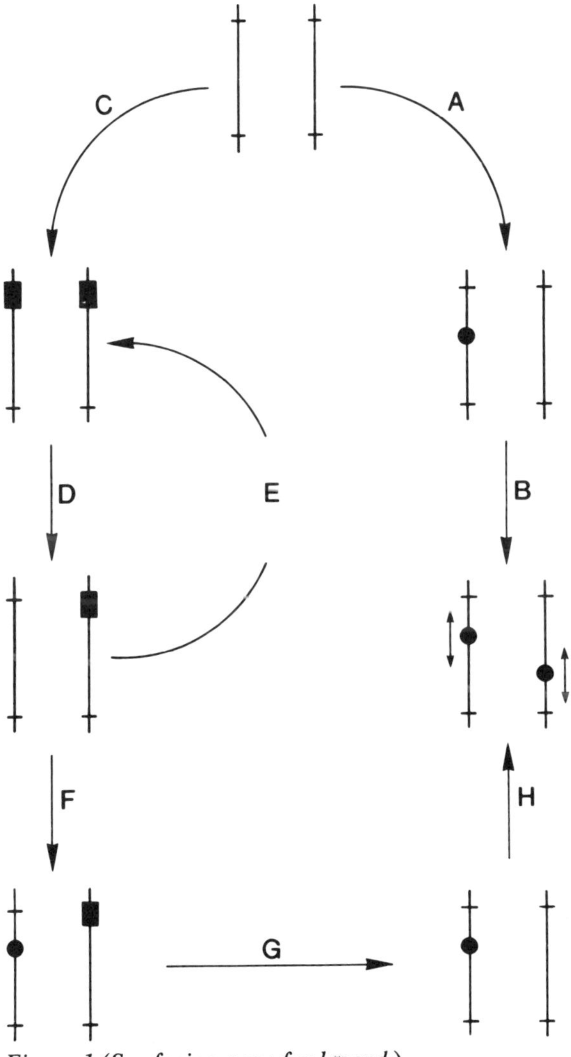

Figure 1 (*See facing page for legend.*)

MRC-5 cells by depriving a growing culture of fetal calf serum for 24 hours and then replacing it. This induces a degree of synchrony in the cells with maximum DNA synthesis 19–20 hours after serum is replaced (F. Watt, pers. comm.). At this time cells were electroporated in the presence of 1.2 mM 5medCTP, which is a higher concentration than that used for CHO cells. We have shown that the frequency of 6-thioguanine resistant (TG^R, $HPRT^-$) colonies is increased at least 20-fold over the controls (Holliday and Ho 1995). Reactivation by 5AC is much more difficult to study in these cells, since they have a finite life span. By the time TG^R colonies have grown and the cells have been picked, treated with 5AC, allowed subsequent expression time, and then plated in HAT (medium containing hypoxanthine, aminopterin, and thymidine which selects for $HPRT^+$), many have become senescent. Nevertheless, we have been able to demonstrate growth in HAT in a proportion of clones treated with 5AC.

These initial results are important because they demonstrate for the first time that the activity of genes in a normal somatic cell can be manipulated by techniques that alter DNA methylation. All previous results have been with established cell lines, which are either partially or fully transformed.

SPECIFIC GENE SILENCING BY METHYLATED OLIGONUCLEOTIDES?

The incorporation of 5medCTP into nascent DNA strands tricks the cells' DNA methyltransferase into methylating previously unmethylated DNA, that is, the parental DNA template. A more specific methylation "imprint" might be achieved by introduction of a methylated single-stranded oligonucleotide. This could anneal to single-stranded DNA at the replication fork, or possibly to single-stranded DNA during transcription. A short hemimethylated region would be formed and the DNA methyltransferase would respond by methylating the parental strand, even if the annealing of the oligonucleotide is transient (Fig. 2).

We have again chosen the CHO APRT gene as a target for methylated oligonucleotide and have used strain D422, which has only one copy of the gene (Bradley and Letovanec 1982; Nalbantoglu et al. 1983). At present we do not know which CpG sites need to be methylated to prevent gene expression, so a number of oligonucleotides containing 5-methylcytosine at CpG doublets have been synthesized, which are complementary to the promoter region. "Phosphate on" is used to phosphorylate the 5′ end of the oligonucleotide, because it is possible that ligation of the oligonucleotides to a normal Okazaki fragment is important for

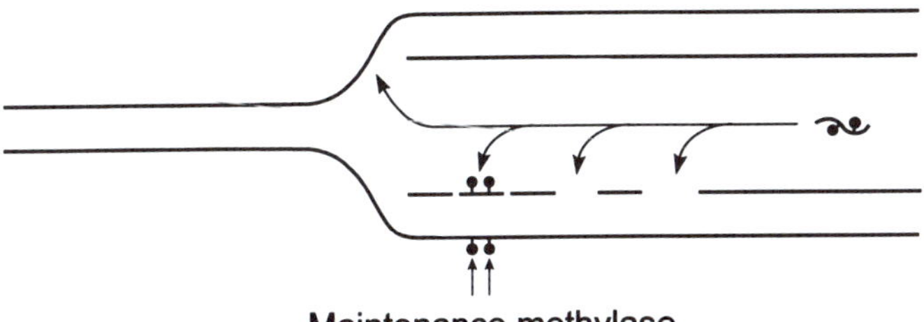

Figure 2 The possible annealing of a methylated single-stranded oligonucleotide to complementary regions of single-stranded chromosomal DNA at the replication fork. Annealing or hybridization could be either to sense or antisense strands, in advance of or between Okazaki fragments. The hemimethylated fragment of DNA may be a substrate for the DNA maintenance methylase, which imposes a heritable "imprint" on parental DNA and thereby silences the target gene. Alternatively, the oligonucleotide might hybridize to single-stranded DNA during transcription of the target gene.

achieving gene silencing. Nonmethylated oligonucleotides are used as controls. Another control is to plate for TG^R ($HPRT^-$) colonies, as this gene should not be targeted by the oligonucleotides. As well as short synthesized oligonucleotides, we have used the PCR product of the promoter region of the APRT gene, which consists of 372 base pairs and 17 CpG sites. These sites are methylated with the SSS1 methylase and the DNA is melted to single-stranded fragments that are used in transfection experiments, with unmethylated DNA as the control. We have also stalled the replication fork with a DNA inhibitor that might facilitate the annealing of the oligonucleotide fragment to the target sequence. Such inhibitors appear to promote de novo methylation at the replication fork (Nyce 1989), but aphidicolin is one of the least active in this respect (J. Nyce, pers. comm.). A number of experimental procedures using aphidicolin are being investigated to optimize silencing of the APRT gene.

In yeast, single-stranded oligonucleotides have been used successfully to transform recipients (Yamamoto et al. 1992 a,b). In this case, genetic changes in the *cyc* gene are selected. It was found that the sense oligonucleotide was about 20 times more effective in transformation than the antisense. In our experiments, we have not documented differences between oligonucleotides with complementary sequence.

In summary, the methylated oligonucleotide experiments carried out with CHO strain D422 are promising, but incomplete. Two advances are necessary. First, to synthesize the appropriate methylated oligonucleotide, it is essential to know whether certain CpG sites in the promoter region must be methylated to silence the APRT gene. This information is not yet available. Second, it is necessary to optimize the conditions of treatment, perhaps by using synchronized cells or different regimes for inhibiting the replication fork.

THE STABILITY OF DNA METHYLATION

All the experimental approaches we have used depend on the stability of the inherited pattern of DNA methylation. If variability in methylation was at all common, then it would be very difficult to obtain the clear phenotypes on which epigenetic analysis depends. The CHO experimental system makes it possible to measure the frequency of de novo methylation of a gene, or the loss of methylation. The best estimate for de novo methylation comes from studies of the TK gene. In strain CHO K1, we have shown in two fluctuation tests that TK^- colonies arise at a frequency of 6×10^{-5} (Holliday and Ho 1990). We also obtained evidence that there is only one active copy of the TK gene in these cells and that the other is already silenced by de novo methylation. The other CHO strain, D422, does not behave in the same way, as spontaneous TK^- cells are much less common. We presume, but have no direct evidence, that this strain has two active TK^+ genes.

Several selective systems can be used to measure the frequency of loss of DNA methylation in CHO cells (see Holliday 1993). These cells are proline auxotrophs (Harris 1984a) and they are also sensitive to cadmium because they lack metallothionein activity. In both cases, one can select directly for gene reactivation. The spontaneous frequencies are very low (10^{-6} or less), which suggests that spontaneous loss of DNA methylation is an infrequent event. This may well mean that loss of a single methylated cytosine is insufficient for reactivation and that several must be lost at the same time. 5AC presumably achieves this effect, since the reactivation frequency can be as high as 10%. In the case of a silent gene, there may be a balance between spontaneous loss of methylation and also de novo methylation. The upshot is that the silent gene is effectively locked into a stable phenotype. In the case of a reactivated gene, silencing may be reimposed if the gene is dispensable and cells grow slightly faster in the absence of transcription and translation. This was shown experimentally using a CHO strain reactivated for cadmium resistance and for proline independence (Cd^R, pro^+) by 5AC. The cells were grown in normal medium for nearly 150 population doublings (PDs). At the end of that time, 100% of cells tested (101 clones) were

Cd^S and pro^- (Holliday and Ho 1990). This provides evidence for gene silencing by de novo methylation and selection. In the case of a 5AC TK^+ induced from a TK^- strain, the TK^+ phenotype was retained over the same number of PDs (99/101 clones tested), presumably because the TK^+ gene is advantageous to the cells.

The genes silenced by electroporation and 5medCTP can also be examined for stability, since the back selection of TK^+, $APRT^+$, and $HPRT^+$ isolates in AAT or HAT medium is possible. The data for TK suggest that the imposed methylation is lost spontaneously at a frequency of about 3×10^{-6} (Holliday 1993). The results from many control experiments in which $HPRT^-$ and $APRT^-$ cells are plated on HAT or AAT medium, respectively, show that the spontaneous reactivation frequency is about 10^{-6}. According to the reasoning just outlined, this may mean that the artificial silencing of these genes is due to the methylation of several CpG doublets, and the likelihood of losing all or most of these in a single step is very low.

CONCLUSIONS AND FUTURE PROSPECTS

The experiments carried out with mammalian cells provide strong evidence that DNA methylation is sufficient to inactivate a gene and that demethylation is sufficient to activate it. These conclusions can be drawn in the absence of biochemical evidence for proteins, which might bind differentially to methylated or nonmethylated promoter sequences. The methylation may bring about binding of a non-sequence-specific protein, which simply prevents access of the transcription machinery. Alternatively, there may be sequence-specific proteins that distinguish between methylated and nonmethylated DNA (for review, see Ehrlich and Ehrlich 1993). In either case, it is justifiable to conclude that methylation causes gene silencing (see Holliday this volume). The experiments also provide further evidence for the stable transmission of DNA methylation. Of course, changes in the pattern of methylation must sometimes occur, but switches of gene activity during normal growth are rare (see above and Holliday 1993).

Gene silencing by 5medCTP is nonspecific, and presumably if the imposed methylation was too extensive, the effect would be lethal. We chose conditions that allow 30–70% survival, and among those survivors, any one gene (such as HPRT) is silenced at a frequency of up to 5×10^{-4}. If there are 20,000 active genes in a cell, then about 10 could be methylated, but not all these need be silenced. Moreover, essential autosomal genes may often be active on both homologous chromosomes, and the probability of both being silenced would be lower. The use of methylated oligonucleotides provides specificity, if there is only one gene target per cell, and that gene is dispensable. One might therefore expect many fewer lethal events, and this would allow, at least in principle, a much higher frequency of gene silencing.

The silencing of a normal housekeeping gene (HPRT) in a primary diploid cell opens up the possibility of silencing other genes of interest, provided some selection procedure is possible. A surprisingly large number of inherited human defects are due to dominant mutations, which implies that the gene product itself is deleterious. The much discussed procedures for gene therapy most frequently relate to the replacement of a mutant gene by a normal gene. In the case of dominant inheritance, a normal gene is already present and it is necessary either to remove the dominant mutation or to silence it. There may be situations where this would become possible in the future. Another important approach would be the removal of histocompatibility surface antigens by silencing the HLA genes. Specific antibodies can be used to select cells lacking such antigens (Turner et al. 1988). By silencing and selection it might be possible to obtain cells for transplantation which are so-called "universal donors."

The field of cancer provides many opportunities for the further study of DNA methylation and gene silencing. At present, enormous effort is put into the sequencing of tumor suppressor genes, such as p53 and Rb, from various cancers. Mutations are commonly detected, but a substantial proportion of tumors lack such mutations. It is possible that some of these have methylated promoters with inactive genes, as has recently been shown for the p16 tumor suppressor gene (Merlo et al. 1995). Another area of great importance is the genomic instability of tumor cells. It is well established that normal cells have a highly conserved pattern of DNA methylation, whereas tumor cells have very variable methylation (see Achten et al. 1991; Behn-Krappa et al. 1991; Kochanek et al. 1990, 1991; Spruck et al. 1993). This provides a source of epigenetic variation, which is probably an essential component for the cell selection that must occur during tumor progression. Very little is known about the silencing of normal genes by DNA methylation in neoplastic cells, or the ectopic expression of genes by demethylation, and the relationship of such events to tumorigenicity. (In this context, ectopic refers to gene expression that is different from the cell in which carcinogenesis is initiated.) The discovery of many oncogenes, tumor suppressor genes, or other genes that regulate the cell cycle opens the way for a concerted investigation of the effects of DNA methylation on such genes and on the cell phenotype. The use of CHO and other cell lines provides real insight into the ways that methylation can change these gene activities in partially or fully transformed cells in vivo.

ACKNOWLEDGMENTS

We thank Mark Meuth for providing CHO strain D422, Fuji Watt for information about synchronizing MRC-5 cells, and Jonathan Nyce for suggesting the use of aphidicolin.

REFERENCES

Achten, S., A. Behn-Krappa, M. Jucker, J. Sprengel, I. Holker, B. Schmitz, H. Tesch, V. Diehl, and W. Doerfler. 1991. Patterns of DNA methylation in selected human genes in different Hodgkin's lymphoma and leukemia cell lines and in normal human lymphocytes. *Cancer Res.* **51:** 3702–3709.

Antequera, F., J. Boyes, and A. Bird. 1990. High levels of *de novo* methylation and altered chromatin structure at CpG islands in cell lines. *Cell* **62:** 503–514.

Behn-Krappa, A., I. Holker, U. Sandaradura de Silva, and W. Doerfler. 1991. Patterns of DNA methylation are indistinguishable in different individuals over a wide range of human DNA sequences. *Genomics* **11:** 1–7.

Bradley, W.E.C. and D. Letovanec. 1982. High-frequency non-random mutational event at the adenine phosphoribosyl transferase (*aprt*) locus of sib-selected CHO variants heterozygous for aprt. *Somatic Cell Mol. Genet.* **8:** 51–60.

Busslinger, M., J. Hurst, and R.A. Flavell. 1983. DNA methylation and the regulation of globin gene expression. *Cell* **34:** 197–206.

Chasin, L.A. 1974. Mutations affecting adenine phosphoribosyl transferase in Chinese hamster cells. *Cell* **2:** 37–41.

Clark, S.J., J. Harrison, and M. Frommer. 1995. CpNpG methylation in mammalian cells. *Nat. Genet.* **10:** 20–27.

Compere, S.J. and R.D. Palmiter. 1981. DNA methylation controls the inducibility of the mouse metallothionein I gene in lymphoid cells. *Cell* **25:** 233–240.

DeMars, R. 1974. Resistance of cultured human fibroblasts and other cells to purine and pyrimidine analogues in relation to mutagenesis detection. *Mutat. Res.* **24:** 335–364.

Ehrlich, M. and K.C. Ehrlich. 1993. Effects of DNA methylation on the binding of vertebrate and plant proteins to DNA. In *DNA methylation: Molecular biology and biological significance* (ed. J.P. Jost and H.P. Saluz), pp. 145–168. Birkhauser, Basel.

Frommer, M., L.E. McDonald, D.S. Millar, C.M. Collis, F. Watt, G. Grigg, P.L. Molloy, and C.L. Paul. 1992. A genomic sequencing protocol which yields a positive display of 5-methylcytosine residues in individual DNA strands. *Proc. Natl. Acad. Sci.* **89:** 1827–1831.

Grigg, G. and S. Clark. 1994. Sequencing 5-methyl cytosine residues in genomic DNA. *BioEssays* **16:** 431–436.

Harris, M. 1982. Induction of thymidine kinase in enzyme deficient Chinese hamster cells. *Cell* **29:** 483–492.

———. 1984a. High frequency induction by 5-aza-cytidine of proline independence in

CHO-K1 cells. *Somatic Cell Mol. Genet.* **10:** 614–624.

———. 1984b. Variants inducible for glutamine synthetase in CHO K1 and V79 cells. *Somatic Cell Mol. Genet.* **10:** 275–281.

Harris, M. 1986. Induction and reversion of asparagine auxotrophs in CHO K1 and V79 cells. *Somatic Cell Mol. Genet.* **12:** 459–466.

Holliday, R. 1984. The significance of DNA methylation in cellular aging. In *Molecular biology of aging* (ed. A.D. Woodhead et al.), pp. 269–283. Plenum Press, New York.

———. 1987. Inheritance of epigenetic defects. *Science* **238:** 163–170.

———. 1991. Mutations and epimutations in mammalian cells. *Mutat. Res.* **250:** 345–363.

———. 1993. Epigenetic inheritance based on DNA methylation. In *DNA methylation: Molecular biology and biological significance* (ed. J.P. Jost and H.P. Saluz), pp. 452–468. Birkhauser, Basel.

Holliday, R. and T. Ho. 1990. Evidence for allelic exclusion in Chinese hamster ovary cells. *New Biol.* **2:** 719–726.

———. 1991. Gene silencing in mammalian cells by uptake of 5-methyl deoxycytidine 5′ phosphate. *Somatic Cell Mol. Genet.* **17:** 537–542.

———. 1995. Evidence for gene silencing by DNA methylation in normal human diploid fibroblasts. *Somatic Cell Mol. Genet.* **21:** 215–218.

Jeggo, P.A. and R. Holliday. 1986. Azacytidine-induced reversion of a DNA repair gene in Chinese hamster ovary cells. *Mol. Cell. Biol.* **6:** 2944–2949.

Jeggo, P.A. and L.M. Kemp. 1983. X-ray sensitive mutants of Chinese hamster ovary cell line. Isolation and cross sensitivity to other DNA-damaging agents. *Mutat. Res.* **112:** 313–327.

Jones, G.E. and P.A. Sargent. 1974. Mutants of cultured Chinese hamster cells deficient in adenine phosphoribosyl transferase. *Cell* **2:** 43–54.

Kochanek, S., A. Radbruch, H. Tesch, D. Renz, and W. Doerfler. 1991. DNA methylation profiles in the human genes for tumor necrosis factors α and β in subpopulations of leukocytes and in leukemias. *Proc. Natl. Acad. Sci.* **88:** 5759–5763.

Kochanek, S., M. Toth, A. Dehmel, D. Renz, and W. Doerfler. 1990. Interindividual concordance of methylation profiles in human genes for tumor necrosis factors α and β. *Proc. Natl. Acad. Sci.* **87:** 8830–8834.

Maynard Smith, J. 1990. Models of a dual inheritance system. *J. Theor. Biol.* **143:** 41–53.

Merlo, A., J.G. Herman, L. Mao, D.J. Lee, E. Gabrielson, P.C. Burger, S.B. Baylin, and D. Sidransky. 1995. 5′CpG island methylation is associated with transcriptional silencing of the tumour suppressor *p16/CDKN2/MTS1* in human cancers. *Nat. Med.* **1:** 686–692.

Meuth, M. 1990. The structure of mutation in mammalian cells. *Biochim. Biophys. Acta* **1032:** 1–17.

Nalbantoglu, J., O. Goncalves, and M. Meuth. 1983. Structure of mutant alleles at the *aprt* locus of Chinese hamster ovary cells. *J. Mol. Biol.* **167:** 575–594.

Nyce, J. 1989. Drug induced DNA hypermethylation and drug resistance in human tumours. *Cancer Res.* **49:** 5829–5836.

———. 1991. Gene silencing in mammalian cells by direct incorporation of electroporated 5-methyl-2′deoxycytidine 5′-phosphate. *Somatic Cell Mol. Genet.* **17:** 543–550.

Siminovitch, L. 1976. On the nature of heritable variation in cultured mammalian cells. *Cell* **7:** 1–11.

Spruck, C.H., III, W.M. Rideout III, and P.A. Jones. 1993. DNA methylation and cancer. In *DNA methylation: Molecular biology and biological significance* (eds. J.P. Jost and H.P. Saluz), pp. 487–509. Birkhauser, Basel.

Turner, D.R., S.A. Grist, M. Janatipour, and A.A. Morley. 1988. Mutations in human lymphocytes commonly involve gene duplication and resemble those seen in cancer cells. *Proc. Natl. Acad. Sci.* **85:** 3189–3192.

Yamamoto, T., R.P. Moerschell, L.P. Wakem, D. Ferguson, and F. Sherman. 1992a. Parameters affecting the frequencies of transformation and cotransformation with synthetic oligonucleotides in yeast. *Yeast* **8:** 935–948.

Yamamoto, R., R.P. Moerschell, L.P. Wakem, S. Komar-Panicucci, and F. Sherman. 1992b. Strand-specificity in the transformation of yeast with synthetic oligonucleotides. *Genetics* **131:** 811–819.

DNA Methyltransferases in Mammalian Development and Genome Defense

Timothy H. Bestor
Department of Genetics and Development
College of Physicians & Surgeons of Columbia University
New York, New York 10032

The mammalian genome is modified by the addition of about 3×10^7 methyl groups, all at the 5 position of cytosine and most at 5′-CpG-3′ dinucleotides. Methylation patterns are transmitted by clonal inheritance (Pfeifer et al. 1990) and increase the information content of the genome (Reinisch et al. 1995); transcription is repressed when CpG sites within promoters are methylated (Iguchi-Ariga and Schaffner 1989). Cytosine methylation is dangerous: 5-methylcytosine (m^5C) is the major endogenous mutagen (deamination results in C→T transition mutations at CpG sites, which account for about one-third of all mutations in humans; Bestor and Coxon 1993), and tumor suppressor genes are frequently inactivated by ectopic de novo methylation of promoter regions (Herman et al. 1994). However, there must be benefits that yield a net selective advantage. This is shown by the retention of cytosine methylation by virtually all organisms with genomes of more than 5×10^8 bp (Bestor 1990) and by the fact that perturbations of methylation patterns are lethal to mouse embryos and to differentiated cells (Li et al. 1992). Severe developmental abnormalities are seen when m^5C levels are reduced in *Arabidopsis* as the result of mutations at uncharacterized loci (Vongs et al. 1993) or by expression of a DNA methyltransferase antisense construct (Finnegan, this volume). Although methylation patterns clearly play an essential role in large-genome eukaryotes, the nature of that role is not understood. It is argued here that an understanding of the regulation of de novo methylation will reveal the biological function of methylation patterns.

HOW ARE METHYLATION PATTERNS ESTABLISHED?

DNA Methyltransferases

The only known component of the system that establishes and maintains methylation patterns is a single form of DNA cytosine-5 methyltransferase that has been purified and characterized (Bestor and Ingram 1983) and cloned and sequenced (Bestor et al. 1988). This enzyme utilizes the same unusual catalytic mechanism as do bacterial cytosine-5 methyltransferases (all enzymes that modify the 5 position of pyrimidines appear to use this mechanism; for review, see Bestor and Verdine 1994), and the carboxy-terminal domain of mammalian DNA methyltransferase is clearly related in sequence to the bacterial enzymes. Key sequence features and functional domains in DNA methyltransferase are depicted in Figure 1.

All of the conserved motifs characteristic of cytosine-5 methyltransferases are easily identifiable within the carboxy-terminal domain of DNA methyltransferase. The sequence divergence between this domain and the most similar bacterial methyltransferases is less than the divergence within the family of bacterial methyltransferases. The carboxy-terminal domain of the DNA methyltransferases from *Arabidopsis* (Finnegan and Dennis 1993), *Xenopus laevis* (J.A. Yoder and T.H. Bestor, unpubl.), chicken (Tajima et al. 1995), and sea urchin (Aniello et al. 1996) are very similar to those of the mammalian enzymes. Conservation within the amino-terminal domain is less, but sufficient to confirm a common origin of all known eukaryotic DNA methyltransferases (Fig. 2).

Joining the carboxy-terminal methyltransferase domain to the large amino-terminal domain is a glycyl- and lysyl-rich tract that thus far has been observed in all eukaryotic DNA cytosine-5 methyltransferases but only in these enzymes; none of the bacterial enzymes contains this tract

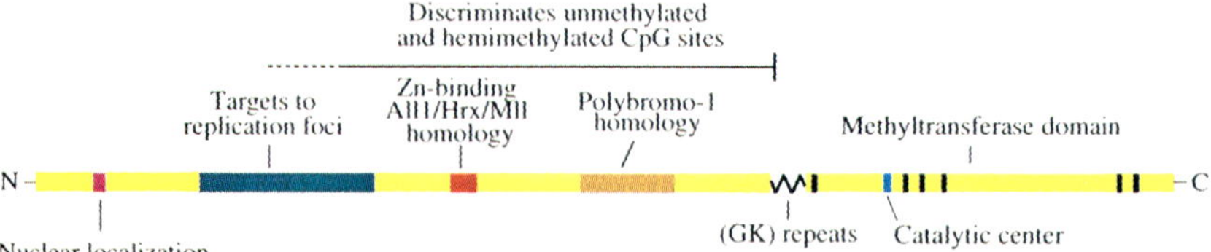

Figure 1 Sequence motifs and functional domains in mammalian DNA methyltransferase. The protein comprises a carboxy-terminal catalytic domain that is closely related to bacterial (cytosine-5)-methyltransferases; this is joined by a lysine/glycine repeat to a large amino-terminal domain that has roles in nuclear import, targeting to replication foci during S phase, and the suppression of de novo methylation. A Zn-binding site is shared with the mammalian homolog of *Drosophila* trithorax (Ma et al. 1993; Bestor 1995), and there is a motif shared with chicken Polybromo-1 (Nicolas and Goodwin 1996). These two proteins are related to *Drosophila* chromosomal proteins that are involved in the clonal propagation of states of gene activity.

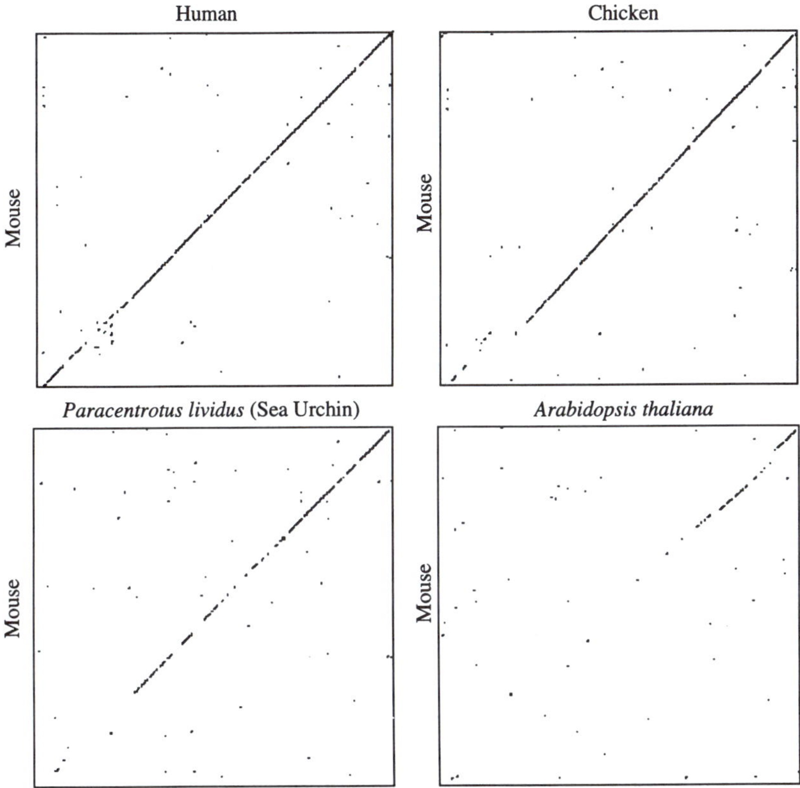

Figure 2 Comparison of eukaryotic DNA methyltransferases. The amino acid sequence of DNA methyltransferase from mouse (Bestor et al. 1988) is compared to that of human (Yen et al. 1992), chicken (Tajima et al. 1995), sea urchin (*Paracentrotus lividus*; GenBank Z50183), and *Arabidopsis thaliana* (Finnegan and Dennis 1993). All are clearly related; similarity decreases with phylogenetic distance. Notice that divergence is least in the carboxy-terminal catalytic domain (upper right in each panel). The Zn-binding site shared with the mammalian trithorax homolog is well-conserved among the animal enzymes but is not discernible in the *Arabidopsis* sequence.

or a domain that might be related to the amino-terminal domain of the eukaryotic enzymes. The similarities in sequence and domain organization suggest that the gene fusion event by which plant and animal DNA methyltransferases arose preceded the divergence of the plant and animal lineages (that is, when the common ancestor was unicellular).

The amino-terminal domain serves several important functions that differentiate the eukaryotic and prokaryotic methyltransferases. The amino-terminal domain contains at least two functional nuclear localization sequences and was the first protein shown to bear a replication focus targeting sequence (Leonhardt et al. 1992); a functionally similar sequence has since been identified in DNA ligase I (Montecucco et al. 1995). The outstanding biochemical difference between the bacterial and mammalian DNA cytosine-5 methyltransferases is the strong preference for hemimethylated substrates displayed by the latter, and controlled proteolysis experiments have shown that the amino-terminal domain is responsible for the suppression of de novo methylation (Bestor 1992). Within this region is an unusual Zn-binding site that is shared with one other protein: the mammalian homolog of *Drosophila* trithorax (trx) that is known as Hrx (Tkachuk et al. 1993) or All-1 (Ma et al. 1993). A second region downstream from the Hrx/All-1 similarity is closely related to a sequence within Polybromo-1 (Nicolas and Goodwin 1996), an avian protein that contains bromo domains, which where first found in Brahma, one of the trithorax group of *Drosophila.* As discussed below, trx and Brahma are chromatin memory proteins of the trithorax group (trxG), and the presence of common structural elements in trxG proteins and DNA methyltransferase formed the first link between the two systems known to be involved in the clonal transmission of states of gene activity in metazoa.

DNA methyltransferases from flowering plants and vertebrates are similar at the sequence level. Although the results of biochemical studies are seldom in agreement, this may be the result of the study of partially purified and often rather severely proteolyzed enzyme preparations. However, with few exceptions, it has been found that the sequence specificity of DNA methyltransferase does not extend past the CpG dinucleotide, and the cues that regulate de novo methylation are unknown. This is in contrast to bacteria, in which genomic methylation patterns are a simple function of the sequence specificity of the methyltransferases present in the cell. It is not known how methylation patterns are established in the genomes of vertebrates and flowering plants. As described in the next section, an understanding of the mechanisms that establish and maintain methylation patterns will also give insights into the biological function of cytosine methylation.

Functions of Genomic Methylation Patterns

There are two salient hypotheses for the role of genomic methylation patterns. The first hypothesis (put forward 20 years ago) suggests that programmed demethylation and de novo methylation play a direct role in gene control during development (Holliday and Pugh 1975; Riggs 1975).

Methylation patterns would be established during gametogenesis or early development, and regulatory factors would mediate the removal of methyl groups from promoters to allow the expression of tissue-specific genes at the appropriate stage of differentiation. Although supported by a large body of indirect and correlative evidence, a definitive test of the causality of cytosine methylation has been elusive, and programmed changes in methylation status have not been proven to be intimately involved in the regulation of any gene. By the same token, a developmental role for cytosine methylation has not been disproved by any course of experiments, and I suggest that cytosine methylation may reinforce the heritability of states of gene expression mediated by chromosomal proteins of the Polycomb and trithorax groups (Bestor and Verdine 1994). This view is supported by the finding of sequence motifs shared by DNA methyltransferase and vertebrate homologs of trxG proteins (Bestor and Verdine 1994), and by the fact that plants that have reduced levels of m^5C at CpG sites show developmental abnormalities and homeotic transformations similar to those that result from mutations in a gene of the *Polycomb* group (Finnegan, this volume). Proteins of the Polycomb and trithorax groups do not establish zones of expression of homeotic genes but are required for their maintenance during early development. The stability of heritable chromatin states could be augmented by a foundation of heritable DNA modification in vertebrates and flowering plants. Some small, short-lived organisms with small genomes such as *Caenorhabditis elegans* and *Drosophila* lack modified bases and can enforce developmental decisions through protein/DNA interactions alone, but as pointed out earlier (Riggs 1989; Bestor 1990), large, long-lived organisms that maintain undifferentiated cell populations for the life of the organism may require much greater stability in transmission of states of gene activity.

The second hypothesis suggests that cytosine methylation is part of a nuclear host defense system that inactivates parasitic sequences such as transposable elements and proviral DNA (Bestor 1990; Bestor and Coxon 1993; Bestor and Tycko 1996). Most such elements are in fact methylated and transcriptionally inert in the genome of mammals, flowering plants, and those fungi whose genomes contain m^5C (Rubin et al. 1994). Treatment of cultured cells or mice with the demethylating drug 5-azacytidine can activate silent retroviruses and endogenous genes that have been silenced by ectopic de novo methylation of regulatory regions (Jaenisch et al. 1985). It is striking that *Drosophila*, whose DNA lacks m^5C, suffers far larger numbers of insertion mutations than do animals whose genomes are methylated (Ashburner 1989). These observations, with support from evolutionary considerations (Bestor 1990), strongly suggest that cytosine methylation is part of a genomic host defense system that suppresses the transcription of parasitic sequence elements. The selective advantage of such a defensive system is obvious, given that a sizable fraction of the genome represents parasitic sequences that are invisible to the immune system and that might inflict intolerable mutational or cytotoxic damage if allowed to proliferate unchecked.

The hypothetical developmental and host defense roles place fundamentally different requirements on the DNA methylating system. The developmental role requires that methylation patterns be established as part of the developmental program via conventional sequence recognition mediated by sequence-specific DNA methyltransferases or specificity factors that interact with the ubiquitous DNA methyltransferase. The host-defense function requires a completely different source of specificity; de novo methylation must be cued by dynamic structural features unique to parasitic sequences in the act of invading the genome, or by events such as transient pairing interactions of repetitive invasive elements. In other words, the developmental role requires that methylation be *directed* to specific sequences as part of the developmental program, whereas the host defense role requires that methylation be *contingent* on the interaction of features characteristic of parasitic sequence elements with the DNA methylating system. It might therefore be possible to deduce the role of methylation patterns from the nature of the biochemical events by which they are established. Although undoubtedly an oversimplification, addressing the question in this way might contribute to the solution of an old and difficult problem.

THE RECOGNITION OF INVASIVE DNA

Invasive DNA *In Flagrante*

The host defense hypothesis requires that the silencing apparatus recognize and inactivate parasitic sequence elements. How could such elements be recognized? Recent data make it clear that nearly all transposition and viral integration intermediates share certain structural features (Craig 1995), and some satellite DNA is thought to undergo amplification by extrachromosomal rolling-circle replication followed by insertion of the array into the genome (Charlesworth et al. 1994). Recognition and de novo methylation of CpG sites in and around features characteristic of integration reactions would ensure the inactivation of the invasive element immediately upon its integration. DNA methyltransferase may have an intrinsic ability to recognize integration intermediates that are characteristic of the above integration events. The de novo sequence specificity of mammalian DNA methyltransferase is strongly dependent on alternative secondary structures in DNA; four-way junctions in cruciform structures formed by inverted repeats in supercoiled plasmids are especially favored targets (Bestor 1987), as are secondary structures in artificial oligonucleotide substrates (Chen et al. 1995). This biochemical property suggests that invasive sequences might be targeted for de novo methylation because of their presentation of alternative secondary structures during integration, as shown in Figure 3.

Three-way junctions share structural features with four-way junctions, and enzymes that recognize four-way junctions (such as T4 endonuclease VII [Jentsch and Kemper 1986] and T7 endonuclease I [Lu et al. 1991]) cleave three-way junctions as well. The nature of the interaction of the methylating system with three-way junctions is of special interest because it may be involved in the etiology of human diseases associated with GGC/GCC triplet repeat expansions. Slippage of replication intermediates can result in extrusion of a segment of GGC/GCC repeats, which form stable three-way junctions despite a lack of perfect complementarity in the extruded segment (Fig. 3) (Chen et al. 1995; Usdin and Woodford 1995). The preference of DNA methyltransferase for junctions may result in de novo methylation of sequences around the three-way junction in slippage structures. If this is true, diseases that result from GGC/GCC triplet repeat expansions may be thought of as autoimmune disorders in which a host defense system attacks an innocuous alteration of an endogenous gene because of its incidental similarity to a parasitic sequence element in the act of integration.

Cytosine Methylation and Pairing Interactions: RIP, MIP, and RIGS

A common characteristic of invasive sequences is their presence in multiple copies, and it has recently become known that repeated sequences can interact so as to trigger their mutual silencing. Fungi and flowering plants have diverse and highly effective means of silencing repeated sequences, and cytosine methylation is associated with silencing in most cases. Repeat-induced point mutation (RIP) imposes methylation and large numbers of C→T transition mutations on repeated sequences during the sexual phase of the fungus *Neurospora crassa* (Selker 1990a,b), whereas methylation induced premeiotically (MIP) inactivates and methylates repeats during the sexual cycle of the fungus *Ascobolus immersus*

(Goyon et al. 1994). Flowering plants can also silence and methylate repeated sequences via repeat-induced gene silencing (RIGS) (Assaad et al. 1993; Matzke and Matzke 1993); the efficiency is such that multicopy transgenes frequently cause mutual silencing of themselves and of ho-

a. Integration/transposition intermediates

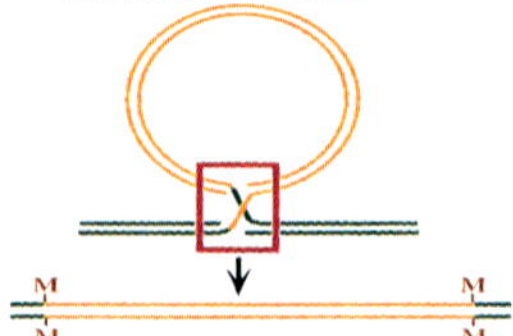

b. Junctions in $(GGC)_n/(GCC)_n$ slippage structures

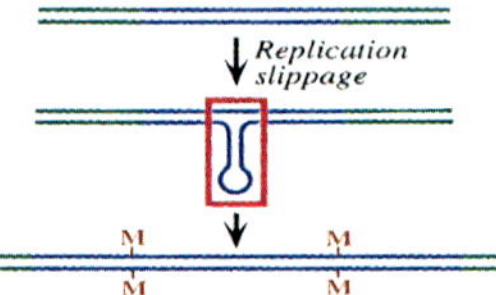

c. Homology/heterology boundaries

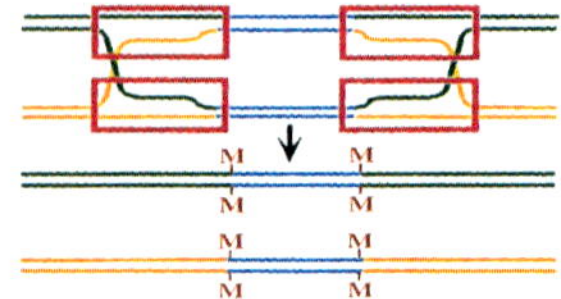

Figure 3 DNA structures that might trigger silencing and de novo methylation. Complementary sequences are depicted in identical colors, and structural features that are proposed to trigger silencing are boxed. A retrovirus integration intermediate is shown in *a*; the step depicted is that at which free 3′ends of the proviral DNA have been linked to 5′ ends of genomic DNA. As recently discussed by Craig (1995), most transposition and retroviral integration events utilize a common mechanism, and integrases and transposases share sequence or structural similarities. Recognition of integration/transposition intermediates by the silencing system could protect the genome from the proliferation of parasitic sequences. (*b*) Formation of three-way junctions in replication slippage structures at (GGC)/(GCC) repeats. Nearly all triplet repeat expansions at (GGC)/(GCC) sequences trigger de novo methylation in and around the repeats, and it is suggested that the three-way junction is the stimulus. (*c*) Detection of invasive sequences via homology/heterology boundaries during strand exchange. Recognition and methylation of the boxed regions would result in mutual silencing of repeated sequences that reside in different sequence contexts. (Reprinted, with permission, from Bestor and Tycko 1996.)

mologous resident sequences. The process can be remarkably efficient; unlinked transgene sequences as small as 300 bp can identify and inactivate each other in a genome of more than 10^9 bp. The silenced state can persist even after the repeated sequences have been separated by segregation in sexual crosses (Matzke and Matzke 1993). Transgene silencing of this type is emerging as a barrier to the improvement of commercially important plant species (Finnegan and McElroy 1994). It is difficult to imagine a selective advantage for RIP, MIP, or RIGS other than host defense against the proliferation of parasitic sequence elements.

How can silencing and methylation be imposed on repeats? Transient pairing interactions are likely to be involved, and two mechanisms can be envisioned. First, strand exchange between repeats at nonallelic positions presents an abrupt loss of homology at the junction with flanking sequences (Fig. 3c), and silencing may be provoked by factors that recognize characteristic structural features at the homology/heterology boundary. This idea was developed by E. Signer to explain copy number-dependent transgene silencing in *Arabidopsis* (Assaad et al. 1993). Second, methylation may propagate through a network of repeated sequences once one or more copies are methylated, as shown in Figure 4.

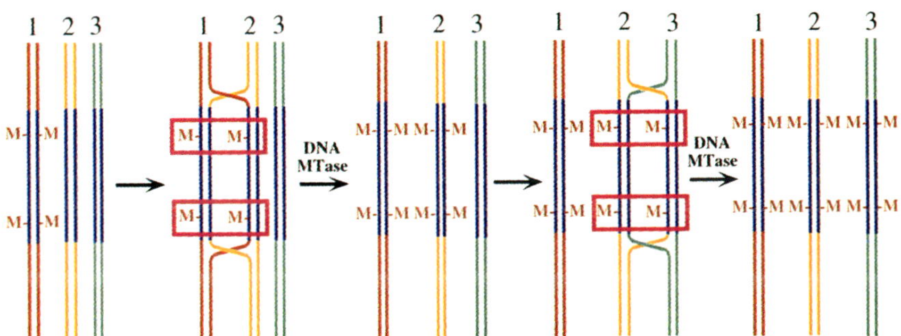

Figure 4 Unidirectional transfer of epigenetic information via paired hemimethylated intermediates. Strand exchange between methylated and unmethylated repeated sequences presents hemimethylated sites, which are the preferred substrate of DNA methyltransferase. Notice that the maintenance activity of DNA methyltransferase causes it to catalyze de novo methylation on these substrates. A common methylation pattern could propagate through unlinked repeats by this mechanism. The loss of allele-specific methylation patterns that has been observed in imprinted genes in certain human diseases may also occur through this mechanism (Moulton et al. 1994). (Reprinted, with permission, from Bestor and Tycko 1996.)

Strand exchange between methylated and unmethylated repeated sequences creates hemimethylated intermediates (Holliday 1987; Flavell 1994; Matzke et al. 1995), which as mentioned previously, are strongly preferred substrates of DNA methyltransferase (Bestor and Ingram 1983). Notice that the maintenance activity of DNA methyltransferase causes it to catalyze de novo methylation on strand-exchange substrates. Strand-exchange reactions (which are thought to occur frequently as part of the double-strand-break-repair pathway; Engels et al. 1994) of the type shown in Figure 4 could cause a common methylation pattern to propagate through a network of repeats, and the probability of pairing with a methylated repeat increases in proportion to the number of methylated repeats. A large number of methylated resident repeats therefore increase the probability that a new, unmethylated copy will be methylated soon after its incorporation into the genome, and a large burden of parasitic sequences may have the paradoxical effect of repressing their own activity and that of homologous invasive sequences. Natural selection may actually favor the retention of large numbers of inactive parasitic sequences for this reason.

Many endogenous genes are present in multiple copies that escape repeat-dependent gene silencing. In *Neurospora*, which seems to be especially aggressive in its response toward repeated sequences, the efficiency of RIP is proportional to the size of the repeats and to their degree of linkage (Selker 1990a,b). There must be special protective mechanisms that shield repeated cellular genes from repeat-induced silencing. A complete intolerance of repeated sequences would also put severe constraints on the evolution of new functions, which depends on the duplication and divergence of existing genes. However, the factors that control sensitivity to repeat-dependent silencing are very poorly understood (Bender and Fink 1995).

Does repeat-dependent silencing or methylation occur in mammals? There is no clear experimental evidence on this point, as exists in fungi and plants, but it can also be said that no objective search for repeat-dependent silencing in mammals seems to have been conducted. Circumstantial evidence suggests that it may occur (Kricker et al. 1992). Much of the m^5C in the mammalian genome is found in repetitive DNA, and most of the repeated and potentially transposable elements in the genome are methylated and transcriptionally inert through most of development (Rubin et al. 1994). Retroviral vectors that transduce reporter genes or therapeutic agents have been observed to undergo silencing after variable periods of expression in animals (Challita and Kohn 1994). Susceptibility to de novo methylation and silencing has limited the usefulness of retroviral vectors in the construction of transgenic mouse lines, and it is not unlikely that silencing phenomena may emerge as barriers to long-term somatic gene therapy in humans. Successful gene transfer may require the development of specialized vectors that evade the silencing response. On the other hand, deleterious sequence elements such as the proviral DNA of HIV could be selectively inactivated by targeted de novo methylation mediated by DNA methyltransferases of the appropriate unique specificity. This approach might be considered a sort of nuclear vaccine and would have the important virtue of stimulation of an existing host defense system.

Unidirectional Transfer of Epigenetic Information at Imprinted Genes

Imprinted mammalian genes have allele-specific methylation patterns that are necessary for maintenance of the imprinted state (Li et al. 1993). Short repetitive sequences are characteristic of imprinted genes (Neumann et al. 1994; Hatada and Mukai 1995); these short repeats may form alternative secondary structures that trigger de novo methylation in the germ line or during early postfertilization development. If the efficiency of de novo methylation differs in the male and female germ lines, a parent-of-origin effect on gene expression results, and in this way a nuclear host defense system may give rise to the genetic phenomenon of imprinting (Barlow 1993).

In the case of asymmetrically methylated alleles at imprinted loci, the mechanism depicted in Figure 3 could mediate the unidirectional transfer of methylation patterns from one allele to the other. Certain pathological human conditions show abnormalities in functional imprinting that appear to involve interchromosomal transfer of allelic methylation patterns. Wilms' tumors frequently show conversion to a uniparental (bipaternal) methylation and expression pattern at the *IGF2/H19* locus on chromosome 11p15.5, which could result from transfer of a methylation pattern from the inactive paternal H19 allele to the normally active maternal allele (Moulton et al. 1994).

Vulnerability of Meiotic Recombination Intermediates to De Novo Methylation

As shown in Figure 5, allelic heteroduplex recombination intermediates present several features that should make them very vulnerable to de novo methylation. First, allelic methylation differences create hemimeth-

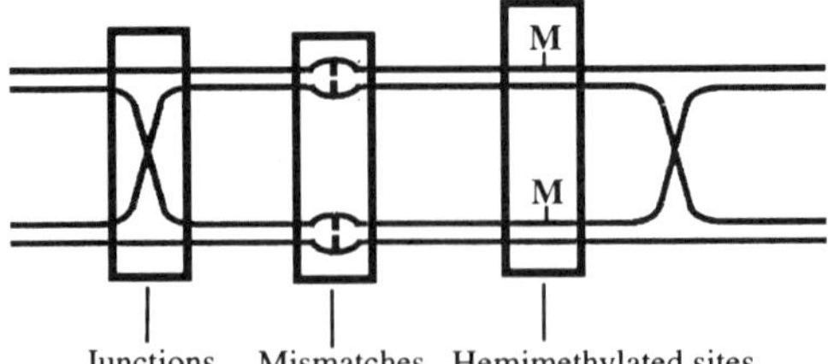

Figure 5 Vulnerability of heteroduplex recombination intermediates to de novo methylation. Crossing-over during meiosis presents DNA methyltransferase with highly preferred targets: four-way junctions, mismatches, and hemimethylated sites. Dysregulated de novo methylation appears to be prevented by sharp down-regulation of DNA methyltransferase during the pachytene stage of meiosis; the mechanism involves the production of a larger, nontranslated DNA methyltransferase mRNA (Jue et al. 1995). (Reprinted, with permission, from Bestor and Tycko 1996.)

ylated sites, which have long been known to be a favored target of DNA methyltransferase (Bestor and Ingram 1983). The maintenance activity of DNA methyltransferase catalyzes de novo methylation under these conditions. Second, Holliday structures and four-way junctions are necessarily present; as described earlier, these structures appear to be favored targets of DNA methyltransferase. Third, mismatches in the vicinity of CpG dinucleotides greatly favor de novo methylation, presumably by lowering the energetic barrier associated with eversion of the target cytosine during the transmethylation reaction (Klimasauskas and Roberts 1995; Yang et al. 1995). Presentation of these vulnerable sites may create a "methylation ratchet" in which methylation levels increase with each meiotic cycle. Are there special conditions that shield recombination intermediates from de novo methylation during meiotic prophase?

Examination of purified male germ cells showed that both DNA methyltransferase protein and the 5.2-kb DNA methyltransferase mRNA found in all proliferating cell types were present in all germ-cell fractions except for pachytene spermatocytes; these spermatocytes contained a 6.2-kb RNA that was not associated with polyribosomes. It is the pachytene stage of meiosis I in which most crossing-over occurs. The ubiquitous 5.2-kb DNA methyltransferase mRNA and DNA methyltransferase protein reappeared at the conclusion of the crossing-over phase of meiosis (Jue et al. 1995). These findings suggest that meiotic recombination intermediates are protected from de novo methylation through down-regulation of DNA methyltransferase via a novel posttranscriptional mechanism that involves the production of a larger, nontranslated RNA transcript.

CONCLUSION: TOWARD AN UNDERSTANDING OF THE FUNCTION OF CYTOSINE METHYLATION

Base methylation increases the information content of the genome, is fully reversible, and can designate specific sequences for interaction with regulatory factors. These features have tempted many to ascribe diverse roles to methylation patterns; if all the postulated roles were to be confirmed, methylation patterns would be a veritable Swiss army knife of the genome. Cytosine methylation has been proposed to reduce the effective size of the genome by masking nonregulatory regions in large-genome eukaryotes (Bestor 1990; Bird 1995), and central roles in DNA repair (Cleaver 1994; Chen et al. 1995) and replication (Taylor 1984) have also been mooted. None of the hypothetical functions of cytosine methylation (and this includes the developmental and host-defense functions) has the support of compelling experimental evidence, and all, some, or none of the hypotheses may be valid. At this time, it seems that the role of cytosine methylation will be understood only when we learn how the cell selects specific sequences for covalent modification.

ACKNOWLEDGMENTS

Work in the author's laboratory was supported by National Institutes of Health grants GM-00616 and CA-60610.

REFERENCES

Aniello, F., A. Locascio, L. Fucci, G. Geraci and M. Branno. 1996. Isolation of cDNA clones encoding DNA methyltransferase of sea urchin *P. lividus*. EMBL accession Z50183.

Ashburner, M. 1989. Drosophila: *A laboratory handbook*, p. 303. Cold Spring Harbor Laboratory Press, Cold Spring Harbor, New York.

Assaad, F.F., K. Tucker, and E. Signer. 1993. Epigenetic repeat-induced gene silencing (RIGS) in *Arabidopsis. Plant Mol. Biol.* **22:** 1067–1085

Barlow, D.P. 1993. Methylation and imprinting: From host defense to gene regulation? *Science* **260:** 309–310.

Bender, J. and G. Fink. 1995. Epigenetic control of endogenous gene family is revealed by a novel blue fluorescent mutant of *Arabidopsis. Cell* **83:** 725–734.

Bestor, T.H. 1987. Supercoiling-dependent sequence specificity of mammalian DNA methyltransferase. *Nucleic Acids Res.* **15:** 3835–3843.

———. 1990. DNA methylation: Evolution of a bacterial immune function into a regulator of genome structure and activity in higher eukaryotes. *Philos. Trans. R. Soc. Lond. B* **326:** 179–187.

———. 1992. Activation of mammalian DNA methyltransferase by cleavage of a Zn-binding regulatory domain. *EMBO J.* **11:** 2611–2618.

———. 1995. The biochemical basis of allele-specific gene expression in genomic imprinting and X inactivation. In *Genomic imprinting: Causes and consequences* (ed. R. Ohlsson et al.), pp. 1109–1117. Cambridge University Press, England.

Bestor, T.H. and A. Coxon. 1993. The pros and cons of cytosine methylation. *Curr. Biol.* **3:** 384–386.

Bestor, T.H. and V. Ingram. 1983. Two species of DNA methyltransferase from murine erythroleukemia cells: Purification, sequence, specificity and mode of interaction with DNA. *Proc. Natl. Acad. Sci.* **80:** 5559–5563.

Bestor, T.H. and B. Tycko. 1996. Creation of genomic methylation patterns. *Nat. Genet.* **12:** 363–367.

Bestor, T.H. and G.L. Verdine. 1994. DNA methyltransferases. *Curr. Opin. Cell Biol.* **6:** 380–389.

Bestor, T., A. Laudano, R. Mattaliano, and V. Ingram. 1988. Cloning and sequencing of a cDNA encoding DNA methyltransferase of mouse cells. *J. Mol. Biol.* **203:** 971–983.

Bird, A.P. 1995 Gene number, noise reduction and biological complexity. *Trends Genet.* **11:** 94–100.

Challita, P.-A. and D.B. Kohn. 1994. Lack of expression from a retroviral vector after transduction of murine hematopoietic stem cells is associated with methylation *in vivo*. *Proc. Natl. Acad. Sci.* **91:** 2567–2571.

Charlesworth, B., P. Sniegowski, and W. Stephan. 1994. The evolutionary dynamics of repetitive DNA in eukaryotes. *Nature* **371:** 215–220.

Chen, X., S.V. Mariappan, P. Catasti, R. Ratliff, R.K. Moyzis, A. Laayoun, S. Smith, E.M. Bradbury, and G. Gupta. 1995. Hairpins are formed by the single DNA strands of the fragile X triplet repeats: Structure and biological implications. *Proc. Natl. Acad. Sci.* **92:** 5199–5203.

Cleaver, J. 1994. It was a very good year for DNA repair. *Cell* **76:** 1–4.

Craig, N.L. 1995. Unity in transposition reactions. *Science* **270:** 253–255.

Engels, W.R., C.R. Preston, and D.M. Johnson-Schlitz. 1994. Long-range cis preference in DNA homology search over the length of the *Drosophila* chromosome. *Science* **263:** 1623–1625.

Finnegan, E.J. and E.S. Dennis. 1993. Isolation and identification by sequence analysis of a putative cytosine methyltransferase from *Arabidopsis thaliana*. *Nucleic Acids Res.* **21:** 2383–2388.

Finnegan, J. and D. McElroy. 1994. Transgene inactivation: Plants fight back! *Bio/technology* **12:** 883–888.

Flavell, R.B. 1994. Inactivation of gene expression in plants as a consequence of specific sequence duplication. *Proc. Natl. Acad. Sci.* **91:** 3490–3496.

Goyon, C., T.I.V. Nogueira, and G. Faugeron. 1994. Perpetuation of cytosine methylation in *Ascobolus immersus* implies a novel type of maintenance methylase. *J. Mol. Biol.* **240:** 42–51.

Hatada, I. and T. Mukai. 1995. Genomic imprinting of p57^{KIP2}, a cyclin-dependent kinase inhibitor, in mouse. *Nat. Genet.* **11:** 204–206.

Herman, J., F. Latif, Y. Weng, M.I. Lerman, B. Zbar, S. Liu, D. Samid, D.-S.R. Duan, J.R. Gnarra, W.M. Linehan, and S.B. Baylin. 1994. Silencing of the VHL tumor-suppressor gene by DNA methylation in renal carcinoma. *Proc. Natl. Acad. Sci.* **91:** 9700–9704.

Holliday, R. 1987. The inheritance of epigenetic defects. *Science* **238:** 163–169.

Holliday, R. and J.E. Pugh. 1975. DNA modification mechanisms and gene activity during development. *Science* **187:** 226–232.

Iguchi-Ariga, S.M.M. and W. Schaffner. 1989. CpG methylation of the cAMP responsive enhancer/promoter TGACGTCA abolishes specific factor binding as well as transcriptional activation. *Genes Dev.* **3:** 612–619.

Jaenisch, R., A. Schnieke, and K. Harbers. 1985. Treatment of mice with 5-azacytidine efficiently activates silent retroviral genomes in different tissues. *Proc. Natl. Acad. Sci.* **82:** 1451–1455.

Jentsch, F. and B. Kemper. 1986. Endonuclease VII resolves Y-junctions in branched DNA *in vitro. EMBO J.* **5:** 181–190.

Jue, K., T.H. Bestor, and J. Trasler. 1995. Regulated synthesis and localization of DNA methyltransferase during spermatogenesis. *Biol. Reprod.* **53:** 561–569.

Klimasuaskas, S. and R.J. Roberts. 1995. M.HhaI binds tightly to substrates containing mismatches at the target base. *Nucleic Acids Res.* **23:** 1388–1395.

Kricker, M.C., J.W. Drake, and M. Radman. 1992. Duplication-targeted DNA methylation and mutagenesis in the evolution of eukaryotic chromosomes. *Proc. Natl. Acad. Sci.* **89:** 1075–1079.

Leonhardt, H., A.W. Page, H.-U. Weier, and T.H. Bestor. 1992. A targeting sequence directs DNA methyltransferase to sites of DNA replication in mammalian nuclei. *Cell* **71:** 865–874.

Li, E., T.H. Bestor, and R. Jaenisch. 1992. Targeted mutation of mouse DNA methyltransferase gene results in embryonic lethality. *Cell* **69:** 915–926.

Li, E., C. Beard, A. Forster, T.H. Bestor, and R. Jaenisch. 1993. DNA methylation, genomic imprinting, and mammalian development. *Cold Spring Harb. Symp. Quant. Biol.* **58:** 297–305.

Lu, M., Q. Guo, W.F. Studier, and N.F. Kallenbach. 1991. Resolution of branched DNA substrates by T7 endonuclease I and its inhibition. *J. Biol. Chem.* **266:** 2531–2536.

Ma, Q., H. Alder, K.K. Nelson, D. Chatterjee, Y. Gu, T. Makamura, E. Canaani, C.M. Croce, L.D. Siracusa, and A.M. Buchberg. 1993. Analysis of the murine *All-1* gene reveals conserved domains with human ALL-1 and identifies a motif shared with DNA methyltransferases. *Proc. Natl. Acad. Sci.* **90:** 6350–6354.

Matzke, M.A. and A.J.M. Matzke. 1993. Genomic imprinting in plants. Parental effects of *trans*-inactivation phenomena. *Annu. Rev. Physiol. Plant Mol. Biol.* **44:** 53–76.

———. 1995. Homology-dependent gene silencing in transgenic plants: What does it really tell us? *Trends Genet.* **11:** 1–3.

Montecucco, A., E. Savini, F. Weighardt, R. Rossi, G. Ciarrocchi, A. Villa, and G. Biamonti. 1995. The N-terminal domain of human ligase I contains the nuclear localization signal and directs the enzyme to sites of DNA replication. *EMBO J.* **14:** 5379–5386.

Moulton, T., T. Crenshaw, Y. Hao, J. Moosidawuwan, N. Lin, F. Dembitzer, T. Hensle, L. Weiss, L. McMorrow, T. Loew, W. Kraus, W. Gerald, and B. Tycko. 1994. Epigenetic lesions at the H19 locus in Wilms' tumor patients. *Nat. Genet.* **7:** 440–447.

Neumann, B., P. Kubicka, and D.P. Barlow. 1994. Characteristics of imprinted genes. *Nat. Genet.* **9:** 451–452.

Nicolas, R.H. and G.H. Goodwin. 1996. Molecular cloning of polybromo, a nuclear protein containing multiple domains including five bromodomains, a truncated HMG-box, and two repeats of a novel domain. *Gene* (in press).

Pfeifer, G., S. Steigerwald, R.S. Hansen, S.M. Gartler, and A.D. Riggs. 1990. Polymerase-chain reaction aided genomic sequencing of an X chromosome linked CpG island: Methylation patterns suggest clonal inheritance, CpG site autonomy, and an explanation of activity state stability. *Proc. Natl. Acad. Sci.* **87:** 8252–8256.

Reinisch, K.M., L. Chen, G.L. Verdine, and W.N. Lipscomb. 1995. The crystal structure of HaeIII methyltransferase covalently complexed to DNA: An extrahelical cytosine and rearranged base pairing. *Cell* **82:** 143–153.

Riggs, A.D. 1975. X inactivation, differentiation, and DNA methylation. *Cytogenet. Cell Genet.* **14:** 9–11.

———. 1989. DNA methylation and cell memory. *Cell Biophys.* **15:** 1–13.

Rubin, C.M., C.A. VandeVoort, R.L. Tiplitz, and C.W. Schmid. 1994. Alu repeated DNAs are differentially methylated in primate germ cells. *Nucleic Acids Res.* **22:** 5121–5127.

Selker, E.U. 1990a. Premeiotic instability of repeated sequences in *Neurospora crassa. Annu. Rev. Genet.* **24:** 579–613.

———. 1990b. DNA methylation and chromatin structure: A view from below. *Trends Genet.* **15:** 103–107.

Tajima, S., H. Tsuda, N. Wakabayashi, A. Asano, S. Mizuno, and K. Nishimori. 1995. Isolation and expression of a chicken DNA methyltransferase cDNA. *J. Biochem.* **117:** 1050–1057.

Taylor, J.H. 1984. Origins of replication and gene regulation. *Mol. Cell. Biochem.* **61:** 99–109.

Tkachuk, D.C.S., S. Kohler, and M.L. Cleary. 1992. Involvement of a homolog of *Drosophila* trithorax by 11q23 chromosomal translocations in acute leukemias. *Cell* **71:** 691–700.

Usdin, K. and K.J. Woodford. 1995. CGG repeats associated with DNA instability and chromosome fragility form structures that block DNA synthesis in vitro. *Nucleic Acids Res.* **23:** 4202–4209.

Vongs, A., T. Katutani, R.A. Martienssen, and E.J. Richards. 1993. *Arabidopsis thaliana* DNA methylation mutants. *Science* **260:** 1926–1928.

Yang, A.S., J.-C. Shen, J.-M. Zingg, S. Mi, and P.A. Jones. 1995. HhaI and HpaII DNA methyltransferases bind DNA mismatches, methylate uracil and block DNA repair. *Nucleic Acids Res.* **23:** 1380–1387.

Yen, R.-W., P.M. Vertino, B.D. Nelkin, J.J. Yu, A. El-Deiry, A. Cumaraswamy, G.G. Lennon, B.J. Trask, P. Celano, and S.B. Baylin. 1992. Isolation and characterization of the cDNA encoding human DNA methyltransferase. *Nucleic Acids Res.* **20:** 2287–2291.

The Mutational Burden of 5-Methylcytosine

Allen S. Yang and Peter A. Jones
Department of Biochemistry and Molecular Biology
USC/Norris Comprehensive Cancer Center
University of Southern California School of Medicine
Los Angeles, California 90033

Atsuko Shibata
Division of Epidemiology
Department of Health Research and Policy
Stanford University School of Medicine
Stanford, California 94305

The importance of the epigenetic modification of cytosine (C) to 5-methylcytosine (m^5C) is discussed in other chapters of this volume. In this chapter, we address the heavy mutational burden induced by the methylation of C at CpG dinucleotides, which is the price that must be paid for having a m^5C epigenetic system. The mutability of m^5C to thymine (T) has presumably led to a depletion of the CpG dinucleotide in the mammalian genome over the course of evolution. Furthermore, m^5C residues, which comprise only about 1% of the human genome, account for about 30% of the base substitution mutations found in genetic disease (Cooper and Yousouffian 1988) and 25% of the mutations found in tumor suppressor genes (Jones et al. 1991; Greenblatt et al. 1994).

The mutability of m^5C was first demonstrated in *Escherichia coli*. Cytosine bases that were methylated in the *E. coli lacI* gene were found to be hot spots for spontaneous base substitution mutations, and the hot spots disappeared when the same sites were unmethylated (Coulondre et al. 1978). It was speculated that the reason for this increase was that whereas C deaminates to uracil (U), m^5C deaminates to T, which is a normal DNA base and therefore inherently more difficult to repair (Duncan and Miller 1980; Shenoy et al. 1987). This mechanism of DNA mutation is unique in that it is not induced by exogenous chemicals and it occurs in nonreplicating DNA. The deamination of m^5C is a first-order chemical process, which is consistent with the fact that deamination occurs on both the transcribed and nontranscribed strand of DNA with equal probability.

MECHANISMS OF MUTATION AT CpG SITES

Chemistry of Deamination

The most likely mechanisms for the induction of C-to-T transition mutations are diagramed in Figure 1. The amino group at the 4 position of an unmethylated C can be hydrolyzed by an addition-elimination reaction catalyzed by the protonation of the N3 position of the C ring creating U (Fig. 1, reaction 1) (Shapiro and Klein 1966). If the U is not excised prior to DNA replication, it is read as T by DNA polymerase and pairs with adenine (A), creating a G:C to A:T transition mutation (Fig. 1, reaction 2).

Cytosine can also be modified by DNA methyltransferase, which adds a methyl group to the 5 position of the C-ring (Fig. 1, reaction 3). Hydrolytic deamination of m^5C will now create T without a U intermediate (Fig. 1, reaction 4). Thymine, a naturally occurring base in DNA, is believed to be repaired less efficiently than U, which can explain the increased mutagenicity of methylation sites (Schmutte et al. 1995).

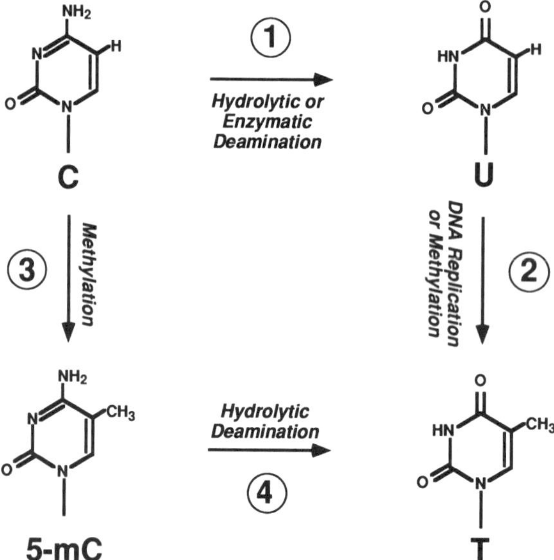

Figure 1 Possible pathways for C-to-T transition mutations at the CpG dinucleotide. (1) Hydrolytic deamination of C can create U, or this deamination can be catalyzed by DNA methyltransferase when the methyl donor, AdoMet, is limiting. (2) Uracil will be recognized as T in subsequent DNA replication or can be methylated to T directly by DNA methyltransferase with low efficiency. (3) Alternatively, DNA methyltransferase can methylate C to m^5C in the presence of AdoMet. (4) Hydrolytic deamination of m^5C creates T without a U intermediate.

The addition of a methyl group to the 5 position of C facilitates deamination by lowering the activation energy for the hydrolytic reaction. In vitro studies have shown that methylation increases the rate of deamination up to fivefold for both the free nucleotide (5-methyl-dCMP) (Lindahl and Nyberg 1974) and for single-stranded DNA (Table 1) (Ehrlich et al. 1986; Frederico et al. 1990). The rate of deamination in double-stranded DNA is slower than in single-stranded DNA (Table 1). The slower rate of deamination in double-stranded DNA may be due to an enthalpy–entropy compensation in base stacking of double-stranded DNA compared to single-stranded DNA. The deamination of C to U or m^5C to T creates energetically unfavorable G:U or G:T mismatches, respectively, in the DNA. A sensitive genetic assay was used to determine the rate of hydrolytic deamination of C and m^5C, and it was determined that the in vitro rate of m^5C deamination was 2- to 3-fold faster than C in double-stranded DNA at 37°C (Table 1).

Deamination of C and m^5C can be facilitated in vitro by acidic conditions and chemicals such as bisulfite, which can selectively deaminate C but not m^5C under the proper conditions. Thus, bisulfite is now being used in genomic sequencing to determine the methylation status of specific C bases in DNA (Clark et al. 1995). It has also been shown that nitric oxide (NO) has the ability to deaminate deoxynucleotides, and mutagenic studies of NO in vivo have demonstrated that this agent increases the rate of C to T transitions in *Salmonella* (Wink et al. 1991). This suggested that NO could play a role in mutagenesis by increasing the deamination of C. However, we have shown that the mutagenicity of NO is not due to an induced deamination of C or m^5C in double-stranded DNA, and that the mutagenic effect of NO in vivo is more likely due to interaction with G (Schmutte et al. 1994). Thus, no exogenous chemical has been shown to be involved in the generation of m^5C mutations in double-stranded DNA in vivo.

Can Enzymatic Deamination Contribute to Mutagenesis?

The bacterial methyltransferase *Hpa*II has been shown in vitro to deaminate C to U directly, when the methyl donor, *S*-adenosyl methionine (AdoMet) is limiting (Fig. 1, reaction 2) (Shen et al. 1992). The catalytic mechanism of cytosine methyltransferases involves nucleophilic attack at the 6 position of the cytosine ring by the cysteine of a proline-cysteine domain common to all cytosine methylases (Wu and Santi 1987). The nucleophilic attack at the 6 position, in order to add a methyl group at the 5 position, can also facilitate deamination of the 4

Table 1 Rate of cytosine and 5-methylcytosine deamination and mutation in vitro and in vivo

Substrate	Rate constant (s^{-1})	Temperature (°C)	$t_{1/2}$ (years)	Reference
In vitro				
mononucleotide[a]				
dCMP	2.6×10^{-7}	95	0.085	Lindahl and Nyberg (1974)
5-methyl-dCMP	1.1×10^{-6}		0.002	Lindahl and Nyberg (1974)
single-stranded DNA				
cytosine	2.1×10^{-10}	37[b]	105	Ehrlich et al. (1986)
5-methylcytosine	9.5×10^{-10}		23	Ehrlich et al. (1986)
cytosine	1×10^{-10}	37	200	Frederico et al. (1990)
double-stranded DNA				
cytosine	7×10^{-13}	37	30,000	Frederico et al. (1990)
cytosine	2.6×10^{-13}	37	85,000	Shen et al. (1994)
5-methylcytosine	5.8×10^{-13}		32,000	Shen et al. (1994)
In vivo germ line				
estimated rate[c]				
5-methylcytosine	1.66×10^{-16}		132,000,000	Cooper and Krawczak (1989)
factor IX gene[d]				
5-methylcytosine	2.9×10^{-17}		770,000,000	Sommer (1995)
Alu element in primate DNA, p53 genes[e]				
5-methylcytosine	2.9×10^{-16}		82,000,000	Yang et al. (1996)

[a]Values for deamination at 95°C. At 95°C, the rates of deamination for dCMP, poly(dC), and denatured *E. coli* DNA (ssDNA) are similar.
[b]Extrapolated values for 37°C from values calculated at 70°C and 95°C.
[c]Estimated rate of CpG mutation in vivo calculated from the in vitro rate.
[d]Derived from a CpG mutation rate of 3.6×10^{-8} per generation for a generation time of 20 years.
[e]Calculated average from rate of CpG germ-line mutation in an *Alu* element in intron 6 of the p53 gene in nine different primate species.

position, creating U. Furthermore, current models of the mechanism of action of cytosine methyltransferases suggest that a conserved domain in cytosine methyltransferases protonates the N3 position of the C ring in order to facilitate nucleophilic attack of the 6 position (Erlanson et al. 1993). This enzymatic protonation would also be expected to promote deamination of C to U, raising the possibility that the methyltransferase enzyme itself may actively deaminate C to U. These studies have been duplicated with the *Eco*RI (Wyszynski et al. 1994), *Hha*I (Yang et al. 1995), and *Sss*I (Bandaru et al. 1995) methylases. However, it is not known whether the mammalian methylase also has the capability of catalyzing deamination (Smith et al. 1992).

We have investigated the importance of AdoMet concentrations in mutagenesis catalyzed by methyltransferase by utilizing site-directed mutagenesis of the AdoMet-binding pocket of *Hpa*II methyltransferase. These enzymes, which can form the reaction intermediate with the target cytosine, but have a decreased ability to methylate the 5 position of the ring, function as cytosine deaminases and are mutagenic in bacteria (Shen et al. 1995). Furthermore, it has been shown that M.*Hha*I and M.*Hpa*II can methylate uracil to thymine (Klimasauskas and Roberts 1995; Yang et al. 1995). This suggests that C, either hydrolytically or enzymatically deaminated to U, may then be methylated to T (Fig. 1, reaction 2). Therefore, a C to U mutation created either by hydrolytic or enzymatic deamination may be fixed to T, escaping uracil DNA glycosylase (UDG) repair in the absence of DNA replication.

Evidence for the presence of an enzymatic deaminase in mammalian cells has been found in mouse cell lines where tandemly linked C-to-T transition mutations at CpG sites were found to occur 6 bases apart at an extremely high frequency. The proximity and tendency for these mutations to occur together argue against a random spontaneous mechanism of mutation but suggest an active mutagen moving processively along the DNA, inducing G:C to A:T mutations at CpG sites (Steinberg and Gorman 1992). Studies in the human globin genes have shown that certain sites which are mutagenic were found to be unmethylated, suggesting that CpG mutation may not always occur through a m^5C intermediate (Perutz 1990).

A major consideration arguing against the possibility of a C to U to T pathway for mutagenesis is the presence of highly active UDG repair systems in all organisms, which would presumably rapidly excise any U formed as a by-product of the methylation reaction. However, we (Yang et al. 1995) and other investigators (Klimasauskas and Roberts 1995) have found that the bacterial methyltransferases form tight binding complexes with G:U mismatches at their target sequences and are capable of shielding the mismatches from UDG (Yang et al. 1995). However, there is no direct evidence as yet implicating enzyme-induced deamination in mutagenesis at CpG sites.

It is also worth mentioning that DNA replication errors, specifically base misincorporation opposite C versus m^5C, may also play a role in the generation of transition mutations at CpG sites. Reverse transcriptase was shown to incorporate dAMP opposite m^5C 4- to 5-fold more efficiently than opposite C in vitro (Shen et al. 1992). However, no significant difference was shown when comparing the copying fidelity for Klenow fragment opposite m^5C and C (Shen et al. 1992).

The rate of spontaneous hydrolytic deamination of C and m^5C in double-stranded DNA is 3 orders of magnitude faster than the rate of CpG mutation calculated in vivo in the germ line (Table 1). Thus, the rate of hydrolytic deamination of m^5C is more than sufficient to account for the frequency of occurrence of mutations at CpG sites. However, mutation of C at a CpG site has been found to be 42-fold more frequent than mutations of C at non-CpG sites (Cooper and Krawczak 1989). Therefore, the increased mutability of m^5C cannot be explained only by the 2- to 3-fold increase in the rate of hydrolytic deamination of m^5C over C. This discrepancy in the calculated 2- to 3-fold difference in the in vitro chemical deamination rate of m^5C and C versus the observed 42-fold difference in the mutation rate of C at CpG sites and non-CpG sites in vivo is believed to be due to the presence of different DNA repair systems for the presumed premutagenic lesions of C and m^5C. The deamination of C to U is repaired by UDG, which is highly efficient; however, deamination of m^5C to T is repaired less efficiently by a G:T mismatch repair system.

ROLE OF MISMATCH REPAIR MUTATIONS AT CpG SITES

The deamination of C to U or m^5C to T creates G:U or G:T mismatches, which are premutagenic lesions that can be repaired or can become fixed as mutations after DNA replication (Fig. 2).

G:U Excision Repair

Uracil, which is not a normal DNA base, is readily excised by UDG in both prokaryotes and eukaryotes (Lindahl et al. 1977; Muller and Caradner 1991). UDG is a small monomeric protein that removes the U base from DNA, but leaves the phosphate-deoxyribose backbone intact. This

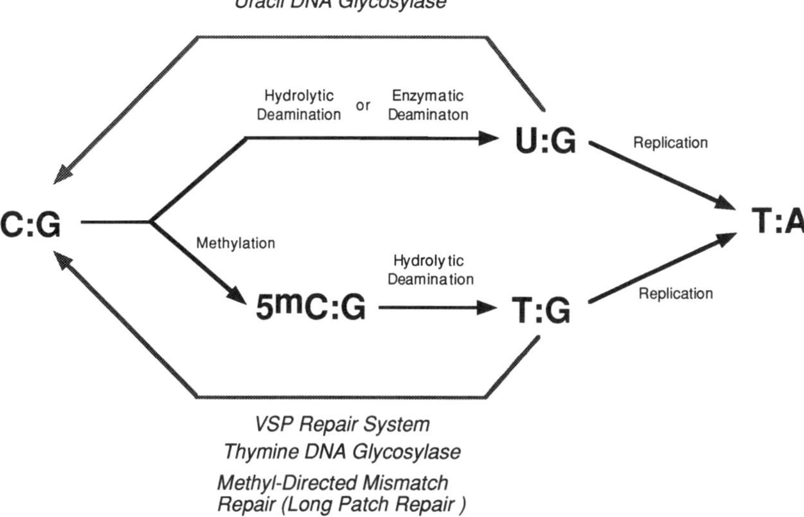

Figure 2 Two possible premutagenic mismatches that can be created by methylation mutations. Deamination of C induced hydrolytically or enzymatically creates a U:G mismatch. Alternatively, C can be methylated to m^{5}C, which can deaminate to T and create a T:G mismatch. These premutagenic mismatches become fixed as C:G to T:A transition mutations after DNA replication. Repair mechanisms exist to recognize and repair both premutagenic mismatches. (VSP) Very short patch.

creates a temporary abasic site (AP) in the DNA (Lindahl et al. 1977), which is then cleaved by the AP-endonuclease and deoxyphosphodiesterase to create a gap which is subsequently filled by DNA polymerase-β and sealed by DNA ligase (Dianov et al. 1992; Palombo et al. 1994). UDG does not remove T from DNA, and the crystal structure of UDG shows that the enzyme has specific mechanisms which prevent excision of T from DNA (Mol et al. 1995; Savva et al. 1995). It was shown that *E. coli* proficient in UDG repair or lacking UDG repair had equivalent frequencies of methylation-induced mutations, suggesting that, at least for *E. coli*, U is an unlikely intermediate for methylation-induced mutation (Duncan and Miller 1980; Wyszynski et al. 1994).

G:T Repair

As mentioned earlier, G:T mismatches are believed to be inherently more difficult to repair than G:U and are created not only by deamination of

m^5C, but also by misincorporation of a G opposite T, or T opposite G, during DNA replication. The G:T mismatch created by the deamination of m^5C should be repaired exclusively in favor of the G by removal of the T to prevent the fixation of mutation. However, G:T mismatches created by replication errors should be repaired in favor of the G or T, depending on which base is in the parent strand. G:T mismatches created from replication errors in *E. coli* are repaired by the long patch repair system (methyl-directed mismatch repair), which uses methylation differences to distinguish the parent from the daughter strand. G:T mismatches created by deamination are corrected by the very short patch (VSP) repair system, which is independent of the long patch repair system and repairs G:T mismatches created by m^5C deamination only at the target sequence of the *dcm* bacterial methylase (Lieb 1985, 1987; Jones et al. 1987). The mechanism of VSP repair involves nicking 5′ of the G:T mismatch, followed by removal of the mismatched T and several adjacent bases by the 5′-3′ exonuclease activity of pol I (Hennecke et al. 1991).

A repair system analogous to the VSP repair system has not yet been found in eukaryotes; however, a 200-kD protein that selectively binds DNA containing G:T mismatches was purified from HeLa cells (Jiricny et al. 1988). Further characterization of this G-T-binding protein (GTBP) showed that it consisted of two proteins 100 kD and 160 kD in size (Jiricny 1994). Peptide sequencing of the 100-kD protein revealed it to be identical to hMSH2, which is part of the human mutS homolog repair system and is analogous to the prokaryotic long patch repair system (Fishel et al. 1993; Leach et al. 1993; Palombo et al. 1994). The MutS protein binds DNA mismatch mutations and recruits other DNA repair proteins. The 160-kD GTBP was found to function as a heterodimer with hMSH2 and is specifically required to repair single-base mismatches or single-base loops (Drummond et al. 1995; Palombo et al. 1995; Papadopoulos et al. 1995). No evidence exists that GTBP or hMSH2 functions specifically in the repair of G:T mismatches created by m^5C deamination, and the fact that it is related to the *E. coli* mutS system implies that it is involved in the repair of G:T mismatches created by replication errors.

In vitro studies of human HeLa cell extracts have shown that base excision repair of G:T mismatches may occur by the use of a thymine DNA glycosylase (TDG) in a mechanism analogous to that catalyzed by UDG (Wiebauer and Jiricny 1989). This TDG has been purified, but the gene has not been cloned (Nedderman and Jiricny 1993). Studies using transformation of DNA containing G:T mismatches into monkey kidney cells have shown that repair of these mismatches is more than 90% efficient in selecting G over T (Brown and Jiricny 1988). Other studies have also shown that repair of G:T mismatches occurs preferentially at methylation sites (Griffin and Karran 1993). These findings suggest a repair mechanism dedicated specifically to protect against loss of m^{5}C, although the exact mechanism has yet to be identified.

G:U Versus G:T Repair

Direct comparison of G:U and G:T mismatch excision repair in vitro at the identical target site with normal and colon tumor extracts has shown at least a 600-fold higher efficiency in the repair of G:U mismatches versus G:T mismatches (Schmutte et al. 1995). In addition, the G:T mismatch repair system seems inefficient, at least in vitro (Schmutte et al. 1995), suggesting that in vitro assays may not be optimal for the study of G:T mismatch repair.

EVOLUTIONARY CONSEQUENCE OF METHYLATION

The evolutionary consequence of CpG methylation and mutation has been the depletion of CpG dinucleotides with a corresponding increase in the number of TpG and CpA dinucleotides in the genome (Bird 1986). Mutation of CpG over the course of evolution has led to a 5-fold depletion of CpGs in the vertebrate genome. Evidence of the role of methylation in causing CpG depletion is seen by comparison of the genomes of organisms containing m^{5}C with organisms such as *Drosophila*, which do not methylate their DNA and are not CpG-depleted (Bird 1986; Gardiner-Garden and Frommer 1987). The entire vertebrate genome, however, is not CpG-depleted, and both m^{5}C and CpG sites are not randomly distributed. Unmethylated CpG-rich clusters, which make up 1–2% of the vertebrate genome, have been described previously (McClelland and Ivarie 1982; Cooper et al. 1983). These CpG-rich regions, referred to as "CpG islands," do not show the strong suppression of CpG seen in the rest of the genome and are usually associated with genes or the promoter regions of genes (Bird 1986). Importantly, CpG islands are not methylated in the germ line (Bird 1986).

The fact that CpG islands are not methylated in the germ line implies that two forces play a role in their generation: selective pressure, which protects against mutation of CpG sites in vital genes, and lack of methylation in the germ line, which relieves CpG islands of the mutagenicity of m^{5}C. Thus, CpG islands are not formed by the creation of CpG-rich regions, but rather are formed by the loss of CpG sites in the bulk of the genome, leaving a non-CpG-depleted region. The key to maintenance of CpG islands is their protection from methylation in the germ line, which can be carried out in two possible ways. Either CpG islands are protected from de novo methylation, or CpG islands are actively demethylated. Evidence exists for both mechanisms (Szyf et al. 1990; Frank et al. 1991).

We have obtained a better understanding of the rate of CpG decay by comparing the sequence of *Alu* repetitive elements within an intron of the p53 tumor suppressor gene shared by primates (Yang et al. 1996). Sequence comparison of these *Alu* elements among the primates revealed an ancestral consensus sequence that was rich in CpG dinucleotides. During the course of evolution, however, mutation led to decay of the CpG sites into TpG and CpA sites, consistent with methylation-induced mutagenesis leading to partial CpG depletion. The results of this study were consistent with other studies and suggest that the current 5-fold suppression of CpG dinucleotide represents a dynamic equilibrium in which the rate of CpG decay is balanced by the rate of CpG dinucleotide formation by random mutation (Cooper and Krawczak 1989; Sved and Bird 1990; Yang et al. 1996). However, there is disagreement on the estimated amount of time needed for a CpG-rich sequence to reach CpG equilibrium. Accurate estimates of the time of *Alu* element insertion by anthropological studies allowed an approximation of the rate of CpG mutation in the primate germ line to be calculated. From these studies, the half-life of m^5C in the primate germ line was estimated to be between 48 and 120 million years (Table 1).

ROLE OF METHYLATION IN GENETIC DISEASE

The evolutionary consequences of CpG methylation and mutability in the past have presumably led to the depletion of CpG dinucleotides and formation of CpG islands. Presently, mutation at CpG sites continues to play a significant role in the formation of new germ-line mutations contributing to genetic disease. Cooper and Krawczak (1990), in a survey of a wide variety of genetic diseases, found 44 of 139 (32%) point mutations were C to T or G to A transitions occurring at CpG dinucleotides.

The most extensively studied mutations have been those in the coagulation factor IX gene, mutation in which is responsible for hemophilia B. Of 216 mutations, 97 (45%) in the factor IX gene are at CpG dinucleotides (Gianelli et al. 1991). Sommer (1995) has compared the relative frequency of CpG mutation to other types of mutations found in genetic disease and has derived an in vivo germ-line rate of CpG mutation (Table 1). The rate of occurrence of transition mutations at CpGs is almost 20-fold higher than the rate of transition mutations at non-CpG sites. This calculation underscores the importance of this endogenous mutational mechanism of m^5C mutation over exogenous carcinogens in the genesis of genetic disease. Li-Fraumeni syndrome is a rare, dominantly inherited syndrome characterized by a wide variety of epithelial and mesenchymal neoplasms that occur in affected individuals. Germ-line mutations in the p53 tumor suppressor gene cause Li-Fraumeni syndrome, and 53% of the 47 point mutations characterized are C to T transitions at CpG dinucleotides (Malkin et al. 1990; Greenblatt et al. 1994).

The predominance of C-to-T transition mutations at CpG dinucleotides in the germ line points out the importance of methylation in mutation of the germ line. Several of the CpG sites in the p53 gene and LDL receptor gene, which are mutated in Li-Fraumeni syndrome or familial hypercholesteremia, have been shown to be methylated in human sperm and somatic tissue (Rideout et al. 1990; Magewu and Jones 1994; Tornaletti and Pfiefer 1995). These results are therefore consistent with the idea that m^5C is an endogenous mutagen in the germ lines of organisms which methylate their DNA.

METHYLATION-INDUCED MUTATIONS IN TUMOR SUPPRESSOR GENES

The role of m^5C in generating mutational hot spots in the germ line has been known for some time (Cooper and Youssoufian 1988). However, the isolation of tumor suppressor genes and the detection of mutations within them in somatic cells has led to the realization that m^5C is a frequent contributor to mutations relevant to human carcinogenesis (Jones et al. 1992).

The p53 gene is the most intensively investigated tumor suppressor gene (Greenblatt et al. 1994). The gene has been found to be mutated in diverse types of human cancer (Hollstein et al. 1991; Greenblatt et al. 1994). The distribution of p53 mutations and the relative occurrence of transition and transversion mutations vary among different cancer types, possibly reflecting different carcinogenic agents in each cancer type. The vast majority of mutations reported in the p53 gene occur in exons 5–8, which contain protein domains that are highly conserved in evolution and are essential for p53 protein function. The spectra of mutations in exons 5–8 of the p53 gene for selected tumor types are shown in Figure 3.

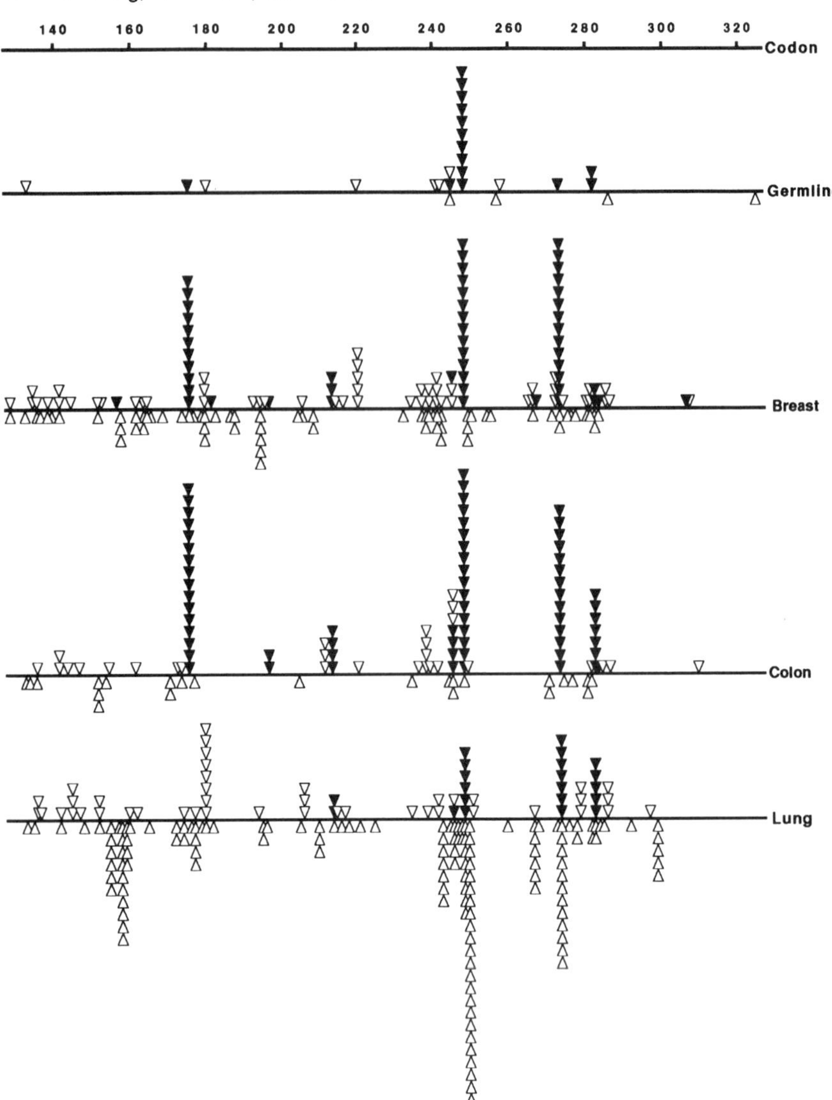

Figure 3 Mutational spectra of the p53 tumor suppressor gene in Li-Fraumeni syndrome and various cancer types. Mutations at various codons are indicated. Each filled triangle represents a CpG transition mutation found in a different individual. Open triangles above the line represent transition mutations not at CpG sites; open triangles below the line represent transversion mutations at any site. Note the predominance of methylation mutations in the germ line (Li-Fraumeni syndrome), breast cancer, and colon cancer, which suggests that endogenous deamination plays an important role in the genesis of these diseases (Shibata 1995). This is in contrast to lung cancer, which has a predominance of transversion mutations presumably reflective of exogenous mutagens found in cigarette smoke or induced by radon.

Although the CpG dinucleotide is suppressed in the genome, C to T transitions at this dinucleotide account for 24% of all the point mutations found in the p53 tumor suppressor gene (Greenblatt et al. 1994). In colon cancer, C to T transitions at CpG sites account for 47% of point mutations reported (Greenblatt et al. 1994). The methylation of C to m^{5}C at CpG sites and the resulting deamination mutation comprise an endogenous process (Fig. 3) (Rideout et al. 1990; Jones 1991). On the other hand, transversion mutations are more frequent in lung cancer (Fig. 3). Exposure of cells to benzo[a]pyrene, one of the numerous carcinogens in tobacco smoke, is known to cause G to T transversions (Mazur and Glickman 1988), and it is not surprising that that is the most common type of point mutation in lung cancer. Thus, the frequency of CpG transition mutations in the p53 gene seems to correlate inversely with the degree of likely direct interaction of exogenous chemical carcinogens with DNA (Fig. 3).

It has been shown that there is a strong strand bias for mutation, with the nontranscribed strand being more susceptible to mutagenesis (Bohr et al. 1987). An examination of the strand bias of mutation at CpG hot spots reveals that there is a bias at some, but not all, mutational hot spots in the p53 gene (Fig. 4). For example, the G to A change most frequently seen at codon 175 (Arg to His) is consistent with deamination of m^{5}C on the noncoding strand. On the other hand, both m^{5}Cs present at the codon 248 hot spot seem equally likely to deaminate to give rise to Arg to Trp and Arg to Gln transitions. This suggests that both the transcribed and nontranscribed strands have an equal possibility of undergoing deamination and that the bias seen at particular hot spots such as codon 175 may be due more to the effect of the amino acid change on the function of the protein than to any inherent strand bias in mutation induction. This is consistent with our studies with *Alu* elements in introns of the p53 gene, which are not under the selection bias of amino acid changes. Here, methylation-induced mutations at CpG sites occurred with equal frequency on the transcribed and nontranscribed strands (Yang et al. 1996). This suggests that CpG mutations are essentially random, and the strand bias previously observed on the *hprt* gene (Skandalis et al. 1994) may have been due to other confounding factors.

CONCLUSION

The methylation of C to m^{5}C in DNA is essential for normal mammalian development (Li et al. 1992). However, this postreplicative epigenetic modification bestows a large mutational burden on organisms that

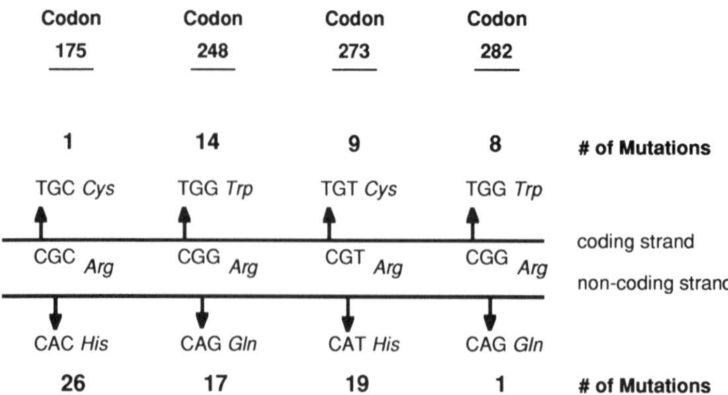

Figure 4 Directionality of CpG mutations in the p53 tumor suppressor gene in colon and breast cancers. Deamination mutations at the palindromic methylated CpG dinucleotide can occur on the coding or noncoding strand. Deamination of the coding strand results in a TpG dinucleotide, and deamination of the noncoding strand results in a CpA. Individual codons appear to have a bias for mutations in the coding (codon 175) or noncoding strand (codon 282); however, these are most likely attributable to the effect of the corresponding amino acid changes on the function of the mutated protein (Shibata 1995).

methylate their genomes. The spontaneous hydrolytic deamination of m^5C to T creates a C-to-T transition mutation at the dinucleotide CpG, which is the target for DNA methyltransferase. In addition, side reactions catalyzed by the methyltransferase may lead directly to premutagenic lesions that are shielded from the DNA repair machinery. Mutation at CpG sites is unique in that it is independent of exogenous carcinogens and is probably a random process that can occur without DNA replication with equal probability on the transcribed and nontranscribed strand of DNA. Despite the slow rate of chemical deamination and the presence of efficient repair mechanisms, C to T transitions at the CpG dinucleotide are the most common point mutations found in genetic disease and cancer. In addition, the mutability of m^5C has led to the depletion of CpG dinucleotides during the course of evolution in the genomes of organisms that methylate their DNA.

REFERENCES

Bandaru, B., M. Wyszynski, and A. S. Bhagwat. 1995. HpaII methyltransferase is mutagenic in *Escherichia coli*. *J. Bacteriol.* **177:** 2950–2952.

Bird, A.P. 1986. CpG-rich islands and the function of DNA methylation. *Nature* **321:** 209–212.

Bohr, V.A., D.H. Phillips, and P.C. Hanawalt. 1987. Heterogeneous DNA damage and repair in the mammalian genome. *Cancer Res.* **47:** 6426–6436.

Brown, T.C. and J. Jiricny. 1988. Different base/base mispairs are corrected with different efficiencies and specificities in monkey kidney cells. *Cell* **54:** 705–711.

Clark, S.J., J. Harrison, and M. Frommer. 1995. CpNpG methylation in mammalian cells. *Nat. Genet.* **10:** 20–27.

Cooper, D.N. and M. Krawczak. 1989. Cytosine methylation and the fate of CpG dinucleotides in vertebrate genomes. *Hum. Genet.* **83:** 181–188.

———. 1990. The mutational spectrum of single base-pair substitutions causing human genetic disease: Patterns and predictions. *Hum. Genet.* **85:** 55–74.

Cooper, D.N. and M. Youssoufian. 1988. The CpG dinucleotide and human genetic disease. *Hum. Genet.* **78:** 151–155.

Cooper, D., M. Taggart, and A. Bird. 1983. Unmethylated domains in vertebrate DNA. *Nucleic Acids Res.* **11:** 647–658.

Coulondre, C., J.H. Miller, P.J. Farabaugh, and W. Gilbert. 1978. Molecular basis of base substitution hotspots in *Escherichia coli. Nature* **274:** 775–780.

Dianov, G., T. Lindahl, and A. Price. 1992. Generation of single-nucleotide repair patches following excision of uracil residues from DNA. *Mol. Cell. Biol.* **12:** 1605–1612.

Drummond, J.T., G.-M. Li, M.-J. Longley, and P. Modrich. 1995. Isolation of an hMSH2-p160 heterodimer that restores DNA mismatch repair to tumor cells. *Science* **268:** 1909–1912.

Duncan, B.K. and J.H. Miller. 1980. Mutagenic deamination of cytosine residues in DNA. *Nature* **287:** 560–561.

Ehrlich, M., K.F. Norris, R.Y.-H. Wang, K.C. Kuo, and C.W. Gehrke. 1986. DNA cytosine methylation and heat-induced deamination. *Biosci. Rep.* **6:** 387–393.

Erlanson, D.A., L. Chen, and G.L. Verdine. 1993. DNA methylation through a locally unpaired intermediate. *J. Am. Chem. Soc.* **115:** 12583–12584.

Fishel, R., M.K. Lescoe, M.R.S. Rao, N.G. Copeland, N.A. Jenkins, J. Garber, M. Kane, and R. Kolodner. 1993. The human mutator gene homolog MSH2 and its applications with hereditary nonpolyposis colon cancer. *Cell* **75:** 1027–1038.

Frank, D., I. Keshet, M. Shani, A. Levine, A. Razin, and H. Cedar. 1991. Demethylation of CpG islands in embryonic stem cells. *Nature* **351:** 239–241.

Frederico, L.A., T.A. Kunkel, and B.R. Shaw. 1990. A sensitive genetic assay for the detection of cytosine deamination: Determination of rate constants and the activation energy. *Biochemistry* **29:** 2532–2537.

Gardiner-Garden, M. and M. Frommer. 1987. CpG islands in vertebrate genomes. *J. Mol. Biol.* **196:** 261–282.

Gianelli, F., P.M. Green, K.A. High, S. Sommer, D.P. Lillicrap, M. Ludwig, K. Olek, P.H. Reitsma, M. Goossens, A. Yoshioka, and G.G. Brownlee. 1991. Haemophilia B: Database of point mutations and short additions and deletions. *Nucleic Acids Res.* **19:** 2193–2219.

Greenblatt, M.S., W.P. Bennett, M. Hollstein, and C.C. Harris. 1994. Mutations in the p53 tumor suppressor gene: Clues to cancer etiology and molecular pathogenesis. *Cancer Res.* **54:** 4855–4878.

Griffin, S. and P. Karran. 1993. Incision at DNA G•T mispairs by extracts of mammalian cells occurs preferentially at cytosine methylation sites and is not targeted by a separate G•T binding reaction. *Biochemistry* **32:** 13032–13039.

Hennecke, F., H. Kolmar, K. Brundl, and H.-J. Fritz. 1991. The vsr gene product of *E. coli* K-12 is a strand- and sequence-specific DNA mismatch endonuclease. *Nature* **353:** 776–778.

Hollstein, M., D. Sidransky, B. Vogelstein, and C.C. Harris. 1991. p53 mutations in human cancer. *Science* **253:** 49–53.

Jiricny, J. 1994. Colon cancer and DNA repair: Have mismatches met their match? *Trends Genet.* **10:** 164–168.

Jiricny, J., M. Hughes, N. Corman, and B.B. Rudkin. 1988. A human 200-kDa protein binds selectively to DNA fragments containing G-T mismatches. *Proc. Natl. Acad. Sci.* **85:** 8860–8864.

Jones, M., R. Wagner, and M. Radman. 1987. Mismatch repair of deaminated 5-methylcytosine. *J. Mol. Biol.* **194:** 155–159.

Jones, P.A., J.D. Buckley, B.E. Henderson, R.K. Ross, and M.C. Pike. 1991. From gene to carcinogen: A rapidly evolving field in molecular epidemiology. *Cancer Res.* **51:** 3617–3620.

Jones, P.A., W.M. Rideout III, J.-C. Shen, C.M. Spruck III, and Y.C. Tsai. 1992. Methylation, mutation and cancer. *BioEssays* **14:** 33–36.

Klimasauskas, S. and R.J. Roberts. 1995. M.HhaI binds tightly to substrates containing mismatches at the target base. *Nucleic Acids Res.* **23:** 1388–1395.

Leach, F.S., N.C. Nicolaides, N. Papadopoulos, B. Liu, J. Jen, R. Parsons, P. Peltomäki, P. Sistonen, L.A. Aaltonen, M. Nyström-Lahti, X.-Y. Guan, J. Zhang, P.S. Meltzer, J.-W. Yu, F.-T. Kao, D.J. Chen, K.M. Cerosaletti, R.E.K. Fournier, S. Todd, T. Lewis, R.J. Leach, S.L. Naylor, J. Weissenbach, J.-P. Mecklin, H. Järvinen, and ten others. 1993. Mutations of a *mutS* homolog in hereditary nonpolyposis colorectal cancer. *Cell* **75:** 1215–1225.

Li, E., T.M. Bestor, and R. Jaenisch. 1992. Targeted mutation of the DNA methyltransferase gene results in embryonic lethality. *Cell* **69:** 915–926.

Lieb, M. 1985. Recombination in the lambda repressor gene: Evidence that very short patch (VSP) mismatch correction restores a specific sequence. *Mol. Gen. Genet.* **199:** 465–470.

———. 1987. Bacterial genes mutL, mutS, and dcm participate in repair of mismatches at 5-methylcytosine sites. *J. Bacteriol.* **169:** 5241–5246.

Lindahl, T. and B. Nyberg. 1974. Heat induced deamination of cytosine residues in deoxyribonucleic acids. *Biochemistry* **13:** 3405–3410.

Lindahl, T., S. Ljungquist, W. Siegert, B. Nyberg, and B. Sperens. 1977. DNA N-glycosidases. *J. Biol. Chem.* **252:** 3287–3294.

Magewu, A.N. and P.A. Jones. 1994. Ubiquitous and tenacious methylation of the CpG site in codon 248 of the p53 gene may explain its frequent appearance as a mutational hot spot in human cancer. *Mol. Cell. Biol.* **14:** 4225–4232.

Malkin, D., F.P. Li, L.C. Strong, J.F. Fraumeni, Jr., C.E. Nelson, D.H. Kim, J. Kassel, M.A. Gryka, F.Z. Bischoff, M.A. Tainsky, and S.H. Friend. 1990. Germ line p53 mutations in a familial syndrome of breast cancer, sarcomas, and other neoplasms. *Science* **250:** 1233–1238.

Mazur, M. and B.W. Glickman. 1988. Sequence specificity of mutations induced by benzo[*a*]pyrene-7,8-diol-9,10-epoxide at endogenous *aprt* gene in CHO cells. *Somatic Cell. Mol. Genet.* **14:** 393–400.

McClelland, M. and R. Ivarie. 1982. Asymmetrical distribution of CpG in an "average" mammalian gene. *Nucleic Acids Res.* **10:** 7865–7877.

Mol, C.D., A.S. Arvai, R.J. Sanderson, G. Slupphaug, B. Kavli, H.E. Krokan, D.W. Mosbaugh, and J.A. Tainer. 1995. Crystal structure of human uracil-DNA glycosylase in complex with a protein inhibitor: Protein mimicry of DNA. *Cell* **82:** 701–708.

Muller, S.J. and S. Caradner. 1991. Isolation and characterization of a human cDNA encoding uracil-DNA glycosylase. *Biochim. Biophys. Acta* **1088:** 197–207.

Neddermann, P. and J. Jiricny. 1993. The purification of a mismatch-specific thymine-DNA glycosylase from HeLa cells. *J. Biol. Chem.* **268:** 21218–21224.

Palombo, F., M. Hughes, J. Jiricny, O. Truong, and J. Hsuan. 1994. Mismatch repair and cancer. *Nature* **367:** 417.

Palombo, F., P. Gallinari, I. Iaccarino, T. Letteri, M. Hughes, A. D'Arrigo, O. Truong, J. Hsuan, and J. Jiricny. 1995. GTBP, a 160-kilodalton protein essential for mismatch-binding activity in human cells. *Science* **268:** 1912–1914.

Papadopoulos, N., N.C. Nicolaides, B. Liu, R. Parsons, C. Lengauer, F. Palombo, A. D'Arrigo, S. Markowitz, J.K.V. Willson, K.W. Kinzler, J. Jiricny, and B. Vogelstein. 1995. Mutations of GTBP in genetically unstable cells. *Science* **268:** 1915–1917.

Perutz, M.F. 1990. Frequency of abnormal human haemoglobins caused by C-T transitions in CpG dinucleotides. *J. Mol. Biol.* **213:** 203–206.

Rideout, W.M.I., G.A. Coetzee, A.F. Olumi, and P.A Jones. 1990. 5-methylcytosine as an endogenous mutagen in the human LDL receptor and p53 genes. *Science* **249:** 1288–1290.

Savva, R., K. McAuley-Hecht, T. Brown, and L. Pearl. 1995. The structural basis of specific base-excision repair by uracil-DNA glycosylase. *Nature* **373:** 487–493.

Schmutte, C., W.M. Rideout, J.-C. Shen, and P.A. Jones. 1994. Mutagenicity of nitric oxide is not caused by deamination of cytosine or 5-methylcytosine in double-stranded DNA. *Carcinogenesis* **15:** 2899–2903.

Schmutte, C.J., A.S. Yang, R.W. Beart, and P.A. Jones. 1995. Base excision repair of U:G mismatches at a mutational hotspot in the p53 gene is more efficient than base excision repair of T:G mismatches in extracts of human colon tumors. *Cancer Res.* **55:** 3742–3746.

Shapiro, R. and R.S. Klein. 1966. The deamination of cytidine and cytosine by acidic buffer solutions. Mutagenic implications. *Biochemistry* **5:** 2358–2362.

Shen, J.-C., W.M. Rideout III, and P.A. Jones. 1992. High frequency of mutagenesis by a DNA methyltransferase. *Cell* **71:** 1073–1080.

———. 1994. The rate of hydrolytic deamination of 5-methylcytosine in double-stranded DNA. *Nucleic Acids Res.* **22:** 972–976.

Shen, J.-C., J.-M. Zingg, A.S. Yang, C. Schmutte, and P.A. Jones. 1995. A mutant HpaII methyltransferase functions as a mutator enzyme. *Nucleic Acids Res.* **23:** 4275–4282.

Shenoy, S., K.C. Ehrlich, and M. Ehrlich. 1987. Repair of thymine-guanine mismatched base-pairs in bacteriophage M13mp18 DNA heteroduplexes. *J. Mol. Biol.* **197:** 617–626.

Shibata, A. 1995. "The implications of p53 mutations in cancer of the bladder and breast: Examples of molecular epidemiology." Ph.D, thesis, University of California, School of Public Health, Los Angeles.

Skandais, A., B.N. Ford, and B.W. Glickman. 1994. Strand bias in mutation involving 5-methylcytosine deamination in the human hprt gene. *Mutat. Res.* **314:** 21–26.

Smith, S.S., B.E. Kaplan, L.C. Sowers, and E.M. Newman. 1992. Mechanism of human methyl-directed DNA methyltransferase and the fidelity of cytosine methylation. *Proc. Natl. Acad. Sci.* **89:** 4744–4748.

Sommer, S.S. 1995. Recent human germ-line mutation: Inferences from patients with hemophilia B. *Trends Genet.* **11:** 141–147.

Steinberg, R.A. and K.B. Gorman. 1992. Linked spontaneous CG-TA mutations at CpG sites in the gene for protein kinase regulatory subunit. *Mol. Cell. Biol.* **12:** 767–772.

Sved, J. and A. Bird. 1990. The expected equilibrium of the CpG dinucleotide in vertebrate genome under a mutation model. *Proc. Natl. Acad. Sci.* **87:** 4692–4696.

Szyf, M., G. Tanigawa, and P. McCarthy. 1990. A DNA signal from the Thy-1 gene defines de novo methylation patterns in embryonic stem cells. *Mol. Cell. Biol.* **10:** 4396–4400.

Tornaletti, S. and G. Pfeifer. 1995. Complete and tissue dependent methylation of CpG sites in the p53 gene: Implications for mutations in human cancers. *Oncogene* **10:** 1493–1499.

Wiebauer, K. and J. Jiricny. 1989. In vitro correction of G•T mispairs to G•C pairs in nuclear extracts from human cells. *Nature* **339:** 234–236.

Wink, D.A., K.S. Kasprzak, C.M. Maragos, R.K. Elespuru, M. Misra, T.M. Dunams, T.A. Cebula, W.H. Koch, A.W. Andrews, J.S. Allen, and L.K. Keefe. 1991. DNA deaminating ability and genotoxicity of nitric oxide and its progenitors. *Science* **254:** 1001–1003.

Wu, J.C. and D.V. Santi. 1987. Kinetic and catalytic mechanism of HhaI methyltransferase. *J. Biol. Chem.* **262:** 4778–4786.

Wyszynski, M., S. Gabbara, and A.S. Bhagwat. 1994. Cytosine deaminations catalyzed by DNA cytosine methyltransferases are unlikely to be the major cause of mutational hot spots at sites of cytosine methylation in *Escherichia coli. Proc. Natl. Acad. Sci.* **91:** 1574–1578.

Yang, A.S., J.-C. Shen, J.-M. Zingg, S. Mi, and P.A. Jones. 1995. HhaI and HpaII DNA methyltransferases bind DNA mismatches, methylate uracil and block DNA repair. *Nucleic Acids Res.* **23:** 1380–1387.

Yang, A.S., M.L. Gonzalgo, J.-M. Zingg, R.P. Millar, J.D. Buckley, and P.A. Jones. 1996. The rate of CpG mutation in an Alu repetitive element within the p53 tumor suppressor gene in the primate germline. *J. Mol. Biol.* **258:** 240–250.

Plant Methyltransferases and Their Targets in the Plant Genome

Roger L.P. Adams, Sriharsa Pradhan,[1] Colin A. Johnson,[2]
Heather Lindsay, Evonne W.L. Shek,
Gareth I. Jenkins, and Nigel A.R. Urwin
Institute of Biomedical and Life Sciences
University of Glasgow, Glasgow G12 8QQ
United Kingdom

In addition to the methylation of cytosine in some CG dinucleotides, plant genomes contain 5-methylcytosine (a cytosine with a methyl group in place of the hydrogen at position 5) in the trinucleotide sequence mCNG, where N is reportedly any of the four common DNA bases (Gruenbaum et al. 1981). 5-Methylcytosine has also been found in nonsymmetrical sequences in transgenes (Meyer et al. 1994), but there is no evidence in plants for the N-4 methylcytosine that is found in some prokaryotes. We have shown that CG and CNG methylation are carried out by different methyltransferases (Pradhan and Adams 1995), and one possible consequence of this is that the two reactions could be independently regulated. This would allow the methylation of CG and CNG sequences to have unrelated functions in the plant cell, although to date there are no data to support this suggestion.

A typical base composition for angiosperm DNA is 40% G+C. If all CG and CNG sequences were to be methylated, then the percentage of cytosines methylated would be 36% (20% in mCG and 20% in mCNG, of which 4% would also be in mCG). Some angiosperms approach this value, and cleavage of pea DNA with methyl-sensitive restriction enzymes indicates that most CG and CWG sequences are methylated (Pradhan et al. 1995). (W stands for A or T and hence the CWG trinucleotide contains CAG on one strand of DNA and CTG on the other.) Yet in the DNA from *Arabidopsis*, only 5% of the cytosines are methylated (Leutwiler et al. 1984). This may reflect the reduced amount of noncoding DNA in *Arabidopsis*, as it has been shown in other species that repetitive sequence DNA is the location of much of the methylcytosine (Deumling 1981; Adams et al. 1983). The obvious conclusion is that in the DNA from *Arabidopsis*, most of the CG and/or CNG sequences are not meth-

Present addresses: [1]New England Biolabs, Beverly, Massachusetts; [2]Department of Anatomy, University of Birmingham, United Kingdom.

 0-87969-490-4/96 $5 + .00

ylated; the possibility must be considered that this may also be true for single-copy, coding DNA from all plants.

Belanger and Hepburn (1990) have shown that in lower plant taxa there is no CG methylation and only limited CNG methylation. This finding can be only partially explained by these cells having a high proportion of coding DNA. The low levels of methylcytosine in these cells are almost certainly a result of the loss of an active methyltransferase (Zhang et al. 1992; Wilkinson et al. 1995; Pinarbarsi et al. 1996).

In this chapter, we do not consider methylation of mitochondrial or chloroplast DNA, a subject that is controversial (Feng and Chiang 1984; Kobayashi et al. 1990; Scott et al. 1991). Rather, we attempt to review the structure of the methyltransferases from pea and their possible targets in the DNA.

DISTINCT CG AND CWG METHYLTRANSFERASES

The CWG methyltransferase has been purified to homogeneity from pea shoot tips. It is a relatively unstable enzyme of 110 kD that methylates cytosines in CAG and CTG sequences (Houlston et al. 1993; Pradhan and Adams 1995). It shows a strong preference for hemimethylated DNA and is found in the nucleus of dividing cells. It does not methylate either the outer cytosine in CCG sequences nor cytosine in CG dinucleotides; neither does it bind to CCG or CG sequences (Pradhan and Adams 1995).

The CG methyltransferase copurifies with the CWG methyltransferases through most purification stages, but the two can be separated by MonoQ chromatography. The native size of the CG methyltransferase from *Pisum sativum* is about 150 kD, but on SDS-PAGE the major band is found at 55 kD with only a minor band at 140 kD. We attribute this to proteolysis of the native enzyme, a problem that also besets attempts to purify mammalian methyltransferases (see, e.g., Xu et al. 1995). Protein bands of about 55 kD are also associated with methyltransferase activity from rice and wheat (Theiss et al. 1987; Giordano et al. 1991). The pea CG methyltransferase does not bind to CWG trinucleotides, but it does bind to CG and to CCG sequences, although it fails to transfer methyl groups to the external cytosine in the latter (Pradhan and Adams 1995).

Finnegan and Dennis (1993) have reported the sequence of an *Arabidopsis* clone putatively encoding a DNA methyltransferase of 1534 amino acids (about 170 kD), and we have a cDNA clone from pea that shows about 70% homology with the *Arabidopsis* sequence and apparently encodes a protein of 1560 amino acids (Pradhan et al., in prep.). Furthermore, Northern analysis indicates the presence of only one mRNA of about 5 kb, a message that is present predominantly in dividing tissues (Pradhan et al., in prep.). In pea, there appears to be a single gene corresponding to this sequence, whereas evidence has been presented for two or more genes in *Arabidopsis* (Finnegan and Dennis 1993).

Like the mammalian enzymes, the plant enzymes consist of a carboxy-terminal region carrying the conserved sequence motifs found in prokaryotic enzymes and a long carboxy-terminal section involved in nuclear localization, etc. (Bestor and Verdine 1994; Adams 1995; Cheng 1995). Cleavage of the predicted pea protein at the linker region between the amino and carboxy-terminal parts (peptide sequence KKKGKCK) would yield a carboxy-terminal protein of 456 amino acids (i.e., ~ 55 kD) that may account for the strong band seen on SDS-PAGE of the purified CG enzyme. This possibility leads us to suggest that the purified enzymes are processed forms of the original protein. It is possible that the initially synthesized protein is processed by alternative pathways to give enzymes with a common carboxyl terminus (Pradhan et al., in prep.).

Kovarik et al. (1994) have found evidence that methylation of CG and CNG sequences is differentially sensitive to various inhibitors. In particular, they have shown that CNG methylation is more sensitive to inhibition by dihydroxypropyladenine than is CG methylation. This inhibitor acts by raising the intracellular concentration of S-adenosyl homocysteine (AdoHcy). AdoHcy is a competitive inhibitor with S-adenosyl methionine (AdoMet) for the transfer of a methyl group to DNA. Therefore, we looked for differential sensitivity of the separated enzymes to inhibition by AdoHcy in vitro. Figure 1 shows the action of this inhibitor on the two activities in the presence of a constant concentration of AdoMet. No difference in sensitivity is apparent, indicating that the interaction of the two enzymes with substrate and inhibitor must be very similar (i.e., the structures of the enzymes at the active site are very similar). It is possible that the in vivo results arise through the modulation of activity by accessory proteins acting on the target DNA.

Kovarik et al. (1994) also find that CNG methylation is refractory to the inhibitory effects of 5-azacytidine. This inhibitor (which contains a nitrogen atom in place of the CH normally present at the 5 position of the pyrimidine ring in cytidine) is incorporated into DNA and results in the formation of a stable, covalent intermediate complex between the methyltransferase and the azacytosine-containing target site in the DNA. As a result of this immobilization of enzyme in a covalent complex with

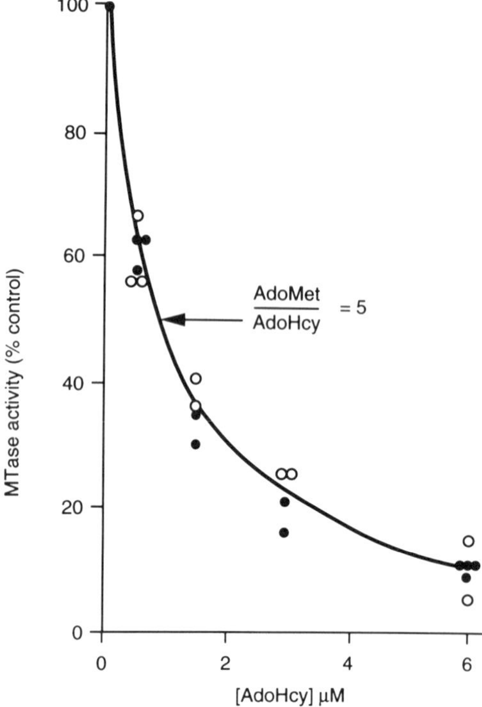

Figure 1 Inhibition of methyltransferases by AdoHcy. DNA methyltransferase was assayed as described previously using substrates specific for the CG or the CWG methyltransferase (Houlston et al. 1993). AdoHcy was present at the indicated concentrations. 50% inhibition was seen for both enzymes when the concentration of AdoHcy was 20% that of AdoMet. This indicates a similar mechanism of action for both enzymes and makes it unlikely that they can be discriminated in vivo by use of agents that affect the AdoMet:AdoHcy ratio.

DNA, the concentration of free enzyme is reduced to a level at which it becomes insufficient to maintain the normal level of DNA methylation. Although the results of Kovarik et al. (1994) might reflect a different mechanism of action for the two enzymes, a similar result would be seen were the CNG enzyme to be in considerable excess over requirements. In such a situation, only a small fraction of the CNG methyltransferase would be immobilized in the covalent complex with azacytosine-containing DNA, leaving adequate levels available to methylate the rest of the newly synthesized DNA. Lower levels of the CG enzyme may lead to its all being immobilized by binding to azacytosine-containing target sites in the DNA.

DISTRIBUTION OF 5-METHYLCYTOSINE

Methylation of CG and CWG

In vertebrates, the presence of high levels of CG methylation is associated with significant deamination of the methylcytosine to thymine, a change that is incompletely or inefficiently repaired (Bird 1986; Shen et al. 1994). Thus, where an mCG dinucleotide pair was initially present in a gene, the deamination process would convert this into a TG/CA dinucleotide pair. This is a common cause of mutation, and the CG dinucleotide is said to be a "hot spot" for mutation (Coulondre et al. 1978; Jones et al. 1992). It is believed that, over evolutionary time, this deamination reaction has resulted in the depletion of CG dinucleotides which, in vertebrates, now occur at only one-fifth their expected frequency (Adams and Eason 1984 and references therein). This phenomenon is known as CG suppression.

In contrast, the considerably higher amount of methylcytosine in higher plant genomes has not been accompanied by such a high level of CG suppression (Gardiner-Garden et al. 1992; Montero et al. 1992), and this may point to a much more effective mismatch repair mechanism in plants. (Mismatch repair is the process whereby mismatches, such as T:G or C:A involving "normal" bases, are corrected. The T:G mismatch commonly arises as a consequence of deamination of methylcytosine to thymine and is specifically corrected to C:G.)

Vertebrates have clusters of unmethylated CG dinucleotides present in the so-called CG islands found near the 5′ end of many genes (Bird 1986). These islands are generally rich in G+C and are important in the control of gene expression. They have avoided CG suppression, and it has been proposed that this results from the fact that they are always unmethylated, even in the germ line. The lack of methylation protects them from the consequences of an ineffective mismatch repair system that might otherwise fail to correct a methylC to T conversion. As mentioned above, plants seem to have a much more efficient mismatch repair system, which is important because plants do not have the well-defined germ-line tissues characteristic of vertebrates, where essential sequences can be protected from the consequences of deamination by being maintained in an unmethylated state (Walbot 1985; Bird 1986).

Plants also have CG islands, although they are not as common as in animal genomes (Antequera and Bird 1988). Additionally, plants have CNG islands, methylation of which can have a dramatic effect on gene expression (Pradhan et al., in prep.).

Upon digestion of genomic DNA with enzymes such as *Pst*I (cleaves at CTGCAG but not when either C is methylated), these islands are cleaved to give a cluster of small fragments known as PTFs (*Pst* tiny fragments). PTFs have been cloned and found to be enriched in CWG sequences but depleted in CG sequences and to have an average base composition of 43% G+C. Occasional *Hpa*II sites occur within or near PTFs; these are unmethylated, implying that these regions of the genome lack all methylation. The distance between these unmethylated *Hpa*II sites ranges from 410 bp to more than 3000 bp (the size of small fragments generated from CG island DNA by *Hpa*II digestion [HTFs] is about 80 bp, whereas PTFs are up to 500 bp in length), and this allows an estimate of the minimum size of CWG islands, which is more than 1 kb. Some CG islands are found in regions of the genome lacking unmethylated CWG sites (i.e., CWG oceans of 20 kb or larger). However, very few unmethylated CWG sites occur in CG oceans; rather, they occur in regions containing occasional unmethylated CCGG sequences that occur at a frequency close to what might be expected on the basis of a random distribution. This conclusion arises from a consideration of the situation where there is no CG suppression, when a random sequence would contain *Hpa*II target sites about every 600 bp. Where CG suppression is extensive, the *Hpa*II target sites would occur, on average, every 2500 bp.

Methylation of CCG

Does mC occur in the sequence mCCG? Clearly, the inner cytosine will be methylated by the CG methyltransferase, as will be the only cytosine in the complementary sequence (CGG). The methylcytosine in mCGG might act as a signal to both the CG and the CNG methyltransferase to indicate that the target site is hemimethylated. This would lead to competition between the two enzymes for the target site. However, we have shown that the CWG methyltransferase from pea does not bind to the sequence CCG/mCGG, whereas this target is recognized by the CG methyltransferase (Pradhan and Adams 1995).

Except for the original observation (Gruenbaum et al. 1981) of a low level of methylation of the 5′ cytosine in CC dinucleotides, most of the evidence for the occurrence of mCCG depends on data obtained by restriction digestion using *Msp*I, an enzyme that is reported to be sensitive to methylation at the outer cytosine but not at the inner cytosine in the sequence CCGG (Waalwijk and Flavell 1978; Walder et al. 1983). There is no satisfactory isoschizomer to use in assessing normal digestion patterns, and the controls used to assess digestion have always been unmethylated plasmid or viral DNA added to the incubation. (The isoschizomer required would be a restriction enzyme that cleaves the same sequence [CCGG] but that is insensitive to methylation of either cytosine.) There is evidence in the literature (Goldsbrough et al. 1982; Keshet and Cedar 1983) that *Msp*I is affected to varying degrees by methylation of the inner cytosine and so may give anomalous results when used to digest CG-methylated genomic DNA. In particular, *Msp*I is effectively blocked by CG methylation in G+C-rich environments (Busslinger et al. 1983; Keshet and Cedar 1983). Figure 2 shows the effect on the digestion by *Msp*I of plasmid DNA that has been methylated at all CGs by M.*Sss*I methyltransferase. The methylated DNA is digested up to 20 times more slowly than is the unmethylated DNA. The rate of digestion depends critically on the source of the enzyme (different suppliers obtain their enzyme from different *Moraxella* species), but in no case have we found the rate of digestion to be more than half that of mock-methylated DNA (i.e., DNA that has been processed in the same way except that the M.*Sss*I has been omitted). Using *excess Msp*I, we can obtain complete digestion of a plasmid that has been fully methylated using M.*Sss*I. We have shown that none of these CCGG sites is methylated at the external cytosine. We have done this by extending the 3′ OH of the external cytosine generated on *Msp*I cleavage using [α-^{32}P]-labeled dGTP. Upon hydrolysis of the resulting DNA to 3′ mononucleotides there is only a single spot of 3′ dCMP on TLC with no evidence for any 3′ methyl dCMP. This confirms that no *Msp*I-sensitive mCCGG sites are generated by treatment with M.*Sss*I and that all the CmCGG sites in the plasmid can be cleaved when sufficient *Msp*I is used. This DNA is not cleaved by *Hpa*II, showing the inner cytosine in all CCGG sites to be methylated.

We have been able to digest pea genomic DNA to small fragments (average size about 600 bp) with *Msp*I (Pradhan and Adams 1995), and Leutwiler et al. (1984) report similar digestion of wheat DNA with *Msp*I. To achieve this digestion requires a long incubation with a very large excess of enzyme. Analysis of the fragments produced by *Msp*I treatment confirms that all cleaved sites are unmethylated at the external cytosine. Thus, the frequently observed failure to digest genomic DNA with *Msp*I may simply reflect the slower digestion rates when the *inner* cytosine (i.e., that in the CG dinucleotide) is methylated. An alternative explanation is that cleavage by *Msp*I is affected by the conformation of DNA with a high methylcytosine content rather than the presence of a methylcytosine specifically in the CCGG target sequence. Either of these possibilities could explain the findings with the highly methylated plant genome. As a result of these studies, we believe the evidence for methylation of the sequence mCCG may be insecure and that some ex-

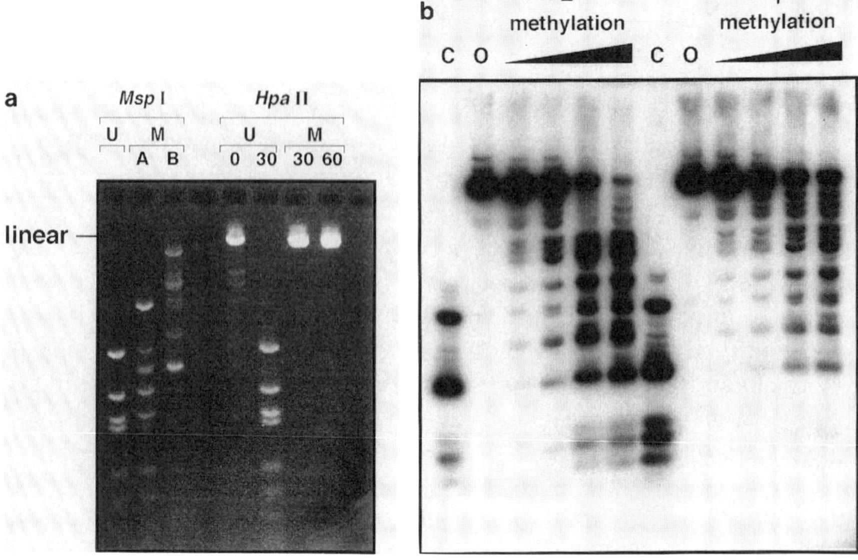

Figure 2 Effect of CG methylation on *Msp*I digestion. (*a*) Plasmid (pBluescript) DNA was mock-methylated or methylated with M.*Sss*I (methylates cytosine in CG dinucleotides) and linearized with *Kpn*I. Treatment with *Hpa*II for 30 or 60 min shows that the methylated plasmid is completely resistant to digestion and hence fully methylated. The mock-methylated plasmid is completely digested by *Hpa*II in 30 min. The plasmid was treated with *Msp*I (2 units per 250 ng) from two different suppliers (A and B). Samples were separated on a 1% agarose/TBE gel. Digestion of the mock-methylated plasmid (U) is complete within 15 min (only the 4-hr digestion is shown) but neither sample of *Msp*I completely digests the methylated plasmid (M) in 4 hr. (*b*) A recombinant plasmid (100 ng pVHCk) was treated similarly but using a third supplier of *Msp*I and zero (0), 5, 15, 30, or 60 digestion units (units x min). The figure shows a Southern blot probed with the entire plasmid labeled by random priming. Analysis of the rates of disappearance of the intact linear plasmid shows a twofold difference. This is a more sensitive assay that analyzes the effect of CG methylation on the site most sensitive to *Msp*I cleavage. C is the plasmid after complete digestion.

periments reported in the literature may have been misinterpreted as a consequence of the low activity of *Msp*I on DNA containing high levels of mCG.

These studies could cast doubt on the use of *Msp*I as a probe for inactive chromatin structures (Levine et al. 1991), as the enzyme responds to the structure of the DNA and not only to the structure of the chromatin. However, in mitigation, we have shown that inactive chromatin structures (as judged by resistance to *Msp*I cleavage) can be present on unmethylated as well as on methylated regions of a transfected plasmid (Kass et al. 1993).

Methyladenine

There is some controversy over the occurrence of methyladenine in plant cells. The significant amounts found in a transgene in wheat cells were attributed to the introduction of a mycoplasma-like organism into the wheat cells at the time of transformation (Chen et al. 1994). However, the presence of methyladenine in plant promoters can have a dramatic effect on their activity (Truss et al. 1992; Torres et al. 1993; and below).

We have studied the expression of reporter genes present on recombinant plasmids one day after their introduction into bean protoplasts (transient expression assay). We have used two different promoters: namely, the bean rubisco small subunit gene promoter (*rbcS2*) and the promoter from the cauliflower mosaic 35S gene (*CaMV35S*). The *CAT* gene (encoding the bacterial chloramphenicol acyltransferase) driven by the bean *rbcS2* promoter is stimulated between six- and sevenfold by prior in vitro methylation using the Dam methyltransferase (an enzyme that methylates the adenines in GATC sequences). A similar enhancement of expression is seen using plasmids grown in Dam$^+$ *Escherichia coli* hosts compared with those grown in Dam$^-$ hosts (Table 1). No stimulation is seen in similar constructs driven by the strong *CaMV35S* promoter, indicating that the effect is mediated by methylation of the *rbcS2* promoter. However, Southern analysis of DNA isolated from a variety of tissues provides no evidence for in vivo methylation of the only *Dam* site in the *rbcS2* promoter present 57 bp upstream of the transcription start site (Fig. 3). Given that lack of adenine methylation is the normal state for all plant DNA (and not just the *rbcS2* promoter), then promoters present in plasmids grown in the usual Dam$^+$ hosts could exhibit abnormally enhanced expression. This may lead to erroneous conclusions concerning promoter strengths and could lead to the masking of the effects of external signals.

METHYLATION AND GENE EXPRESSION

It is clear from the discovery of two distinct methyltransferases, and the finding that some genes appear to have CWG islands rather than CG islands, that the control over the expression of these two classes of genes may be distinct. A comparison of 40 different plant promoters indicates

Table 1 Effect of adenine methylation on gene expression

E. coli host	In vitro Dam methylation	CAT activity (% control)	
		rbcS2	*CaMV35S*
Dam^-	–	100	100
Dam^-	+	670	104
Dam^+	–	620	120

Plasmids were grown in *E. coli* Xl1-Blue (Dam^+, Dcm^+) or GM2163 (Dam^-, Dcm^-). Dam^+ cells contain the adenine methylase that transfers methyl groups to GATC sequences; Dcm^+ cells contain the cytosine methylase that transfers methyl groups to CCWGG sequences. The plasmids contained the *CAT* reporter gene linked to either the bean *rbcS2* promoter or the *CaMV35S* promoter. In vitro methylation was with the Dam methyltransferase (New England Biolabs), and the efficiency of methylation was checked by exposure to *Dpn*I (cuts only methylated GATC sequences) and *Mbo*I (cuts only unmethylated GATC sequences). 100% CAT activity represents 15,000 ^{14}C cpm (*35S* promoter) or 700 ^{14}C cpm (*rbcS* promoter) in acylchloramphenicol as assayed by the method of Seed and Sheen (1988).

that, although the concentration of CG dinucleotides is fairly constant (between 6.5% and 12% of cytosines are in CG dinucleotides), the fraction of cytosine in CWG trinucleotides varies between 2% and 17% (Pradhan et al., in prep.). A few promoters do not fall into this general pattern in that they have a much lower or a much higher proportion of cytosines in CG dinucleotides. We cannot generalize at this time, but this may be related to the strength of the promoters or to the types of signal to which the promoter responds. When studied as part of a transgene in transient expression assays, the strong, universally expressed *CaMV35S* promoter that has a high CG content is relatively insensitive to CWG methylation. In contrast, the weak, light-responsive *rbcS* promoter that is CG-deficient is very sensitive to CWG methylation. In contrast, methylation of all the CGs in the recombinant plasmid DNAs leads to complete inhibition of expression from both promoters.

There are many reports in the literature showing an inverse correlation between CG methylation and gene expression, but the link between the two is generally not obvious. Transient expression experiments have the advantage of at least clearly showing that it is methylation that inhibits transcription rather than the reverse. In animal systems, as with plants, blanket methylation of all CGs in the transfected recombinant DNA leads to inhibition of gene expression; however, the observed inhibition may be independent of the promoter and may be mediated by the formation and spreading of inactive chromatin structures (Kass et al. 1993).

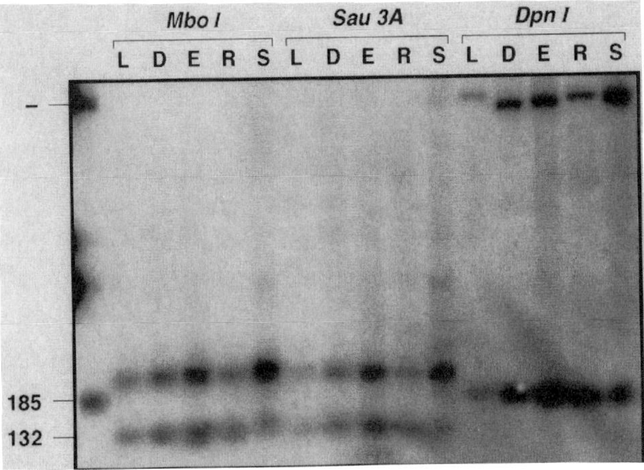

Figure 3 Absence of adenine methylation in the bean *rbcS2* promoter. The bean *rbcS2* promoter contains the sequence GATC at –57 relative to the transcription start site. Genomic DNA from various bean tissues was digested with *Kpn*I and *Hin*dIII to release the pertinent 185-bp fragment extending to just after the transcription start site, and also with one of the indicated restriction enzymes, and separated on a 1% polyacrylamide gel. The Southern transfer was probed with the 185-bp fragment that had been labeled by random priming. This probe illuminates not only the 185-bp fragment from the *rbcS2* promoter, but also a much larger fragment from a related *rbcS* promoter; both are seen in the first lane. The same two bands are seen in DNA treated with *Dpn*I, which cuts GATC *only* when it is methylated at the adenine. Both bands disappear (and are replaced by smaller bands; the smallest of which is only 53 bp long and is not seen on the gel) when the DNA is cleaved with *Mbo*I, which cuts GATC independently of its methylation status, or *Sau3*A, which does not cut methylated GATC sequences. DNA was isolated from leaves from light-grown plants (L), dark-adapted plants (D), or etiolated plants (E), or from roots (R) or seed (S).

CONCLUSIONS

We are beginning to discover more about plant DNA methylation, but it is still a mystery why plants have so much methylcytosine in their genomes. Although we know there are separate CG and CWG methyltransferases, we do not know what controls their production or their activity. If we assume that the small genome of *Arabidopsis* is largely expressed and unmethylated, the corollary would be that most methylcytosine in more typical higher plants is present in redundant (multicopy) and inactive genes, in introns, or in "junk" DNA. Certainly, most individual endogenous genes that have been investigated are unmethylated, but these have been chosen for study just because they are functional and active. It is frequently found that when extra copies of genes are introduced into plant cells, both the introduced and the endogenous genes become silenced in a process that is associated with their methylation.

Vertebrates have developed a mechanism of reprogramming the pattern of methylation with each generation; a mechanism that involves periods of demethylation and de novo methylation. Plants propagate both by genetic and by vegetative mechanisms, and many cells in a mature plant can give rise to a new organism (i.e., they are totipotent). We know nothing of the mechanisms controlling DNA methylation under these circumstances, yet it must be that plants have a highly developed mechanism of demethylation to prevent the irreversible methylation and inactivation of essential genes, a process that would be inimical to the processes of both genetic and vegetative propagation. As with methylation, there may be distinct enzymes involved in the demethylation of mCG and mCWG sequences.

ACKNOWLEDGMENTS

We thank the British Commonwealth Association for a scholarship for S.P., the SERC for a grant to R.L.P.A., and the University of Glasgow and Professor Lindsay for the provision of facilities.

REFERENCES

Adams, R.L.P. 1995. Eukaryotic DNA methyltransferases—Structure and function. *BioEssays* **17:**139–145.

Adams, R.L.P. and R. Eason. 1984. Increased G + C content of DNA stabilizes methyl CpG dinucleotides. *Nucleic Acids Res.* **12:** 5869–5877.

Adams, R.L.P., R.H. Burdon, and J. Fulton. 1983. Methylation of satellite DNA. *Biochem. Biophys. Res. Commun.* **113:** 695–702.

Antequera, F. and A.P. Bird. 1988. Unmethylated CpG islands associated with genes in higher plant DNA. *EMBO J.* **7:** 2295–2299.

Belanger, F.C. and A.G. Hepburn. 1990. The evolution of CpNpG methylation in plants. *J. Mol. Evol.* **30:** 26–35.

Bestor, T.H. and G.L. Verdine. 1994. DNA methyltransferases. *Curr. Opin. Cell Biol.* **6:** 380–389.

Bird, A.P. 1986. CpG-rich islands and the function of DNA methylation. *Nature* **321:** 209–213.

Busslinger, M., E. deBoer, S. Wright, F.G. Grosveld, and R.A. Flavell. 1983. The sequence GGC^{m}CGG is resistant to *Msp*I cleavage. *Nucleic Acids Res.* **11:** 3559–3569.

Chen, D.F., P.J. Dale, J.S. Heslop-Harrison, J.W. Snape, W. Harwood, S. Bean, and P.M. Mullineaux. 1994. Stability of transgenes and presence of N^{6} methyladenine DNA in transformed cells. *Plant J.* **5:** 429–436.

Cheng, X. 1995. Structure and function of DNA methyltransferases. *Annu. Rev. Biophys.*

Biomol. Struct **24:** 293–318.

Coulondre, C., J.H. Miller, P.J. Farabaugh, and W. Gilbert. 1978. Molecular basis of base substitution hotspots in *Escherichia coli*. *Nature* **274:** 775–780.

Deumling, B. 1981. Sequence arrangement of a highly methylated satellite DNA of a plant, *Scilla:* A tandemly repeated inverted repeat. *Proc. Natl. Acad. Sci.* **78:** 338–342.

Feng, T.-Y. and K.-S. Chiang. 1984. The persistence of maternal inheritance in *Chlamydomonas* despite hypomethylation of chloroplast DNA induced by inhibitors. *Proc. Natl. Acad. Sci.* **81:** 3438–3442.

Finnegan, E.J. and E.S. Dennis. 1993. Isolation and identification by sequence homology of a putative cytosine methyltransferase from *Arabidopsis thaliana*. *Nucleic Acids Res.* **21:** 2383–2388.

Gardiner-Garden, M., J.A. Sved, and M. Frommer. 1992. Methylation sites in angiosperm genes. *J. Mol. Evol.* **34:** 219–230.

Giordano, M., M.E. Mattachini, R. Cella, and G. Pedrali-Noy. 1991. Purification and properties of a novel DNA methyltransferase from cultured rice cells. *Biochem. Biophys. Res. Commun.* **177:** 711–719.

Goldsbrough, P.B., T.H.N. Ellis, and G.P. Lomonossoff. 1982. Sequence variation and methylation of the flax 5S RNA genes. *Nucleic Acids Res.* **10:** 4501–4514.

Gruenbaum, Y., T Naveh-Many, H. Cedar, and A. Razin. 1981. Sequence specificity of methylation in higher plant DNA. *Nature* **292:** 860–862.

Houlston, C.E., H. Lindsay, S. Pradhan, and R.L.P. Adams. 1993. DNA substrate specificity of pea DNA methylase. *Biochem. J.* **293:** 617–624.

Jones, P.A., W.M. Rideout III, J.-C. Shen, C.H. Spruck, and Y.C. Tsai. 1992. Methylation, mutation and cancer. *BioEssays* **14:** 33–36.

Kass, S.U., J.P. Goddard, and R.L.P. Adams. 1993. Inactive chromatin spreads from a focus of methylation. *Mol. Cell. Biol.* **13:** 7372–7379.

Keshet, E. and H. Cedar. 1983. Effect of CpG methylation on *Msp*I. *Nucleic Acids Res.* **11:** 3571–3580.

Kobayashi, H., J, Ngernprasirtsiri, and T. Akazawa. 1990. Transcriptional regulation and DNA methylation in plastids during transitional conversion of chloroplasts to chromoplasts. *EMBO J.* **9:** 307–313.

Kovarik, A., B. Koukalova, A. Holy, and M. Bezdek. 1994. Sequence-specific hypomethylation of the tobacco genome induced with dihydroxypropyladenine, ethionine and 5-azacytidine. *FEBS Lett.* **353:** 309–311.

Leutwiler, L.S., B.R. Hough-Evans, and E.M. Meyerowitz. 1984. The DNA of *Arabidopsis thaliana*. *Mol. Gen. Genet.* **194:** 15–23.

Levine, A., G.L.Cantoni, and A. Razin. 1991. Inhibition of promoter activity by methylation: Possible involvement of protein mediators. *Proc. Natl. Acad. Sci.* **89:** 10119–10123.

Meyer, P., I. Niederhof, and M. ten Lohuis. 1994. Evidence for cytosine methylation of non-symmetrical sequences in transgenic *Petunia hybrida*. *EMBO J.* **13:** 2084–2088.

Montero, L.M., J. Filipski, P. Gil, J. Capel, J.M. Martinez-Zapater, and J. Salinas. 1992. The distribution of 5-methylcytosine in the nuclear genome of plants. *Nucleic Acids Res.* **20:** 3207–3210.

Pinarbarsi, E., J. Elliott, and D.P. Hornby. 1996. Activation of a yeast pseudo DNA methyltransferase by deletion of a single amino acid. *J. Mol. Biol.* (in press).

Pradhan, S. and R.L.P. Adams. 1995. Distinct CG and CNG DNA methyltransferases in *Pisum sativum*. *Plant J.* **7:** 471–481.

Scott, N.S., J.M. Timmis, and M.A. Aycliffe. 1991. Chloroplast genes in higher plants. *Aust. J. Plant Physiol.* **18:** 533–541.

Seed, B. and J.-Y. Sheen. 1988. A simple phase-extraction assay for chloramphenicol acyltransferase activity. *Gene* **67:** 271–277.

Shen, J.-C., W.M. Rideout III, and P. Jones. 1994. The rate of hydrolytic deamination of 5-methylcytosine in double-stranded DNA. *Nucleic Acids Res.* **22:** 972–976.

Theiss, G., R. Schleicher, R. Schimpff-Weiland, and H. Follmann. 1987. DNA methylation in wheat: Purification and properties of DNA methyltransferase. *Eur. J. Biochem.* **167:** 89–96.

Torres, J.Y., A. Block, K. Hahlbrock, and I.E. Somssich. 1993. Influence of bacterial strain genotype on transient expression of plasmid DNA in plant protoplasts. *Plant J.* **4:** 587–592.

Truss, M., J. Bartsch, G. Chalepakis, and M. Beato. 1992. Artificial steroid hormone response element generated by Dam-methylation. *Nucleic Acids Res.* **20:** 1483–1486.

Waalwijk, C. and R.A. Flavell. 1978. *Msp*I, an isoschizomer of *Hpa*II which cleaves both unmethylated and methylated *Hpa*II sites. *Nucleic Acids Res.* **5:** 3231–3236.

Walbot, V. 1985. On the life strategies of plants and animals. *Trends Genet.* **1:** 165–169.

Walder, R.Y., C.J. Laugtimm, R. Catterjee, and J.A. Walder. 1983. Cloning of the *Msp*I modification enzyme. *J. Biol. Chem.* **258:** 1235–1241.

Wilkinson, C.R.M., R. Bartlett, P. Nurse, and A.P. Bird. 1995. The fission yeast gene *pmt1*$^+$ encodes a DNA methyltransferase homologue. *Nucleic Acids Res.* **23:** 203–210.

Xu, G., J. Flynn, J.F. Glickman, and N.O. Reich. 1995. Purification and stabilization of mouse DNA methyltransferase. *Biochem. Biophys. Res. Commun.* **207:** 544–551.

Zhang, Y., M. Nelson, and J.L. van Etten. 1992. A single amino acid change restores DNA cytosine methyltransferase activity in a cloned chlorella virus pseudogene. *Nucleic Acids Res.* **20:** 1637–1642.

Mechanism of DNA Demethylation in Vertebrates and Its Biological Significance

Jean-Pierre Jost
Friedrich Miescher Institut
CH-4002, Basel
Switzerland

The aim of this chapter is to present some general features of site-specific and genome-wide demethylation. Demethylation of DNA is part of the process responsible for the formation of specific methylation patterns. The establishment of a CpG methylation pattern during embryonic development involves at least three components: DNA methyltransferase, a demethylation system (passive or active), and sequence-specific *cis-trans* controlling elements located on or near the CpG methylation sites. The *cis-trans* controlling elements either enhance or inhibit the methylation of specific sequences. Whether a given CpG site is methylated or not depends on the mole ratio of activities of the different factors involved and of their respective affinities for a specified CpG site on the DNA. This concept of titratable factors is also valid for the *cis*-acting control elements. For example, the introduction of DNA-binding sites for specific proteins into the cell could possibly titrate out protein factors and engender hypermethylation of the DNA (Parker et al. 1986; Mehtali et al. 1990; Assaad et al. 1993).

The DNA-methyltransferase level could either be controlled by its synthesis at the correct time during the S phase, or it could lag behind the S phase, resulting in a lower level of DNA methylation (Adams et al. 1990). Alternatively, the level of DNA-methyltransferase activity could be regulated by small-molecular-weight peptides (E.S. Lyon et al. 1982; S.B. Lyon et al. 1987). The selective distribution of DNA methyltransferase between nuclei and the cytoplasm could favor either methylation or no methylation of DNA (Carlson et al. 1992; Leonhart et al. 1992). DNA methyltransferase could also be progressively diluted in the cell, resulting, upon cell divisions, in a decrease in DNA methylation (Sano et al. 1984; Rocamora and Mezquita 1989). Alternatively, DNA methyltransferase could be regulated by Fos, Jun, and Ras, since its promoter has AP-1 (binding sites of Fos/Jun) control elements (Szyf 1994; Rouleau et al. 1995). Recently it was shown, by two different groups

Epigenetic Mechanisms of Gene Regulation

(Brandeis et al. 1994; Macleod et al. 1994), that Sp1-binding sites flanking the CpG islands provided protection against methylation of the CpG islands in the mouse adenosine phosphoribosyl transferase gene.

At the moment, we know very little about the factors that regulate the DNA demodifying machinery. However, it has been shown that in G8 mouse myoblasts (muscle precursor cells) there is, upon differentiation, a sharp increase in 5-methylcytosine DNA/glycosylase (this enzyme removes the methylated cytosine from DNA, leaving the deoxyribose intact), and in developing chicken embryos there is a very abrupt decrease in 5-methylcytosine/DNA glycosylase activity after the 16th day of embryonic development (Jost 1993; Jost and Jost 1994).

PASSIVE VERSUS ACTIVE DNA DEMETHYLATION

As illustrated in Figure 1, there are two possible ways the demethylation of DNA could take place: passive and active demethylation.

Passive Demethylation

For passive DNA demethylation, at least two cycles of replication are necessary to obtain an almost complete demethylation of a given CpG site. According to this model, the DNA methyltransferase is not able to methylate specified CpG sites on the DNA during the S phase. This inhibition could possibly have several causes; either a protein binds to the hemimethylated DNA, preventing the action of the DNA methyltransferase, or there could be a combination of the binding of a protein to the hemimethylated site with a lag in the appearance of the DNA methyltransferase during the replication of DNA. In each case, a 50% demethylation of DNA is achieved after the first cycle of replication and a 75% demethylation after the second (Razin et al. 1984a).

Active DNA Demethylation

For active DNA demethylation, there are in essence four possibilities, three of which are shown in Figure 1A–C. The most probable reaction for which there is some evidence is depicted in Figure 1A. It involves a 5-methylcytosine/DNA glycosylase that removes methylcytosine from the hemimethylated DNA generated after one cycle of replication. One major feature of 5-methylcytosine/DNA glycosylase is its preferential reaction with hemimethylated DNA, which is 6–10 times faster than with bifilarly methylated DNA (Jost et al. 1995). The reaction specificity of

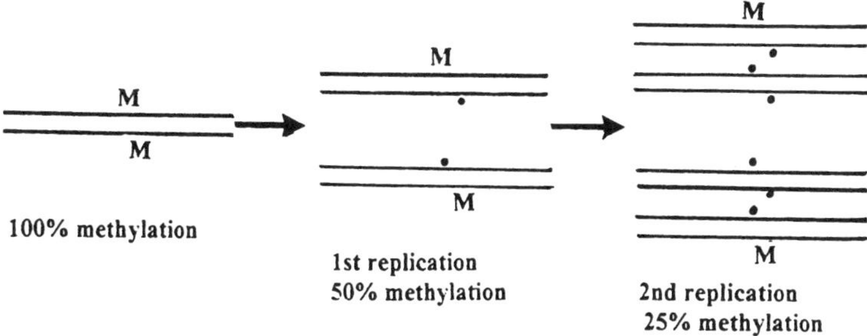

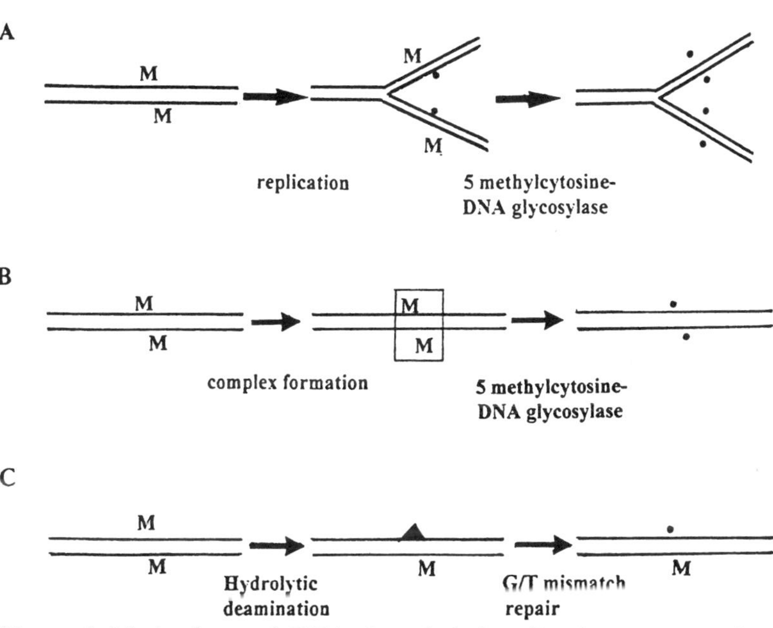

Figure 1 Mechanisms of DNA demethylation. Passive versus active DNA demethylation. (M) Methyl group; (*dot*) no methyl, i.e., cytosine.

the purified 5-methylcytosine/DNA glycosylase is discussed in detail below in the section on "Properties of a highly purified 5-methylcytosine/DNA glycosylase."

Figure 1B shows that there is also a possibility that methylcytosine is removed from bifilarly methylated DNA. A 5-methylcytosine/DNA glycosylase that has a preference for bifilarly methylated DNA has been recently presented by Vairapandi and Duker (1993). In a HeLa nuclear system this enzyme removes methylcytosine in the presence of Mg^{++}. Figure 1C shows a further possibility, which is hydrolytic deamination (spontaneous deamination, without enzymatic activity) of 5-methylcytosine followed by conversion to thymine, creating a G/T mismatch, which is the substrate of an enzyme called mismatch-specific-thymine DNA glycosylase that has been purified to homogeneity (Neddermann and Jiricny 1993). However, it is doubtful that this reaction is efficient enough to account for the active demethylation of DNA. Using a new and very sensitive assay, Shen et al. (1994) have determined the rate constants for the spontaneous hydrolytic deamination of 5-methylcytosine and cytosine in double-stranded DNA at 37°C. The rate constants were 5.8×10^{-13} s^{-1} and 2.6×10^{-13} s^{-1}, respectively. These rate constants may be sufficient to account for the appearance of spontaneous mutations, but they seem insufficient to explain the demethylation of a specific gene, which can occur within hours or days. A fourth possibility (not shown) is a small DNA patch repair pathway, but for this we have no evidence.

SITE-SPECIFIC DNA DEMETHYLATION

General reviews showing maps of the site-specific methylation of different genes have been published previously (Yeivin and Razin 1993; Razin et al. 1994a).

Most investigators have used methylation-sensitive restriction enzymes (Bird and Southern 1978) for the detection of methylation at specific sites. This method has the advantage of being simple and precise, but it is limited by the number of CpG sites that can be directly studied. One variation of the above method is a combination with polymerase chain reaction (PCR) (Singer-Sam et al. 1990). This improved procedure drastically decreases the amount of DNA required for one experiment, and the test can be performed on the DNA of a single cell. The major disadvantage of this very precise method is its dependence on CpGs situated within restriction sites of methyl-sensitive restriction enzymes.

Since the pattern of methylation is lost during the cloning of genomic DNA, it was necessary to develop methods where the state of methylation of all CpGs could be studied from total uncloned genomic DNA. A new generation of techniques, called genomic sequencing, was initiated by Church and Gilbert (1984). Their procedure was further improved by Saluz and Jost (1986, 1989, 1990), but soon this method was replaced by ligation-mediated PCR (Pfeifer et al. 1989). Now the most popular technique involves the bisulfite reaction of C combined with PCR (Frommer et al. 1992; Clark et al. 1994; Raizis et al. 1995). These above procedures have the major advantage that every single CpG of any known gene can be studied.

By reviewing data on the very large number of genes whose methylation has been studied by means of restriction enzymes and other methods, some general conclusions can be drawn.

1. There is an inverse correlation between gene methylation and gene expression, and there is a positive correlation between DNA demethylation and sites on the chromatin that are deoxyribonuclease hypersensitive.
2. Not all sites on a given gene are demethylated at the same time.
3. The promoter region and/or the 3′ end of the gene may be the focal point of demethylation. At these sites, demethylation starts and progresses in both directions. Only in a few cases does the demethylation of DNA precede the onset of gene transcription. Usually it occurs in parallel with, or a little after, the onset of transcription.
4. Demethylation of the coding region of a gene may proceed at a slower pace than that of the promoter region. Demethylation of the gene body usually happens long after the onset of gene transcription and may represent a combination of two independent processes—active and passive demethylation. Several lines of evidence indicate that the inhibitory effect of methylcytosine on transcription is indirect and that the inhibition of transcription depends on the presence of specific proteins capable of binding to methylated DNA (Boyes and Bird 1991; Levine et al. 1991, 1992; Johnson et al. 1995). The dynamic aspect of these interactions of protein with methylated DNA may be rather complex and may involve signal transduction pathways. For example, in the case of the avian vitellogenin gene, following a single injection of estradiol into immature roosters there is first a rapid drop (within a few hours) in the binding activity of methylated DNA-binding protein 2 that has homology with histone H1 (MDBP-2-H1) (Jost et al. 1991; Jost and Hofsteenge 1992), followed by the onset of DNA-strand-specific DNA demethylation and gene transcription. The binding of MDBP-2-H1 to methylated DNA depends on this protein state of phosphorylation. Estradiol triggers the dephosphorylation of MDBP-2-H1, which then does not bind to methylated DNA (Bruhat and Jost 1995). These observations and

others led to the general conclusion that the temporal fate of each methylation site depends, as mentioned above (Fig. 1), on the stoichiometry and relative binding activity of these factors to DNA at a given moment.

GENOME-WIDE DNA DEMETHYLATION

Description of the Phenomenon

Genome-wide or global demethylation is a phenomenon in which a large portion of the genome becomes transiently hypomethylated during differentiation. Examples in tissue culture are differentiating mouse teratocarcinoma (cancerous embryonic cells) and erythroleukemia cells (abnormal precursor of red blood cells), the induction of Epstein-Barr virus in producer cells (Bestor et al. 1984; Razin et al. 1984b; Young and Tilghman 1984; Szyf et al. 1985), and, in vivo, spermatogenesis in the chicken (Rocamora and Mezquita 1989).

When mouse teratocarcinoma cells are induced to differentiate in vitro, they undergo a genome-wide demethylation amounting to as much as 30% of all mCpGs in the whole genome (random demethylation). In vivo, a similar genome-wide demethylation is observed in mouse extraembryonic membranes, yolk sac, and placenta. It is interesting to note that the teratocarcinoma cells are derived from the same extraembryonic tissues. Similarly, murine erythroleukemia cells (MEL) lose between 4% and 30% of their methyl groups during differentiation triggered by chemical substances (Razin et al. 1986, 1988). In differentiating erythroleukemia cells, the loss of methylcytosine is transient, whereas in teratocarcinoma cells, the level of demethylation depends on the length of treatment with retinoic acid. In a third system, where the induction of Epstein-Barr virus was observed in the presence of phorbol esters and sodium butyrate (Szyf et al. 1985), massive hypomethylation (up to 30% of all CpGs) of the DNA in producer cells occurred at a time when DNA synthesis was inhibited, suggesting that an active demethylation of DNA was taking place. In G8 mouse myoblasts, it was recently observed (Jost and Jost 1994) that incubation in a differentiation-inducing medium (low serum concentration) resulted in a transient genome-wide demethylation (~30–40% of CpGs). This genome-wide demethylation was correlated with a large increase in base-excision repair activity of 5-methylcytosine. Moreover, preceding the genome-wide demethylation, a drop in DNA methyltransferase activity was also observed, creating, most likely, hemimethylated sites that could be the substrate for 5-methylcytosine/DNA glycosylase (Jost and Jost 1994). One well-documented case of in vivo demethylation is spermatogenesis in chicken (Rocamora and Mezquita 1989). DNA was separated in different fractions according to their reassociation kinetics, and it was observed that the undermethylation in meiotic and postmeiotic cells not only was a property of repetitive sequences, but also was found in unique sequences. A large proportion of the slowly renaturing, intermediately renaturing, and rapidly renaturing DNA contained 27%, 32%, and 31% less methylcytosine in meiotic and postmeiotic cells than in the corresponding fractions of premeiotic cells. In this case, DNA methyltransferase is also lower in meiotic and postmeiotic cells containing the hypomethylated DNA.

In the mouse haploid genome there are approximately 2.5×10^7 CpGs, of which 70–90% are normally methylated. A change of 1% in methylcytosine content reflects a change in the methylation status of about 2.5×10^5 CpG sites per haploid genome (Razin et al. 1984a). This means that a 30% demethylation in mouse myoblasts, for example, corresponds to 7.5×10^6 methylated CpG sites per haploid genome. Taking an average of 500 methylated CpGs per gene, this means that 15,000 different genes could be demethylated per haploid genome. This is probably more than the number of genes turned on during the differentiation of myoblasts; therefore, genome-wide DNA demethylation could possibly have another function (DNA repair?) besides gene regulation during development.

Timing of Genome-wide DNA Demethylation

There are no general rules for the timing of genome-wide demethylation. In F9 teratocarcinoma cells differentiating in tissue culture, the percentage of DNA demethylated under similar experimental conditions depends on how long the cells have been treated with retinoic acid. For example, Razin et al. (1984b) observed a 30% demethylation of DNA after 13 days of treatment of the cells, whereas Bestor et al. (1984) found 9% demethylation after 4 days of retinoic acid treatment. In sharp contrast, in mouse erythroleukemia cells and differentiating G8 mouse myoblasts (muscle precursor cells), genome-wide demethylation is transient (Razin et al. 1988; Jost and Jost 1994). In the first case it occurs before the commitment of the cells to differentiate, whereas in the second case it is a relatively late event, occurring after expression of muscle-specific genes. In differentiating myoblasts it is unlikely that genome-wide DNA demethylation is related to cell determination.

Mechanisms of Genome-wide Demethylation

There are at least two separate mechanisms that could explain genome-wide demethylation. First, passive demethylation, where the maintenance methyltransferase is inhibited during several cycles of replication, and second, active DNA demethylation, which may require only one cycle of replication for the complete demethylation of DNA. Both mechanisms may be used, as they are not mutually exclusive.

In the case of differentiating mouse erythroleukemia cells, it has been found that the transient demethylation of DNA may occur by a combination of a decrease in DNA-methyltransferase activity (Adams et al. 1990) and an enzymatic replacement of methylcytosine by cytosine. Similarly, during the differentiation of mouse G8 myoblasts, there is a drop in the activity of DNA methyltransferase preceding the increase in 5-methylcytosine/DNA glycosylase activity (Jost and Jost 1994). In contrast, the progressive genome-wide demethylation of DNA observed in mouse F9 teratocarcinoma cells required several cycles of replication, suggesting that passive demethylation of DNA is occurring in these cells.

Biological Significance of Genome-wide Demethylation

Is there a causal relationship between genome-wide DNA demethylation and cellular differentiation? This crucial question was addressed by Razin et al. (1988), who used as a model the differentiating mouse erythroleukemia cells treated with hexamethylenbiacetamide (HMBA). Exposure of the cells to 3-deazaadenosine, an adenosine analog, in combination with homocysteine resulted in the intracellular accumulation of 3-deazaadenosylhomocysteine, which caused the inhibition of HMBA-induced hypomethylation of DNA. This inhibition of genome-wide demethylation was also correlated with a corresponding inhibition of the differentiation of erythroleukemia cells. Furthermore, Razin et al. (1988) showed that exposure of cells to cycloheximide (an inhibitor of protein synthesis) during the first day of treatment with HMBA resulted in a complete inhibition of differentiation without affecting genome-wide demethylation. These results suggest that DNA demethylation may be necessary, but not sufficient, for differentiation. In addition, the machinery for demethylation is already present in the cells prior to stimulation with HMBA. In cases where demethylation is slow to appear and occurs only after the appearance of marker enzymes of differentiation, it is, as mentioned above, doubtful that genome-wide DNA demethylation is the cause of differentiation.

If global demethylation is a consequence of cell differentiation, what is its purpose? One can envision the following two possibilities: First, genome-wide demethylation could help the DNA repair machinery to repair the group of genes that have recently undergone the transition from heterochromatin to euchromatin (active chromatin). Some experimental data support the idea that a hypomethylated genome is more rapidly repaired than a normally methylated genome. For example, while studying the repair of a UV-damaged dihydrofolate reductase gene (DHFR) in Chinese hamster ovary cells (CHO), Ho et al. (1989) found that 24 hours post UV irradiation, hypomethylated cells were repaired twice as quickly as normally methylated cells. They concluded that methylation may play an indirect role in the regulation of DNA repair through an effect on chromatin structure or transcriptional activity. Furthermore, it is well documented that in bacteria and mammalian cells there is a methylation-directed DNA repair activity (Hare and Taylor 1985; Dohet et al. 1986; Modrich 1989; Cooper et al. 1993; Grilley et al. 1993) where DNA methylation determines which strand is to be repaired. In this case the newly replicated DNA, which has not yet been methylated, is the preferential substrate for the repair of any replication error. The second possibility is that genome-wide demethylation is related to the periodic changes in chromatin organization that are associated with a general rearrangement of repair patches together with DNA excision repair in mammalian cells (Mathis and Althaus 1986).

EVIDENCE FOR THE PRESENCE OF AN ENZYMATIC SYSTEM REPLACING METHYLCYTOSINE WITH CYTOSINE

Indirect Evidence for the Presence of an Active DNA Demethylating System

Wilks et al. (1984) showed that a specific site in the promoter region of the avian vitellogenin gene became demethylated after induction by estradiol in the absence of DNA synthesis. In addition, the same group showed that these sites went through a state of hemimethylation during which very little DNA was synthesized (Saluz et al. 1986). In differentiating erythroleukemia cells, it was also shown that a transient and genome-wide demethylation occurred in the absence of DNA synthesis (Razin et al. 1984b). Using a different approach, Paroush et al. (1990) showed that demethylation of a methylated, transfected α-actin gene in L-8 myoblasts occurred in two stages, with a hemimethylated intermediate formed within a few hours posttransfection. The removal of the 5-methylcytosine on the opposite strand took place after a delay of 48 hours in the absence of DNA synthesis (Paroush et al. 1990).

Evidence that demethylation occurred after the completion of DNA replication during meiotic prophase in the mouse was provided by Trasler et al. (1990) and Frank et al. (1991), who showed that the methylated CpG islands in transiently transfected F9 cells became rapidly demethylated (within 48 hours), and demethylation seemed to be specific for cells derived from embryonic carcinomas. Shemer et al. (1991) found that complete demethylation of the apolipoprotein A1 gene during embryonic development occurred between the 8- and 16-cell stages. The same group also demonstrated that for the β-globin gene, Vk, insulin II, and Zeta gene, total demethylation took place after only one cycle of replication.

Direct Evidence for the Presence of an Active Demethylating System

Direct evidence for active demethylation was first reported by Gjerset and Martin (1982), who found that in a cell-free extract prepared from erythroleukemia cells, there is an enzymatic removal of acid-soluble labeled material from DNA double-labeled with ^{32}P and tritiated CH_3 by use of *Hpa*II methylase in the presence of tritiated *S*-adenosyl methionine. Unfortunately, the authors did not analyze the product of the reaction, and no enzyme has been purified from this system. Further evidence was presented by Razin et al. (1986) using erythroleukemia cells induced to differentiate and labeled in vivo with 5-bromo deoxyuridine and [^{3}H]cytidine. Upon separation of the DNA on CsCl gradients, they showed that demethylation of the DNA during differentiation is achieved by an enzymatic replacement of 5-methylcytosine by cytosine (Razin et al. 1986).

More recently, we succeeded in reconstituting an in vitro nuclear cell-free system in which methylcytosine is replaced by cytosine. The cell-free system was prepared from nuclei of differentiating mouse myoblasts and 12-day-old chicken embryos (Jost 1993; Jost and Jost 1994). One of the key enzymes of this system, a 5-methylcytosine/DNA glycosylase, was purified 30,000-fold from 12-day-old chicken embryos (Jost et al. 1995). Similarly, Vairapandi and Duker (1993) have detected 5-methylcytosine/DNA glycosylase activity in nuclear extracts from HeLa cells. The enzyme that they characterized in the crude nuclear extracts had a different substrate specificity from the 5-methylcytosine/DNA glycosylase purified from chicken embryo and mouse myoblasts. The enzyme from HeLa cells cleaves preferentially bifilarly methylated DNA in the presence of Mg^{++}, whereas the enzyme isolated from chicken embryos and mouse myoblasts requires hemimethylated DNA in the absence of Mg^{++}. (The assay is done in 25 mM EDTA or EGTA, or both together.) A direct comparison of the two enzymes is not possible.

PROPERTIES OF A HIGHLY PURIFIED 5-METHYLCYTOSINE/DNA GLYCOSYLASE

An enzyme that was first called mCpG endonuclease was purified from chicken embryos to homogeneity (Jost et al. 1995). During the purification, while several possible enzymatic activities were being tested, it turned out that mCpG endonuclease activity was composed of two different enzymes: a methylcytosine/DNA glycosylase and an AP/endonuclease (apyrimidic-site endonuclease). During the early stage of purification, it was not possible to detect the release of 5-methylcytosine from the DNA substrate, since it was rapidly converted into thymine by a 5-methylcytosine deaminase. The reaction product of the highly purified enzyme (one single band from an SDS-polyacrylamide gel) is 5-methylcytosine, and the apyrimidic site can be cleaved by alkali hydrolysis. The purified enzyme has a molecular mass of 52.5 kD on an SDS-polyacrylamide gel. The optimal pH is between 6.5 and 7.5, and the pI is between 5.5 and 7.5. The catalytic properties are 8×10^{-8} M with a V_{max} of 4×10^{-11} mole/hr/µg of protein. The 5-methylcytosine/DNA glycosylase is very specific for mCpG and does not react with mCpC, mCpT, mCpA, mApG, or mCpApG. Since the enzyme reacts with hemimethylated DNA six to ten times faster than with bifilarly methylated DNA, the in vivo methylcytosine/DNA glycosylase may have a preference for newly replicated DNA. However, the small but significant activity of 5-methylcytosine/DNA glycosylase on bifilarly methylated DNA also suggests that the same enzyme may, under certain conditions, demethylate symmetrically methylated DNA. As tested in vitro with a synthetic polynucleotide, the reaction of 5-methylcytosine/DNA glycosylase with methylated DNA is clearly distributive, with some sites being better substrates than others (Jost et al. 1995). All of our attempts to detect 5-methylcytosine/DNA glycosylase, using a hemimethylated oligonucleotide as a substrate, from HeLa nuclear extracts have failed. Moreover, Steinberg (1995) has shown that the removal of 5-methylcytosine from poly(dG-5-methyl-dC) by HeLa cell nuclear extracts was not by a DNA glycosylase. One interesting aspect of the purification of 5-methylcytosine/DNA glycosylase is its copurification with a mismatch-specific thymine/DNA glycosylase activity. A single peptide band cut out of an SDS-polyacrylamide gel, when denatured and renatured, gave activity for both enzymes. The same preparation also had a thymine and uracyl mismatch glycosylase activity, but no activity on a single-stranded DNA substrate containing one uracyl residue. Two im-

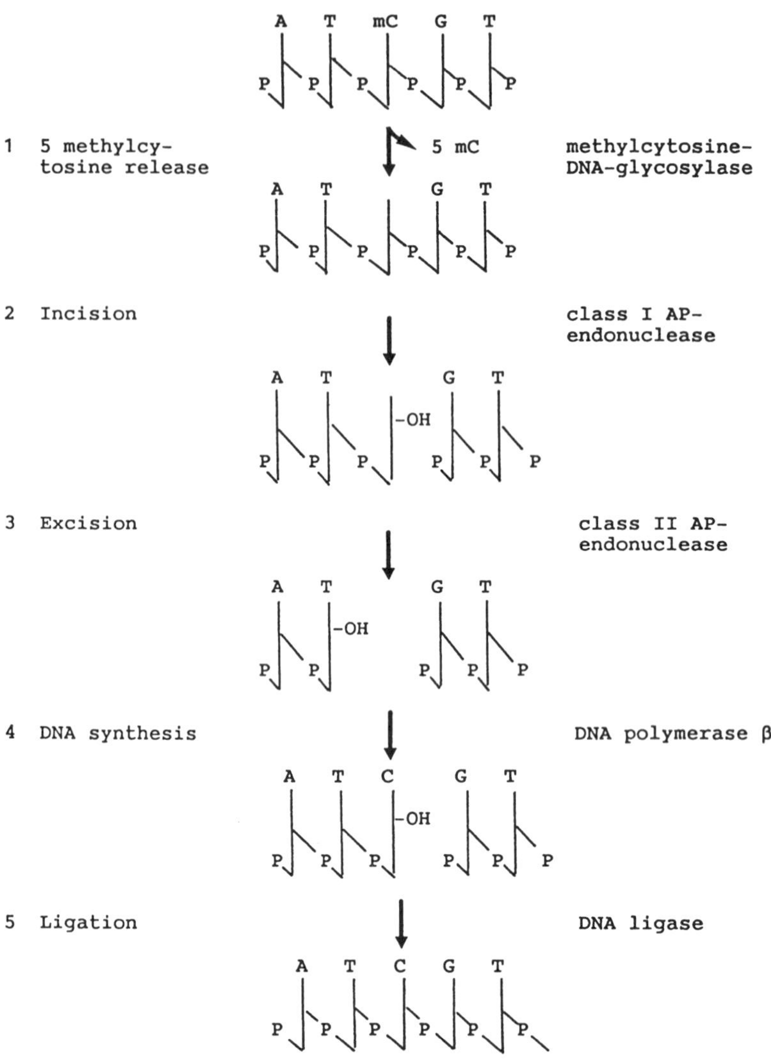

Figure 2 Pathway of active demethylation through the action of 5-methylcytosine/DNA glycosylase. (AP) Apyrimidic AP-endonuclease I, which hydrolyzes 3′ from the AP site, whereas AP-endonuclease II hydrolyzes 5′ from the AP site.

portant questions still need to be answered. The first concerns whether these multienzymatic activities are measured from a single peptide. The second question concerns the possibility of having a close relationship between DNA demethylation and DNA repair. As already mentioned above, it is known that a hypomethylation of the whole genome accelerates DNA repair (Ho et al. 1989), so it is conceivable that there is a link between the methylcytosine/DNA glycosylase and DNA repair. Recently, the group of Jiricny has purified to homogeneity a mismatch-specific thymine/DNA glycosylase from HeLa cells (Neddermann and Jiricny 1993). However, there are two major differences between this enzyme and the 5-methylcytosine/DNA glycosylase. On affinity columns the binding of the 5-methylcytosine/DNA glycosylase has no selectivity for methylated or unmethylated DNA, whereas the mismatch-specific thymine/DNA glycosylase binds only to DNA containing a G/T mismatch.

CONCLUSION

In conclusion, demethylation of DNA by either a passive or an active mechanism is likely to be important for the selective activation of genes. Passive demethylation occurs through the inhibition of, or the lack of, DNA-methyltransferase activity during the S phase of the cell cycle. The most likely pathway reponsible for the active demodification of DNA is presented in Figure 2, where 5-methylcytosine/DNA glycosylase plays a key role. The other steps leading toward the replacement of methylcytosine by cytosine have been inferred from other experiments in vitro where the reaction products were analyzed and/or the reactions were tested with selective inhibitors. Now it remains to be proven that 5-methylcytosine/DNA glycosylase is indeed the enzyme responsible for the in vivo active demethylation of DNA. As a conclusive proof, it will be necessary to carry out gene knock-out experiments in mice by homologous recombination. In addition, we cannot exclude the possibility that one purpose of genome wide demethylation is to facilitate the genome-wide DNA repair that may occur during cellular differentiation.

ACKNOWLEDGMENTS

I thank Drs. A. Bruhat and E. Oakeley for critical reading of the manuscript, and I am grateful to Mrs. Y.-C. Jost for typing it.

REFERENCES

Adams, R.L.P., A. Hanley, and A. Rinaldi. 1990. DNA demethylation in erythroleukemia cells. *FEBS Lett.* **269:** 29–31.

Assaad, F.F., K.L. Tucker, and E. Signer. 1993. Epigenetic repeat induced gene silencing (RIGS) in *Arabidopsis. Plant Mol. Biol.* **22:** 1067–1085.

Bestor, T.H., S.B. Hellewell, and V.M. Ingram. 1984. Differentiation of two mouse cell lines is associated with hypomethylation of their genome. *Mol. Cell. Biol.* **4:** 1800–1806.

Bird, A.P. and E.M. Southern. 1978. Use of restriction enzymes to study eucaryotic DNA methylation. *J. Mol. Biol.* **118:** 27–47.

Boyes, J. and A. Bird. 1991. DNA methylation inhibits transcription indirectly via methyl CpG binding proteins. *Cell* **64:** 1123–1134.

Brandeis, M., D. Frank, I. Keshet, Z. Siegfried, M. Mendelson, A. Nemes, V. Temper, A. Razin, and H. Cedar. 1994. Sp1 elements protect a CpG island from *de novo* methylation. *Nature* **371:** 435–438.

Bruhat, A. and J.-P. Jost. 1995. *In vivo* estradiol dependent dephosphorylation of MDBP-2-H1 correlates with the loss of *in vitro* preferential binding to methylated DNA. *Proc. Natl. Acad. Sci.* **92:** 3678–3682.

Carlson, L.L., A.W. Page, and T.H. Bestor. 1992. Properties and localization of DNA methyltransferase in preimplantation mouse embryos: Implications for genomic imprinting. *Genes Dev.* **6:** 2536–2541.

Church, G.M. and W. Gilbert. 1984. Genomic sequencing. *Proc. Natl. Acad. Sci.* **81:** 1991–1995.

Clark, S.J., J. Harrison, C.L. Paul, and M. Frommer. 1994. High sensitivity mapping of methylated cytosines. *Nucleic Acids Res.* **22:** 2990–2997.

Cooper, D.L., R.S. Lahue, and P. Modrich. 1993. Methyl-directed mismatch repair is bidirectional. *J. Biol. Chem.* **268:** 11823–11829.

Dohet, C., R. Wagner, and M. Radman. 1986. Methyl-directed repair of frame shift mutations in heteroduplex DNA. *Proc. Natl. Acad. Sci.* **83:** 3395–3397.

Frank, D., N. Keshet, M. Shani, A. Levine, A. Razin, and H. Cedar. 1991. Demethylation of CpG islands in embryonic cells. *Nature* **351:** 239–241.

Frommer, M., L.E. McDonald, D.S. Millar, C.M. Collis, F. Watt, G.W. Grigg, P.L. Molloy, and C.L. Paul. 1992. A genomic sequencing protocol which yields a positive display of 5-methylcytosine residues in individual DNA strands. *Proc. Natl. Acad. Sci.* **89:** 1827–1831.

Gjerset, R.A. and D.W. Martin. 1982. Presence of a DNA demethylating activity in the nucleus of murine erythroleukemia cells. *J. Biol. Chem.* **257:** 8581–8583.

Grilley, M., J. Griffith, and P. Modrich. 1993. Bidirectional excision in methyl-directed mismatch repair. *J. Biol. Chem.* **268:** 11830–11837.

Hare, J.T. and J.H. Taylor. 1985. One role for DNA methylation in vertebrate cells is strand discrimination in mismatch repair. *Proc. Natl. Acad. Sci.* **82:** 7350–7354.

Ho, L., V.A. Bohr, and P.C. Hanawalt. 1989. Demethylation enhances removal of pyrimidine dimers from the overall genome and from specific DNA sequences in Chinese hamster ovary cells. *Mol. Cell. Biol.* **9:** 1594–1603.

Johnson, C.A., J.P. Goddard, and R.L.P. Adams. 1995. The effect of histone H1 and DNA methylation on transcription. *Biochem. J.* **305:** 791–798.

Jost, J.-P. 1993. Nuclear extracts of chicken embryos promote an active demethylation of DNA by excision repair of 5-methyldeoxycytidine. *Proc. Natl. Acad. Sci.* **90:** 4684–

4688.

Jost, J.-P. and J. Hofsteenge. 1992. The repressor MDBP-2-H1 is a member of histone H1 family that binds preferentially to methylated DNA *in vitro* and *in vivo*. *Proc. Natl. Acad. Sci.* **89:** 9499–9503.

Jost, J.-P. and Y.-C. Jost. 1994. Transient DNA demethylation in differentiating mouse myoblasts correlates with higher activity of 5 methyldeoxycitidine excision repair. *J. Biol. Chem.* **269:** 10040–10043.

Jost, J.-P., H.-P. Saluz, and A. Pawlak. 1991. Estradiol down regulates the binding activity of an avian vitellogenin gene repressor (MDBP-2) and triggers a gradual demethylation of the mCpG of its binding site. *Nucleic Acids Res.* **19:** 5771–5775.

Jost, J.-P., M. Siegmann, L. Sun, and R. Leung. 1995. Purification and characterization of a 5 methylcytosine-DNA glycosylase from chicken embryos. *J. Biol. Chem.* **270:** 9734–9739.

Leonhart, H., A.W. Page, H.U. Weier, and T.H. Bestor. 1992. A targeting sequence directs DNA methyltransferase to sites of DNA replication in mammalian nuclei. *Cell* **71:** 865–873.

Levine, A., G.L. Cantoni, and A. Razin. 1991. Inhibition of promoter activity by methylation: Possible involvement of protein mediators. *Proc. Natl. Acad. Sci.* **88:** 6515–6518.

———. 1992. Methylation in the preinitiation domain suppresses gene transcription by an indirect mechanism. *Proc. Natl. Acad. Sci.* **89:** 10119–10123.

Lyon, E.S., P. McPhie, and W.B. Jakoby. 1982. Methinin: A peptide inhibitor of methylation. *Biochem. Biophys. Res. Commun.* **108:** 846–850.

Lyon, S.B., L. Buonocore, and M. Miller. 1987. Naturally occurring methylation inhibitor: DNA hypomethylation and hemoglobin synthesis in human K562 cells. *Mol. Cell. Biol.* **7:** 1759–1763.

Macleod, D., J. Charlton, J. Mullins, and A.P. Bird. 1994. Sp1 sites in the mouse Aprt gene promoter are required to prevent methylation of the CpG islands. *Genes Dev.* **8:** 2282–2292.

Mathis, G. and F.R. Althaus. 1986. Periodic changes of chromatin organization associated with rearrangement of repair patches accompany DNA excision repair in mammalian cells. *J. Biol. Chem.* **261:** 5758–5765.

Mehtali, M., M. LeMeur, and R. Lathe. 1990. The methylation-free status of a housekeeping transgene is lost at high copy number. *Gene* **91:** 179–184.

Modrich, P. 1989. Methyl-directed DNA mismatch corrections. *J. Biol. Chem.* **264:** 6597–6600.

Neddermann, P. and J. Jiricny. 1993. The purification of a mismatch specific thymine-DNA glycosylase from HeLa cells. *J. Biol. Chem.* **268:** 21218–21224.

Parker, M.I., J.B. deHaan, and W. Gevers. 1986. DNA hypomethylation in sodium butyrate-treated WI-38 fibroblasts. *J. Biol. Chem.* **261:** 2786–2790.

Paroush, Z., I. Keshet, J. Yisraeli, and H. Cedar. 1990. Dynamics of demethylation and activation of the actin gene in myoblasts. *Cell* **63:** 1229–1237.

Pfeifer, G.P., S.D. Steigerwald, P.R. Mueller, B. Wold, and A.D. Riggs. 1989. Genomic sequencing and methylation analysis by ligation mediated PCR. *Science* **246:** 810–813.

Raizis, A.M., F. Schmitt, and J.-P. Jost. 1995. A bisulphite method of 5-methylcytosine mapping that minimizes template degradation. *Anal. Biochem.* **226:** 161–166.

Razin, A., H. Cedar, and A.D. Riggs. (eds). 1984a. *DNA methylation, biochemistry and biological significance*, pp. 392. Springer-Verlag, Berlin.

Razin, A., A. Levine, T. Kafri, S. Agostini, T. Gomi, and G.L. Cantoni. 1988. Relationship between transient DNA hypomethylation and erythroid differentiation of murine erythroleukemia cells. *Proc. Natl. Acad. Sci.* **85:** 9003–9006.

Razin, A., C. Webb, M. Szyf, J. Yisraelli, A. Rosenthal, T. Naveh-Many, and N. Sciaky-Gallili. 1984b. Variations in DNA methylation during mouse cell differentiation *in vivo* and *in vitro*. *Proc. Natl. Acad. Sci.* **81:** 2275– 2279.

Razin, A., M. Szyf, T. Kafri, M. Roll, H. Giloh, S. Scarpa, D. Carotti, and G.L. Cantoni. 1986. Replacement of 5-methylcytosine by cytosine: A possible mechanism for transient DNA demethylation during differentiation. *Proc. Natl. Acad. Sci.* **83:** 2827–2831.

Rocamora, N. and C. Mezquita. 1989. Chicken spermatogenesis is accompanied by a genome wide loss of DNA methylation. *FEBS Lett.* **247:** 415–418.

Rouleau, J., A.R. MacLeod, and M. Szyf. 1995. Regulation of the DNA methyltransferase by Ras-AP-1 signaling pathway. *J. Biol. Chem.* **270:** 1595–1601.

Saluz, H.-P. and J.-P. Jost. 1986. Optimized genomic sequencing as a tool for the study of cytosine methylation in the regulatory region of the chicken vitellogenin II gene. *Gene* **42:** 151–157.

———. 1989. Genomic footprinting with Taq polymerase. *Proc. Natl. Acad. Sci.* **86:** 2602– 2606.

———. 1990. A laboratory guide for *in vitro* studies of DNA methylation and protein-DNA interaction. *Biomethods* **3:** 129–214.

Saluz, H.-P., J. Jiricny, and J.-P. Jost. 1986. Genomic sequencing reveals a positive correlation between the kinetics of strand specific DNA demethylation of the overlapping estradiol/glucocorticoid receptor binding site and the rate of avian vitellogenin mRNA synthesis. *Proc. Natl. Acad. Sci.* **83:** 7167–7171.

Sano, H., C. Grabowy, and R. Sager. 1984. Loss of chloroplast DNA methylation during dedifferentiation of *Chlamydomonas reinhardi* gametes. *Mol. Cell. Biol.* **4:** 2103–2108.

Shemer, R., T. Kafri, A. O'Connell, S. Eisenberg, J.L. Breslow, and A. Razin. 1991. Methylation changes in the apolipoprotein AI gene during embryonic development of the mouse. *Proc. Natl. Acad. Sci.* **88:** 11300–11304.

Shen, J.C., W.M. Rideout III, and P.A. Jones. 1994. The rate of hydrolytic deamination of 5-methylcytosine in double stranded DNA. *Nucleic Acids Res.* **22:** 972–976.

Singer-Sam, T.P., N. Yang, R.L. Mori, J.M. Tanguay, J.C. LeBon, and A.D. Riggs. 1990. DNA methylation in the 5′ region of the mouse PGK-1 gene and a quantitative PCR assay for methylation in nucleic acids methylation. *UCLA Symp. Mol. Cell. Biol. New Ser.* **1:** 28.

Steinberg, R.A. 1995. Enzymic removal of 5-methylcytosine from poly(dG-5-methyl-dC) by HeLa cell nuclear extracts is not by a DNA glycosylase. *Nucleic Acids Res.* **23:** 1621– 1624.

Szyf, M. 1994. DNA methylation properties: Consequences for pharmacology. *Trends Pharmacol. Sci.* **15:** 233–238.

Szyf, M., L. Eliasson, V. Mann, G. Klein, and A. Razin. 1985. Cellular and viral DNA hypomethylation associated with induction of Epstein-Barr virus lytic cycle. *Proc. Natl. Acad. Sci.* **82:** 8090–8094.

Trasler, J.M., L.E. Kake, P.A. Johnson, A.A. Alcivar, C.F. Millette, and H.B. Hecht. 1990. DNA methylation and demethylation events during meiotic prophase in the mouse testis. *Mol. Cell. Biol.* **10:** 1828–1834.

Vairapandi, M. and N.J. Duker. 1993. Enzymatic removal of 5-methylcytosine from DNA by a human DNA glycosylase. *Nucleic Acids Res.* **21:** 5323–5327.

Wilks, A., M. Seldran, and J.-P. Jost. 1984. An estrogen dependent demethylation at the 5′ end of the avian vitellogenin gene is independent of DNA synthesis. *Nucleic Acids Res.* **12:** 1163–1177.

Yeivin, A. and A. Razin. 1993. Gene methylation patterns and expression. In *DNA methylation molecular biology and biological significance* (ed. J.-P. Jost and H.-P. Saluz), pp. 523–568. Birkhäuser Verlag, Basel.

Young, P.Y., and S.M. Tilghman. 1984. Induction of α fetoprotein synthesis in differentiating F9 teratocarcinoma cells is accompanied by a genome wide loss of DNA methylation. *Mol. Cell. Biol.* **4:** 898–907.

The Role of DNA Methylation in Plant Development

E. Jean Finnegan
CSIRO, Division of Plant Industry
Canberra, ACT
Australia, 2601

DNA methylation has been implicated in a wide range of biological processes including gene regulation (Razin and Cedar 1991; Finnegan et al. 1993), determination of chromatin structure (Lewis and Bird 1991), timing of DNA replication (Lewis and Bird 1991), genomic imprinting (Sasaki et al. 1993), increasing the spontaneous mutation rate (Coulondre et al. 1978; Duncan and Miller 1980), as the basis for epigenetic phenomena (Holliday 1993), and as a defense mechanism to protect against foreign or invading DNA (Bird 1986; Bestor 1990; Doerfler 1991). Despite this, the existence of organisms such as yeast and *Drosophila* that lack DNA methylation has led to debate about the role of DNA methylation. There can be no doubt that DNA methylation plays a fundamental role in mouse development following the establishment of a targeted mutation of the DNA methyltransferase gene in the germ line of transgenic mice. Embryos that were homozygous for the mutation had methylation levels reduced to about 30% of wild-type levels. These embryos were stunted and delayed in development compared to their normal siblings and spontaneously aborted in mid-gestation. Although the underlying cause for the embryo-lethal phenotype has not been determined, it has been proposed that inappropriate gene expression resulting from the reduced level of DNA methylation is involved (Li et al. 1992). In contrast, mutants of *Neurospora crassa* that abolish all detectable methylation showed only minor changes in vigor and growth morphology, suggesting that DNA methylation is not absolutely essential for development of this organism (Foss et al. 1993). This raises the question as to whether DNA methylation is essential for plant development.

Until recently, there was no direct evidence demonstrating that DNA methylation is important for normal plant development despite correlative evidence linking DNA methylation to repression of transcription

(Finnegan et al. 1993). Two different approaches, mutagenesis and transformation with a methyltransferase antisense, that reduced DNA methylation in *Arabidopsis* have now clearly demonstrated that DNA methylation is also fundamentally important in plant development. I briefly describe the features of DNA methylation in plants, introduce the DNA methyltransferase family of *Arabidopsis*, and present evidence for the role of DNA methylation in plant development before discussing the possible interaction between DNA methylation and other heritable mechanisms of gene regulation.

PLANT DNA METHYLATION

Methylation of cytosine residues at the carbon 5 position is the most common modification of DNA in plant cells. The level of cytosine methylation in plants is generally higher than that found in mammals, where about 3–8% of cytosines are methylated (Shapiro 1975). In plants, 20–30% of cytosines are modified, on average, although there is considerable interspecies variation in the level of cytosine methylation, ranging from 4–6% in *Arabidopsis thaliana*, which has a small genome with relatively little repetitive DNA (Leutwiler et al. 1984), to 37% in *Helianthus* (Ergle and Katterman 1961). The predominant sites for methylation in plant DNA are cytosines found in CpG dinucleotides or CpNpG trinucleotides; CpApG and CpTpG appear to be the most commonly modified trinucleotides in plants (Gruenbaum et al. 1981; Messeguer et al. 1991). Frequent methylation of CpNpG trinucleotides, which are only rarely modified in mammalian DNA (Clark et al. 1995), is one reason for the large difference in methylation levels observed in plant and animal DNA. In addition, CpG dinucleotides occur at a much lower frequency in mammalian than in plant DNA (McClelland 1983; Gardiner-Garden et al. 1992). Both CpG and CpNpG motifs show strand symmetry, and it has been noted that when a cytosine of a CpG dinucleotide is methylated, the corresponding cytosine of the opposing strand is also generally methylated (Bird 1978; Cedar et al. 1979). Because the frequency of methylation in cytosines of CpT and CpA dinucleotides is equivalent, it is probable that methylation at CpNpG trinucleotides is also symmetrical (Gruenbaum et al. 1981). Methylation patterns are transmitted through mitosis by the action of maintenance DNA methyltransferases which methylate cytosines on the newly synthesized strand of hemimethylated CpG (and CpNpG) motifs. Methylation of nonsymmetrical cytosines has been demonstrated by genomic sequencing of transgenes that have become inactivated and heavily methylated (Meyer and Heidmann 1994). It is not known whether methylation of asymmetrical sites is restricted to "foreign" DNA, or whether it also occurs in heavily methylated endogenous sequences; this issue will only be resolved when more endogenous sequences have been subjected to genomic sequencing.

Methylated residues are not distributed at random throughout the plant genome; for example, in cotton, unique sequences that comprise about 38% of the genome have only 4% methylated cytosines (Gusienov et al. 1975). Clusters of unmethylated CpG residues can be found in a number of plant species, even though on average approximately 55–80% of CpG dinucleotides are methylated (Gruenbaum et al. 1981; Antequera and Bird 1988; Messeguer et al. 1991; Gardiner-Garden and Frommer 1992). These clusters are identified as small fragments ranging in size from about 25 bp to about 250 bp in *Hpa*II-digested DNA and are reminiscent of *Hpa*II tiny fragments characteristic of CpG islands in vertebrate genomes (Cooper et al. 1983). The transcription start site of angiosperm housekeeping genes is not always surrounded by a CpG island, as is the case in mammalian and avian genomes (Gardiner-Garden and Frommer 1992).

DNA METHYLTRANSFERASES IN *ARABIDOPSIS*

A cDNA encoding a putative DNA methyltransferase, *METI*, that is structurally similar to other eukaryote DNA methyltransferases (Bestor et al. 1988; Yen et al. 1992; Tajima et al. 1995) has been isolated from *Arabidopsis* (Finnegan and Dennis 1993). The carboxy-terminal domain of this putative methyltransferase has 50% identity with the mouse enzyme and is structurally similar to the prokaryote cytosine-5 methyltransferases. Eight of the ten conserved motifs identified in prokaryote enzymes (Som et al. 1987; Lauster et al. 1989; Pósfai et al. 1989) occur in the plant protein, including the putative binding site for the methyl donor, S-adenosyl methionine (Ingrosso et al. 1989), and the universally conserved proline-cysteine doublet that forms the active site of the enzyme (Wu and Santi 1987; Chen et al. 1991; Wyszynski et al. 1992). The amino-terminal domain is less well conserved between the plant and the mouse enzyme; homology in this domain includes the sequences that target the mouse enzyme to the replication fork (Leonhardt et al. 1992), suggesting that the *Arabidopsis* enzyme is a maintenance methyltransferase. Southern analyses of *Arabidopsis* genomic DNA indicate that there are at least two methyltransferase genes (Finnegan and Dennis

1993, E.J. Finnegan, unpubl.); mapping data have demonstrated that this family is not clustered in the *Arabidopsis* genome (E.J. Finnegan and R.K. Genger, unpubl.). Transformation of *Arabidopsis* with an antisense construct encoding the methyltransferase domain of *METI* reduced methylation of CpG dinucleotides by up to 90%; methylation of CpCpG nucleotides was also affected, but there was little effect on the level of methylation of CpA/TpG sites (Finnegan et al. 1996). These data suggest that members of the methyltransferase gene family differ in target specificity; for example, METI may methylate both CpG and CpCpG sequences. Alternatively, the antisense construct may interact with transcripts from more than one member of the gene family. Two methyltransferases that differ in target specificity have been partially purified from pea shoots. Both enzymes carried out maintenance and de novo methylation and whereas one enzyme methylated CpApG and CpTpG trinucleotides, the other preferentially methylated CpG dinucleotides but also methylated CpCpG trinucleotides at low efficiency (Pradhan and Adams 1995).

Methylation-deficient *Arabidopsis* mutants have been identified in a mutagenized M2 population by screening for plants with reduced methylation in centromeric repeat DNA (Vongs et al. 1993). Three independent mutants (*ddm*) were identified, at least two of which lie in the same complementation group. Methylation of cytosines in both CpG and CpNpG motifs in homozygous *ddm1* mutants was reduced to about 30% of wild type, but methyltransferase activity in mutant extracts was comparable to wild type. The *ddm1* locus is not closely linked to the mapped putative methyltransferase genes (E.J. Richards, pers comm.); taken together, the biochemical data and the genetic mapping data indicate that *ddm1* does not encode a DNA methyltransferase. The intracellular pool of S-adenosyl methionine in homozygous *ddm1* plants was also comparable to wild type, suggesting that hypomethylation did not result from a reduction in the availability of the methyl donor. It has been suggested that the *ddm1* gene product serves as a regulator of DNA methylation by interacting directly with DNA methyltransferases, perhaps affecting the localization of the enzyme. Alternatively, *ddm1* mutations could alter chromosome structure to prevent methylation of some chromosomal domains or, by altering the timing of replication, could exclude some genomic sequences from methyltransferase-rich replication foci (Kakutani et al. 1995). Factors that regulate DNA methylation have been implicated in mouse, where increased intranuclear localization of the methyltransferase in preimplantation embryos does not coincide with increased DNA methylation (Carlson et al. 1992).

DNA METHYLATION IN DEVELOPMENT

Decreasing Methylation Results in Morphological Abnormalities

The role of DNA methylation in development has been investigated in plants that have substantially reduced levels of methylation. As discussed above, methylation-deficient *Arabidopsis* lines have been identified following mutagenesis. Homozygous *ddm1* plants initially showed no phenotypic abnormalities (Vongs et al. 1993) but, after several generations of self-pollination, progeny developed morphological abnormalities. These included rounded rosette leaves, increased cauline leaf number, and slightly delayed flowering in conditions that promote early flowering of wild-type plants. A number of more dramatic morphological changes were noted but remain to be characterized (E.J. Richards, pers. comm.). Although the relationship between the alterations in phenotype and DNA hypomethylation is not currently understood, it has been suggested that reducing DNA methylation leads to aberrant gene expression, which alters development (Kakutani et al. 1995).

Arabidopsis plants transformed with a fragment of the putative DNA methyltransferase cDNA have DNA methylation reduced to as little as 10% of the wild-type level (Finnegan et al. 1996). These plants displayed a range of dramatic phenotypic abnormalities, including reduced apical dominance resulting in a bushy appearance, reduced size, small rounded leaves (Fig. 1A) or leaves that curl toward the upper surface (Fig. 1B), altered root development, and homeotic-like transformation of floral organs (Fig. 1C–F). The floral homeotic transformation affected organs in one or more whorls (Fig. 1D, E) or resulted in the development of complete flowers in place of a floral organ or in the internode between floral organs (Fig. 1F). Some of the homeotic transformations resemble those caused by loss of function of previously described floral organ identity, cadastral or meristem identity genes (Jack et al. 1993; Ma 1994). Homeotic transformations caused by mutation of floral genes were mimicked experimentally by ectopic expression of different floral homeotic genes (Mizukami and Ma 1992; Jack et al. 1994), indicating that, in antisense plants, transformation could be caused by either loss- or gain-of-function "mutations." Ectopic expression of two floral homeotic genes was observed in the antisense plants, suggesting that the floral abnormalities were caused by ectopic expression rather than mutation (Finnegan et al. 1996).

The whole plant and floral phenotypes of methyltransferase antisense plants were very variable and included some or all of these characteristics. Similar phenotypic changes were observed in progeny of indepen-

Figure 1 Abnormal *Arabidopsis* plants and flowers from progeny of two independent methyltransferase antisense plants. (*A*) Sibling plants from family #10. Both plants carry the methyltransferase antisense, but the plant on the left looks normal, whereas the plant on the right has reduced apical dominance, is smaller, and has rounded leaves. (*B*) Two plants from family #22-6. The plant on the left is smaller than normal and has reduced apical dominance, whereas its sibling on the right is of more normal size but has leaves that curl to the upper surface. (*C*) Wild-type flower. *Arabidopsis* flowers have organs arranged in four concentric whorls with four sepals in the outer whorl; then four petals in the second whorl; six stamens, the male reproductive organs, in the third whorl; and a single female reproductive organ, the gynoecium consisting of two fused carpels, in the fourth whorl. (*D*) Flower from a plant in the T3 generation of family #10. The organs in the outer two whorls are normal, but this flower has additional stamens and the carpels of the gynoecium are unfused. This flower is similar to those formed on plants homozygous for a mutation in the floral cadastral gene *superman* or to flowers formed by plants transformed with a 35S-*apetala*3 chimeric gene. (*E*) Flower from a plant in the T3 generation of family #10. This flower has four normal sepals but has petals or staminoid petals in all the inner whorls. It is a phenocopy of flowers seen on plants homozygous for the mutations in both the *agamous* and *superman* genes. (*F*) Flower from a plant in the T3 generation of family #10. This flower has extra stamens and a reduction in carpel tissue. There are two secondary flowers formed within the main flower. This flower appears to be a loss-of-function mutation for the genes *superman* and *apetala1*.

dent transgenic plants with reduced methylation. This indicates that the developmental abnormalities were due to decreased DNA methylation, not to mutation caused by insertion of the transgene. The pleiotropic phenotype of methyltransferase antisense plants suggests that the expression of many genes was affected. The severity of the phenotype was correlated with the extent of demethylation such that plants with the lowest level of methylation were the most severely affected; for example, where methylation was reduced to 10–20% of wild type, all progeny showed abnormalities in root, shoot, and flower development. Almost all flowers on these plants were affected, and most were female-sterile. Plants with higher levels of methylation (70–80% of wild type) were generally larger, and only the late-formed flowers were abnormal.

The phenotype became more severe in selfed progeny, even though there was no evidence for decreasing DNA methylation in successive generations (Finnegan et al. 1996). The progressive phenotype seen in antisense plants is consistent with the development of phenotypic abnormalities after several generations of selfing *ddm1* homozygotes (Kakutani et al. 1995). The increasing severity of the phenotypic abnormalities could be a consequence of the accumulation of different demethylated sites in sexual progeny and/or of destabilization of chromatin structure in repeated cycles of cell division (see below).

Reduced Methylation Promotes Flowering in Short Days

The initiation of flowering in many angiosperms is affected by environmental factors including day length, light quality, and temperature. In many ecotypes and late-flowering mutants of *Arabidopsis*, flowering can be promoted by a prolonged period of low temperature, a response termed vernalization. The vernalization response has been well characterized, and it has been shown that nondividing cells do not respond to cold treatment, suggesting that cell division and probably DNA replication are required for cells to be receptive to cold (Wellensiek 1964). There is no evidence for a transmissable signal in response to low-temperature treatment, because vernalization only results when cells that will ultimately give rise to the floral meristem are exposed to low temperatures (Schwabe 1954; Metzger 1988). There is a strong mitotic memory for the cold treatment, as cells from root explants that are cold-treated regenerate into plants that flower early (Burn et al. 1993). Each generation requires vernalization, indicating that the memory of cold treatment is not transmitted through meiosis (Lang 1965). It has been proposed that the vernalization response is mediated by demethylation of a gene(s) critical for the initiation of flowering (Burn et al. 1993; Dennis et al. 1996); indeed, prolonged exposure to low temperatures decreases DNA methylation (Burn et al. 1993; Dennis et al. 1996).

The vernalization response of late-flowering mutants of *Arabidopsis* was mimicked by treatment of imbibed seeds with the demethylating agent, 5-azacytidine. The reduction in time to flower was not as great as that caused by prolonged cold treatment, suggesting that demethylation can partially substitute for cold treatment (Burn et al. 1993). Similarly, reduction of methylation by a methyltransferase antisense reduced flowering time from 62 days (22.7 leaves) to 31 days (10.4 leaves) when plants were grown in short days, conditions that delay flowering of wild-type plants. Under the same growth conditions, cold-treated wild-type and antisense plants flowered at 21 days (6.3 leaves) and 23 days (6.9 leaves), respectively (Dennis et al. 1996; R.K. Genger, unpubl.). These observations are consistent with the hypothesis that vernalization is mediated by demethylation of a gene(s) critical for the initiation of flowering, but given the other developmental abnormalities observed in antisense transgenic plants, it is premature to conclude promotion of flowering, in each case, occurs by the same mechanism. This hypothesis can only be tested definitively when the genes involved have been identified and cloned.

De Novo Methylation in *Arabidopsis*

Methylation patterns can be altered by demethylation or by de novo methylation of previously unmethylated cytosines. Eukaryote DNA methyltransferases do not have well-defined target sequences, and the signals that direct methylation of some, but not all, cytosines in CpG (and CpNpG) motifs are not currently understood. De novo methylation of multicopy transgene sequences has been reported in *Arabidopsis* (Kilby et al. 1992; Assaad et al. 1993), suggesting that repeated DNA may be a strong signal for DNA methylation in *Arabidopsis*. However, when the *METI* methyltransferase antisense was segregated by self-pollination of a hemizygous plant, progeny that did not inherit the antisense construct still had reduced levels of DNA methylation. The level of methylation remained below wild type in selfed progeny of the antisense-null plants, indicating that de novo methylation in *Arabidopsis* is very slow (Finnegan et al. 1996). A similar observation was made when homozygous *ddm1* plants were crossed to wild type; methylation in the F_1 progeny was intermediate between that of the parents, even though *ddm1* is recessive rather than semidominant. Repeated crossing of heterozygotes resulted in the gradual restoration of methylation, consistent with dilution of the undermethylated DNA segments rather than de novo methylation (Vongs et al. 1993). The inability of *Arabidopsis* to rapidly restore DNA methylation, following substantial demethylation, even after passage through meiosis, suggests that plants may not undergo the global demethylation followed by remethylation that is characteristic of early mammalian embryo development (for review, see Razin and Cedar 1993). The significance of the slow rate of de novo methylation on *Arabidopsis* development is unknown but presumably reflects a strong preference for hemimethylated substrates by the DNA methyltransferases.

CONCLUSIONS

Although the developmental abnormalities observed in plants with reduced DNA methylation have clearly established that methylation plays a role in plant development, the molecular mechanisms involved are unknown. Decreasing DNA methylation in *Arabidopsis* resulted in a pleiotropic phenotype including homeotic transformation of floral organs, suggesting that the expression of many genes has been affected. Aberrant expression of two floral homeotic genes was observed in plants with reduced levels of methylation. DNA methylation represses transcription directly by interfering with protein/DNA interactions and indirectly by altering chromatin structure. This raises the question as to whether DNA methylation has replaced the chromatin structural elements that maintain patterns of gene expression in *Drosophila*, or whether DNA methylation stabilizes patterns of gene activity established by other factors (Bestor et al. 1994; Pirotta 1995).

Homeotic transformations were observed in *Drosophila* carrying mutation in genes encoding members of the polycomb (PcG) or trithorax (trxG) group of proteins. PcG proteins, which are involved in long-term gene repression, affect higher-order chromatin structure. Chromatin complexes formed by these proteins are stably transmitted through mitosis, providing a cellular memory of gene expression patterns that are established by other developmental regulators early in embryo development. Although PcG proteins were initially identified by their effect on homeotic genes, they are now believed to regulate expression of a wide range of genes involved in body segmentation, pattern formation, and development of the nervous system (Paro 1993; Moerhle and Paro 1994; Pirrotta 1995). trxG proteins maintain an active transcription state, but the mechanism of action is less well understood (Pirrotta 1995).

Two lines of evidence point to DNA methylation being an addition to, rather than a replacement for, chromatin-based gene regulatory mechanisms. The *CLF* gene of *Arabidopsis* encodes a protein with homology to a member of the PcG proteins of *Drosophila*. *CLF* mutants have a pleiotropic phenotype, some facets of which are very similar to the phenotype of methyltransferase antisense plants, including curled leaves, homeotic-like transformations of floral organs, and early flowering; the expression of the floral homeotic gene, *AGAMOUS*, is no longer restricted to defined regions of the floral bud but occurs in the outer whorls of the bud, leaves, and stems in the *clf* mutant (P. Puangsomlee et al., pers. comm.). The similarity between the phenotype of *clf* and methyltransferase antisense plants suggests that DNA methylation and homologs of the PcG/trxG system may act together to stabilize patterns of gene expression in *Arabidopsis*. An interaction of this type would account for both gain- and loss-of-function mutations.

Homology between the mammalian DNA methyltransferases and a human homolog (ALL-1) of trithorax (Ma et al. 1993) provides another link between the two systems involved in heritable gene repression during development. A conserved cysteine-rich region may cause these proteins to sense the methylation status of DNA and to respond differently to unmethylated, hemimethylated, and symmetrically methylated DNA. Interestingly, the *Drosophila* trithorax protein, which interacts with DNA lacking methylation, does not contain the cysteine-rich region present in its mammalian homologs (Bestor and Verdine 1994). The *Arabidopsis* methyltransferase also lacks the cysteine-rich region but, despite this, it is clear that this enzyme has a strong preference for hemimethylated substrates (Finnegan and Dennis 1993; Finnegan et al. 1996).

In conclusion, there is strong evidence to suggest that, in addition to a role in defense, DNA methylation has evolved to play a fundamental role in plant development. There are a number of mechanisms by which methylation could regulate genes critical to normal development, including repression of transcription by methylation of promoter sequences, binding of methyl-binding proteins that indirectly suppress transcription (Meehan et al. 1989; Boyes and Bird 1991), condensation of methylated DNA that limits access of transcription machinery, and interaction with other gene regulatory mechanisms, such as homologs of *Drosophila* polycomb and trithorax group proteins.

ACKNOWLEDGMENTS

I thank Ruth Genger, Eric Richards, Justin Goodrich, George Coupland, and Preeya Puangsomlee for communication of results prior to publication, and Richard Brettell and Michael Graham for critically reading the manuscript.

REFERENCES

Antequera, F. and A.P. Bird. 1988. Unmethylated CpG islands associated with genes in higher plant DNA. *EMBO J.* **7:** 2295–2299.

Assaad, F.F., K.L. Tucker, and E.R. Signer. 1993. Epigenetic repeat-induced silencing (RIGS) in *Arabidopsis. Plant Mol. Biol.* **22:** 1067–1085.

Bestor, T.H. 1990. DNA methylation: Evolution of a bacterial immune function into a regulator of gene expression and genome structure in higher eukaryotes. *Philos. Trans. R. Soc. Lond. B* **326:** 179–187.

Bestor, T.H. and G.L. Verdine. 1994. DNA methyltransferases. *Curr. Opin. Cell Biol.* **6:** 380–389.

Bestor, T.H., V.L. Chandler, and A.P. Feinberg. 1994. Epigenetic effects in eukaryote gene expression. *Dev. Genet.* **15:** 458–462.

Bestor, T., A. Laudano, R. Mattaliano, and V. Ingram. 1988. Cloning and sequencing of a cDNA encoding DNA methyltransferase of mouse cells. *J. Mol. Biol.* **203:** 971–983.

Bird, A.P. 1978. Use of restriction enzymes to study eukaryotic DNA methylation: II. The symmetry of methylated sites supports semi-conservative copying of the methylation pattern. *J. Mol. Biol.* **118:** 49–60.

———. 1986. CpG-rich islands and the function of DNA methylation. *Nature* **321:** 209–213.

Boyes, J. and A. Bird. 1991. DNA methylation inhibits transcription indirectly via a methyl-CpG binding protein. *Cell* **64:** 1123–1134.

Burn, J.E., D.J. Bagnall, J.M. Metzger, E.S. Dennis, and W.J. Peacock. 1993. DNA methylation, vernalization, and the initiation of flowering. *Proc. Natl. Acad. Sci.* **90:** 287–283.

Carlson, L.L., A.W. Page, and T.H. Bestor. 1992. Properties and localization of DNA methyltransferase in preimplantation mouse embryos: Implications for genomic imprinting. *Genes Dev.* **6:** 2536–2541.

Cedar, H., A. Solage, G. Glaser, and A. Razin. 1979. Direct detection of methylated cytosine in DNA by use of the restriction enzyme MspI. *Nucleic Acids Res.* **6:** 2125–2132.

Chen, L., A.M. MacMillan, W. Chang, K. Ezaz-Nikpay, W.S. Lane, and G.L. Verdine. 1991. Direct identification of the active-site nucleophile in a DNA (cytosine-5)-methyltransferase. *Biochemistry* **30:** 11018–11025.

Clark, S.J., J. Harrison, and M. Frommer. 1995. CpNpG methylation in mammalian cells. *Nat. Genet.* **10:** 20–27.

Cooper, D.N., M.H. Taggart, and A.P. Bird. 1983. Unmethylated domains in vertebrate DNA. *Nucleic Acids Res.* **11:** 647–657.

Coulondre, C., J.H. Miller, P.J. Farabaugh, and W. Gilbert. 1978. Molecular basis of base substitution hotspots in *Escherichia coli. Nature* **274:** 775–780.

Dennis, E.S., E.J. Finnegan, P. Bilodeau, A. Chaudhury, R. Genger, C.A. Helliwell, C.C.

Sheldon, and W.J. Peacock. 1996. Vernalization and the initiation of flowering. *Semin. Dev. Biol.* (in press).

Doerfler, W. 1991. Patterns of DNA methylation—Evolutionary vestiges of foreign DNA inactivation as a host defence. *Biol. Chem. Hoppe-Seyler* **372:** 557–564.

Duncan, B.K. and J.H. Miller. 1980. Mutagenic deamination of cytosine residues in DNA. *Nature* **187:** 560–561.

Ergle, D.R. and F.R.H. Katterman. 1961. Deoxyribonucleic acid of cotton. *Plant Physiol.* **36:** 811–815.

Finnegan, E.J. and E.S. Dennis. 1993. Isolation and identification by sequence homology of a putative cytosine methyltransferase from *Arabidopsis thaliana. Nucleic Acids Res.* **21:** 2383–2388.

Finnegan, E.J., R.I.S. Brettell, and E.S. Dennis. 1993. The role of DNA methylation in plant development. In *DNA methylation: Molecular biology and biological significance* (ed. J.-P. Jost and H.-P. Saluz), pp. 218–261. Birkhauser Verlag, Basel.

Finnegan, E.J., W.J. Peacock, and E.S. Dennis. 1996. Reduced DNA methylation in *Arabidopsis thaliana* results in abnormal plant development and floral homeotic-like transformations. (in press).

Foss, H.M., C.J. Roberts, K.M. Claeys, and E.U. Selker. 1993. Abnormal chromosome behavior in *Neurospora* mutants defective in DNA methylation. *Science* **262:** 1737–1741.

Gardiner-Garden, M. and M. Frommer. 1992. Significant CpG-rich regions in angiosperm genes. *J. Mol. Biol.* **34:** 231–245.

Gardiner-Garden, M., J.A. Sved, and M. Frommer. 1992. Methylation sites in angiosperm genes. *J. Mol. Evol.* **34:** 219–230.

Gruenbaum, Y., T. Naveh-Many, H. Cedar, and A. Razin. 1981. Sequence specificity of methylation in higher plant DNA. *Nature* **292:** 860–862.

Guseinov, V.A., G.I. Kiryanov, and B.F. Vanyushin. 1975. Intragenome distribution of 5-methylcytosine in DNA of healthy and wilt-infected cotton plants (*Gossypium hirsutum* L.). *Mol. Biol. Rep.* **2:** 59–63.

Holliday, R. 1993. Epigenetic inheritance based on DNA methylation. In *DNA methylation: Molecular biology and biological significance* (ed. J.P. Jost and H.P. Saluz), pp. 452–468. Birkhauser Verlag, Basel.

Ingrosso, D., A.V. Fowler, J. Bleibaum, and S. Clarke. 1989. Sequence of the D-aspartyl/L-isoaspartyl protein methyltransferase from human erythrocytes. *J. Biol. Chem.* **264:** 20131–20139.

Jack, T., G.L. Fox, and E.M. Meyerowitz. 1994. Arabidopsis homeotic gene *APETALA3* ectopic expression: Transcriptional and posttranscriptional regulation determine floral organ identity. *Cell* **76:** 703–716.

Jack, T., L.E. Sieburth, and E.M. Meyerowitz. 1993. Genes that control flower development in *Arabidopsis. Semin. Dev. Biol.* **4:** 51–63.

Kakutani, T., J. Jeddeloh, and E.J. Richards. 1995. Characterization of an *Arabidopsis thaliana* DNA hypomethylation mutant. *Nucleic Acids Res.* **23:** 130–137.

Kilby, N.J., H.M.O. Leyser, and I.J. Furner. 1992. Promoter methylation and progressive transgene inactivation in *Arabidopsis. Plant Mol. Biol.* **20:** 103–112.

Lang, A. 1965. Physiology of flower initiation. In *Encyclopedia of plant physiology* (ed. W. Ruhland), pp. 1489–1536. Springer Verlag, Berlin.

Lauster, R., T.A. Trautner, and M. Noyer-Weidner. 1989. Cytosine-specific type II DNA methyltransferases. *J. Mol. Biol.* **206:** 305–312.

Leonhardt, H., A.W. Page, H.-U. Weier, and T.H. Bestor. 1992. A targeting sequence directs DNA methyltransferase to sites of DNA replication in mammalian nuclei. *Cell* **71:** 865–873.

Leutwiler, L.S., B.R. Hough-Evans, and E.M. Meyerowitz. 1984. The DNA of *Arabidopsis thaliana*. *Mol. Gen. Genet.* **194:** 15–23.

Lewis, J. and A. Bird. 1991. DNA methylation and chromatin structure. *FEBS Lett.* **285:** 155–159.

Li, E., T.H. Bestor, and R. Jaenisch. 1992. Targeted mutation of the DNA methyltransferase gene results in embryonic lethality. *Cell* **69:** 915–926.

Ma, H. 1994. The unfolding drama of flower development: Recent results from genetic and molecular analyses. *Genes Dev.* **8:** 745–756.

Ma, Q., H. Alder, K.K. Nelson, D. Chatterjee, Y. Gu, T. Makamura, E. Canaani, C.M. Croce, L.D. Siracusa, and A.M. Buchberg. 1993. Analysis of the murine *All-1* gene reveals conserved domains with human *ALL-1* and identifies a motif shared with DNA methyltransferases. *Proc. Natl. Acad. Sci.* **90:** 6350–6354.

McClelland, M. 1983. The frequency and distribution of methylatable DNA sequences in leguminous plant protein coding genes. *J. Mol. Evol.* **19:** 346–354.

Meehan, R.R., J.D. Lewis, S. McKay, E.L. Kleiner, and A.P. Bird. 1989. Identification of a mammalian protein that binds specifically to DNA containing methylated CpGs. *Cell* **58:** 499–507.

Messeguer, R., M.W. Ganal, J.C. Steffens, and S.D. Tanksley. 1991. Characterisation of the level, target sites and inheritance of cytosine methylation in tomato nuclear DNA. *Plant Mol. Biol.* **16:** 753–770.

Metzger, J.D. 1988. Localization of the site of perception of thermoinductive temperatures in *Thlaspi arvense* L. *Plant Physiol.* **88:** 424–428.

Meyer, P. and I. Heidmann. 1994. Epigenetic variants of a transgenic petunia line show hypermethylation in transgene DNA: An indication for specific recognition of foreign DNA in transgenic plants. *Mol. Gen. Genet.* **243:** 390–399.

Mizukami, Y. and H. Ma. 1992. Ectopic expression of the floral homeotic gene *AGAMOUS* in transgenic Arabidopsis plants alters floral organ identity. *Cell* **71:** 119–131.

Moehrle, A. and R. Paro. 1994. Spreading the silence: Epigenetic transcriptional regulation during *Drosophila* development. *Dev. Genet.* **15:** 478–484.

Paro, R. 1993. Mechanisms of heritable gene repression during development of *Drosophila*. *Curr. Opin. Cell Biol.* **5:** 999–1005.

Pirrotta, V. 1995. Chromatin complexes regulating gene expression in *Drosophila*. *Curr. Opin. Genet. Dev.* **5:** 466–472.

Pósfai, J., A.S. Bhagwat, G. Pósfai, and R.J. Roberts. 1989. Predictive motifs derived from cytosine methyltransferases. *Nucleic Acids Res.* **17:** 2421–2435.

Pradhan, S. and R.L.P. Adams. 1995. Distinct CG and CNG DNA methyltransferases in *Pisum sativum*. *Plant J.* **7:** 471–481.

Razin, A. and H. Cedar. 1991. DNA methylation and gene expression. *Microbiol. Rev.* **55:** 451–458.

———. 1993. DNA methylation and embryogenesis. In *DNA methylation: Molecular biology and biological significance* (ed. J.P. Jost and H.P. Saluz), pp. 343–357. Birkhauser Verlag, Basel.

Sasaki, H., N.D. Allen, and M.A. Surani. 1993. DNA methylation and genomic imprinting in mammals. In *DNA methylation: Molecular biology and biological significance* (ed. J.P. Jost and H.P. Saluz), pp. 469–488. Birkhauser Verlag, Basel.

Schwabe, W.W. 1954. Factors controlling flowering in the chrysanthemum. *J. Exp. Biol.* **5:** 389–400.

Shapiro, H.S. 1975. Content of 6-methylaminopurine and 5-methylcytosine in DNA. In *Handbook of biochemistry, selected data for molecular biology: Nucleic Acids* (ed. G.D. Fasman), pp. 282–283. CRC Press, Boca Raton, Florida.

Som, S., A.S. Baghwat, and S. Friedman. 1987. Nucleotide sequence and expression of the gene encoding the *Eco*RII modification enzyme. *Nucleic Acids Res.* **15:** 313–326.

Tajima, S., H. Tsuda, N. Wakabayashi, A. Asano, S. Mizuno, and K. Nishimori. 1995. Isolation and expression of a chicken DNA methyltransferase cDNA. *J. Biochem.* **117:** 1050–1057.

Vongs, A., T. Kakutani, R.A. Martienssen, and E.J. Richards. 1993. *Arabidopsis thaliana* DNA methylation deficient mutants. *Science* **260:** 1926–1928.

Wellensiek, S.J. 1964. Dividing cells as the prerequisite for vernalization. *Plant Physiol.* **39:** 832–835.

Wu, J.C. and D.V. Santi. 1987. Kinetic and catalytic mechanism of *Hha*I methyltransferase. *J. Biol. Chem.* **262:** 4778–4786.

Wyszynski, M.W., S. Gabbara, and S. Bhagwat. 1992. Substitutions of a cysteine conserved among DNA cytosine methylases result in a variety of phenotypes. *Nucleic Acids Res.* **20:** 319–326.

Yen, R.W.C., P.M. Vertino, B.D. Nelkin, J.J. Yu, W. Eldeiry, A. Cumaraswamy, G.G. Lennon, B.J. Trask, P. Celano, and S.B. Baylin. 1992. Isolation and characterization of the cDNA encoding human DNA methyltransferase. *Nucleic Acids Res.* **20:** 2287–2291.

Methylation-related Epigenetic Signals in Bacterial DNA

Josep Casadesús and Joaquín Torreblanca
Departamento de Genética
Universidad de Sevilla
Seville, Spain

The term "epigenetic signals" may seem extraneous to the life-style of bacteria, because the classic concept of epigenetics has been applied to the changes in gene expression that govern differentiation and development (Waddington 1956). However, because differentiation is often associated with changes in DNA or chromatin structure (Holliday and Pugh 1975; Riggs 1975), DNA modification has provided an attractive model for the study of epigenetic regulation.

In prokaryotes, the most common DNA modification is methylation: For instance, DNAs from enteric bacteria contain minor amounts of 6-methyladenine and 5-methylcytosine (Marinus 1987). Base modifications are formed by two classes of methylases: (1) those associated with restriction/modification systems (Bickle and Krüger 1993) and (2) two methyltransferases that do not have a restriction enzyme counterpart: the Dam methylase (DNA adenine methyltransferase) and the Dcm methylase (DNA cytosine methyltransferase). Both enzymes transfer a methyl group from *S*-adenosylmethionine to a base located in specific, double-stranded DNA sequences (Noyer-Weidner and Trautner 1993). The target for Dam methylase is 5′-GATC-3′ (Lacks and Greenberg 1977; Hattman et al. 1978). The Dcm enzyme methylates the inner cytosine of the pentanucleotides 5′-CCAGG-3′ and 5′-CCTGG-3′ (May and Hattman 1975).

While the biological role of the Dcm enzyme remains largely unknown (Marinus 1987; Noyer-Weidner and Trautner 1993; Gläsner et al. 1995), Dam methyltransferase is known to participate in the control of many cellular processes: DNA replication, chromosome segregation, mismatch repair, and transcriptional regulation of certain genes (for reviews, see Marinus 1987; Messer and Noyer-Weidner 1988; Barras and Marinus 1989; Noyer-Weidner and Trautner 1993). Thus, the bacterial cell seems to make wide and versatile use of methylation as a source of epigenetic information. However, despite its seemingly important roles, Dam methylation is a dispensable function in *Escherichia coli* (Marinus

Epigenetic Mechanisms of Gene Regulation

and Morris 1973) and is not found in all eubacteria (Barbeyron et al. 1984). Therefore, its evolutionary significance must be discussed in the framework of the advantages that DNA methylation may provide to the few eubacteria that have acquired it. Furthermore, it should be noted that DNA methylation is not the only epigenetic mechanism that operates in bacteria. A number of phenomena reminiscent of epigenetic gene silencing and unlikely to be methylation-related have also been described previously (Schaefler 1967; Hotchkiss and Gabor 1980; Lynch and Wang 1995; Schnetz 1995; Sledjeski and Gottesman 1995). This field of epigenetic phenomena unrelated to DNA methylation is not discussed in this chapter.

METHYLATION AND HEMIMETHYLATION OF DAM TARGETS: CELL-CYCLE-COUPLED EPIGENETIC SIGNALING

Bulk DNA from enteric bacteria contains 5′-GATC-3′ sites at the expected frequency of 1/256 bp (Marinus 1987). Most of these GATC sites are fully methylated. Upon replication, hemimethylated GATC sites are formed, but their existence is transient: In the daughter molecules, remethylation of the newly synthesized strand follows replication at a distance of some 3 kb (Marinus 1987). Although it is short-lived, hemimethylated DNA serves as a signal for several biological processes, some of which are briefly described in the following sections.

Methyl-directed Mismatch Repair

Mismatched base pairs inevitably arise during DNA replication, because proofreading by DNA polymerase does not eliminate all replication errors. If uncorrected, mismatches give rise to mutations after one replication round. If corrected randomly, mutations appear in half of the cases. Thus, efficient mismatch repair requires a procedure to distinguish between the template strand and the newly synthesized, error-prone strand. This information is provided by the transient lack of methylation in the newly synthesized strain (Wagner and Meselson 1976; Modrich 1991). Whenever the mismatch repair apparatus identifies a mismatch, the MutH endonuclease initiates repair in the nonmethylated strand, which is thus identified as the one that contains the mispaired base (Welsh et al. 1987).

Because of the instructive role of hemimethylation, mismatch repair is performed randomly in Dam$^-$ mutants. As a consequence, they are hypermutable (Marinus and Morris 1975). Hypermutability is also observed in mutants that overproduce Dam methylase, because they undergo faster DNA remethylation (Herman and Modrich 1981; Pukkila et al. 1983). The similar effects of these opposite mutations illustrate the importance of methylation as a signal to indicate *when* and *where* a specific DNA/protein interaction must occur (Messer and Noyer-Weidner 1988).

Regulation of Promoter Activity

The DNA methylation/hemimethylation cycle at GATC sites can be used to trigger transcription of certain promoters during a specific stage of the cell cycle. The simplest and clearest examples are provided by promoters activated by hemimethylation, such as the promoter of the transposase gene (*tnp*) of insertion element IS*10* (Roberts et al. 1985). The IS*10 tnp* promoter contains a GATC site in its –10 module; when this Dam site is fully methylated, the promoter is inactive. Passage of the replication fork activates transcription; later, DNA remethylation brings the promoter back to the repressed state. Thus, the IS*10 tnp* promoter is only active while hemimethylated, thereby limiting transposase synthesis to a brief period of the cell cycle (Roberts et al. 1985). Other genes whose promoters are activated by hemimethylation are the cell division gene *sulA* and the biosynthetic genes *trpR, trpS,* and *glnS* (Marinus 1985; Sternberg 1985; Marinus 1987; Plumbridge 1987). Except for *sulA*, which encodes a cell division inhibitor, the logic of coupling transcription of these genes to DNA replication is not obvious (Marinus 1987; Barras and Marinus 1989).

One example of the opposite situation (repression by hemimethylation) is provided by a promoter of the *dnaA* gene, which is only transcribed if fully methylated. However, as discussed below, transcription of this promoter during a short, specific period of the cell cycle requires the participation of a mechanism that delays remethylation (Campbell and Kleckner 1990).

To date, most genes regulated by Dam methylation have been discovered by reverse genetics: The presence of GATC sites in or near the promoter has provided evidence for regulation by Dam methylation (Marinus 1987; Wang and Church 1992; Hale et al. 1994). However, Dam-regulated genes can also be detected by genetic screenings like that shown in Figure 1. In this screening, host promoters are fused with a *lacZY* operon, which serves as reporter gene. The fusions are generated using a transposable element that inserts randomly in the host genome; thus, isolates carrying gene fusions in random, unknown loci can be collected. If a fusion is driven by a promoter regulated by Dam methylation, its activity can be expected to change upon transfer to a Dam⁻ host (see Fig. 1 for details). A caveat is that this screening relies on the use of unmethylated DNA; therefore, it may give rise to artifacts whenever hemimethylation and unmethylation are not equivalent signals. Another problem is that the screening may not detect slight changes in gene activity. Despite its potential limitations, the screening has proved useful to detect a novel locus repressed by Dam methylation, located in the *Salmonella typhimurium* virulence plasmid (J. Torreblanca and J. Casadesús, unpubl.).

Outside enteric bacteria, a methylase that recognizes a target slightly different from Dam (5′-GAnTC-3′) has been proposed to control the expression of genes that coordinate the life cycle of the dimorphic bacterium *Caulobacter crescentus* (Stephens et al. 1995).

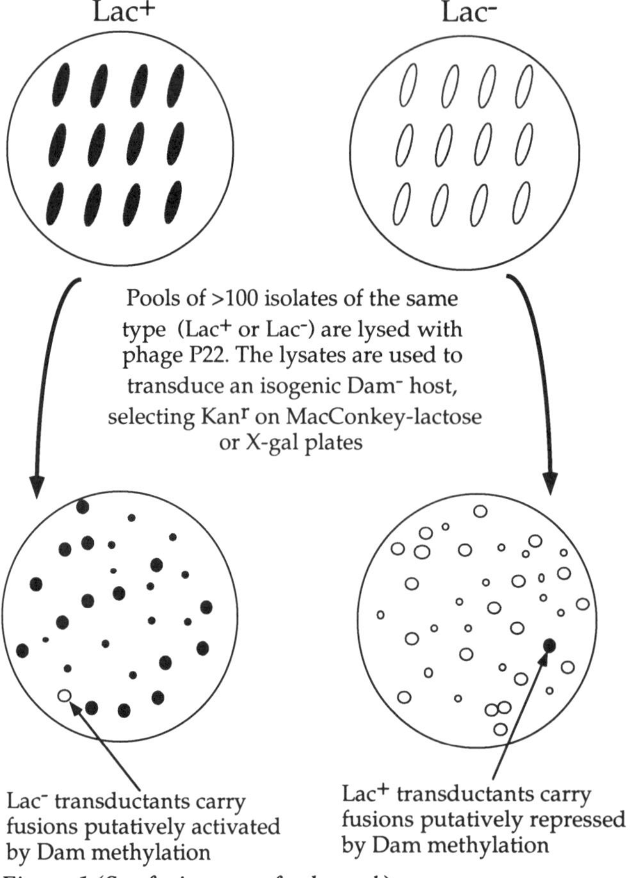

Figure 1 (See facing page for legend.)

REGULATION OF REMETHYLATION BY TRANSIENT TARGET SEQUESTRATION

Although the general rule is that remethylation follows DNA replication rather closely, certain GATC sites remain hemimethylated for a longer fraction of the cell cycle. Current evidence indicates that maintenance of the hemimethylated state requires sequestration by specific proteins.

Initiation of Chromosome Replication

The *oriC* region of the *E. coli* chromosome contains 11 GATC sites in 245 bp, a density tenfold higher than expected from a random distribution (Meijer et al. 1979). Chromosome replication can only start if the origin is fully methylated (Messer et al. 1985; Boye and Løbner-Olesen 1990). Upon DNA replication, the GATC sites of *oriC* are not immediately remethylated; sequestration by an unknown mechanism allows these targets to remain hemimethylated for a substantial fraction of the cell cycle (Campbell and Kleckner 1990).

Figure 1 Search for Dam-regulated genes in *S. typhimurium.* Transcriptional *lac* fusions are generated upon mutagenesis with a Mu-derived transposable element (Mu*d*J). This element carries a kanamycin resistance gene and a promoterless *lac* operon; insertion in the appropriate orientation generates fusions in which the transposon-borne *lac* operon is transcribed from a host promoter. Lac^+ isolates carry active fusions; Lac^- isolates carry either fusions in genes that are not expressed or inserts that do not generate fusions (e.g., inserted in the wrong orientation, with the *lacZY* genes oriented opposite to the direction of transcription). Lac^+ and Lac^- colonies are distinguished by plating on indicator media (e.g., on MacConkey-lactose plates, Lac^+ isolates form red colonies and Lac^- isolates form white colonies, whereas on X-gal plates, Lac^+ isolates form blue colonies and Lac^- isolates form white colonies). Fusions are obtained in a Dam^+ strain and classified according to their Lac phenotype (Lac^+ or Lac^-). Fusions that change their Lac phenotype upon transfer to a Dam^- host can be expected to be transcribed from a Dam-regulated promoter. Transfer is achieved by transduction with phage P22. To avoid the tedious analysis of individual fusions, isolates of the same type (Lac^+ or Lac^-) are "pooled" to generate a single P22 lysate. Candidates carrying Dam-regulated fusions are detected among the Kan^r transductants obtained on a Dam^- recipient: (1) Lac^- isolates that turn Lac^+ in a Dam^- background carry fusions putatively repressed by Dam methylation; (2) Lac^+ isolates that turn Lac^- in a Dam^- background carry fusions putatively activated by Dam methylation. (Kan^r) Kanamycin resistant; (x-gal) 5-bromo-4-chloro-3-indolyl-β-D-galactopyranoside.

Sequestration of *oriC* is coordinated with DnaA synthesis. Binding of DnaA protein to an active (fully methylated) *oriC* is required to initiate chromosome replication. The *dnaA* gene is transcribed from two promoters; one of them (*dnaA2*) contains three GATC sites and is only active if these Dam sites are fully methylated (Braun and Wright 1986; Kücherer et al. 1986). Methylation of the *dnaA2* promoter is regulated by a sequestration mechanism analogous (and probably identical) to that operating at *oriC* (Campbell and Kleckner 1990). Since *oriC* and *dnaA* are separated by only 50 kb, both regions become hemimethylated almost simultaneously and are probably sequestered by a common mechanism involving a protein called SeqA (Lu et al. 1994).

Chromosome Segregation

Initial observations made by Ogden et al. (1988) indicated that the *E. coli oriC* region bound to the cell envelope when *oriC* was hemimethylated (but not when it was fully methylated or unmethylated). The interaction of *oriC* with the outer membrane relies on two GATC-rich tracts and only occurs when they are hemimethylated (Herrick et al. 1994). The same authors have identified a gene that encodes a protein (HobH) involved in the *oriC*/membrane interaction (Herrick et al. 1994).

METHYLATION-BLOCKING FACTORS: TARGET SEQUESTRATION BEYOND CELL DIVISION

In the former section, we introduced the concept of target sequestration, which indicates that remethylation rates can vary among individual GATC targets. However, in the *oriC* and *dnaA* examples, target sequestration is limited to a certain portion of the cell cycle. Because these signals do not change the status of a genetic locus beyond cell division, they are not in a strict sense epigenetic. Situations in which a given locus can be inherited in alternative methylation states are discussed in the following sections.

Regulation of the *pap* Operon of *E. coli*

E. coli pathogenic strains synthesize various types of pili that are involved in binding to specific host receptors. For instance, the pyelonephritis-associated-pilus (*pap*) operon encodes pili typical of strains that cause kidney infections (O'Hanley et al. 1985). Expression of the *pap* operon is controlled at the transcriptional level by a complex epigenetic mechanism that involves Dam methylation (van der Woude et al. 1992; Low 1994). Transcription of the *pap* promoter is regulated by binding of the leucine-responsive regulatory protein (Lrp), an *E. coli* protein that plays a major role in global metabolic control (Calvo and Matthews 1994). Lrp binds to two upstream regulatory regions separated by more than 100 bp (Braaten et al. 1991). Each Lrp-binding region contains one GATC site (designated GATC-1 and GATC-2; see Fig. 2). These sites can be methylated only in the absence of Lrp (Nou et al. 1993).

In the OFF phase, Lrp binds cooperatively and with high affinity around the GATC-2 site, possibly preventing RNA polymerase binding to the *pap* promoter. In this situation, the GATC-1 site is free and thus

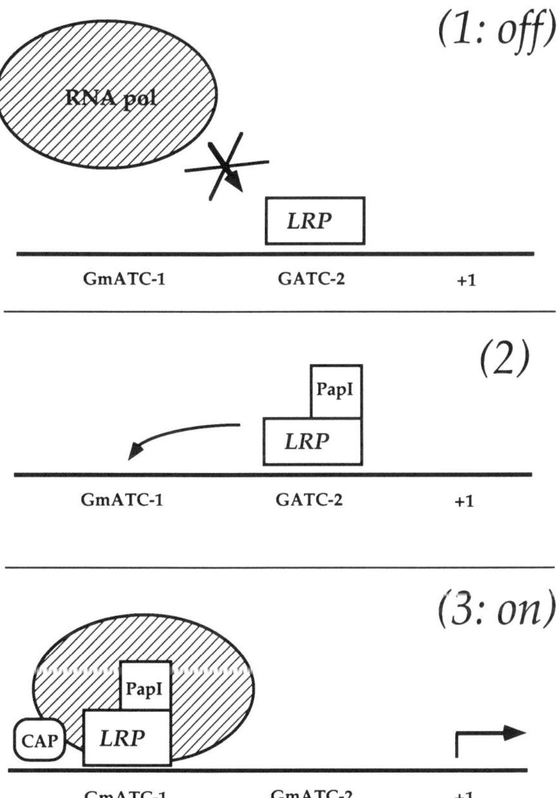

Figure 2 Model for transcriptional regulation of the *pap* operon, adapted from publications by D.A. Low and colleagues. The position "+1" indicates the transcription start point. (RNA pol) RNA polymerase; (LRP) leucine-responsive protein; (mA) 6-methyladenine; (CAP) cAMP-binding protein.

can be fully methylated (Fig. 2). Transition to the ON stage is triggered by binding of PapI, a *pap*-encoded regulatory protein, to the Lrp moiety bound to *pap* DNA (Braaten et al. 1992; Kaltenbach et al. 1995). PapI reduces the affinity of Lrp for the GATC-2 region and increases its affinity for sites around GATC-1. Binding of Lrp near GATC-1, together with binding of the transcriptional activator CAP at a site 50 bp upstream of GATC-1, results in transcription activation (Braaten et al. 1994).

Dam methylation plays dual roles in Pap phase variation. Methylation of GATC-1 in phase OFF cells blocks phase variation by inhibiting binding of Lrp-PapI to the GATC-1 region (Fig. 2). Every round of DNA replication offers an opportunity for phase variation, because the complex Lrp-PapI can bind with low affinity to the hemimethylated GATC-1 site; thus, the hemimethylated state is a time window for switching to the ON phase. Switching also requires full methylation of the GATC-2 site; current data suggest that methylation protects this site from binding of free Lrp and/or from rebinding of Lrp bound at GATC-1. Thus, methylation of GATC-2 may lock cells in the ON state (D.A. Low, pers. comm.).

Although the ultimate logic of *pap* regulation is still obscure, the involvement of two general regulators of *E. coli* such as Lrp and CAP suggests that synthesis of Pap fimbriae responds to environmental stimuli (Low 1994). Moreover, it shows that transduction of environmental signals can change the methylation state of certain GATC sites.

Undermethylation as a Physiological Signal

Before analysis of the *pap* operon, GATC sites were believed to be only hemimethylated transiently, and unmethylated DNA was viewed as a nonphysiological condition (Barras and Marinus 1989). Aside from its intrinsic elegance, David Low's model for *pap* regulation has introduced the important concept that undermethylated GATC targets can also exist as physiological signals (van der Woude et al. 1992; Low 1994). Furthermore, recent studies have shown that the *pap* operon is not a rare exception: certain GATC sites of the *E. coli* chromosome can be stably hemimethylated or unmethylated (Ringquist and Smith 1992; Wang and Church 1992; Hale et al. 1994). Interestingly, a number of unmethylated sites are found in the upstream regulatory regions of genes (Wang and Church 1992), and the methylation state of certain Dam targets seems to be under the influence of environmental conditions. For instance, leucine, the carbon source, and the bacterial growth state seem to regulate the methylation patterns of a subset of the GATC sites analyzed (Hale et al. 1994).

Because Dam methylase is a fast and efficient enzyme, GATC targets are unlikely to remain unmethylated unless they are protected from remethylation (Marinus 1987; Wang and Church 1992). Various mechanisms could provide such protection: (1) binding of proteins involved in gene regulation or nucleoid organization; (2) conformational steric hindrance of DNA (e.g., acquisition of DNA conformations resistant to methylation); (3) demethylation by DNA repair. Unmethylated sites may tend to stay unmethylated because the Dam methylase prefers hemimethylated DNA over unmethylated DNA (Herman and Modrich 1982). Thus, temporary protection of a Dam target may be sufficient to cause (to a certain extent) stable unmethylation.

A SPECULATION ON THE ROLE OF EPIGENETIC SIGNALING IN METABOLIC ADAPTATION

Because bacteria cannot exert any control over their environment and are not equipped with homeostatic mechanisms, their survival depends on adaptability (Neidhardt et al. 1990). As a consequence, they employ complex, overlapping mechanisms for the accurate transduction of environmental signals (Neidhardt 1987). Although a general involvement of Dam methylation in adaptation is far from being proved, the unexpected flexibility of methylation patterns suggests that they might serve as regulatory signals. The presence of Dam sites in binding regions for global transcriptional regulators such as Lrp and CAP is notable, because DNA methylation can strongly influence protein binding (Bölker and Kahmann 1989; Nou et al. 1993). Thus, it is tempting to speculate that the cell may establish distinct patterns of metabolic activity by dictating the methylation state of Dam targets located in the binding sites of transcriptional regulators. A caveat is that protection of a GATC site can block remethylation: Thus, undermethylation of a given GATC site may be the epigenetic mark itself or may merely reflect the existence of a protection mechanism.

Current evidence suggests that, in most cases, transcriptional control by Dam methylation is exerted in combination with other regulatory mechanisms (Braaten et al. 1992; Hale et al. 1994; J. Torreblanca and J. Casadesús, unpubl.). However, an ancillary role of epigenetic signaling would not diminish its adaptive value. Because DNA methylation patterns can be inherited, the daughter cells may use them as a short-term memory of the metabolic state in which the former generation thrived and divided.

CONCLUSIONS

DNA adenine methylation regulates many cellular processes, indicating *when* and *where* a specific DNA/protein interaction must occur (or, in the terms used by Messer and Noyer-Weidner [1988], for "timing and targeting"). Full methylation of *oriC* and *dnaA* dictates the start of every replication round. Upon chromosome replication, hemimethylated DNA provides signals for mismatch repair, transient activation of certain promoters, binding of the daughter chromosomes to the outer membrane, and temporary sequestration of *oriC*.

The normal cycle of hemimethylation and remethylation can be disrupted by protection mechanisms that prevent remethylation of specific targets, thus generating GATC sites that are stably hemimethylated or unmethylated (Wang and Church 1992; Hale et al. 1994). Maintenance of methylation patterns beyond cell division is a genuine mode of epigenetic inheritance, whose importance in bacterial physiology remains to be established. To date, the paradigm of epigenetic regulation in bacteria is the *pap* operon of *E. coli*, in which Dam methylation and Lrp binding are the major mechanisms that regulate transcription (Braaten et al. 1994). The example of *pap* is unlikely to be unique: Both physical and genetic screeenings suggest that novel examples of loci regulated by DNA adenine methylation will be identified in the near future (Ringquist and Smith 1992; Wang and Church 1992; Hale et al. 1994; J. Torreblanca and J. Casadesús, unpubl.). On the other hand, the presence of undermethylated sites within the binding regions of regulatory proteins such as Lrp and CAP strongly suggests that DNA methylation may play a role in global regulation (Wang and Church 1992; Hale et al. 1994). Thus, one major difference between eukaryotic and prokaryotic epigenetic signals may be that the latter are used for adaptation, rather than for differentiation. Concomitant with this adaptive role, it is not suprising that prokaryotic epigenetic marks are unstable (and thus elusive).

ACKNOWLEDGMENTS

Work in our laboratory is supported by grant PB93-649 from the Dirección General de Investigación Científica y Técnica of the Government of Spain. We are grateful to Andrés Aguilera, Bruce Braaten, Sebastián Chávez, David Low, Rob Martienssen, Art Riggs, Enzo Russo, and Marjan van der Woude for critical reading of the manuscript.

REFERENCES

Barbeyron, T., K. Kean, and P. Forterre. 1984. DNA adenine methylation of GATC sequences appeared recently in the *Escherichia coli* lineage. *J. Bacteriol.* **160:** 586–590.

Barras, F. and M.G. Marinus. 1989. The great GATC: DNA methylation in *E. coli. Trends Genet.* **5:** 139–143.

Bickle, T.A., and D.H. Krüger. 1993. Biology of DNA restriction. *Microbiol. Rev.* **57:** 434–450.

Bölker, M. and R. Kahmann. 1989. The *Escherichia coli* regulatory protein OxyR discriminates between methylated and unmethylated states of the phage Mu *mom* promoter. *EMBO J.* **8:** 2403–2410.

Boye, E. and A. Løbner-Olesen. 1990. The role of *dam* methyltransferase in the control of DNA replication in *E. coli. Cell* **62:** 981–989.

Braaten, B.A., L.B. Blyn, B.S. Skinner, and D.A. Low. 1991. Evidence for a methylation-blocking factor (*mbf*) locus involved in *pap* pilus expression and phase variation in *Escherichia coli. J. Bacteriol.* **173:** 1789–1800.

Braaten, B.A., X. Nou, L.S. Kaltenbach, and D.A. Low. 1994. Methylation patterns in *pap* regulatory DNA control pyelonephritis-associated pili phase variation in *E. coli. Cell* **76:** 577–588.

Braaten, B. A., J.W. Platko, M.W. van der Woude, B.H. Simons, F.K. de Graaf, J.M. Calvo, and D.A. Low. 1992. Leucine-responsive regulatory protein controls the expression of both the *pap* and *fan* pili operons in *Escherichia coli. Proc. Natl. Acad. Sci.* **89:** 4250–4254.

Braun, R.E. and A. Wright. 1986. DNA methylation differentially enhances the expression of one of the two *E. coli dnaA* promoters in vivo and in vitro. *Mol. Gen. Genet.* **202:** 246–250.

Calvo, J.M. and R.G. Mattews. 1994. The leucine-responsive regulatory protein, a global regulator of metabolism in *Escherichia coli. Microbiol. Rev.* **58:** 466–490.

Campbell, J.L. and N. Kleckner. 1990. *E. coli oriC* and the *dnaA* gene promoter are sequestered from *dam* methyltransferase following the passage of the chromosomal replication fork. *Cell* **62:** 967–979.

Gläsner, W., R. Merkl, V. Schellenberger, and H.J. Fritz. 1995. Substrate preferences of Vsr DNA mismatch repair endonuclease and their consequences for the evolution of the *Escherichia coli* genome. *J. Mol. Biol.* **245:** 1–7.

Hale, W.B., M.W. van der Woude, and D.A. Low. 1994. Analysis of nonmethylated GATC sites in the *Escherichia coli* chromosome and identification of sites that are differentially methylated in response to environmental stimuli. *J. Bacteriol.* **176:** 3438–3441.

Hattman, S., J.E. Brooks, and M. Masurekar. 1978. Sequence specificity of the P1-modification methylase (M.*Eco*P1) and the DNA methylase (M.*Eco* dam) controlled by the *E. coli dam* gene. *J. Mol. Biol.* **126:** 367–380.

Herman, G.E. and P. Modrich. 1981. *Escherichia coli* K-12 clones that overproduce *dam* methylase are hypermutable. *J. Bacteriol.* **145:** 644–646.

———. 1982. *Escherichia coli dam* methylase. *J. Biol. Chem.* **257:** 2605–2612.

Herrick, J., R. Kern, S. Guha, A. Landoulsi, O. Fayet, A. Malki, and M. Kohiyama. 1994. Parental strand recognition of the DNA replication origin by the outer membrane in *Escherichia coli. EMBO J.* **13:** 4695–4703.

Holliday, R. and J.E. Pugh. 1975. DNA modification mechanisms and gene activity during development. *Science* **187:** 226–232.

Hotchkiss, R.D. and M.H. Gabor. 1980. Biparental products of bacterial protoplast fusion showing unequal parental chromosome expression. *Proc. Natl. Acad. Sci.* **77:** 3553–3557.

Kaltenbach, L.S., B.A. Braaten, and D.A. Low. 1995. Specific binding of PapI to Lrp-pap DNA complexes. *J. Bacteriol.* **177:** 6449–6455.

Kücherer, C., H. Lother, R. Kölling, M.A. Schauzu, and W. Messer. 1986. Regulation of transcription of the chromosomal *dnaA* gene of *Escherichia coli. Mol. Gen. Genet.* **205:** 115–121.

Lacks, S. and B. Greenberg. 1977. Complementary specificity of restriction endonucleases of *Diplococcus pneumoniae* with respect to DNA methylation. *J. Mol. Biol.* **114:** 153–168.

Low, D.A. 1994. Methylation-dependent and Lrp-dependent fimbrial gene regulation in *Escherichia coli.* In *Molecular genetics of bacterial pathogenesis* (ed. V.L. Miller et al.), pp. 423–436. ASM Press, Washington, D.C.

Lu, M., J.L. Campbell, E. Boye, and N. Kleckner. 1994. SeqA: A negative modulator of replication initiation in *E. coli. Cell* **77:** 413–426.

Lynch, A.S. and J.C. Wang. 1995. SopB protein-mediated silencing of genes linked to the *sopC* locus of *Escherichia coli* F plasmid. *Proc. Natl. Acad. Sci.* **92:** 1896–1900.

Marinus, M.G. 1985. DNA methylation influences *trpR* promoter activity in *Escherichia coli* K-12. *Mol. Gen. Genet.* **200:** 185–186.

———. 1987. DNA methylation in *Escherichia coli. Annu. Rev. Genet.* **21:** 113–131.

Marinus, M.G. and N.R. Morris. 1973. Isolation of deoxyribonucleic acid methylase mutants of *Escherichia coli* K-12. *J. Bacteriol.* **114:** 1143–1150.

———. 1975. Pleiotropic effects of a DNA adenine methylation mutation *dam-3* in *Escherichia coli* K-12. *Mutat. Res.* **28:** 15–26.

May, M.S. and S. Hattman. 1975. Analysis of bacteriophage deoxyribonucleic acid sequences methylated by host- and R-factor- controlled enzymes. *J. Bacteriol.* **123:** 768–770.

Meijer, M., E. Beck, F.G. Hansen, H. E. Bergmans, W. Messer, K. von Meyenburg, and H. Schaller. 1979. Nucleotide sequence of the origin of replication of the *E. coli* K-12 chromosome. *Proc. Natl. Acad. Sci.* **76:** 580–584.

Messer, W. and M. Noyer-Weidner. 1988. Timing and targeting: The biological functions of Dam methylation in *E. coli. Cell* **54:** 735–737.

Messer, W., U. Bellekes, and H. Lother. 1985. Effect of *dam* methylation on the activity of the *E. coli* origin, *oriC. EMBO J.* **4:** 1327–1332.

Modrich, P. 1991. Mechanisms and biological effects of mismatch repair. *Annu. Rev. Genet.* **25:** 229–253.

Neidhardt, F.C. 1987. Multigene systems and regulons. In Escherichia coli *and* Salmonella typhimurium: *Cellular and molecular biology* (ed. F.C. Neidhardt et al.), pp. 1313–1317. American Society for Microbiology, Washington, D.C.

Neidhardt, F.C., J.L. Ingraham, and M. Schaechter. 1990. *Physiology of the bacterial cell.* Sinauer, Sunderland, Massachusetts.

Niki, H., A. Jaffé, R. Imamura, T. Ogura, and S. Hiraga. 1991. The new gene *mukB* codes for a 77 kDa protein with coiled-coil domains involved in chromosome partitioning of *E. coli. EMBO J.* **10:** 183–193.

Nou, X, B. Skinner, B. Braaten, L. Blyn, D. Hirsch, and D. Low. 1993. Regulation of pyelonephritis-associated pili phase-variation in *Escherichia coli*: Binding of the PapI and the Lrp regulatory proteins is controlled by DNA methylation. *Mol. Microbiol.* **7:**

545–553.

Noyer-Weidner, M. and T.A. Trautner. 1993. Methylation of DNA in prokaryotes. In *DNA methylation: Molecular biology and biological significance* (ed. J.P. Jost and H.P. Saluz), pp. 39–108. Birkhäuser Verlag, Basel, Switzerland.

Ogden, G.B., M.J. Pratt, and M. Schaechter. 1988. The replicative origin of the *E. coli* chromosome binds to cell membranes only when hemimethylated. *Cell* **54:** 127–135.

O'Hanley, P., D. Lark, S. Falkow, and G. Schoolnik. 1985. Molecular basis of *Escherichia coli* colonization of the upper urinary tract in BALB/c mice. *J. Clin. Invest.* **75:** 347–360.

Plumbridge, J. 1987. The role of *dam* methylation in controlling gene expression. *Biochimie* **69:** 439–443.

Pukkila, P.J., J. Peterson, G. Herman, P. Modrich, and M. Meselson. 1983. Effects of high levels of DNA adenine methylation on methyl-directed mismatch repair in *Escherichia coli*. *Genetics* **104:** 571–582.

Riggs, A.D. 1975. X-inactivation, differentiation and DNA methylation. *Cytogenet. Cell Genet.* **14:** 9–11.

Ringquist, S. and C.L. Smith. 1992. The *Escherichia coli* chromosome contains specific, unmethylated *dam* and *dcm* sites. *Proc. Natl. Acad. Sci.* **89:** 4539–4543.

Roberts, D., B.C. Hoopes, W.R. McClure, and N. Kleckner. 1985. IS10 transposition is regulated by DNA adenine methylation. *Cell* **43:** 117–130.

Schaefler, S. 1967. Inducible system for the utilization of β-glucosides in *Escherichia coli*. I. Active transport and utilization of β-glucosides. *J. Bacteriol.* **93:** 254–263.

Schnetz, K. 1995. Silencing of *Escherichia coli bgl* promoter by flanking sequence elements. *EMBO J.* **14:** 2545–2550.

Sledjeski, D. and S. Gottesman. 1995. A small RNA acts as an antisilencer of the H-NS-silenced *rcsA* gene of *Escherichia coli*. *Proc. Natl. Acad. Sci.* **92:** 2003–2007.

Stephens, C.M., G. Zweiger, and L. Shapiro. 1995. Coordinate cell cycle control of a *Caulobacter* DNA methyltransferase and the flagellar genetic hierarchy. *J. Bacteriol.* **177:** 1662–1669.

Sternberg, N. 1985. Evidence that adenine methylation influences DNA-protein interactions in *E. coli*. *J. Bacteriol.* **164:** 490–493.

van der Woude, M.W., B.A. Braaten, and D.A. Low. 1992. Evidence for global regulatory control of pilus expression in *Escherichia coli* by Lrp and DNA methylation: Model building based on the analysis of *pap*. *Mol. Microbiol.* **6:** 2429–2435.

Waddington, C.H. 1956. *Principles of embryology*. Allen & Unwin, London, England.

Wagner, R. and M. Meselson. 1976. Repair tracts in mismatched DNA heteroduplexes. *Proc. Natl. Acad. Sci.* **73:** 4135–4139.

Wang, M.X. and G.M. Church. 1992. A whole genome approach to *in vivo* DNA-protein interactions in *E. coli*. *Nature* **360:** 606–610.

Welsh, K.M., A.L. Lu, S. Clark, and P. Modrich. 1987. Isolation and characterization of the *Escherichia coli mutH* gene product. *J. Biol. Chem.* **262:** 15625–15629.

DNA Modification and Restriction: Selfish Behavior of an Epigenetic System

Ichizo Kobayashi
Department of Molecular Biology
Institute of Medical Science
University of Tokyo, Tokyo 108, Japan

Type II restriction endonucleases (R) are known to make double-strand breaks within or near specific recognition sequences on duplex DNA. Cognate modification enzymes (M) can methylate these sequences and protect them from cleavage (Wilson and Murray 1991). The tight association of a cognate restriction enzyme gene with a modification gene has been termed the (type II) restriction-modification (RM) system.

Although the RM systems that have been described are individually highly specific for one or a few recognition sequences, collectively the sequences recognized are quite diverse. Type II RM systems are ubiquitous in prokaryotes and in archaebacteria but are absent from eukaryotes (Wilson and Murray 1991).

Type II restriction enzymes cleave foreign DNA such as viral and plasmid DNA when this DNA has not been modified by the appropriate modification enzyme. In this way, cells are protected from invasion by foreign DNA. Thus, it has been widely believed that the evolution and maintenance of type II restriction modification systems have been driven by the cell's need to protect itself from infection by foreign DNA (the cellular defense hypothesis). However, there are several unresolved issues that cannot be explained satisfactorily by this cellular defense hypothesis. Here, I present experimental evidence and theoretical arguments for an alternative hypothesis (the selfish gene hypothesis) that the evolution of at least some RM gene pairs is driven by their "selfishness" in the genetic and evolutionary sense of the term.

WEAK POINTS OF THE CELLULAR DEFENSE HYPOTHESIS

The phenomena associated with type II RM systems that have not been explained (or have not been well explained) by the cellular defense hypothesis are listed in Table 1. Alternative explanations for many of these phenomena are considered below.

Table 1 Cellular-defense hypothesis versus selfish-gene hypothesis for type II RM systems

Phenomena	Cellular-defense hypothesis	Selfish-gene hypothesis
Evolution and maintenance		
widespread occurrence in the prokaryotes	defense against invading phage and plasmids by cleavage of their DNA	resistance to displacement by host-killing, especially in exclusive competition with similar genes (Fig. 1B,C)
defense against phage only short-lived	colonization selection (defense effective during formation of a colony from a bacterial cell in the presence of phage)	phage defense secondary
individual specificity and collective diversity in sequence recognition	frequency-dependent selection for effective defense against phage among bacteria	frequency-dependent selection arising from competition for recognition sequences among RM elements (Fig. 3B,C)
virtual absence of sequence homology among R genes	many independent origins	early specialization to specific recognition sequences followed by sequence divergence in R genes

Gene organization		
restriction and modification activities on separate molecules		allows postsegregational cleavage of the host chromosome by restriction enzyme molecule (Fig. 1)
tight linkage of R and M genes		simultaneous loss essential to post-segregational killing (Fig. 1)
horizontal transfer and independence from the rest of the genome		mobility in harmony with the selfish nature (as in viruses)
Putative anti-RM systems		
solo methylases		protection of bacterial chromosome against some restriction enzymes
paucity of some RM sites in some bacterial chromosomes		protection of host chromosomes against some RM systems
methyl-specific restriction enzyme	defense against phage	suicidal defense against invasion of some RM systems by cleaving chromosome at methylated sites
homologous recombination	repair of phage DNA cleavage by restriction enzyme (recombination machinery of lambda and a lambdoid prophage)	repair of host chromosomes (and destruction of invading DNA) after cleavage by restriction enzyme (RecA RecBCD recombination machinery of *E. coli*)

See text for details and references.

Laboratory studies have demonstrated that defense against bacteriophages by RM systems is only short-lived (Korona and Levin 1993). Conceivably, a novel RM system could help bacteria to colonize a new habitat full of phages. However, this advantage would be quickly lost because some invading viral DNA occasionally becomes modified and escapes restriction, thus affording protection against restriction for itself and its descendants (Korona and Levin 1993).

It is also unclear whether the cellular defense hypothesis can account for the individual high specificity and collective wide diversity in the sequence recognition of type II RM systems. According to the cellular defense hypothesis, this diversity is a result of frequency-dependent selection against phage infection (Levin 1988): Rare RM types are favored because only a small number of phage are modified for the recognition sequence.

With few exceptions, restriction enzyme genes do not display a high degree of sequence homology (Wilson and Murray 1991; Siksnys et al. 1995). This suggests their independent evolution. The cognate restriction enzyme and methylation enzyme usually do not share sequence homology or a common strategy for sequence recognition (Janulaitis et al. 1992). Thus, there is redundancy compared with other RM systems, such as type I and type III RM systems, where only one molecule determines sequence specificity. In addition, although the cognate restriction and modification genes are tightly linked, the gene products are not known to associate physically. Thus, the evolution of RM systems was not straightforward.

There are examples suggesting horizontal transfer of RM genes between different bacterial genera (Wilson and Murray 1991). The *Eco*RI RM unit appears to have been inserted into a plasmid, which suggests that the *Eco*RI RM unit can be easily transferred. This mobility has not been explained by the cellular defense mechanism. Finally, the absence of type II RM systems from the eukaryotic world remains a mystery.

SELFISH BEHAVIOR OF THE RM SYSTEMS

Experimental evidence for the selfish behavior of RM systems can be summarized as follows.

Stabilization of Plasmid DNA That Encodes a Type II RM System

Plasmids that carry the *Pae*R7 RM gene pair have been shown not to be easily lost from host bacteria even in the absence of selection (Naito et al. 1995). This stabilization is lost when a restriction-minus derivative is used. This type of restriction-dependent plasmid stabilization has been observed with other type II RM gene pairs such as *Eco*RI (Kulakauskas et al. 1995; Kusano et al. 1995; Naito et al. 1995), *Bsp*6I (Kulakauskas et al. 1995), and *Eco*RV (Y. Nakayama and I. Kobayashi, unpubl.).

Host-killing Model for Stabilization

On the basis of the above results and supporting data (see below), the following host-killing model for plasmid stabilization was postulated (Fig. 1B) (Naito et al. 1995).

In a cell carrying an RM plasmid, essentially all of the numerous recognition sites on the host chromosome are modified by the methylation enzyme and protected from the restriction enzyme (Fig. 1B, top). When a cell loses the RM plasmid, its descendants retain fewer and fewer molecules of the modification and restriction enzymes after each cell division cycle. Eventually, the capacity of modification enzymes to modify the recognition sites becomes inadequate. One of these unmodified sites is then cleaved by the remaining pool of restriction enzymes with consequent cell death (Fig. 1B, bottom). The net result of this would be the stable maintenance of the plasmid genes in the population of viable cells.

To test this model, we used the *Pae*R7 RM genes cloned into a thermosensitive replicon. In this way, plasmid loss could be thermally induced (Kusano et al. 1995; Naito et al. 1995). When the temperature was shifted to the nonpermissive temperature for the replicon, the increase in the viable cell count stopped, and this was found to correlate with the stop of increase in the number of plasmid-carrying cells. Degradation of the host chromosome followed. A similar loss of cell viability was observed at the nonpermissive temperature for cells bearing thermosensitive replicons encoding *Eco*RI (Kusano et al. 1995; Naito et al. 1995) and *Eco*RV RM genes (Y. Nakayama and I. Kobayashi, unpubl.).

Resistance of a Plasmid Carrying an RM Gene Pair to Replacement by Another, Incompatible, Plasmid

We originally showed that a plasmid carrying the *Pae*R7 RM genes could not be readily displaced by a second, incoming plasmid when both plasmids were members of the same incompatibility group (Naito et al. 1995). This could not be accounted for by restriction of the incoming

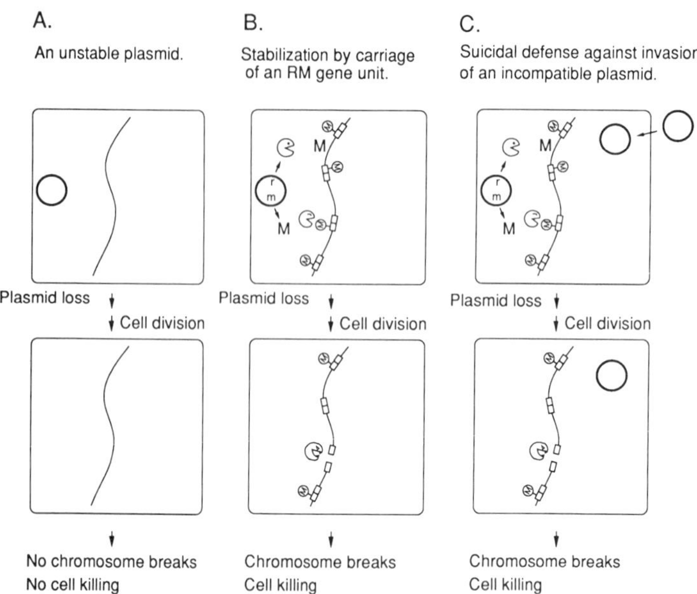

Figure 1 Stabilization of a plasmid as a result of an encoded RM gene unit. (*A*) One plasmid without an RM gene pair. Loss of the plasmid does not lead to cell killing. (*B*) One plasmid with an RM gene pair. Loss of the plasmid leads to dilution of the modification enzyme and cell killing by the restriction enzyme through chromosome cleavage. The RM plasmid is maintained at a high frequency in the population of the viable cells. (*C*) Invasion by a second plasmid with an incompatible replication apparatus. Loss of the resident RM plasmid leads to cell killing as in *B*. This is suicidal defense against invasion by the second plasmid (Fig. 2). (Large box) Bacterial host cell. (Circle) Plasmid. (r) Restriction gene on the plasmid. (Face) Restriction endonuclease. (m) Cognate modification gene. (M) Modification methyltransferase. (Central curved line) Bacterial chromosome. (Small squares on line) Recognition sites of the RM system on the chromosome. (Circled M) Methyl group on the recognition site.

plasmid DNA, because the incoming plasmid did not contain *Pae*R7 restriction enzyme sites. Rather, further experiments suggested that the incoming plasmid was indeed capable of displacing the RM plasmid from the cell, but that the resulting cell and, consequently, the invading plasmid failed to survive because of cleavage of the cellular DNA by the imbalanced restriction enzyme activity (Fig. 1C).

This "host-killing" strategy programmed by the RM gene pairs prevents plasmids carrying the RM genes from being displaced by other replicons of the same incompatibility group (Fig. 2A,B). This situation is similar to the "altruistic" suicide of some cells upon infection with some

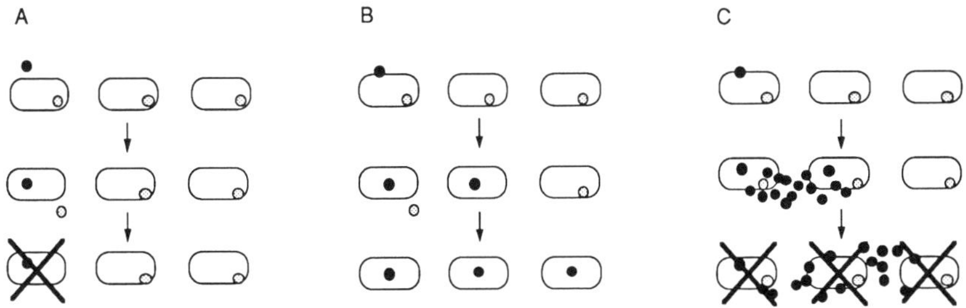

Figure 2 Suicidal defense against invasion programmed by a selfish gene. (*A*) Cell killing by a resident suicide gene is triggered by invasion of plasmids, viruses, or other selfish genes in general. The selfish gene can, by programming such altruistic deaths, maintain its presence in the population of viable cells. (*B*) Invasion and no suicide. The resident gene is replaced by the invading gene (plasmid) in the first cell. The copies of the resident gene are replaced by the copies of the invading gene through secondary infections or through competition. (*C*) Virus infection and no suicide. The copies of the resident gene are all lost together with the host cells as a result of secondary infections. (Gray circle) Resident selfish suicide gene such as an RM gene unit. (Black circle) Invading selfish gene. (Cross) Cell death.

viruses (Yarmolinsky 1995). The driving force behind the evolution of such suicide genes is likely to be prevention of secondary infection of neighboring cells by the viral progeny of the infected cell, which are very likely carrying copies of the same suicide gene (Fig. 2A,C).

Comparison with "Postsegregational Killing" by Plasmids

The host-killing mechanism for stable maintenance that we describe here is reminiscent of the postsegregational killing mechanisms of plasmid maintenance (Ogura and Hiraga 1983; Gerdes et al. 1986; Jensen and Gerdes 1995). In these cases, a killer and anti-killer gene pair on a plasmid constitute what has been termed an "addictive module" (Lehnherr et al. 1993; Yarmolinsky 1995). It renders the bacterial host addicted to the continual presence of the "dispensable" genetic element that it harbors (Yarmolinsky 1995). Plasmid loss leads to an imbalance in the ratio of killer to anti-killer gene products and consequent cell killing through a variety of mechanisms (Gerdes et al. 1986; Jensen and Gerdes 1995).

Some type II RM genes, such as those in *Bacillus* and *Haemophilus*, are carried by the chromosome rather than the plasmid. The same cell-killing mechanism would be expected to operate in the case of chromosomal RM genes when they are "threatened" by homologous recombination with homologous invading DNA. This was effectively shown to be the case for the *Pae*R7 RM gene pair when it was placed on the chromosome (Y. Nakayama and I. Kobayashi, unpubl.).

SELFISH-GENE HYPOTHESIS FOR THE EVOLUTION AND MAINTENANCE OF RM SYSTEMS

The RM gene pair may be referred to as a "selfish gene" entity in the genetic sense of the term. In general, two homologous alleles in diploid eukaryotic cells segregate in a one-to-one ratio at meiosis. This Mendelian law of segregation is violated by some genes in a process called "meiotic drive." These selfish genes are preferentially transmitted over the other, non-self alleles (Lyttle 1993). In particular, the action of maternal-effect selfish genes that causes postfertilization killing (Beeman et al. 1992) appears to be formally similar to that of the RM genes described here; i.e., loss of the selfish gene leads to killing of the progeny by the residual gene product.

These meiotic-drive selfish genes are expected to spread in a population, even in the absence of any additional selective advantage (Wade and Beeman 1994). It has been predicted that there will be a faster increase in the frequency of such genes in populations that are divided into small subgroups where there is competition for limited resources (Bull et al. 1992; Wade and Beeman 1994).

The host-killing strategy of an RM gene unit makes any DNA that carries the RM gene unit an effective competitor when faced with other DNA that can exclude it (Fig. 2A,B). This has been demonstrated to be the case for two incompatible plasmids driven by the same replication machinery (Fig. 1C) and for two allelic chromosomal gene units as described above. It is possible that this competitive advantage drove the original spreading of RM systems and continues to assure their maintenance.

Some features of the organization and behavior of the RM gene units can be readily explained by the selfish-gene concept. The tight linkage of R and M genes is in contrast to the physical separation of their protein products. This can be explained by the mechanism of postsegregational host killing as presented in Figure 1B. Their tight linkage allows simultaneous loss of R and M genes whether they are carried on plasmids or on the chromosome. Their physical separation allows execution of host killing by the restriction enzyme.

The selfish-gene concept for the RM gene pairs implies a certain degree of independence from the rest of the genome and consequent movement of RM gene pairs between genomes. A gene that behaves selfishly by killing the host would benefit from moving from one genome to another. Inversely, a gene that can move between host genomes would benefit if it behaved selfishly and killed the host cell. Thus, the movement of the RM genes (Wilson and Murray 1991), which has not been explained by the cellular defense hypothesis, makes sense under the selfish-gene hypothesis. The RM gene unit appears similar to some mobile genetic elements such as bacteriocins and viruses and could be related to them by evolution.

INDIVIDUAL SPECIFICITY AND COLLECTIVE DIVERSITY IN SEQUENCE RECOGNITION

Two RM Systems That Recognize the Same Sequence Are Mutually Exclusive

The next series of experiments deals with the interactions between two RM systems present in the same host cell. If these two RM systems (*Pae*R7 and *Eco*RI, for example) recognize different sequences (5′-CTCGAG and 5′-GAATTC, respectively), loss of even one of them leads to killing of the host cell (Fig. 3B). Thus, the cell is forced to maintain these two RM systems. The host killing by an RM gene pair (*Eco*RI, for example) will not work when the second RM gene pair shares the same sequence specificity (an isoschizomer of *Eco*RI, for example, *Rsr*I). Methylation of the recognition site by the modification enzyme of one pair protects it from cleavage by the restriction enzyme of the other pair (Fig. 3C). Thus, two RM systems of the same sequence specificity (*Eco*RI and *Rsr*I) cannot simultaneously enjoy stabilization. These predictions were confirmed by Kusano et al. (1995).

We may call this phenomenon "incompatibility" between RM systems, since it brings to mind the incompatibility that exists between plasmid replication units. In this way, each of the recognition sequences would define one incompatibility group. For example, *Eco*RI, *Rsr*I, and presumably their corresponding isoschizomer RMs, which recognize the sequence 5′-GAATTC, thus form one incompatibility group with respect to stabilization. (The actual modes of interference of one RM system by another RM system may be quite complex because of the recognition sequence overlap, the different modes of methylation, and the differential sensitivity of restriction enzymes to various modes of methylation.)

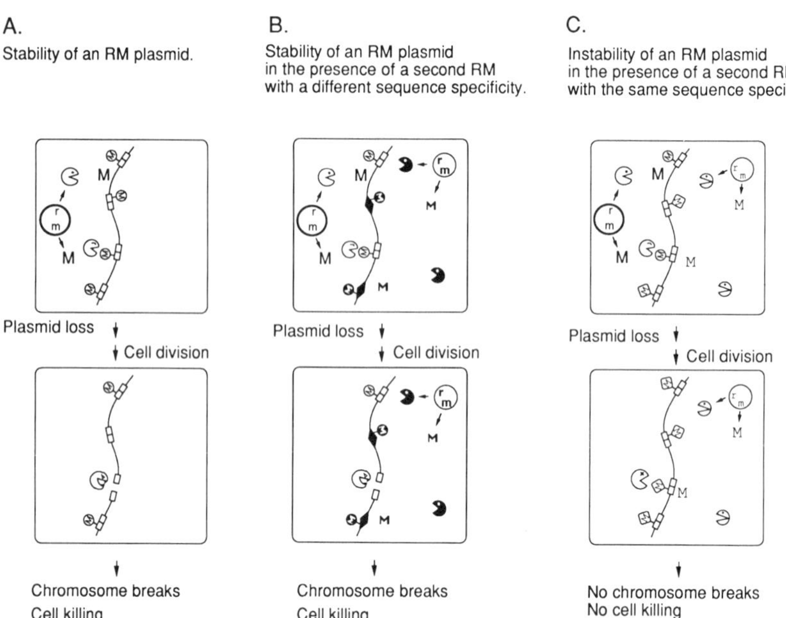

Figure 3 Relationship between two RM gene units residing within the same host. (*A*) One plasmid with an RM gene pair. Loss of the plasmid (and, therefore, the RM gene unit present on it) leads to dilution of the modification enzyme and cell killing by the restriction enzyme. The RM plasmid (and the RM gene unit that it carries) is maintained at a high frequency in the population of viable cells (as in Fig. 1B). (*B*) Two plasmids with RM gene pairs of a different sequence specificity. Loss of one of the plasmids leads to cell killing, as in *A*. (*C*) Two plasmids with RM gene pairs of the same sequence specificity. Loss of one of the plasmids does not lead to cell killing through chromosome cleavage because the other RM system modifies and protects the recognition sites on the host chromosome. Symbols are as in Fig. 1. A circle on the right indicates the second plasmid with the second RM gene unit. Their gene products, their chromosomal recognition sites, and methyl groups are shown.

Origin of Diversity and Specificity in Sequence Recognition

Incompatibility defined on the basis of the recognition sequence (Fig. 3) can explain the nature of RM sequence recognition; i.e., the individual high specificity combined with collective wide diversity. Because this type of incompatibility implies exclusive competition for specific sequences in genomes by RM systems, this would result in the specialization of each selfish-gene unit in only one of many diverse sequences. In general, competitive exclusion drives the adaptation of individual species to one of many small ecological niches in an exclusive way. We may assume that the ecological niche of an RM system is its recognition sequence.

The independent evolution of numerous restriction enzymes, as inferred from the virtual absence of sequence homology between them, can be explained, at least partly, by "adaptive radiation" at very early periods; i.e., specialization to specific recognition sequences by the above diversifying selection. A different sequence for each RM unit would serve to isolate RM systems and ensure their independent evolution.

ANTI-RM ADAPTATION BY THE HOST

Our selfish-gene hypothesis raises the possibility that bacteria may have evolved systems to cope with the parasitism of RM systems. The following may represent such anti-RM adaptations.

Methyl-specific Restriction

Methyl-specific restriction systems, such as McrA, McrBC, and Mrr in *Escherichia coli*, require methylation of specific sequences in order to cut DNA (Noyer-Weidner et al. 1986; Raleigh and Wilson 1986). These restriction systems would abort the invasion of RM units by cleaving the host chromosomal DNA at cognate restriction sites after de novo methylation by the invading RM unit methylase. (These systems would also ensure destruction of the methylated plasmid before expression of the methylase gene if the plasmid happened to carry the recognition sequence in an appropriate configuration.)

Solo Methylases

Some bacteria code for DNA methylase genes, such as *dam* and *dcm*, that are not associated with a restriction enzyme gene. These methylases could play a role in defending their chromosomal DNA from cleavage by restriction enzymes coded for by foreign DNA. These methylases may have evolved because they serve as a kind of vaccination against parasitism by certain RM systems.

Homologous Recombination

Various modes of homologous recombination can be understood as responses to DNA double-strand breaks (Kobayashi 1992). The parasitism of RM systems may have been important in their evolution.

The RecBCD enzyme of *E. coli* (Myers and Stahl 1994) degrades duplex DNA from a double-stranded end. The chi sequence, 5′-GCTGGTGG, blocks exonucleolytic action of RecBCD enzyme and locally enhances RecA-mediated homologous recombination. The chi sequence is present more frequently than expected on a random basis on the *E. coli* chromosome. This RecA/RecBCD system could destroy invading non-self DNA such as bacteriophage DNAs after restriction cleavage. On the other hand, the RecA/RecBCD system would allow recombinational restoration of its own chi-marked chromosomal DNA after restriction cleavage.

Homologous recombination mechanisms coded by lambdoid bacteriophages promote double-strand break repair (Takahashi and Kobayashi 1990; Yokochi et al. 1995). These mechanisms likely represent a system developed by bacteriophages to protect their DNA against restriction.

Paucity of Restriction Sites among Bacterial Chromosomes

Some DNA restriction sites are less abundant than expected on the chromosomes of some bacteria (Karlin et al. 1992). These authors mentioned "Bacterial DNA may also experience some selection against restriction sites, either because of occasional failure of methylation modification..." This may represent an adaptive mechanism developed by bacteria to prevent parasitism by RM systems.

SOME EXTRAPOLATIONS

Potential Selfishness of Other Nucleases and Transposons

There is a group of endonuclease genes that can introduce a double-strand break into a specific sequence and then insert itself between the break in a process called homing (Lambowitz and Belfort 1993). This bears resemblance to the RM gene unit: The occupied state of a site corresponds to the methylated (or self) state and the unoccupied state to the unmethylated (or non-self) state. Their precise excision regenerates the recognition sequence and should lead to cleavage of the chromosome and cell killing by the endonuclease. As in the case of the RM system, this cell killing would maintain them in the population of viable cells.

Some transposons make a double-strand break at their resident site (donor suicide). The chromosome break could kill the host under conditions that threaten the maintenance of the transposon. In this way, they might behave as a suicide gene, which is selfish and altruistic at the same time, in a manner similar to the RM units (Fig. 2A).

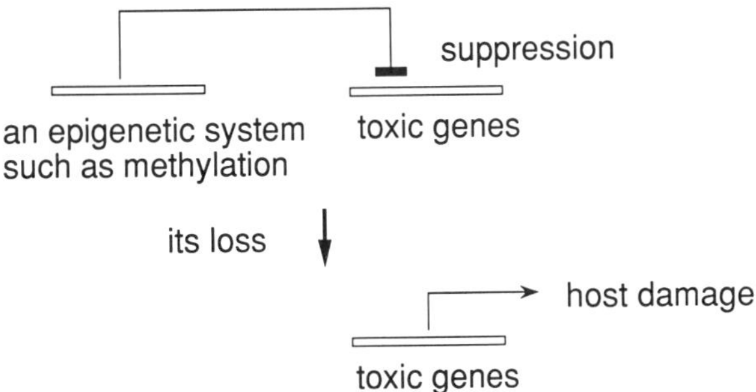

Figure 4 Selfish nature of epigenetic systems. An epigenetic system somehow suppresses a toxic gene. Removal of the epigenetic system leads to expression of the toxic gene and damages the host. The epigenetic system forces its maintenance on host cells through addiction. The black bar indicates that the epigenetic system somehow inhibits the action of the gene.

Potential Selfishness of Other Methylases and the Evolution of Gene Number

The methylated state of a gene is inherited via replication by maintenance methylases (Holliday 1993). Methylation of a gene or a site often represses its function. A methylase (or, more generally, an epigenetic system) should be able to force its maintenance on the host if it represses a killer gene (or many slightly toxic genes) or site (or many slightly toxic sites) (Fig. 4). Loss of the methylase gene would kill the host (or decrease its fitness) by allowing expression of the killer gene or site.

Such selfishness may explain the presence of methylation in some organisms and its absence from others, such as *Saccharomyces*. Excision repair functions in *S. cerevisiae* can recognize and repair methylated adenines created artificially by the *E. coli dam* gene (Malone 1986). The presence of such repair systems may explain why this organism has been refractory to infection by methylation systems.

The increase in the total gene number in a genome correlates with the increase in the extent of methylation during the evolution of vertebrates (Antequera and Bird 1993). I hypothesize that the continual creation of genes has been the strategy used by hosts to cope with the selfish spreading of methylation and the consequent suppression of gene expression.

A Scenario for the Evolution of RM Systems and a Reexamination of the Cellular Defense Hypothesis

One may ask at this stage, "What then was the ancestor of restriction enzymes?" I presume that there was a gene whose product killed the host by a chromosomal double-strand break at or close to its own insertion site under conditions that threatened its maintenance (Fig. 2A). It would have been similar to the homing endonucleases and the donor-suicide transposons. Such a cell-death strategy would have ensured their maintenance in the population of viable cells (Fig. 2A).

Methylases, which belong to another group of selfish-gene entities, may have evolved to further the propagation of these selfish endonucleases. Methylation would have been useful in the suppression of endonuclease gene expression and/or DNA cleavage in the first stage of their symbiosis. Together they would have forced their maintenance on the host (Fig. 4). Once the cutting sites were marked with methylation, these sites no longer needed to be close to the endonuclease gene. After this stage, the convergence of their recognition sequences probably occurred: The recognition sequence of the endonuclease became shorter (and, therefore, more abundant), and the recognition sequence of the methylase became longer (and, therefore, less abundant).

Extrapolation of our competition experiments (Fig. 3B,C) suggests the presence of selective pressure for shortening (or degeneration, in general) of the recognition sequence of an RM system. Imagine a competition between an RM unit recognizing a sequence (8 bp, for example) and a derivative RM unit recognizing a shorter sequence (7 bp, for example) that is included in the longer sequence and undergoes methylation at the same position(s). If the two units, RM7 and RM8, are in the same bacteria, the RM7 would have a selective advantage over the RM8 for the following reason. The hypothetical M7 methylase would recognize and methylate the 7-bp restriction sites as well as the 8-bp restriction sites. Therefore, both the 7-bp and 8-bp sites would be protected from the R7 restriction enzyme and the R8 restriction enzyme. On the other hand, the M8 methylase would methylate only the 8-bp restriction site, as the 7-bp restriction site does not constitute its entire recognition sequence. Because the RM7 methylase would protect all the restriction sites (R7 and R8), the loss of the RM8 unit would not kill the cell and it would be eventually selected against for this reason. On the contrary, the loss of the RM7 unit would kill the cell through the mechanism shown in Figure 3B.

There would thus be selective pressure for the recognition sequences of an RM system to become shorter during evolution. During the process of shortening, the diversification of the recognition sequence driven by competitive exclusion could have taken place among the siblings of the same parental RM system. (In order to change a recognition sequence of an RM system, it is necessary to change the specificity of both the restriction and modification enzymes. Thus, R and M genes may have changed partners during this type of evolution of the recognition sequence.)

Why are most RM sites 4–6 bp long instead of even shorter? The abundance of RM sites on the chromosome of their natural host does not vary very much (McClelland 1988). There may be some selective pressure against further shortening and increase in the number of recognition sites. One possibility may be the cost of methylation, because methylation is known to inhibit gene activity. Another explanation may be the cost of host killing by restriction. In either case, this phenomenon may be explained as a sort of parasite virulence in coevolution with the host.

In the later stages of evolution, some of the RM systems may have developed to counteract invading viral DNAs or incoming homologous DNAs by cleaving them. In this scenario, some aspects of the cellular defense hypothesis would fit into the selfish-gene hypothesis. Further work will be necessary to distinguish between these two hypotheses for each of the type II RM systems as well as other RM systems.

Meiosis, a Eukaryotic Counterpart of the RM Systems?

One of the remaining mysteries associated with the RM systems is their virtual absence from the eukaryotic world. Meiotic recombination is a process specific to eukaryotes and, at least in some organisms, proceeds as the repair of DNA double-strand breaks, just as recombination does in some bacteriophages (the double-strand-break-repair models) (Lichten and Goldman 1995). Earlier I proposed that meiosis might play a role similar to that of the RM systems: It may defend chromosomes against invasion by selfish DNAs such as transposons (Kobayashi et al. 1995). In particular, the homing endonuclease genes present in the organella genome may have been the major menace. Its insertion into the genome of the early eukaryotes may have generated introns.

According to this hypothesis, when a chromosomal locus suffers insertion, an endonuclease somehow senses the insertion and introduces a double-strand break at the insertion site. This break stimulates homologous recombination with a homolog, which results in two (or one) chromosomes free from the insertion with frequent crossing-over of flanking sequences.

In the case of a heterozygous diploid, the effect is more than elimination of the inserted transposon. Let us assume that a transposon has inserted into the sequence B in a diploid of genotype ABC/abc. The gene conversion during the recombination event would replace the site B, chosen by the transposon, by another allelic site (b). The crossing-over would replace the region (AC) with other regions (Ac and aC) different from the original and thus less favorable for transposon insertion. Therefore, I hypothesize that the resulting chromosomes would be less likely to be infected again with the same transposon.

This hypothesis, "the patrol model," is related to the hypotheses that homologous recombination and meiosis have evolved to eliminate deleterious mutations (Kondrashov 1993), to cope with parasites (Hamilton et al. 1990), and to fight against meiotic drive (Haig and Grafen 1991).

CONCLUSIONS

Loss of some type II RM gene units leads to killing of the host as a result of cleavage of the genomic DNA. Such killing forces the maintenance of these systems on the host. If RM genes are treated as selfish-gene entities, this would answer several questions associated with their evolution and maintenance.

The selfish-gene systems could also help the host by eliciting cell death, a process that can be important for the survival of the host genome at various stages in the population. For example, it could program suicidal defense against infection. As long as these death processes increase the fitness of the host, the relationship between the host genome and the selfish gene may be regarded as a form of symbiosis rather than one-sided parasitism. Here is one answer to the question of why and how symbiosis can evolve from parasitism.

Any epigenetic system that can repress toxic genes should be able to behave in a similar way. Loss of the epigenetic system could lead to damage or a general decrease in fitness of the host through the action of the toxic genes. This type of addiction may have been one of the reasons that epigenetic systems, functioning as selfish-gene entities, became part of the genome in the first place.

ACKNOWLEDGMENTS

I am grateful to Michael Yarmolinsky, Allan Campbell, Katsutoshi Mise, and Noriko Takahashi Kobayashi for comments on the manuscript, and to Lisa Raleigh, Howard Ochman, Richard Roberts, Jim Crow, and Hajime Ishikawa for discussions. The work in our group was carried out in collaboration with Taku Naito, Kohji Kusano, Youichi Nakayama, and Naofumi Handa and has been supported by Monbusho (Department of

Education, Science, Sports and Culture) and Koseisho (Department of Health and Welfare) of the Japanese government and by Nissan Science Foundation.

REFERENCES

Antequera, F. and A. Bird. 1993. Number of CpG islands and genes in human and mouse. *Proc. Natl. Acad. Sci.* **90:** 11995–11999.

Beeman, R.W., K.S. Friesen, and R.E. Denell. 1992. Maternal-effect selfish genes in flour beetles. *Science* **256:** 89–92.

Bull, J.J., I.J. Molineux, and J.H. Werren. 1992. Selfish genes. *Science* **65:** 65–66.

Gerdes, K., P.B. Rasmussen, and S. Molin. 1986. Unique type of plasmid maintenance function: Postsegregational killing of plasmid-free cells. *Proc. Natl. Acad. Sci.* **83:** 3316–3120.

Haig, D. and A. Grafen. 1991. Genetic scrambling as a defense against meiotic drive. *J. Theor. Biol.* **153:** 531–558.

Hamilton, W.D., R. Axelrod, and R. Tanese. 1990. Sexual reproduction as an adaptation to resist parasites. *Proc. Natl. Acad. Sci.* **87:** 3566–3573.

Holliday, R. 1993. Epigenetic inheritance based on DNA methylation. In *DNA methylation: Molecular biology and biological significance* (ed. J.P. Jost and H.P. Saluz), pp. 452–468. Birkhauser, Basel.

Janulaitis, A., R. Vaisvila, A. Timinskas, S. Klimasauskas, and V. Butkus. 1992. Cloning and sequence analysis of the genes coding for Eco57I type IV restriction-modification enzymes. *Nucleic Acids Res.* **20:** 6051–6056.

Jensen, R.B. and K. Gerdes. 1995. Programmed cell death in bacteria: Proteic plasmid stabilization systems. *Mol. Microbiol.* **17:** 205–210.

Karlin, S., C. Burge, and A.M. Campbell. 1992. Statistical analyses of counts and distributions of restriction sites in DNA sequences. *Nucleic Acids Res.* **20:** 1363–1370.

Kobayashi, I. 1992. Mechanisms for gene conversion and homologous recombination: The double-strand break repair model and the successive half crossing-over model. *Adv. Biophys.* **28:** 81–133.

Kobayashi, I., K. Sakagami, K. Kusano, F. Youhci, N. Takahashi-Kobayashi, and H. Yoshikura. 1995. Homologous interaction as a gene destruction process: Hypotheses for homology-driven non-homologous recombination and double-strand break repair. In *Modification of gene expression and non-Mendelian inheritance: Proceedings of the U.S.–Japanese joint meeting* (ed. K. Oono and F. Takaiwa), pp. 55–76. National Institute of Agrobiological Resources, Ministry of Agriculture, Forestry, and Fisheries, Tsukuba Science City, Ibaraki, Japan

Kondrashov, A. 1993. Classification of hypotheses on the advantage of amphimixis. *J. Hered.* **84:** 372–387.

Korona, R. and B.R. Levin. 1993. Phage-mediated selection for restriction-modification. *Evolution* **47:** 565–575.

Kulakauskas, S., A. Lubys, and S.D. Ehrlich. 1995. DNA restriction-modification systems mediate plasmid maintenance. *J. Bacteriol.* **177:** 3451–3453.

Kusano, K., T. Naito, N. Handa, and I. Kobayashi. 1995. Restriction-modification systems as genomic parasites in competition for specific sequences. *Proc. Natl. Acad. Sci.* **92:** 11095–11099.

Lambowitz, A.M. and M. Belfort. 1993. Introns as mobile genetic elements. *Annu. Rev. Biochem.* **62:** 587–622.

Levin, B.R. 1988. Frequency-dependent selection in bacterial populations. *Philos. Trans. R. Soc. London B Biol. Sci.* **319:** 459–472.

Lehnherr, H., E. Maguin, S. Jafri, and M.B. Yarmolinsky. 1993. Plasmid addiction genes of bacteriophage P1: Doc, which causes cell death on curing of prophage, and phd, which prevents host death when prophage is retained. *J. Mol. Biol.* **233:** 414–428.

Lichten, M. and A.S.H. Goldman. 1995. Meiotic recombination hotspots. *Annu. Rev. Genet.* **29:** 423–444.

Lyttle, T.W. 1993. Cheaters sometimes prosper: Distortion of Mendelian segregation by meiotic drive. *Trends Genet.* **9:** 205–210.

Malone, R.E. 1986. Excision repair functions in *Saccharomyces cerevisiae* recognize and repair methylation of adenine by the *Escherichia coli dam* gene. *Mol. Cell. Biol.* **6:** 3555–3558.

McClelland, M. 1988. Recognition sequences of type II restriction systems are constrained by the G + C content of host genomes. *Nucleic Acids Res.* **16:** 2283–2294.

Myers, R.S. and F.W. Stahl 1994. χ and the RecBCD enzyme of *Escherichia coli. Annu. Rev. Genet.* **28:** 49–70.

Naito, T., K. Kusano, and I. Kobayashi. 1995. Selfish behavior of restriction-modification systems. *Science* **267:** 897–899.

Noyer-Weidner, M., R. Diaz, and L. Reiners. 1986. Cytosine-specific DNA modification interferes with plasmid establishment in *Escherichia coli* K-12: Involvement of *rglB*. *Mol. Gen. Genet.* **205:** 469–475.

Ogura, T. and S. Hiraga. 1983. Mini-F plasmid genes that couple host cell division to plasmid proliferation. *Proc. Natl. Acad. Sci.* **80:** 4784–4788.

Raleigh, E.A. and G. Wilson. 1986. *Escherichia coli* K-12 restricts DNA containing 5-methylcytosine. *Proc. Natl. Acad. Sci.* **83:** 9070–9074.

Siksnys, V., A. Timinskas, S. Klimasauskas, V. Butkus, and A. Janulaitis. 1995. Sequence similarity among type-II restriction endonucleases, related by their recognized 6-bp target and tetranucleotide-overhang cleavage. *Gene* **157:** 311–314.

Takahashi, N. and I. Kobayashi. 1990. Evidence for the double-strand break repair model of bacteriophage λ recombination. *Proc. Natl. Acad. Sci.* **87:** 2790–2794.

Wade, M.J. and R.W. Beeman. 1994. The population dynamics of maternal-effect selfish genes. *Genetics* **138:** 1309–1314.

Wilson, G.G. and N.E. Murray. 1991. Restriction and modification systems. *Annu. Rev. Genet.* **25:** 585–627.

Yarmolinsky, M. 1995. Programmed cell death in bacterial populations. *Science* **267:** 836–837.

Yokochi, T., K. Kusano, and I. Kobayashi. 1995. Evidence for conservative (two-progeny) DNA double-strand break repair. *Genetics* **139:** 5–17.

Organization and Control of Imprinted Genes: The Common Features

Justin F-X. Ainscough and M. Azim Surani
Wellcome/CRC Institute
Cambridge, CB2 1QR
United Kingdom

During mammalian development, the two parental genomes are not functionally equivalent. This was first demonstrated in the early 1980s by the pronuclear transfer experiments of Surani and Solter, which demonstrated that both parental genomes are essential for normal development (McGrath and Solter 1984; Surani et al. 1984). Embryos that contain two paternal genomes (androgenetic) show very poor embryonic development. Gynogenetic embryos, which contain two maternal genomes, show exceptionally poor extraembryonic tissue development. In both instances, lethality occurs by mid-gestation. These results suggest that the parental genomes play complementary roles, involving differential (monoallelic) expression of essential genes in embryonic development. In recent years, a variety of techniques have been used to identify an increasing number of mouse and human genes that are differentially expressed depending on their parental origin. For example, polymorphisms in a given gene may be used to assess the parental origin of an expressed allele. Such genes, which "remember" which parent they are inherited from, have been termed imprinted genes.

Other work with mice containing uniparental duplications of subchromosomal regions (both copies derived from one parent) showed that duplication of some domains results in a phenotype which is often embryonic-lethal (Cattanach and Kirk 1985). This indicates that essential genes in these regions are imprinted. Ten such domains have been identified, located on mouse chromosomes 2, 6, 7, 11, 12, and 17. Some imprinted genes are clustered within these domains and may be controlled by common regulatory elements. Other imprinted genes may have their own imprinting regulatory elements and work as individual units. To date, more than 15 imprinted genes have been identified in mouse and/or human, which are summarized in Table 1. It is noteworthy that all imprinted genes mapped so far are located within the large imprinted chromosomal regions determined by Cattanach and coworkers (Beechey and Cattanach 1995).

Epigenetic Mechanisms of Gene Regulation

Table 1 Imprinted genes in mouse and man

		Expressed allele		Reference	
Gene	Mouse chromosome	mouse	human	mouse	human
H19	7	maternal	maternal	Bartolomei et al. (1991)	Rachmilewitz et al. (1992)
Igf2	7	paternal	paternal	DeChiara et al. (1991)	Giannoukakis et al. (1993)
Ins2	7	paternal		Giddings et al. (1994)	
Mash2	7	maternal		Guillemot et al. (1995)	
$p57^{Kip2}$	7	maternal		Hatada and Mukai (1995)	
Snrpn	7	paternal	paternal	Cattanach et al. (1992)	Leff et al. (1992)
IPW	7		paternal		Wevrick et al. (1994)
PAR1	7		paternal		Sutcliffe et al. (1994)
PAR 5	7		paternal		Sutcliffe et al. (1994)
Znf127	7	paternal	paternal	Cattanach et al. (1995)	Cattanach et al. (1995)
Mas	17	paternal		Villar and Pedersen (1994)	
Igf2r	17	maternal	biallelic	Barlow et al. (1991)	Ogawa et al. (1993)
Peg1	6	paternal		Kaneko-Ishino et al. (1995)	
Peg3	7	paternal		Kuroiwa et al. (1996)	
U2afbp-rs	11	paternal		Hatada et al. (1993); Hayashizaki et al. (1994)	
Xist	X	paternal		Kay et al. (1993)	

The genes are presented in groups according to their chromosomal locations. In the mouse *H19*-$p57^{Kip2}$ are located on distal chromosome 7, *Snrpn-Znf127* are on central chromosome 7, and *Peg3* is on proximal chromosome 7. *Mas* and *Igf2r* are clustered on chromosome 17 in mouse. Lowercase type refers to mouse homologs, uppercase type denotes human homologs. The functions of *H19, IPW, PAR1, PAR5, Znf127, Peg1, Peg3,* and *Xist* are unclear. *Igf2* and *Ins2* are fetal growth factors, and *Igf2r,* among other roles, is involved in regulating the availability of *Igf2*. *Mash2* is a basic-helix-loop-helix transcription factor that is essential for extraembryonic development, and $p57^{Kip2}$ is a cyclin-dependent kinase inhibitor. *Snrpn* and *U2afbp-rs* are closely related to known splicing factors. *Mas* encodes a mitogenic G-protein-coupled cell-surface receptor.

The analysis of imprinted genes is of considerable interest because of their possible role in a number of human genetic disorders. For example, *H19*, which encodes a nontranslated RNA, and insulin-like growth factor II (*IGF2*) are implicated in Beckwith-Weidemann syndrome (BWS) and a number of childhood tumors, and small nuclear ribonucleoprotein polypeptide N (*SNRPN*), among other imprinted genes, lies in the critical region for Prader-Willi syndrome (PWS) (Cattanach et al. 1992; Ozcelik et al. 1992).

The purpose of this chapter is to examine what common features, if any, are associated with imprinted genes. First, current information on the structural features of imprinted genes is compared to that of genes which are biallelically expressed. Second, the evidence for imprinting control elements in specific regions is discussed, with reference to two genomic domains on chromosome 7 in the mouse. Finally, possible models to explain how genes may be imprinted are discussed.

HOW ARE TWO COPIES OF A GENE IN ONE NUCLEUS DIFFERENTIALLY REGULATED?

For a parental imprint to be established, some form of epigenetic modifications must distinguish the DNA inherited from the mother and father. This mark is presumably established in the germ line when the two parental genomes are separate and may be propagated by further post-fertilization epigenetic events. Possible candidates for epigenetic marks include DNA methylation, or differences in chromatin structure induced by local structural features and relative accessibility to imprinting factors. The current knowledge on how some genes are imprinted is vague, but a number of intriguing associated features are becoming apparent.

Methylation

Before any endogenous imprinted genes were identified, a number of transgenes were observed to be active only after passage through the germ line of one sex (for review, see Reik et al. 1990; see also Chaillet et al. 1995). Some of these were studied in detail and were found to have different methylation patterns depending on their parental origin. In general, the active allele was unmethylated and the inactive allele was methylated.

Modification of DNA by cytosine methylation at CpG dinucleotides has properties that fulfill the main requirements of imprinting. DNA methylation affects gene expression, is heritable, and is reversible. In the mammalian genome, CpGs occur at well below the expected frequency, and most of these are methylated. However, about 2% of CpGs are clustered into islands of around 1 kb in length, where they occur at the expected frequency (Bird 1986). The majority of CpG islands are found at the 5′ end of genes and are not methylated regardless of the gene's transcriptional activity.

Imprinted Genes Have Differentially Methylated Regions

Contrary to the general rule that CpG islands are unmethylated, CpG islands associated with imprinted genes become comparatively methylated on one allele. In the case of the *H19* gene, the inactive paternal allele is highly methylated whereas the active maternal allele is unmethylated around the promoter and first exon (Bartolomei et al. 1993; Ferguson-Smith et al. 1993). A difference in sensitivity to DNase I was also noted in this region, suggesting that the active maternal allele has a more open chromatin conformation. The *Igf2* gene, a fetal-specific growth factor located 70 kb upstream of *H19* on mouse distal chromosome 7, shows more subtle differential methylation in a region 3 kb upstream of promoter 1 (Sasaki et al. 1992) and also around exon 5 to 6 (Feil et al. 1994). Like *H19*, the paternal allele of *Igf2* is slightly more methylated than the maternal allele, but unlike *H19*, the paternal allele is active. Both alleles are equally sensitive to DNase I, and low-level transcription has been observed from the maternal allele, suggesting that its promoters are potentially active but are restrained by an unidentified mechanism.

The maternally expressed insulin-like growth factor-II/cation-independent mannose 6-phosphate receptor (*Igf2r*) gene, located on mouse chromosome 17, contains two regions that show differential methylation between parental alleles (Stoger et al. 1993). One of these (region 1) contains the transcriptional start site and is methylated on the silent paternal chromosome, whereas region 2, located in an intron, is methylated on the active maternal allele. It is interesting to note that the human *IGF2R* gene is also differentially methylated at the equivalent intronic region, but this gene is usually biallelically expressed (Smrzka et al. 1995). This suggests that features in addition to a methylation imprint may be required to generate imprinted gene expression.

Other imprinted genes that have been shown to contain regions of parentally determined methylation include *U2afbp-rs*, a gene with significant similarity to U2 small nuclear ribonucleoprotein auxiliary factor, and *SNRPN*. *U2afbp-rs* was isolated on the basis of differential methylation, using a methylation-sensitive technique called restriction landmark genomic scanning (RLGS) (Hatada et al. 1993; Hayashizaki et al. 1994). *U2afbp-rs* is expressed from the paternal allele and hypermethylated on the inactive maternal allele. The paternally expressed *SNRPN* gene, which is located in the critical region for PWS in humans, contains sites within intron 5 that are paternally methylated (Glenn et al. 1993), whereas the CpG island 5′ to the gene is maternally methylated (Sutcliffe et al. 1994). The methylation status of many of the other imprinted genes has yet to be determined.

In summary, no simple correlations exist between genomic imprinting and DNA methylation, because researchers have identified both paternal and maternal hypermethylation associated with transcriptionally active and transcriptionally quiescent genes. However, with the exception of the *Igf2* gene (see above), for which the 5′ CpG island is unmethylated on both parental alleles, a correlation does appear to exist between 5′ CpG methylation and inactivity of imprinted genes.

Is Methylation the Primary Imprint?

Differences in methylation of CpG-rich regions associated with imprinted genes suggest that methylation might play a crucial role in their specification. However, it is unclear whether methylation is involved in establishing the initial imprinting mark or only in maintaining differential gene expression from one cell generation to the next. As mentioned earlier, a primary imprinting mark is likely to be established in the germ line, before the two parental genomes are exposed to the same cellular environment. Tremblay and coworkers recently demonstrated that three CpGs in a region 5′ to the *H19* gene are paternally methylated (Tremblay et al. 1995), and this methylation is maintained from the sperm through early embryogenesis, escaping the genome-wide demethylation event that occurs in the morula (Monk et al. 1987). Similarly, sites in region 2 of the mouse *Igf2r* gene are inherited from the female gamete in a methylated state and remain methylated during preimplantation development (Stoger et al. 1993). However, more detailed analysis of the early postfertilization embryo suggests that methylation of these sites may not be continuous (see Sherner and Razin, this volume). Propagation of germ-line methylated CpGs in islands of other imprinted genes has also been reported (Brandeis et al. 1993; Razin and Cedar 1994), including the mouse *Xist* gene, which lies in the X-chromosome inactivation center (Ariel et al. 1995; Zuccotti and Monk 1995). Therefore, methylation of some CpGs is established in the germ line and may be propagated through early development, but it remains to be seen whether this germ-line methylation constitutes the primary imprint signal.

Whether it is the primary imprint or not, DNA methylation is essential for maintenance of imprinting. This was clearly demonstrated by deletion of the mouse DNA methyltransferase gene using gene targeting in embryonic stem (ES) cells (Li et al. 1992, 1993). Lack of DNA methyltransferase resulted in embryonic lethality before day 10 of gestation. In the mutant embryos, both copies of *H19* were active and both copies of *Igf2* and *Igf2r* were silent, demonstrating that for *H19*, methylation has a negative effect on transcription, and for *Igf2* and *Igf2r*, it acts positively. More recently, it was shown that in methyltransferase-deficient male mice, expression of *Xist* is activated, correlating with demethylation of the 5′ end of the gene (Beard et al. 1995). One possible explanation for these effects is that methylation of active genes prevents binding of proteins that might repress transcription. Conversely, methylation of inactive genes might prevent binding of proteins that activate transcription. The resulting transcriptional status of the gene would then be dependent on whether binding of a positive or a negative regulator of transcription is affected by methylation (see Fig. 1a).

It has recently been shown that one protein, MeCP2, has a high affinity for methylated CpGs and is able to bind to a single methylated site (Meehan et al. 1992), whereas another, MeCP1, also binds methylated sequences but requires multiple methylated CpGs (Antequera et al. 1989; Boyes and Bird 1992). Another protein that binds methylated DNA is MDBP-2-H1. This is a member of the histone H1 family which, when phosphorylated, has a high affinity for the methylated form of the avian vitellogenin II gene promoter region (Bruhat and Jost 1995). Binding of these, and other such proteins, may provide a physical block to transcriptional regulators, positive or negative (Fig. 1b).

In addition to proteins whose binding is affected by methylation, other proteins have been found to modify the methylation status of DNA. Brandeis and coworkers (1994) demonstrated that the transcription factor Sp1 plays a key role in protecting the CpG island of the adenine phosphoribosyltransferase (*APRT*) gene from de novo methylation during early embryogenesis. This may be an active process in which local demethylation is induced by Sp1 binding (Fig. 1c).

The relative activity of DNA-binding proteins (which influence, or are influenced by, DNA methylation) on the two parental alleles of a gene may have a substantial effect on the activity of that gene. Differential protein activities may occur through a variety of mechanisms, including subnuclear compartmentalization, higher-order chromatin structure, or localized modifications to the proteins, such as phosphorylation.

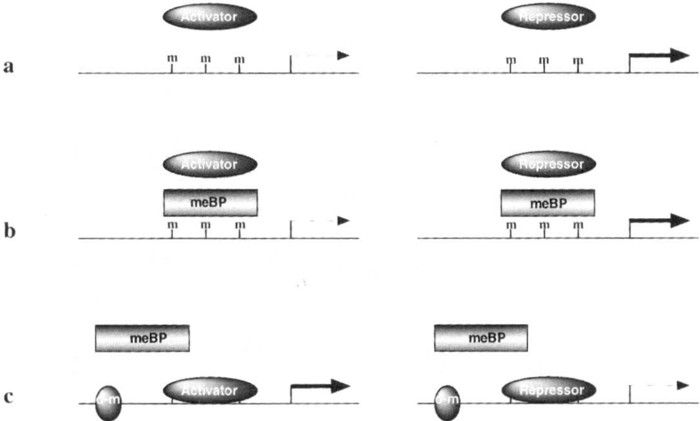

Figure 1 Methylation and gene regulation. Schematic representation of the possible interactions between DNA methylation (m), methyl-binding proteins (meBP), and demethylating factors (d-m) in influencing the transcriptional activity of a gene. (*a*) Methylation of CpG dinucleotides prevents binding of the transcriptional activator/repressor proteins directly. Factors that actively demethylate DNA are not functional. Alternatively, as in (*b*), methyl-binding proteins associated with methylated DNA may present a physical block to transcriptional activator/repressor proteins. (*c*) Binding of active demethylating factors prevents binding of methyl-binding proteins and makes the DNA available to transcriptional activator/repressor proteins. Bold arrow signifies transcription, dashed arrow signifies no (or basal level) transcription.

Repeats and CpG Islands

In a number of cases, the CpG islands of imprinted genes contain repeated sequences of variable size. For example, the differentially methylated region 2 of *Igf2r* contains a number of tandem direct repeats ranging in size from 25 bp to 75 bp (Neumann et al. 1995). Neumann and coworkers also studied the sequences of a number of other imprinted genes and found similar direct tandem repeats in CpG-rich regions of *Igf2, U2afbp-rs*, and *SNRPN*. Sequence analysis of the upstream region of the *H19* gene has revealed the presence of a 6-bp repeated sequence located near CpG dinucleotides that are methylated on the paternal allele (Tremblay et al. 1995). A structurally similar repeated region was reported in the *p57*^{Kip2} gene, a maternally expressed cyclin-dependent kinase inhibitor, again localizing to a region that demonstrates parental-specific differential methylation (Hatada and Mukai 1995). The mouse *Xist* and human *XIST* genes also contain a number of short tandem repeats throughout their length (Brockdorff et al. 1992; Brown et al. 1992).

One transgene, *RSVIgmyc*, also contains repeats associated with CpG-rich DNA and has been reported to imprint wherever it is inserted in the mouse genome (Chaillet et al. 1995). This suggests that it contains all *cis*-acting signal(s) needed for expression and methylation imprinting.

Although many imprinted genes contain repeated sequences close to regions of parental-specific differential methylation, there is no apparent sequence homology between the repeats. This suggests that they are not sequence-specific binding sites for imprinted gene regulatory proteins but may be involved in modifying local chromatin structure. Methylation, or other epigenetic modifications of the DNA and chromatin components, may influence this, thus affecting the availability of the DNA to essential transcription factors.

Repetitive elements in genomic DNA have been proposed to induce homology-dependent gene silencing by formation of heterochromatin (Dorer and Henikoff 1994). In this study, it was clearly shown that an increase in the local copy number of a *white* transgene led to suppression of gene activity in *Drosophila*, and this suppression was influenced by known modifiers of position-effect variegation, hence suggesting a role for heterochromatin formation. Similar effects have been reported in fungi (Pandit and Russo 1992) and in plants (for review, see Matzke and Matzke 1995), where increase in copy number leads to suppression of gene activity. Similar mechanisms involving repeat-induced modifications to local chromatin structure in imprinted genes of mammals may be of considerable interest.

Imprinted Domains May Contain Genes Whose Functional Product Is RNA

An unusual feature of some imprinted genes is that they have no apparent conserved open reading frame, although in other aspects they look like conventional mRNAs. They are therefore suspected to function as RNAs rather than being translated. This group includes *Xist*, which is paternally expressed in extraembryonic mouse tissues and monoallelically expressed in the embryo (not dependent on parental origin) (Kay et al. 1993). The human homolog of this noncoding RNA (*XIST*) has been localized to the X-inactivation center (XIC), which is the region of the X chromosome where the inactivation process commences (see Avner; X Chromosome Inactivation and Epigenetic Mechanisms [Riggs and Porter]; both this volume). X-Inactivation involves transcriptional repression of most genes on the chromosome and is associated with chromatin compaction, large-scale DNA methylation, reversion to late replication (for review, see Singer-Sam and Riggs 1993), and histone deacetylation

(Jeppesen and Turner 1993). In fact, *Xist* is the only gene known to be expressed exclusively from the inactive X chromosome, and since its expression precedes inactivation, *Xist* may have a primary role in initiating the inactivation program (Rastan 1994).

Another unusual feature of *Xist* is that it remains localized to the nucleus, which strengthens the hypothesis that it functions as an RNA, since mRNAs must be transported to the cytoplasm to be translated. In fact, Lawrence and coworkers recently demonstrated that the RNA remains in contact with the inactive X chromosome, coating it in its entirety (Clemson et al. 1996). They propose that the RNA plays a structural role in the subsequent condensation and inactivation of the chromosome.

The RNA product of the *H19* gene, on the other hand, is largely localized to the cytoplasm, specifically in 28S-30S cytoplasmic particles of unknown function (Brannan et al. 1990). This gene is highly transcribed but, like *Xist*, no conserved open reading frames have been identified between the mouse and human homologs. However, sequence conservation between the two species suggests that patterns of primary and secondary structure are conserved (Tilghman et al. 1992). Similar conservation of secondary structures has also been noted for the mouse *Xist* and human *XIST* genes (Pfeifer and Tilghman 1994).

H19 and *Xist* are both expressed in chromosomal domains where other genes are inactivated. For *Xist* this extends over the entire X chromosome. For *H19* the inactivated region is much more localized, extending to the *Igf2* and *Ins2* genes, but not further upstream, since the imprinted *Mash2* gene is expressed off the same parental chromosome as *H19*.

It is also possible, but not clear at the present time, that the same is true for the human *IPW* gene, an *i*mprinted gene in the *P*rader-*W*illi syndrome region (see below and Fig. 2b). As with *H19*, the RNA product of *IPW* is found predominantly in the cytoplasm. It is expressed from the paternal allele, which is the same as for other imprinted genes in the critical region for PWS (Wevrick et al. 1994). Thus, *IPW* expression does not appear to correlate with inactivation of other genes in *cis*. However, *IPW* also lies relatively close (50 kb) to the critical region for another human genetic disorder, Angelman syndrome (AS), for which maternally expressed genes have been proposed, although none have been found to date. Therefore, it is possible that expression of *IPW* plays a role in inactivation of the maternally expressed AS gene(s) on the paternally derived chromosome but is prevented from inactivating genes in the PWS region by an unknown mechanism. Two other paternally expressed

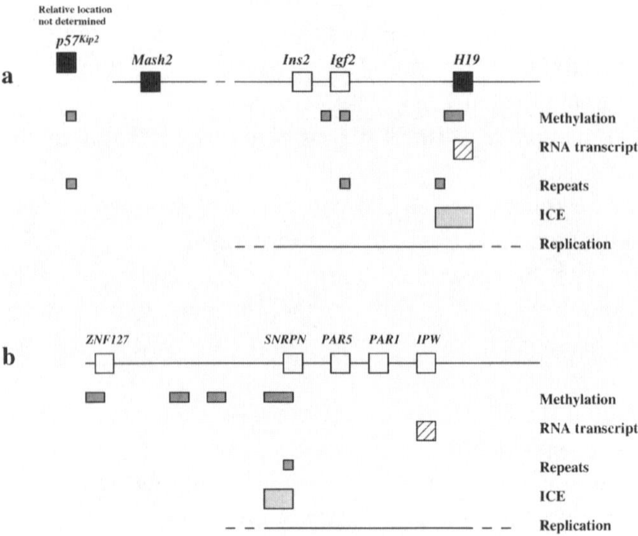

Figure 2 Parental origin-specific epigenetic modifications in imprinted domains. A number of general features have been identified which localize to, and appear to be consistent between, imprinted domains. These include parental origin-specific hypermethylation (Methylation), paternal early replication (Replication), repeated elements in CpG-rich regions (Repeats), and the presence of genes whose functional product may be an RNA (RNA transcript). The localization of these features in two imprinted domains is shown (*a*) in the mouse *H19*/*Igf2* domain and (*b*) in the human *SNRPN* domain. In both regions, a putative imprinting control element (ICE) has been localized that may influence the chromatin over large distances. White boxes denote paternally expressed genes, black boxes denote maternally expressed genes. The location of $p57^{Kip2}$ in *a* has not been determined. Nonimprinted genes located in the *SNRPN* domain are not shown. Not drawn to scale.

genes, *PAR1* and *PAR5*, identified in the PWS region are also of potential interest in this context since sequence analysis did not reveal any significant open reading frames (Sutcliffe et al. 1994). This analysis is, however, incomplete.

The possibility that RNA molecules have a structural role in the nucleus is not a new concept, since inhibition of RNA synthesis is associated with chromatin collapse (see, e.g., Davis et al. 1993). In addition to the imprinted genes whose functional products appear to be RNA molecules, much of the mammalian genome is highly repeated DNA of unknown function, which is transcribed but never translated. It is therefore possible that these high-copy RNA molecules are structurally involved in the architectural organization of chromatin.

Replication Timing

The regulation of replication of regions that contain imprinted genes is different for the two parental copies. Imprinted genes replicate asynchronously in the cell cycle, not synchronously as generally observed for biallelically expressed genes. Interestingly, it is always the paternal allele that replicates first, regardless of the transcriptional status of genes in the region (Kitsberg et al. 1993). For example, *H19, Igf2*, and *Ins2* (in both mouse and human) fall into such a replication domain even though *H19* is maternally expressed and the other two genes are paternally expressed, or biallelically expressed in the case of *Ins2* in the pancreas. Thus, a differential replication program of the two parental alleles may provide a mark to a large genomic domain (which seems to be larger than the expected average of a single replicon) allowing the cells transcriptional (and epigenetic DNA modification) machinery to distinguish between them and control them appropriately.

It is interesting to note that one region has been identified which shows the opposite pattern of asynchrony; that is the intergenic region between the γ-aminobutyric acid receptor subunit β3 (*GABRB3*) and α5 (*GABRA5*) genes in humans which replicates the maternal copy first (Knoll et al. 1994). The allele specificity is lost in patients with uniparental duplications of chromosome 15 (LaSalle and Lalande 1995). This region lies telomeric to the imprinted domain associated with AS/PWS, but as yet no imprinted genes have been identified here. The dramatic switch between paternal and maternal early replication near the *GABRB3* gene may represent a boundary for the end of an imprinted replication domain.

IS THERE EVIDENCE FOR REGIONAL CONTROL OF GENOMIC IMPRINTING?

Distal Chromosome 7

To date, five imprinted genes have been localized to mouse distal chromosome 7, four of which lie in a region spanning approximately 300 kb (Fig. 2a). *H19* and *Igf2* are separated by 70 kb (Zemel et al. 1992; Koide et al. 1994) and lie in the same transcriptional orientation. They are expressed in an almost identical array of tissues of both mesodermal and endodermal origin at the same time in development (Lee et al. 1990; Poirier et al. 1991). However, they are expressed from different parental alleles (*Igf2* is paternally expressed and *H19* is maternally expressed) (Bartolomei et al. 1991; DeChiara et al. 1991). *Ins2* is located about 20 kb upstream of *Igf2*, also in the same transcriptional orientation, and is paternally expressed in the yolk sac but shows biallelic expression in the pancreas (Giddings et al. 1994). *Mash2* (mammalian achaete scute homolog) is located further upstream, and its expression (maternal) is restricted to the spongiotrophoblast (Guillemot et al. 1995). The most recently identified gene in this region is the maternally expressed *p57^{Kip2}* gene (Hatada and Mukai 1995), but its relative location is not yet characterized. Is clustering of these genes essential for them to maintain their imprinted status?

The similarity in expression of *H19* and *Igf2* led to the suggestion that they may share common regulatory elements, with competition for these elements resulting in their differential expression (Bartolomei et al. 1993). Thus, when the unmethylated maternal allele of *H19* interacts with the enhancers, expression of *Igf2* from the maternally derived chromosome is prevented. However, when the *H19* gene is methylated, as on the paternally derived chromosome, interaction with the enhancers is blocked, leaving them free to activate *Igf2* expression. So far, the only enhancers identified in this region are downstream from the *H19* gene, and their action is restricted to tissues of endodermal origin (Yoo-Warren et al. 1988) and the sclerotome and its derivatives (J.D. Brenton, unpubl.). No expression is directed in the myotome or its derivatives. However, a YAC transgene containing the entire *Igf2/H19* domain appears to contain the necessary regulatory elements to drive full expression of an *Igf2-LacZ* fusion when stably integrated into the mouse genome by microinjection of fertilized eggs (Fig. 3) (J. Ainscough, unpubl.). This suggests that the enhancers needed for expression in the myotome and its derivatives are located within this domain. We have previously identified a small region between *H19* and *Igf2* that is hypersensitive to DNase I enzyme, unmethylated, GC-rich, and conserved between mammals (Koide et al. 1994). This region is a strong candidate for the mesodermal enhancer(s) and/or an imprinting control element (ICE). Investigations into whether this genomic domain actually contains the necessary signals for imprinting of these two genes are currently under way.

Experiments using a 30-kb rat *Igf2* transgene, in which the coding sequences were replaced with *LacZ*, failed to give appropriate tissue-specific expression or imprinting in five out of six expressing lines (Lee et al. 1993). One line, however, did show a subset of the expected pattern of expression and was appropriately imprinted. This transgene insert was mapped to a region on chromosome 19 not expected to contain imprinted genes, based on genetic studies. It is possible that the *Igf2* region may contain some information important for its own imprinting. Bartolomei and coworkers reported that two of their original seven *H19* transgenes,

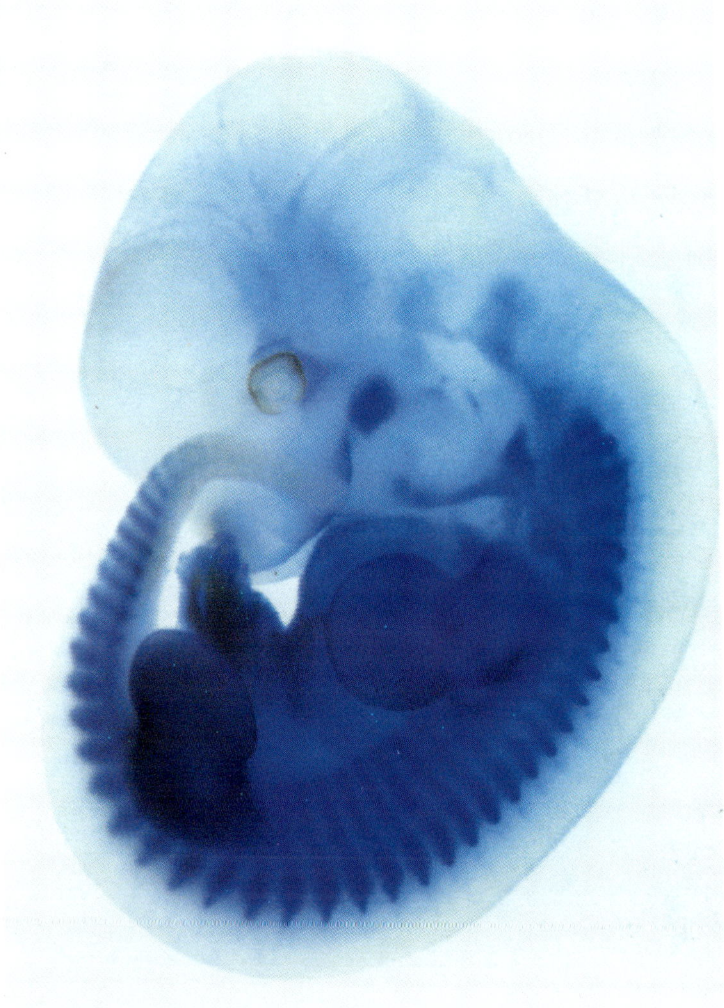

Figure 3 A YAC transgene contains full information for proper *Igf2* expression. Transgenic mice containing a 130-kb YAC show appropriate expression of a mouse *Igf2/LacZ* gene fusion. Whole-mount LacZ staining of an embryo at day 12 of development.

containing the upstream region and the downstream enhancers, displayed paternal-specific methylation and maternal expression, suggesting that the *H19* region may contain an imprinting signal (Bartolomei et al. 1993). Due to the low numbers of transgenic lines reported so far, this is by no means conclusive.

Recent experiments by Leighton and coworkers have gone some way toward addressing the possible coordinate expression and imprinting of *H19* and *Igf2*. Deletion of the *H19* gene and approximately 10 kb of upstream sequence by gene targeting in mouse ES cells was shown to induce expression of maternally inherited *Igf2* and *Ins2* (Leighton et al. 1995b). In addition, the upstream region of the *Igf2* gene adopted a paternal-type methylation pattern. No effect was seen following paternal transmission of the deletion, which was expected, since *H19* is normally paternally silent. This result demonstrates that the *H19* gene, its RNA product, or the upstream sequence plays a role in controlling the imprinting of other genes in the locality. However, analysis of RNA from a range of neonatal tissues showed that the increase in levels of *Igf2* expression was not uniform between samples. In the gut and liver, a 1.3- to 1.4-fold increase was observed instead of the expected 2-fold increase seen in heart, yolk sac, muscle, and kidney, as predicted by the competition model. This suggests that additional regulatory elements are probably involved in controlling the expression of *Igf2* from the maternal allele in gut and liver. In the brain, a 1.5-fold increase was detected, which is difficult to reconcile with the current knowledge of *Igf2* expression, since both alleles are normally active in the choroid plexus and leptomeninges, the only tissues in the brain where *Igf2* expression is seen (DeChiara et al. 1991). This suggests that at least one of the alleles is being up-regulated in these tissues, or that *Igf2* is being expressed in additional tissues.

Deletion of the downstream *H19* enhancers clearly showed that these elements are in fact utilized by both *H19* and *Igf2*, since maternal transmission led to loss of *H19* expression in cells of endodermal origin, whereas paternal transmission resulted in a similar loss of *Igf2* expression in the same cell types (Leighton et al. 1995a). Some reduction of *Ins2* expression was noted in the yolk sac, but no effect was seen on *Mash2* following either maternal or paternal inheritance.

The imprinting of *H19* and *Igf2* therefore seems to be mechanistically linked and regulated, at least in some tissues, by a common ICE. It also seems that the imprinting of *Ins2* is influenced by this same region, since expression in the yolk sac is disturbed when the *H19* genomic region is deleted. However, this could simply be due to the relative strength of the enhancers in this region inadvertently driving *Ins2* expression in the yolk sac, the imprinting of this gene being consequential to what is happening to these enhancers. No effect was seen on the imprinting of *Mash2* by deletion of the putative imprinting element near *H19* or by deletion of the downstream enhancers (Leighton et al. 1995a). It is not yet known whether there is any direct effect on the imprinting of $p57^{Kip2}$. It is possible that these genes are imprinted independently or are contained in an imprinting domain adjacent to the *H19/Igf2* region, their sizes being determined by domain organization analogous to replicons. It should be remembered, however, that in these experiments, small deletions are generated at the locus by gene targeting, so the possibility of long-range control elements interacting with the genes cannot be addressed. Experiments with YAC transgenes, mentioned above, will allow a critical evaluation of the containment of control elements required for proper imprinting.

Central Chromosome 7

There is at least one other imprinted domain on chromosome 7 in the mouse, which is analogous to 15q11-13 in humans. This contains the critical regions for the human genetic disorders AS and PWS, the phenotypes of which are only manifested following maternal and paternal transmission, respectively. This suggests that the PWS gene(s) is paternally expressed and the AS gene(s) is maternally expressed. Five genes in this region are now known to be imprinted in man, and are interspersed with nonimprinted genes. The imprinted genes are *SNRPN, IPW, ZNF127, PAR1,* and *PAR5*, which lie in a region extending over 1.5 Mb, shown in Figure 2b. All of these are expressed exclusively from the paternal allele. The mouse *Snrpn* gene has also been shown to be imprinted in the same direction as the human gene (Leff et al. 1992). Sutcliffe and coworkers demonstrated that *SNRPN*, *PAR1*, and *PAR5* expression was lost in three PWS patients following paternal transmission of small deletions covering the differentially methylated CpG island upstream, and exon α, of *SNRPN* (Sutcliffe et al. 1994). These three genes lie within a 200-kb region. These investigators also noted that the methylation pattern of the region was affected by these deletions, in one family extending as far as *ZNF127,* over 1 Mb centromeric to the deletion (Reis et al. 1994). These findings led to the proposal that there may be a genetic element (ICE) that controls the expression of nearby genes from the paternal chromosome; deletion or inactivation of this element may result in PWS. Analysis of replication timing in the 500- to 650-kb region around *SNRPN* showed that deletion of sequences including this putative controlling element also results in loss of late replication control in lymphocytes when paternally inherited (Gunaratne et al. 1995). Further analysis of microdeletions in this region demonstrated an association with AS, even though the putative AS gene(s) is thought to map hundreds of kilobases distally (Buiting et al. 1995). It is therefore possible that within the AS/PWS imprinted domain there is a single ICE (Fig. 2b) that affects the chromatin structure and methylation status, thus controlling gene expression in the region.

SUMMARY

It is unclear whether genomic imprinting occurs at the level of the individual gene or affects large regions containing a cluster of genes. There is evidence that both scenarios may occur, and the possibility of highly complex levels of control is very real. The primary imprinting mark that specifies the parental origin of imprinted regions could be one of several epigenetic modifications, including DNA methylation, histone acetylation, and ubiquitination. DNA methylation has been studied extensively, but little is known about the other modifications. In addition, differential replication and usage of scaffold attachment regions or boundary elements, as well as the binding of germ-line-specific factors, could all provide a mark that could be maintained throughout numerous cell divisions, then reset after passage through the germ line.

In recent years, it has become clear that the nucleus is a highly organized organelle, containing a number of specialized structures that allow compartmentalization of nuclear functions. This may facilitate an additional level of control that allows two near-identical chromosomal regions to be differentially regulated. For example, the two parental chromosomes may maintain their identity by localizing to separate nuclear compartments. Restriction of specific transcription factors to only one of these compartments could allow gene activity from only one chromosome. It is important to note that distinguishing epigenetic marks may still be required to direct such compartmentalization.

There are a number of similarities between the two domains on chromosome 7 that contain imprinted genes. These are summarized in Figure 2. Both regions contain areas of differentially methylated CpG-rich DNA with associated repeated elements; they are differentially replicated and contain genes encoding mRNAs that are apparently not translated. Most importantly, deletions of comparatively small regions in these domains have been identified that affect the methylation status and expression of genes up to hundreds of kilobases away. These regions may therefore house the elusive ICEs responsible for regional control of chromatin structure and gene expression.

It has been demonstrated that chromosomal regions containing imprinted genes are differentially replicated, with the paternal copy replicating earlier than its maternal homolog. Whether this is memory of replication patterns set up in the male and female germ lines is yet to be tested. If it is, the proposed ICEs may contain replication origins that retain the replicative capacity they started out with in the germ line, resulting in the differential replication timing observed in imprinted domains. Nonimprinted gene regions, conversely, may have lost the initial differential replication during early development, rendering the two alleles indistinguishable until the replication status is reset in the germ line.

Some imprinted genes may be located close to ICEs that control large domains. For example, *SNRPN* lies close to a region that affects the imprinting status of genes at least 300 kb away. The DNA around *H19* may also possess such a signal that controls the imprinting of *H19, Igf2*, and possibly other genes in this domain. The relative proximity of these genes to the imprinting elements may simply be fortuitous, making this small subset of imprinted genes appear "self-imprinting" in transgenic experiments. Alternatively, the relative ordering and transcriptional orientation of imprinted genes with respect to imprinting elements may have functional significance similar to the β-globin locus control region, which controls the temporal and spatial expression of five linked genes (Dillon and Grosveld 1993).

Assuming that imprinted domains are marked by such processes, what actually makes the genes present in these regions imprintable? Could it be that any gene which happens to find itself in one of these domains becomes imprinted? This seems unlikely, since nonimprinted genes have been identified interspersed between imprinted genes. From the information now available, it appears that there may be some features, such as the presence of repetitive DNA and CpG islands susceptible to differential methylation, that are essential for a gene to be responsive to higher-order imprinting signals.

ACKNOWLEDGMENTS

We thank members of our laboratory for helpful discussions, and the Wellcome Trust (grant 036481) for financial support.

REFERENCES

Antequera, F., D. MacLeod, and A. Bird. 1989. Specific protection of methylated CpGs in mammalian nuclei. *Cell* **58:** 509–517.

Ariel, M., E. Robinson, J. McCarrey, and H. Cedar. 1995. Gamete specific methylation correlates with imprinting of the murine *Xist* gene. *Nat. Genet.* **9:** 312–315.

Barlow, D.P., R. Stoger, B.G. Herrmann, K. Saito, and N. Schweifer. 1991. The mouse insulin-like growth factor type-2 receptor is imprinted and closely linked to the *Tme* locus. *Nature* **349:** 84–87.

Bartolomei, M.S., S. Zemel, and S.M. Tilghman. 1991. Parental imprinting of the mouse *H19* gene. *Nature* **351:** 153–155.

Bartolomei, M.S., A.L. Webber, M.E. Brunkow, and S.M. Tilghman. 1993. Epigenetic mechanisms underlying the imprinting of the mouse *H19* gene. *Genes Dev.* **7:** 1663–1673.

Beard, C., E. Li, and R. Jaenisch. 1995. Loss of methylation activates *Xist* in somatic but not in embryonic cells. *Genes Dev.* **9:** 2325–2334.

Beechey, C. and B. Cattanach. 1995. Genetic imprinting map. *Mouse Genome* **93:** 89–92.

Bird, A. 1986. CpG-rich islands and the function of DNA methylation. *Nature* **321:** 209–213.

Boyes, J. and A. Bird. 1992. Repression of genes by DNA methylation depends on CpG density and promoter strength: Evidence for involvement of a methyl-CpG binding protein. *EMBO J.* **11:** 327–333.

Brandeis, M., T. Kafri, M. Ariel, J.R. Chaillet, J. McCarrey, A. Razin, and H. Cedar. 1993. The ontogeny of allele-specific methylation associated with imprinted genes in the mouse. *EMBO J.* **12:** 3669–3677.

Brandeis, M., D. Frank, I. Keshet, Z. Siegfried, M. Mendelsohn, A. Nemes, V. Temper, A. Razin, and H. Cedar. 1994. Sp1 elements protect a CpG island from *de novo* methylation. *Nature* **371:** 435–438.

Brannan, C.I., E.C. Dees, R.S. Ingram, and S.M. Tilghman. 1990. The product of the *H19* gene may function as an RNA. *Mol. Cell. Biol.* **10:** 28–36.

Brockdorff, N., A. Ashworth, G. Kay, V. McCabe, D. Norris, P. Cooper, S. Swift, and S. Rastan. 1992. The product of the mouse *Xist* gene is a 15 kb inactive X-specific transcript containing no conserved ORF and located in the nucleus. *Cell* **71:** 515–526.

Brown, C., B. Hendrich, J. Rupert, R. Lafreniere, Y. Xing, J. Lawrence, and H. Willard. 1992. The human *XIST* gene: Analysis of a 17 kb inactive X-specific RNA that contains conserved repeats and is highly localised within the nucleus. *Cell* **71:** 527–542.

Bruhat, A. and J.-P. Jost. 1995. *In vivo* estradiol-dependent dephosphorylation of the repressor MDBP-2-H1 correlates with the loss of *in vitro* preferential binding to methylated DNA. *Proc. Natl. Acad. Sci.* **92:** 3678–3682.

Buiting, K., S. Saitoh, S. Gross, B. Dittrich, S. Schwartz, R.D. Nicholls, and B. Horsthemke. 1995. Inherited microdeletions in the Angelman and Prader-Willi syndromes define an imprinting centre on human chromosome 15. *Nat. Genet.* **9:** 395–400.

Cattanach, B.M. and M. Kirk. 1985. Differential activity of maternally and paternally derived chromosome regions in mice. *Nature* **315:** 496–498.

Cattanach, B., J. Barr, and J. Jones. 1995. Use of chromosome rearrangements for investigations into imprinting in the mouse. In *Genomic imprinting. Causes and consequences* (ed. R. Ohlsson, et al.), pp. 327–341. Cambridge University Press, England.

Cattanach, B.M., J.A. Barr, E.P. Evans, M. Burtenshaw, C.V. Beechey, S.E. Leff, C.I. Brannan, N.G. Copeland, N.A. Jenkins, and J. Jones. 1992. A candidate mouse model for Prader-Willi syndrome which shows an absence of *Snrpn* expression. *Nat. Genet.* **2:** 270–274.

Chaillet, J., D. Bader, and P. Leder. 1995. Regulation of genomic imprinting by gametic and embryonic processes. *Genes Dev.* **9:** 1177–1187.

Clemson, C.M., J.A. McNeil, H.F. Willard, and J.B. Lawrence. 1996. *XIST* RNA paints the inactive X chromosome at interphase: Evidence for a novel RNA involved in nuclear/chromosome structure. *J. Cell Biol.* **132:** 259–275.

Davis, L., M. Cadrin, D. Brown, and N. Chaly. 1993. Reversible disassembly of transcription domains in lymphocyte nuclei during inhibition of RNA synthesis by DRB. *Biol. Cell* **78:** 163–180.

DeChiara, T.M., E.J. Robertson, and A. Efstratiadis. 1991. Parental imprinting of the mouse insulin-like growth factor II gene. *Cell* **64:** 849–859.

Dillon, N. and F. Grosveld. 1993. Transcriptional regulation of multigene loci: Multilevel control. *Trends Genet.* **9:** 134–137.

Dorer, D.R. and S. Henikoff. 1994. Expansions of transgene repeats cause heterochromatin formation and gene silencing in *Drosophila*. *Cell* **77:** 993–1002.

Feil, R., J. Walter, N.D. Allen, and W. Reik. 1994. Developmental control of allelic methylation in the imprinted mouse *Igf2* and *H19* genes. *Development* **120:** 2933–2943.

Ferguson-Smith, A.C., H. Sasaki, B.M. Cattanach, and M.A. Surani. 1993. Parental-origin-specific modification of the mouse *H19* gene. *Nature* **362:** 751–755.

Giannoukakis, N., C. Deal, J. Paquette, C. Goodyer, and C. Polychronakos. 1993. Parental genomic imprinting of the human *IGF-2* gene. *Nat. Genet.* **4:** 98–101.

Giddings, S.J., C.D. King, K.W. Harman, J.F. Flood, and L.R. Carnaghi. 1994. Allele specific inactivation of insulin 1 and 2, in the mouse yolk sac, indicates imprinting. *Nat. Genet.* **6:** 310–313.

Glenn, C., K. Porter, M. Jong, R. Nicholls, and D. Driscoll. 1993. Functional imprinting and epigenetic modification of the human *SNRPN* gene. *Hum. Mol. Genet.* **2:** 2001–2005.

Guillemot, F., T. Caspary, S.M. Tilghman, N.G. Copeland, D.J. Gilbert, N.A. Jenkins, D.J. Anderson, A.L. Joyner, J. Rossant, and A. Nagy. 1995. Genomic imprinting of *Mash2*, a mouse gene required for trophoblast development. *Nat. Genet.* **9:** 235–242.

Gunaratne, P., M. Nakao, D. Ledbetter, J. Sutcliffe, and A. Chinault. 1995. Tissue-specific and allele-specific replication timing control in the imprinted human Prader-Willi syndrome region. *Genes Dev.* **9:** 808–820.

Hatada, I. and T. Mukai. 1995. Genomic imprinting of *p57KIP2*, a cyclin-dependent kinase inhibitor, in mouse. *Nat. Genet.* **11:** 204–206.

Hatada, I., T. Sugama, and T. Mukai. 1993. A new imprinted gene cloned by a methylation-sensitive genome scanning method. *Nucleic Acids Res.* **21:** 5577–5582.

Hayashizaki, Y., H. Shibata, S. Hirotsune, H. Sugino, Y. Okazaki, N. Sasaki, K. Hirose, H. Imoto, H. Okuizumi, M. Muramatsu, H. Komatsubara, T. Shiroishi, K. Moriwaki, M. Katsuki, N. Hatano, H. Sasaki, T. Ueda, N. Mise, N. Takagi, C. Plass, and V.M. Chapman. 1994. Identification of an imprinted U2af binding protein related sequence on mouse chromosome ll using the RLGS method. *Nat. Genet.* **6.** 33–40.

Jeppesen, P. and B.M. Turner. 1993. The inactive X chromosome in female mammals is distinguished by a lack of histone H4 acetylation, a cytogenetic marker for gene expression. *Cell* **74:** 281–289.

Kaneko-Ishino, T., Y. Kuroiwa, N. Miyoshi, T. Kohda, R. Suzuki, M. Yokoyama, S. Viville, S. Barton, F. Ishino, and M.A. Surani. 1995. *Pegl/Mest* imprinted gene on chromosome 6 identified by cDNA subtraction hybridization. *Nat. Genet.* **11:** 52–59.

Kay, G.F., G.D. Penny, D. Patel, A. Ashworth, N. Brockdorff, and S. Rastan. 1993. Expression of *Xist* during mouse development suggests a role in the initiation of X chromosome inactivation. *Cell* **72:** 171–182.

Kitsberg, D., S. Selig, M. Brandeis, I. Simon, I. Keshet, D.J. Driscoll, R.D. Nicholls, and H. Cedar. 1993. Allele-specific replication timing of imprinted gene regions. *Nature* **364:** 459–463.

Knoll, J.H.M., S.-D. Cheng, and M. Lalande. 1994. Allele specificity of DNA replication timing in the Angelman/Prader-Willi syndrome imprinted chromosomal region. *Nat. Genet.* **6:** 41–46.

Koide, T., J. Ainscough, M. Wijgerde, and M.A. Surani. 1994. Comparative analysis of *Igf-2/H19* imprinted domain: Identification of a highly conserved intergenic DNase I hypersensitive region. *Genomics* **24:** 1–8.

Kuroiwa, Y., T. Kaneko-Ishino, F. Kagitani, T. Kohda, L.-L. Li, M. Tada, R. Suzuki, M. Yokayama, T. Shiroishi, S. Wakana, S.C. Barton, F. Ishino, and M.A. Surani. 1996. *Peg3* imprinted gene on proximal chromosome 7 encodes for a zinc finger protein. *Nat. Genet.* **12:** 186–190.

LaSalle, J. and M. Lalande. 1995. Domain organisation of allele-specific replication within the *GABRB3* gene cluster requires a biparental 15q11-13 contribution. *Nat. Genet.* **9:** 386–394.

Lee, J., J. Pintar, and A. Efstratiadis. 1990. Pattern of the insulin-like growth factor II gene expression during early mouse development. *Development* **110:** 151–159.

Lee, J.E., U. Tantravahi, A.L. Boyle, and A. Efstratiadis. 1993. Parental imprinting of an *Igf-2* transgene. *Mol. Reprod. Dev.* **35:** 382–390.

Leff, S.E., C.I. Brannan, M.L. Reed, T. Ozcelik, U. Francke, N.G. Copeland, and N.A. Jenkins. 1992. Maternal imprinting of the mouse *Snrpn* gene and conserved linkage homology with the human Prader-Willi syndrome region. *Nat. Genet.* **2:** 259–264.

Leighton, P., J. Saam, R. Ingram, C. Stewart, and S. Tilghman. 1995a. An enhancer deletion affects both *H19* and *Igf2* expression. *Genes Dev.* **9:** 2079–2089.

Leighton, P.A., R.S. Ingram, J. Eggenschwiler, A. Efstratiadis, and S.M. Tilghman. 1995b. Disruption of imprinting caused by deletion of the H19 gene region in mice. *Nature* **375:** 34–39.

Li, E., C. Beard, and R. Jaenisch. 1993. Role of DNA methylation in genomic imprinting. *Nature* **366:** 362–365.

Li, E., T.H. Bestor, and R. Jaenisch. 1992. Targeted mutation of the DNA methyltransferase gene result in embryonic lethality. *Cell* **69:** 915–926.

Matzke, M. and A. Matzke. 1995. How and why do plants inactivate homologous (trans)genes. *Plant Physiol.* **107:** 679–685.

McGrath, J. and D. Solter. 1984. Completion of mouse embryogenesis requires both the maternal and paternal genomes. *Cell* **37:** 179–183.

Meehan, R., J. Lewis, and A. Bird. 1992. Characterization of MeCP2, a vertebrate DNA binding protein with affinity for methylated DNA. *Nucleic Acids Res.* **20:** 5085–5092.

Monk, M., M. Boubelik, and S. Lehnert. 1987. Temporal and regional changes in DNA methylation in the embryonic, extraembryonic and germ cell lineages during mouse embryo development. *Development* **99:** 371–382.

Neumann, B., P. Kubicka, and D. Barlow. 1995. Characteristics of imprinted genes. *Nat. Genet.* **9:** 12–13.

Ogawa, O., L.A. McNoe, R. Eccles, I.M. Morrison, and A.E. Reeve. 1993. Human insulin-like growth factor type I and type II receptors are not imprinted. *Hum. Mol. Genet.* **2:** 2163–2165.

Ozcelik, T., S. Leff, W. Robinson, T. Donlon, M. Lalande, E. Sanjines, A. Schinzel, and

U. Francke. 1992. Small nuclear ribonucleoprotein polypeptide N (*SNRPN*), an expressed gene in the Prader-Willi syndrome critical region. *Nat. Genet.* **2:** 265–269.

Pandit, N.N. and V.E.A. Russo. 1992. Reversible inactivation of a foreign gene, hph, during the asexual cycle in *Neurospora crassa* transformants. *Mol. Gen. Genet.* **234:** 412–422.

Pfeifer, K. and S.M. Tilghman. 1994. Allele-specific gene expression in mammals: The curious case of the imprinted RNAs. *Genes Dev.* **8:** 1867–1874.

Poirier, F., C.-T.J. Chan, P. Timmons, E.J. Robertson, M.J. Evans, and P.W.J. Rigby. 1991. The murine *H19* gene is activated during embryonic stem cell differentiation *in vitro* and at the time of implantation in the developing embryo. *Development* **113:** 1105–1114.

Rachmilewitz, J., R. Goshen, I. Ariel, T. Schneider, N. de Groot, and A. Hochberg. 1992. Parental imprinting of the human *H19* gene. *FEBS Lett.* **309:** 25–28.

Rastan, S. 1994. X chromosome inactivation and the *Xist* gene. *Curr. Opin. Genet. Dev.* **4:** 292–297.

Razin, A. and H. Cedar. 1994. DNA methylation and genomic imprinting. *Cell* **77:** 473–476.

Reik, W., S.K. Howlett, and M.A. Surani. 1990. Imprinting by DNA methylation: From transgenes to endogenous gene sequences. *Dev. Suppl.* pp. 99–106.

Reis, A., B. Dittrich, V. Greger, K. Buiting, M. Lalande, G. Gillesen-Kaesbach, M. Anvret, and B. Horsthemke. 1994. Imprinting mutations suggested by abnormal DNA methylation patterns in familial Angelman and Prader-Willi syndromes. *Am. J. Hum. Genet.* **54:** 741–747.

Sasaki, H., P.A. Jones, J.R. Chaillet, A.C. Ferguson-Smith, S.C. Barton, W. Reik, and M.A. Surani. 1992. Parental imprinting: Potentially active chromatin of the repressed maternal allele of the mouse insulin-like growth factor II (*Igf2*) gene. *Genes Dev.* **6:** 1843–1856.

Singer-Sam, J. and A. Riggs. 1993. X chromosome inactivation and DNA methylation. *Experientia* (suppl.) **64:** 358–384.

Smrzka, O.W., I. Fae, R. Stoger, R. Kurzbauer, G.F. Fischer, T. Henn, A. Weith, and D.P. Barlow. 1995. Conservation of a maternal-specific methylation signal at the human *IGF2R* locus. *Hum. Mol. Genet.* **4:** 1945–1952.

Stoger, R., P. Kubicka, C.G. Liu, T. Kafri, A. Razin, H. Cedar, and D.P. Barlow. 1993. Maternal-specific methylation of the imprinted mouse *Igf2r* locus identifies the expressed locus as carrying the imprinting signal. *Cell* **73:** 61–71.

Surani, M.A.H., S.C. Barton, and M.L. Norris. 1984. Development of reconstituted mouse eggs suggests imprinting of the genome during gametogenesis. *Nature* **308:** 548–550.

Sutcliffe, J.S., M. Nakao, S. Christian, K.H. Orstavik, N. Tommerup, D.H. Ledbetter, and A.L. Beaudet. 1994. Deletions of a differentially methylated CpG island at the *SNRPN* gene define a putative imprinting control region. *Nat. Genet.* **8:** 52–58.

Tilghman, S., M. Brunkow, C. Brannan, C. Dees, M. Bartolomei, and K. Phillips. 1992. The mouse *H19* gene: Its structure and function in mouse development. In *Nuclear processes and oncogenes* (ed. P. Sharp), pp. 188–199. Academic Press, San Diego.

Tremblay, K.D., J.R. Saam, R.S. Ingram, S.M. Tilghman, and M.S. Bartolomei. 1995. A paternal-specific methylation imprint marks the alleles of the mouse *H19* gene. *Nat. Genet.* **9:** 407–413.

Villar, A.J. and R.A. Pedersen. 1994. Parental imprinting of the *Mas* protooncogene in mouse. *Nat. Genet.* **8:** 373–379.

Wevrick, R., J.A. Kerns, and U. Francke. 1994. Identification of a novel paternally expressed gene in the Prader-Willi syndrome region. *Hum. Mol. Genet.* **3:** 1877–1882.

Yoo-Warren, H., V. Pachnis, R.S. Ingram, and S.M. Tilghman. 1988. Two regulatory domains flank the mouse *H19* gene. *Mol. Cell. Biol.* **8:** 4707–4715.

Zemel, S., M.S. Bartolomei, and S.M. Tilghman. 1992. Physical linkage of two mammalian imprinted genes, *H19* and *Igf2*. *Nat. Genet.* **2:** 61–65.

Zuccotti, M. and M. Monk. 1995. Methylation of the mouse *Xist* gene in sperm and eggs correlates with imprinted *Xist* expression and paternal X-inactivation. *Nat. Genet.* **9:** 316–320.

Genomic Imprinting and Modifier Genes in the Mouse

Jörn Walter, Torsten Krüger, and Sabine Engemann
Max-Planck-Institut für Molekulare Genetik
14195 Berlin, Germany

Nicholas D. Allen, Gavin Kelsey, Robert Feil, Thierry Forné, and Wolf Reik
Department of Development and Genetics
The Babraham Institute
Cambridge CB2 4AT
United Kingdom

Imprinted genes provide an excellent experimental system to examine the epigenetic control of gene expression. This is, first, because an active and inactive allele of the same gene are present in the same cell, so that the differentiation between expressed and nonexpressed status involves *cis*-acting genetic elements, and, second, because there must exist a long-lasting memory of the origin of alleles (from sperm or egg) that is sustained through numerous rounds of replication. Here we examine the epigenetic features of the imprinted insulin-like growth factor 2 (Igf2) gene and postulate that some of the methylation and chromatin features in this gene are influenced or controlled by methylation and chromatin modifier genes. In the second part of the chapter, we describe a genetic system for the identification and potential isolation of such modifiers of methylation and chromatin.

In mouse and human, the Igf2 gene is almost exclusively expressed from the paternal chromosome (DeChiara et al. 1991; Giannoukakis et al. 1993; Ohlsson et al. 1993). For a number of reasons, this particular gene is important for the study of mechanisms and consequences of imprinting. First, Igf2 has a very strong effect on fetal growth, one of the key phenotypes controlled by imprinted genes. Second, there are at least two other (maternally expressed) imprinted genes, Igf2r and H19, that interact with the Igf2 pathway (Wang et al. 1994; Leighton et al. 1995a). Third, the Igf2 region of imprinted genes is implicated in a number of human diseases, most notably the fetal overgrowth and tumor syndrome, Beckwith-Wiedemann syndrome (BWS) (Ogawa et al. 1993a; Ohlsson et al. 1993; Reik et al. 1994, 1995b; Weksberg 1994). Finally, the Igf2 gene

Epigenetic Mechanisms of Gene Regulation

is surrounded by other imprinted genes that form a cluster of at least five, but perhaps more, imprinted genes on distal chromosome 7 in the mouse and on chromosome 11p15.5 in the human (Fig. 1). It is not known why imprinted genes are clustered, but there may be functional, mechanistic, and evolutionary reasons. However, conceivably, the control of imprinting relies on both regional (replication-, locus-) control elements and gene-specific control elements (enhancers, silencers, etc.).

EPIGENETIC MODIFICATIONS IN THE IGF2 GENE AND THEIR DEVELOPMENTAL CONTROL

Structure, Function, and Regulation of the Igf2 Gene

The mouse Igf2 gene contains 8 exons and 4 promoters (Fig. 2). Therefore, its structure is very similar to that of the human and rat Igf2 genes. Current work in our laboratory has shown exons ψ1 and ψ2, which were previously regarded as pseudoexons, are actually transcribed (T. Moore et al., unpubl.). Depending on the promoters used, transcripts comprise various sets of untranslated leader exons which are all spliced onto three coding exons. The different transcripts generated are under developmental and tissue-specific control and, in addition, have different translational capabilities. However, only one polypeptide is translated and after proteolytic cleavage a mature Igf2 polypeptide of 67 amino acids is produced. Transcription mainly occurs in fetal tissues, with particularly strong expression in mesodermal, endodermal, and extraembryonic tissues. Transcriptional silencing occurs in the immediate postnatal period, with the exception of the liver (in the human) and the choroid plexus and

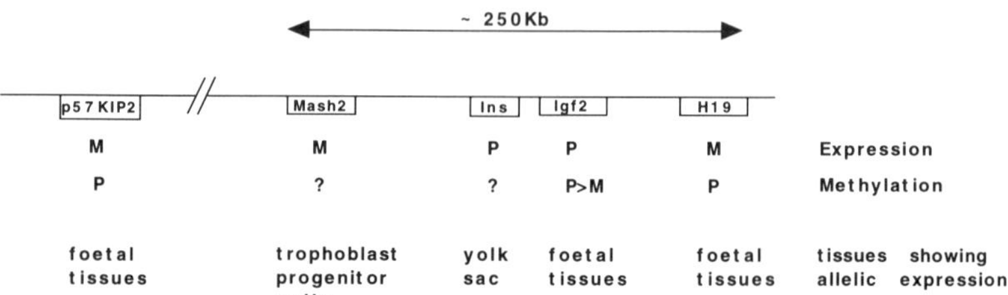

Figure 1 Location of imprinted genes (*boxes*) on distal chromosome 7 of the mouse. The physical distance between Mash2 and H19 genes is given according to Guillemot et al. (1995). A summary of allele-specific methylation and expression (M = maternal, P = paternal) and the tissue specificity of imprinting in the mouse is given below. The interrupted line between Mash2 and p57Kip2 indicates that the physical distance has not been determined. The synteny in this region is conserved on human chromosome 11p15.5.

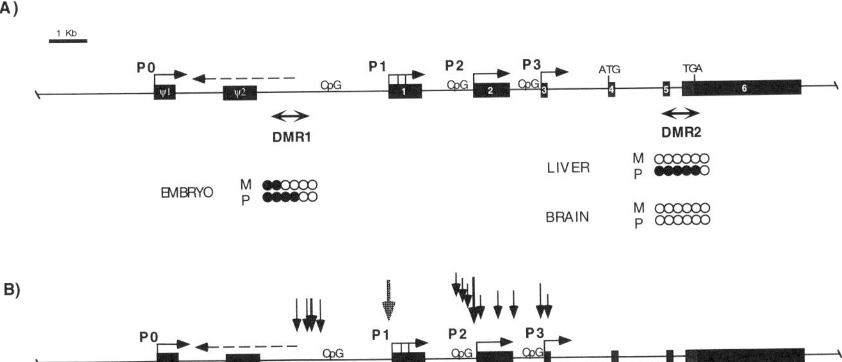

Figure 2 Structure of the murine Igf2 gene and summary of DNA methylation patterns in liver and brain from d16 fetuses (*A*) and DNase I sensitivity studies (*B*). Black boxes depict the locations of exons. The first two exons ψ1 and ψ2 were previously regarded as pseudoexons. Promoters are indicated by P0,1, 2, and 3 and transcriptional start points by horizontal arrows above the exons. The translated region within exons 4, 5, and 6 is highlighted by gray shading and start (ATG) and stop (TGA) codons. The antisense transcript that overlaps with ψ2 is shown with a dashed arrow. The relative positions of CpG islands are indicated by CpG. (*A*) Location of DMRs 1 and 2 and their allele-specific methylation patterns in certain tissues. The relative extent of methylation on maternal (M) and paternal alleles (P) is indicated by the number of filled (methylated) or open (unmethylated) circles. (*B*) Positions of DNase I sensitive sites within the Igf2 gene are shown by vertical arrows. Thickness of these arrows refers to relative accessibility of the chromatin. The gray shaded arrow above promoter P1 shows the position of a hypersensitive site only found in newborn liver on both parental alleles.

leptomeninges in the nervous system, where expression is maintained in adults. The Igf2 polypeptide exerts its main known effects in a paracrine and endocrine fashion and mainly signals through the Igf1-receptor. Deletion of the Igf2 gene in the mouse results in fetal growth reduction to 60% of the birth weight (DeChiara et al. 1990). Transgenic overexpression of the polypeptide can result in an increase of birth weight by 30% (F.L. Sun et al., unpubl.).

The maternal allele is repressed in all tissues, with the exception of the blastocyst stage of very early development (Latham et al. 1994; Szabo and Mann 1995a) and the choroid plexus in the nervous system, where both parental alleles are expressed. In the human, expression from promoter 1 also shows some relaxation of imprinting: Although there is paternal expression in early gestation fetuses (Taniguchi et al. 1995), at later stages, and in particular in postnatal liver, expression is biallelic (Vu and Hoffman 1994; Ekström et al. 1995). Biallelic expression is, however, also observed in pathological situations, most notably in the human fetal overgrowth syndrome, in BWS, and in various cancers.

Methylation

Two regions of allele-specific DNA methylation have been detected in and around the mouse and human Igf2 genes. These are called differentially methylated regions 1 (DMR1) and 2 (DMR2) (Fig. 2). DMR1 is about 3 kb upstream of promoter P1, and DMR2 comprises the last intron and 5′ end of the last exon of the gene (Sasaki et al. 1992; Feil et al. 1994, 1995). Both regions are small clusters of CpGs (not quite islands) that are methylated on the expressed paternal chromosome. Both regions are also associated with clusters of direct repeats, which have been proposed to have some function in the control of imprinting (Neumann et al. 1995). Bisulfite sequencing, used to examine the methylation state of all C residues on individual chromosomes, revealed that the methylation of DMR1 is highly mosaic: On average, the methylation on paternal chromosomes is about 50%, whereas it is about 30% on maternal chromosomes. Hence, regional methylation density – rather than absolute methylation levels of individual sites – may be important for its function. Bisulfite sequencing in DMR2 showed a different picture, as there is very little maternal methylation. Where paternal methylation is present (see below), it is quite extensive and non-mosaic (J. Walter et al., unpubl.). In contrast to CpG methylation in the DMRs, the CpG islands associated with promoters 2 and 3 and one just 3′ of DMR1 (Fig. 2) remain constitutively unmethylated on both parental alleles. The CpG island adjacent to DMR1 may coincide with the promoter for an antisense transcript which is also paternally expressed (T. Moore et al., unpubl.).

The allelic methylation patterns in the DMRs are under significant developmental and tissue-specific control. Sites in DMR1 are methylated in both oocyte and sperm (Brandeis et al. 1993; Feil et al. 1994) and are rapidly demethylated following fertilization (Razin amd Shemer 1995; Shemer et al. 1996). Allelic methylation of the paternal copy first appears at the morula stage and is similar in all embryonic and adult tissues, with the exception of the choroid plexus where high levels of methylation are found on both parental copies (Feil et al. 1994). In contrast, methylation in DMR2 is highly tissue-specific. It is known to be methylated in sperm, but the patterns in the oocyte and during preimplantation development have not been elucidated. In fetal and adult stages, methylation of the paternal allele is extensive in liver and intestine, but low in brain and kidney (see Fig. 2) (Feil et al. 1994; J. Walter et al., unpubl.). A thorough analysis is currently being carried out. The first results leave us with the impression that extensive paternal methylation of DMR2 may predominate in endoderm-derived tissues. This is particularly interesting in light of the endoderm-specific enhancers in the neighboring H19 gene (see below). The human DMR2 region also shows tissue variation in its paternal methylation pattern (Schneid et al. 1993; W. Reik et al., unpubl.). We propose that both DMRs are negative control elements (e.g., silencers), whose function can be modulated by DNA methylation (see below for more extensive discussion).

Other imprinted genes do not necessarily show the same pattern of allele-specific methylation. For example, in the H19 (upstream region), Igf2r (region 2), and U2afbp genes, much more robust and constitutive allele-specific methylation patterns have been described (Razin and Shemer 1995; Tremblay et al. 1995). Some of these paternal and maternal methylation differences are apparently inherited from the gametes and can be present on the active maternal allele (Igf2r) or the repressed maternal or paternal alleles (U2afbp and H19, respectively). The H19 and Igf2r genes also contain regions of methylation associated with the inactive allele which are established later in development.

Chromatin

Chromatin analyses using DNase I and specific restriction enzymes have been carried out (Fig. 2B) (Sasaki et al. 1992; Feil et al. 1995). Wherever hypersensitive sites are found, they are present on both parental chromosomes (subtle differences cannot be excluded), ruling out extensive heterochromatinization of the repressed maternal allele. Other features of chromatin and gene activity, e.g., matrix binding sites, nucleosome positioning, and transcription factor binding, have so far not been investigated. Hypersensitive sites are chiefly associated with the promoters and, in addition, with the DMR1 (Fig. 2B). In this context, a proper CpG island is situated just 3′ of the DMR1, and this may well be a promoter region for an antisense transcript that originates in this region and is transcribed in the 5′ direction (Fig. 2) (Rivkin et al. 1993; T. Moore, unpubl.). Interestingly, hypersensitive sites in this region are present before Igf2 is transcriptionally active; that is, in embryonic stem (ES) cells. There is significant developmental and tissue-specific control in other hypersensitive sites. The hypersensitive site in promoter P1, for example, is strongly present in fetal liver, but absent in other fetal tissues. This again highlights the possibility that in the liver—which is partly endodermal—a distinct Igf2 control mechanism is operating in contrast to tissues of mesodermal origin. The significance of strong hypersensitive sites in maternal allele promoters is unclear: Whether this means that the promoters are "open" and therefore potentially ready for action, or whether promoters actually initiate transcription, but that transcription terminates prematurely, still remains to be established. However, it is intriguing that features of imprinting (differential methylation) show close temporal and spatial coregulation with more general epigenetic features of tissue-specific control (hypersensitive sites). We therefore propose that imprinting and the control of tissue-specific gene activity are intricately linked.

Imprinting of Igf2: Functional Analysis

The functional significance of the epigenetic features of Igf2 remains to be examined. The only study to test whether *cis*-acting sequences in the Igf2 gene or its immediate vicinity are sufficient for imprinting control is inconclusive: Some transgenic lines with the same cosmid-based construct of the rat gene showed paternal expression, but others did not (Lee et al. 1993; Efstratiadis 1994). However, now there is good evidence that sequences at some distance from the Igf2 gene (namely about 90 kb away) which contain the imprinted and maternally expressed H19 gene have major effects on the control of Igf2 imprinting. We examine this evidence first. Then we return to the Igf2 gene and ask how the epigenetic features in Igf2 may interact with distant control elements (in H19 and elsewhere).

The H19 gene is remarkable, not only because it does not appear to encode a polypeptide (Brannan et al. 1990), but also because in addition to its reciprocal imprinting, its tissue and developmental stage-specific expression closely parallels that of the Igf2 gene (Bartolomei et al. 1991). These observations have given rise to a number of speculations concerning the possible coregulation, or common imprinting, of the two genes (Bartolomei and Tilghman 1992). The most explicit model posits the existence of an enhancer competition (Leighton et al. 1995a), with a set of endoderm-specific enhancers 3′ to H19. These are only able to access either the H19 promoter, or the Igf2 promoters, on the same chromosome (Fig. 3A). Moreover, the H19 gene carries its own imprinting signal ("imprinting box"), which causes the paternal copy to be inactive and the maternal copy to be active (Tremblay et al. 1995). A number of observations seem to prove the correctness of the most important parts of this model. Yet other observations indicate that the model, as it stands, cannot account for all aspects of Igf2 imprinting.

First of all, the H19 gene possesses a region of allele-specific methylation 5′ to the promoter, where paternal methylation is already detected in the sperm, and the gene is unmethylated in the oocyte, which is maintained throughout development (Tremblay et al. 1995). This methylation could serve as a primary germ-line-specific imprinting signal. Second, H19 transgenes containing these sequences (as well as the 3′ enhancers) show appropriate imprinting with maternal expression, although this imprinting appears to be faithful on certain genetic backgrounds of inbred mice only (Bartolomei et al. 1993). Third and most important, when the maternal copy of H19 is either deleted (Leighton et al. 1995a) (but the enhancers are still there), or inactivated by methylation (Moulton et al. 1994; Steenmann et al. 1994; Reik et al. 1995b), the maternal copy of Igf2 is expressed (Fig. 3B). Fourth, when the H19 3′ enhancers are deleted on the maternal chromosome, H19 is not transcribed, whereas when they are deleted on the paternal chromosome, Igf2 is not transcribed (in endoderm tissues) (Fig. 3C) (Leighton et al. 1995b).

Let us consider the observations that are more difficult to reconcile with the simple model. First, there are instances where Igf2 and H19 seem to be expressed from the same chromosome. In the fetal mouse choroid plexus, for example, Igf2 expression is biallelic, whereas H19 is expressed from the maternal chromosome in its usual fashion (Svensson et al. 1995). Similarly, in some BWS patients with translocations centromeric of Igf2 in the maternal chromosome, Igf2 expression is biallelic but H19 expression is maternal (Fig. 3D; W. Reik, unpubl.). The same is true for some BWS patients without translocations (E. Maher and W. Reik, unpubl.). In addition, in the early mouse embryo following implantation, Igf2 expression becomes monoallelic very quickly, whereas H19 expression remains biallelic (Szabo and Mann 1995b). Similarly, H19 is expressed biallelically in first-trimester human placenta where Igf2 expression is monoallelic (Jinno et al. 1995).

Whereas maternal deletion of H19 results in activation of the silent Igf2 allele, this activation is incomplete in some tissues, most notably in those derived from endoderm (Leighton et al. 1995a). These are the only tissues in which there is extensive methylation of DMR2 on the paternal allele. The key features in Igf2 not explained by the simple enhancer competition model are the allele-specific methylation and the presence of hypersensitive sites in the promoters of the inactive Igf2 gene. Allele-specific methylation may be an entirely passive phenomenon that is simply related to whether or not the allele is expressed. We consider this unlikely, because for both DMR1 and 2, there are situations where differential methylation and differential expression are not linked. For example, DMR1 is differentially methylated in adult tissues when Igf2 is not expressed, and in the DMR2 there is no differential methylation in some fetal tissues that show monoallelic expression. It seems more likely that allelic methylation is actively controlled in some way, and that it participates directly in allele-specific expression; for example, because it involves (regulates) elements having a negative effect (silencers) on transcription. In addition, the fact that there is hypersensitivity in the maternal Igf2 promoters and that these promoters are unmethylated suggests to us that the promoters may be actively engaged with the (H19 or other) enhancers, but that successful transcription is blocked (presumably by the DMR elements). This could even be a block to transcriptional

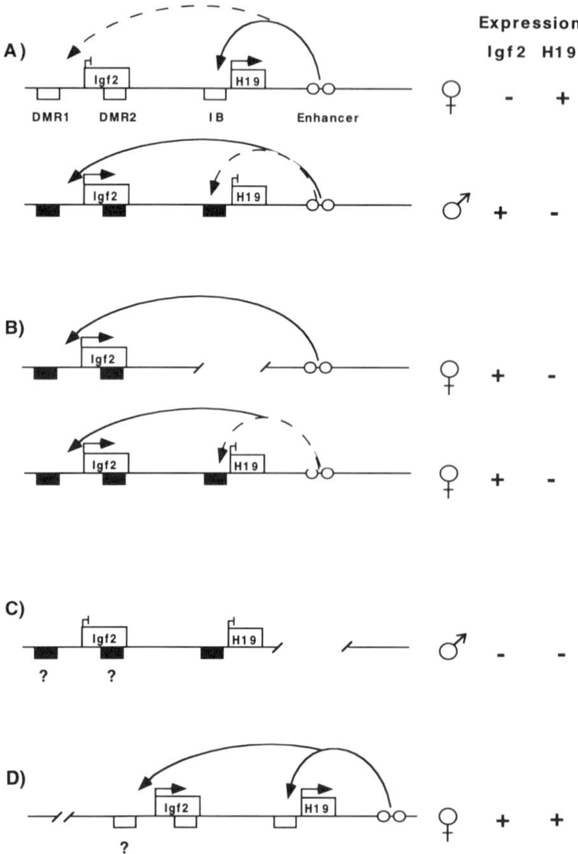

Figure 3 (*See facing page for legend.*)

elongation, rather than initiation, which is something that needs to be tested.

The enhancer competition model could easily be modified to accept that competition is not exclusive, but that elements within the Igf2 promoter provide an independent or additional level of control, which ensures a more stringently regulated system. This is why Igf2 promoters are potentially active on the maternal chromosome, yet are prevented from full activity by unmethylated DMR elements. Consequently, in situations when DMR elements on the maternal chromosome become methylated, both genes should be active.

How is differential methylation in Igf2 brought about? The first possibility is that the DMRs are bona fide germ-line imprints. This is apparently not the case for DMR1 (as described above). However, it cannot be ruled out for DMR2. Thus, the early genesis of these patterns needs to be clarified. Still, it is clear that they can be modified at later stages and that apparently interactions with the H19 gene (enhancers) or gene product (among other factors) could be involved in this modification. First, in adult choroid plexus where Igf2 expression is biallelic and H19 is not expressed, DMR1 is highly methylated on both alleles (Feil et al. 1994).

Figure 3 Models for the control of allele-specific expression of the Igf2 and H19 genes. Both genes are shown as boxes in their relative positions on the chromosome (*thin line*). Transcriptionally active alleles are marked by horizontal arrows. The locations of differentially methylated regions (DMR1 and 2) and the imprinting box of H19 (IB) are given below the line; methylation is indicated by filled boxes. Potential interactions between the H19 enhancer elements (*open circles*) and promoter regions are indicated by curved arrows. (*A*) Wild-type situation on both parental alleles. The dashed arrows represent our extended view of the enhancer competition model between H19 downstream enhancers and Igf2 and H19 promoter regions. The transcriptional activity of maternal and possibly also paternal chromosomes may not be determined exclusively by epigenetic changes in the H19 imprinting box (and the promoter) but may also be influenced by methylation in the DMRs 1 and 2. (*B*) Methylation and expression of maternal Igf2 alleles in which the H19 gene and upstream region (imprinting box) have been deleted (*upper* part) or the maternal H19 gene is methylated (*lower* part), e.g., in BWS patients or Wilms' tumors. (*C*) Paternal chromosomes of mice on which the enhancer region has been deleted show a strong decrease of the Igf2 gene expression in certain tissues (e.g., liver); the methylation of DMRs 1 and 2 in this situation has not been analyzed (indicated by ?). (*D*) Activity of H19 and Igf2 genes on the maternal allele in some BWS patients with translocations centromeric of Igf2 (indicated by the broken line). The breakpoint is approximately 300 kb centromeric (upstream) of Igf2.

Second, in mice with the maternal deletion of H19, both DMR1 and DMR2 become methylated on the maternal chromosome (T. Forné and T. Koüger unpubl.). Third, in BWS patients with maternal H19 methylation, the maternal copy of DMR2 also becomes methylated. It is possible that this is a reflection of the alteration of the transcriptional status of Igf2, but we consider this less likely for the reasons discussed above. In consequence, we are tempted to speculate that the H19 gene may have some direct or indirect role in influencing chromatin or methylation patterns. With respect to the Igf2 DMRs, the prediction is that some aspect of H19 expression decreases methylation levels in (at least) the maternal Igf2 copy. Formally at least, the H19 gene can accordingly be described as a methylation modifier, and an imprinted one at that.

MODIFIERS OF DNA METHYLATION AND IMPRINTING

We are beginning to understand the *cis*-acting DNA sequences important for imprinting, but the identification of *trans*-acting factors lags behind. In fact, the only *trans*-acting factor known to be involved is the DNA methyltransferase. In DNA-methyltransferase-deficient mice, imprinting of H19, Igf2, and Igf2r is altered in the predicted direction, thus establishing that DNA methylation is indeed crucial for the control of imprinting (Li et al. 1993). However, since the DNA methyltransferase seems to have very little sequence or developmental specificity, other factors are required that modify methylation patterns ("methylation modifiers"). For an imprinted gene, one of the simplest scenarios illustrating some of these requirements is as follows:

Consider an imprinted gene where the germ-line-specific methylation difference, say maternal methylation, is present and maintained during development in tissues where monoallelic expression occurs. Thus, de novo methylation, or the prevention thereof, in oogenesis or spermatogenesis, respectively, has to be regulated. The next challenge for the methylated maternal allele is to remain methylated during preimplantation development despite genome-wide demethylation during this period. This could be achieved by factors that protect against the attack of demethylases. This is followed by an effort by the paternal allele to stay unmethylated despite genome-wide de novo methylation at early gastrulation stages. Furthermore, both the protection against demethylation and the protection against de novo methylation could be developmental stage- and lineage-specific. Therefore, specificity has to be conferred by modifiers for sequence and/or methylation state and for developmental stages and tissues. Some of the factors conferring this specificity may well be shared with the general machinery controlling gene expression. Yet other factors may be more specific to the imprinting mechanism. It is therefore important to distinguish between these different types of modifiers when devising experimental systems for the identification of methylation modifiers.

If there are indeed genes that modify imprints (through methylation or other chromatin mechanisms), there should be genetic systems where polymorphisms in such modifiers could lead to variations in the way imprinting functions. Despite some efforts to find such polymorphic systems in the mouse, there is no real evidence for this so far. Only one observation indicates the possibility of polymorphic imprinting: the rescue of T-associated maternal effect (Tme) lethality (caused by lack of expression of Igf2 receptor) by mating to males of some *M.m.musculus* strains (Forjet and Gregorova 1992). Nevertheless, there is no direct proof that imprinting is altered in this cross, and the genetics of the system have turned out to be very complex (Forjet et al. 1995). It has been suggested that polymorphic imprinting might exist in the human, because allele-specific expression of IGF2R and WT1 has been observed in some but not all individuals (Kalscheuer et al. 1993; Ogawa et al. 1993b; Polychronakos et al. 1993; Jinno et al. 1994). However, the reasons for this are unclear and may be unrelated to the genetic control of imprinting.

MODIFIERS OF TRANSGENE METHYLATION

Given the difficulty in identifying modifiers that affect endogenously imprinted genes, an alternative approach is to use transgenic systems to identify genes that modify methylation, including (potentially) methylation in imprinted genes. Examples of such modifiers of methylation have been described for various transgenic mouse systems (Sapienza et al. 1989; Allen et al. 1990; Reik et al. 1990; Chaillet et al. 1991; Engler et al. 1991). The key observation is that methylation patterns of such transgenes depend on the genetic background of the inbred strains used. The inference is that different alleles of modifiers are present in different inbred strains. This provides the basis for the genetic mapping and subsequent isolation of the respective genes. So far one strain-specific modifier of transgene methylation, Ssm1, has been mapped successfully to distal chromosome 4 (Engler et al. 1991). Ssm1 modifies transgene pHRD in a position-independent and parent-of-origin-independent manner.

Two other transgenic model systems are more closely related to the control of imprinting. First, the imprinted and maternally methylated

RSV-myc transgene is susceptible to strain-specific modifiers (Chaillet et al. 1995). Specifically, undermethylation of the paternal allele is present in a FVB/N strain background but is lost in a C57BL/6J background, and a dominant FVB/N allele thus seems to prevent methylation of the paternal transgene. This may pinpoint a genetic factor specifically involved in protecting imprinted alleles against de novo methylation, one of the requirements described above.

The second system is that of transgene TKZ751, in which the modifier gene itself appears to be imprinted. We have previously argued that this indicates that the modifier may also act on an endogenous imprinted locus (Reik et al. 1995a). Briefly, this is because genes interacting with imprinted genes are under selective pressure to become imprinted. The fact that the imprinted H19 gene seems to behave as a methylation modifier clearly strengthens this contention. We have begun a detailed genetic and molecular analysis of the TKZ751 transgenic system (Allen et al. 1990). The structure of the TKZ construct and detailed genetic analyses have been reported previously. In contrast to the situation of the RSV-myc (Chaillet et al. 1995) or pHRD (Engler et al. 1991) transgenes, the control of methylation of the TKZ751 transgene is position-dependent, suggesting that the strain-specific modifier action is intimately linked to structures at the site of integration. We have cloned and sequenced parts of the chromosomal regions directly flanking the transgene locus and found that the transgene copies are inserted into clusters of L1 repeat elements (S. Engemann, unpubl.). This specific chromosomal scenario, i.e., the integration into clustered repetitive elements, may be of significance for control by strain-specific modifiers.

As with various other transgenic systems, the degree of TKZ751 methylation is dependent on the mouse strain background. However, the control of TKZ751 methylation shows some unique aspects, which makes it an interesting system with respect to modifier activity. On a BALB/c background, a modifier gene(s) operates which increases transgene methylation, whereas in a DBA/2 background, methylation remains low. The analysis of F_1 progeny from reciprocal crosses between BALB/c and DBA/2 strains has revealed that the BALB/c-specific effect of increasing methylation only occurs when the BALB/c modifier is maternally inherited (Fig. 4A). Further genetic analysis has demonstrated that the BALB/c-specific modifier effect cannot be caused by a maternal oocyte-specific factor, but rather that the activity of the BALB/c-specific modifier itself is controlled by parental imprinting (Allen and Mooslehner 1992). In contrast, the parental origin of the transgene itself does not influence its methylation status. However, for reasons stated above, we

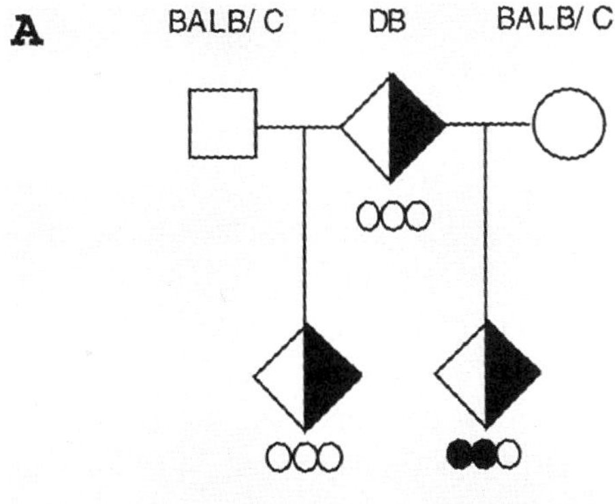

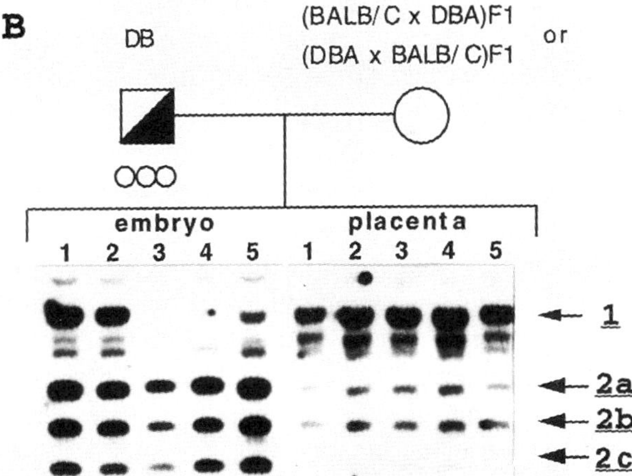

Figure 4 Genetic control of methylation of the TKZ751 transgene locus. Transgenic animals are shown as half-filled symbols: circles = females, squares = males, and diamonds = both sexes. Degrees of methylation of the transgene are given by the amount of filled (methylation) or open (unmethylated) circles below the genetic symbols. (*A*) Methylation of the unmodified transgene (*open circles*) in a DB (DBA/2J) line shows a parental-origin effect when crossed to BALB/c mice. Partial methylation only occurs in crosses to BALB/c females. (*B*) Schematic presentation of the backcross performed for modifier mapping. The methylation phenotypes of progeny from one litter are shown below as determined by *Pst*I-*Mlu*I digestion of chromosomal DNA hybridized with the transgene probe. *Mlu*I digestion is sensitive to methylation. Marker positions: 1 is the *Mlu*I uncleaved transgene copy (only *Pst*I-digested); 2a–c are digestion products of the transgene by *Mlu*I. The methylation profiles of extraembryonic (placenta) and embryonic tissues of individual offspring are shown in parallel. Note that there are large variations of methylation phenotype in embryonic tissues but not in placentae.

assume that the endogenous targets for the modifier are likely to include imprinted genes.

The mechanism of the modifier action remains unclear, but preliminary analysis of transgene methylation in early stages of embryonic development has suggested that the methylation occurs as early as the blastocyst stage (Allen and Mooslehner 1992). The fact that methylation of the transgene is uniform in all embryonic tissues analyzed (N. Allen and J. Walter, unpubl.), and does not show cellular mosaicism, also indicates that the modifier acts at very early developmental stages. However, similar to the situation observed for the pHRD transgene, the influence of strain-specific modifier alleles seems to be confined to embryonic tissues. In all extraembryonic tissues analyzed, methylation has been found to be identical. This is even true in embryos for which extreme methylation phenotypes (hypo- versus hypermethylated) are observed (Fig. 4B).

As a first step toward the molecular understanding of the regulation of TKZ751 modification, we have initiated the genetic mapping of the modifier gene(s) involved. For this purpose, we have set up a backcross between (BALB/c x DBA/2) F_1 females and DBA/2 males carrying the unmethylated transgene. The degrees of transgene methylation from about 150 progeny of this backcross have been determined by Southern blot hybridizations, as shown in Figure 4B. The transgene methylation profiles of the offspring cover a wide spectrum of low, medium, and high methylation phenotypes, suggesting that various combinatorial effects of modifier genes (enhancers and suppressors) operate. DNAs of 65 animals (in which more than 50% of the transgene copies were methylated) have been chosen for genetic mapping using microsatellite markers covering the entire genome at an average spacing of about 15 cM. The mapping data suggest that a main BALB/c-specific modifier of TKZ751 is located on the proximal part of chromosome 17, with 80% of the progeny being heterozygous for the microsatellite marker D17MIT80. The genetic data show that the BALB/c modifier(s) is only active when maternally derived; i.e., the gene(s) appears to be imprinted and maternally expressed. It is interesting to note that the Igf2r gene and the Mas protooncogene are located in close vicinity to D17MIT80 (~2–3 cM apart) and are imprinted and maternally and paternally expressed, respectively (Barlow et al. 1991; Villar and Pederson 1994). Conclusive evidence for the presence of a BALB/c modifier in this region and for the question of whether this gene is maternally expressed (and paternally imprinted) will come from the methylation analysis of progeny derived from DBA/2 females into which the putative BALB/c modifier allele of chromosome 17 has been introduced.

The segregation of methylation phenotypes in the backcross suggests the possibility of several modifier loci contributing to control of methylation. In addition to the chromosome-17 region, our initial mapping indicates that minor additional effects might be contributed by genes on distal chromosome 4 (in the region of Ssm1) and a few other chromosomal regions. Yet these associations have not reached statistical significance. The expansion of our mapping analysis to 300–400 animals will give us further insight into the location and contribution of other modifier loci involved in the control of TKZ751 methylation.

Although the molecular nature of modifier genes remains unclear, the genetic mapping and the methylation results indicate that the control of TKZ751 methylation involves various major and minor modifier genes. It is possible that one molecular basis of TKZ751 methylation control involves the formation of heterochromatin-like structures in the transgene. In view of some similarities between strain-specific methylation variation in mammals and position-effect variegation (PEV) in *Drosophila* (Henikoff 1992; Bestor et al. 1994), the cloning and subsequent testing of mammalian homologs of known modifiers (enhancers and suppressors) of PEV might turn out to be a fruitful complementation to the genetic mapping approaches and eventual positional cloning efforts.

As mentioned above, some observations indicate that imprinting of endogenous genes, e.g., the IGF2R and WT1 genes in the human, might be influenced by genetic background factors. The identification of individual strain-specific "transgene" modifier genes, like Ssm1 or the TKZ751 modifier on chromosome 17, will allow us to address the question of whether such modifiers are also involved in the epigenetic control of endogenous chromosomal sequences and, in particular, imprinted genes.

CONCLUSIONS

Regional regulation of imprinted genes in clusters is an important feature of imprinting control. Enhancer sharing, as exemplified by Igf2–H19, is one of the mechanisms contributing to regional control. We propose that the differentially methylated regions in the Igf2 gene are additional *cis*-acting control elements. The function of these regions, which is modulated by methylation, would be to control full expression or repression of the Igf2 alleles in conjunction with H19 enhancer interactions. Allelic methylation in these regions may be influenced by germ-line imprints and modifier genes, including, directly or indirectly, the H19 gene. Thus, genes like H19 may control regional allelic methylation and could act as genetic components of imprinting centers.

It may not be surprising, therefore, that some modifiers of transgene methylation, like those acting on the transgene locus TKZ751, are themselves imprinted. Although initial mapping experiments indicate the existence of multiple genetic loci interacting additively or substitutively to control the precise methylation pattern of the transgene, one major locus, which may be maternally expressed, has been mapped to proximal chromosome 17, close to the imprinted region containing Igf2r and Mas genes. Imprinted or non-imprinted modifiers of methylation or chromatin may be an integral part of imprinted gene clusters.

ACKNOWLEDGMENTS

We thank the editors of this book for their helpful comments on the manuscript. We are also grateful to Tom Moore for his help on the manuscript. We thank him and Joachim Oswald for permission to include some of their unpublished results. This work was supported by funding from the Biotechnology and Biological Sciences Research Council; Action Research; Ministry of Agriculture, Fisheries and Food; and the European Community.

REFERENCES

Allen, N.A. and K.A. Mooslehner. 1992. Imprinting, transgene methylation and genotype-specific modification. *Semin. Dev. Biol.* **3:** 87–98.

Allen, N.D., M.L. Norris, and M.A. Surani. 1990. Epigenetic control of transgene expression and imprinting by genotype-specific modifiers. *Cell* **61:** 853–861.

Barlow, D.P., R. Stoeger, B.G. Herrmann, K. Saito, and N. Schweifer. 1991. The mouse insulin-like growth factor type-2 receptor is imprinted and closely linked to the *Tme* locus. *Nature* **349:** 84–87.

Bartolomei, M.S. and S.M. Tilghman. 1992. Parental imprinting of mouse chromosome 7. *Semin. Dev. Biol.* **3:** 107–117.

Bartolomei, M.S., S. Zemel, and S.M. Tilghman. 1991. Parental imprinting of the H19 gene. *Nature* **351:** 153–155.

Bartolomei, M.S., A.L. Webber, M.E. Brunkow, and S.M. Tilghman. 1993. Epigenetic mechanisms underlying the imprinting of the mouse H19 gene. *Genes Dev.* **7:** 1663–1673.

Bestor, T.H., V.L. Chandler, and A.P. Feinberg. 1994. Epigenetic effects in eukaryotic gene expression. *Dev. Genet.* **15:** 458–462.

Brandeis, M., T. Kafri, M. Ariel, J.R. Chaillet, J. McCarrey, A. Razin, and H. Cedar. 1993. The ontogeny of allele-specific methylation associated with imprinted genes in the mouse. *EMBO J.* **12:** 3669–3677.

Brannan, C.I., E.C. Dees, R.S. Ingram, and S.M. Tilghman. 1990. The product of the H19 gene may function as an RNA. *Mol. Cell. Biol.* **10:** 28–36.

Chaillet, J.R., D.S. Bader, and P. Leder. 1995. Regulation of genomic imprinting by gametic and embryonic processes. *Genes Dev.* **9:** 1177–1187.

Chaillet, J.R., T.F. Vogt, D.R. Beier, and P. Leder. 1991. Parental-specific methylation of an imprinted transgene is established during gametogenesis and progressively changes during embryogenesis. *Cell* **66:** 77–83.

DeChiara, T.M., A. Efstratiadis, and E.J. Robertson. 1990. A growth-deficiency phenotype in heterozygous mice carrying an insulin-like growth factor II gene disrupted by targeting. *Nature* **345:** 78–80.

DeChiara, T.M., E.J. Robertson, and A. Efstratiadis. 1991. Parental imprinting of the mouse insulin-like growth factor II gene. *Cell* **64:** 849–859.

Efstratiadis, A. 1994. Parental imprinting of autosomal mammalian genes. *Curr. Opin. Genet. Dev.* **4:** 265–280.

Ekström, T.J., H.M. Cui, A. Nystrom, E.M. Rutanen, and R. Ohlsson. 1995. Monoallelic expression of igf2 at the human fetal maternal boundary. *Mol. Reprod. Dev.* **41:** 177–183.

Engler, P., D. Haasch, C.A. Pinkert, L. Doglio, M. Glymour, R. Brinster, and U. Storb. 1991. A strain-specific modifier on mouse chromosome 4 controls the methylation of independent transgene loci. *Cell* **65:** 939–947.

Feil, R., J. Walter, N.A. Allen, and W. Reik. 1994. Developmental control of allelic methylation in the imprinted mouse Igf2 and H19 genes. *Development* **120:** 2933–2943.

Feil, R., M.-A. Handel, N.D. Allen, and W. Reik. 1995. Chromatin structure and imprinting: Developmental control of DNase-I sensitivity in the mouse insulin-like growth factor 2 gene. *Dev. Genet.* **17:** 240–248.

Forejt, J., and S. Gregorová. 1992. Genetic analysis of genomic imprinting: An Imprintor-1 gene controls inactivation of the paternal copy of the mouse Tme locus. *Cell* **70:** 443–450.

Forjet, J., S. Gregorová, M. Landíková, J. Capková, and L.M. Silver. 1995. Genetic variations in parental imprinting on mouse chromosome 17. In *Genomic imprinting—Causes and consequences* (ed. R. Ohlsson et al.), pp. 29–45. Cambridge University Press, England.

Giannoukakis, N., C. Deal, J. Paquette, C.G. Goodyer, and C. Polychronakos. 1993. Parental genomic imprinting of the human IGF2 gene. *Nat. Genet.* **4:** 98–101.

Guillemot, F., T. Caspary, S.M. Tilghman, N.G. Copeland, D.J. Gilbert, N.A. Jenkins, D.J. Anderson, A.L. Joyner, J. Rossant, and A. Nagy. 1995. Genomic imprinting of mash2, a mouse gene required for trophoblast development. *Nat. Genet.* **9:** 235–242.

Henikoff, S. 1992. Position effect and related phenomena. *Curr. Opin. Genet. Dev.* **2:** 907–912.

Jinno, Y., K.K. Yun, K. Nishiwaki, T. Kubota, O. Ogawa, A.E. Reeve, and N. Niikawa. 1994. Mosaic and polymorphic imprinting of the wt1 gene in humans. *Nat. Genet.* **6:** 305–309.

Jinno, Y., Y. Ikeda, K. Yun, M. Maw, H. Masuzaki, H. Fukuda, K. Inuzuka, A. Fujishita, Y. Ohtani, T. Okimoto, T. Ishimaru, and N. Niikawa. 1995. Establishment of functional imprinting of the H19 gene in human developing placentae. *Nat. Genet.* **10:** 318–324.

Kalscheuer, V.M., E.C. Mariman, M.T. Schepens, H. Rehder, and H.-H. Ropers. 1993. The insulin-like growth factor type-2 receptor gene is imprinted in the mouse but not in humans. *Nat. Genet.* **5:** 74–78.

Latham, K.E., A.S. Doherty, C.D. Scott, and R.M. Schultz. 1994. Igf2r and igf2 gene-expression in androgenetic, gynogenetic, and parthenogenetic preimplantation mouse embryos—Absence of regulation by genomic imprinting. *Genes Dev.* **8:** 290–299.

Lee, J.E., U. Tantravahi, A.L. Boyle, and A. Efstratiadis. 1993. Parental imprinting of an Igf-2 transgene. *Mol. Reprod. Dev.* **35:** 382–390.

Leighton, P.A., R.S. Ingram, J. Eggenschwiler, A. Efstratiadis, and S.M. Tilghman. 1995a. Disruption of imprinting caused by deletion of the h19 gene region in mice. *Nature* **375:** 34–39.

Leighton, P.A., J.R. Saam, R.S. Ingram, C.L. Stewart, and S.M. Tilghman. 1995b. An enhancer deletion affects both h19 and igf2 expression. *Genes Dev.* **9:** 2079–2089.

Li, E., C. Beard, and R. Jaenisch. 1993. Role for DNA methylation in genomic imprinting. *Nature* **366:** 362–365.

Liu, J.Q., A.I. Kahri, P. Heikkila, V. Ilvesmaki, and R. Voutilainen. 1995. H19 and insulin-like growth factor-ii gene-expression in adrenal-tumors and cultured adrenal-cells. *J. Clin. Endocrinol. Metabol.* **80:** 492–496.

Moulton, T., T. Crenshaw, Y. Hao, J. Moosikasuwan, N. Lin, F. Dembitzer, T. Hensle, L. Weiss, L. McMorrow, T. Loew, T. Kraus, W. Gerald, and B. Tycko. 1994. Epigenetic lesions at the H19 locus in Wilms' tumour patients. *Nat. Genet.* **7:** 440–447.

Neumann, B., P. Kubicka, and D.P. Barlow. 1995. Characteristics of imprinted genes. *Nat. Genet.* **9:** 12–13.

Ogawa, O., L.A. McNoe, M.R. Eccles, I.M. Morison, and A.E. Reeve. 1993a. Human insulin-like growth factor type I and type II receptors are not imprinted. *Hum. Mol. Genet.* **2:** 2163–2165.

Ogawa, O., M.R. Eccles, J. Szeto, L.A. McNoe, K. Yun, M.A. Maw, P.J. Smith, and A.E. Reeve. 1993b. Relaxation of insulin-like growth factor II gene imprinting implicated in Wilms' tumour. *Nature* **362:** 749–751.

Ohlsson, R., A. Nystrom, O.S. Pfeifer, V. Tohonen, F. Hedborg, P. Schofield, F. Flam, and T.J. Ekstrom. 1993. IGF2 is parentally imprinted during human embryogenesis and in the Beckwith-Wiedemann syndrome. *Nat. Genet.* **4:** 94–97.

Polychronakos, C. 1993. Parental imprinting of the genes for IGF-II and its receptor. *Adv. Exp. Med. Biol.* **343:** 189–203.

Razin, A. and R. Shemer. 1995. DNA methylation in early development. *Hum. Mol. Genet.* **4:** 1751–1755.

Reik, W., S.K. Howlett, and M.A. Surani. 1990. Imprinting by DNA methylation: From transgenes to endogenous gene sequences. *Dev. Suppl.*, pp. 99–106.

Reik, W., R. Feil, N.D. Allen, T. Moore, and J. Walter. 1995a. Imprinted genes, allelic methylation, and imprinted modifier genes of methylation. In *Genomic imprinting—Causes and consequences* (ed. R. Ohlsson and M. Ritzen), pp. 157–170. Cambridge University Press, England.

Reik, W., K.W. Brown, H. Schneid, Y. Lebouc, W. Bickmore, and E.R. Maher. 1995b. Imprinting mutations in the Beckwith-Wiedemann syndrome suggested by an altered imprinting pattern in the igf2-h19 domain. *Hum. Mol. Genet.* **4:** 2379–2385.

Reik, W., K.W. Brown, R.E. Slatter, P. Sartori, M. Elliott, and E.R. Maher. 1994. Allelic methylation of h19 and igf2 in the Beckwith-Wiedemann syndrome. *Hum. Mol. Genet.* **3:** 1297–1301.

Rivkin, M., K.M. Rosen, and K.L. Villa. 1993. Identification of an antisense transcript from the IGF-II locus in mouse. *Mol. Reprod. Dev.* **35:** 394–397.

Sapienza, C., J. Paquette, T.H. Tran, and A. Peterson. 1989. Epigenetic and genetic factors affect transgene methylation imprinting. *Development* **107:** 165–168.

Sasaki, H., P.A. Jones, J.R. Chaillet, S.A. Ferguson, S.C. Barton, W. Reik, and M.A. Surani. 1992. Parental imprinting: Potentially active chromatin of the repressed maternal allele of the mouse insulin-like growth factor II (Igf2) gene. *Genes Dev.* **6:** 1843–1856.

Shemer, R., Y. Birger, W.L. Dean, W. Reik, A.D. Riggs, and A. Razin. 1996. Dynamic methylation adjustment and counting as part of imprinting mechanisms. *Proc. Natl. Acad. Sci.* (in press).

Steenman, M.J., S. Rainier, C.J. Dobry, P. Grundy, I.L. Horon, and A.P. Feinberg. 1994. Loss of imprinting of IGF2 is linked to reduced expression and abnormal methylation of H19 in Wilms' tumour [published erratum appears in *Nat. Genet.* 1994:203]. *Nat. Genet.* **7:** 433–439.

Svensson, K., C. Walsh, R. Fundele, and R. Ohlsson. 1995. H19 is imprinted in the choroid plexus and leptomeninges of the mouse foetus. *Mech. Dev.* **51:** 31–37.

Szabo, P.E. and J.R. Mann. 1995a. Biallelic expression of imprinted genes in the mouse germ line: Implications for erasure, establishment, and mechanism of genomic imprinting. *Genes Dev.* **9:** 1857–1868.

Szabo, P. and J.R. Mann. 1995b. Allelic-specific expression and total expression levels of imprinted genes during early mouse development: Implications for imprinting mechanisms. *Genes Dev.* **9:** 3097–3108.

Taniguchi, T., A.E. Schofield, J.L. Scarlett, I.M. Morison, M.J. Sullivan, and A.E. Reeve. 1995. Altered specificity of IGF2 promoter imprinting during fetal development and onset of Wilms tumour. *Oncogene* **11:** 751–756.

Tremblay, K.D., J.R. Saam, R.S. Ingram, S.M. Tilghman, and M.S. Bartolomei. 1995. A paternal-specific methylation imprint marks the alleles of the mouse h19 gene. *Nat. Genet.* **9:** 407–413.

Villar, A.J. and R.A. Pederson. 1994. Parental imprinting of the *Mas* protooncogene in mouse. *Nat. Genet.* **8:** 373–379.

Vu, H. and A.R. Hoffman. 1994. Promoter-specific imprinting of the human insulin-like growth factor-II gene. *Nature* **371:** 714–715.

Wang, Z.Q., M.R. Fung, D.P. Barlow, and E.F. Wagner. 1994. Regulation of embryonic growth and lysosomal targeting by the imprinted Igf2/Mpr gene. *Nature* **372:** 464–467.

Weksberg, R. 1994. Wiedemann-Beckwith syndrome—Genomic imprinting revisited. *Am. J. Med. Genet.* **52:** 235–236.

Establishment of Imprinted Methylation Patterns during Development

Ruth Shemer and Aharon Razin
Department of Cellular Biochemistry
Hebrew University Hadassah Medical School
Jerusalem, Israel 91120

Genomic imprinting is a classic example of epigenetic control of gene expression in mammals. It represents a process that marks the parental origin of certain genes, resulting in their allele-specific expression. Imprinting, which is implicated in the inequality of the maternal and paternal genomes in mammals, results in the developmental failure of isoparental embryos to develop properly (Surani 1993). An imbalance of specific parental chromosomes in the embryo or aberrant expression of the imprinted genes results in a number of genetic disorders in man and may also contribute to tumor development (Feinberg 1993; Nicholls 1993).

Several model explanations have been proposed for the evolutionary acquisition of genomic imprinting as a developmental control mechanism in mammals (Moore and Haig 1991; Barlow 1993; Varmuza and Mann 1994). One of these models (Moore and Haig 1991) suggests that imprinting might have evolved in mammals because of the conflicting "interests" of maternal and paternal genes within the embryo. This model proposes that paternally expressed genes promote embryonic growth, whereas maternal genes act to restrain the use of maternal resources. A suitable example that corroborates this argument is the paternally expressed Igf2 gene and its counterpart, maternally expressed receptor (Igf2r). The different contribution to the embryonic phenotype of paternal and maternal genes had in fact been demonstrated in parthenogenetic and androgenetic embryos. Although the development of extraembryonic tissues in parthenogenetic embryos harboring two copies of the maternal genome is impaired, the development of the embryo proper is normal. In contrast, the embryo proper in androgenones carrying two copies of the paternal genome does not develop properly whereas extraembryonic tissues develop normally (Surani 1993).

DNA METHYLATION AND IMPRINTING

A large body of evidence strongly suggests that DNA methylation is involved in the imprinting process. First, all imprinted genes studied in the mouse and man show regions that are differentially methylated in an allele-specific manner, and the monoallelic expression of imprinted genes correlates with this allele-specific methylation (Fig. 1). Differential methylation is established by a gradual process that starts in the gametes, continues throughout embryonic development, and involves allele-specific demethylation as well as monoallelic de novo methylation (Fig. 2). In this chapter, we discuss the features of DNA methylation and emphasize its involvement in the imprinting process. The imprinting process can be divided into two steps. The first step is to mark the alleles according to the parental source. The second step is to silence one allele, thereby establishing a parent-specific expression pattern of the imprinted genes.

DNA methylation is known to affect protein/DNA interactions, to control gene expression, and to be clonally inherited. It is therefore a suitable candidate to play a role at all stages of the imprinting process. DNA methylation, being an epigenetic mark on the DNA, can serve as a signal to mark the alleles. This mark can be maintained throughout development by a maintenance methyltransferase (Gruenbaum et al. 1982) and erased during gametogenesis by a global demethylation process (Kafri et al. 1992). This clonally inherited mark can provide the specificity required for the establishment, during early embryonic devel-

Figure 1 Methylation, expression, and chromosomal location of imprinted genes. (*Filled boxes*) Methylated; (*open boxes*) unmethylated. The genes in man are designated by uppercase letters. Genes in the mouse are designated by lowercase letters. Where genes of both mouse and man are mentioned, the uppercase form is used. The methylation status is only indicated in regions that were determined. (N.D.) Not determined. Arrow indicates transcription start. Superscripts designate reference numbers as follows: (1) Bartolomei et al. (1993); (2) Zhang et al. (1993); (3) Feil et al. (1994); (4) Reik et al. (1995); (5) Hatada and Mukai (1995); (6) Shemer et al. (1996); (7) Glenn et al. (1993a); (8) C.C. Glenn et al. (in prep.); (9) Glenn et al. (1993b); (10) Stoger et al. (1993); (11) Hatada et al. (1995); (12) Norris et al. (1994); (13) Guillemot et al. (1995); (14) Giddings et al. (1994); (15) Villar and Pedersen (1994); (16) Kaneko-Ishino et al. (1995); (17) Kuroiwa et al. (1996); (18) Smrzka et al. (1995); (19) Wervick et al. (1994); (20) Sutcliffe et al. (1994). *The Xist gene is paternally expressed in the yolk sac but randomly expressed in other female tissues. The Ins 1 and Ins 2 genes are paternally expressed only in the yolk sac.

opment, of final differential methylation patterns as they are observed in the adult (Fig. 1). In addition to a possible role in marking the alleles, DNA methylation can take part in the control of allele-specific expression. Involvement of DNA methylation in the control of gene activity has

GENE	Differentially methylated regions		Differential expression		Chromosomal location	
	Mouse	Human	Mouse	Human	Mouse	Human
H19[1,2]	p m 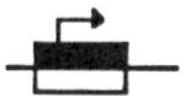	p m 	Maternal	Maternal	7 distal	11p15.5
IGF2[3,4]	m 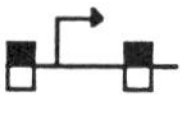	p m 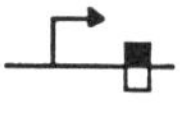	Paternal	Paternal	7 distal	11p15.5
P57/KIP2[5]	p m 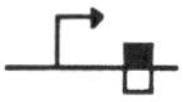	N.D.	Maternal	Maternal	7 distal	11p15.5
SNRPN[6,7,8,20]	p m 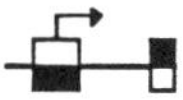	p m 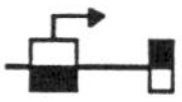	Paternal	Paternal	7 central	15q11q13
ZNF127[9]	N.D.	p m 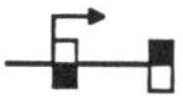	Paternal	Paternal	7 central	15q11q13
IGF2R[10,18]	p m 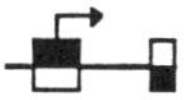	p m 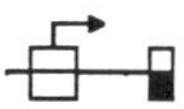	Maternal	Biparental	17 proximal	6q25-q27
U2af1rs1[11]	p m 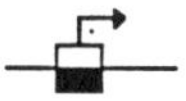	N.D.	Paternal	N.D.	11 proximal	N.D.
Xist[12]	p m 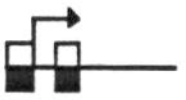	N.D.	Random/ Paternal*	Random/ Paternal*	X	X
Mash 2[13]	N.D.	N.D.	Maternal	N.D.	7 distal	11p15.5
Ins 2[14]	N.D.	N.D.	Paternal*	N.D.	7 distal	11p15.5
Ins 1[14]	N.D.	N.D.	Paternal*	N.D.	6 proximal	N.D.
Mas[15]	N.D.	N.D.	Paternal	N.D.	17 proximal	6q25-q27
Peg1/Mest[16]	N.D.	N.D.	Paternal	N.D.	6 proximal	N.D.
Peg3[17]	N.D.	N.D.	Paternal	N.D.	7 proximal	N.D.
IPW[19]	N.D.	N.D.	N.D.	Paternal	N.D.	15q11q13
PAR 1[20]	N.D.	N.D.	N.D.	Paternal	N.D.	15q11q13
PAR 5[20]	N.D.	N.D.	N.D.	Paternal	N.D.	15q11q13

Figure 1 (*See facing page for legend.*)

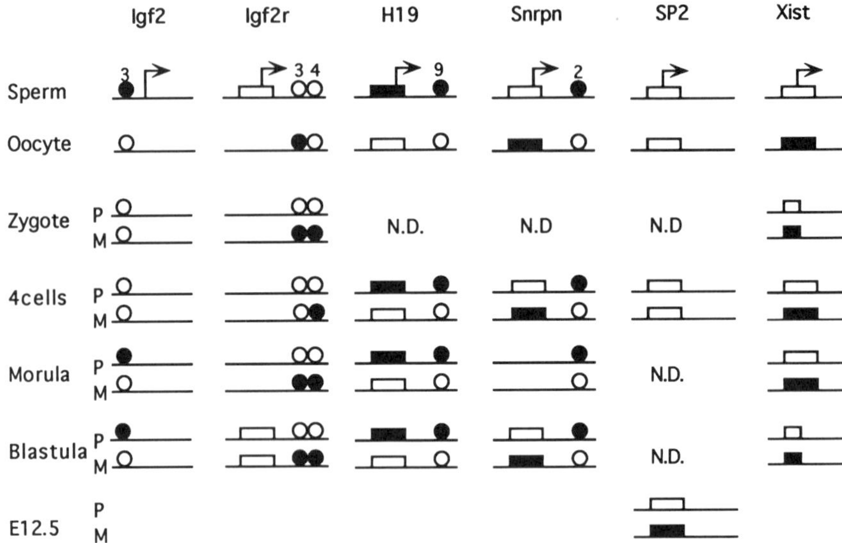

Figure 2 Establishment of differentially methylated patterns in mouse imprinted genes. A compilation of the data available for the status of various differentially methylated sites in a number of imprinted genes at various stages of embryo development. The sites can be classified in three categories: (1) sites that become differentially methylated in the gametes or in the zygote prior to syngamy; (2) sites that undergo demethylation postfertilization and allele-specific de novo methylation before implantation; and (3) sites that undergo allelic methylation in the postimplantation embryo. The first category is represented by site 4 of the intronic differentially methylated region 2 of the Igf2r gene (Stoger et al. 1993), sites in the promoter region of H19 (Bartolomei et al. 1993), the downstream site 9 of the H19 gene, site 2 in exon 5 of Snrpn (Shemer et al. 1996), and sites in the promoter region of the Xist gene (Ariel et al. 1995; Zuccotti and Monk 1995). Sites such as Igf2 site 3 (Brandeis et al. 1993) and Igf2r site 3 (Shemer et al. 1996) belong to the second category. In the third category are the promoter regions of Igf2r, which become methylated on the paternal allele with 4 day neonate (Stoger et al. 1993) and of SP2 (U2af1rs1), which become methylated on the maternal allele by 12.5 dpc (Hatada et al. 1995). The sites are positioned relative to the transcription start (*arrows*). Individual sites (*open circles*, unmethylated; *filled circles*, methylated) are marked by their corresponding designated numbers in the above referred papers. Clusters of 2 or more sites (promoter regions) are represented by rectangles (*open*, unmethylated; *filled*, methylated). (P) Paternal allele; (M) maternal allele; (N.D.) not determined. Arrows indicate transcription start and apply to all developmental stages.

been shown in many in vivo and in vitro systems (Razin and Kafri 1994); therefore, the differentially methylated regions found in all imprinted genes may play a role in differential expression of the imprinted genes. It should be understood, however, that the two suggested roles for DNA methylation in establishing the imprint and controlling differential expression are not necessarily linked. Moreover, if methylation does play a dual role, each role may involve different regions of the imprinted gene. This is probably true in the case of the mouse Igf2r gene and the human and mouse Snrpn gene. In these two genes, the promoter region is methylated on the silent allele, and a downstream region is methylated on the active allele. It is possible that the methylation of the promoter region in these genes is associated with the activity of the gene, and the downstream region serves as the imprinting element harboring the imprinting signal. However, the possibility that the downstream region is involved in expression cannot be ruled out. If that is indeed the case, methylation of this region on the active allele may interfere with the binding of a repressor protein that binds freely to the undermethylated allele, leading to its silencing. The repressor model is supported by several lines of evidence. For example, the 3′exonic region of the Igf2 gene is paternally methylated, and the level of methylation correlates positively with the activity of the gene. The expression of this gene is tissue-specific, being active in fetal liver where it is heavily methylated and repressed in fetal brain where it is unmethylated (Feil et al. 1994). More direct evidence supporting this model comes from the analysis of expression of imprinted genes in DNA-methyltransferase-deficient mice (Li et al. 1993). In these mice, the activity of Igf2r and Igf2 was completely shut off, suggesting that undermethylation results in silencing of these genes. More importantly, the change in the expression pattern of imprinted genes in the methyltransferase-deficient mice strongly suggests that DNA methylation plays an active role in the control of imprinted gene expression.

ALLELE-SPECIFIC METHYLATION PATTERNS IN IMPRINTED GENES

Whereas some imprinted genes, such as H19, are differentially methylated across the entire gene, other imprinted genes are differentially methylated only in discrete regions (Fig. 1). The mouse Igf2 gene contains a region located 3 kb upstream of the transcription start that is preferentially methylated on the active paternal allele (Sasaki et al. 1992; Brandeis et al. 1993), in addition to a paternally methylated region spanning exon 5, intron 5, and the 5′part of exon 6 (Feil et al. 1994). Similarly, a 3-kb intronic region located 27 kb downstream from the promoter region of the mouse Igf2r gene (Stoger et al. 1993) and an exonic region (exon 5) of the mouse Snrpn gene (Shemer et al. 1996) show preferential methylation on the active allele. In addition, two human paternally expressed genes, SNRPN and ZNF127 (Glenn et al. 1993b), have differentially methylated regions that are methylated on the active paternal allele. In these and other imprinted genes such as H19, Xist, and U2af1-rs1(sp2), the promoter region is methylated, as expected, on the silenced allele (Fig. 1).

Two kinds of modification are thus observed. In the first, the activity of the allele is correlated with the unmethylated state of the gene's promoter region, whereas in the second, a non-promoter region is methylated on the active allele. It is important to note that many of the differentially methylated regions in imprinted genes are located within CpG islands. This observation is especially interesting in light of the fact that CpG islands are invariably unmethylated, with the exception of X-linked genes which are methylated on the inactive X chromosome. This unusual methylation of CpG islands in imprinted genes in itself may suggest that methylation plays a role in recognizing imprinted genes.

DIFFERENTIAL METHYLATION PATTERNS IN THE GAMETES

It was suggested above that methylation may serve as an imprinting signal. For that to be the case, the first methylation event should take place when the male and female genomes are in separate entities; namely, during the final steps of gamete maturation or, at the latest, in the zygote prior to syngamy. This is the case with several imprinted genes, Xist, H19, and Igf2r (see Fig. 2). Sites at the 5′end of the Xist gene are modified in the oocyte and unmodified in the sperm; modification of the maternal allele is maintained in the early embryo (Ariel et al. 1995; Zucotti and Monk 1995). Similarly, modification of the 5′end of H19 is established in the gametes and maintained in the embryo (Tremblay et al. 1995). Two sites in the downstream differentially methylated region of Igf2r become methylated in the mature oocyte and are unmethylated in sperm (Stoger et al. 1993). Another site in this region becomes methylated on the maternal allele in the zygote prior to syngamy (Fig. 2).

If methylation does indeed serve as the imprinting signal, it should be maintained throughout embryonic development and erased during gametogenesis. For the vast majority of differentially methylated sites this is what occurs. Primordial germ cells (PGCs) emerge from the epiblast region of the early embryo and are first seen in the base of the allantois at 7.5 days post coitum (dpc) (McCarrey 1993). By 11.5 dpc, the PGCs colonize the developing genital ridges where the imprinted genes are found completely unmethylated (Brandeis et al. 1993). Since imprinted genes in cells of the epiblast region of the early embryo show differential methylation, the unmethylated status of imprinted genes in the 11.5 dpc PGCs suggests that demethylation takes place at some unknown point in the eclipse period when the PGCs migrate from the base of the allantois to the genital ridge. This is in accord with the observation of monoallelic expression of Snrpn in the PGCs at the base of the allantois and biallelic expression of this gene, as well as Igf2 and Igf2r, in postmigrating PGCs (Szabo and Mann 1995a,b).

The Xist gene, which is thought to control X-inactivation, represents an exception to the rule. Sites in the Xist gene are observed to be modified in 12.5–13.5 dpc primordial germ cells (Ariel et al. 1995). This methylation is maintained until 18.5 dpc in both sexes. By 21.5 dpc, however, these sites undergo specific demethylation in the developing spermatogonia, whereas they remain modified throughout oogenesis. The continuous methylation of these sites in the Xist gene throughout oogenesis is in accord with the reactivation of the X chromosome at about 13.5 dpc during oogenesis (Kratzer and Chapman 1981) when Xist is inactive (McCarrey and Dilworth 1992).

The methylation of imprinted genes starts in the gametes, but by no means is this methylation complete. This may explain the biallelic expression of Igf2 and Igf2r seen in the gametes and early embryo (Latham et al. 1994; Szabo and Mann 1995a,b). Differential expression of the imprinted genes may therefore require established differential methylation patterns.

Although there is no direct proof that the gamete-specific methylations of imprinted genes serve as imprinting signals, it is likely, considering that some parent-specific methylation in the gametes is copied and survives the genome-wide demethylation that takes place in the precavitation embryo (Kafri et al. 1993).

ESTABLISHMENT OF DIFFERENTIALLY METHYLATED PATTERNS IN THE NORMAL PREIMPLANTATION EMBRYO

Although the vast majority of sites in imprinted genes that are methylated in the gametes do not escape the global demethylation that takes place in the precavitation embryo, a few specific sites acquire parent-specific methylation in the gametes and maintain their methylation during early development (Fig. 2). As mentioned above, these sites fulfill the requirements of an imprinting signal. It is clear, however, that the establishment of the final differential methylation pattern involves dynamic changes in methylation during embryogenesis. It is possible that these dynamic changes in methylation are guided by the initial methylation pattern in the gametes (Fig. 2).

The allele-specific de novo methylation that takes place in the preimplantation embryo is unique to imprinted genes, since all nonimprinted genes tested are unmethylated at that embryonic stage. Thus, demethylation precedes the global de novo methylation of nonimprinted genes that takes place at the pregastrula stage between 5.5 and 6.5 dpc (Kafri et al., 1992). It should be mentioned that the final differential methylation patterns of imprinted genes as they are observed in the adult may be completed relatively late in development, as is the case for the Igf2r promoter region (Stoger et al. 1993). The stepwise methylation of this gene coincides with its mode of expression. The Igf2r gene is biallelic when expressed in the early embryo; monoallelic expression is exhibited only in the 8.5 dpc embryo.

A deeper insight into the mechanism involved in the establishment of differential methylation patterns of imprinted genes has recently been obtained by experiments using preimplantation embryos lacking complementary sets of chromosomes (Shemer et al. 1996). The methylation status of several sites in differentially methylated regions of imprinted genes has been shown to be similar in diploid parthenogones and androgenones and in tetraploid parthenogenetic preimplantation embryos as well as in parthenogenic and androgenic embryonic stem cells (Feil et al. 1994; Shemer et al. 1996). The level of methylation in all these different cells approached the 50% methylation level characteristic of normal embryos. These observations clearly point to a methylation adjustment mechanism that senses gene dosage by monoallelic demethylation or de novo methylation. Although tetraploid embryos show the same methylation adjustments as in diploids resulting in 50% methylation, haploid parthenogenetic embryos do not show methylation adjustment at all. Haploid embryos maintain the maternal methylation status that was established in the oocyte. The observed methylation adjustment is therefore achieved by a mechanism that counts one allele per genome and thus is similar to the counting mechanism seen in the X-inactivation phenomenon. Genes on the X chromosome in mammalian female cells are known to be inactive on one of the two alleles, and this inactivaiton is associated with methylation (Riggs and Pfiefer 1992). Whereas this inactivation and methylation occurs randomly on one of the X chromosomes in the postimplantation female embryo at about 6.5 dpc, the paternal X chromosome in extraembryonic tissues is specifically inactivated between 3.5 and 4.5 dpc (Takagi and Sasaki 1975). As mentioned above, the Xist gene, which is located at the X inactivation center (Xic), is monoallelically transcribed from the inactive X and is thought to inactivate the X chromosome *in cis*. Accordingly, in extraembryonic tissue, Xist is transcribed exclusively from the paternal X.

ESTABLISHING THE IMPRINT

The primary imprint may be established by differential methylation in the gamete at discrete loci. This differential methylation is maintained throughout early embryo development and guides the gradual establishment of the final differential methylation pattern. Alternatively, if methylation is not the primary signal, gamete-specific protein factors may recognize the imprinted genes and bind specifically to discrete sequences during late stages of gametogenesis. These proteins may mark the alleles until differentially methylated patterns are established. Although a specific protein may be the initial mark, it is difficult to imagine how this mark could then be clonally inherited. Therefore, according to this model, differential methylation can play a role in stabilizing the imprint and regulating differential expression, rather than being the primary signal.

As argued above, it is more likely that differential methylation serves as an imprinting marker, but no direct evidence exists to prove this point. Similarly, no direct evidence exists to show that methylation directs allele-specific expression of the imprinted genes. In the mouse U2af1-rs1 (sp2) gene, for example, allele-specific expression is already observed in the two-cell stage. This differential expression is not accompanied by allele-specific methylation. Therefore, at least in this case, the paternal methylation of the upstream region of this gene probably plays a role in the maintenance of differential expression rather than in establishment of paternal expression of the gene. It is possible that at this early stage in development specific protein factors of gametic origin still bind to the DNA and guide the differential expression of the gene (Hatada et al. 1995). However, it should also be noted that methylation was analyzed only on the upstream region, and other differentially methylated regions might already exist at an earlier stage in development.

REGIONAL CONTROL ELEMENTS INVOLVED IN ESTABLISHING THE IMPRINT

In addition to the local control elements in the immediate domain of the gene, the control of genomic imprinting may also involve regional control elements. This is suggested by the existence of large chromosomal regions in the mouse and human genomes that show differential expression of genes in these regions. In fact, the great majority of the known imprinted genes are clustered in reciprocally imprinted regions. Such a cluster is found in the distal part of mouse chromosome 7 (Zemel et al. 1992; Guillemot et al. 1995). In this cluster, four imprinted genes have been identified in a region spanning about 350 kb. The genes are Mash 2 (Guillemot et al. 1995), H19 (Bartolomei et al. 1991), and Ins2 and Igf2 (DeChiara et al. 1991; Giddings et al. 1994; see Avner, this volume). Another such cluster is found in the PWS/AS region on human chromosome 15 spanning about 300 kb. In this region, the following genes have been identified: SNRPN (Reed and Leff 1994), IPW (Wevrick et al. 1994), PAR-5, and PAR-1 (Sutcliff et al. 1994). This region in the human genome has a syngeneic domain on the centromeric part of mouse chromosome 7 that includes the mouse Snrpn gene.

The idea of regional control elements is supported by the analysis of replication timing and recombination rates of imprinted domains. Experimental evidence suggests that large chromosomal domains that include imprinted genes show asynchronous replication, with the paternal allele replicating earlier (Kitsberg et al. 1993; Knoll et al. 1994). In addition, imprinted regions were shown to recombine with very high frequency during male meiosis and with very low frequency during female meiosis (Paldi et al. 1995). This asynchronous replication timing and differential recombination frequency presumably reflect parent-of-origin-specific differences in the chromatin structure. For example, the paternal copy may be euchromatic in nature, whereas the maternal copy may be more heterochromatic. This may be analogous to the phenomenon of position-effect variegation observed in *Drosophila* (Hendrich and Willard 1995). A chromatin model has also been suggested to explain imprinting of Igf2/H19 genes (Banerjee and Smallwood 1995).

Recently an "imprinting center" model has been suggested, based on two clinically distinct syndromes, Prader-Willi syndrome (PWS) and Angelman syndrome (AS) located on human chromosome 15. In patients with these disorders, monoparental deletions of this region overlapping the SNRPN gene, or near it, are accompanied by altered methylation and expression of genes at loci distal to the chromosomal rearrangement (Glenn et al. 1993b; Reis et al. 1994; Sutcliffe et al. 1994; Buiting et al. 1995). These observations suggested that an imprinting control element (IC) is located near the SNRPN gene (Lalande 1994; Nicholls 1994). Mice deleted for H19 failed to imprint nearby genes, suggesting a role of H19 in the IC model (Leighton et al. 1995). If such an imprinting center exists, it is possible that imprinted genes in the controlled domain are regulated by intertalk between the postulated IC and the imprinting boxes located within the genes. Suitable candidates for imprinting boxes are region 2 in the Igf2r gene, exon 5 in Snrpn, and exon 6 in Igf2. These gene regions are differentially methylated and may harbor the imprinting signal. This regulation *in cis* may be analogous to the situation observed in the inactivation of X chromosome in female mammals, where the inactivation spreads from a single X-inactivation center. The inactive X chromosome has a more condensed chromatin structure, is late-replicating, and is methylated; most genes on the inactive X are silenced.

Further studies will be required to substantiate the possibility that methylation is the imprinting mark used to determine the differential expression of imprinted genes and to clarify the involvement of the IC in the establishment of the differential methylation of imprinted genes.

CONCLUSIONS

All imprinted genes tested for methylation show differential methylation. Most genes are differentially methylated at discrete regions, at the 5′ and 3′ parts of the gene. In many cases, a 3′ region is methylated on the expressed allele, and the 5′ end of the gene is methylated on the inactive allele. An exception to this rule exists in the RNA-producing genes, H19 and Xist, which are methylated throughout on the inactive allele.

The differential methylation patterns are established gradually during embryo development. This process starts with specific sites that are positioned in differentially methylated domains, generally at the 3′ end of the gene. These sites inherit their methylation status from the gametes and may guide the allele-specific methylation of other sites. The formation of the differential methylation patterns is completed, sometime after birth, by establishing the differentially methylated status of the 5′ end of the gene.

It is therefore postulated that the methylation at the 3′ end which is inherited from the gametes serves as the imprinting signal. Once allele-specific expression is established, the methylation at the 5′ end is formed to secure differential expression of the gene.

If this model proves to be correct, many questions remain to be solved. The most interesting question is, How is the imprinting signal recognized and the information translated into allele-specific expression? It is clear that *trans*-acting protein factors or chromatin structure is involved in such a complex process. These factors must be identified and characterized before we can understand the imprinting phenomenon.

ACKNOWLEDGMENTS

The work done in the authors' laboratory was supported by grants from the National Institutes of Health and the Council for Tobacco Research.

REFERENCES

Ariel, M., E. Robinson, J.R. McCarrey, and H. Cedar. 1995. Gamete-specific methylation correlates with imprinting of the murine *Xist* gene. *Nat. Genet.* **9:** 312–315.

Banerjee, S. and A. Smallwood. 1995. A chromatin model of IGF2/H19 imprinting. *Nat. Genet.* **11:** 237–238.

Barlow, D.P. 1993. Methylation and imprinting: From host defense to gene regulation? *Science* **260:** 153–155.

Bartolomei, M.S., S. Zemel, and S.M. Tilghman. 1991. Parental imprinting of the mouse H19 gene. *Nature* **351:** 153–155.

Bartolomei, M.S., A.L. Webber, M.E. Brunkow, and S.M. Tilghman. 1993. Epigenetic mechanisms underlying the imprinting of the mouse H19 gene. *Genes Dev.* **7:** 1663–1673.

Brandeis, M., T. Kafri, M. Ariel, J.R. Chaillet, J. McCarrey, A. Razin, and H. Cedar. 1993. The ontogeny of allele-specific methylation associated with imprinted genes in the mouse. *EMBO J.* **12:** 3669–3677.

Buiting, K., S. Saitoh, S. Gross, B. Dittrich, S. Schwartz, D.R. Nicholls, and B. Horsthemke. 1995. Inherited microdeletions in the Angelman and Prader-Willi syndromes define an imprinting centre on human chromosome 15. *Nat. Genet.* **9:** 2079–2089.

DeChiara, T.M., E.J. Robertson, and A. Efstratiadis. 1991. Parental imprinting of the mouse insulin-like growth factor II gene. *Cell* **64:** 849–859.

Feil, R., J. Walter, N. Allen, and W. Reik. 1994. Developmental control of allelic methylation in the imprinted mouse Igf2 and H19 genes. *Development* **120:** 2933–2943.

Feinberg, A. 1993. Genomic imprinting and gene activation in cancer. *Nat. Genet.* **4:** 110–113.

Giddings, S.J., C.D. King, K.W. Harman, J.F. Flood, and L.R. Carnaghi. 1994. Allele specific inactivation of insulin 1 and 2, in the mouse yolk sac, indicates imprinting. *Nat. Genet.* **6:** 310–313.

Glenn, C.C., K.A. Porter, M.T.C. Jong, R.D. Nicholls, and D.J. Driscoll. 1993a. Functional imprinting and epigenetic modification of the human SNRPN gene. *Hum. Mol. Genet.* **2:** 2001–2005.

Glenn, C.C., R.D. Nicholls, W.P. Robinson, S. Saitoh, N. Niikawa, A. Schinzel, B. Horsthemke, and D.J. Driskoll. 1993b. Modification of 15q11-q13 DNA methylation imprints in unique Angelman and Prader-Willi patients. *Hum. Mol. Genet.* **2:** 1377–1382.

Gruenbaum, Y., H. Cedar, and A. Razin. 1982. Substrate and sequence specificity of a eukaryotic DNA methylase. *Nature* **195:** 620–622.

Guillemot, F., T. Caspary, S.M. Tilghman, N.G. Copeland, D.J. Gilbert, N.A. Jenkins, D.J. Anderson, A.L. Joyner, J. Rossant, and A. Nagy. 1995. Genomic imprinting of *Mash2*, a mouse gene required for trophoblast development. *Nat. Genet.* **9:** 235–241.

Hatada, I. and T. Mukai. 1995. Genomic imprinting of $p57^{KIP2}$, a cyclin-dependent kinase inhibitor, in mouse. *Nat. Genet.* **11:** 204–206.

Hatada, I., K. Kitagawa, T. Yamnoka, X. Wang, Y. Arai, K. Hashido, S. Ohishi, J.

Masuda, J. Ogata, and T. Mukai. 1995. Allele-specific methylation and expression of an imprinted U2af1-rs1(SP2) gene. *Nucleic Acids Res.* **23:** 36–41.

Hendrich, D.B. and F.H. Willard. 1995. Epigenetic regulation of gene expression: The effect of altered chromatin structure from yeast to mammals. *Hum. Mol. Genet.* **4:** 1765–1777.

Kafri, T., X. Gao, and A. Razin. 1993. Mechanistic aspects of genome-wide demethylation in the preimplantation mouse embryo. *Proc. Natl. Acad. Sci.* **90:** 10558–10562.

Kafri, T., M. Ariel, M. Brandeis, R. Shemer, L. Urven, J. McCarrey, H. Cedar, and A. Razin. 1992. Developmental pattern of gene specific DNA methylation of the mouse embryo and germ line. *Genes Dev.* **6:** 705–714.

Kaneko-Ishino, T., Y. Kuroiwa, N. Miyoshi, T. Kohda, R. Suzuki, M. Yokoyama, S. Viville, S.C. Barton, F. Ishino, and A.M. Surani. 1995. Pegl/Mest imprinted gene on chromosome 6 identified by cDNA subtraction hybridization. *Nat. Genet.* **11:** 52–59.

Kitsberg, D., S. Selig, M. Brandeis, I. Simon, I. Keshet, D.J. Driscoll, R.D. Nicholls, and H. Cedar. 1993. Allele-specific replication timing of imprinted gene regions. *Nature* **364:** 459–463.

Knoll, J.H.M., S.D. Cheng, and M. Lalande. 1994. Allele specific of DNA replication timing in the Angelman/Prader-Willi syndrome imprinted chromosomal region. *Nat. Genet.* **6:** 41–45.

Kratzer, P.G. and V.M. Chapman. 1981. X-chromosome reactivation in oocyte of *Mus caroli. Proc. Natl. Acad. Sci.* **78:** 3093–3097.

Kuroiwa, Y., T. Kaneko-Ishino, F. Kagitani, T. Kohda, L.-L. Li, M. Tada, R. Suzuki, M. Yokoyama, T. Shiroishi, S. Wakana, S.C. Barton, F. Ishino, and A.M. Surani. 1996. Peg3 imprinted gene on proximal chromosome 7 encodes for a zinc finger protein. *Nat. Genet.* **12:** 186–190.

Lalande, M. 1994. In and around SNRPN. *Nat. Genet.* **8:** 5–7.

Latham, K.E., A.S. Doherty, C.D. Scott, and R.M. Schultz. 1994. Igf2r and Igf2 gene expression in androgenetic, gynogenetic, and parthenogenetic preimplantation mouse embryos: Absence of regulation by genomic imprinting. *Genes Dev.* **8:** 290–299.

Leighton, P.A., J.R. Saam, R.S. Ingram, C.L. Stewart, and S.M. Tilghman. 1995. An enhancer deletion affects both H19 and Igf2 expression. *Genes Dev.* **9:** 2079–2089.

Li, E., C. Beard, and R. Jaenisch. 1993. Role for DNA methylation in genomic imprinting. *Nature* **366:** 2282–2292.

McCarrey, J.R. 1993. Development of the germ cell. In *The testis: Cell and molecular biology of the testis* (ed. C. Desjardinsand and L.L. Ewing), vol. 5, chap. 3. Oxford University Press, New York.

McCarrey, J.R. and D.D. Dilworth. 1992. Expression of Xist in mouse germ cells correlates with X-chromosome inactivation. *Nat. Genet.* **3:** 200–203.

Moore, T. and D. Haig. 1991. Genomic imprinting in mammalian development: A parental tug of war. *Trends Genet.* **7:** 45–49.

Nicholls, R.D. 1993. Genomic imprinting and candidate genes in the Prader-Willi and Angelman syndromes. *Curr. Opin. Genet. Dev.* **3:** 445–456.

———. 1994. New insights reveal complex mechanisms involved in genomic imprinting. *Am. J. Hum. Genet.* **54:** 733–740.

Norris, D.P., D. Patel, G.F. Kay, G.D. Penny, N. Brockdorff, S.A. Sheardown, and S. Rastan. 1994. Evidence that random and imprinted *Xist* expression is controlled by preemptive methylation. *Cell* **77:** 41–51.

Paldi, A., G. Gyapay, and J. Jami. 1995. Imprinted chromosomal regions of the human genome display sex-specifc meiotic recombination frequencies. *Curr. Biol.* **5:** 1030–1035.

Razin, A. and T. Kafri. 1994. DNA methylation from embryo to adult. *Prog. Nucleic Acids Res. Mol. Biol.* **48:** 53–81.

Reed, L.M. and E.S. Leff. 1994. Maternal imprinting of human *SNRPN*, a gene deleted in Prader-Willi syndrome. *Nat. Genet.* **6:** 163–167.

Reik, W., K.W. Brown, H. Schneid, Y. LeBouc, W. Bickmore, and E.R. Maher. 1995. Imprinting mutations in the Beckwith-Wiedemann syndrome suggested by an altered imprinting pattern in the Igf2-H19 domain. *Hum. Mol. Genet.* **4:** 2379–2385.

Reis, A., B. Dittrich, V. Greger, K. Buiting, M. Lalande, G. Gillessen-Kaesbach, and M. Anvert. 1994. Imprinting mutations suggested by abnormal DNA methylation patterns in familial Angelman and Prader-Willi syndromes. *Am. J. Hum. Genet.* **54:** 741–747.

Riggs, A.D. and G.D. Pfeifer. 1992. X-chromosome inactivation and cell memory. *Trends Genet.* **8:** 169–174.

Sasaki, H., P.A. Jones, J.R. Chaillet, A.C. Ferguson-Smith, S.R. Barton, W. Reik, and M.A. Surani. 1992. Parental imprinting: Potentially active chromatin of the repressed maternal allele of the mouse insulin-like growth factor II (Igf2). *Genes Dev.* **6:** 1843–1856.

Shemer, R., Y. Birger, W.L. Dean, W. Reik, A.D. Riggs, and A. Razin. 1996. Dynamic methylation adjustment and counting as part of imprinting mechanisms. *Proc. Natl. Acad. Sci.* (in press).

Smrzka, O.W., I. Fae, R. Stoger, R. Kurzbauer, G.F. Fischer, T. Henn, A. Weith, and D.P. Barlow. 1995. Conservation of a maternal-specific methylation signal at the human IGF2R locus. *Hum. Mol. Genet.* **4:** 1945–1952.

Stoger, R., P. Kubicka, C.-G. Liu, T. Kafri, A. Razin, H. Cedar, and D.P. Barlow. 1993. Maternal specific methylation of the imprinted mouse Igf2r locus identifies the expressed locus as carrying the imprinted signal. *Cell* **73:** 61–71.

Surani, M.A. 1993. Silence of the genes. *Nature* **366:** 302–304.

Sutcliffe, J.S., M. Nakao, S. Christian, K.H. Orstavik, N. Tommerup, D.H. Ledbetter, and A.L. Beaudet. 1994. Deletions of a differentially methylated CpG island at the *SNRPN* gene define a putative imprinting control region. *Nat. Genet.* **8:** 52–58.

Szabo, P.E. and J.R. Mann. 1995a. Allele-specific expression and total expression levels of imprinted genes during early mouse development: Implications for imprinting mechanisms. *Genes Dev.* **9:** 1–12.

———. 1995b. Biallelic expression of imprinted genes in the mouse germ line: Implications for erasure, establishment, and mechanisms of genomic imprinting. *Genes Dev.* **9:** 1857–1868.

Takagi, N. and M. Sasaki. 1975. Preferential inactivation of the paternally derived X-chromosome in the extraembryonic membranes of the mouse. *Nature* **256:** 640–642.

Tremblay, K.D., J.R. Saam, R.S. Ingram, S.M. Tilghman, and M.S. Bartolomei. 1995. A paternal-specific methylation imprint marks the alleles of the mouse H19 gene. *Nat. Genet.* **9:** 407–413.

Varmuza, S. and M. Mann. 1994. Genomic imprinting—Defusing the ovarian time bomb. *Trends Genet.* **10:** 118–123.

Villar, A.J. and R.A. Pedersen. 1994. Parental imprinting of the Mas protooncogene in mouse. *Nat. Genet.* **8:** 373–379.

Wevrick, R., J.A. Kerns, and U. Francke. 1994. Identification of a novel paternally expressed gene in the Prader-Willi syndrome region. *Hum. Mol. Genet.* **3:** 1877–1882.

Zemel, S., M.S. Bartolomei, and S.M. Tilghman. 1992. Physical linkage of two mammalian imprinted genes, H19 and insulin-like growth factor 2. *Nat. Genet.* **2:** 61–65.

Zhang, Y., T. Shields, T. Crenshaw, Y. Hao, T. Moulton, and B. Tycko. 1993. Imprinting of human H19: Allele-specific CpG methylation. Loss of the active allele in Wilms tumor and potential for somatic allele switching. *Am. J. Hum. Genet.* **53:** 113–124.

Zuccotti, M. and M. Monk. 1995. Methylation of the mouse *Xist* gene in sperm and eggs correlates with imprinted *Xist* expression and paternal X-inactivation. *Nat. Genet.* **9:** 316–320.

X-Chromosome Inactivation and Epigenetic Mechanisms

Arthur D. Riggs and Thomas N. Porter
Biology Department
Beckman Research Institute of the City of Hope
Duarte, California 91010

In each female mammalian cell, one of the two X chromosomes is genetically silent and heterochromatin-like, as a result of a process called X-chromosome inactivation (XCI). Transcriptional silencing by XCI is efficient, stable, and somatically heritable, yet genes on the inactive X chromosome (the X_i) retain the ability to function in the next generation. In mammals, XCI was the first recognized epigenetic phenomenon that, by definition, does not involve irreversible alteration of base sequence. Instead of sequence changes, epigenetic phenomena such as XCI are usually assumed to depend on heritable chromatin structure for the mitotic or meiotic transfer of information from parent to progeny. Ohno et al. (1959) first correctly deduced that the Barr body, a dark-staining, condensed object seen in interphase nuclei, was a single X chromosome. Soon thereafter, Lyon (1961) cogently explained some unusual mouse coat-color genetics as the result of epigenetic inactivation of one X chromosome. We now know that mammals are cellular mosaics, with approximately half of somatic cells expressing only paternally derived X-linked genes and half expressing only maternally derived X-linked genes. XCI thus causes monoallelic expression of X-linked genes in diploid cells, a feature shared with imprinted autosomal genes (see chapters on genomic imprinting by Ainscough and Surani; Walter et al.; both this volume). In this review we focus on the question, How is this accomplished at the molecular level? Although this question cannot yet be fully answered, we review more than 30 years of research on XCI that has led to a considerable body of experimental data relevant to epigenetic mechanisms and heritable chromatin structure.

XCI and genomic imprinting share several features or properties in addition to the ones mentioned above, and these include: (1) control by *cis*-acting elements that regulate domains (or the entire X chromosome) rather than individual genes; (2) cytosine methylation that is essential for maintenance and perhaps establishment; (3) CpG islands that are methylated in the inactive X chromosome and in the elements hypothesized to

Epigenetic Mechanisms of Gene Regulation

be primary imprint centers; (4) earlier replication in S phase of the active allele relative to the inactive one; and (5) a counting mechanism and methylation adjustment that can be seen in unusual cases (see Hendrich and Willard 1995; Shemer et al. 1996). Because of these similarities, it is very likely that XCI and genomic imprinting are evolutionarily related and share common mechanisms.

FUNDAMENTALS OF X-CHROMOSOME INACTIVATION

Figure 1 schematically illustrates X-chromosome inactivation, which is a chromosome-wide phenomenon unique to mammals. By the cleavage stage after fertilization, at least some genes on both X chromosomes are active. Biallelic expression continues until differentiation from totipotency, at which time cytodifferentiation signals trigger the genetic silencing of most genes on one X chromosome. The resulting monoallelic expression thus compensates for gene-dosage differences between XX female and XY male cells. For a comprehensive summary and discussion of the earlier literature, a review by Grant and Chapman (1988) is recommended. XCI also has been the subject of several more recent reviews (Riggs and Pfeifer 1992; Lyon 1993; Singer-Sam and Riggs 1993; Gartler and Goldman 1994; Rastan 1994; Hendrich and Willard 1995). Because of space limitations, most original papers are not cited if they were published prior to 1990 and included in the above reviews.

Cytodifferentiation and X-chromosome inactivation take place earliest in the extraembryonic tissues such as the trophectoderm and primitive endoderm, at 3.5 and 4.5 days post coitus, respectively. In these lineages, XCI is not random; instead, the paternal X chromosome is preferentially silenced, providing a clear example of genomic imprinting. In the embryonic lineage that forms the adult mouse, XCI does not take place until late blastocyst, at about the time of placental implantation at 5.5–6.5 days post coitus. By this time, the imprinting mark that differentiates the paternal X from the maternal X has been erased; thus, inactivation is random with respect to paternal source in the progenitors of all somatic cells. Once chromosomal differentiation is established at the late blastocyst stage, the X_i becomes a somatically heritable entity. The same X is maintained inactive in all progeny of a given cell, and, as a result, mammalian females are cellular mosaics with respect to X-linked genes. Macroscopic monoclonal patches are seen in most tissues, including the skin (Lyon 1961) and bladder epithelium (Tsai et al. 1995). The stability of the inactive X is truly remarkable (for reviews, see Grant and Chapman 1988; Gartler and Goldman 1994). For example, the reactivation of

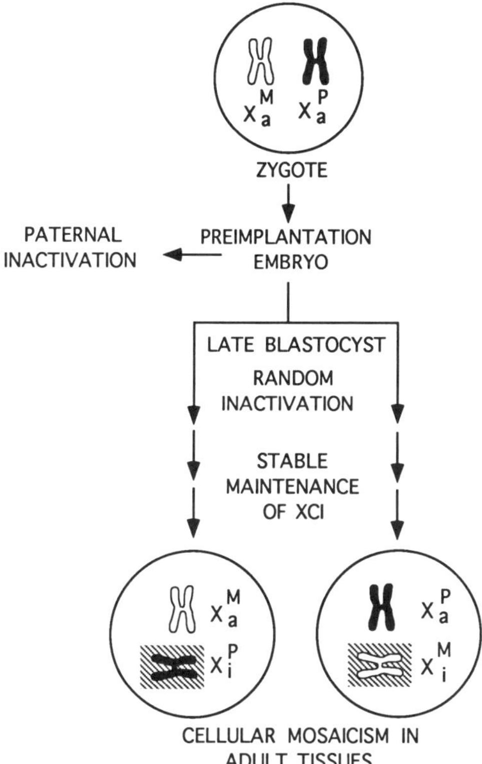

Figure 1 Schematic overview of X-chromosome inactivation (XCI). X^M and X^P depict maternally and paternally derived X chromosomes, respectively. X_a and X_i depict active and inactive X chromosomes, respectively. The X^M is white; the X^P is black. The shaded box represents the heterochromatinization which correlates with genetic inactivation.

the HPRT gene on the X_i is undetectable (10^{-8} or less) in normal human cells in culture, and there is no evidence for reactivation in the course of a human lifetime, during which roughly 10^6 cells are dividing every second. In transformed cells or in cross-species hybrid cells, the human X_i is significantly less stable, yet the spontaneous reactivation rate is 10^{-6} or less.

Table 1 lists some of the differences between the active and inactive X chromosomes (the X_a and X_i, respectively). Cytologically and functionally, the X_a is very similar to autosomes and seems to be a normal euchromatic chromosome. Since the X_i is genetically silent and appears similar to heterochromatin, heterochromatinization is said to occur as part of the inactivation process, and the X_i is considered to be facultative

Table 1 Differences between the X_a and X_i

Property or feature	X_a/X_i differences[a]
Transcriptional silence	most genes on X_i are inactive, a few are not
Appearance in interphase	X_i is more condensed, like heterochromatin
Chromatin structure	little known; positioned nucleosomes likely on PGK promoter of X_i; general DNase I sensitivity is similar
Mitotic chromosomes	histone H4 is relatively hypoacetylated on X_i; transcription factors absent from X_a and X_i
Xist/XIST	only transcribed from X_i
Replication timing	X_a replicates in 1st half of S, X_i in 2nd half
Cytosine methylation	many CpG islands are heavily methylated only on X_i; Xist promoter is opposite, methylated on X_a

[a]See text for more details and references.

or conditional heterochromatin. The X_a and X_i replicate asynchronously; in somatic cells the X_i replicates later in S phase than the X_a.

With regard to epigenetics and heritable chromatin structure, the two most difficult to explain, and thus most fascinating, features of XCI are (1) the differential treatment of identical alleles in the same nucleus and (2) the *cis*-limited spreading of inactivation that encompasses an entire chromosome of more than 10^8 bp (5% of the human genome).

The differential treatment of sequence-identical genes is seen clearly in inbred mice, in which XCI occurs normally. The X_i is in the same nucleus with the X_a, and yet ubiquitously expressed housekeeping genes such as phosphoglycerate kinase (*PGK1*, human; *Pgk-1*, mouse) and hypoxyanthine phosphoribosyltransferase (*HPRT*, human; *Hprt*, mouse) are actively transcribed only from the X_a. The cognate gene on the X_i is not detectably expressed. Active expression from the X_a means that all necessary transcription factors for full expression are present in the nucleus. Does the chromatin structure of the X_i completely restrict access of these factors to the promoters and control elements on the X_i? If these factors are not excluded, then what mechanism ensures the silencing of genes on the X_i? Whatever the reason, it is clear that the X_i displays a type of immunity to the effects of transcription factors.

The second important feature of XCI is that inactivation is *cis-limited.* To be subject to inactivation, DNA must be in physical continuity with a locus called the X-inactivation center (XIC), located at human Xq13 (Brown et al. 1991b). A similar locus (Xic or Xce) has been mapped on the mouse X chromosome (see Avner, this volume). X-linked genes separated from the X inactivation center are not inactivated, and autosomal genes translocated to the X chromosome are subject to inactivation. The inactive chromatin configuration is able to "spread" into autosomal chromatin, although the extent of spreading is limited, variable, and probably not as stable as it would be for normal X chromatin (Schanz and Steinbach 1989). Recently, a gene mapping within the XIC region has been identified which, in contrast to all other known X-linked genes, is expressed only from the inactive X chromosome (Brown et al. 1991a). For this reason, the gene has been named *XIST* (*Xist* in mouse), for X-inactivation-specific transcript. The transcript is large (15–17 kb) and seems not to be translated. Although its mechanism of action is completely unknown, the *XIST/Xist* gene is likely to be involved in, and necessary for, the establishment of XCI (Jani et al. 1995; Penny et al. 1996), but it seems not to be necessary for the maintenance of XCI (Brown and Willard 1994).

DNA METHYLATION AND X-CHROMOSOME INACTIVATION

It is generally accepted that methylation of CpG islands (regions of higher than average density of CpG dinucleotides) is essential for the efficient maintenance of XCI. Much of the evidence for this is not new and has been comprehensively reviewed by Grant and Chapman (1988) and Singer and Riggs (1993). The following observations should be noted: (1) CpG islands on the X_i tend to be heavily methylated whereas the same islands on the X_a are completely unmethylated. (2) X_i-derived DNA functions less well for DNA-mediated transfection than does X_a-derived DNA. (3) DNA cytosine methylation inhibitors such as 5-azacytidine or 5-azadeoxycytidine are the only known exogenous agents that efficiently cause reactivation of X-linked genes, as much as 10^4-fold over the spontaneous rate. (4) Demethylation, followed by changes in chromatin nuclease sensitivity, precedes transcription during the reactivation of HPRT by 5-azacytidine (Sasaki et al. 1992). (5) In marsupials, X-linked CpG islands at genes so far tested are not heavily methylated, and these genes are not kept stably inactivated either in culture or in vivo. (6) Transgenic embryos with a severe deficiency in DNA methyltransferase are not able to maintain genomic imprinting, nor are they able to properly control the expression of Xist (Beard et al. 1995), an imprinted, X-linked gene critical gene for the establishment of X inactivation (Penny et al. 1996).

Reactivation of X-linked genes by methylation inhibitors is piecemeal, not chromosome-wide, suggestive of regional or domain-wide control. XCI can be reversed by fusing mouse lymphocytes with teratocarcinoma cells (Takagi et al. 1983) or by fusing transformed mouse somatic cells with human chorionic villi cells (Luo et al. 1995). In these cases, reactivation is thought to be chromosome-wide.

Methylation and Chromatin Structure

With regard to methylation and chromatin structure at an X-linked gene, the best-studied promoter is that of human PGK1. Studies done at nucleotide-level resolution by ligation-mediated polymerase chain reaction (PCR) have established that the active promoter on the X_a is completely unmethylated and shows clear footprints for several transcription sites (Pfeifer et al. 1990a; Pfeifer and Riggs 1991; Riggs and Pfeifer 1992). In contrast, the inactive PGK1 promoter on the X_i is covered by two positioned nucleosomes or nucleosome-sized complexes and is methylated at 60 of 61 CpG sites in an 800-bp region. The large number of CpG sites is consistent with the promoter of PGK1 being a typical CpG island. A cardinal feature of autosomal CpG islands is that, although they consist of numerous potentially methylatable sites, they are maintained in a methylation-free state (Antequera and Bird 1993). The heavy methylation of the PGK1 island on the X_i dramatically contrasts with autosomal CpG islands and exemplifies what seems to be a general feature of XCI—methylation of X-linked CpG islands (Norris et al. 1991). Differential methylation of CpG islands is also a candidate for the primary mark that influences transcriptional control of imprinted genes (Li and Jaenisch 1993; Leighton et al. 1996).

Does methylation maintain genetic silence directly by inhibiting the binding of transcription factors or indirectly by affecting some aspect of chromatin structure? Little is known about the effect of methylation on chromosome structure other than its influence on the binding of many protein factors. As reviewed by Ehrlich and Ehrlich (1993), the binding of more than 20 proteins, mostly transcription factors, is now known to be directly affected by methylation of their cognate DNA-binding sites. Usually DNA binding is inhibited, but sometimes it is enhanced. In addition, there is evidence that methylation of CpG-rich promoters, including X-linked PGK, results in inhibition of in vitro transcription by an indirect mechanism, probably involving proteins such as MeCP1 or MeCP2, which bind in a sequence-independent manner to methylated DNA (Boyes and Bird 1992).

Is Methylation a Primary or Secondary Event in XCI?

The complete answer to this question is not yet known. Many accept that methylation is involved in, and may be sufficient for, the maintenance of XCI but question whether methylation is involved in the initial steps for establishment of XCI in the early embryo. The two primary reasons for considering methylation as only a locking mechanism are: (1) Methylation is less extensive at CpG islands in marsupials (which are assumed to have a more primitive X inactivation system) and also in the extraembryonic tissues of placental mammals, yet XCI and asynchronous replication occur; and (2) some methylation-sensitive restriction sites in the first exon of the mouse Hprt gene become methylated well after XCI takes place (Lock et al. 1987).

Despite these data, it is nevertheless premature to draw any conclusions regarding the role of methylation in the initial establishment of XCI. First, concerning gene-specific control, most methylation is likely to be irrelevant, representing a background that tends to obscure methylation at key sites. There are reported cases where methylation at a single CpG site is crucial for the maintenance of transcriptional silence (Graessmann et al. 1994; Robertson et al. 1995) and the elimination of factor binding (see Gaston and Fried 1995). Second, the time of XCI establishment is a time of major changes in genome-wide methylation, proceeding from a very hypomethylated state in the early blastocyst to heavy methylation in the late blastocyst (see Shemer and Razin, this volume). Third, in contrast to the earlier study of intron sites by Lock et al. (1987), more recent studies of the promoter regions of the X-linked genes Pgk-1, G6pd, and Xist have shown that methylation takes place concomitant with genetic silencing (Singer-Sam et al. 1990; Norris et al. 1991; Grant et al. 1992; Beard et al. 1995).

Recent studies on methylation of the Xist gene offer the most insight into the mechanism of establishment of XCI. The Xist gene is differentially methylated in the gametes; sperm DNA is unmethylated and oocyte DNA is methylated (Kay et al. 1994; Norris et al. 1994; Ariel et al. 1995). Thus, the embryo starts with methylation-marked Xist genes. In the extraembryonic tissues, the unmethylated, paternal Xist gene is preferentially transcribed, and the methylated, maternal Xist gene is transcriptionally silent (Kay et al. 1994; Norris et al. 1994). Recalling that Xist is expressed only from the Xi, this result is in accord with methylation-mediated silencing, the correlation seen for most autosomal and X-linked genes. In the embryonic lineages, which differentiate later in development, the methylation imprint is lost, and the paternal and maternal X chromosomes are equally likely to be inactivated. Beard et al. (1995) studied the expression of Xist in transgenic mouse embryos with two levels of DNA methyltransferase deficiencies, moderate and severe. Homozygous deficiency of DNA methyltransferase is embryonic-lethal, but the embryos do survive long enough to permit studies of Xist expression. The key finding of Beard et al. (1995) is that the Xist gene in severely methyltransferase-deficient embryos is hypomethylated and expressed even in males, which have only an X_a and normally have a methylated and transcriptionally silent Xist gene. The involvement of methylation in controlling the expression of Xist in the embryo seems even more likely in light of the finding that only severe, not moderate, methylation-deficient embryos express Xist and are hypomethylated at the Xist promoter.

An emerging generalization is that sites vary in their resistance to demethylation and that the sites most resistant to the extensive demethylation that occurs during the morula and early blastocyst stages of development are the best candidates for primary imprint sites. The most attractive hypothesis regarding imprinted genes is that methylation at this limited number of primary sites survives through the morula stage to later guide the establishment of the final methylation pattern of the gene or larger chromosomal domain. This concept is also relevant for the maintenance of XCI in germ cells. Primordial germ cells possess an X_i, but the methylation level of premeiotic and early-stage-meiotic germ cells is very low (Driscoll and Migeon 1990; Ghazi et al. 1992; Singer-Sam et al. 1992).

Where and What Are the Key Elements?

Before methylation-pattern correlations can be used to argue for or against the involvement of methylation in a developmental process, the key DNA elements controlling the process must first be identified. Often this is difficult. In this regard, it should be noted that although Xist maps to the Xce, the mouse X-inactivation center, mapping resolution is limited, and Courtier et al. (1995) have reported that a region lying 15 kb distal to Xist contains several sites that show hypermethylation specifically associated with the X_a. Furthermore, analysis of this region in various Xce strains has revealed a correlation between the strength of the Xce allele carried and the methylation status of this region. It thus seems likely that genes or elements other than Xist determine the probability of inactivation (see Avner, this volume).

METHYLATION-INDEPENDENT MECHANISMS

There is no direct evidence that methylation-independent mechanisms are necessary to explain XCI, but methylation probably does not explain all of the features that govern epigenetic regulation of XCI and imprinted genes. The most compelling argument for the existence of methylation-independent mechanisms is that both *Drosophila* and yeast display cell memory and, for yeast mating type, chromatin-state heritability and factor immunity; yet neither of these organisms has any detectable 5-methylcytosine. For XCI, two experimental results suggest methylation-independent mechanisms: (1) Marsupials have little or no methylation at some X-linked CpG islands, and (2) primordial germ cells, which possess an Xi, are hypomethylated. For these reasons, several models have been considered as candidates for explaining X-chromosome inactivation. These have included nuclear compartmentalization, replication timing, and heritable chromatin structure (see other chapter by Riggs and Porter; this volume). In contrast to *Drosophila* position-effect variegation (PEV; see Henikoff, this volume), feedback circuits using diffusible, *trans*-acting factors are not usually considered as a candidate mechanism for XCI because the differential treatment of homologs seems to require a *cis*-acting mechanism.

Replication timing has been proposed (see Riggs and Pfeifer 1992) as the ancestral mechanism for XCI, since it is the most obviously conserved feature (Graves 1987). The X_a and X_i replicate asynchronously, in different parts of S phase, so they could be temporally separated into a different environment during DNA synthesis and the re-formation of mature chromatin. This model suggests that X-linked replication origins are likely to be key elements, but no X-linked replication origin has yet been identified; thus, directly relevant experimental evidence is completely lacking. It is intriguing, however, that imprinted autosomal genes also show asynchronous replication (Kitsberg et al. 1993). Nuclear compartmentalization, one of the first models proposed for XCI, continues to be attractive. Nuclear compartmentalization could aid maintenance by keeping the X_a and X_i in different environments with respect to transcription factors and chromosomal proteins. This model must be given serious consideration, because it has been found that Xist RNA is preferentially localized to the subnuclear region occupied by the human X_i (Clemson et al. 1996).

As commonly used, the meaning of "heritable chromatin structure" is vague enough to include models for XCI that propose "heterochromatinization" as the first step of XCI (see, e.g., Gartler et al. 1985). This class of models depends on different nucleoprotein complexes being established on the X_a and X_i, and the differences somehow surviving DNA replication (and metaphase condensation). Various mechanisms for accomplishing this are discussed by Kass and Wolffe (this volume; see also other chapter by Riggs and Porter, this volume). Even though nucleosomes and proteins of the basal transcription machinery are stripped from the DNA during replication, adequate amounts may stay in the general vicinity of their previous location to "seed" the assembly of similar structures within a domain.

Promoters and other control elements of the X_i are presumed to be packaged in chromatin so as to be inaccessible or immune to the effects of transcription factors. For example, PGK1 is expressed at moderately high levels in virtually all cells, yet transcription from the X_i is undetectable—less than 0.1% of the X_a (LeBon et al. 1995). Thus, it is unlikely that the complex patterns of active and inactive nucleoprotein complexes and/or chromosomal domains on the X chromosomes are created completely de novo each cell generation. More likely, some kind of chromatin structural information passes to progeny cells and assists the re-formation of the differential X_a/X_i states and the complex pattern of active and inactive genes that determine cell type.

In order for chromatin-based information to pass to progeny cells, it must survive not only DNA replication, but also mitosis. Metaphase chromosomes have an extremely compacted chromatin fiber, approximately 10,000-fold condensed relative to extended DNA. What information remains on compacted metaphase chromosomes that can help progeny cells faithfully maintain XCI and monoallelic expression from imprinted genes? To investigate this question for X-linked genes, metaphase cells were isolated and assayed for in vivo footprints (Hershkovitz and Riggs 1995). It was found that footprints for eight transcription factors on the constitutively active promoter of the human PGK1 gene are absent from the chromatin of metaphase cells. Similar results have been obtained using DNase I and permeabilized cells (M. Hershkovitz and A.D. Riggs, unpubl.). In a study of autosomal genes, Martinez-Balbas et al. (1995) did genomic footprinting of HeLa cells arrested at mitosis by nocodazole and found that several transcription factors seen in interphase cells on the induced promoter of the heat-shock gene hsp70 are absent from metaphase chromatin. They also showed that the transcription factors are still present in the cell, but they are not bound to mitotic chromatin. Thus, the multiprotein transcriptional complexes on the constitutive PGK1 gene, the inducible hsp70 gene, and presumably a myriad of other transcription complexes in the genome must form de novo each cell generation.

The absence of transcription factors from the promoters so far examined by in vivo footprinting does not mean, however, that all nucleoprotein structures are stripped from metaphase chromatin. Immunocytochemical studies indicate that of eight transcription factors studied, one, AP-2, is detectable on metaphase chromosomes of HeLa cells (Martinez-Balbas et al. 1995). Raff et al. (1994) found that GAGA factor, a transcription factor that binds to GA/AT-rich sequences occurring in *Drosophila* heterochromatin, can be detected by immunostaining on chromosomes at all stages of the cell cycle, indicating that it is possible for sequence-specific proteins to remain on tightly condensed chromatin. In vivo footprinting studies have indicated that for the matrix attachment site near the human β-interferon gene, the structure of metaphase chromatin is not significantly different from that of interphase chromatin (Hershkovitz and Riggs 1996). Matrix attachment sites thus are promising candidates for study as key elements for the epigenetic inheritance of chromatin.

X_a/X_i DIFFERENCES OTHER THAN METHYLATION

Histone Acetylation

A very strong positive correlation has emerged between histone acetylation and the active chromatin state (see Kass and Wolffe, this volume). This correlation holds for the differences between the chromatin of the X_a and the facultative heterochromatin of the X_i. Using immunofluorescent staining of metaphase mammalian cells with antibodies specific for acetylated histone H4, Jeppesen and Turner (1993) found that the X_a staining pattern is similar to that of autosomes, but the X_i is barely visible, comparable with heterochromatin. Since histone acetylation is preserved in metaphase chromosomes and affects nucleosomal and chromatin function, it is an attractive candidate as a component of the information transfer system(s) that guides chromatin decondensation and reactivation in progeny cells. It should be noted, however, that inhibitors of deacetylation, such as sodium butyrate or trichostatin, which increase histone acetylation levels, have not been reported to reactivate the X_i. Our attempts to reactivate HPRT in Chinese hamster/human hybrid cells by treatment with trichostatin have been negative (M. Hershkovitz and A.D. Riggs, unpubl.).

Condensation and Accessibility to Enzyme and Protein Factors

A key difference between active euchromatin and inactive heterochromatin is that heterochromatin appears more condensed in the light microscope during telophase and interphase. The cytological evidence interpreted as evidence for condensation is usually based on differential staining, e.g., the Barr body. Active chromatin has long been known to have a two- to fourfold increased general sensitivity to DNase I as well as localized hypersensitive sites. For X-linked genes, cytogenetic, Southern blot, and other studies have demonstrated somewhat increased sensitivity of X_a genes to DNase I in interphase chromatin and in metaphase chromatin (Gazit and Cedar 1982; Kuo et al. 1982; Riley et al. 1986; Yang and Casky 1987; Lin and Chinault 1988; Martinez-Balbas et al. 1995). However, despite the seemingly obvious fact that heterochromatin is more condensed and should be less accessible to regulatory factors, there is a body of data that argues against this notion.

Schlossher et al. (1994) used a sensitive, quantitative variation on ligation-mediated PCR (Steigerwald et al. 1990) to study chromatin accessibility in *Drosophila* and came to the conclusion that inactive chromatin silenced by PEV is as accessible to restriction enzymes as is active chromatin. They also found that mutations affecting PEV and polycomb silencing have no effect on enzyme accessibility. Other studies support these experiments. Wines et al (1996) found that PEV has little effect on overall chromatin accessibility to the *Escherichia coli* Dam methylase, and McCall and Bender (1996) found that T7 RNA polymerase has access to PEV-silenced genes. With regard to XCI, Cremer et al. (1993), in a study of nuclear compartments, used in situ hybridization methods and careful analysis of serial sections to compare the volumes occupied by the X_a and X_i in interphase nuclei. The surprising result is that although the shape of the X_i is significantly different from that of the X_a, the volume it occupies is not. Even though DNase I differences are observed between the X_a and X_i in mammalian cells, the differences are only about twofold, as measured by Southern blot experiments (Riley et al. 1986; Yang and Casky 1987; Lin and Chinault 1988). In vivo footprinting experiments of PGK1 (Pfeifer and Riggs 1991), which involved treatment of permeabilized cells and had an excellent internal control, indicated little difference between the X_a and X_i in susceptibility to DNase I nicking. These experiments, which have been confirmed for mouse cells (L. Wang and A.D. Riggs, unpubl.), are consistent with and confirm the Southern blot experiments. Interphase chromatin has been thought to be coiled into a 30-nm solenoid which is then compacted by looping to form a 250-nm fiber. Further compaction takes place during mitosis to form the 700-nm chromatid of a metaphase chromosome. The solenoid model has been called into question by recent data, and a looser three-dimensional zigzag model has been proposed

(Leuba et al. 1994; Woodcock 1994). A tight solenoid would be expected to exclude protein factors and enzymes such as DNase I, but perhaps the zigzag structure would not.

CONCLUSIONS AND FUTURE DIRECTIONS

Most X-linked genes are monoallelically expressed in diploid cells, as are at least 15 imprinted genes (see Ainscough and Surani, this volume). In addition to monoallelic expression, X-linked and imprinted genes share two other features: asynchronous replication and differential methylation of CpG islands. In this review, we have focused on the features of XCI that are relevant to the differential treatment of sequence-identical genes and the stable, epigenetic inheritance of alternate chromatin states. In addition to the older body of evidence strongly suggesting that DNA methylation is essential for the maintenance of XCI, recent studies on the Xist gene in DNA-methyltransferase-deficient mice (Beard et al. 1995) have provided strong evidence that methylation is involved in maintaining the silence of this gene, which is a *cis*-acting gene necessary for the establishment of XCI in the early embryo. However, the mechanism of action of the Xist gene is completely unknown, as are details of the establishment mechanism(s). Regarding maintenance of XCI, the clearest difference between the X_a and the X_i is the degree of methylation in the PGK1 promoter, where full methylation of 60 CpG sites is found on the X_i and no methylation is found on the X_a. De novo methylation of the X_i CpG island and promoter is known to occur at about 5% per cell generation, so this observation, along with the assumption that multiple key sites in the promoter are affected by methylation, essentially explains the stability of the methylated, inactive state (Pfeifer et al. 1990b). Even if occasionally methylation is lost at a particular site due to a mistake in maintenance, de novo methylation will repair this loss, and unmethylated sites will not accumulate. The question remains, however: How is the promoter-containing CpG island of the X_a maintained free from methylation? The involvement of Sp1 should be considered, as it has been reported to keep neighboring DNA free of methylation (Brandeis et al. 1994; Macleod et al. 1994), and it binds, as well, to the active PGK1 promoter (Pfeifer and Riggs 1991). In this context, it should be noted that Sp1 binding is either not affected or is increased by methylation of its binding site, yet Sp1 is not bound to the methylated PGK promoter of the X_i either in interphase chromatin or at metaphase (Pfeifer and Riggs 1991; Hershkovitz and Riggs 1995). In this case, the action of methylation on Sp1 binding seems necessarily to be indirect. Why is Sp1 on the X_a but not on the X_i?

Better understanding of chromatin structure at both the nucleosomal and higher-order level is needed to more fully comprehend the details of XCI. Chromatin repression and restricted accessibility are now recognized as important components of genetic control mechanisms (Kornberg and Lorch 1995) and seem likely to be key factors for control of genes that are monoallelically expressed, such as X-linked and imprinted genes. It often is assumed that monoallelic immunity to factors can be a result of differentially condensed chromatin structure that limits access to one allele. However, as reviewed in the previous section, accessibility may be similar between active and inactive chromatin, and additional studies are needed to address this point. Other areas that need study because of their likely importance for understanding XCI and imprinting are (1) histone acetylation (see Kass and Wolffe, this volume), (2) metaphase chromosome structure, and (3) replication origins. The degree of histone acetylation is a known difference between the X_a and X_i and also is known to affect the control of yeast mating type (see Grunstein et al. 1995; Holmes et al., this volume). Nucleoprotein complexes that survive on metaphase chromosomes are likely to be involved in the transfer to progeny of information needed to maintain monoallelic expression of X-linked and imprinted genes (see discussion in this chapter and in Hershkovitz and Riggs 1995). X-linked replication origins should be investigated because replication of the X_i and X_a at different stages of S phase has been proposed as the ancestral mechanism for XCI (Riggs and Pfeifer 1992), and also because origin of replication complexes are known to be involved in the chromatin-based silencing of yeast mating-type genes (see Pillus and Grunstein 1995; Holmes et al., this volume).

ACKNOWLEDGMENTS

This work was supported by a National Institutes of Health grant (GM-50575) to A.D.R.

REFERENCES

Antequera, F. and A. Bird. 1993. CpG Islands. In *DNA methylation: Molecular biology and significance* (ed. J.P. Jost and H.P. Saluz), pp. 169–185. Birkhauser Verlag, Berlin.

Ariel, M., E. Robinson, J.R. McCarrey, and H. Cedar. 1995. Gamete-specific methylation correlates with imprinting of the murine Xist gene. *Nat. Genet.* **9:** 312–315.

Beard, C., E. Li, and R. Jaenisch. 1995. Loss of methylation activates *Xist* in somatic but not in embryonic cells. *Genes Dev.* **9:** 2325–2334.

Boyes, J. and A. Bird. 1992. Repression of genes by methylation depends on CpG density and promoter strength: Evidence for involvement of a methyl-CpG binding protein. *EMBO J.* **11:** 327–333.

Brandeis, M., D. Frank, I. Keshet, Z. Siegfried, M. Mendelsohn, A. Nemes, V. Temper, A. Razin, and H. Cedar. 1994. Sp1 elements protect a CpG island from de novo methylation. *Nature* **371:** 435–438.

Brown, C.J. and H.F. Willard. 1994. The human X-inactivation centre is not required for maintenance of X-chromosome inactivation. *Nature* **368:** 154–156.

Brown, C.J., A. Ballabio, J.L. Rupert, R.G. Lafreniere, M. Grompe, R. Tonlorenzi, and H.F. Willard. 1991a. A gene from the region of the human X inactivation centre is expressed exclusively from the inactive X chromosome. *Nature* **349:** 38–44.

Brown, C.J., R.G. Lafreniere, V.E. Powers, G. Sebastio, A. Ballabio, A.L. Pettigrew, D.H. Ledbetter, E. Levy, I.W. Craig, and H.F. Willard. 1991b. Localization of the X inactivation centre on the human X chromosome in Xq13. *Nature* **349:** 82–84.

Clemson, C.M., J.A. McNeil, H.F. Willard, and J.B. Lawrence. 1996. XIST RNA paints the inactive X chromosome at interphase: Evidence for a novel RNA involved in nuclear chromosome structure. *J. Cell Biol.* **132:** 259–275.

Courtier, B., E. Heard, and P. Avner. 1995. Xce haplotypes show modified methylation in a region of the active X chromosome lying 3′ to Xist. *Proc. Natl. Acad. Sci.* **92:** 3531–3535.

Cremer, T., A. Kurz, R. Zirbel, S. Dietzel, B. Rinke, E. Schrock, M.R. Speicher, U. Mathieu, A. Jauch, P. Emmerich, H. Scherthan, T. Ried, C. Cremer, and P. Lichter. 1993. Role of chromosome territories in the functional compartmentalization of the cell nucleus. *Cold Spring Harbor Symp. Quant. Biol.* **58:** 777–792.

Driscoll, D.J. and B.R. Migeon. 1990. Sex difference in methylation of single-copy genes in human meiotic germ cells: Implications for X chromosome inactivation, parental imprinting, and origin of CpG mutations. *Somat. Cell Mol. Genet.* **16:** 267–282.

Ehrlich, M. and K.C. Ehrlich. 1993. Effect of DNA methylation on the binding of vertebrate and plant proteins to DNA. In *DNA methylation: Molecular biology and biological significance* (ed. J.P. Jost and H.P. Saluz), pp. 145–168. Birkhauser Verlag, Berlin.

Gartler, S.M. and M.A. Goldman. 1994. Reactivation of inactive X-linked genes. *Dev. Genet.* **15:** 504–514.

Gartler, S.M., K.A. Dyer, J.A.M. Graves, and M. Rocchi. 1985. A two step model for mammalian X chromosome inactivation. In *Biochemistry and biology of DNA methylation* (ed. G.L. Cantoni and A. Razin), pp 223–235. A.R. Liss, New York.

Gaston, K. and M. Fried. 1995. CpG methylation has differential effects on the binding of YY1 and ETS proteins to the bi-directional promoter of the Surf-1 and Surf-2 genes. *Nucleic Acids Res.* **23:** 901–909.

Gazit, B. and H. Cedar. 1982. Active genes are sensitive to deoxyribonuclease during metaphase. *Science* **217:** 648–650.

Ghazi, H., F. Gonzales, and P.A. Jones. 1992. Methylation of CpG-island-containing genes in human sperm, fetal and adult tissues. *Gene* **114:** 203–210.

Graessmann, A., G. Sandberg, E. Guhl, and M. Graessmann. 1994. Methylation of single sites within the herpes simplex virus tk coding region and the simian virus 40 T-antigen intron causes gene inactivation. *Mol. Cell. Biol.* **14:** 2004–2010.

Grant, M., M. Zuccotti, and M. Monk. 1992. Methylation of CpG sites of two X-linked genes coincides with X-inactivation in the female mouse embryo but not in the germ line. *Nat. Genet.* **2:** 161–166.

Grant, S.G. and V.M. Chapman. 1988. Mechanisms of X-chromosome inactivation. *Annu. Rev. Genet.* **22:** 199–233.

Graves, J.A.M. 1987. The evolution of mammalian sex chromosomes and dosage compensation: Clues from marsupials and monotremes. *Trends Genet.* **3:** 252–256.

Grunstein, M., A. Hecht, G. Fisher-Adams, J. Wan, R.K. Mann, S.T. Strahl-Bolsinger, T. Laroche, and S. Gasser. 1995. The regulation of euchromatin and heterochromatin by histones in yeast. *J. Cell Sci. Suppl.* **19:** 29–36.

Hendrich, B.D. and H.F. Willard. 1995. Epigenetic regulation of gene expression: The effect of altered chromatin structure from yeast to mammals. *Hum. Mol. Genet.* **4:** 1765–1777.

Hershkovitz, M. and A.D. Riggs. 1995. Metaphase chromosome analysis by ligation-mediated PCR: Heritable chromatin structure and a comparison of active and inactive X chromosomes. *Proc. Natl. Acad. Sci.* **92:** 2379–2383.

———. 1996. Ligation-mediated PCR for chromatin-structure analysis of interphase and metaphase chromosomes. *Methods* (in press).

Jani, M. M., B.S. Torchia, G.S. Pai, and B.R. Migeon. 1995. Molecular characterization of tiny ring X chromosomes from females with functional X chromosome disomy and lack of cis X inactivation. *Genomics* **27:** 182–188.

Jeppesen, P. and B.M. Turner. 1993. The inactive X chromosome in female mammals is distinguished by a lack of histone H4 acetylation, a cytogenetic marker for gene expression. *Cell* **74:** 281–289.

Kay, G.F., S.C. Barton, M.A. Surani, and S. Rastan. 1994. Imprinting and X chromosome counting mechanisms determine Xist expression in early mouse development. *Cell* **77:** 639–650.

Kitsberg, D., S. Selig, M. Brandeis, I. Simon, I. Keshet, D.J. Driscoll, R.D. Nicholls, and H. Cedar. 1993. Allele-specific replication timing of imprinted gene regions. *Nature* **364:** 459–463.

Kornberg, R.D. and Y. Lorch. 1995. Interplay between chromatin structure and transcription. *Curr. Opin. Cell Biol.* **7:** 371–375.

Kuo, M.T., B. Iyer, and R.J. Schwarz. 1982. Condensation of chromatin into chromosomes preserves an open configuration but alters the DNase I hypersensitive cleavage sites of the transcribed gene. *Nucleic Acids Res.* **10:** 4565–4579.

LeBon, J.M., P.P.L. Tam, J. Singer-Sam, A.D. Riggs, and S.S. Tan. 1995. Mouse endogenous X-linked genes do not show lineage-specific delayed inactivation during development. *Genet. Res.* **65:** 223–227.

Leighton, P.A., J.R. Saam, R.S. Ingram, and S.M. Tilghman. 1996. Genomic imprinting in mice: Its function and mechanism. *Biol. Reprod.* **54:** 273–278.

Leuba, S.H., G. Yang, C. Robert, B. Samori, K. Van Holde, J. Zlatanova, and C. Bustamante. 1994. Three-dimensional structure of extended chromatin fibers as revealed by tapping-mode scanning force microscopy. *Proc. Natl. Acad. Sci.* **91:** 11621–11625.

Li, E.B.C. and R. Jaenisch. 1993. Role of DNA methylation in genomic imprinting. *Nature* **366:** 362–365.

Lin, D. and A.C. Chinault. 1988. Comparative study of DNase I sensitivity at the X-linked human HPRT locus. *Somat. Cell Mol. Genet.* **14:** 261–272.

Lock, L.F., N. Takagi, and G.R. Martin. 1987. Methylation of the Hprt gene on the inactive X occurs after chromosome inactivation. *Cell* **48:** 39–46.

Luo, S., B.S. Torchia, and B.R. Migeon. 1995. XIST expression is repressed when X inactivation is reversed in human placental cells: A model for study of XIST regulation. *Somat. Cell Mol. Genet.* **21:** 51–60.

Lyon, M.F. 1961. Gene action in the X-chromosome of the mouse. *Nature* **190:** 372–373.

———. 1993. Epigenetic inheritance in mammals. *Trends Genet.* **9:** 123–128.

Macleod, D., J. Charlton, J. Mullins, and A.P. Bird. 1994. Sp1 sites in the mouse aprt gene promoter are required to prevent methylation of the CpG island. *Genes Dev.* **8:** 2282–2292.

Martinez-Balbas, M.A., A. Dey, S.K. Rabindran, K. Ozato, and C. Wu. 1995. Displacement of sequence-specific transcription factors from mitotic chromatin. *Cell* **83:** 29–38.

McCall, K. and W. Bender. 1996. Probes for chromatin accessibility in the *Drosophila* bithorax complex respond differently to *Polycomb*-mediated repression. *EMBO J.* **15:** 569–580.

Norris, D.P., N. Brockdorff, and S. Rastan. 1991. Methylation status of CpG-rich islands on active and inactive mouse X chromosomes. *Mamm. Genome* **1:** 78–83.

Norris, D.P., D. Patel, G.F. Kay, G.D. Penny, N. Brockdorff, S.A. Sheardown, and S. Rastan. 1994. Evidence that random and imprinted Xist expression is controlled by preemptive methylation. *Cell* **77:** 41–51.

Ohno, S., W.D. Kaplan, and R. Kinosita. 1959. Formation of the sex chromatin by a single X-chromosome in liver cells of *Rattus norvegicus. Exp. Cell Res.* **18:** 415–418.

Penny, G.D., G.F. Kay, S.A. Sheardown, S. Rastan, and N. Brockdorff. 1996. Requirement for *Xist* in X chromosome inactivation. *Nature* **379:** 131–137.

Pfeifer, G.P. and A.D. Riggs. 1991. Chromatin differences between active and inactive X chromosomes revealed by genomic footprinting of permeabilized cells using DNase I and ligation-mediated PCR. *Genes Dev.* **5:** 1102–1113.

Pfeifer, G.P., R.L. Tanguay, S.D. Steigerwald, and A.D. Riggs. 1990a. In vivo footprint and methylation analysis by PCR-aided genomic sequencing: Comparison of active and inactive X chromosomal DNA at the CpG island and promoter of human PGK1. *Genes Dev.* **4:** 1277–1287.

Pfeifer, G.P., S.D. Steigerwald, R.S. Hansen, S.M. Gartler, and A.D. Riggs. 1990b. Polymerase chain reaction-aided genomic sequencing of an X chromosome-linked CpG island: Methylation patterns suggest clonal inheritance, CpG site autonomy, and an explanation of activity state stability. *Proc. Natl. Acad. Sci.* **87:** 8252–8256.

Pillus, L. and M. Grunstein. 1995. Chromatin structure and epigenetic regulation in yeast. In *Chromatin structure and gene expression* (ed. C.R. Elgin), pp. 123–146. Oxford University Press, New York.

Raff, J.W., R. Kellum, and B. Alberts. 1994. The *Drosophila* GAGA transcription factor is associated with specific regions of heterochromatin throughout the cell cycle. *EMBO J.* **13:** 5977–5983.

Rastan, S. 1994. X chromosome inactivation and the Xist gene. *Curr. Opin. Genet. Dev.* **4:** 292–297.

Riggs, A.D. and G.P. Pfeifer. 1992. X chromosome inactivation and cell memory. *Trends Genet.* **8:** 169–174.

Riley, D.E., M.A. Goldman, and S.M. Gartler. 1986. Chromatin structure of active and inactive human X-linked phosphoglycerate kinase gene. *Somat. Cell. Mol. Genet.* **12:** 73–80.

Robertson, K.D., S.D. Hayward, P.D. Ling, D. Samid, and R.F. Ambinder. 1995. Transcriptional activation of the Epstein-Barr virus latency C promoter after 5-azacytidine treatment: Evidence that demethylation at a single CpG site is crucial. *Mol. Cell. Biol.* **15:** 6150–6159.

Sasaki, T., R.S. Hansen, and S.M. Gartler. 1992. Hemimethylation and hypersensitivity are early events in transcriptional reactivation of human inactive X-linked genes in a hamster x human somatic cell hybrid. *Mol. Cell Biol.* **12:** 3819–3826.

Schanz, S. and P. Steinbach. 1989. Investigation of the "variable spreading" of X inactivation into a translocated autosome. *Hum. Genet.* **82:** 244–248.

Schlossher, J., H. Eggert, R. Paro, and S. Cremer. 1994. Gene inactivation in *Drosophila* mediated by the polycomb gene product or by position-effect variegation does not involve major changes in the chromatin fibre. *Mol. Gen. Genet.* **243:** 453–462.

Shemer, R., Y. Birger, W.L. Dean, W. Reik, A.D. Riggs, and A. Razin. 1996. Dynamic methylation adjustment and counting as part of imprinting mechanisms. *Proc. Natl. Acad. Sci.* (in press).

Singer-Sam, J. and A.D. Riggs. 1993. X chromosome inactivation and DNA methylation. In *DNA methylation: Molecular biology and biological significance* (ed. J. Jost and H. Saluz), pp 358–384. Birkhauser Verlag, Berlin.

Singer-Sam, J., V. Chapman, J.M. LeBon, and A.D. Riggs. 1992. Parental imprinting studied by allele-specific primer extension after PCR: Paternal X chromosome-linked genes are transcribed prior to preferential paternal X chromosome inactivation. *Proc. Natl. Acad. Sci.* **89:** 10469–10473.

Singer-Sam, J., M. Grant, J.M. LeBon, K. Okuyama, V. Chapman, M. Monk, and A.D. Riggs. 1990. Use of a HpaII-polymerase chain reaction assay to study DNA methylation in the Pgk-1 CpG island of mouse embryos at the time of X-chromosome inactivation. *Mol. Cell Biol.* **10:** 4987–4989.

Steigerwald, S.D., G.P. Pfeifer, and A.D. Riggs. 1990. Ligation-mediated PCR improves the sensitivity of methylation analysis by restriction enzymes and detection of specific DNA strand breaks. *Nucleic Acids Res.* **18:** 1435–1439.

Takagi, N., M.A. Yoshida, O. Sugawara, and M. Sasaki. 1983. Reversal of X-inactivation in female mouse somatic cells hybridized with murine teratocarcinoma stem cells in vitro. *Cell* **34:** 1053–1062.

Tsai, Y.C., A.R. Simoneau, C.H.R. Spruck, P.W. Nichols, K. Steven, J.D. Buckley, and P.A. Jones. 1995. Mosaicism in human epithelium: Macroscopic monoclonal patches cover the urothelium. *J. Urol.* **153:** 1697–1700.

Wines, D.R., P.B. Talbert, D.V. Clark, and S. Henikoff. 1996. Introduction of a DNA methyltransferase into *Drosophila* to probe chromatin structure in vivo. *Chromosoma* **104:** 332–340.

Woodcock, C.L. 1994. Chromatin fibers observed in situ in frozen hydrated sections. *J. Cell. Biol.* **125:** 11–19.

Yang, T.P. and C.T. Casky. 1987. Nuclease sensitivity of the mouse HPRT gene promoter region: Differential sensitivity on the active and inactive X chromosome. *Mol. Cell. Biol.* **7:** 2994–2998.

X-Chromosome Inactivation: *XCE* and the Candidate Region for the X-Inactivation Center

Philip Avner
Unité de Génétique Moléculaire Murine
Institut Pasteur
F-75724 Paris Cedex 15, France

X-Chromosome inactivation is characterized by the transcriptional silencing of one of the two X chromosomes present in somatic cells of the female. Transcriptional silencing of the X chromosome carried by male germ cells having actively entered meiosis is also observed, but it remains to be proved whether the mechanism underlying X-inactivation in the two cases is identical (Lyon 1983).

The process of X-inactivation has been somewhat arbitarily divided into three phases: initiation, spreading, and maintenance. Most theories of initiation suggest it is triggered by a developmental cue at strictly programmed specific stages of embryogenesis in the female and spermatogenesis in the male. In the female mouse, X-inactivation first occurs in the trophoectoderm and extraembryonic endoderm at 3.5–4.5 days of gestation. This is followed by the X-inactivation of the tissues of the embryo itself, which occurs at the egg cylinder stage, between days 5.5 and 6.0 of gestation, around the time of uterine implantation (for review, see Epstein 1983; Lyon 1983). The developmental cue that triggers inactivation is thought to be associated with the presence of a "blocking signal" that is produced in limited quantity and binds to a single X chromosome per diploid cell, protecting it from inactivation. Implicit in this model is the idea that prior to, or during, the initial phases of inactivation some sort of counting mechanism is capable of sensing the number of X chromosomes in relation to the autosomal chromosome complement. There is no evidence that this "sensing" is necessarily direct, and it has been suggested that indirect sensing through mechanisms such as cell volume, which is normally proportional to chromosome ploidy, may be involved.

There is strong evidence for the existence of a unique locus known as the X-inactivation center or *Xic*, whose presence on the X chromosome (i.e., *in cis*) is necessary for inactivation (Fig. 1). An X chromosome car-

Epigenetic Mechanisms of Gene Regulation

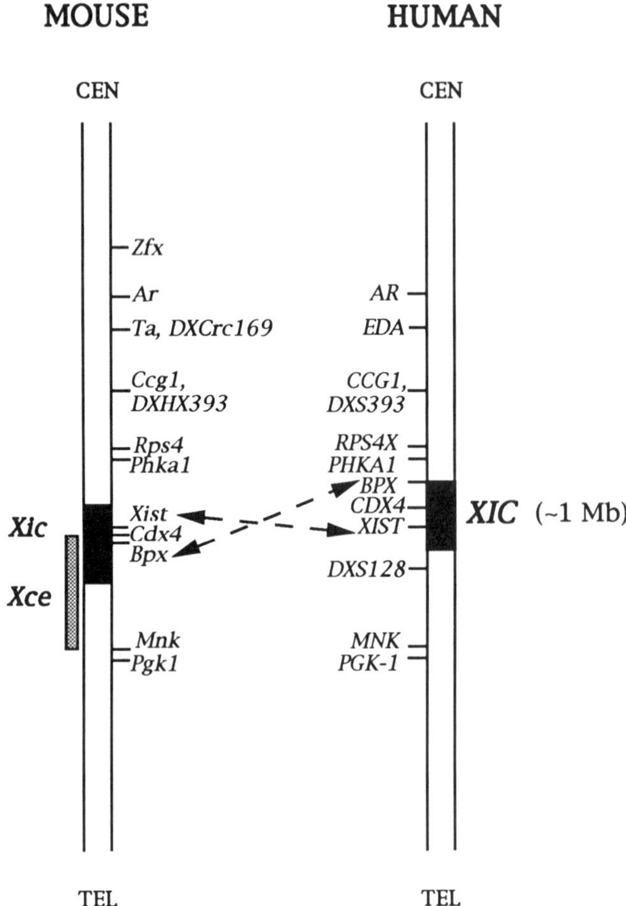

Figure 1 Map of the XIC/*Xic* candidate regions on the human and mouse X chromosomes. The solid areas on both X chromosomes represent the XIC candidate regions as defined in human, spanning 680–1200 kb. The location of the *Xce* locus is indicated by the shaded bar on the left. The 5′→3′ orientations of the XIST/*Xist* genes are indicated by arrows. Locus acronyms include: (Zfx) zinc finger protein; (AR/*Ar*) androgen receptor; (EDA) ectodermal dysplasia; (Ta) tabby; (CCG1/*Ccg1*) cell-cycle progression gene; (RPS4X/*Rps4x*) ribosomal protein S4; (PHKA/Phkα) phosphorylase kinase; (XIST/*Xist*) X-inactive-specific transcript; (MNK/*Mnk*) Menkes; (PGK1/*Pgk1*) phosphoglycerate kinase. DXS393/DXHS393 and DXS128 are anonymous sequences isolated from human.

rying an *Xic* is potentially able to undergo X-inactivation, whereas one from which *Xic* is missing cannot. The existence of this locus and its localization have been deduced from studies of the X-inactivation status of cells carrying structurally altered X chromosomes (Russell 1983; Therman and Sarto 1983). In man the candidate region extends over approximately 1 Mb of the proximal long arm within band Xq13 (Brown et al. 1991b; Lafrenière et al 1993; Leppig et al. 1993). In the mouse the candidate region is bordered toward the centromere-containing end of the X chromosome by the T16(X:16)H translocation localizing to band XD (Rastan 1983) and toward the telomere-containing end of the chromosome by the deletion breakpoint carried by the HD3 E.S. cell line (Rastan and Robertson 1985). The candidate region thus defined in the mouse is considerably larger than that in man, but the overall linkage conservation of the region of the human X chromosome corresponding to Xq11-Xq23 in the mouse has led to suggestions that the minimal region defined in man can be directly extrapolated to the mouse (but see below).

The *Xist* (X-inactive specific transcript) locus lies within the candidate region as defined in both species (Borsani et al. 1991; Brockdorff et al. 1991; Brown et al. 1991a). It is the only sequence so far characterized that is transcribed exclusively from the inactive X chromosome. Its unique pattern of transcriptional activity and its position within the *Xic* candidate region have led to suggestions that *Xist* and *Xic* are synonomous. The human and mouse *Xist* sequences are characterized by a relatively low level of sequence homology (Brockdorff et al. 1992; Brown et al. 1992), and this is in agreement with the difficulties that have been encountered in identifying a marsupial homolog of the *Xist* sequence. The degree of conservation of the human and murine *Xist* sequences can be visualized by reference to the β-globin locus. The level of homology observed for human and mouse *Xist* lies somewhere between the level of homology shown by the human and murine exonic β-globin sequences and that shown by the corresponding intronic sequences (Hendrich et al. 1993). The level of evolutionary conservation between the human and mouse *Xist* sequences is similar to that observed between the human and mouse *H19* sequences (Hendrich et al. 1993).

The *H19* gene, which is subject to imprinting, is expressed in most endodermal and mesodermal cell types of the developing embryo. It lies close to the *Igf2* locus on mouse chromosome 7. It has been suggested that *H19* transcription plays a role in the reciprocal imprinting of the neighboring *Igf2* locus such that the paternal copy of *Igf2* and the maternal copy of *H19* are expressed in development (Bartolomei et al. 1991; De Chiara et al. 1991).

The lack of open reading frames of any significant length in the *Xist*/XIST[1] sequence have similarly led to the mechanism and action of *Xist* on the X-inactivation being compared to that of *H19* on the imprinting of the neighboring *Igf2* locus (Brannan et al. 1990). In both cases it has been suggested that the effects may be mediated at the RNA level or may be a consequence of the act of transcription itself (Brockdorff et al. 1992; Brown et al. 1992). An alternative possibility has been opened up by the findings of Clemson et al. (1996), who noted that XIST RNA remains in the nucleus and often encircles the heterochromatic Barr body. The apparent close association of the XIST RNA with the nuclear matrix fraction suggests that it may constitute an architectural element, possibly a component of nonchromatin nuclear structure that specifically associates with the territory of the inactive X chromosome.

A further genetic element, the X-controlling element (*Xce*), influences in an as-yet-undefined manner the randomness of the X-inactivation process in the mouse (Cattanach and Williams 1972). Three main alleles have been defined at the *Xce* locus: Xce^a, Xce^b, and Xce^c. In Xce^a/Xce^b heterozygotes, the X carrying the Xce^a allele is more likely to be inactivated than that carrying the Xce^b allele; similarly in Xce^b/Xce^c heterozygotes, the X carrying the Xce^b allele is more likely to be inactivated in most cells. Most extreme nonrandomness is, however, seen in Xce^a/Xce^c heterozygotes. *Mus spretus* is thought to contain a fourth allele, Xce^d, although this allele has been less extensively characterized (Cattanach et al. 1989).

DETAILED CHARACTERIZATION OF THE *XIC* REGION IN THE MOUSE

Detailed analysis of the *Xist* locus itself in man and mouse initially showed that the 5′→3′ orientation of this locus relative to the centromere in the two species differs, implicating the presence of a small inversion (Fig. 1) (Brockdorff et al. 1992; Lafrenière et al. 1993; Rougeulle et al. 1994). The recent discovery of two other genes, *Cdx4* (mouse caudal homolog 4) (Horn and Ashworth 1995) and *Bpx* (brain protein X-linked) (Rougeulle and Avner 1996), both of which map telomeric to the 3′ end of the murine *Xist* locus yet lie centromeric to the 3′ end of XIST in man, suggests that this inversion between man and mouse extends over at least 300 kb in the mouse and 600 kb in man. The large inverted duplica-

[1]Mouse loci are denoted by an italicized name starting with a capital letter. The equivalent human locus is all capitals and nonitalicized. *Xist* therefore refers to the mouse Xist locus and XIST to its human counterpart.

tion previously identified by Lafrenière et al. (1993) as lying between the *PHKA1* gene and the neighboring locus *DXS227* may well represent the proximal boundary of the rearranged chromosomal segment in man. The identification of further conserved transcription units and, in particular, those lying close to the 5′ end of the murine *Xist* locus will be necessary to allow the extent and complexity of this chromosomal rearrangement to be completely defined.

Interestingly, the chromosomal inversion involving *Xist* is accompanied by the loss of a considerable amount of chromosomal material within the translocated region. *Bpx*, which codes for an X-linked brain protein, physically maps, for instance, at least 600 kb from the 3′ end of XIST in man but only 300 kb from the mouse *Xist* locus. The major part of this loss of genetic material seems to have occurred between *Xist* and *Cdx4* rather than between *Cdx4* and *Bpx*, as Horn and Ashworth (1995) have reported that the *Cdx4* gene which lies 100 kb distal to Xist in the mouse lies 400 kb proximal to XIST on the human X chromosome. The loss of genetic material from around *Xist*/XIST may be tolerated because of the apparently relatively gene-poor nature of this region. In this respect, both the finding of an *S19* pseudogene just upstream of the 5′ promoter region of the murine *Xist* locus (C. Rougeulle and P. Avner, in prep.) and the apparently high and nonrandom frequency of transgene insertion sites such as *Mintd* within band XD may be revealing (Taketo et al. 1992; Wu et al. 1992; Dandolo et al. 1993).

Since both *Cdx4* and *Bpx* are, unlike *Xist*, subject to normal X-inactivation, the boundary elements separating an eventual *Xist* domain from that subject to X-inactivation must lie within the 100-kb region separating *Xist* from *Cdx4*. The genomic sequencing of this region in the mouse, which has recently allowed the identification and preliminary characterization of a novel sequence from this region that is expressed in the adult exclusively in the testis (M.-C. Simmler et al.; D. Cunningham et al.; both in prep.), should facilitate the identification of such elements. That part of the human XIST–CDX4 region that is deleted in the mouse can presumably be excluded from consideration as part of the candidate *Xic*/XIC region.

MAPPING OF *XCE* IN RELATION TO *XIST* AND *XIC* IN THE MOUSE

The *Xce* locus was initially defined by Cattanach and his colleagues in a series of outstanding papers as lying between the Tabby and the Mottled loci *(ATP7a)*, a genetic region coincident with, but considerably more extensive than, the *Xic* candidate region itself (Cattanach and Williams

1972). The observation that *Xist* expression, when analyzed by Northern blot, was markedly lower in the *M. spretus* female than in the C57Bl/10 female led to suggestions by Rastan and colleagues that *Xist* and *Xce* may be one and the same locus and that *Xce* alleles may correspond to different *Xist* alleles whose promoters have different affinities for a presumptive blocking factor (Brockdorff et al. 1991). Experiments in our laboratory using the more sensitive semi-quantitative reverse transcriptase polymerase chain reaction (RT-PCR) technique suggested that there was a 20-fold difference in the level of Xist transcript in female C3H/He (Xce^a) compared to C57Bl/6(Xce^b) and a 5-fold greater level of Xist transcript compared with the C3H.Pgk1a strain (Xce^c). However, when the same technique was applied to a pair of congenic strains (129 and 129.Pgk1a) differing solely by a chromosomal segment extending around the *Xce* locus, these differences in *Xist* expression levels could not be confirmed (B. Courtier et al., unpubl.; but see also Buzin et al. 1994). These observations highlight the difficulties in interpreting in terms of *Xce* molecular observations obtained using solely the standard *Xce* tester mouse stocks with their highly diverse genetic backgrounds.

To refine the genetic localization of *Xce* and to explore the relationship between the *Xce* and *Xist* loci, a series of three microsatellite markers lying close to, and on either side of, *Xist* were isolated (Simmler et al. 1993). These microsatellite markers, which are informative for mouse strains carrying the a, b, and c alleles at the *Xce* locus, show linkage disequilibrium with the *Xce* locus. Analysis of a first series of mice recombinant for the chromosomal span lying between the Tabby and Mottled genes, and which were of known haplotype for both *Xce* and the microsatellite markers, confirmed that the *Xce* locus must be tightly linked to the *Xist* locus itself. It did not, however, allow the precise localization of *Xce* with respect to *Xist*. Genetic analysis of the standard *Xce* tester strain, Ju/Ct, clearly suggested, however, that *Xce* and *Xist* are separable loci with *Xce* lying distal to the 3′ end of the *Xist* gene (Simmler et al. 1993).

Recently, H.F. Willard and his colleagues (pers. comm.) have suggested that a mutation within the human XIST 5′ promoter region is associated with the presence of bias in X-inactivation within some human females. This mutation is not an obvious human homolog of the *XCE* locus, notwithstanding its obvious interest, since the XCE locus, if it exists in man, would, on the basis of the mouse data, be expected to lie 3′ to XIST rather than 5′. Further data concerning the extent of the inversion involving XIST in man will be necessary to exclude this possibility completely.

METHYLATION PATTERNS IN THE VICINITY OF *XIST*

Although *Xist* is not associated with a classic CpG island site, sequences 5′ to *Xist* (Norris et al. 1994) have been shown to be hypermethylated on the active X chromosome and hypomethylated on the inactive X chromosome. This pattern is the opposite of that normally associated with gene-associated CpG islands on the inactive X chromosome. In both cases, however, hypermethylation is associated with the transcriptionally silenced gene copy, whether that be of a housekeeping gene such as *Hprt* or of *Xist* itself. Certain of the CpG sites associated with the promoter region of *Xist* have been shown to be differentially methylated in eggs and sperm and may be associated with imprinted *Xist* expression in the early embryo and paternal X-inactivation (Ariel et al. 1995; Zuccotti and Monk 1995). More recently, sequences 3′ to *Xist* have similarly been shown to be hypermethylated when associated with the inactive Xist sequence on the active X chromosome (Heard et al. 1993; Courtier et al. 1995).

METHYLATION PATTERN VARIATION IN THE *XIST* REGION WITH *XCE* TYPE

The methylation pattern of the region 3′ to *Xist* has been explored in some detail, since as discussed above, it may be of interest in terms of *Xce* function. The hypermethylated region specifically associated with the active X chromosome we have identified lies some 15 kb distal to *Xist* and extends over 2.4 kb. It contains both a *Sal*I site (referred to as *Sal**) and a neighboring nest of *Hpa*II sites lying within a tandem repeat. Analysis of this region in various *Xce* strains has suggested a correlation between the strength of the *Xce* allele present and the methylation status of the region (Courtier et al. 1995). Thus, DNA from males of standard strains of the Xce^a and Xce^b type show higher hypermethylation than Xce^c or Xce^d animals at sites within this region. The significance of this observation was reinforced by the use of a pair of strains on the 129 background congenic for the *Pgk1* region. These animals, which should differ by only a short differential X chromosome segment linked to the *Pgk1* locus, would be expected to also differ for the closely linked *Xist* and *Xce* loci. The clear differences in the methylation profiles of the *Sal** and *Hpa*II sites within the repeat between the congenic strains suggests the presence of one or more regulatory elements present within the differential X chromosome segment that influences the methylation profile of the active X chromosome.

Repeat sequences have recently been implicated in imprinting phenomena (Neumann et al. 1995), genetic diseases through triplet expansion (Bates and Lehrach 1994; Richards and Sutherland 1994; Sutherland and Richards 1995), and direct gene control. One example of the latter is the recent findings concerning the human insulin gene linked polymorphic region (ILPR), a tandem repeat, which is composed of multiple 14- to 15-mer repeat units. The ILPR, which is situated within the proximal promoter of the insulin gene, is itself implicated in susceptibility to type 1 diabetes with susceptibility being associated with short allele length (Owerbach and Gabbay 1993). The ILPR has been shown not only to be a major S1 nuclease hypersensitive site capable of adopting an altered DNA structure (Hammond-Kosack et al. 1992), but also to be capable of influencing transcriptional activity of the insulin promoter, with a long ILPR having greater activity than the short ILPR allele (Kennedy et al. 1995). Another example of a similar finding refers to the human androgen receptor (hAR) gene (Kazemi-Esfarjani et al. 1995), although in this case using an hAR cDNA expression vector, an inverse correlation between the size of a (CAG) trinucleotide repeat and the degree of gene expression was observed.

Since hypermethylation of the 3′ region of *Xist* is observed in adult male cells, it obviously occurs in the absence of X-inactivation itself. In embryonic stem (ES) cells, for instance, where X-chromosome inactivation is observed in female cells upon differentiation, the *Sal** site is unmethylated in both female and male ES cells in the undifferentiated state and becomes methylated upon differentiation. One possibility is that the observed methylation is part of an X-autosome interaction occurring in the absence of a second copy of the X chromosome. Indeed, we have hypothesized (Courtier et al. 1995) that the observed methylation of the Xa in this region of the X chromosome may be part of some form of sensing of the chromosome content of the cell involved in the choice of both which and how many X chromosomes will be inactivated. Implicit in this model is the idea that the choice between the two chromosomes precedes X-inactivation itself. The methylation effects are seen in both hemizygote males that carry a single copy of the mouse X chromosome and females that are heterozygous for the X chromosome; i.e., that carry two dissimilar X chromosomes, but not apparently in female homozygotes carrying two identical X chromosomes. This may imply a degree of *trans*-sensing and discrimination between the two chromosomes, involving either the autosomes and/or the X chromosomes themselves. Speculation as to the nature of such *trans*-sensing has ranged widely, with certain models suggesting a limited supply of an autosomal gene product that binds to *Xic* and others suggesting a single nuclear site to which the active chromosome can attach. This phenomenon could involve the protection through a blocking mechanism of one chromosome—the future active X chromosome or Xa—in an open chromatin conformation. In female cells, inactivation of the other, unprotected X chromosome would then occur at a later stage. Methylation analysis of additional sites lying within the *Xic* region will be necessary to determine to what extent there is a clearly definable *Xce* methylation-sensitive domain around the *Xist* sequence.

IN VIVO AND IN VITRO EXPLORATION OF THE ROLE OF *XIC* CANDIDATE SEQUENCES

To clarify the role of the *Xic*/XIC and of *Xist*/XIST in X-inactivation, a number of distinct experimental approaches have been adopted. These can schematically be regrouped into approaches aimed at exploiting "loss of function" experimental systems and those using "gain of function" approaches.

Loss of Function

Willard and his colleagues (Brown and Willard 1994) have exploited a series of somatic cell hybrids deleted for various parts of the inactive X chromosome including the XIC that were obtained using the "pushmi–pullyu" technique, from a hybrid originally containing an inactive human X chromosome. In this elegant but complicated technique, human female cell lines containing an inactive X chromosome are fused with a mouse cell line containing a temperature-sensitive defect in the *Ube1* gene that can be complemented by either the active or inactive human X chromosome, since the human gene escapes X-inactivation. The UBE1 gene, which is situated on the Xp or human X chromosome short arm, is used for cell selection in tandem with the HPRT gene situated on distal Xq or the human long arm and which *is subject* to X-inactivation. Hybrids that have lost the human active X chromosome but retained the inactive X chromosome are obtained after selection for a functional UBE1 gene by growth at the nonpermissive temperature in the presence of medium containing azaguanine or thioguanine.

Subsequently, to obtain breakpoints within the inactive human X chromosome retained in these hybrids, reactivants of the HPRT gene are obtained either spontaneously or after demethylation by azacytidine and selection on HAT medium. In a third stage corresponding to the essential pushmi–pullyu step, chromosomes having broken between UBE1 and

HPRT are then selected for by simultaneous double selection aimed at securing UBE1 + HPRT clones; i.e., clones that have retained part of Xp but lost various portions of the rest of the human X chromosome.

Using such hybrids, it has been demonstrated that the inactivated state can apparently be maintained in the absence of the XIC, suggesting that neither XIST nor indeed the XIC is necessary for the maintenance of X-inactivation in adult somatic cells (Brown and Willard 1994). Given both the in vitro nature of the experimental system used and the extent of the experimental manipulation imposed on the hybrid cells, analysis of in vivo systems to validate the conclusions appears necessary. Apparent confirmation has been obtained in studies of tumor material from two female patients having myelodysplasia concomitant with the presence of an isodicentric X chromosome with a breakpoint in Xq13 and the loss of the XIST gene. These studies similarly indicate that XIST is not necessary for the maintenance of X-inactivation (Rack et al. 1994).

Ring chromosomes provide another experimental system for the study of X-inactivation. The presence of small X-ring chromosomes has been associated with an absence of X-inactivation leading to abnormal dosage of X-linked genes on the ring and abnormal phenotype and/or mental retardation. Most such X-ring chromosomes that do not become inactive do not contain either XIC or XIST, and this explains their non-inactivation (Wolff et al. 1994). Recent data from Migeon et al. (1993) suggest, however, that in a minority of cases, the presence within the ring chromosome of the XIST gene itself may not be sufficient for X-inactivation, suggesting the eventual need for additional genetic elements. Although the relatively limited amount of chimerism seen in certain of the X-ring-chromosome patients makes it likely that in such cases X-inactivation indeed occurred after ring formation, it must be borne in mind that the interpretation of the ring chromosome data, and indeed most of the human data on the XIC, depends crucially on the chromosomal rearrangement having occurred prior to the X-inactivation step itself.

This remark applies less to the mouse, as female mouse ES cells contain two active X chromosomes. X-Inactivation occurs on differentiation in vivo or in vitro. Deletions of the *Xic* region can therefore be targeted using homologous recombination into the region of interest in such undifferentiated ES cells prior to the onset of X-inactivation.

Recently, Penny et al. (1996) used such an approach to target a deletion into one of the two *Xist* alleles of a female XX ES cell line and investigated the effects of this mutation following differentiation in vitro and in vivo. Mutation appeared to disrupt the X-inactivation of the targeted X chromosome, which failed to inactivate while the untargeted X chromosome continued to inactivate normally. These results clearly reinforce the hypothesis that *Xist* is required *in cis* for X-inactivation to occur and have been interpreted by the authors as suggesting that both counting of the number of copies of the X chromosome present in the cell relative to the number of autosomes and the choice of which of the two X chromosomes present in the cell will be inactivated may not involve *Xist* itself.

Xist may therefore be *necessary* but not *sufficient* for the initiation of X-inactivation. Indeed, not only the results of Penny et al. (1996), but also both our own results concerning the *Xce* locus and those of Migeon et al. (1993) concerning the human X-ring chromosomes can be interpreted as suggesting that the initiation of X-chromosome inactivation requires more than *Xist* itself.

Gain of Function

A second approach to identifying elements that may be concerned in X-inactivation initiation and to defining the minimal chromosomal domain that is necessary depends on the study of biological material carrying additional copies of parts of the *Xic*/XIC region to be examined. Fontes and his colleagues have exploited the case of a male patient carrying a single X chromosome with a duplication of the XIC region including XIST to explore whether the presence of an additional XIC region was sufficient to confer X-inactivation in the male (Muscatelli et al. 1992). No signs of X-inactivation involving the duplicated region were detected when analysis at the level of replication timing, XIST expression, and the activity of other genes in the duplicated region were examined. Two major hypotheses can be envisaged to explain this result: First, two copies of the XIC on two different X chromosomes are required for X-inactivation to occur. Second, it is the absence of the counting elements or other elements acting upstream of the XIC that is responsible for the absence of phenotype.

In the mouse, transgenic animals can be created to test in a similar manner whether sequences from the *Xic* region that include *Xist* are sufficient for the initiation of inactivation. Heard et al. (1996) created a series of transgenic mouse strains by direct pronuclear injection into the fertilized egg of a 460-kb *Xist*-containing YAC. In each case, a single copy of the YAC or a fragment derived from this YAC had inserted into an autosomal location. The transgenic lines were analyzed for *Xist* expression, expression of two other genes carried by the YAC, *Cdx4* and

Bpx that are normally silent on the inactive X chromosome, and a *LacZ* reporter gene that had been inserted into one of the arms of the YAC. No *Xist* expression was detected and clear evidence for the non-inactivation of the reporter gene and of the *Bpx* and *Cdx4* genes carried by the YAC was obtained. Since identical results were obtained with both male and female transgenic mice carrying either a single copy or two copies of the transgene, the lack of transgene inactivation cannot be due to the absence of either the counting mechanism or another element capable of activity *in trans*. Such transgenic data are compatible with the idea that essential sequences for *Xist* expression and X-inactivation are absent from the YAC array used. Such sequences may include the *Xce* locus and also other *cis* elements necessary for correctly constituting the *Xic* chromosomal domain.

One possibility that must be considered with such in vivo experiments is whether embryonic lethality associated with YAC transgene insertion into an autosome might lead to the loss of precisely those animals in which the YAC is capable of undergoing inactivation, thereby inducing the partial monosomy of neighboring autosomal sequences. Several lines of argument, including the overall efficiency of transgenesis which was very similar to that obtained by other investigators using pronuclear microinjection of YACs, make this possibility unlikely (for discussion, see Heard et al. 1996).

Matsuura et al. (1996), using a similar approach to Heard and her collaborators, have derived two transgenic lines using a 350-kb YAC clone encompassing *Xist*. One of the transgenic lines carries four copies of the YAC integrated into an autosomal site, and the second line carries four copies integrated into a site on the long arm of the Y chromosome. *Xist* expression was not observed in mice carrying the autosomal transgene, in agreement with the above results obtained by Heard et al. (1996). Some *Xist* expression was, however, observed from the Y chromosome transgenic array commensurate with full expression of a single *Xist* gene. Unfortunately, data as to the status of the other expressed genes could not be obtained for *Cdx4* due to fragmentation of the YAC.

The finding by Matsuura et al. (1996) of *Xist* expression from the YAC copies on the Y chromosome is compatible with the idea that *Xist* itself may be capable of responding to the heterochromatinized state present on the Y chromosome long arm or, alternatively, that such sequences can, albeit apparently inefficiently, allow both *Xist* expression and inactivation of the YAC assembly. Further exploration of additional YAC transgenes will be necessary to clarify the length of sequence necessary for reproducible *Xist* expression and X-inactivation. Neither of the hypotheses advanced to explain the currently obtained YAC transgenesis results invalidates the data obtained by gene targeting as to the necessary nature of *Xist* for X-inactivation to proceed.

CONCLUSION AND PERSPECTIVES

Increasing evidence suggests that the gene silencing associated with X-inactivation involves complex interactions leading to the induction of an altered chromatin state. The mechanisms and the cast of players underpinning this epigenetic process remain for the most part unknown. The experimental confirmation of a role for *Xist* in the inactivation process represents an important step forward, as does our increasing knowledge of its immediate genetic environment.

Despite the superficial similarities between the *H19* and *Xist* molecules, it remains to be established to what extent common mechanisms prevail in systems such as imprinting and X-inactivation. Nature's formidable capacity to reach similar operational endpoints by diverse means should never be underestimated. Epigenetic gene regulation in nonmammalian organisms is marked by great diversity in the genes and proteins involved, even though the basic endpoint—an altered chromatin structure—is the same. This clearly contrasts with the organizational and functional evolutionary conservation of genes, such as the homeobox genes, concerned in laying down the body plan during embryogenesis in different species where a basic repertoire of genes appears to be used. This has important potential implications in assessing the relevance of data obtained in nonmammalian systems. Many of the genes involved in sex determination and dosage compensation in *Drosophila*, such as scute, are also involved in neurogenesis and play important roles in the development of both the embryonic and peripheral nervous systems of larvae and adult flies. Does this have any implications for the mammalian systems? Could *Bpx*, which encodes a *Nap1*-related nucleosomal protein and lies within the *Xic*, actually be involved in the inactivation process itself?

With respect to such questions, it is important to stress that imprinting and X-inactivation clearly differ from the systems of dosage compensation found in other organisms by their ability to identify and control individual chromosomal homologs within the same cell. It may well be this requirement for marking and recognizing homologs that lie within the same cell, a requirement that has been restricted during evolution to only a handful of species, that will underlie any communality between the X-inactivation and imprinting systems.

ACKNOWLEDGMENTS

The parts of this review relating to work in the author's laboratory were supported by grants from the Centre National de la Recherche Scientifique (URA 1968), the Groupemente de recherches et d'etudes sur les genomes (no. 68/94), and the Association Française contre les Myopathies.

REFERENCES

Ariel, M., E. Robinson, J.R. McCarrey, and H. Cedar. 1995. Gamete-specific methylation correlates with imprinting of the murine *Xist* gene. *Nat. Genet.* **9:** 312–315.

Bates, G. and H. Lehrach. 1994. Trinucleotide repeat expansions and human genetic disease. *BioEssays* **16:** 277–284.

Bartolomei, M.S., S. Zemel, and S.M. Tilghman. 1991. Parental imprinting of the mouse *H19* gene. *Nature* **351:** 153–155.

Borsani, G., R. Tonlorenzi, M.C. Simmler, L. Dandolo, D. Arnaud, V. Capra, M. Grompe, A. Pizzuti, D. Muzny, C. Lawrence, H.F. Willard, P. Avner, and A. Ballabio. 1991. Characterization of a murine gene expressed from the inactive X chromosome. *Nature* **351:** 329–331.

Brannan, C.I., E. Claire Dees, R. Ingram, and S.M. Tilghman. 1990. The product of the *H19* gene may function as an RNA. *Mol. Cell. Biol.* **10:** 28–36.

Brockdorff, N., A. Ashworth, G.F. Kay, V.M. McCabe, D. Norris, P.J. Cooper, S. Swift, and S. Rastan. 1992. The product of the mouse Xist gene is a 15kb inactive X-specific transcript containing no conserved ORF and located in the nucleus. *Cell* **71:** 515–526.

Brockdorff, N., A. Ashworth, G.F. Kay, P. Cooper, S. Smith, V.M. McCabe, D. Norris, G.D. Penny, D. Patel, and S. Rastan. 1991. Conservation of position and exclusive expression of mouse Xist from the inactive X chromosome. *Nature* **351:** 329–331.

Brown, C.J, and H.F. Willard. 1994. The human X-inactivation centre is not required for maintenance of X-chromosome inactivation. *Nature* **368:** 154–156.

Brown, C.J., A. Ballabio, J.L. Rupert, R.G. Lafrenière, M. Grompe, R. Tonlorenzi, and H.F. Willard. 1991. A gene from the region of the human X inactivation centre is expressed exclusively from the inactive X chromosome. *Nature* **349:** 38–44.

Brown, C.J., B.D. Hendrich, J.L. Rupert, R.G. Lafrenière, Y. Xing, J. Lawrence, and H.F. Willard. 1992. The human XIST gene: Analysis of a 17kb inactive X-specific RNA that contains conserved repeats and is highly localised within the nucleus. *Cell* **71:** 527–542.

Brown, C.J., R.G. Lafrenière, V.E. Powers, G. Sebastio, A. Ballabio, A.L. Pettigrew, D.H. Ledbetter, E. Levy, I.W. Craig, and H.F. Willard. 1991b. Molecular localization of the X inactivation center on the human X chromosome in Xq13. *Nature* **349:** 82–84.

Buzin, C.H., J.R. Mann, and J. Singer-Sam. 1994. Quantitative RT-PCR assays show Xist RNA levels are low in mouse female adult tissue, embryos and embryoid bodies. *Development* **120:** 3529–3536.

Cattanach, B. and C.E. Williams. 1972. Evidence of non-random X-chromosome activity in the mouse. *Genet. Res.* **19:** 229–240.

Cattanach, B.M., C. Rasberry, E.P. Evans, and M. Burtenshaw. 1989. *Mouse News Lett.* **83:** 165.

Clemson, C.M. J.A. Mc Neil, H.F. Willard, and J.B. Laurence. 1996. Xist RNA paints the inactive X chromosome at interphase: Evidence for a novel RNA involved in nuclear/chromosome structure. *J. Cell Biol.* **132:** 259–275.

Courtier, B., E. Heard, and P. Avner. 1995. Xce haplotypes show modified methylation in a region of the active X chromosome lying 3′ to Xist. *Proc. Natl. Acad. Sci.* **92:** 3531–3535.

Dandolo, L, C.L. Stewart, M.G. Mattei, and P.R. Avner. 1993. Inactivation of an X-linked transgene in murine extraembryonic and adult tissues. *Development* **118:** 641–649.

De Chiara, T.M., E.J. Robertson, and A. Efstratiadis. 1991. Parental imprinting of the mouse insulin-like growth factor II gene. *Cell* **64:** 849–859.

Epstein, C.J. 1983. The X chromosome in development. In *Cytogenetics of the mammalian X chromosome. Part A. Basic mechanisms of X chromosome behaviour* (ed. AA. Sandberg), pp. 51–65. Alan Liss, New York.

Hammond-Kosack, M.C.U, B. Dobrinski, R. Lurz, K. Docherty, and M.W. Kilpatrick. 1992. The human insulin gene polymorphic region exhibits an altered DNA structure. *Nucleic Acids Res.* **20:** 231–236.

Heard, E., M.C. Simmler, Z. Larin, C. Rougeulle, B. Courtier, H. Lehrach, and P. Avner. 1993. Physical mapping and YAC contig analysis of the region surrounding Xist on the mouse X chromosome. *Genomics* **15:** 559–569.

Heard, E., C. Kress, F. Mongelaard, B. Courtier, C. Rougeulle, A. Ashworth, C. Vourc'h, C. Babinet, and P. Avner. 1996. Transgenic mice carrying on *Xist*-containing YAC. *Hum. Mol. Genet.* **5:** 441–451.

Hendrich, B.D., C.J. Brown, and H.F. Willard. 1993. Evolutionary conservation of possible functional domains of the human and murine Xist genes. *Hum. Mol. Genet.* **2:** 663–672.

Horn, J.H. and A. Ashworth. 1995. A member of the Caudal family of homeobox genes maps to the X-inactivation centre region of the mouse and human X chromosomes. *Hum. Mol. Genet.* **4:** 1041–1047.

Kazemi-Esfarjani, P, M.A Triforo, and L. Pinsky. 1995. Evidence for a repressive function of the long polyglutamine tract in the human androgen receptor: Possible pathogenetic relevance for the (CAG)n-expanded neuropathies. *Hum. Mol. Genet.* **4:** 523–527.

Kennedy, G.C., M.S. German, and W.J. Rutter. 1995. The minisatellite in the diabetes susceptibility locus IDDM2 regulates insulin transcription. *Nat. Genet.* **9:** 293–298.

Lafrenière, R.G., C.J. Brown, S. Rider, J. Chelly, P. Taillon-Miller, A.C. Chinault, A.P. Monaco, and H.F. Willard. 1993. 2.6 Mb YAC contig of the human X inactivation center region in Xq13: Physical linkage of the RPS4X, PHKA1, XIST and DXS128E genes. *Hum. Mol. Genet.* **2:** 1105–1115.

Leppig, K.A., C.J. Brown, S.L. Bressler, K. Gutashaw, R.A. Pagon, H.F. Willard, and C.M. Disteche. 1993. Mapping of the distal boundary of the X-inactivation center in a rearranged X chromosome from a female expressing XIST. *Hum. Mol. Genet.* **2:** 82–84.

Lyon, M.F. 1983. The X chromosome and their levels of activation. In *Cytogenetics of the mammalian X chromosome. Part A. Basic mechanisms of X chromosome behaviour* (ed. A.A. Sandberg), pp. 187–204. Alan Liss, New York.

Matsuura, S., V. Episkopou, and S.D.M. Brown. 1996. Xist expression from an Xist YAC transgene carried on the mouse Y chromosome. *Hum. Mol. Genet.* **5:** 451–459.

Migeon, B.R., S. Luo, B.A. Stasiowoki, M. Jani, J. Axelman, D.L. Van Dyke, L. Weiss, P.A. Jacobs, T.L. Yang-Feng, and J.E. Wiley. 1993. Deficient transcription of XIST from tiny ring X chromosomes in females with severe phenotypes. *Proc. Natl. Acad. Sci.* **90:** 12025–12029.

Muscatelli, F., D. Lena, M.G. Mattei, and M. Fontes. 1992. A male with two contiguous inactivation centers on a single X chromosome: Study of X inactivation and XIST expression. *Hum. Mol. Genet.* **1:** 115–119.

Neumann, B., P. Kubicka, and D.P Barlow. 1995. Characteristics of imprinted genes. *Nat. Genet.* **9:** 12–13.

Norris, D.P., D. Patel, G.F. Kay, G.D. Penny, N. Brockdorff, S.A. Sheardown, and S. Rastan. 1994. Evidence that random and imprimed expression is controlled by preemptive methylation. *Cell* **77:** 41–51.

Owerbach, D. and K.H. Gabbay. 1993. Localisation of a type 1 diabetes susceptibility loss to the variable tandem repeat region flanking the insulin gene. *Diabetes* **42:** 1708–1714.

Penny, G.D., G.F. Kay, S.A. Sheardown, S. Rastan, and N. Brockdorff. 1996. The Xist gene is required in cis for X chromosome inactivation. *Nature* **379:** 131–137.

Rack, K.A., J. Chelly, R.J. Gibbons, S. Rider, D. Benjamin, R.G. Lafrenière, D. Oscier, R.W. Hendriks, I.W. Craig, H.F. Willard, A.P. Monaco, and V.J. Buckle. 1994. Absence of the XIST gene from late replicating isodicentric X chromosomes in leukemia. *Hum. Mol. Genet.* **3:** 1053–1059.

Rastan, S. 1983. Non-random X-chromosome inactivation in mouse X-autosome translocation embryos—location of the X inactivation centre. *J. Embryol. Exp. Morphol.* **78:** 1–22.

Rastan, S. and E.J. Robertson. 1985. X-chromosome deletions in embryo-derived (EK) cell lines associated with lack of X-chromosome inactivation. *J. Embryol. Exp. Morphol.* **90:** 379–388.

Richards, R.I and G.R. Sutherland. 1994. Simple repeat DNA is not replicated simply. *Nat. Genet.* **6:** 114–116.

Rougeulle, C. and P. Avner. 1996. Cloning and characterization of a murine brain specific gene Bpx and its human homolog lying within the Xic candidate region. *Hum. Mol. Genet.* **5:** 41–49.

Rougeulle, C., L. Colleaux, B. Dujon, and P. Avner. 1994. Generation and characterization of an ordered lambda clone array for the 460-kb region surrounding the murine Xist sequence. *Mamm. Genome* **5:** 416–423.

Russell, L.B. 1983. X-autosome translocations in the mouse. Their characterization and use as tools to investigate gene inactivation. In *Cytogenetics of the mammalian X chromosome. Part A. Basic mechanisms of X chromosome behaviour* (ed. A.A. Sandberg), pp. 205–250. Alan Liss, New York.

Simmler, M.-C., B.M. Cattanach, C. Rasberry, C. Rougeulle, and P. Avner. 1993. Mapping the murine Xce locus with (CA)n repeats. *Mamm. Genome* **4:** 523–530.

Sutherland G.R and R.I. Richards. 1995. Simple tandem repeats and human genetic disease. *Proc. Natl. Acad. Sci.* **92:** 3636–3641.

Taketo, M., T.A. Howard, and M.F. Seldin. 1992. Mapping of recombinant retrovirus integration sites that cause expression of the viral genome in murine embryonal carcinoma cells. *Mamm. Genome* **2:** 240–245.

Therman, E. and G.E. Sarto. 1983. Inactivation center on the human X chromosome. In *Cytogenetics of the mammalian X chromosome. Part A. Basic mechanisms of X chromosome behaviour* (ed. A.A. Sandberg), pp. 315–325. Alan Liss, New York.

Wolff, D.J., C.J. Brown, S. Schwartz, A.M.V. Duncan, U. Surti, and H.F. Willard. 1994. Small marker X chromosomes lack the X inactivation center implications for karyotype/phenotype correlations. *Am. J. Hum. Genet.* **55:** 87–95.

Wu, H., R. Fassler, A. Schnieke, D. Barker, K.H. Lee, V. Chapman, U. Francke, and R. Jaenisch. 1992. An X-linked human collagen transgene escapes X inactivation in a subset of cells. *Development* **116:** 687–695.

Zuccotti, M. and M. Monk. 1995. Methylation of the mouse Xist gene in sperm and eggs correlates with imprinted Xist expression and paternal X-inactivation. *Nat. Genet.* **9:** 316–320.

Epigenetic Silencing and Activation of a Maize *r* Gene

Jerry L. Kermicle
Laboratory of Genetics
University of Wisconsin
Madison, Wisconsin 53706

INTRODUCTION

The *r* Gene Family

The *r* (*red color*) family of maize genes codes for transcriptional activators of the basic helix-loop-helix class (Ludwig et al. 1989) that regulate genes in the anthocyanin biosynthetic pathway (for review, see Dooner et al. 1991). Members of the *r* gene family, which also includes *b* (*booster*) and *lc*/*sn* (*leaf color* and *scutellar node* color), show high sequence homology in their coding regions, and their products are functionally interchangeable (Goff et al. 1990; Ludwig et al. 1990). However, sequences in their control regions are divergent, and these differences govern the distribution of anthocyanin pigmentation plant part by plant part (Styles et al. 1973; Radicella et al. 1992). Because the pigment is conspicuous, subtle differences between genes and alleles have been discerned. Genes of the *r* family thus serve as sensitive reporters of their own regulation. Perhaps for this reason, a number of epigenetic phenomena involving this gene family have been discovered (for review, see Brink 1973; Patterson and Chandler 1995). Those involving the *r* locus are discussed here; those involving *b* are discussed separately (Chandler et al., this volume).

Among strains of maize collected from diverse geographic sources, the *r* locus on chromosome 10 sometimes is represented by one *r* gene, sometimes by two or more, thus comprising a gene complex. When multiple, each *r* gene constitutes a separate complementation group, with different combinations of *r* genes resulting in different pigmentation patterns. For example, the *R-r:standard* complex comprises a simple *r* gene, *P*, which confers plant color, and a compound gene, *S*, which confers seed color (Walker et al. 1995). In some strains the additional *r* gene designated as *lc* or *sn* resides at an insertional duplication two map units distal from the *r* complex on chromosome 10L.

 0-87969-490-4/96 $5 + .00

Paramutation and Imprinting

When maintained homozygous, most alleles of *r* are stable, including those that pigment the aleurone layer of kernels (*R* class). However, *R* genes indigenous to regions outside the Andean Highlands of South America, such as *S* in *R-r:standard*, confer a low level of pigmentation after being heterozygous with most Andean *R* complexes, such as *R-stippled* (*R-st*) (Brink 1956). Brink (1958) termed this alteration "paramutation." Andean complexes that incite the change are said to be paramutagenic; sensitive genes are paramutable; altered *R* is paramutated, designated *R'*. A minority of Andean *R* genes are paramutationally neutral (Van der Walt and Brink 1969), as are others worldwide that do not confer seed color (*r* class).

Seed pigmentation by *R-r* represents a narrow window in the maize life cycle for detecting whether paramutation has occurred. Darkly mottled kernels are produced both when *R-st*/*R-st* plants and plants homozygous for the null allele (*r*) are pollinated with *R-r*/*R-r*. Only when *R-r*/*R-st* F_1 plants are grown and then crossed to *r*/*r* as female are the effects of paramutation clearly evident (Fig. 1A). Thus, paramutation occurs during a period of phenotypic eclipse between fertilization and the production of kernel progeny from testcrosses on *r*/*r*.

Curiously, the effect of paramutation is not seen when the testcross is made in the reciprocal direction, with *r*/*r* as male and *R-r*/*R-st* as female. In this case, *R-r*-carrying kernels typically are solidly pigmented (Fig. 1A) as though paramutation had not occurred. However, progeny grown

Figure 1 (*A*) Testcross kernels illustrating *R-r:standard* action before and after being heterozygous with *R-st* (paramutation; *left* vs. *right* columns) and when transmitted to the endosperm via ovule and pollen (imprinting; *upper* vs. *lower* rows). Anthocyanin pigments the outer layer of the endosperm, whose genotype is indicated. *R-r:standard* is abbreviated *R*. An *R* allele that has experienced prior heterozygosity with *R-st* is indicated by *R'*. In the genotype given for the triploid endosperms, alleles of maternal origin are on the left of the slash and the paternal allele is on the right. All the kernels were produced by crossing *R*- or *R'*-carrying plants with colorless (*r*/*r*) tester strains. (*B*) Pairs of testcross kernels resulting from crossing three first-generation paramutant progeny plants (*R'*/*r*) to *r*/*r*. The plant-to-plant variation seen in this generation traces to three *R'* gametes from a single *R*/*R-st* plant. (*C*) Allelic structure and phenotypic effects of the alleles used to illustrate paramutation. *R-r:standard* is adapted from Walker et al. (1995); *R-st* from Eggleston et al. (1995). Coding and promoter regions of *r* are magnified relative to regions between genes (*thin lines*).

from such kernels and tested on *r*/*r* females exhibit paramutation. Pigmentation has been reduced to the same level on average as when plants grown from overtly paramutated kernels produced in the reciprocal cross are tested similarly (Cooper 1964). Where had the window of opportunity for observing paramutation closed when *R-r*/*R-st* was female?

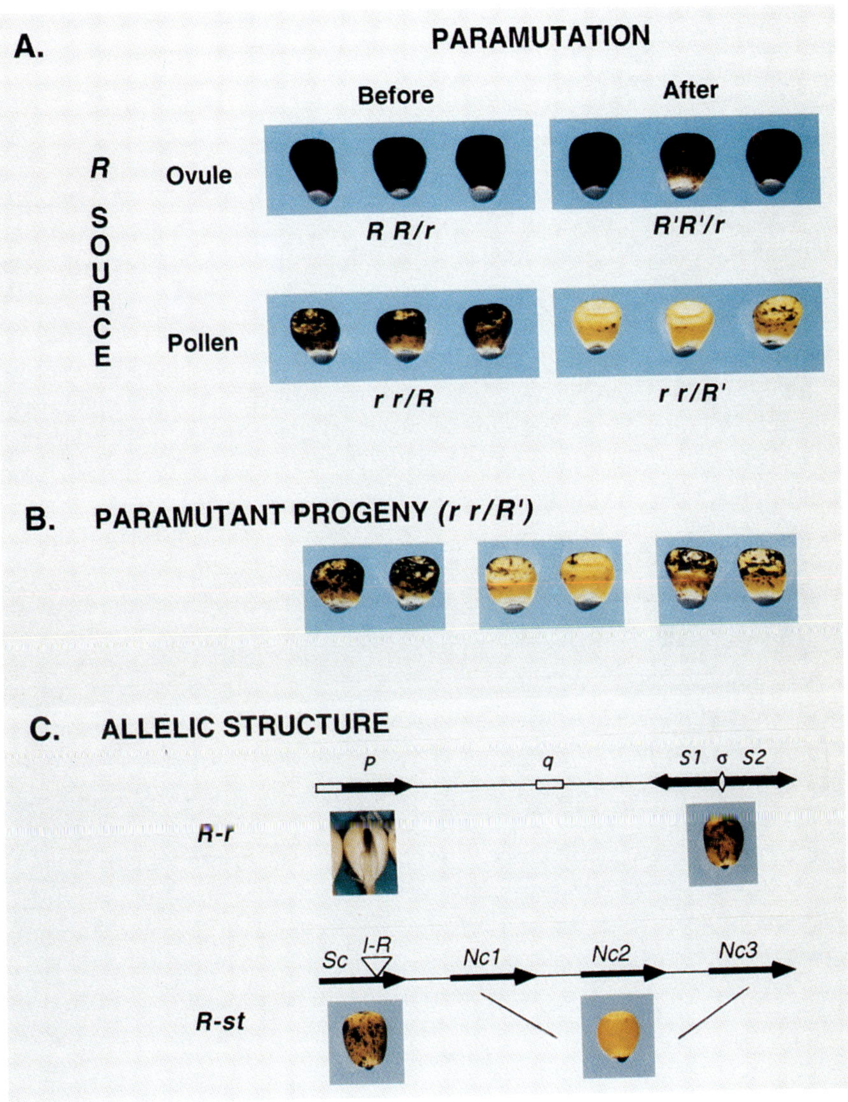

Figure 1 (See facing page for legend.)

The answer lies with the origin of the aleurone. It originates not as part of the embryo proper but as part of the extraembryonic endosperm, a triploid tissue derived by a separate fertilization of the ovule's binucleate central cell. In terms of genotype, aleurone resulting from *r*/*r* female x *R-r*/*R-st* male crosses is *r*/*r* // *R'*, whereas that produced in the reciprocal cross is *R'*/*R'* // *r* (the double slash separates alleles of maternal and paternal origins).

The difference between *r*/*r* // *R'* and *R'*/*R'* // *r* phenotypes proved not to be due to differences in *R* and *r* dosage, but to the mode of *R'* transmission (Kermicle 1970b, 1978). Maternally derived *R'* is activated, substantially masking the effect of prior paramutation. Like silencing through paramutation, activation of *R'* by parental imprinting is chromosomally based. Hence, *r* imprinting falls within Brink's (1960) broadened sense of paramutation to include systems of inheritance which, although chromosomally based, do not conform to Mendelian assumptions concerning genetic stability. Jorgensen (1994) recently has emphasized the usefulness of broad-sense "paragenetics" as a term identifying a subclass of heterogeneous epigenetic phenomena.

This review relates the molecular organization of pigmenting genes in *r* complexes to chromosome segments associated with paramutation. Characterization of *r* paragenetic changes also has begun. This information promises to illuminate the relationship between paramutation and imprinting and opens the way to consider how these changes come about.

R PARAMUTATION: A CASE OF EPIGENETIC SILENCING

Genetic Properties

In paramutationally active heterozygotes, virtually 100% of *R* gametes can be altered (Brown 1968). The extent of alteration, however, depends on the particular heterozygote and conditions of the test. Andean *R* complexes and their derivatives differ in paramutagenic strength (McWhirter and Brink 1962; Van der Walt and Brink 1969), whereas paramutable genes from diverse geographic origins differ in sensitivity (Bray and Brink 1966). Paramutation in active heterozygotes progresses over generations (Mikula 1961; McWhirter and Brink 1962) and increases with dosage of the paramutagenic complex in trisomic plants (Kermicle et al. 1995). When the sensitive gene was carried in rearranged chromosomes (Kester and Brink 1966; Brink 1969) or when F_1 seeds were irradiated (Shih and Brink 1969), paramutation decreased detectably. It was reduced markedly when seeds were mutagenized with alkylating agents (Axtell and Brink 1967; Brink et al. 1968). Environmental manipulation of seedlings also influences the extent of paramutation (Mikula 1995). The effectiveness of various treatments administered to seeds and young plants suggests that events critical to paramutation occur early in plant development.

The paramutant phenotype is heritable but not stable. In combination with a neutral allele or when hemizygous, *R'* reverts partially in the first generation, and recovery continues over generations in an erratic, apparently stochastic manner (Brink et al. 1968; Styles and Brink 1969). In long-term homozygous lineages, the level of expression appears to stabilize. When two *R'/R'* lineages were followed for ten progeny generations, a marked partial reversion occurred in the first generation, a small increment of additional reversion occurred in the second, and no further change was detected in generations three through ten (Kermicle [cited in Brink 1964] and unpubl.). Because *R'* itself is weakly paramutagenic (Brown and Brink 1960), mutual paramutation between the two alleles in *R'R'* homozygotes may result in a state of conditional stability.

Levels of paramutation vary among gametes produced by individual plants (Fig. 1B), resulting in progeny lineages with different phenotypic effects. By comparing samples of pollen from individual tassel branches, Sastry et al. (1965) showed a major component of this variation to derive from different initial levels of paramutation. Mosaicism was particularly evident in heterozygotes involving weakly paramutagenic alleles, where gametes from occasional branches escaped paramutation. First-generation progeny (*R'/r*) recovering from paramutation also were detectably mosaic. The results of Sastry et al. illuminate numerous experiments showing that generations of paramutation or reversion increase phenotypic variation in *R*, whereas generations of homozygosity decrease it.

The foregoing synopsis of *r* paramutation is based largely on behavior of the paramutable source *R-r:standard* (*R-r*), the paramutagenic source *R-stippled* (*R-st,*), and their derivatives. Genetic and molecular analyses revealed both to be complex, composed of more than one *r* gene. Recombination occasionally fractionates these complexes. In a pioneering study using *R-r/R-st* heterozygotes, Ashman (1965a) identified crossovers that were recombinant for pigmenting properties of the parent complexes. These hybrid complexes had different levels of paramutagenicity, and this study pointed the way for analyzing the organization of paramutational properties (Ashman 1965b). Unequal recombination also occurs in *R-r/R-r* and *R-st/R-st* homozygotes, generating deletions and higher-order duplications. The following discussion relies heavily on the characterization of deletions obtained in this manner.

Organization of Paramutability in *R-r:standard*

R-r:standard belongs to the colored-seed, red-plant class of *r* complexes. Unequal recombination in *R-r* homozygotes results in loss of plant or seed color (Stadler and Neuffer 1953; Dooner and Kermicle 1971). Molecular analyses on these derivatives performed in Stephen Dellaporta's laboratory confirmed the presence of duplicate *r* sequences (Robbins et al. 1991; Walker et al. 1995) and revealed the structure of *R-r* given in Figure 1C. The centromere-proximal *r* gene, known as *P*, confers plant color. Centrally located and 190 kb distal to *P* is *q*, a truncated *r* gene containing a *P*-like promoter but no *r* coding sequences. A tandemly inverted duplication involving two *r* coding sequences, *S1* and *S2*, is located another 15 kb distal. *S1* and *S2* flank a 400-bp junction region denoted as σ. It comprises terminal and subterminal repeats of *doppia*, a member of the CACTA family of transposable elements, plus a rearranged segment that includes sequences homologous to the promoter region of *P*. The σ region presumably is responsible for regulating *S1* and *S2* expression in the aleurone layer of seeds.

In *R-r*/*R-st* heterozygotes, sensitivity to paramutation regularly cosegregates with seed (*S*) pigmenting function. Brown (1968) identified 64 darkly pigmented *R-r* progeny kernels among 122,000 *R-r* gametes of *R-r*/*R-st* heterozygotes. These gametes appeared, therefore, to have escaped paramutation. When retested against *R-st,* however, all 64 were shown to retain full sensitivity in the next generation.

Figure 2 summarizes qualitatively the effect on paramutability of deleting various regions within the *R-r* complex. Deletions extending from the promoter region of *P* to a corresponding region in *q* cause loss of plant color (*R-g* derivatives). *R-g* seed color and sensitivity to *R-st* remained unchanged relative to *R-r* (Brown 1966). Removal of the region immediately distal to *S2* from *R-g* produced similar results (Kermicle et al. 1995). In this case, the deletions extend from *S2* to *lc*, the duplicated *r* sequence displaced two centimorgans distal to the principal *r* locus (Dooner and Kermicle 1976).

In contrast, each of 20 losses of seed color (*r-r* derivatives) was paramutationally inactive (Brown 1966). Because *r-r* alleles confer a colorless seed phenotype, paramutation of these alleles was evaluated by their inability to acquire paramutagenicity from *R-st.* Subsequent molecular analysis showed such *r-r* derivatives to be of two classes (Robbins et al. 1991). A majority result from unequal crossing-over between *P* and *S2*, deleting the entire intervening region. A minority involve deletion confined to the *S* complex, typically removing the shared promoter region and a portion of one or both *S* coding regions. Walker et

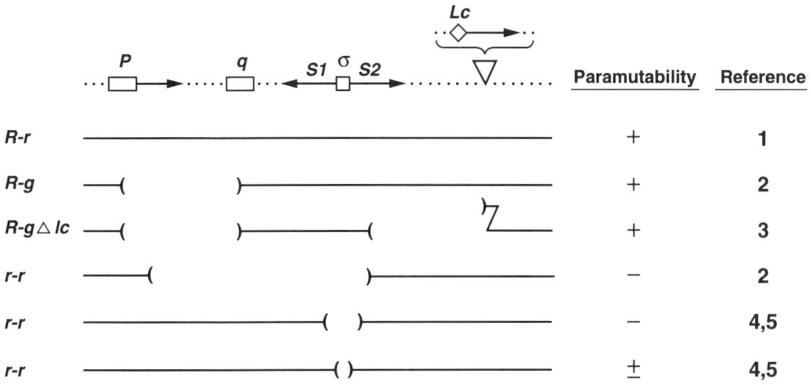

	Paramutability	Reference
R-r	+	1
R-g	+	2
R-g△*lc*	+	3
r-r	−	2
r-r	−	4,5
r-r	±	4,5

Figure 2 Deletion analysis of *R-r* sensitivity to paramutation (paramutability). In the heading, *r* coding regions are shown by arrows, promoter regions by the open square, diamond, and rectangles. The inverted triangle represents a displaced duplication that includes the *Lc* gene. The deletions occurred by unequal crossing-over in *R-r* homozygotes (plus or minus the *r*-duplicate gene *Lc*) or from spontaneous deletion within the inverted *S* duplication. References: (1) Brink 1956; (2) Brown 1966; (3) Kermicle et al. 1995; (4) Walker et al. 1995; (5) Table 1 of this paper.

al. (1995) characterized five *r-r* deletions of this class. Table 1 reports an initial evaluation of their potential to acquire paramutagenicity during three generations of heterozygosity with *R-st*. Only the smallest deletion (*r-r: N1-3-1*) acquired a significant level of paramutagenicity. It reduced pigmentation of the test allele by 0.14 units, compared with 0.42 units for *R-r'*. This 0.4-kb deletion is nearly coextensive with σ, the presumed promoter region of *S1* and *S2*. Hence, this region and that delimited by the next smallest deletion (paramutationally neutral) appear to be particularly involved in mediating paramutation.

The presence of *doppia* within this region raises the presumption of its involvement in paramutation. Since *doppia* sequences are repetitive in the maize genome, *S* might be subject to a general system of transposon or repetitive DNA regulation. In plants, repetitive sequences typically are sequestered by cytosine 5-methylation (Antequera and Bird 1988; Bennetzen et al. 1994). Regulation of the multicopy maize transposon families *Spm*/*En* and *Mu* by methylation has been characterized in detail (Fedoroff; Martienssen; both this volume). In both systems, autonomous elements influence methylation of nonautonomous receptor elements *in trans*. Remarkably, this influence is not limited to elements in allelic positions, and multiple receptor elements can respond coordinately.

Table 1 Acquisition of paramutagenicity by *R-r* and its *r-r* derivatives deleted within the seed color (*S*) gene

Test allele	Reduction in pigmentation[a]		Size of *S* deletion (kb)
R-r control	+0.02[b]	n.s.[d]	0
R-r'	+0.42[c]	**	0
r-r:n35	−0.01	n.s.	4.3
r-r:n101	−0.06	n.s.	1.1
r-r:n156	−0.02	n.s.	3.4
r-r:N1-3-1	+0.14	*	0.4
r-r:N1-3-2	+0.18	n.s.	2.0

[a]*R-r* and the five *r-r* strains were crossed three generations to *R-st* and then, together with control *R-r*, to a colorless-seed, green-plant stock (*r-g*/*r-g*). The resulting *R-r'*/*r-g*, *r-r*/*r-g*, and *R-r*/*r-g* heterozygotes were crossed to a stabilized *R-g'*/*R-g'* strain to compare the effects on *R-g'* of *R-r'*, *r-r*, and *R-r* with paramutagenically neutral *r-g*. Effects on kernel pigmentation were quantified visually by matching kernels produced from testcrosses on *r-g*/*r-g* against a standard set of six which, together with colorless, define seven pigmentation classes (Brink et al. 1960). Values given are the average testcross scores for *R-g'*/*r-g* plants minus that of *R-g'*-carrying sibs that were heterozygous for one of the alleles under test (*R-r'*, *r-r*, or control *R-r*.)

[b]Average of six test lines.

[c]Average of five test lines.

[d](n.s.) Not statistically significant; (*) $p < 0.05$; (**) $p < 0.01$.

Within the maize genome, *doppia* is present upstream of the *Rhoades* allele of *pl* (Cone et al. 1993). (*pl* also codes for a regulator of anthocyanin biosynthesis, but it does not share coding sequence similarity with *r*.) Intriguingly, *Pl:Rhoades* recently was reported to be unstable, changing spontaneously to a paramutagenic state (Hollick et al. 1995; Chandler et al., this volume). At present, however, there is no direct evidence linking paramutation at *pl* and *r* through *doppia*.

Organization of Paramutagenicity in *R-st*

A combination of genetic and molecular evidence has revealed the overall structure of *R-st* illustrated in Figure 1C. *R-st* is unstable both somatically and meiotically. Reversion of *R-st* to full or "self" colored (*R-sc*, Ashman 1960) results when the transposable element *I-R (Inhibitor of R)* excises from the strong aleurone pigmenting gene *Sc* (McWhirter and Brink 1962; Williams et al. 1984). Self-colored and other classes of variants also originate through intralocus recombination, indicating complexity of the allele (Ashman 1965a, 1970; Kermicle 1970a; Gavazzi 1977). Molecular analyses reveal four *r*-homologous genes in *R-st* (Eggleston et al. 1995). Unequal crossover derivatives from *R-st*/*R-st* homozygotes retained either three, two, or one *r* gene. Because the series is complete, all the *r* gene segments should lie in the same orientation. Previously, Alleman and Kermicle (1993) showed that *Sc* in the single *r* derivative allele *R-sc:124* was oriented with its promoter toward the centromere. This is the same orientation as the *P* and *S2* genes in *R-r* (Fig. 1C). Molecular characterization is not yet sufficiently detailed to identify the duplication junctions in *R-st*. A different segment may have been duplicated than in *R-r:standard*. Thus, whereas the *R-r* duplication includes the neighboring gene *Inhibitor of striate*, this gene maps distal to the *R-st* complex (Dooner and Kermicle 1974; Kermicle and Axtell 1981; Eggleston et al. 1995).

Do the four *r*-homologous DNA segments in *R-st* reported by Eggleston et al. (1995) encompass functionally independent *r* genes? Through recombination, Ashman (1965a) showed the kernel phenotype of *R-st* to comprise two distinct parts: (1) clonal spots of intense pigment and (2) faint, incompletely penetrant background pigmentation distributed in a mottled pattern and described as near-colorless (Nc). By examining revertant and recombinant alleles, Eggleston et al. (1995) found the intense spotting to be associated with excision of a 3.3-kb insert (*I-R*) from the 3′ end of the proximal *r* gene, *Sc*. The three remaining *r* segments were associated with the near-colorless phenotype and were designated *Nc1*, *Nc2*, and *Nc3*. Many of the crossovers that served to separate *Sc* and *Nc* genes occurred within *r* coding regions, producing chimeric genes. In each case, the 5′ portion determined the phenotype; no differences in phenotype attributable to the 3′ ends of various *r* genes (*Sc*, *Nc1*, *Nc2*, *Nc3*, *P*, *S2*, and *Lc*) have been reported.

Recombinant complexes and those resulting from excision of the transposable element *I-R* have been useful in mapping paramutagenic regions within *R-st*. Derivatives that lose *I-R*, but retain the four *r* segments intact, retain full paramutagenicity (McWhirter and Brink 1962; Kermicle et al. 1995). In contrast, when unequal crossing-over in *R-st* homozygotes resulted in deletion of all but a single *r* gene, the resulting products were not detectably paramutagenic. Reduction in the number of *r* gene segments from four through three to two resulted in a progressive decline in paramutagenicity, both when lost in the order *Nc1* through *Nc3* and when lost in reverse order. A parallel series, graded in level of paramutagenicity, was identified among recombinant products of *R-r*/

R-st heterozygotes (Ashman 1965a). One such red-plant, near-colorless aleurone isolate was characterized subsequently as *P* plus *Nc3*. When heterozygous with an *Sc::I-R* plus *Nc1* allele from the first series, unequal crossing-over produced derivatives having increased copy number. Both of the two-copy parents were weakly paramutagenic. The three-*r*-copy resynthesized derivatives were moderately strong, and the four-*r*-copy derivatives were as strongly paramutagenic as *R-st* (Kermicle et al. 1995).

Figure 3 summarizes the results of the *R-st* deletion and resynthesis experiments diagrammatically. The nonparamutagenicity of single-copy alleles shows that some function essential to paramutagenicity resides within the deleted regions. Four-*r*-copy complexes are strongly paramutagenic, and the duplicate *r* gene *lc* (*leaf color*) seems to be able to substitute for one *Nc* gene. All three-*r*-copy derivatives are moderately strong. It is notable that the deletions involved in the different classes of three-copy derivatives overlap. Hence, no particular region of the *Nc* complex is essential for paramutagenicity. The two-copy derivatives are only weakly paramutagenic, indicating that multiple segments are required for the moderately strong level characteristic of three-copy complexes.

The multigenic basis of paramutagenicity in *R-st* could reflect additive, independent action of the different *Nc*-containing segments or some form of complementary interaction. If the effect is additive, increasing the dosage of weakly paramutagenic two-copy derivatives should increase paramutagenicity. Marked increases occurred for both two-copy derivatives (third and fourth from last entries in Fig. 3) tested by doubling the dosage through the use of trisomic-*10* plants (Kermicle et al. 1995). For weakly paramutagenic *R-st*Δ*lc*, two doses overlapped the effect of *R-st* itself, suggesting equivalence of two *r* copies in each of two chromosomes to four in one. Furthermore, the effect of one each of the two-copy derivatives was intermediate between two doses of either one, indicating an absence of complementation. The results of these experiments are consistent with a cumulative effect on paramutagenicity of the *Sc*- and *Nc*-marked segments in *R-st*.

The *r* genes in *R-st* are not contiguous but are separated by from 10 kb to 25 kb of intervening sequences (W.B. Eggleston, pers. comm.). Therefore, change in *r* copy number by unequal crossing-over is accompanied by a loss or gain of regions between *r* genes. Thus, it is possible that paramutagenicity is mediated by the intergenic regions rather than by the *r* genes themselves. To make this distinction, crossovers within *r* genes involving nonparamutagenic *r* alleles and the *r* genes of

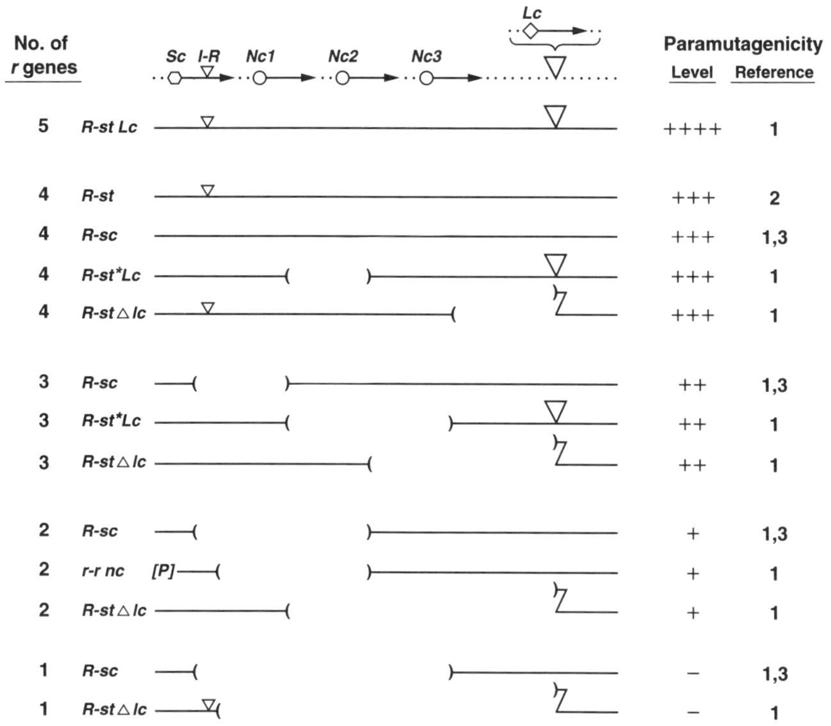

Lc

Sc I-R Nc1 Nc2 Nc3

No. of r genes		Paramutagenicity Level	Reference
5	R-st Lc	++++	1
4	R-st	+++	2
4	R-sc	+++	1,3
4	R-st*Lc	+++	1
4	R-st△lc	+++	1
3	R-sc	++	1,3
3	R-st*Lc	++	1
3	R-st△lc	++	1
2	R-sc	+	1,3
2	r-r nc [P]	+	1
2	R-st△lc	+	1
1	R-sc	–	1,3
1	R-st△lc	–	1

Figure 3 Deletion analysis of *R-st* paramutagenicity. *r* coding regions and promoter symbols are as in Fig. 2. The small inverted triangle represents the transposable element *I-R*. The deletions occurred from either loss of the transposable element *I-R* (producing *R-sc*), from unequal crossing-over in *R-st* and *R-st Lc* homozygotes (producing *R-sc* and *R-st*Δ*lc*), from recombination between *R-r* and *R-st* (producing *r-r nc*), or from recombination between partially deleted products (producing reconstituted *R-st*Lc*). References: (1) Kermicle et al. 1995; (2) Brink 1956; (3) McWhirter and Brink 1962.

R-st are useful. Crossovers between the *Sc* gene of *R-st* with non-paramutagenic *R-navajo* and with the *P* gene of *R-r:std* have been identified and characterized. Exchange between the segment extending from the promoter to the site of *I-R* insertion in *Sc* with the homologous region in *R-nj* or *P* imposes unstable expression on the tissues pigmented by *R-nj* or *P*, and transfers the near-colorless genes from *R-st* to the recombinant product. In both series of recombinants, paramutagenicity was reduced by an amount roughly equivalent to the reduction from four-*r*-copy to three-*r*-copy *R-sc* derivatives of *R-st* (Satyanarayana 1970; Williams 1972). Hence, this increment of paramutagenicity appears to map 5′ to the *I-R* insertion site in *Sc*. Comparable intragenic recombinants between nonparamutagenic *r* genes and the *Nc* genes in *R-st* have not been identified.

Are Paramutable and Paramutagenic Components Related to Each Other?

Early studies suggested that regions at *r* controlling paramutability and paramutagenicity not only were distinct from pigmentation genes, but also were different from one another (Brink 1964; Van der Walt and Brink 1969). In part, this view was supported by the nonoverlapping geographic distribution of the two classes of *r* complexes. Subsequent findings point to a closer relationship between them, however.

The *S* gene of *R-r* and the *Nc* genes of *R-st* share a number of genetic features. First, both pigment the aleurone layer of seeds. Second, the distribution of pigmented aleurone cells due to *Nc* resembles the mottled distribution characteristic of *S* when inherited from the male. Pigmented and colorless cells are intermingled. Variable proportions of cells in the resulting patches are pigmented, unlike the solidly colored spots resulting from excision of the *I-R* transposon from the *Sc* gene in *R-st*. Third, within a given series of recombinant products, the higher the paramutagenic level, the lower the level of (*Nc*) pigmentation (Satyanarayana 1970; Eggleston et al. 1995; Kermicle et al. 1995). These evidences support the view that the *Nc* genes in *R-st* are themselves paramutable (Satyanarayana 1970). *S* of *R-r* behaves in parallel in that *R-r* becomes paramutagenic as the level of *S* pigmentation is reduced by paramutation. A more detailed analysis of the *Nc* genes is needed to determine whether their promoter regions are homologous with *S*. Even if homologous, this relationship need not underlie paramutagenicity of other *r* complexes. In *R-marbled*, for example, (*Nc*) phenotype evidently is absent (Weyers 1961).

PARENTAL IMPRINTING: A CANDIDATE FOR EPIGENETIC ACTIVATION

Reciprocal crosses of *R-r/R-st* with colorless *r/r* produce contrasting endosperm phenotypes (see Introduction). When *R-r/R-st* is male, the effect of paramutation is seen as a reduction from intensely to weakly mottled *r/r // R′* aleurone. When *R-r/R-st* is female, the resulting *R′/R′ // r* aleurone of most kernels is solid-colored, with occasional mottling (Fig. 1A). The difference between *r/r // R′* and *R′/R′ // r* could reflect an *R′* to *r* dosage inequality resulting from the double contribution of genes from the ovule's central cell to endosperm tissue. Alternatively, this difference in pigmentation could result from genomic imprinting or other sorts of maternal effects. When parental dosage was equalized by introducing two *r*-carrying chromosomes through pollen, *r*/*r* // *R*′/*R*′ vs. *R*′/*R*′ // *r*/*r* aleurones retained the mottled versus full-color distinction of *r*/*r* // *R*′ versus *R*′/*R*′ // *r* (Kermicle 1970b). Other sorts of maternal effects could be discounted, and the difference between maternal solid color and paternal mottling was demonstrated to be caused by chromosomally based parental imprinting (Kermicle 1978).

There are two ways in which parental imprinting might influence *R* gene expression in aleurone tissue. The *R* allele that is inherited from the male parent might be repressed by imprinting, or the *R* allele that is inherited from the female parent might be activated by imprinting. In animal systems, it is often assumed that imprinting involves gene repression, although the evidence is not always clear. One way to distinguish these possibilities is first to identify genetic modifiers that affect the imprinting process, and then to determine in which parent the modifiers act. Genetic evidence supporting maternal activation of *R* derives from analysis of one such modifier gene and is described below.

When a variant form of a modifier gene known as *mdr* is inherited maternally along with *R*′, the aleurone phenotype is no longer solid-colored (Kermicle 1978). Instead it is mottled, resembling *R*′ inherited from a male. Thus, the variant form of *mdr* results in failure of imprinting. The variant form was viewed as a loss of wild-type function, and hence the modifier gene was named after its assumed wild-type function, *maternal derepression of r*. Viewed otherwise, however, the variant allele (*mdr*) could be interpreted as a gain of function relative to wild type (*Mdr*), repressing maternal as well as paternal *R*. In the former view, wild type is expected to be dominant to variant; in the latter view, variant is expected to be dominant to wild type. A ready assessment of dominance was precluded by *Mdr*'s critical action in the female gametophyte, which is haploid (Brink et al. 1970; Kermicle and Alleman 1990). To test dominance, disomic *Mdr*/*mdr* female gametophytes were synthesized from plants trisomic for the region of chromosome 4 bearing the *mdr* locus. When transmitted together with *R*′, the aleurone was solid-colored, indicating dominance of the wild-type allele (Kermicle 1995).

Because paternally and maternally derived *R* phenotypes differ more markedly following paramutation (Fig. 1A), imprinting and paramutation appear to be associated phenomena. Furthermore, whereas genes known to imprint are sensitive to paramutation, the phenotypes of both genes of paramutagenic complexes tested depended on dosage rather than parentage. However, in two circumstances, *r* imprinting and paramutation are effectively uncoupled. The *R-cherry* complex is sensitive to paramutation; but unlike *R-r:standard*, it showed little if any response to transmission through the female (J.L. Kermicle, unpubl.). Second, in an *mdr* genetic background, female- as well as male-transmitted *R-r′* exhibits a paramutant phenotype. Thus, neither an imprintable *r* allele nor a genetic background in which imprinting is observed appears necessary for paramutation to occur.

DNA METHYLATION IN PARAMUTATION AND IMPRINTING

The nature of the molecular alteration that distinguishes *R′* from *R* is a central question of paramutation, as is the nature of the activation in maternally imprinted alleles. For that reason, a summary of observations regarding cytosine methylation of *r* sequences is given, although studies are still in progress (Eggleston et al. 1995; M. Alleman; W. Eggleston; both unpubl.). In the following summary, altered digestibility by methylation-sensitive restriction enzymes has been interpreted as reflecting altered DNA methylation, although other interpretations are possible. (For a general review of methylation in eukaryotes, see Martienssen and Richards 1995.)

1. Genomic restriction analysis using methylation-insensitive enzymes has not revealed DNA structural differences attributable to paramutation and imprinting. Using methylation-sensitive enzymes, differences also have not been detected in the 3′ portion of *r* coding regions. However, differences are found using probes that span the inferred transcription start site.
2. Promoter region DNA from the *S* (seed color) and *P* (plant color) genes of *R-r*/*R-r* seedlings is normally partially methylated, but the pattern of methylation is changed following paramutation.
3. Promoter region DNA from the *Nc* (near colorless) genes in *R-st*/*R-st* seedlings is extensively methylated, whereas the *Sc* promoter is partially methylated. In derivative alleles with fewer *Nc* genes, no change was detected in *Nc* methylation, but the methylation of *Sc* became progressively reduced. No corresponding difference in *Sc*-mediated pigmentation was observed.
4. When inherited from the male, *S* gene methylation levels in DNA extracted from aleurones resembled that from seedlings. When inherited from the female, the *S* gene promoter was less methylated in aleurone DNA, correlating with the increased level of pigmentation. This difference occurred both in parental and paramutated *R-r* stocks. Maternally inherited *P* gene sequences were also less methylated in

aleurone DNA, although the *P* gene does not contribute to aleurone pigmentation.

5. Methylation of maternally inherited *S* and *P* genes is increased in DNA from *mdr/mdr // Mdr* aleurones, correlating with reduced levels of seed pigmentation.
6. Pigmentation conferred by the *r* duplicate gene *Sn* correlated inversely with *Sn* promoter methylation (Ronchi et al. 1995). Methylation increased after paramutation but decreased after treatment with 5-azacytidine.

The discovery of base modification attending both paramutation and imprinting provides a means of envisioning *r* paragenetic alterations. Such modifications may prove to be primary or secondary, and other mechanisms of epigenetic inheritance are not excluded. It serves minimally as a means to organize the available evidence. The key features to be accounted for include the potential for bidirectional change, the quantitative nature of the change, and the metastability and heritability of altered forms.

R derivatives generated by paramutation range continuously from virtually colorless through intermediate grades toward the parental level. Initially, the quantitative variation was difficult to explain short of postulating repetitive, extragenic inhibitory units (Brink 1964). Alternatively, McWhirter (1968) speculated that differential methylation of cytosine residues within *r* genes could account for the variation. The number of potentially methylated sites and combinations thereof could yield an overlapping array of *R* products. Because DNA methyltransferases prefer hemimethylated duplexes as substrates, the symmetry of CG and CNG sequences provides a likely basis for somatic propagation and sexual transmission of given methylation patterns. Mechanisms leading to change in methylation are less obvious. A number of influences likely are at play. A residual level of de novo methylation (possibly nonspecific) and paramutation or other means of site-specific methylation promote modification. Conditions associated with demethylation include hemizygosity and ovule transmission.

PARAMUTATION AS A COMBINATION OF *CIS* AND *TRANS* INACTIVATION PHENOMENA

cis-Acting features of *r* paramutation include (1) the negative effect of *Nc* copy number on pigmentation in *R-st*, (2) the paramutability of *R-r*'s plant color only when coupled with seed color (Brown 1966), and (3) the presence of transposon terminal repeats in the 5′ region of the *S* gene.

Because *cis*-arranged homologous sequences are known to recombine in plant somatic cells (see, e.g., Swoboda et al. 1994), direct DNA-DNA interaction should be considered as a means of mediating these changes. One possible mechanism involves cross-hybridization between strands of homologous duplexes to form a double Holliday structure (Flavell 1994; Matzke et al. 1994). The hybrid strands constitute heteroduplexes in terms of methylation differences present in the parent duplexes (Fig. 4). DNA methyltransferase then acts on the resulting hemimethylated substrates to fix methylation in the hybridized region. This means of epigenetic, or, in the present context, paragenetic, transformation could spread and stabilize methylation within tandem repeats.

Homologous DNA-DNA interactions presumably occur less efficiently between sequences present *in trans*. This expectation is based on the rank of mitotic recombination frequencies whereby adjacent *cis* copies >

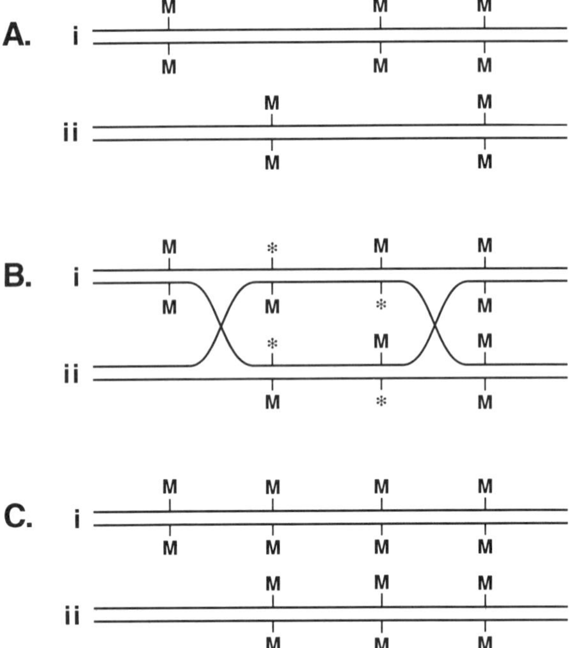

Figure 4 Transfer of cytosine methylation between homologous DNA duplexes, as proposed by Matzke et al. (1994) and Flavell (1994). (*A*) Differentially methylated duplexes *i* and *ii*. (*B*) Local cross-hybridization between *i* and *ii* forming hemizygous methylation sites (*), a preferred substrate for DNA methyltransferase. (*C*) Molecules after resolution. Gene conversion or crossing-over need not have occurred.

displaced *cis* copies > *trans* copies. The gradient is best established in yeast, *Ascobolus*, and *Drosophila*, all eukaryotes having moderately small genomes. These categories may differ still more in organisms containing large amounts of repetitive DNA, such as maize.

Distinctly *trans* features of paramutation include the reduction in *S* gene action in *R-r/R-st* heterozygotes, the acquired paramutagenicity of *S*, and the conditional stability of *S* in *R/R* and *R′/R′* homozygotes. Such *trans* interactions may be mediated by *r*-encoded products, either RNA, or protein acting as an autogenous regulator. The involvement of RNA is particularly attractive because of the sequence-specific recognition properties of nucleic acids. For example, aberrant transcripts might accumulate in the nucleus and hybridize with homologous DNAs. A suggestive analogy involves the potato spindle tuber RNA virus (Wassenegger et al. 1994). A cDNA copy of the viroid integrated in the tobacco genome became methylated, but only following autonomous viroid RNA-RNA replication. Flanking T-DNA and tobacco genomic sequences were not affected. The mechanism of RNA-dependent methylation was not evident.

Some *trans* paramutational changes progress gradually, over the course of several plant generations. Change in a single generation can be imperceptible. For changes that occur infrequently per generation, the possibility of direct DNA-DNA interaction is not unthinkable. One example is the gradual enhancement of nonparamutant *R* expression when maintained heterozygous with a paramutationally neutral allele (Styles and Brink 1966). In the absence of a homologous template for copy correction, any spontaneous loss of methylation might become fixed. Another example is the acquisition of paramutagenicity by the deletion derivative *r-r:N1-3-1*. Mediation of paramutagenicity at the product level seems unlikely in this case, since deletion removed the presumed promoter region of the *S* complex.

Observations concerning *r* paramutation are consistent with the view that gene silencing via methylation may be a means of protection against foreign DNA segments such as transposable elements (Kricker et al. 1992). Some eukaryotic transposable elements overreplicate and transpose preferentially to nearby chromosomal sites. De novo methylation of either copy might be transferred to the other by *cis*-dependent mechanisms, become fixed, and then spread to *trans* copies, silencing expression. Insertion of a transposon near or within a gene, such as *doppia* within *S*, could bring the gene under the influence not only of the transposon, but also of the host's system for regulating the transposon. This process would be not host-sequence-specific, but general.

SUMMARY

The seed color gene (*S*) of the maize *R-r* complex undergoes both epigenetic silencing (paramutation) and activation (parental imprinting). Formal genetic analysis serves to identify conditions promoting these changes and to describe the heritability and potential variability of the altered forms. Progress is being made in understanding the organization of *r* complexes that respond to and promote paragenetic alterations. Further work is needed to disclose the nature of the primary alterations, the mechanisms bringing them about, and the basis for their propagation.

ACKNOWLEDGMENTS

I thank Mary Alleman and William Eggleston for permission to cite unpublished findings; Ron Okagaki for discussions; Millard Susman, William Eggleston, and Rob Martienssen for detailed critiques of the manuscript; and Sherry LaRue for technical assistance concerning it. This research is supported by Department of Energy grant FG02-86ER13539 and U.S. Department of Agriculture grant 91-37301-6379. This is paper No. 3452 from the Laboratory of Genetics.

REFERENCES

Alleman, M. and J.L. Kermicle. 1993. Somatic variegation and germinal mutability reflect the position of transposable element *Dissociation* within the maize *R* gene. *Genetics* **135:** 189–203.

Antequera, F. and A.P. Bird. 1988. Unmethylated CpG islands associated with genes in higher plant DNA. *EMBO J.* **7:** 2295–2299.

Ashman, R.B. 1960. Stippled aleurone in maize. *Genetics* **45:** 19–34.

———. 1965a. Mutants from maize plants heterozygous *R-r R-st* and their association with crossing over. *Genetics* **51:** 305–312.

———. 1965b. Paramutagenic action of mutants from maize plants heterozygous *R-r R-st*. *Genetics* **52:** 835–841.

———. 1970. The compound structure of the *R-st* allele in maize. *Genetics* **64:** 239–245.

Axtell, J.D. and R.A. Brink. 1967. Chemically induced paramutation at the *R* locus in maize. *Proc. Natl. Acad. Sci.* **58:** 181–187.

Bennetzen, J.L., K. Schrick, P.S. Springer, W.E. Brown, and P. San-Miguel. 1994. Active maize genes are unmodified and flanked by diverse classes of modified, highly repetitive DNA. *Genome* **37:** 565–576.

Bray, R.A. and R.A. Brink. 1966. Mutation and paramutation at the *R* locus in maize. *Genetics* **54:** 137–149.

Brink, R.A. 1956. A genetic change associated with the *R* locus in maize which is directed and potentially reversible. *Genetics* **41:** 872–889.

———. 1958. Paramutation at the *R* locus in maize. *Cold Spring Harbor Symp. Quant.*

Biol. **23:** 379–391.

———. 1960. Paramutation and chromosome organization. *Q. Rev. Biol.* **35:** 120–137.

———. 1964. Genetic repression of *R* action in maize. In *Role of the chromosomes in development* (ed. M. Locke), pp. 183–230. Academic Press, New York.

———. 1969. Abnormal chromosome *10* and *R* paramutation in maize. *Mutat. Res.* **8:** 285–302.

———. 1973. Paramutation. *Annu. Rev. Genet.* **7:** 129–152.

Brink, R.A., J.L. Kermicle, and N.K. Ziebur. 1970. Derepression in the female gametophyte in relation to paramutant *R* expression in maize endosperms, embryos, and seedlings. *Genetics* **66:** 87–96.

Brink, R.A., E.D. Styles, and J.D. Axtell. 1968. Paramutation: Directed genetic change. *Science* **159:** 161–170.

Brink, R.A., D.F. Brown, J.L. Kermicle, and W.H. Weyers. 1960. Locus dependence of the paramutant *R* phenotype in maize. *Genetics* **45:** 1297–1312.

Brown, D.F. 1966. Paramutability of *R-g* and *r-r* mutant genes derived from an *R-r* allele in maize. *Genetics* **54:** 899–910.

———. 1968. A test to find non-paramutable variants of the *R-r* gene in maize. *Proc. Int. Congr. Genet.* **12:** 254.

Brown, D.F. and R.A. Brink. 1960. Paramutagenic action of paramutant *R-r* and *R-g* alleles in maize. *Genetics* **45:** 1313–1316.

Cone, K.C., S.M. Cacciolone, C.A. Moehlenkamp, T. Weber, B.J. Drummond, L.A. Tagliani, B.A. Bowen, and G.H. Perrot. 1993. Role of the regulatory gene *pl* in the photocontrol of maize anthocyanin pigmentation. *Plant Cell* **5:** 1807–1816.

Cooper, H.B., Jr. 1964. "Paramutation and reversion of the *R-r* allele in maize." Ph.D. thesis, University of Wisconsin, Madison.

Dooner, H.K. and J.L. Kermicle. 1971. Structure of the *R-r* tandem duplication in maize. *Genetics* **67:** 427–436.

———. 1974. Reconstitution of the *R-r* compound allele in maize. *Genetics* **78:** 691–701.

———. 1976. Displaced and tandem duplications in the long arm of chromosome *10* in maize. *Genetics* **82:** 309–322.

Dooner, H.K., T.P. Robbins, and R.A. Jorgensen. 1991. Genetic and developmental control of anthocyanin biosynthesis. *Annu. Rev. Genet.* **25:** 173–199.

Eggleston, W.B., M. Alleman, and J.L. Kermicle. 1995. Molecular organization and germinal instability of *R-stippled* maize. *Genetics* **141:** 361–372.

Flavell, R. B. 1994. Inactivation of gene expression in plants as a consequence of specific sequence duplication. *Proc. Natl. Acad. Sci.* **91:** 3490–3496.

Gavazzi, G. 1977. The genetic complexity of the *R* locus in maize. *Stadler Genet. Symp.* **9:** 37–61.

Goff, S.A., T.M. Klein, B.A. Roth, M.E. Fromm, K.C. Cone, J.P. Radicella, and V.L. Chandler. 1990. Transactivation of anthocyanin biosynthetic genes following transfer of *B* regulatory genes into maize tissues. *EMBO J.* **9:** 2517–2522.

Hollick, J.B., G.I. Patterson, E.H. Coe, Jr., K.C. Cone, and V.L. Chandler. 1995. Allelic interactions heritably alter the activity of a metastable maize *pl* allele. *Genetics* **141:** 709–719.

Jorgensen, R. 1994. Developmental significance of epigenetic imposition on the plant genome: A paragenetic function for chromosomes. *Dev. Genet.* **15:** 523–532.

Kermicle, J.L. 1970a. Somatic and meiotic instability of *R-stippled*, an aleurone spotting factor in maize. *Genetics* **64:** 247–258.

———. 1970b. Dependence of the *R*-mottled aleurone phenotype in maize on mode of sexual transmission. *Genetics* **66:** 69–85.

———. 1978. Imprinting of gene action in maize endosperm. In *Maize breeding and genetics* (ed. D.B. Walden), pp. 357–371. Wiley, New York.

———. 1995. Location, time of action, and dominance relations of an imprintor of *R*-mottled maize. In *Modification of gene expression and non-Mendelian inheritance* (ed. K. Oono and F. Takaiwa), pp. 119–134. National Institute of Agrobiological Resources, Tsukuba, Japan.

Kermicle, J.L. and M. Alleman. 1990. Genetic imprinting in maize in relation to the angiosperm life cycle. *Dev. Suppl.*, pp. 9–14.

Kermicle, J.L. and J.D. Axtell. 1981. Modification of chlorophyll striping by the *R* region. *Maydica* **26:** 185–197.

Kermicle, J.L., W.B. Eggleston, and M. Alleman. 1995. Organization of paramutagenicity in *R-stippled* maize. *Genetics* **141:** 361–372.

Kester, D.E. and R.A. Brink. 1966. Variation in paramutability of *R-r* in maize following extraction of the allele from a linked reciprocal translocation. *Genetics* **51:** 1401–1407.

Kricker, M.C., J.W. Drake, and M. Radman. 1992. Duplication-targeted DNA methylation and mutagenesis in the evolution of eukaryotic chromosomes. *Proc. Natl. Acad. Sci.* **89:** 1075–1079.

Ludwig, S.R., B. Bowen, L. Beach, and S.R. Wessler. 1990. A regulatory gene as a novel visible marker for maize transformation. *Science* **247:** 449–450.

Ludwig, S.R., L.F. Habera, S.L. Dellaporta, and S.R. Wessler. 1989. *Lc*, a member of the maize *R* gene family responsible for tissue-specific anthocyanin production, encodes a protein similar to transcriptional activators and contains the *myc*-homology region. *Proc. Natl. Acad. Sci.* **86:** 7092–7096.

Martienssen, R.A. and E.J. Richards. 1995. DNA methylation in eukaryotes. *Curr. Opin. Genet. Dev.* **5:** 234–242.

Matzke, A.J.M., F. Neuhuber, Y.-D. Park, P.F. Ambros, and M.A. Matzke. 1994. Homology-dependent gene silencing in trangenic plants: Epistatic silencing loci contain multiple copies of methylated transgenes. *Mol. Gen. Genet.* **244:** 219–229.

McWhirter, K.S. 1968. Methylation of DNA as the molecular basis for paramutation. *Maize Genet. Coop. Newsl.* **42:** 160–161.

McWhirter, K.S. and R.A. Brink. 1962. Continuous variation in level of paramutation at the *R* locus in maize. *Genetics* **47:** 1053–1074.

Mikula, B.C. 1961. Progressive conversion of *R*-locus expression in maize. *Proc. Natl. Acad. Sci.* **47:** 566–571.

———. 1995. Environmental programming of heritable epigenetic changes in paramutant *r*-gene expression using temperature and light at a specific stage of early development in maize seedlings. *Genetics* **140:** 1379–1387.

Patterson, G.I. and V.I. Chandler. 1995. Paramutation in maize and related allelic interactions. In *Gene silencing in higher plants and related phenomena in other eukaryotes* (ed. P. Meyer), pp. 121–141. Springer-Verlag, Berlin.

Radicella, J.P., D. Brown, L.A. Tolar, and V.L. Chandler. 1992. Allelic diversity of the maize *b* regulatory gene: Different leader and promoter sequences of two *b* alleles determine distinct tissue specificities of anthocyanin production. *Genes Dev.* **6:** 2152–2164.

Robbins, T.P., E.L. Walker, J.L. Kermicle, M. Alleman, and S.L. Dellaporta. 1991.

Meiotic instability of the *R-r* complex arising from displaced intragenic exchange and intrachromosomal rearrangement. *Genetics* **129:** 271–283.

Ronchi, A., K. Petroni, and C. Tonelli. 1995. The reduced expression of endogenous duplications (REED) in the maize *R* gene family is mediated by DNA methylation. *EMBO J.* **14:** 5318–5328.

Sastry, G.R.K., H. B. Cooper, Jr., and R.A. Brink. 1965. Paramutation and somatic mosaicism in maize. *Genetics* **52:** 407–424.

Satyanarayana, K.V. 1970. "Organization of the pigmenting and paramutagenic determinants of the *R-stippled* gene in maize." Ph.D. thesis, University of Wisconsin, Madison.

Shih, K.L. and R.A. Brink. 1969. Effects of X-irradiation on aleurone pigmenting potential of standard *R-r* and a paramutant form of *R-r* in maize. *Genetics* **61:** 179–189.

Stadler, L.J. and M. Neuffer. 1953. Problems of *R* gene structure. II. Separation of *R-r* elements (P) and (S) by unequal crossing over. *Science* **117:** 471–472.

Styles, E.D. and R.A. Brink. 1966. The metastable nature of paramutable *R* alleles in maize. I. Heritable enhancement in level of standard *R-r* action. *Genetics* **54:** 433–439.

———. 1969. The metastable nature of paramutable *R* alleles in maize. IV. Parallel enhancement of *R* action in heterozygotes with *r* and in hemizygotes. *Genetics* **61:** 801–811.

Styles, E.D., O. Ceska, and K.-T. Seah. 1973. Developmental differences in action of *R* and *B* alleles in maize. *Can. J. Genet. Cytol.* **15:** 59–72.

Swoboda, P., S. Gal, B. Hohn, and H. Pachta. 1994. Homologous recombination in whole plants. *EMBO J.* **13:** 484–489.

Van der Walt, W.J. and R.A. Brink. 1969. Geographic distribution of paramutable and paramutagenic *R* alleles in maize. *Genetics* **61:** 677–695.

Walker, E.L., T.P. Robbins, T.E. Bureau, J.L. Kermicle, and S.L. Dellaporta. 1995. Transposon-mediated chromosomal rearrangements and gene duplications in the formation of the maize *R-r* complex. *EMBO J.* **14:** 2350–2363.

Wassenegger, M., S. Heimes, L. Riedel, and S.L. Sänger. 1994. RNA-directed *de novo* methylation of genomic sequences in plants. *Cell* **76:** 567–576.

Weyers, W.H. 1961. Expression and stability of the marbled allele in maize. *Genetics* **46:** 1061–1067.

Williams, W.M. 1972. "Variability of the *R-stippled* gene in maize." Ph.D. thesis, University of Wisconsin, Madison.

Williams, W.M., K.V. Satyanarayana, and J.L. Kermicle. 1984. *R-stippled* maize as a transposable element system. *Genetics* **107:** 477–488.

b and *pl* Paramutation in Maize: Heritable Transcription States Programmed during Development

Vicki L. Chandler, Kenneth M. Kubo,[1] and Jay B. Hollick
Institute of Molecular Biology
University of Oregon
Eugene, Oregon 97403

As discussed in the accompanying chapters, most epigenetic regulatory mechanisms result in mitotically stable, but meiotically unstable, changes. The regulatory mechanism(s) associated with paramutation in flowering plants is exceptional because the changes initiated during somatic development are meiotically heritable; in a sense, blurring the distinction of epigenetic versus genetic alterations. Specifically, paramutation is an interaction between two alleles that leads to a directed alteration in transcription that is both mitotically and meiotically heritable. That such meiotically heritable changes appear to be confined to plants (Jorgensen 1993) suggests either that plants have evolved distinct regulatory mechanisms or that the basic differences in reproductive biology allow such somatic changes to be sexually transmitted (Patterson and Chandler 1995a). An understanding of paramutation should reveal how homologous chromosomes communicate in somatic cells and how transcription states that are programmed early in development can be stably transmitted through numerous cell divisions and into the next generation.

Paramutation was first described at the *r* locus in maize (Brink 1958) and has been studied at two other maize genes, *b* (Coe 1966) and *pl* (Hollick et al. 1995). All three loci encode transcription factors that regulate the purple anthocyanin pigment biosynthetic pathway. The *r* and *b* genes encode functionally equivalent basic-helix-loop-helix (bHLH) proteins, with different alleles at each gene expressed under distinct tissue-specific and developmental controls (Goff et al. 1990; Ludwig and Wessler 1990;

[1]Present address: Department of Molecular and Cellular Biology, University of Arizona, Tucson, Arizona 85721-1016.

 0-87969-490-4/96 $5 + .00

Ludwig et al. 1990; Radicella et al. 1992). All *b* alleles examined have only a single coding region, whereas most *r* alleles contain several coding regions. The *pl* gene encodes a myb-related transcription factor (Cone et al. 1993a). Most *pl* alleles are expressed throughout the seedling and mature plant, but the functionally equivalent *c1* gene (Paz-Ares et al. 1987) is expressed in the seed and embryo (Coe 1985). All characterized *c1* and *pl* alleles contain a single coding region. Transcriptional activation of the anthocyanin pathway requires the functional interaction of one bHLH factor and one MYB factor (Goff et al. 1992). This pathway is an excellent system for studying gene expression because the amount of pigment is a very sensitive indicator of the levels of the regulatory proteins, producing distinct phenotypes throughout development. This review focuses on *b* paramutation and *pl* paramutation, which share many common features. Because *r* paramutation is reviewed by Kermicle (this volume), in this chapter it is only briefly compared with *b* and *pl* paramutation.

When discussing paramutation, it is useful to define three types of alleles: paramutagenic, paramutable, and neutral. Paramutagenic alleles induce a change in sensitive paramutable alleles, whereas neutral alleles neither induce nor respond to paramutation. The paramutable *B-I* (*B-Intense*) (Table 1) and *Pl-Rh* (*Pl-Rhoades*) alleles are unstable, spontaneously producing the paramutagenic derivatives, *B'* and *Pl'-mah* (*Pl'-mahogany*) at variable frequencies that usually range between 1% and 10% of the progeny (Coe 1966; Patterson et al. 1993; Hollick et al.

Table 1 b and *pl* alleles

Allele	Phenotype	Paramutation activity
B-Intense (*B-I*)[a]	dark purple plant	paramutable
B'[a]	weak purple plant	paramutagenic
b-v[a]	green plant	paramutable
B-Peru[b]	weak plant color; strong purple seed color	neutral
Pl-Rhoades (*Pl-Rh*)[c]	dark purple plant; dark purple anthers	paramutable
Pl'-mahogany (*Pl'-mah*)[c]	mahogany red plant color; variegated anthers	paramutagenic
Pl-Blotched (*Pl-Bh*)[d]	variegated plant color; variegated anthers	neutral

Maize nomenclature convention is to represent dominant alleles with uppercase italics and recessive alleles with lowercase italics. (a) Coe 1966; (b) Patterson et al. 1995; (c) Hollick et al. 1995; (d) Cocciolone and Cone 1992.

1995). Paramutation involves the alteration of a darkly pigmented paramutable *B-I* or *Pl-Rh* allele to a lightly pigmented *B'* or *Pl'-mah* paramutagenic allele in heterozygous *B-I/B'* or *Pl-Rh/Pl'-mah* plants, such that only *B'* or *Pl'-mah* is transmitted. Linked genetic markers demonstrate that chromosomes entering these crosses carrying *B-I* or *Pl-Rh* exit carrying *B'* or *Pl'-mah* (Coe 1966; Hollick et al. 1995; Patterson and Chandler 1995b). Newly derived *B'* and *Pl'-mah* alleles continue to promote paramutation in subsequent crosses to naive (previously unexposed) *B-I* and *Pl-Rh* alleles. Paramutation is allele-specific; most *b* and *pl* alleles are neutral (Coe 1966; Patterson et al. 1993). There is no evidence for interaction between the loci, as paramutagenic *b* alleles have no effect on paramutable *pl* alleles and vice versa (J. Hollick and V. Chandler, unpubl.)

The paramutagenic alleles at *b* and *pl* very efficiently alter the paramutable alleles, with the reduced-pigment phenotype observable in young F_1 plants. The phenotype of heterozygous *B-I/B'* and *Pl-Rh/Pl'-mah* plants is lightly pigmented like *B'* and *Pl'-mah*. In addition, both paramutable alleles spontaneously change to paramutagenic alleles at a high frequency. These features of *b* and *pl* paramutation are not shared by *r* paramutation. During *r* paramutation, the paramutable allele (*R-r*) is altered to a lighter pigmented state (*R-r'*), but it is not changed into a strongly paramutagenic allele. In fact, the strongly paramutagenic alleles, such as R^{st} (*R-stippled*) and R^{mb} (*R-marbled*), do not derive spontaneously from any known paramutable allele and are quite distinct from paramutable *r* alleles in both phenotype and structure (Eggleston et al. 1995; Walker et al. 1995).

A major difference between *b* paramutation and *pl* paramutation is the stability of the paramutagenic state. The *B'* state is very stable (Coe 1966; Patterson et al. 1995), whereas *Pl'-mah* can revert to *Pl-Rh* (Hollick et al. 1995). Several reviews summarize the early literature on *r* and *b* paramutation and other examples of similar phenomena in plants (Brink 1964, 1973; Coe 1966; Brink et al. 1968; Patterson and Chandler 1995a). Herein, we review genetic and molecular characterizations which established that the altered pigment levels are caused by stable, but reversible, changes in transcriptional activity, and we discuss when these changes occur during development. Molecular comparisons of paramutable and paramutagenic *b* and *pl* alleles, and genetic experiments to identify *cis*-acting sequences and other genes required for paramutation, are summarized. Epigenetic and genetic models to explain paramutation are presented and discussed in relation to epigenetic phenomena in other organisms.

PROGRAMMING OF TRANSCRIPTION STATES DURING DEVELOPMENT

Paramutation Affects Transcription

Paramutation at either *b* or *pl* results in a decrease in epidermal and subepidermal pigment in numerous plant tissues. Reduced pigment is readily seen in seedling leaf sheaths (first 1–3 true leaves), juvenile leaf sheaths, mature plant leaf sheaths, husks, cob, culm, and tassel. RNA gel blots revealed a decrease in steady-state *b* and *pl* mRNA upon paramutation, and nuclear run-on transcription assays established that this decrease is accounted for by a decrease in transcription rate (Patterson et al. 1993; G. Patterson and V. Chandler, unpubl.). The size, splicing, and 5′ ends of *b* mRNA are not altered upon paramutation (Patterson et al. 1993). Both dramatic (10- to 20-fold) and subtle (~2-fold) changes in transcriptional activity are observed as visible changes in pigment levels (Patterson and Chandler 1995b; G. Patterson and V. Chandler, unpubl.). This sensitive visual assay showed that in the F_1, and following meiotic segregation, a "new" *B′* allele (*B-I* in the immediately preceding generation) was slightly darker than an "old" *B′* allele (*B′* in prior generations) (Patterson and Chandler 1995b). The new *B′* allele had about 2-fold higher levels of *b* mRNA than old *B′*. Thus, although all *B′* alleles have notably less *b* transcription and less pigment than *B-I*, different transcriptional states of *B′* can exist that are both mitotically and meiotically heritable. *Pl′-mah* also exists in different phenotypic states (Fig. 1) and, although transcription rates have not been compared for all these classes, they too likely reflect discrete transcriptional states.

A unique feature of *b* and *pl* paramutation is that paramutagenic alleles derive spontaneously from sensitive paramutable alleles. These spontaneous events occur during development and are often visualized as lighter pigmented sectors within a plant. When such sectors appear in the ear or tassel, lightly pigmented paramutagenic derivatives are observed in progeny. Our current hypothesis is that the spontaneous events that generate paramutagenic alleles are mechanistically related to classic paramutation that occurs in the F_1 between paramutagenic and paramutable alleles. Thus, we propose a broader definition for paramutation: a mitotically and meiotically heritable change in the transcriptional activity of a gene that can occur spontaneously or be induced by a paramutagenic allele.

Paramutation Occurs Early in Plant Development

The visual and cell-autonomous nature of anthocyanin synthesis, combined with the expression of the *b* and *pl* alleles in most plant parts, has

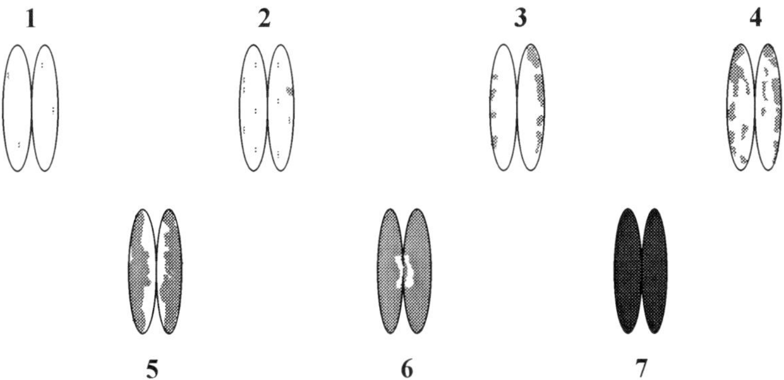

Figure 1 Schematic representation of *Pl'-mah* anther phenotypes. Anthers classified on the basis of visual anthocyanin pigments were assigned a particular anther color score (ACS) on a scale of 1–7 (Hollick et al. 1995). ACS of 1 has either no pigment or very few individual pigmented cells, and ACS of 7 is defined by the fully pigmented *Pl-Rh* state. Discrimination of the intervening ACS classes is based on the frequency and distribution of pigmented cells within the anther (for details, see Hollick et al. 1995).

been used to examine clonal relationships during corn development (Poethig 1987). The shoot meristem is an organized group of dividing cells that replenishes itself and gives rise to differentiated plant structures repeatedly throughout development. Because plant cells do not migrate, a particular cell's position within a plant reflects its developmental history. By simply looking at where pigment changes occur in plants undergoing paramutation, we can learn when in development the altered transcription states are established.

All pigmented tissues of *Pl'-mah/Pl-Rh* plants, from the seedling to the anthers, have less color than the *Pl-Rh* parent. That the reduced-pigment phenotype is observed in the first leaves of such F_1 plants suggests that paramutation has occurred during early development. The analogous change in *B'/B-I* plants also occurs early, but the absolute timing is slightly more variable, as the first leaf sheaths of different F_1 seedlings have pigment levels ranging from that of normal *B-I* seedlings to that of *B'* seedlings. Pigment in *B'/B-I* plants is nonetheless always significantly reduced relative to *B-I* in the second leaf and in the rest of the plant (Patterson and Chandler 1995b). The well-defined clonal lineages in maize combined with the non-sectored phenotype suggest that the altered transcription state is established very early during embryogenesis, prior to the point where multiple cells contribute clonal descendants to the aerial plant structures.

Spontaneous paramutation is often observed as the appearance of weakly pigmented progeny from two darkly pigmented parents (Coe 1966; Patterson et al. 1995). Spontaneous events are also visualized as lightly pigmented clonal sectors appearing in darkly pigmented plants (Coe 1966; K. Kubo et al., unpubl.). There is remarkable uniformity of the pigment level within each whole plant or independent sector, although the amount of pigment may be variable between different derivatives. Whole-plant events are most readily observed with *pl* alleles in the tassel where all the anthers are pigmented to the same extent (Fig. 1) (Hollick et al. 1995). Assuming that paramutation results in mitotically heritable changes in transcription states (as suggested by the clonal sectors that are observed), then whole-tassel events are a reflection of paramutation having occurred either very late in the previous generation or very early in embryo development. This is because the tassel comprises multiple cell lineages derived from the early proembryo (Poethig et al. 1986). Sectors contained within the ear and tassel demonstrate that the heritable change can also occur quite late, after ear and tassel primordia have been determined. Keep in mind that even though these states must be programmed early in development, transcription does not occur until terminal differentiation of the pigment-forming cells takes place. The variability of phenotypes exhibited between individuals and the uniformity of a given phenotype within a plant, or sector, suggest that paramutation can program quantitatively distinct transcriptional competency states, which are then faithfully maintained throughout somatic development and sexually transmitted to progeny. The observation that sectors can occur throughout the plant suggests that these events are not limited to a particular developmental time.

The fact that paramutation of a sensitive allele (*B-I* or *Pl-Rh*) invariably occurs when heterozygous with a paramutagenic partner (*B'* or *Pl'-mah*) suggests that an allelic communication allows efficient paramutation of the sensitive partner. The question then becomes how long can the alleles coexist in the same nucleus before this allele-based event takes place. Coe (1966) addressed this issue by inducing hemizygous sectors via X-ray-induced chromosome loss. His irradiations of *B-I*/*B'* developing embryos showed that loss of the chromosome arm containing *B'* resulted in plants with *B-I*-like pigmentation throughout or resulted in plants with large *B-I*-like sectors (1/2–1/8 plant sectors). The hemizygosity of the sectors was confirmed using a linked marker. Unfortunately, no progeny were obtained from these plants, so it was not possible to confirm that the *B-I* phenotype was meiotically heritable. However, these sectors suggest that the two alleles coexist for at least a few cell divisions without an irreversible change in the paramutable allele.

Coe's experiments demonstrate that *B-I* has not been permanently altered to *B'* in early embryogenesis. These experiments do not distinguish between *B-I* remaining unchanged versus the possibility that change is followed by reversion in the absence of the *B'* allele. Our phenotypic analyses indicate that the change has occurred prior to or during the formation of the first leaves. One possibility is that the transcriptional competency state is established during shoot meristem formation, such that all the cells that derive from the meristem during subsequent development will inherit that program. Alternatively, the state could be established later as cells differentiate, with the same state faithfully established numerous times during a single plant's reiterative development. An interesting puzzle is how one state is uniformly established within a plant, yet different plants establish distinct states. Further genetic mosaic analysis throughout development should reveal when paramutation occurs in heterozygous plants.

In contrast to the consistency of the phenotype within a given *Pl'-mah* plant (all anthers have the same *a*nther *c*olor *s*core, ACS), each anther of *Pl'-mah* plants has a variegated-pigment phenotype (Fig. 1). This variegation appears to represent nonclonal decisions to produce pigment. Within a plant, all the anthers contain a similar number of pigmented cells, suggesting that during development there is a characteristic probability that any particular anther cell will make pigment. This probability can differ between plants; anthers with higher ACS values have more pigmented cells.

Interestingly, the frequency and pattern of pigmented cells within an anther indicate both the extent to which paramutation has occurred and the probability with which a given transcriptional competency state will be meiotically inherited (Hollick et al. 1995). When plants with an ACS of 1–4 are crossed with *Pl-Rh*, they almost invariably transmit only highly paramutagenic *Pl'-mah* alleles. Progeny predominantly have phenotypes distributed among lower ACS values than that of their paramutagenic parent. No darkly pigmented *Pl-Rh*-like plants are recovered. This invariable meiotic transmission of a somatic transcription state is similar to that seen with distinct *B'* states (Patterson and Chandler 1995b; K. Kubo and V. Chandler, unpubl.). In contrast, when plants with an ACS value of 5 or 6 are crossed with *Pl-Rh*, they transmit a range of *Pl'-mah* alleles such that all the progeny together represent the full range of ACS values of 1–7. In such crosses, *Pl-Rh*-like plants are frequently recovered that maintain typical *Pl-Rh*/*Pl-Rh* transmission characteristics. We have not observed an analogous intermediate state (ACS 5–6) for *B'* alleles.

A fundamental difference between *Pl'-mah* (ACS 1–4) and *Pl'-mah* (ACS 5–6) anthers may account for their different transmission patterns. Two pigmented cell layers, each derived from distinct clonal lineages (Dawe and Freeling 1990), are found in the mature anther. This is easily observed, because the outer cells are arranged in longitudinal arrays and the inner cells are transversely oriented. The inner cells are devoid of pigment in *Pl'-mah* (ACS 1–4) individuals, whereas increasing numbers of these cells are pigmented in *Pl'-mah* (ACS 5–6) individuals. The significance of this observation is that the inner cells are clonally related with the gametes such that the genotype of these cells is often transmitted to progeny (Dawe and Freeling 1990). The stochastic pigmentation that occurs in the inner cell layer may thus reflect the situation in the gamete population; some will transmit a *Pl-Rh*-like transcriptional competency state, whereas others will maintain a *Pl'-mah* state. The correlation with increased numbers of unpigmented inner cells and more *Pl'-mah* progeny strengthens the interpretation that mitotically observed transcription states can be meiotically heritable.

STRUCTURAL FEATURES OF *b* AND *pl* ALLELES

The structures of the paramutable and paramutagenic alleles were examined to investigate the molecular basis for the different transcription states and the ability to cause or respond to paramutation. The alleles involved in *b* paramutation are molecularly simple, consisting of only a single coding region. Although *B-I* and *B'* adopt significantly different transcription states, no obvious differences in DNA structure are detectable between these alleles (extensive restriction mapping has been carried out from −11,000 to +13,000 relative to the transcription start at +1 [Patterson et al. 1993, 1995]). DNA sequence analysis of the proximal promoter region (−1487 to +730) revealed no differences between *B-I* and *B'* (Patterson et al. 1993). In addition, the changes in *B-I* and *B'* transcription states do not correlate with changes in methylation, as the methylation patterns of *B-I* and *B'* are indistinguishable. In total, more than 60 sites were compared using about 20 methylation-sensitive restriction enzymes (Patterson et al. 1993). Furthermore, in the region from −3500 to +5500, an estimated 20% of the cytosines in canonical methylation sites were assayed. For a given allele, some sites were unmethylated and other sites were partially methylated (10–50%); the remaining sites were completely methylated. Nevertheless, in virtually every instance, no consistent differences in methylation pattern were found between *B-I* and *B'*. At one site where a difference in methylation was initially observed, examination of a larger population of *B-I* and *B'* plants (~60 individuals) showed only a weak, and probably insignificant, correlation with genotype (Patterson et al. 1993). These results contrast with examples where changes in gene expression are correlated with differences in methylation. In these cases, the changes in methylation are extensive enough to be readily detected by examination of only a few restriction sites. The absence of extensive methylation changes during *b* paramutation is not surprising, as increased methylation changes are often associated with silenced genes, and *B'* transcription, although substantially less than *B-I*, is still fairly robust. If changes in methylation or DNA sequence do occur during paramutation, they must be very distant from the *b* coding region or involve subtle alterations.

An initial comparison of the *Pl-Rh* and *Pl'-mah* alleles has not revealed any differences in DNA structure (J. Hollick et al., unpubl.). Most *pl* alleles have numerous sequence differences relative to *Pl-Rh* and *Pl'-mah* (Cone et al. 1993b). A notable exception is *Pl-Bh* (*Pl-Blotched*), an allele of *pl* that is virtually identical in DNA sequence: Only ten nucleotide differences have been found within a 7.0-kb region, the closest of which are 1.2 kb upstream or 0.63 kb downstream from the coding region (Cocciolone and Cone 1993; Hoekenga and Cone 1995). Despite this similarity in DNA sequence, *Pl-Bh* differs dramatically from *Pl-Rh* in pigment phenotype (Cocciolone and Cone 1993). *Pl-Bh* produces blotchy kernel pigmentation, variegated plant color, and variegated anthers. In fact, the *Pl-Bh* anther phenotype is very similar to that of *Pl'-mah*. However, *Pl-Bh* does not participate in paramutation (J. Hollick and V. Chandler, unpubl.). *Pl-Rh* and *Pl-Bh* show differences in methylation at several sites that were examined near the 3' end of the gene (Cocciolone and Cone 1993). There are no methylation differences between *Pl-Rh* and *Pl'-mah* at these same sites (J. Hollick and V. Chandler, unpubl.).

GENETIC ANALYSIS OF PARAMUTATION

The ability to cause or respond to paramutation is tightly linked to the *b* and *pl* loci. Using a small population (44 plants), Coe (1966) demonstrated that the ability of *B'* to cause paramutation is linked to *B'*. Recently, much larger populations were used in recombination experiments to investigate whether the sequences required for paramutation could be separated from the sequences controlling *B-I* and *B'* tissue-specific expression (Patterson et al. 1995; K. Kubo and V. Chandler, unpubl.). More than 10,000 plants were screened for these recombinant alleles, and none were obtained. On the basis of these findings, the sequences mediating paramutation (at >0.95 probability) are located <0.1 cM from *B'* and <0.05 cM from *B-I* (Patterson et al. 1995; K. Kubo and V. Chandler, unpubl.). Similar experiments have established that the sequences which mediate *pl* paramutation are tightly linked to *pl* (Hollick et at. 1995; J. Hollick and V. Chandler, unpubl.).

To map the sequences required for *b* paramutation, *B-I* and *B'* were used to create intragenic recombinants with *B-Peru*, a *b* allele that has distinct tissue-specific transcription and is neutral with respect to paramutation. Like other *b* alleles that do not undergo paramutation, *B-Peru* contains numerous sequence differences relative to *B'* and *B-I*, notably in the 5'-flanking region (Radicella et al. 1992; Patterson et al. 1995). Recombination events within the transcribed region of *b* generated a series of hybrid alleles. For one series, the promoter and 5'-flanking sequences of *B-I* or *B'* were linked with the coding and 3'-flanking sequences of *B-Peru*. A complementary series was also created, in which the 5' sequences of *B-Peru* were joined to the coding and 3'-flanking sequences of *B-I* or *B'*. These recombinant *b* alleles were functional for activating the anthocyanin pathway, and phenotypic analysis of these hybrid *b* alleles demonstrated that sequences required for *B-Peru*, *B-I*, and *B'* tissue-specific pigment expression are located in the 5' regions of each allele. Furthermore, the abilities to cause and respond to paramutation also map to the 5' regions of *B'* and *B-I*, respectively. On the basis of these results and the tight genetic linkage of paramutation to *b*, we are attracted to the view that paramutation requires the regulatory sequences specifying *B-I* and *B'* transcription.

Three lines of evidence suggest that the B protein is not required for paramutation. First, recombinants containing the 5' region of *B-Peru* coupled to the coding sequences of *B-I* or *B'* do not participate in paramutation (Patterson et al. 1995). Second, the B proteins encoded by *B-I* and *B'* have equivalent function with respect to activating the anthocyanin pathway (Patterson et al. 1993). Third, insertions in the coding region of *b* that disrupt B protein function (T. Shroyer et al., unpubl.) do not affect paramutation (Coe 1966).

Mutational analysis has revealed *trans*-acting components that may play a role in paramutation. A dominant mutation, not tightly linked to *b*, has been isolated that reduces the frequency of paramutation (K. Kubo and V. Chandler, unpubl.). This mutation was isolated in an ethyl methanesulfonate mutagenesis as an exceptional *B'/B-I* dark plant (all of its sibs had the expected light pigment). This mutation does not make *B-I* resistant to paramutation: dark *B'/B-I* plants carrying this mutation produce only *B'* plants when crossed with *B'*, demonstrating that *B-I* is still sensitive to paramutation. *B'* plants with this mutation that are crossed with *B-I* give rise to about 30–35% dark plants, whereas sibling *B'* plants lacking this mutation produce only lightly pigmented *B'* progeny. Thus, we favor the view that this mutation reduces the efficiency of *B'* to induce paramutation, with dark plants escaping paramutation in about 30% of the progeny from F_1 plants with the mutant background, as compared to only 0.002% escapes in wild-type backgrounds (K. Kubo and V. Chandler, unpubl.).

We have isolated a recessive mutation that may affect the stability of *B'* and that is unlinked to *b*. This mutation was isolated from stocks that contain active *Mutator* transposable elements (Chandler and Hardeman 1992). When this mutation is homozygous, *B'* plants are darkly pigmented with pale *B'*-like sectors (K. Kubo and V. Chandler, unpubl.). The presence of these sectors suggests that the mutation may be caused by a transposable element insertion. Prior to isolation of this mutation, *B'* had never been observed to change to a dark allele. It is tempting to speculate that this mutation is affecting a *trans*-acting component required for the reduced transcription of the *B'* state. Loss of this component by mutation might destabilize the *B'* state, causing *B'* to change to a more highly transcribed *B-I* state. Further characterization of this mutation will further define its potential role in paramutation.

MODELS

To provide a framework for thinking about the mechanistic basis of paramutation, genetic and epigenetic models are discussed below.

Genetic Model

The stability of *B'* in diverse genetic backgrounds suggests that the change from *B-I* to *B'* might involve a DNA sequence alteration. One simple hypothesis is that *B-I* contains a unique DNA sequence that promotes higher transcription rates. This sequence is spontaneously lost from the *B-I* allele at a high frequency, generating *B'*. When *B'* is crossed to *B-I*, the heterology triggers a unidirectional mitotic gene conversion, resulting in only *B'* alleles. The phenotype of the F_1 plants and the appearance of light sectors in *B-I* plants suggest the event happens in somatic cells. The sequence transferred could be upstream of the sequenced region, smaller than we can detect by restriction mapping, or more distant from the locus than the restriction sites examined. If gene conversion is operating, it must occur unidirectionally, at a high frequency, and without crossing-over (Patterson et al. 1995), as has been seen in yeast (Haber 1983; Gimble and Thorner 1992).

One aspect of paramutation that genetic models do not easily explain is the pigment differences between the "new" and "old" *B'* alleles and the quantitatively variable states observed with *Pl'-mah*. It is possible that the alteration happens stepwise, with heritable intermediate states, as has been proposed for paramutation of *R-r* (Brink et al. 1968). For example, Fragile X syndrome is associated with an amplification of the number of CGG triplets in the FMR1 gene (Sutcliffe et al. 1992). Different extents of amplification could produce different states. Genetic models are difficult to reconcile with the reversion of *Pl'-mah* to *Pl-Rh* at high frequencies (Hollick et al. 1995) and with the reversible changes between *B'* and *B-I* states in certain mutant backgrounds. For these reasons, we favor an epigenetic basis for paramutation.

Epigenetic Model

Paramutation may occur at a given locus via epigenetic changes in transcription protein complexes bound to the participating alleles. Studies of developmental repression of homeotic genes provide a useful paradigm, in that the transient presence of transcriptional repressors establishes transcription states that are heritably maintained through mitosis by Polycomb group proteins (Simon et al. 1992; Moehrle and Paro 1994; see Paro and Harte, this volume).

In a *B-I/B-I* cross, most progeny are *B-I*, but about 10% are *B'*. *B'* might occur spontaneously when the chromatin proteins or transcription complexes rearrange into a more stable complex that leads to less activation of transcription. In Figure 2a, this is schematically represented as enabling a repressor to down-modulate *b* transcription in the *B'* state. Since the *B'* state is much more stable than the *B-I* state (spontaneous changes from *B'* to *B-I* have not been seen in most backgrounds), *B'* may represent the normal transcription state dictated by the DNA sequence, and *B-I* may be a rare state maintained by human selection. The stable alteration of *B-I* by *B'* in the F_1 (Fig. 2b) can be explained by the interaction of proteins bound to both alleles modifying the unstable *B-I* complex, resulting in a *B'* state that is inherited because this state is dictated by the DNA sequence. A model in which protein factors bound to *B'* act *in trans* to reduce the transcription of *B-I* is similar to models for *trans*-inactivation and transvection in *Drosophila* (Tartof and Henikoff 1991; see Pirotta; Henikoff; both this volume) and for *trans*-inactivation

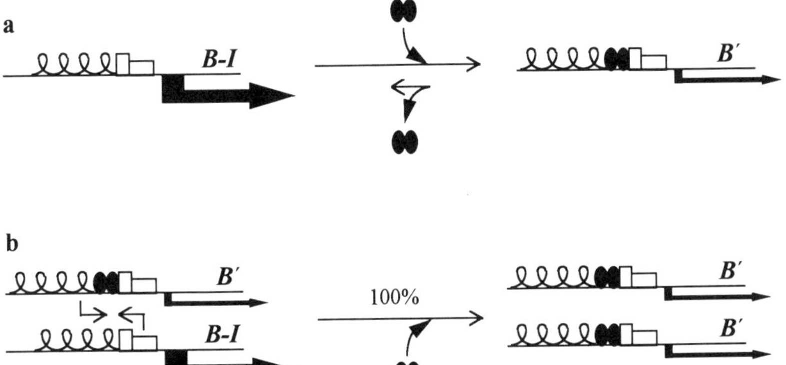

Figure 2 Schematic model for epigenetic changes during *b* paramutation. Size of filled arrows associated with *B-I* and *B′* indicates the amount of transcription. Details of the model are described in the text. (*a*) Diagram of protein complexes for the *B-I* and *B′* states. *B-I* can spontaneously change to *B′* by acquisition of additional proteins, whereas reestablishment of *B-I* is depicted as a much rarer event. (*b*) Paramutation in *B′/B-I*. As diagrammed, *B′* efficiently changes *B-I* to *B′* by protein-protein interactions between the two alleles.

in plants (Matzke and Matzke 1993). It is possible that paramutation is mechanistically similar to *trans*-dominant suppression in *Drosophila* (Tartof and Henikoff 1991) but is meiotically heritable in plants. Mitotically heritable states are frequently transmitted to progeny when germ cells differentiate from somatic cells late in plant development (Jorgensen 1993), whereas animals set aside germ cells early in embryogenesis. DNA methylation may play a role in the stable maintenance of transcription states. However, the fact that transcription states are heritable through mitosis in yeast and *Drosophila* in the absence of DNA methylation makes it clear that eukaryotic cells have mechanisms in addition to DNA methylation for stably maintaining distinct transcription states. For the *B-I* transcription state to be observed in about 90% of the progeny, some aspect of chromatin structure, the transcription complex, or DNA methylation must be preserved through numerous mitotic cell divisions and through the significant reworking of chromatin and DNA methylation that are likely to occur during gametogenesis.

It has been postulated that paramutation occurs during meiosis (Coe 1966), although all current evidence favors a premeiotic change (Patterson and Chandler 1995b). There is no physical evidence for homolog synapsis in somatic plant cells, but very little is known about the genome's organization at any developmental stage. Our model requires only transient interactions between alleles, because once the *B'* state is established, it is stably maintained. Studies of homology-dependent gene silencing in plants strongly suggest that the genome has mechanisms for detecting very small duplications (see Matzke and Matzke; Vaucheret; both this volume), which suggests the operation of a homology scanning mechanism.

This model can account for the different quantitative states observed with *pl*, and for the new and old *B'* states, by invoking multiple *cis*-acting sequences, with the extent of occupancy of these sites influencing the transcription level. An interesting aspect of this quantitative variation is that even small 2-fold differences in transcription rate are relatively stable through mitosis and meiosis (Patterson et al. 1993; Patterson and Chandler 1995b). Potentially, a threshold concentration of factors is set to a particular level on a paramutagenic allele. The level set influences the occupancy of the regulatory sites, which in turn influences the transcription state established as the plant develops. This state is then communicated to the other allele, such that the same state is established on the paramutable allele. There is variation between plants in what level is set, but once established, the levels are maintained in somatic cells. However, it is clear from examining the variegated *Pl'-mah* anthers that there is a stochastic nature to the transcription state eventually expressed in individual anther cells. We propose that the threshold establishes the probability that a given cell will adopt a transcription state competent to express *pl*. In the context of the model in Figure 2, the higher the level of factors, the lower the probability of expression, resulting in less-pigmented cells and a lower ACS. It is our hope that the recently isolated mutant that influences *B'* stability will provide insight into such postulated factors.

Predictions of this epigenetic model are that differences in chromatin structure or transcription factor occupancy may be observed between *B-I* and *B'*, or *Pl-Rh* and *Pl'-mah*. Sequences required for paramutation and transcription should overlap, and *trans*-acting mutations affecting paramutation might occur in loci that encode chromatin proteins or transcription factors. We hope to distinguish between genetic and epigenetic models by identifying the minimum *cis*-acting sequences required for paramutation and by identifying *trans*-acting modifier genes.

ACKNOWLEDGMENTS

The authors' research on paramutation is supported by an American Cancer Society grant (NP-875) to V.L.C., and National Science Foundation postdoctoral fellowships to K.M.K (BIR-9104373) and J.B.H. (BIR-9303601). We thank F. Bryan Pickett for useful comments on this review.

REFERENCES

Brink, R.A. 1958. Basis of a genetic change which invariably occurs in certain maize heterozygotes. *Science* **127:** 1182–1183.

———. 1964. Genetic repression of *R* action in maize. In *The role of chromosomes in development: The 23rd Symposium of The Society for the Study of Development and Growth* (ed. M. Locke), pp. 183–230. Academic Press, New York.

———. 1973. Paramutation. *Annu. Rev. Genet.* **7:** 129–152.

Brink, R.A., E.D. Styles, and J.D. Axtell. 1968. Paramutation: Directed genetic change. *Science* **159:** 161–170.

Chandler, V.L and K.J. Hardeman. 1992. The *Mu* elements of *Zea mays. Adv. Genet.* **30:** 77–122.

Cocciolone, S.M. and K.C. Cone. 1993. *Pl-Bh*, an anthocyanin regulatory gene of maize that leads to variegated pigmentation. *Genetics* **135:** 575–588.

Coe, E.H., Jr. 1966. The properties, origin, and mechanism of conversion-type inheritance at the *B* locus in maize. *Genetics* **53:** 1035–1063.

———. 1985. Phenotypes in corn: Control of pathways by alleles, time and place. *UCLA Symp. Mol. Cell. Biol. New Ser.* pp. 509–521.

Cone, K.C., S.M. Cocciolone, F.A. Burr, and B. Burr. 1993a. Maize anthocyanin regulatory gene *pl* is a duplicate of *c1* that functions in the plant. *Plant Cell* **5:** 1795–1805.

Cone, K.C., S.M. Cocciolone, C.A. Moehlenkamp, T. Weber, B.J. Drummond, L.A. Tagliani, B.A. Bowen, and G.H. Perrot. 1993b. Role of the regulatory gene *pl* in the photocontrol of maize anthocyanin pigmentation. *Plant Cell* **5:** 1807–1816.

Dawe, R.K. and M. Freeling. 1990. Clonal analysis of the cell lineages in the male flower of maize. *Dev. Biol.* **142:** 233–245.

Eggleston, W., M. Alleman, and J.L. Kermicle. 1995. Molecular organization and germinal instability of *R-stippled* maize. *Genetics* **141:** 347–360.

Gimble, F.S. and J. Thorner. 1992. Homing of a DNA endonuclease gene by meiotic gene conversion in *Saccharomyces cerevisiae. Nature* **357:** 301–306.

Goff, S.A., K.C. Cone, and V.L. Chandler. 1992. Functional analysis of the transcriptional activator encoded by the maize *B* gene: Evidence for direct functional interaction between two classes of regulatory proteins. *Genes Dev.* **6:** 864–875.

Goff, S.A., T.M. Klein, B.A. Roth, M.E. Fromm, K.C. Cone, J.P. Radicella, and V.L. Chandler. 1990. Transactivation of anthocyanin biosynthetic genes following transfer of *B*-regulatory genes into maize tissues. *EMBO J.* **9:** 2517–2522.

Haber, J.E. 1983. Mating type genes of *Saccharomyces cerevisiae*. In *Mobile genetic elements* (ed. J. Shapiro), pp. 560–621. Academic Press, New York.

Hoekenga, O.A. and K.C. Cone. 1995. Ten nucleotide differences exist between *Pl* and *Pl-Bh* within a 7 kb region. *Maize Genet. Coop. Newsl.* **69:** 47.

Hollick, J.B., G.I. Patterson, E.H. Coe, Jr., K.C. Cone, and V.L. Chandler. 1995. Allelic interactions heritably alter the activity of a metastable maize *pl* allele. *Genetics* **141:** 709–719.

Jorgensen, R. 1993. The germinal inheritance of epigenetic information in plants. *Philos. Trans. R. Soc. Lond.* **339:** 173–181.

Ludwig, S.R. and S.R. Wessler. 1990. Maize *R* gene family: Tissue-specific helix-loop-helix proteins. *Cell* **62:** 849–851.

Ludwig, S.R., B. Bowen, L. Beach, and S.R. Wessler. 1990. A regulatory gene as a novel visible marker for maize transformation. *Science* **247:** 449–450.

Matzke, M. and A.J.M. Matzke. 1993. Genome imprinting in plants: Parental effects and trans-inactivation phenomena. *Annu. Rev. Plant. Physiol. Plant Mol. Biol.* **44:** 53–76.

Moehrle, A. and R. Paro. 1994. Spreading the silence: Epigenetic transcriptional regulation during *Drosophila* development. *Dev. Genet.* **15:** 478–484.

Patterson, G.I. and V.L. Chandler. 1995a. Paramutation in maize and related allelic interactions. *Curr. Top. Microbiol. Immunol.* **197:** 121–141.

———. 1995b. Timing of *b* paramutation. *Maydica* **40:** 35–41.

Patterson, G.I., C.J. Thorpe, and V.L. Chandler. 1993. Paramutation, an allelic interaction, is associated with a stable and heritable reduction of transcription of the maize *b* regulatory gene. *Genetics* **135:** 881–894.

Patterson, G.I., K.M. Kubo, T. Shroyer, and V.L. Chandler. 1995. Sequences required for paramutation of the maize *b* gene map to a region containing the promoter and upstream sequences. *Genetics* **140:** 1389–1406.

Paz-Ares, J., D. Ghosal, U. Wienand, P.A. Peterson, and H. Saedler. 1987. The regulatory *c1* locus of *Zea mays* encodes a protein with homology to *myb* proto-oncogene products and with structural similarities to transcriptional activators. *EMBO J.* **6:** 3553–3558.

Poethig, R.S. 1987. Clonal analysis of cell lineage patterns in plant development. *Am. J. Bot.* **74:** 581–594.

Poethig, R.S., E.H. Coe Jr., and M.M. Johri. 1986. Cell lineage patterns in maize embryogenesis: A clonal analysis. *Dev. Biol.* **117:** 392–404.

Radicella, J.P., D. Brown, L.A. Tolar, and V.L. Chandler. 1992. Allelic diversity of the maize *B* regulatory gene—Different leader and promoter sequences of two *B* alleles determine distinct tissue specificities of anthocyanin production. *Genes Dev.* **6:** 2152–2164.

Simon, J., A. Chiang, and W. Bender. 1992. Ten different *Polycomb* group genes are required for spatial control of the AbdA and AbdB homeotic products. *Development* **114:** 493–505.

Sutcliffe, J.S., D.L. Nelson, F. Zhang, M. Pierretti, C.T. Caskey, D. Saxe, and S.T. Warren. 1992. DNA methylation represses FMR-1 transcription in Fragile X syndrome. *Hum. Mol. Genet.* **1:** 397–400.

Tartof, K.D. and S. Henikoff. 1991. Trans-sensing effects from *Drosophila* to humans. *Cell* **65:** 201–203.

Walker, E.L., T.P. Robbins, T.E. Bureau, J. Kermicle, and S.L. Dellaporta. 1995. Transposon-mediated chromosomal rearrangements and gene duplications in the formation of the maize *R-r* complex. *EMBO J.* **14:** 2350–2363.

DNA Methylation of Flower Color Transgenes in *Petunia hybrida*

Peter Meyer
Centre for Plant Biochemistry and Biotechnology
Department of Biology
The University of Leeds
Leeds LS2 9JT, United Kingdom

Inactivation of transgene constructs that have been introduced into plants is a frequently reported phenomenon. Many such silencing events are associated with DNA methylation, although it is still a matter of debate whether de novo methylation is the cause or the consequence of gene inactivation.

A correlation between DNA methylation and gene inactivation is not limited to transgenes but is also observed for transposable elements (Chandler and Walbot 1986; Schwarz and Dennis 1986) and for some (Spena et al. 1983), but not all (Nick et al. 1986), endogenous genes. Interestingly, changes in methylation patterns that correlate with changes in tissue-specific expression have been found in distant upstream regions of some genes (Langdale et al. 1986), indicating that methylation-mediated control of gene expression does not exclusively affect promoter regions.

We can assign at least two distinct molecular functions to hypermethylated sequences. First, it is conceivable that methylation patterns influence the binding of transcription factors to regulatory promoter elements. For a number of viral or human cellular promoters, position-specific methylation has been shown to inhibit or decrease promoter activity (Muiznieks and Doerfler 1994), supporting the assumption of a direct regulatory role for DNA methylation in transcription. Second, DNA methylation within a larger area may alter the local chromatin structure. This would also prevent transcription factors from accessing promoters within such regions due to the inaccessibility of the genomic region harboring the promoter, rather than the modification of a specific binding. Strong evidence for this indirect role of de novo methylation in the regulation of gene expression comes from the observation that de novo methylation alters chromatin structure (Keshet et al. 1986; Antequera et al. 1990) and from the analysis of methyl-CpG-binding proteins in mammals that act as mediators of repression and bind to heterochromatic regions (Lewis et al. 1992).

Epigenetic Mechanisms of Gene Regulation

A crucial question that arises from these models for the control of transgene activity is whether we can define spatial, structural, or sequence characteristics that regulate the induction of de novo methylation within a transgene construct integrated into the host genome. This chapter covers predominantly recent work employing a visible marker system to identify certain parameters involved with transgene-specific de novo methylation. A cDNA of the A1 gene was linked to the 35S promoter and introduced into petunia protoplasts via a PEG-mediated transformation technique. Expression of the maize A1 gene in the white-flowering petunia line Rl01 invokes a new pigmentation pathway that results in the accumulation of brick-red floral pigments (Meyer et al. 1987). The activity of the A1 gene can easily be monitored in individual petal cells of flowers that are constantly produced by transgenic petunia plants. Transgenic lines are propagated via stem cuttings, which allows continuous monitoring of transgene activity over several years and under various environmental conditions (Fig. 1).

THE INFLUENCE OF TRANSGENE COPY NUMBER ON TRANSGENE METHYLATION

Analysis of primary A1 transformants revealed an inverse correlation between the number of integrated transgene copies and the expression of the transgene (Linn et al. 1990). Most single-copy transformants had uniformly red flower pigmentation, suggesting that the A1 gene was expressed. A CCGG recognition sequence for the methylation-sensitive enzymes *Hpa*II and *Msp*I at the boundary between the promoter and the A1 cDNA was found unmethylated in these plants. Other primary transformants contained multiple truncated transgenes that were mainly integrated at different loci, but occasionally also linked as direct or inverted repeats at one locus. Plants containing multicopy transgenes displayed either no pigmentation at all, or a variegated expression phenotype, with A1 expression in some floral cells. The *Hpa*II site of such multiple integration lines was hypermethylated, suggesting that repeated copies provide a target for de novo methylation, reminiscent of repeat-directed methylation mechanisms in *Neurospora crassa* (Bull and Wootton 1984; Selker et al. 1987) and *Ascobolus immersus* (Faugeron et al. 1990), and specific methylation of multiple transgene copies during the asexual cycle of *N. crassa* (Pandit and Russo 1992; Russo et al., this volume).

An inverse correlation between transgene copy number and activity is not specific for the A1 gene, but has also been reported for other trans-

Figure 1 Different phenotypes of transgenic petunia flowers that carry an A1-expression construct. Activity of the A1 transgene can be monitored in the floral tissue, which provides a simple monitoring system to differentiate among uniform expression (*red* flowers), variegated expression (*mottled* flower), and inactivation (*white* flower).

genes introduced into *Petunia hybrida* (Jones et al. 1987; Renckens et al. 1992). In tobacco, the preferential methylation of multiple sequences has been detected in transformants that were retransformed with a vector containing sequences homologous to the transgene already present in the transgenic line (Matzke et al. 1989). In *Arabidopsis,* it has been shown that hypermethylated and inactivated transgenes that were inserted as concatemers could be reactivated after intrachromosomal deletion events reduced the number of transgene copies (Assaad and Signer 1992; Mittelsten Scheid et al. 1994), confirming the importance of repeats for de novo methylation.

In accordance with observations made for *Ascobolus* (Rossignol and Faugeron 1995), tandem repeats in plants appear to be very efficient targets for de novo methylation. This suggests the activity of a DNA-

DNA pairing mechanism that either induces the initiation of hypermethylation or enhances the transfer and spread of methylation patterns among homologous sequences. A direct pairing between homologous DNA copies would also explain why individual transgenes differ in their potential to inactivate homologous ectopic copies (Matzke et al. 1989; Vaucheret 1993), if one assumes that DNA pairing is influenced by the location of homologous sequences within the interphase nucleus. If homologous transgenes are located in close proximity within the interphase nucleus, this should enhance physical pairing and lead to a more efficient pairing-dependent transfer of methylation patterns. A transgene that inactivates homologous sequences very efficiently has been located close to a telomere (Matzke et al. 1994), indicating that a telomeric location may stimulate pairing with homologous sequences.

POSITION EFFECTS

Although the integration of single transgene copies is an important prerequisite to escape homology-dependent methylation, it is not sufficient to prevent de novo methylation of transgenes. The analysis of three transgenic lines containing single copies of the A1 gene construct revealed the importance of the chromosomal integration region for the stability of transgene expression (Pröls and Meyer 1992). In the two lines 16 and 24, the transgene was inserted into a hypermethylated chromosomal environment consisting of a highly repetitive region in line 16 and a low-copy genomic region in line 24. In the first flowers of both primary transformants, the A1 gene was uniformly expressed, but its activity became significantly reduced in subsequent flowers, until the two lines showed either no pigmentation (line 16) or only a weak coloration (line 24). Because new flowers are continuously produced from meristematic cells, these results implied that an epigenetic modification occurred during the production of new cells. Analysis of the junctions between the transgene DNA and the genomic integration site indicated that hypermethylation patterns had spread into the transgene. Another primary transformant, line 17, displayed uniform pigmentation in all flowers. Genomic analysis revealed that the A1 construct had integrated into a unique genomic region that was stably unmethylated.

These data suggest that hypermethylation patterns which are imposed on defined genomic regions can be slowly extended into adjacent transgenes, thus reducing their expression. It is unclear which features make a particular genomic region a target for hypermethylation. Repetitive regions of the plant genome are frequently hypermethylated, but the detection of a stable hypermethylation pattern in a low-copy genomic region suggests that there are alternative determinants of hypermethylation. These could include specific sequence elements, secondary structures, or a particular location of the DNA region within the nucleus. If methylation patterns can spread into adjacent regions, the cell must have evolved ways to protect endogenous genes from this effect. An elucidation of the regulation of the initiation and spread of methylation should therefore provide insight into the control of transgene expression.

ENVIRONMENTAL INDUCTION OF TRANSGENE METHYLATION

Plants are frequently exposed to rapid and significant changes of environmental conditions that can modify expression patterns of endogenous genes (Sachs and Ho 1986) and transposable elements (Walbot 1988, 1992). Tissue culture, which can be considered an environmental stress, also generates genetic variation and changes in methylation patterns of endogenous sequences (Brown 1989; Kaeppler and Phillips 1993). Several reports confirm that changing environmental conditions can reduce or inactivate the expression of transgenes. Heat treatment has been reported to induce inactivation of a phosphinothricin resistance marker in cell suspension cultures of *Medicago sativa* (Walter et al. 1992). Another report (Renckens et al. 1992) describes the inactivation of a negative selectable marker in tissue culture. Constitutive expression of the marker gene would have inhibited cell growth under certain culture conditions. Cells that survived negative selection displayed hypermethylation of the marker. A single-copy A1 cDNA driven by the 35S promoter in *P. hybrida* became frequently methylated and inactivated when plants were grown in the field, whereas plants kept in the greenhouse had predominantly pigmented flowers (Meyer et al. 1992). At least in one case analyzed, methylation occurred specifically within the transgene, as adjacent genomic sequences remained hypomethylated (Meyer and Heidmann 1994). Treatment of tissue-culture material of the same line with propionic acid or butyric acid, which was applied as a stress factor, also induced promoter methylation and variegated expression of the A1 transgene (ten Lohuis et al. 1995a). No changes occurred in the methylation states of several endogenous sequences tested, again indicating that the transgene was a specific target for stress-induced methylation.

Modulation of methylation patterns generates new epigenetic states, which may enhance the plant's potential to adapt to drastic changes in the environment. However, such a mechanism should cause considerable damage if it interferes with the expression of essential endogenous genes. Housekeeping genes and genes that are under strict developmental or tissue-specific control should therefore have developed ways to escape from the influence of environmental stress. The location of a gene, its chromosomal environment, or its sequence context may mediate the protection of endogenous genes from methylation. This assumption is supported by the identification of insulating elements that confer position-independent gene activity. Such scaffold or matrix attachment regions are thought to form the boundaries of DNA loops that act as independent expression domains. They have proven to mediate efficient transgene stabilization in animal systems (Grosveld et al. 1987; Stief et al. 1989), and similar effects have been reported for individual plant transgenes (Breyne et al. 1992; Schöffl et al. 1993; Mlynárová et al. 1994), but evidence for a general applicability of insulator elements is still lacking. As an alternative to the embedding of transgenes in insulator elements, transgene methylation may be prevented when the transgenes are targeted into trancriptionally active regions. However, it is not clear whether this strategy will be sufficient to prevent transgene methylation in plants, considering that *Agrobacterium*-mediated transformation often results in transgenes integrated into transcriptionally active regions (Mayerhofer et al. 1991).

SPECIFIC RECOGNITION OF FOREIGN SEQUENCES?

The specific de novo methylation of transgenes, which can also be observed for transposable elements (Chandler and Walbot 1986; Schwarz and Dennis 1986), suggests that plants contain a recognition mechanism for DNA elements that have inserted into new genomic locations. Such a mechanism would act in combination with the previously described homology-dependent methylation mechanism and would also affect single-copy insertions if these are recognizable as foreign DNA. This implies that methylation has evolved from a bacterial immune function into a defense mechanism against foreign DNA (Bestor 1990). Such a mechanism may be induced by the illegitimate recombination process of integration, or it may recognize elements that integrate into new chromosomal environments because their sequence composition deviates from that of the neighboring region (Meyer and Heidmann 1994). It has been shown that the nuclear genomes of angiosperms are mosaics of about 200-kb-long, compositionally homogeneous DNA segments called isochores (Salinas et al. 1988), and that functional genes and their chromosomal environment contain characteristic isochores with defined overall GC contents (Matassi et al. 1989). As the AT content differs for bacterial, monocotyledonous, and dicotyledonous genes, it is conceivable that genes derived from different species are identified because they have a different isochore composition. The maize A1 gene has a low AT-content compared to its dicotyledonous homologs and is frequently methylated in the dicotyledonous *P. hybrida* (Meyer and Heidmann 1994). The A1 homolog of the dicotyledonous *Gerbera* species encodes the same enzyme function, but its sequence resembles a dicotyledonous-specific isochore composition with a higher AT content. When the A1 cDNAs of *Gerbera* and maize, respectively, were linked to the 35S promoter and transferred into *P. hybrida* via *Agrobacterium*-mediated transfer, only the *Gerbera* A1 gene remained mainly unmethylated (Elomaa et al. 1995), which might reflect an escape of the transgene from a cellular recognition mechanism. However, additional data are required to assess the significance of sequence composition for de novo methylation.

CIS AND *TRANS* EFFECTS OF DNA METHYLATION PATTERNS

DNA methylation patterns not only affect the expression of a transgene, but they can also influence the activity of adjacent genes (ten Lohuis et al. 1995b) and homologous sequences located at allelic (Meyer et al. 1993) or ectopic positions (Matzke et al. 1994). When a repetitive sequence element (RPS) was placed adjacent to a glucuronidase (GUS) expression marker, transformants with the RPS construct were more often variegated than were control transformants that did not contain the RPS. RPS is a 1.6-kb-long repetitive sequence that was isolated from the petunia genome. The multiple endogenous RPS sequences in tobacco and petunia are predominantly hypermethylated. Restriction-site mapping of the RPSs of the integrated transgene construct showed that the RPSs were completely methylated at a particular *Hha*I site in all transformants tested. It appears that, due to the presence of specific signals within the RPS region or interaction of the RPS with other endogenous hypermethylated homologs, the RPS functions as an initiation region for de novo methylation and induces variegated expression in adjacent sequences (ten Lohuis et al. 1995b).

Not only can methylation patterns spread along the chromosome, but they also can be transferred to homologous regions of other chromosomes. When the hypermethylated allele of a single-copy transgene was combined with a hypomethylated allele of the same locus in a cross, the methylation pattern was transferred to the hypomethylated allele in a semidominant way (Meyer et al. 1993), because in a large number of plants, A1 expression was inhibited. This effect was reminiscent of paramutation effects described in various plant species (Hagemann and Berg 1977). A similar transfer can also occur between certain homologous transgenes that are located at different positions in the genome. The susceptibility of a transgene to adopt a methylation pattern via interaction with a homologous transgene at the same locus is enhanced if the locus is metastable and thus prone to lose expression (Matzke et al. 1994). Methylation states that have been established via allelic or ectopic interactions are largely preserved in subsequent generations, even when the transgenes segregate away from each other. It is conceivable that allelic and ectopic *trans*-inactivation are mediated by transient somatic pairing of the two homologous copies, and that the DNA methylation patterns serve as memory imprints of this interaction (Matzke and Matzke 1990).

The *cis* and the *trans* effects of methylation patterns resemble gene inactivation processes of *Drosophila*, which are mediated by chromatin components, since *Drosophila* does not possess a DNA methylation system. Position-effect variegation (PEV) (Muller 1930), gene regulation caused by positioning genes next to a heterochromatic region, has been interpreted as a spreading of heterochromatin into adjacent sequences (Henikoff 1990 and this volume). As with hypermethylation in plants, inactive chromatin states can also be transferred among different chromosomes. The *Drosophila brown* locus shows dominant PEV, in which a heterochromatinized *brown* allele imposes its heterochromatic state on a homologous allele that becomes *trans*-inactivated (Dreesen et al. 1991). There is evidence that repeated sequences induce the formation of heterochromatin, as arrays of three or more transgenes produced PEV-like inactivation effects that could be modulated by known modifiers of PEV (Dorer and Henikoff 1994).

LINKS BETWEEN DNA METHYLATION AND CHANGES IN CHROMATIN STRUCTURE

Hypermethylation can be associated with a repressed chromatin state (Weintraub 1985; Keshet et al. 1986; Antequera et al. 1990), which indicates that changes in methylation could be either the cause or the consequence of structural modifications mediating *cis* and *tran*s inactivation. Structural changes might also induce the methylation of C residues outside the common symmetrical sequence contexts CG and CXG. The symmetry of the target sequences CG and CNG provides the information for the faithful propagation of methylation patterns after replication. Because hemimethylated DNA is the preferred substrate for DNA methyltransferase, it is assumed that the enzyme recognizes 5-methylcytosine (m^5C) residues in the template strand of semiconservatively replicated DNA and that it methylates C residues in the synthesized strand, if the template strand carries a m^5C residue at a symmetrical position (Holliday and Pugh 1975; Riggs 1975). Nonsymmetrical methylation patterns were found in a hypermethylated A1 transgene, where they occurred clustered in certain regions. It was not always the same C residue, but one or several C residues within a particular region that were affected (Meyer et al. 1994). When a hypermethylated allele was compared with a hypomethylated variant of the same locus, the local chromatin showed a reduced accessibility for endonucleases, implying that hypermethylation involved a condensed chromatin structure (Meyer 1995). These results imply that, similar to RNA methylation patterns (Kane and Beemon 1985), the induction and conservation of nonsymmetrical methylation patterns are probably not encoded in the sequence, but more likely are due to secondary structure effects.

Nonsymmetrical methylation patterns have been found in chromosomal replication origins of mammalian cells that have replicated their DNA. The function of the methylation patterns is not known, but it has been proposed that they are involved in binding of the origins to the nuclear envelope; that they determine the early replication of the labeled origins; or that they serve as a tag to prevent redundant initiation of replication at loci that have already been replicated (Tasheva and Roufa 1994).

In plants, nonsymmetrical methylation has so far been detected in transgenes (Ingelbrecht et al. 1994; Meyer et al. 1994), and non CG or CXG methylation has been widely reported in maize genes and transposons at restriction sites such as *Sst*I. Dense nonsymmetrical methylation patterns also occur in silenced repeats in *Neurospora* (Selker et al. 1993) and *Ascobolus* (Goyon et al. 1994), again pointing out common features of silencing mechanisms in plants and filamentous fungi.

CONCLUSION

Transgene technology has helped us to identify various interesting aspects of the regulation of gene expression. There are at least three major topics that need to be elucidated before we can understand and control de novo methylation in transgenic plants: the specific methylation of repeated sequences, the role of chromosomal location and sequence composition that allow a gene to escape de novo methylation, and, finally, the cellular factors responsible for the formation and spread of methylation patterns. The striking similarities of de novo methylation phenomena in plants and filamentous fungi should encourage a closer collaboration among scientists in the two areas. Filamentous fungi offer a powerful genetic system that may complement and speed up the analysis of the above-mentioned features in plants.

ACKNOWLEDGMENT

I thank Dr. Celia Knight for critical reading of the manuscript.

REFERENCES

Antequera, F., J. Boyes, and A. Bird. 1990. High levels of de novo methylation and altered chromatin structure at CpG islands in cell lines. *Cell* **62:** 503–514.

Assaad, F.F. and E.R. Signer. 1992. Somatic and germinal recombination of a direct repeat in *Arabidopsis. Genetics* **132:** 553–566.

Bestor, T.H. 1990. DNA methylation: Evolution of a bacterial immune function into a regulator of gene expression and genome structure in higher eukaryotes. *Philos. Trans. R. Soc. Lond. B Biol. Sci.* **326:** 179–187.

Breyne, P., M. Van Montague, A. Depicker, and G. Gheysen. 1992. Characterization of a plant scaffold attachment region in a DNA fragment that normalizes transgene expression in tobacco. *Plant Cell* **4:** 463–471.

Brown, P.T.H. 1989. DNA methylation in plants and its role in tissue culture. *Genome* **31:** 717–729.

Bull, J.H. and J.C. Wootton. 1984. Heavily methylated amplified DNA in transformants of *Neurospora crassa. Nature* **310:** 701–704

Chandler, V.L. and V. Walbot. 1986. DNA modification of a maize transposable element correlates with loss of activity. *Proc. Natl. Acad. Sci.* **83:** 1767–1771.

Dorer, R.D. and S. Henikoff. 1994. Expansion of transgene repeats cause heterochromatin formation and gene silencing in *Drosophila. Cell* **77:** 993–1002.

Dreesen, T.D., S. Henikoff, and K. Loughney. 1991. A pairing-sensitive element that mediates *trans*-inactivation is associated with the *Drosophila* brown gene. *Genes Dev.* **5:** 331–340.

Elomaa, P., Y. Helariutta, R.J. Griesbach, M. Kotilainen, P. Seppänen, and T.H. Teeri. 1995. Transgene inactivation in *Petunia hybrida* is influenced by the properties of the foreign gene. *Mol. Gen. Genet.* **248:** 649–656.

Faugeron, G., L. Rhounim, and J.-L. Rossignol. 1990. How does the cell count the number of ectopic copies of a gene in the premeiotic inactivation process acting in *Ascobulus immersus? Genetics* **124:** 585–591.

Goyon, C., T.I.V. Nogueira, and G. Faugeron. 1994. Perpetuation of cytosine methylation in *Ascobolus immersus* implies a novel type of maintenance methylase. *J. Mol. Biol.* **240:** 42–51

Grosveld, F., G. Blom van Assendelft, D.R. Greaves, and G. Kollias. 1987. Position-independent, high-level expression of the human beta-globin gene in transgenic mice. *Cell* **51:** 975–985.

Hagemann, R. and W. Berg. 1977. Vergleichende Analyse der Paramutationssysteme bei höheren Pflanzen. *Biol. Zentbl.* **96:** 257–301.

Henikoff, S. 1990. Position-effect variegation after 60 years. *Trends Genet.* **6:** 422–425.

Holliday, R. and J.E. Pugh. 1975. DNA modification mechanisms and gene activity during development. *Science* **187:** 226–232.

Ingelbrecht, I., H. Van Houdt, M. Van Montague, and A. Depicker 1994. Posttranscriptional silencing of reporter transgenes in tobacco correlates with DNA methylation. *Proc. Natl. Acad. Sci.* **91:** 10502–10506.

Jones, J.D.G., D.E. Gilbert, K.L. Grady, and R.A. Jorgensen. 1987. T-DNA structure and gene expression in petunia plants transformed by *Agrobacterium tumefaciens* C58 derivatives. *Mol. Gen. Genet.* **207:** 478–485.

Kaeppler, S.M. and R.L. Phillips. 1993. DNA methylation and tissue culture–induced variation in plants. *In Vitro Cell. Dev. Biol.* **29P:** 125–130.

Kane, S.E. and K. Beemon. 1985. Precise localization of m6A in Rous sarcoma virus RNA reveals clustering of methylation sites: Implication for RNA processing. *Mol. Cell. Biol.* **5:** 2298–2306.

Keshet, I., J. Lieman-Hurwitz, and H. Cedar. 1986. DNA methylation affects the formation of active chromatin. *Cell* **44:** 535–543.

Langdale, J.A., W.C. Taylor, and T. Nelson. 1986. Cell-specific accumulation of maize phosphoenolpyruvate carboxylase is correlated with demethylation at a specific site >3kb upstream of the gene. *Mol. Gen. Genet.* **225:** 49–55.

Lewis, J.D., R.R. Meehan, W.J. Henzel, I. Maurer-Fogy, P. Jeppesen, F. Klein, and A. Bird. 1992. Purification, sequence, and cellular localization of a novel chromosomal protein that binds to methylated DNA. *Cell* **69:** 905–914.

Linn, F., I. Heidmann, H. Saedler, and P. Meyer. 1990. Epigenetic changes in the expression of the maize A1 gene in *Petunia hybrida:* Role of numbers of integrated gene copies and state of methylation. *Mol. Gen. Genet.* **222:** 329–336.

Matassi, G., L.M. Monterero, J. Salinas, and G. Bernardi. 1989. The isochore organization and the compositional distribution of homologous coding sequences in the nuclear genome of plants. *Nucleic Acids Res.* **13:** 5273–5290.

Matzke, A.J.M., F. Neuhuber, Y.D. Park, P.F. Ambros, and M.A. Matzke. 1994. Homology-dependent gene silencing in transgenic plants—Epistatic silencing loci contain multiple copies of methylated transgenes. *Mol. Gen. Genet.* **244:** 219–229.

Matzke, M.A. and E.J.M. Matzke. 1990. Gene interactions and epigenetic variation in transgenic plants. *Dev. Genet.* **11:** 214–223.

Matzke, M.A., M. Priming, J. Trnovsky, and A.J.M. Matzke. 1989. Reversible methylation and inactivation of marker genes in sequentially transformed tobacco plants. *EMBO J.* **8:** 643–649.

Mayerhofer, R., Z. Koncz-Kalman, C. Nawrath, G. Bakkeren, A. Crameri, K. Angelis, G.P. Redei, J. Schell, B. Hohn, and C. Koncz. 1991. T-DNA integration: A mode of illegitimate recombination in plants. *EMBO J.* **10:** 697–704.

Meyer, P. 1995. DNA methylation and transgene silencing in *Petunia hybrida.* In *Gene silencing in higher plants and related phenomena in other eukaryotes* (ed. P. Meyer), vol. 197, pp. 15–28. Springer-Verlag, Berlin.

Meyer, P. and I. Heidmann. 1994. Epigenetic variants of a transgenic petunia line show hypermethylation in transgene DNA: An indication for specific recognition of foreign DNA in transgenic plants. *Mol. Gen. Genet.* **243:** 390–399.

Meyer, P., I. Heidmann, and I. Niedenhof. 1993. Differences in DNA-methylation are associated with a paramutation phenomenon in transgenic petunia. *Plant J.* **4:** 89–100.

Meyer, P., I. Niedenhof, and M. ten Lohuis. 1994. Evidence for cytosine methylation of non-symmetrical sequences in transgenic *Petunia hybrida. EMBO J.* **13:** 2084–2088.

Meyer, P., I. Heidmann, G. Forkmann, and H. Saedler. 1987. A new petunia flower colour generated by transformation of a mutant with a maize gene. *Nature* 330: 677–678.

Meyer, P., F. Linn, I. Heidmann, H. Meyer, I. Niedenhof, and H. Saedler. 1992. Endogenous and environmental factors influence 35S promoter methylation of a maize A1 gene construct in transgenic petunia and its colour phenotype. *Mol. Gen. Genet.* **231:** 345–352.

Mittelsten Scheid, O., K. Afsar, and J. Paszkowski. 1994. Gene inactivation in *Arabidopsis thaliana* is not accompanied by an accumulation of repeat-induced point mutations. *Mol. Gen. Genet.* **244:** 325–330.

Mlynáová, L., A. Loonen, J. Heldens, R.C. Jansen, P. Keizer, W.J. Stiekema, and J.-P. Nap. 1994. Reduced position effect in mature transgenic plants conferred by the chicken lysozyme matrix-associated region. *Plant Cell* **6:** 417–426.

Muiznieks, I. and W. Doerfler. 1994. The impact of 5′-CG-3′ methylation on the activity of different eukaryotic promoters: A comparative study. *FEBS Lett.* **344:** 251–254.

Muller, H.J. 1930. Types of visible variations induced by X-rays in *Drosophila. J. Genet.* **22:** 299–334.

Nick, H., B. Bowen, R.J. Ferl, and W. Gilbert. 1986. Detection of cytosine methylation in the maize alcohol dehydrogenase gene by genomic sequencing. *Nature* **319:** 243–246.

Pandit, N.N. and V.E.A. Russo. 1992. Reversible inactivation of a foreign gene, hph, during the asexual cycle in *Neurospora crassa. Mol. Gen. Genet.* **234:** 412–422

Pröls, F. and P. Meyer. 1992. The methylation patterns of chromosomal integration regions influence gene activity of transferred DNA in *Petunia hybrida. Plant J.* **2:** 465–475.

Renckens, S., H. De Greve, M. Van Montagu, and J.-P. Hernalsteens. 1992. Petunia plants escape from negative selection against a transgene by silencing the foreign DNA via methylation. *Mol. Gen. Genet.* **233:** 53–64.

Riggs, A.D. 1975. X inactivation, differentiation and DNA methylation. *Cyotogenet. Cell Genet.* **14:** 9–25.

Rossignol, J.-L. and G. Faugeron. 1995. MIP: An epigenetic gene silencing process in *Ascobolus immersus.* In *Gene silencing in higher plants and related phenomena in other eukaryotes* (ed. P. Meyer), vol. 197, pp. 179–191. Springer-Verlag, Berlin.

Sachs, M.M. and T.-H.D. Ho. 1986. Alteration of gene expression during environmental stress in plants. *Annu. Rev. Plant Physiol.* **37:** 363–376.

Salinas, J., G. Matassi, L.M. Montero, and G. Bernardi. 1988. Compositional compartmentalization and compositional patterns in the nuclear genomes of plants. *Nucleic Acids. Res.* **16:** 4269–4285.

Schöffl, F., G. Schröder, M. Kliem, and M. Rieping. 1993. A SAR sequence containing 395bp DNA fragment mediates enhanced, gene-dosage-correlated expression of a chimeric heat shock gene in transgenic tobacco plants. *Transgenic Res.* **2:** 93–100.

Schwarz, D. and E. Dennis. 1986. Transposase activity of the Ac controlling element in maize is regulated by its degree of methylation. *Mol. Gen. Genet.* **205:** 476–482.

Selker, E.U., D.Y. Fritz, and M.J. Singer.1993. Dense non-symmetrical DNA methylation resulting from repeat-induced point mutation (RIP) in *Neurospora. Science* **262:**

1724–1728.

Selker, E.U., E.B. Cambareri, B.C. Jensen, and K.R. Haack. 1987. Rearrangement of duplicated DNA in specialized cells of Neurospora. *Cell* **51:** 741–752.

Spena, A., A. Viotti, and V. Pirrotta. 1983. Two adjacent genomic zein sequences: Structure, organization and tissue-specific restriction pattern. *J. Mol. Biol.* **169:** 799–811.

Stief, A., D.M. Winter, W.H. Straetling, and A.E. Sippel. 1989. A nuclear DNA attachment element mediates elevated and position-independent gene activity. *Nature* **341:** 343–345.

Tasheva, E.S. and D.J. Roufa. 1994. Densely methylated DNA islands in mammalian chromosomal replication origins. *Mol. Cell. Biol.* **14:** 5636–5644.

ten Lohuis, M., H. Galliano, I. Heidmann, and P. Meyer. 1995a. Treatment with propionic and butyric acid enhances expression variegation and promoter methylation in plant transgenes. *Biol. Chem. Hoppe-Seyler* **376:** 311–320.

ten Lohuis, M., A. Müller, I. Heidmann, I. Niedenhof, and P. Meyer. 1995b. A repetitive DNA fragment carrying a hot spot for de novo DNA methylation enhances expression variegation in tobacco and petunia. *Plant Journal* **8:** 919–932.

Vaucheret, H. 1993. Identification of a general silencer for 19S and 35S promoters in a transgenic tobacco plant: 90 bp of homology in the promoter sequence are sufficient for *trans*-inactivation. *C.R. Acad. Sci.* **316:** 1471–1483.

Walbot, V. 1988. Reactivation of the mutator transposable element system following gamma irradiation of seed. *Mol. Gen. Genet.* **212:** 259–264.

———. 1992. Reactivation of mutator transposable elements of maize by ultraviolet light. *Mol. Gen. Genet.* **234:** 353–360.

Walter, C., J. Broer, D. Hillemann, and A. Puehler. 1992. High frequency, heat treatment–induced inactivation of the phosphinothricin resistance gene in transgenic single cell suspension cultures of *Medicago sativa*. *Mol. Gen. Genet.* **235:** 189–196.

Weintraub, H. 1985. Assembly and propagation of repressed and derepressed chromosomal states. *Cell* **42:** 705–711.

Position-effect Variegation in *Drosophila:* Recent Progress

Steven Henikoff
Howard Hughes Medical Institute and
Basic Sciences Division
Fred Hutchinson Cancer Research Center
Seattle, Washington 98104

Chromosomes of higher eukaryotes are cytologically differentiated into euchromatin, which decondenses at telophase, and heterochromatin, which remains visibly condensed (Heitz 1929). In *Drosophila,* heterochromatin forms at blocks of repetitive DNA, both simple sequence repeats (Mayfield and Ellison 1975; Lohe et al. 1993) and moderately repetitive sequences apparently derived from transposable elements (Carmena and Gonzalez 1995; Le et al. 1995; Pimpinelli et al. 1995). When euchromatin and heterochromatin are juxtaposed, mosaic gene silencing can occur, a phenomenon known as position-effect variegation (PEV). Although PEV has been familiar to geneticists for decades, mechanistic questions have only recently been addressed: What differences between euchromatin and heterochromatin are responsible for silencing? How do *trans*-acting factors participate in silencing? How are genes silenced from a distance? Do repeats play a direct role? Answering these questions is becoming increasingly important, given the mounting evidence that some of the same questions arise for PEV-like phenomena in a wide variety of systems.

The focus of this brief critical review is recent work that seems most relevant to epigenetic phenomena, especially those involving repeats. Other aspects of PEV are discussed in excellent comprehensive reviews by Lewis (1950), Spofford (1976), and Weiler and Wakimoto (1995).

DIFFERENT TYPES OF PEV

Genes are subject to PEV in different situations that represent different vantage points from which interactions between euchromatin and heterochromatin can be observed. The most common situation is when a euchromatic gene is moved next to heterochromatin by chromosomal rearrangement, i.e., classic PEV. Figure 1a illustrates classic PEV on the famous *white* eye pigment gene. Similar effects are seen when transgenes

 0-87969-490-4/96 $5 + .00

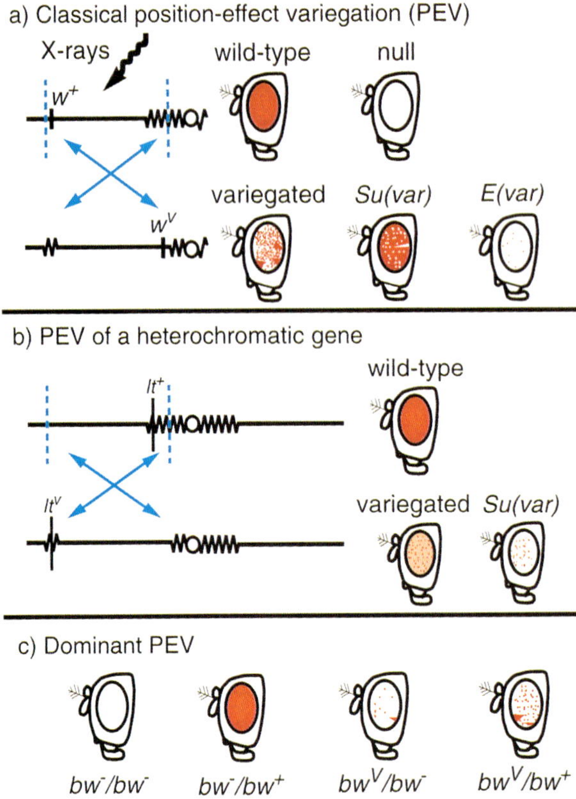

Figure 1 Three different categories of PEV. (*a*) X-rays have induced an inversion, *white-mottled-4,* thus placing the wild-type *white* (w^+) gene within about 25 kb of pericentric heterochromatin, giving rise to a mutant allele with variegating eyes, w^v. *Su(var)*s suppress and *E(var)*s enhance the variegating phenotype. Wiggly lines represent heterochromatin and straight lines represent euchromatin. (*b*) Heterochromatic genes show PEV in chromosomal rearrangements. The *light* (lt^+) gene displays variegated eyes when moved to distal euchromatin. Variegation of *light* (lt^v) can be enhanced by *Su(var)*s and suppressed by *E(var)*s. (*c*) Dominance is seen for a variegating allele of the *brown* (*bw*) gene. Heterozygotes between *brown-Variegated* (bw^V) and wild-type *brown,* bw^V/bw^+, have variegated eyes, although somewhat suppressed relative to bw^V/bw^-. (Reprinted, with permission, from Henikoff 1995.)

are mobilized and land in heterochromatin (Wallrath and Elgin 1995; Zhang and Spradling 1995). *trans*-Acting modifiers of PEV dominantly enhance or suppress classic PEV in a consistent manner but typically have no effects on other gene-silencing phenomena in *Drosophila*, such as telomere position effects (Talbert et al. 1994; Wallrath and Elgin 1995).

Most genes are located in euchromatin and variegate when moved to heterochromatin. However, a relative handful of genes are located within heterochromatin, and these variegate when they are moved to distal euchromatin. Effects of *trans*-acting modifiers on PEV of heterochromatic genes are generally opposite to the effects seen for classic PEV. For example, the heterochromatic *light* gene variegates when the block of heterochromatin in which it lies is moved far away from the bulk of heterochromatin, and this effect can be enhanced by PEV suppressors (see Fig. 1b) (Hearn et al. 1991). Different rules pertain to PEV of euchromatic genes and PEV of heterochromatic genes: Heterochromatic genes were found to be sensitive to the size of the heterochromatic block that they were embedded in, whereas the size of the heterochromatic block did not determine the extent of variegation of an adjacent euchromatic gene (Howe et al. 1995).

Dominant PEV reveals a different mode of silencing by heterochromatin. Classic PEV alleles of the *brown* eye pigment gene are dominant over wild-type *brown*, unlike PEV alleles of the *white* gene, which are recessive to wild-type *white* (Fig. 1c). This dominant effect requires somatic pairing of homologs: It is thought that pairing causes contact between the *cis*-inactivated copy of *brown* and the *trans* copy, resulting in *trans*-inactivation (Dreesen et al. 1991). Expression of the *brown* gene and its susceptibility to *trans*-inactivation are tightly linked and have been conserved for at least about 60 million years, suggesting an intimate relationship (Martin-Morris and Henikoff 1995).

PEV AND CHROMATIN STRUCTURE

Genes subject to PEV provide a potential means of examining the molecular basis for differences between euchromatin and heterochromatin. Surprisingly, a molecular corollary to the striking cytological distinction has not been easy to detect. Chromatin sensitivity to DNase I digestion was found to differ only slightly for the *white* gene between PEV-enhanced and PEV-suppressed states that are readily distinguished cytologically in polytene chromosomes (Hayashi et al. 1990). Restriction endonuclease probes (Schlossherr et al. 1994) or endogenous nucleases (Locke 1994) also failed to detect differences at *white*. Although one might worry that nuclease treatments disrupt higher-order structure, an in vivo assay using *Escherichia coli dam* methylase in live flies likewise failed to detect consistent differences in accessibility between euchromatic and heterochromatic sequences (Wines et al. 1996).

Despite these negative results for classic PEV, a study of transgene

PEV revealed that the heterochromatic state can indeed be associated with chromatin structural alterations (Wallrath and Elgin 1995). When transgenes encoding the 26-kD heat shock protein were inserted into heterochromatin, they became relatively refractory to restriction endonuclease cleavage and showed a sharpened nucleosomal pattern characteristic of heterochromatin in general. An interesting possibility is that transgenes completely embedded in heterochromatin are more susceptible to chromatin structural alterations than euchromatic genes that are only moved close to a junction with heterochromatin (Wallrath and Elgin 1995).

MODIFIERS OF PEV

Genes subject to PEV also provide a potential means of identifying *trans*-acting components of heterochromatin by the recovery of modifier mutations. Most modifiers are dosage dependent, referred to as *Su(var)*s (*Suppressors-of-variegation*) and *E(var)*s (*Enhancers-of-variegation*). There may be hundreds of dosage-dependent *Su(var)*s (Henikoff 1979; Wustmann et al. 1989), of which a subset show effects when the gene is in one dose (heterozygous with a null mutation) and in three doses (carrying a duplication) (Locke et al. 1988). Although it has been argued that these fully dosage-sensitive modifiers may encode structural components of heterochromatin (Locke et al. 1988), alternative possibilities must be considered in the absence of biochemical or cytological evidence. It is possible that most modifier genes act indirectly: Known indirect effects on PEV phenotypes are caused by environmental differences in temperature and pH during development (Michailidis et al. 1988), and so mechanistic interpretations should be regarded with caution.

Several modifier genes have been cloned, but only in the case of *Su(var)205*, which encodes the heterochromatin protein 1 (HP1), is there direct evidence for a structural role. Anti-HP1 antibody most strongly decorates the diffuse heterochromatic chromocenter of the otherwise banded polytene chromosomes (James and Elgin 1986), consistent with a role in packaging heterochromatin. HP1 shares a 37-amino-acid "chromodomain" motif with the Polycomb protein, a component of a multiprotein complex involved in developmental silencing of homeotic genes (Paro and Hogness 1991; see Paro and Harte; Pirrotta; both this volume), which suggests that these proteins have similar roles as subunits of their respective complexes. More direct evidence comes from the finding that a chimeric protein produced by substitution of the Polycomb chromodomain for its counterpart in HP1 binds both heterochromatic regions and Polycomb-binding sites (Platero et al. 1995). As befits a structural constituent of heterochromatin, HP1 is required at about the time visible heterochromatin forms, and null mutations are defective in chromosome segregation, evidently because of the failure of heterochromatin to fully condense (Kellum and Alberts 1995).

*E(var)*s appear to constitute a diverse class of genes that act independently on different PEV alleles (Bishop 1992). One *E(var)* encodes a protein component of a chromatin insulator (Gerasimova et al. 1995), which can protect a sensitive reporter gene promoter from nearby enhancers and/or silencers. Others encode members of the trithorax group (trxG) (Dorn et al. 1993; Farkas et al. 1994), which act oppositely to Polycomb group (PcG) proteins in developmental silencing of homeotic gene complexes (Moehrle and Paro 1994). The striking parallels between *Su(var)* and PcG proteins and between *E(var)* and trxG proteins has led to the idea that *E(var)*- and trxG-encoded complexes package active chromatin and so oppose gene silencing mediated by *Su(var)*- and PcG-encoded complexes (Locke et al. 1988; Moehrle and Paro 1994).

THE "OOZING" MODEL

The most widely discussed mechanism for PEV falls into the class of molecular models termed "oozing" (Tartof et al. 1984; Ptashne 1986). Oozing was proposed to explain the genetic phenomenon of spreading, which is the silencing of closely linked genes in a polar manner. Spreading has been observed phenotypically for two genes adjacent to a heterochromatic breakpoint, in which case the gene closer to the breakpoint is more frequently inactive. Spreading also has been observed cytologically in polytene chromosomes as the variable loss of banding for regions in which affected genes lie (Hartmann-Goldstein 1967; Hayashi et al. 1990; Belyaeva and Zhimulev 1991). These regions resemble heterochromatin in that they assume a diffuse unbanded appearance. By both phenotypic and cytological criteria, spreading decreases in extent when PEV is suppressed and increases when PEV is enhanced. Oozing models are straightforward extensions of these phenotypic and cytological observations to the molecular level, presumably involving heterochromatin-specific proteins such as HP1.

To account for polarity of the spreading effect, oozing of heterochromatic complexes must start in constitutive heterochromatin, move through the heterochromatin-euchromatin junction, then end downstream from the farthest affected gene in that cell. To account for variegation, oozing would sometimes stop short of each gene. Specific models

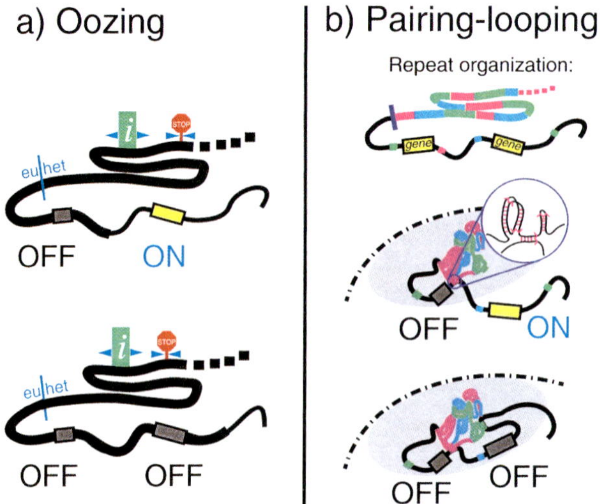

Figure 2 Two models for PEV. (*a*) Oozing. Heterochromatic protein complexes (*thick bars*) that ooze from initiation sites in heterochromatin (*i*) can pass through a heterochromatin/euchromatin junction (eu/het) in the absence of a terminator (stop) site, causing polar spreading into two nearby genes (*open boxes*). (*b*) Pairing-looping. Looping into the heterochromatic compartment (*gray oval*) on the nuclear envelope (*dashed and dotted lines*) results from pairing between middle repetitive elements that are dispersed in euchromatin and concentrated in heterochromatin. Different repeat families are represented by different patterns, extended in the top cartoon and folded into paired structures below. Paired associations between euchromatic and heterochromatic repeats (shown in magnification) lead to inactivation of the more proximal gene only or of both genes. Hypothetical pairing interactions are schematically illustrated as ladder rungs. (Reprinted, with permission, from Henikoff 1995.)

(Tartof et al. 1984; Grigliatti 1991; Moehrle and Paro 1994) hypothesize ooze initiation centers, and (optionally) ooze terminators (see Fig. 2a). The availability of more proteins such as HP1 leads to more extensive oozing into neighboring euchromatin. If the specific site of termination differs from cell to cell, then oozing of heterochromatin complexes along the chromosome can account for the polar spreading effect.

Unfortunately, the oozing model fails to explain other basic features of PEV. For example, there seems to be a "fit" between the euchromatin and the heterochromatin comprising a PEV junction. X-ray mutagenesis screens of the *brown* gene provide an illustration. In one study, none of 21 PEV alleles was found to have translocated the chromosomal *brown* gene to X heterochromatin, and only 2 were translocated to the Y chromosome (Slatis 1955). However, in an X-ray mutagenesis screen of a P transposon carrying *brown*, a very different distribution of heterochromatic breaks was found: Of the 3 PEV alleles recovered, 2 were translocated to X heterochromatin, and the other was translocated to the Y chromosome (Dreesen et al. 1991). Such examples indicate that heterochromatin is not uniform in its ability to induce PEV, a feature that depends on the euchromatic location of the gene under study. The oozing model does not explain how a gene region discriminates between heterochromatic complexes oozing from different heterochromatic regions.

Oozing also fails to explain PEV of heterochromatic genes, because this requires that heterochromatin retreats back past the junction with euchromatin into a gene such as *light*. However, breaks that cause *light* PEV can be extremely far away from the gene, translocating not only *light*, but also the large heterochromatic block in which *light* lies (Wakimoto and Hearn 1990). If any ooze initiators lie within the translocated block between *light* and the breakpoint, then it is hard to understand how heterochromatin can retreat past these initiators to induce PEV.

Oozing is inconsistent with discontinuous compaction, in which a heterochromatic block appears to be separated from the polytene chromocenter (Belyaeva and Zhimulev 1991). Discontinuous compaction has been confirmed for a polytene chromosome puff that lies between the separated blocks (Belyaeva et al. 1993): This site of active transcription is seen to be flanked by heterochromatin that stains with anti-HP1 antibody, as if the putative ooze has skipped the puff.

Initiator elements at oozing centers have been thought to correspond to specific sequences in heterochromatin. However, it appears that heterochromatin can form at any sequence that is locally repetitive, as predicted more than 50 years ago (Pontecorvo 1944). PEV phenotypes were seen for repeat arrays of three or more copies of a 10-kb P transposon carrying the mini-*white* gene, even when located far from natural heterochromatin (Dorer and Henikoff 1994). Variegation associated with arrays was indistinguishable from classic *white* PEV and was sensitive to the effects of PEV modifiers, indicating the formation of heterochromatin. Reversals of transposons within tandem arrays strengthened PEV, suggesting that paired structures form between neighboring repeats to nucleate condensation: Paired hairpin structures between adjacent reverse repeats should form readily by zippering from a single point in between, whereas loop structures between tandem repeats require the association of sequences a full repeat unit apart.

According to the oozing model, distal sequence changes should have no effect; that is, it should not be possible to attract or repel the ooze from a distance. However, this notion is contradicted by the finding that alterations at a single site strongly regulate *brown* gene PEV (Sabl and Henikoff 1996). Starting with a classic PEV allele consisting of an X-ray-induced translocation that places pericentric heterochromatin adjacent to the *brown* gene on a P transposon, changes were obtained in the number and orientation of copies. Increasing transposon copy number enhanced PEV, and reversing the transposon suppressed PEV. If oozing from the heterochromatic breakpoint were responsible for the PEV phenotype, then more copies must inexplicably attract the ooze, whereas a reversed orientation must repel.

THE ELIMINATION MODEL

In polytene chromosomes, DNA of heterochromatin is differentially underrepresented, and satellite sequences are essentially absent. In polyploid and polytene ovarian tissues, DNA of at least some heterochromatin is found to have unexpected properties suggestive of rearrangement, modification, and loss (Spradling 1993; Glaser and Spradling 1994). These observations have raised the possibility that DNA structural modifications can cause PEV, perhaps by an elimination mechanism (Spradling et al. 1993). However, in several cases, silenced genes are not eliminated from position-affected cells. For example, PEV silencing of a salivary gland chromosome puff did not affect gene copy number determined for DNA extracted from affected glands (Umbetova et al. 1991). For genes such as *white* and *brown*, where PEV occurs in tissue that is likely to be diploid, there is even less reason to favor a gene-loss mechanism, because position-affected genes appear to be normally represented in diploid tissue (Zhang and Spradling 1995). Thus, even though underrepresentation can accompany PEV, DNA elimination cannot be a general explanation of the phenomenon.

THE PAIRING-LOOPING MODEL

The repetitive sequence organization of heterochromatin suggests a different mechanism for PEV. DNA in heterochromatin consists of simple sequence satellites interspersed with blocks of middle repetitive sequences derived from transposable elements (Miklos et al. 1988; Carmena and Gonzalez 1995; Le et al. 1995; Pimpinelli et al. 1995). These middle repetitive sequences typically have counterparts scattered throughout euchromatin. Euchromatic repetitive elements might provide sites that are able to mediate homologous pairing associations between heterochromatin and euchromatin (Sabl and Henikoff 1996). As a result, rearrangements that move repetitive elements present in heterochromatin close to homologous elements in euchromatin would form paired structures, with the sequences in between looped out (Fig. 2b). Homologous interactions would draw euchromatic elements to the surface of the heterochromatic compartment on the nuclear envelope, and this would prevent access of diffusible factors needed for expression of affected genes. According to this pairing-looping model, cytological differences arise from pairing-mediated networks present in heterochromatin but not in euchromatin. For example, cytological compaction in polytene chromosomes would result from network formation stabilized by heterochromatic protein complexes that disrupt the regular band-interband pattern characteristic of the precisely paired euchromatic arms.

Silencing of genes in a polar manner would result from the far greater likelihood for a proximal repetitive element to pair stably with its homolog(s) in heterochromatin than for a distal element. For two nearby genes, when the more proximal gene is looped into the heterochromatic compartment, the more distal gene is not necessarily held down, and it escapes silencing; however, when the more distal gene is looped in, the more proximal gene is held down on both sides and unable to escape silencing (Fig. 2b). Discontinuous compaction around a puff (Belyaeva et al. 1993) would result from the unusually large distance between the heterochromatic breakpoint and the compacted distal region. This would sometimes allow intervening chromatin to loop out far enough from the heterochromatic compartment to avoid being drawn in.

The pairing-looping model unites classic PEV with PEV of heterochromatic genes, previously proposed to involve loss of association of genes with the heterochromatic compartment (Devlin et al. 1990; Wakimoto and Hearn 1990). Associations with repeats would silence euchromatic genes by sequestering them from euchromatic transcriptional components and would activate heterochromatic genes, which require a heterochromatic environment in order to be transcribed. This environment appears to depend on the total amount of heterochromatin in the vicinity of the reporter gene (Howe et al. 1995). Moreover, a heterochromatic environment can be affected by nonadjacent but evidently associated heterochromatin. For both the *rolled* heterochromatic gene (Eberl et al. 1993) and the bw^D heterochromatic element, which silences the euchromatic *brown* gene (Talbert et al. 1994; Henikoff et al. 1995), moving the gene and its accompanying heterochromatin closer to major heterochromatic blocks *in cis* leads to a stronger heterochromatic effect on a reporter gene, and vice versa. In the case of the bw^D element, this heterochromatin-distance effect has a cytologically visible counterpart, whereby enhancement of silencing caused by movement closer to centric heterochromatin correlates with looping into the chromocenter of polytene chromosomes.

The pairing-looping model suggests a different role for heterochromatin-associated proteins than the oozing model. If each association between members of a repeat pair is complexed by HP1 and other *Su(var)*-encoded components of heterochromatin, then overall condensation is sensitive to protein concentration (Fig. 3). These proteins would recognize paired hairpin or loop structures characteristic of closely linked repeats and would stabilize them by binding (Dorer and Henikoff 1994). More of these proteins available in the nucleus will increase the total number of paired segments bound and so increase the stability of homologous associations in general. Paired structures in the immediate vicinity of a position-affected gene will be more stable, and therefore more likely to be drawn into the heterochromatic compartment. General transcriptional factors encoded by *E(var)*s that act to remodel chromatin might counteract the formation and stabilization of paired structures. This model resembles a looping model for PcG-mediated silencing (Pirrotta and Rastelli 1994), except that the binding specificity is for paired structures rather than specific protein-binding sites.

CONCLUSIONS: PAIRED SEQUENCES AS NUCLEAR BRACES?

The proposed involvement of pairing in heterochromatin formation and PEV silencing might be relevant to transgene-silencing phenomena observed in other higher eukaryotes. In mammals, variable expression of transgenes in arrays has been hypothesized to occur by the formation of locally paired structures and condensation into heterochromatin (Dorer and Henikoff 1994). In plants (Matzke et al. 1994) and fungi (Pandit and Russo 1992), many examples of transgene silencing are thought to reveal an efficient homology-scanning mechanism (see Matzke and Matzke; Russo et al.; both this volume). Pairing of homologous sequences may be a general feature of chromatin. If so, how does pairing occur?

A possible answer comes from recent models of the structure of the 30-nm chromatin fiber, a prominent chromosomal structure. For many years, the 30-nm fiber was thought to consist of regular folding of beaded nucleosomes into a solenoid. However, Woodcock et al. (1993) have argued that the 30-nm fiber follows a highly irregular trajectory be-

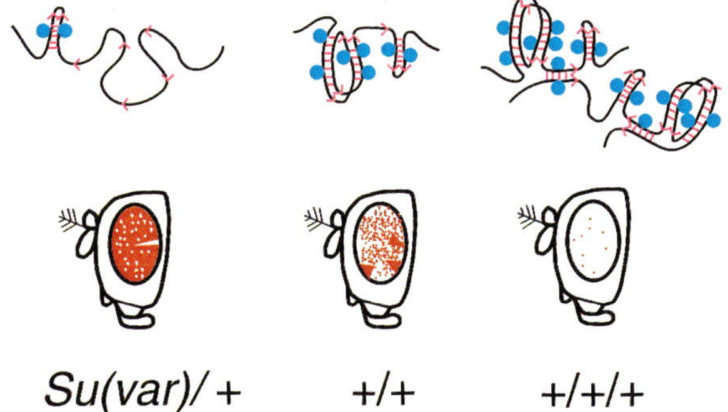

Figure 3 Assembly of heterochromatin by the pairing-looping model. HP1 or other proteins that package heterochromatin (*circles*) bind to paired structures that form transiently at locally repetitive sequences (indicated by *arrowheads* on thin lines), causing stabilization of the structures and formation of networks between paired arrays nearby, thus silencing the eye pigment reporter gene. The frequency of PEV silencing of the reporter depends on the dosage of heterochromatin packaging protein encoded by a *Su(var)*$^+$ gene. Increasing the dose of the *Su(var)*$^+$ gene from one to two to three copies provides increasing levels of protein for stabilization of paired repeats and formation of networks. Larger networks should become heterochromatic more readily, as is the case for transposon repeat arrays, which show a more strongly mutant phenotype as the numbers of copies increase (Dorer and Henikoff 1994; Sabl and Henikoff 1996). Larger networks would associate more readily with other blocks of heterochromatin (Eberl et al. 1993), increasing the frequency with which a gene is silenced by mislocalization to a chromocenter (Talbert et al. 1994; Henikoff et al. 1995). (Reprinted, with permission, from Henikoff 1995.)

cause of differences in the spacings between nucleosomes, and studies of chromatin using newer technologies (Horowitz et al. 1994; Leuba et al. 1994) support this view. To the extent that nucleosomal spacing is ultimately a consequence of sequence-specific constraints, then zippering of homologs might be a natural result. That is, homologous 30-nm fibers that have initiated pairing, perhaps via a DNA-DNA interaction (Kleckner and Weiner 1993), will zipper up because identical irregularities should follow identical and adjacent paths, just as adjacent strands of curly hair curl together rather than individually. The idea that pairing is the default state is consistent with the recovery of mutants that improve pairing: For example, a missense mutation in flies causes mature polyploid ovarian nuclei that lack visible structures to be polytene, with banded chromosomes indicative of precisely aligned chromatids and homologs (Steinhauer and Kalfayan 1992).

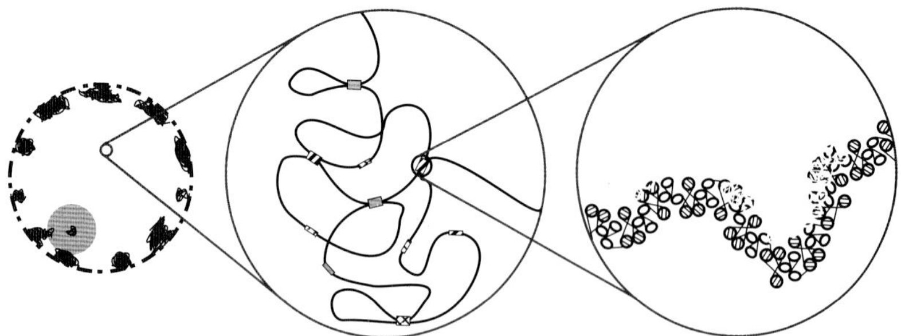

Figure 4 Model for higher-order structure of euchromatin. Magnification of a single hypothetical 30-nm fiber in an interphase nucleus (*left*), showing associations between a subset of dispersed repeats (various boxes in middle diagram). Further magnification of a paired association (*right*) shows the hypothetical alignment of two repeat members as 30-nm fibers with identical trajectories, based on proposed sequence-specific variations in linker length, inspired by the model of Woodcock et al. (1993).

Is there a general housekeeping function for pairing interactions? If pairing of closely spaced repeats holds together heterochromatin on the nuclear envelope at interphase, then perhaps pairing of dispersed repeats provides structural support for euchromatin in the nuclear interior. Structural support has been a proposed function of the nuclear matrix. However, in polytene nuclei, the giant chromosomes appear tightly packed (Hochstrasser et al. 1986), so that any supporting matrix must be insubstantial by comparison. Rather, we might attribute a support role to the 30-nm fiber itself, both in polytene chromosomes, where "ectopic" chromatin fibers are often seen that connect separate but apparently homologous sites (Barr and Ellison 1972; Pardue 1986), and in diploid nuclei. By this view, Velcro-like pairing between dispersed repeats would act as braces for interphase chromatin fibers to provide lateral support (Fig. 4). In diploid nuclei, condensation of chromatin at metaphase would disrupt these paired associations, which could re-form during the following telophase. The abundance of dispersed repeats belonging to many different families should assure that a sufficient number of pairings occur rapidly, favoring pairs in spatial proximity, without requiring that any particular repeat pair comes together at every telophase. In contrast, the precise alignment of chromatin fibers in polytene chromosomes would result from the zippering that can occur when replicating chromosomes are not subject to mitotic stresses, displacing all cross-connections except for those seen as ectopic fibers.

Epigenetic silencing phenomena might reveal heritable alterations in this nuclear support system.

ACKNOWLEDGMENTS

I thank Amy Csink, Doug Dorer, Linda Martin-Morris, Joy Sabl, and Paul Talbert for their intellectual and experimental contributions to the pairing-looping model. Work in my laboratory is supported by the Howard Hughes Medical Institute.

REFERENCES

Barr, H.J. and J.R. Ellison. 1972. Ectopic pairing of chromosome regions containing chemically similar DNA. *Chromosoma* **39:** 53–61.

Belyaeva, E.S., and I.F. Zhimulev. 1991. Cytogenetic and molecular aspects of position effect variegation in *Drosophila*. III. Continuous and discontinuous compaction of chromosomal material as a result of position effect variegation. *Chromosoma* **100:** 453–466.

Belyaeva, E.S., O.V. Demakova, G.H. Umbetova, and I.F. Zhimulev. 1993. Cytogenetic and molecular aspects of position-effect variegation in *Drosophila melanogaster*. V. Heterochromatin-associated protein HP1 appears in euchromatic chromosomal regions that are inactivated as a result of position-effect variegation. *Chromosoma* **102:** 583–590.

Bishop, C.P. 1992. Evidence for intrinsic differences in the formation of chromatin domains in *Drosophila melanogaster*. *Genetics* **132:** 1063–1069.

Carmena, M. and C. Gonzalez. 1995. Transposable elements map in a conserved pattern of distribution extending from beta-heterochromatin to centromeres in *Drosophila melanogaster*. *Chromosoma* **103:** 676–684.

Devlin, R.H., B. Bingham, and B.T. Wakimoto. 1990. The organization and expression of the *light* gene, a heterochromatic gene of *Drosophila melanogaster*. *Genetics* **125:** 129–140.

Dorer, D.R. and S. Henikoff. 1994. Expansions of transgene repeats cause heterochromatin formation and gene silencing in *Drosophila*. *Cell* **77:** 993–1002.

Dorn, R., V. Krauss, G. Reuter, and H. Saumweber. 1993. The enhancer of position-effect variegation of *Drosophila, E(var)3-93D,* codes for a chromatin protein containing a conserved domain common to several transcriptional regulators. *Proc. Natl. Acad. Sci.* **90:** 11376–11380.

Dreesen, T.D., S. Henikoff, and K. Loughney. 1991. A pairing-sensitive element that mediates *trans*-inactivation is associated with the *Drosophila brown* gene. *Genes Dev.* **5:** 331–340.

Eberl, D.F., B.J. Duyf, and A.J. Hilliker. 1993. The role of heterochromatin in the expression of a heterochromatic gene, the *rolled* locus of *Drosophila melanogaster*. *Genetics* **134:** 277–292.

Farkas, G., J. Gausz, M. Galloni, G. Reuter, H. Gyurkovics, and F. Karch. 1994. The *Trithorax-like* gene encodes the *Drosophila* GAGA factor. *Nature* **371:** 806–808.

Gerasimova, T.I., D.A. Gdula, D.V. Gerasimov, O. Siminova, and V.G. Corces. 1995. A

Drosophila protein that imparts directionality on a chromatin insulator is an enhancer of position-effect variegation. *Cell* **82:** 587–597.

Glaser, R.L. and A.C. Spradling. 1994. Unusual properties of genomic DNA molecules spanning the euchromatic-heterochromatic junction of a *Drosophila* minichromosome. *Nucleic Acids Res.* **22:** 5068–5075.

Grigliatti, T. 1991. Position-effect variegation—An assay for nonhistone chromosomal proteins and chromatin assembly and modifying factors. In *Functional organization of the nucleus: A laboratory guide* (ed. B.A. Hamkalo and S.C.R. Elgin), pp. 587–627. Academic Press, San Diego.

Hartmann-Goldstein, I.J. 1967. On the relationship between heterochromatization and variegation in *Drosophila* with special reference to temperature-sensitive periods. *Genet. Res.* **10:** 143–159.

Hayashi, S., A. Ruddell, D. Sinclair, and T. Grigliatti. 1990. Chromosomal structure is altered by mutations that suppress or enhance position effect variegation. *Chromosoma* **99:** 391–400.

Hearn, M.G., A. Hedrick, T.A. Grigliatti, and B.T. Wakimoto. 1991. The effect of modifiers of position-effect variegation on the variegation of heterochromatic genes of *Drosophila melanogaster*. *Genetics* **128:** 785–797.

Heitz, E. 1929. Heterochromatin, Chromocentren, Chromomeren. *Ber. Dtsch. Bot. Ges.* **47:** 274–284.

Henikoff, S. 1979. Position effects and variegation enhancers in an autosomal region of *Drosophila melanogaster*. *Genetics* **93:** 105–115.

———. 1995. A pairing-looping model for position-effect variegation in *Drosophila*. In *Proceedings of the 21st Stadler Genetics Symposium* (ed. J.P. Gustafson and R.B. Flavell), pp. 211–242. Plenum Press, New York.

Henikoff, S., J.M. Jackson, and P.B. Talbert. 1995. Distance and pairing effects on the *brown*^{Dominant} heterochromatic element in *Drosophila*. *Genetics* **140:** 1007–1017.

Hochstrasser, M., D. Mathog, Y. Gruenbaum, H. Saumweber, and J.W. Sedat. 1986. Spatial organization of chromosomes in the salivary gland nuclei of *Drosophila melanogaster*. *J. Cell Biol.* **102:** 112–123.

Horowitz, R.A., D.A. Agard, J.W. Sedat, and C.L. Woodcock. 1994. The three-dimensional architecture of chromatin in situ: Electron tomography reveals fibers composed of a continuously variable zig-zag nucleosomal ribbon. *J. Cell Biol.* **125:** 1–10.

Howe, M., P. Dimitri, M. Berloco, and B.T. Wakimoto. 1995. *Cis*-effects of heterochromatin on heterochromatic and euchromatic gene activity in *Drosophila melanogaster*. *Genetics* **140:** 1033–1045.

James, T.C. and S.C.R. Elgin. 1986. Identification of a nonhistone chromosomal protein associated with heterochromatin in *Drosophila melanogaster* and its gene. *Mol. Cell. Biol.* **6:** 3862–3872.

Kellum, R. and B.M. Alberts. 1995. Heterochromatin protein 1 is required for correct chromosome segregation in *Drosophila* embryos. *J. Cell Sci.* **108:** 1419–1431.

Kleckner, N. and B.M. Weiner. 1993. Potential advantages of unstable interactions for pairing of chromosomes in meiotic, somatic, and premeiotic cells. *Cold Spring Harbor Symp. Quant. Biol.* **58:** 553–565.

Le, M.-H., D. Duricka, and G.H. Karpen. 1995. Islands of complex DNA are widespread in *Drosophila* centric heterochromatin. *Genetics* **141:** 283–303.

Leuba, S.H., G. Yang, C. Robert, B. Samori, K. van Holde, J. Zlatanova, and C. Bustamante. 1994. Three-dimensional structure of extended chromatin fibers as revealed by tapping-mode scanning force microscopy. *Proc. Natl. Acad. Sci.* **91:** 11621–11625.

Lewis, E.B. 1950. The phenomenon of position effect. *Adv. Genet.* **3:** 73–115.

Locke, J. 1994. Examination of DNA sequences undergoing chromatin conformation changes at a variegating breakpoint in *Drosophila melanogaster*. *Genetica* **92:** 33–41.

Locke, J., M.A. Kotarski, and K.D. Tartof. 1988. Dosage-dependent modifiers of position effect·variegation in *Drosophila* and a mass action model that explains their effect. *Genetics* **120:** 181–198.

Lohe, A.R., A.J. Hilliker, and P.A. Roberts. 1993. Mapping simple repeated DNA sequences in heterochromatin of *Drosophila melanogaster*. *Genetics* **134:** 1149–1174.

Martin-Morris, L.E. and S. Henikoff. 1995. Conservation of *brown* gene *trans*-inactivation in *Drosophila*. *Genetics* **140:** 193–199.

Matzke, M.A., A.J.M. Matzke, and O. Mittelsten-Scheid. 1994. Inactivation of repeated genes—DNA-DNA interaction? In *Homologous recombination in plants* (ed. J. Paszkowski), pp. 271–307. Kluwer, Amsterdam.

Mayfield, J.E. and J.R. Ellison. 1975. The organization of interphase chromatin in *Drosophilidae*. The self adhesion of chromatin containing the same DNA sequences. *Chromosoma* **52:** 37–48.

Michailidis, J., N.D. Murray, and J.A.M. Graves. 1988. A correlation between development time and variegated position effect in *Drosophila melanogaster*. *Genet. Res.* **52:** 119–123.

Miklos, G.L.G., M.-T. Yamamoto, J. Davies, and V. Pirrotta. 1988. Microcloning reveals a high frequency of repetitive sequences characteristic of chromosome 4 and the beta-heterochromatin of *Drosophila melanogaster*. *Proc. Natl. Acad. Sci.* **85:** 2051–2055.

Moehrle, A. and R. Paro. 1994. Spreading the silence: Epigenetic transcriptional regulation during *Drosophila* development. *Dev. Genet.* **15:** 478–484.

Pandit, N.N. and V.E. Russo. 1992. Reversible inactivation of a foreign gene, hph, during the asexual cycle in *Neurospora crassa* transformants. *Mol. Gen. Genet.* **234:** 412–422.

Pardue, M.L. 1986. In situ hybridization to DNA of chromosomes and nuclei. In Drosophila, *a practical approach*. (ed. D.B. Roberts), pp. 111–137. IRL Press, Oxford.

Paro, R. and D.S. Hogness. 1991. The Polycomb protein shares a homologous domain with a heterochromatin-associated protein of *Drosophila*. *Proc. Natl. Acad. Sci.* **88:** 263–267.

Pimpinelli, S., M. Berloco, L. Fanti, P. Dimitri, S. Bonaccorsi, E. Marchetti, R. Caizzi, C. Caggese, and M. Gatti. 1995. Transposable elements are stable structural components of *Drosophila* heterochromatin. *Proc. Natl. Acad. Sci.* **92:** 3804–3808.

Pirrotta, V. and L. Rastelli. 1994. *white* gene expression, repressive chromatin domains and homeotic gene regulation in *Drosophila*. *BioEssays* **16:** 549–556.

Platero, J.S., T. Hartnett, and J.C. Eissenberg. 1995. Functional analysis of the chromo domain of HP1. *EMBO J.* **14:** 3977–3986.

Pontecorvo, G. 1944. Structure of heterochromatin. *Nature* **153:** 365–367.

Ptashne, M. 1986. Gene regulation by proteins acting nearby and at a distance. *Nature* **322:** 697–701.

Sabl, J.F. and S. Henikoff. 1996. Copy number and orientation determine the susceptibility of a gene to silencing by nearby heterochromatin in *Drosophila*. *Genetics* **142:** 447–458.

Schlossherr, J., H. Eggert, R. Paro, S. Cremer, and R.S. Jack. 1994. Gene inactivation in *Drosophila* mediated by the *Polycomb* gene product or by position-effect variegation does not involve major changes in the accessiblity of the chromatin fiber. *Mol. Gen. Genet.* **243:** 453–462.

Slatis, H.M. 1955. Position effects at the brown locus in *Drosophila melanogaster. Genetics* **40:** 5–23.

Spofford, J.B. 1976. Position-effect variegation in *Drosophila*. In *Genetics and biology of Drosophila* (ed. M. Ashburner and E. Novitski), vol. 1c, pp. 955–1019. Academic Press, London.

Spradling, A.C. 1993. Position effect variegation and genomic instability. *Cold Spring Harbor Symp. Quant. Biol.* **58:** 585–596.

Spradling, A.C., G. Karpen, R. Glaser, and P. Zhang. 1993. Evolutionary conservation of developmental mechanisms: DNA elimination in *Drosophila*. In *Evolutionary conservation of developmental mechanisms* (ed. A.C. Spadling), pp. 39–53. Wiley-Liss, New York.

Steinhauer, W.R. and L.J. Kalfayan. 1992. A specific ovarian tumor protein isoform is required for efficient differentiation of germ cells in *Drosophila* oogenesis. *Genes Dev.* **6:** 233–243.

Talbert, P.B., C.D.S. LeCiel, and S. Henikoff. 1994. Modification of the *Drosophila* heterochromatic mutation $brown^{Dominant}$ by linkage alterations. *Genetics* **136:** 559–571.

Tartof, K.D., C. Hobbs, and M. Jones. 1984. A structural basis for variegating position effects. *Cell* **37:** 869–878.

Umbetova, G.H., E.S. Belyaeva, E.M. Baricheva, and I.F. Zhimulev. 1991. Cytogenetic and molecular aspects of position effect variegation in *Drosophila melanogaster*. IV. Underreplication of chromosomal material as a result of gene inactivation. *Chromosoma* **101:** 55–61.

Wakimoto, B.T. and M.G. Hearn. 1990. The effects of chromosome rearrangements on the expression of heterochromatic genes in chromosome *2L* of *Drosophila melanogaster. Genetics* **125:** 141–154.

Wallrath, L.L. and S.C.R. Elgin. 1995. Position effect variegation in *Drosophila* is associated with an altered chromatin structure. *Genes Dev.* **9:** 1263–1277.

Weiler, K.S. and B.T. Wakimoto. 1995. Heterochromatin and gene expression in *Drosophila. Annu. Rev. Genet.* **29:** 577–605.

Wines, D.R., P.B. Talbert, D.V. Clark, and S. Henikoff. 1996. Introduction of a DNA methyltransferase in *Drosophila* to probe chromatin structure *in vivo. Chromosoma* **104:** 332-340.

Woodcock, C.L., S.A. Grigoryev, R.A. Horowitz, and N. Whitaker. 1993. A chromatin folding model that incorporates linker variability produces fibers resembling the native structures. *Proc. Natl. Acad. Sci.* **90:** 9021–9025.

Wustmann, G., J. Szidonya, H. Taubert, and G. Reuter. 1989. The genetics of position-effect variegation modifying loci in *Drosophila melanogaster. Mol. Gen. Genet.* **217:** 520–527.

Zhang, P. and A.C. Spradling. 1995. The *Drosophila* salivary gland chromocenter contains highly polytenized subdomains of mitotic heterochromatin. *Genetics* **139:** 659–670.

Control and Function of DNA Methylation in *Neurospora crassa*

Ann T. Hagemann and Eric U. Selker
Institute of Molecular Biology
University of Oregon
Eugene, Oregon 97403-1229

Genomic DNA of the filamentous fungus *Neurospora crassa* is largely devoid of methylation. Methylation at the 5 position of cytosine is heavy in certain areas of the genome, however. Overall, approximately 1.5% of the cytosines in *N. crassa* DNA are methylated (Russell et al. 1987; Foss et al. 1993; Selker et al. 1993b), but in certain regions, nearly 100% of the cytosines are methylated (Selker and Stevens 1985; Selker et al. 1993a). The total number of such methylated patches in the *N. crassa* genome is probably in the hundreds, based on the total amount of methylation (1.5%); the size and density of the two best-characterized, natural, methylated patches (1.6 kb for ζ–η, Selker et al. 1993b; and ~2 kb for ψ–63, B. Margolin and E. Selker, unpubl.); and the size of the genome (~4 x 10^7 bp, Orbach et al. 1988b). Genetic analyses have revealed that mutations in a number of genes can reduce the level of DNA methylation in *Neurospora* (Foss et al. 1993; Roberts and Selker 1995; H. Foss and E. Selker, unpubl.), and one mutant apparently lacks all DNA methylation in vegetative tissue (*dim-2*, for *d*efective *i*n *m*ethylation; Foss et al. 1993).

The cytosine residues that are methylated in *Neurospora* are not present exclusively, or even predominantly, in CpG dinucleotides (Bull and Wootton 1984; Selker and Stevens 1985; Selker et al. 1993a). This is unlike the situation in higher eukaryotes and some other fungi that have been studied (Selker 1993). Non-CpG methylation cannot be propagated by the attractive mechanism proposed by Riggs (1975) and Holliday and Pugh (1975). In the proposed mechanism, an unmethylated CpG that, following replication, is base-paired with a methylated CpG has a methyl group added to the cytosine, thus propagating the methylation pattern. Methylcytosines at nonsymmetrical sites cannot be maintained in this manner. Thus, *Neurospora*, and other organisms with methylation at nonsymmetrical sites, e.g., *Ascobolus immersus* (Goyon and Faugeron 1989), must have an alternative way of keeping cytosines methylated.

What causes certain patches of *Neurospora* DNA to be methylated? In general, a methylated region can be stripped of methylation by replicating it in *Escherichia coli* or by amplification in vitro, and it will once again become methylated when reintroduced into *Neurospora* (Selker et al. 1987a; Cambareri et al. 1991). The methylated DNA can also be moved to another location where it becomes methylated and also frequently causes methylation of neighboring sequences (Selker and Stevens 1987; Selker et al. 1993b; V. Miao and E. Selker, unpubl.). Therefore, it appears that information in the DNA sequence is sufficient to attract methylation to a site in the genome. This is called "de novo methylation" because it does not require preexisting methylation, in contrast to "maintenance methylation," discussed above, which does. The information in the DNA sequence that directs de novo methylation is referred to here as the "methylation signal." The nature of this signal remains largely unknown.

Most, or perhaps all, methylation signals in wild-type *Neurospora* were created by the process of repeat-induced point mutation (RIP; for review, see Selker 1990). RIP creates multiple C-G to T-A transition mutations in duplicated sequences if the strain carrying them is put through the sexual cycle. *Neurospora*'s sexual cycle begins with formation of specialized organs that are fertilized by nuclei of the opposite mating type. This results in a dikaryon containing separate nuclei of the two mating types. These nuclei proliferate within the developing fruiting body, and, following premeiotic DNA synthesis, nuclei of opposite mating type undergo karyogamy (fusion) and meiosis to form haploid ascospores. RIP occurs within the nuclei of the dikaryon, after fertilization but before karyogamy, and affects the duplicated sequence in only the nuclei carrying the duplication (Selker et al. 1987b). RIP can be heavy; in some cases nearly 50% of the cytosines in a sequence are mutated (Cambareri et al. 1989; Singer et al. 1995). RIP has not been detected in vegetative cells, nor has it been demonstrated to occur outside of the genus *Neurospora*. Frequently, a sequence mutated by RIP will thereafter be methylated and will be spontaneously remethylated following removal of the methylation by growth in the presence of the methylation inhibitor 5-azacytidine (5AC) or by replication outside of *Neurospora*. Therefore, RIP of a sequence commonly produces a de novo methylation signal (Cambareri et al. 1991; Selker et al. 1993b; Singer et al. 1995).

WHAT IS THE RELATIONSHIP BETWEEN RIP AND METHYLATION?

The production of de novo methylation signals as a result of RIP acting on a normally unmethylated sequence indicates that methylation is a consequence of RIP. Whether methylation is also required for RIP to occur is still an open question. One reason for wondering whether methylation might be required for RIP is that two alternative mechanisms, illustrated in Figure 1, have been proposed to explain how RIP makes mutations (Selker 1990). One of these involves methylation of cytosines in the duplicated sequence, followed by deamination of the 5-methylcytosines to thymines (Fig. 1A). An alternative mechanism involves deamination of cytosines to uracils (Fig. 1B). For a mutation created by either of these two scenarios to be made permanent, it must be either replicated or subjected to mismatch repair favoring the mutation. Experiments are under way to determine if RIP can occur in the absence of methylation.

A process called MIP (methylation induced premeiotically), which occurs in the fungus *A. immersus*, is related to RIP in that it inactivates duplicated sequences during the sexual cycle (Rossignol and Faugeron 1994). MIP causes methylation without mutations. Thus, it is economical to suppose that RIP evolved from MIP by adding the ability to mutate. This view would be consistent with a role for methylation in the RIP mechanism.

WHAT CONSTITUTES A DE NOVO METHYLATION SIGNAL?

Neurospora methylation signals have been studied in several ways. One approach utilized a collection of sequences altered by RIP, some methylated and some not, and attempted to distinguish features of the methylated sequences that were absent from the unmethylated ones (Singer et

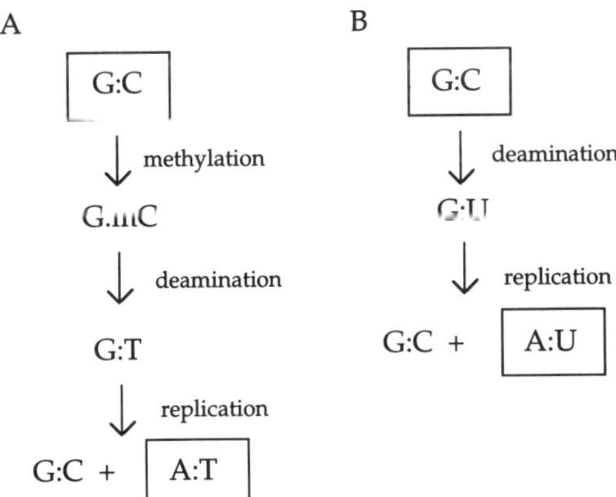

Figure 1 Two alternative mechanisms for RIP have been proposed, only one of which (*A*) involves methylation.

al. 1995). Another approach involved deletion analyses of a methylated region to determine the minimum effective methylation signal (Selker et al. 1993b; V. Miao and E. Selker, unpubl.). Several conclusions came from these studies: (1) There is no simple sequence, analogous to a restriction recognition sequence or a promoter element, that is shared by all methylated regions and absent from unmethylated ones. (2) A-T-rich sequences are more likely to be methylated, but A-T-richness per se is not the signal. It is possible, however, that A-T-richness in a particular context or pattern, such as a high density of TpA dinucleotides, is the signal. The reason for suspecting A-T-richness or TpA involvement is that RIP converts C to T and most often mutates cytosines in CpA dinucleotides, creating TpA dinucleotides. A-T- or TpA-richness may be influencing methylation only indirectly, for example, by altering the DNA structure. (3) The signal is redundant; several small (<300 bp), nonoverlapping fragments of a methylated patch are capable of acting alone as methylation signals. (4) Different fragments containing signals promote different levels of methylation, suggesting either several signals with incremental effects or signals of differing strengths. (5) Spreading of methylation into adjacent sequences, known to lack methylation signals, sometimes occurs.

Studies of methylation signals that involve engineered constructs are complicated by frequent methylation of transgenic DNA. *Neurospora* often integrates transforming DNA ectopically in multiple, linked copies, and the integrated DNA is often methylated (Bull and Wootton 1984; Orbach et al. 1988a; Pandit and Russo 1992; Romano and Macino 1992; Selker et al. 1993b). Therefore, an early concern about the methylation induced by the first methylated region identified in *Neurospora*, a native sequence called ζ–η, was that the methylation was induced by the sequence duplication; ζ–η is a diverged, tandem duplication of 5S rRNA pseudogenes (Selker and Stevens 1985). This possibility was ruled out by introducing the two halves of the duplication separately (Selker et al. 1993b), as well as by the deletion analysis mentioned above (V. Miao and E. Selker, unpubl.); in both cases it was shown that a duplication is not required for sequences from the ζ–η region to act as methylation signals.

The above experiments demonstrated that a duplication is not required for induction of methylation, but they did not address the possibility that a duplication might be sufficient to induce methylation. There are a number of clues, however, that repeats are not sufficient. For example, the *am* gene of *Neurospora*, when transformed without vector sequences, is not methylated in multicopy arrays (Selker et al. 1993b; B.

Margolin and E. Selker, unpubl.). Direct and inverted duplications containing two copies of the *Neurospora am* gene (2 kb), separated by 2.3 kb (including the *hph* gene), are not methylated when integrated at any of a variety of genomic locations (Irelan et al. 1994; J. Irelan and E. Selker, unpubl.). There is an indication that the *E. coli hph* gene and the *N. crassa* **a** mating-type region, together about 4 kb, are not methylated when present as an uninterrupted, tandem repeat (Chang and Staben 1994). It has also been reported that the *Neurospora eth-1* gene did not become methylated in multicopy transformants (Mautino et al. 1996). These examples indicate that repeated DNA is not sufficient to induce methylation, but it is still possible that repeats play a critical role in determining methylation status, in combination with other factors. The features of transforming DNA that cause it to be methylated may include its sequence, copy number, arrangement, and chromosomal context.

Although methylation of transforming DNA occurs de novo, it can be distinguished from that directed by the de novo methylation signals discussed above. DNA carrying a de novo methylation signal triggers methylation when present as a single copy in any genomic location. To avoid confusion due to spurious methylation of transforming DNA, a potential methylation signal can be tested as a single copy inserted into the *am* locus by gene replacement (Miao et al. 1994). Several *Neurospora* DNA sequences that act as methylation signals at their native loci also act as signals when integrated in *am*. In contrast, a number of other sequences, both native ones that are normally unmethylated, and ones foreign to *Neurospora*, are unmethylated in *am* (Miao et al. 1994; V. Miao et al., unpubl.).

WHAT IS THE ROLE OF DNA METHYLATION IN *NEUROSPORA*?

The biological role for methylation in *Neurospora* has not yet been determined, but we know that methylation can inhibit gene expression in *Neurospora*, as it does in higher eukaryotes. Several known cases of gene regulation by methylation involve engineered constructs. Methylation of some alleles of the *am* gene created by RIP was associated with loss of transcripts detectable by Northern analysis. When methylation was prevented, however, either by treatment with 5AC or by crossing the *am* alleles into the methylation-deficient *dim-2* mutant background, transcripts were detected (M. Rountree and E. Selker, unpubl.). These *am* alleles were not functional; therefore, the effect of methylation could not be assayed genetically. A situation in which the effect of methylation can be detected genetically exists for the *E. coli hph* gene, which encodes resistance to hygromycin, in a construct where it is flanked on both sides by *am* genes and is integrated at an unknown site in the *N. crassa* genome (Irelan et al. 1994). This construct was initially unmethylated, but, after going through the sexual cycle, the duplicated *am* genes were mutated by RIP. In a small fraction of these Am^{-} progeny, methylation of the *hph* gene occurred, presumably spreading from the adjacent, mutated sequences, to inactivate *hph*. This effect was alleviated by treating with the drug 5AC, which caused loss of methylation and restored hygromycin resistance (Irelan et al. 1994; J. Irelan and E. Selker, unpubl.).

One other example of a role for methylation in gene regulation is seen with an unstable allele of the *am* gene that was selected in the laboratory (Kinsey and Helber 1989). This allele was created by insertion of Tad (a transposon from the strain Adiopodoume) upstream of the *am* coding sequence. The *am* gene is active when the transposon is methylated and presumably transcriptionally silent, whereas the *am* gene is inactive when the transposon is unmethylated and presumably transcriptionally active (Cambareri et al. 1996).

It had been reported earlier that *Neurospora* mutants defective in DNA methylation are prone to becoming aneuploid, suggesting a role for methylation in chromosome segregation or replication control (Foss et al. 1993). It has since been shown that this effect is not in fact due to the lack of methylation, but is likely due to other mutations in these methylation-defective strains (Foss et al. 1995). Because the *dim-2* mutant lacks all detectable methylation in vegetative tissues yet is viable and has no obvious phenotypes, it is unlikely that methylation plays a crucial biological role for *Neurospora* growing vegetatively under laboratory conditions.

Another phenomenon found in *Neurospora* that may be related to DNA methylation is transgene silencing or "quelling." Introduced genes may be inactivated in vegetative cells upon transformation. This can affect either a foreign gene (Pandit and Russo 1992), or a native gene when homology is present on the transforming DNA (Romano and Macino 1992). This phenomenon may be related to methylation of ectopic transforming sequences, discussed above. However, inactivation of native genes following transformation with homologous sequences has been shown to occur efficiently in the *dim-2* mutant, which lacks methylation in vegetative cells (Cogoni et al. 1996). Therefore, methylation is not responsible for this type of silencing. Whether methylation is responsible for the silencing of foreign sequences is uncertain and is addressed for a particular foreign sequence, *hph*, by Russo et al. (this volume).

EVIDENCE FOR MAINTENANCE METHYLATION IN *NEUROSPORA*

In most cases, sequences that are methylated in *Neurospora* recover their methylation after it is lost as a result of treatment with 5AC. Surprisingly, two alleles of *am* that cannot recover their methylation were discovered among six methylated *am* alleles generated by RIP (Singer et al. 1995). The methylation of these two alleles was stable through vegetative growth following the cross in which they were created by RIP, but it was not reestablished, even after many replication cycles, following removal of the methylation by 5AC. Thus, these two alleles do not carry signals for de novo methylation in vegetative cells. Retention of the original methylation during vegetative growth must therefore have been due to maintenance methylation, rather than to de novo methylation. This result was verified by removing the methylation using another method. Specifically, the *am* gene was isolated from these strains, amplified in *E. coli*, and reintroduced at the native *am* locus. The replaced versions did not become remethylated.

Because the two alleles of *am* that display maintenance methylation were products of the sexual cycle and are not methylated de novo during vegetative growth, it is likely that their methylation arose as part of the RIP process. Alternatively, the methylation may have arisen at some other time during the sexual cycle. This raises the question of whether control of methylation during the sexual phase of the *Neurospora* life cycle is different from that during vegetative growth.

Prior to the discovery of these *am* alleles, *Neurospora* was thought to lack maintenance methylation, since the methylation of transformed sequences appeared to be independent of the methylation status of the transforming DNA (B. Margolin et al., unpubl.). The new evidence of maintenance methylation therefore suggests that only particular sequences can support maintenance methylation in this organism; i.e., in *Neurospora*, maintenance methylation may require more than just preexisting methylation. A region's propensity for propagating methylation could be influenced by DNA sequence and structure. Perhaps maintenance methylation depends on a feature that is also found in de novo methylation signals. The methylated *am* alleles that lack signals for de novo methylation have fewer mutations than the other methylated alleles (Singer et al. 1995). Thus, a quantitative rather than qualitative difference between the sequences may distinguish those that can induce methylation in vegetative tissue and those that can simply maintain preexisting methylation. De novo signals may also function as maintenance "signals," but this feature would be obscured by their ability to become methylated de novo. Figure 2 illustrates the possibility that the

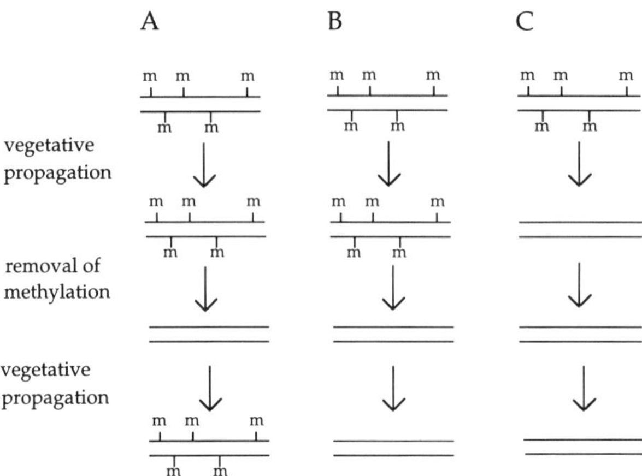

Figure 2 Three possible types of sequences that may be distinguished by the *Neurospora* methylation machinery. Although methylation is heterogeneous within a population of nuclei in *Neurospora* (Selker and Stevens 1985; Selker et al. 1993a), for simplicity, only one pattern of methylation (m) is shown. The three types of sequences are drawn following the establishment of methylation (e.g., in the sexual cycle) at the top, and their methylation states following vegetative propagation, as well as following demethylation and further vegetative propagation, are indicated below. Only DNA carrying a de novo signal (*A*) is remethylated after its methylation is removed. Some sequences can propagate methylation for many replication cycles but cannot direct their own remethylation after the methylation is removed (*B*). Other sequences neither trigger methylation de novo nor serve as substrates for maintenance methylation (*C*). This type of DNA sequence may comprise most of the *Neurospora* genome, although very few sequences have been tested for the presence of maintenance signals.

Neurospora methylation machinery distinguishes three classes of sequences: those that direct de novo methylation (A), those supporting maintenance but not de novo methylation (B), and those incapable of supporting either type of methylation (C).

SUMMARY

Whereas most of the *Neurospora* genome is unmethylated, in certain regions nearly every cytosine can be methylated; the methylation is not confined to symmetric sequences (e.g., CpG). De novo methylation depends on a signal, which is defined as information present in the sequence that attracts methylation to the area. The nature of the signal is not fully understood, but it is not a simple sequence motif. Maintenance methylation, which is propagated from preexisting methylation, seems to also require a signal, in the sense that not all sequences are subject to it. Methylation signals are commonly created by RIP of a duplicated sequence, but it is not yet known whether the RIP mechanism requires methylation. Methylation can inhibit gene expression in *Neurospora*, but, at least in vegetative tissue, it is not essential, and it has not been shown to play a major role in the biology of this organism.

ACKNOWLEDGMENTS

We thank J. Irelan, B. Margolin, and H. Foss for thoughtful comments on the manuscript, and M. Rountree, H. Foss, B. Margolin, V. Miao, J. Irelan, and E. Cambareri for sharing unpublished results. This work was supported by National Institutes of Health grant GM-35690 and a National Science Foundation research training grant, Genetic Mechanisms of Evolution, BIR-9014265.

REFERENCES

Bull, J.H. and J.C. Wootton. 1984. Heavily methylated amplified DNA in transformants of *Neurospora crassa*. *Nature* **310:** 701–704.

Cambareri, E.B., M.J. Singer, and E.U. Selker. 1991. Recurrence of repeat-induced point mutation (RIP) in *Neurospora crassa*. *Genetics* **127:** 699– 710.

Cambareri, E.B., B.C. Jensen, E. Schabtach, and E.U. Selker. 1989. Repeat-induced G-C to A-T mutations in *Neurospora*. *Science* **244:** 1571–1575.

Cambareri, E.B., H.M. Foss, M.R. Rountree, E.U. Selker, and J.A. Kinsey. 1996. Epigenetic control of a transposon-inactivated gene in *Neurospora* is dependent on DNA methylation. *Genetics* **143:** 137–146.

Chang, S. and C. Staben. 1994. Directed replacement of *mt A* by *mt a-1* effects a mating type switch in *Neurospora crassa*. *Genetics* **138:** 75–81.

Cogoni, C., J.T. Irelan, M. Schumacher, T.J. Schmidhauser, E.U. Selker, and G. Macino. 1996. Transgene silencing of the *al-1* gene in vegetative cells of *Neurospora* is mediated by a cytoplasmic effector and does not depend on DNA:DNA interactions or DNA methylation. *EMBO J.* **15:** 3153–3163.

Foss, H.M., C.J. Roberts, K.M. Claeys, and E.U. Selker. 1993. Abnormal chromosome behavior in *Neurospora* mutants defective in DNA methylation. *Science* **262:** 1737–1741.

———. 1995. Abnormal chromosome behavior in *Neurospora* mutants defective in DNA methylation (letter). *Science* **267:** 316.

Goyon, C. and G. Faugeron. 1989. Targeted transformation of *Ascobolus immersus* and de novo methylation of the resulting duplicated DNA sequences. *Mol. Cell. Biol.* **9:** 2818–2827.

Holliday, R. and J.E. Pugh. 1975. DNA modification mechanisms and gene activity during development. *Science* **187:** 226–232.

Irelan, J.T., A.T. Hagemann, and E.U. Selker. 1994. High frequency repeat-induced point mutation (RIP) is not associated with efficient recombination in *Neurospora*. *Genetics* **138:** 1093–1103.

Kinsey, J.A. and J. Helber. 1989. Isolation of a transposable element from *Neurospora crassa*. *Proc. Natl. Acad. Sci.* **86:** 1929–1933.

Mautino, M.R., J.L. Barra, and A.L. Rosa. 1996. *eth-1*, the *Neurospora crassa* locus encoding S-adenosylmethionine synthetase: Molecular cloning, sequence analysis and *in vivo* overexpression. *Genetics* **142:** 789–800.

Miao, V.P.W., M.J. Singer, M.R. Rountree, and E.U. Selker. 1994. A targeted replacement system for identification of signals for de novo methylation in *Neurospora crassa. Mol. Cell. Biol.* **14:** 7059–7067.

Orbach, M.J., W.P. Schneider, and C. Yanofsky. 1988a. Cloning of methylated transforming DNA from *Neurospora crassa* in *Escherichia coli. Mol. Cell. Biol.* **8:** 2211–2213.

Orbach, M.J., D. Vollrath, R.W. Davis, and C. Yanofsky. 1988b. An electrophoretic karyotype of *Neurospora crassa. Mol. Cell. Biol.* **8:** 1469– 1473.

Pandit, N.N. and V.E.A. Russo. 1992. Reversible inactivation of a foreign gene, *hph*, during the asexual cycle in *Neurospora crassa* transformants. *Mol. Gen. Genet. 234:* 412–422.

Riggs, A.D. 1975. X-inactivation, differentiation and DNA methylation. *Cytogenet. Cell Genet.* **14:** 9–25.

Roberts, C.J. and E.U. Selker. 1995. Mutations affecting the biosynthesis of S-adenosylmethionine cause reduction of DNA methylation in *Neurospora crassa. Nucleic Acids Res.* **23:** 4818–4826.

Romano, N. and G. Macino. 1992. Quelling: Transient inactivation of gene expression in *Neurospora crassa* by transformation with homologous sequences. *Mol. Microbiol.* **6:** 3343–3353.

Rossignol, J.-L. and G. Faugeron. 1994. Gene inactivation triggered by recognition between DNA repeats. *Experientia* **50:** 307–317.

Russell, P.J., K.D. Rodland, E.M. Rachlin, and J.A. McCloskey. 1987. Differential DNA methylation during the vegetative life cycle of *Neurospora crassa. J. Bacteriol.* **169:** 2902–2905.

Selker, E.U. 1990. Premeiotic instability of repeated sequences in *Neurospora crassa. Annu. Rev. Genet.* **24:** 579–613.

———. 1993. Control of DNA methylation in fungi. In *DNA methylation: Molecular biology and biological significance* (ed. J.P. Jost and H.P. Saluz), pp. 212–217. Birkhäuser Verlag, Boston.

Selker, E.U. and J.N. Stevens. 1985. DNA methylation at asymmetric sites is associated with numerous transition mutations. *Proc. Natl. Acad. Sci.* **82:** 8114–8118.

———. 1987. Signal for DNA methylation associated with tandem duplication in *Neurospora crassa. Mol. Cell. Biol.* **7:** 1032–1038.

Selker, E.U., D.Y. Fritz, and M.J. Singer. 1993a. Dense non-symmetrical DNA methylation resulting from repeat-induced point mutation (RIP) in *Neurospora. Science* **262:** 1724–1728.

Selker, E.U., B.C. Jensen, and G.A. Richardson. 1987a. A portable signal causing faithful DNA methylation *de novo* in *Neurospora crassa. Science* **238:** 48–53.

Selker, E.U., E.B. Cambareri, B.C. Jensen, and K.R. Haack. 1987b. Rearrangement of duplicated DNA in specialized cells of *Neurospora. Cell* **51:** 741–752.

Selker, E.U., G.A. Richardson, P.W. Garrett-Engele, M.J. Singer, and V. Miao, 1993b. Dissection of the signal for DNA methylation in the ζ–η region of *Neurospora. Cold Spring Harbor Symp. Quant. Biol.* **58:** 323–329.

Singer, M.J., B.A. Marcotte, and E.U. Selker. 1995. DNA methylation associated with repeat-induced point mutation in *Neurospora crassa. Mol. Cell. Biol.* **15:** 5586–5597.

Silencing of the Gene *hph* in *Neurospora crassa*

Vincenzo E.A. Russo, Yeon-Su Lee, and Antonio C. Codón
Max-Planck-Institut für Molekulare Genetik
D14195 Berlin, Germany

Neurospora crassa is a eukaryotic organism belonging to the group of fungi called Ascomycetes. In nature, it grows on decaying or burnt vegetation in tropical or subtropical areas. Since Beadle and Tatum (1941) isolated the first biochemical mutants and suggested the "one gene–one enzyme hypothesis," *Neurospora* has been a very popular model for biochemical genetics, molecular genetics, and for development (Russo and Pandit 1992).

Neurospora has several advantages for epigenetical studies: (1) it grows very fast; (2) it can grow on minimal defined medium, and (3) it produces many germinal cells such as conidia. A typical culture obtained from 20 ml of agar medium contains about 10^9 conidia. Conidia contain on average two or three haploid nuclei. The mycelium is coenocytic, as the cells of the filaments forming the mycelium have dozens of nuclei. The organism is haploid except for a transient phase that precedes meiosis during the sexual cycle, and the haploid genome contains about 4×10^7 base pairs on seven different chromosomes (Russo and Pandit 1992). Today, thousands of mutants of *N. crassa* are known and more than 620 genes have been mapped, and about 200 genes have been cloned (J. Kinsay [Director of Fungal Genetics Stock Center, Kansas City, Kansas], pers. comm.).

Neurospora is a heterothallic organism in that it has two distinct mating types, although these are morphogenetically indistinguishable. In the asexual life cycle there are three major morphological structures: the germinal cells called *conidia*; *hyphae*, which branch forming a mat called the mycelium; and *aerial hyphae*, which produce conidia. The whole asexual life cycle of *Neurospora* lasts a few days. The sexual life cycle takes a few weeks and results in a dozen different major morphological structures (Russo and Pandit 1992).

N. crassa protoplasts can be easily transformed with an efficiency of 10^3–10^4 transformants per microgram of DNA (Vollmer and Yanofsky 1986). The transformants arise by integration of incoming DNA into the

Epigenetic Mechanisms of Gene Regulation

chromosome, although hardly anything is known about the integration process. Research is being conducted to find out whether the transforming DNA integrates preferentially in some chromosomes, if it integrates in any particular regions of a chromosome (e.g., near the centromere or the telomere), or if there are preferred nucleotide sequences for the integration of the transforming DNA. None of the enzymes involved in the integration process is known. It was during our studies to elucidate this transformation process that we stumbled onto the phenomenon of gene silencing, unknown until then in *Neurospora*.

THE PHENOMENON OF *HPH* SILENCING

Transformants of *Neurospora* often contain multiple copies of the transforming DNA (Bull and Wootton 1984). To study the integration process, the first question we asked was: Are the multiple copies inserted into the same nucleus? Germinated conidia containing several nuclei are routinely used for transformation. Therefore, multiple insertions could be either in one single nucleus or in several nuclei. Our laboratory discovered that up to eight integrations can occur in one and the same nucleus (Pandit and Russo 1992), and integrations in both nuclei in the same transformant are rare.

To obtain these results, we used a special strain, the heterokaryon (*al*-1, *arg*-10)*(*al*-2, *arg*-1), as the recipient in the transformation. The two arginine markers ensured that on minimal medium only the heterokaryotic conidia grew. These then contained at least one nucleus with the wild-type form of the *arg*-1 gene, and another nucleus with the *arg*-10 wild-type gene. The albino mutations were chosen to be able to distinguish if a given mycelium contained one or the other, or both, types of nuclei by visual inspection. The presence of nuclei with only the *al*-1 mutation in a mycelium confers upon it a yellow color, due to low production of carotenoids; nuclei with only the *al*-2 mutation result in a white color (no carotenoids) of the mycelium, and the heterokaryon containing both types of nuclei in the same mycelium is orange (Pandit and Russo 1992).

The plasmid pCSN44 used for transformation is shown in Figure 1. It contains the bacterial *hph* gene that confers resistance to the antibiotic hygromycin B (Hyg). On the basis of Southern blot analysis, this plasmid does not have any sequence homology with the *Neurospora* genome. After transformation, we observed that 13 out of 36 transformants had only one plasmid inserted into their genomes, which rendered their co-

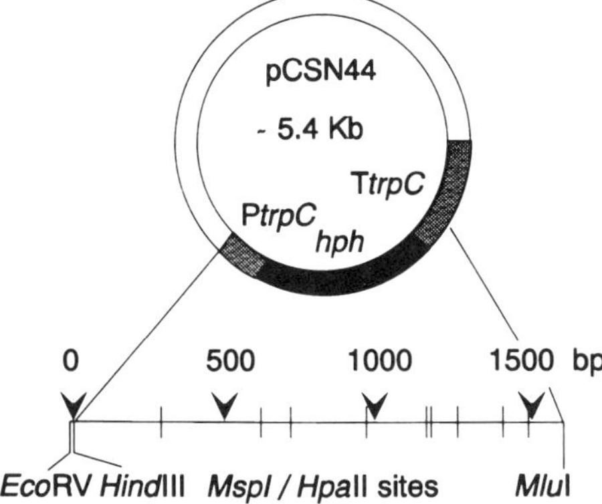

Figure 1 Map of plasmid pCSN44. P*trp*C is the promoter region, *hph* is the coding region, and T*trp*C is the termination region. P*trp*C and T*trp*C are the promoter and the termination regions of the tryptophan gene of *Aspergillus nidulans*. Once attached to the coding region of the bacterial *hph* gene, a gene is created that is active in *Neurospora*. The DNA stretch from the *Eco*RV to the *Mlu*I restriction sites was used as a probe for the Southern blot of Fig. 4. The *Hpa*II/*Msp*I sites of this DNA stretch are shown.

nidia 100% resistant to hygromycin B. The other 23 transformants contained between 2 and 10 inserted plasmids, and half of these transformants had a special phenotype: The expression of the *hph* gene could be silenced under certain conditions.

This phenomenon of silencing can be illustrated using the transformant HTA5 as an example (Fig. 2). The primary transformant HTA5 contains eight plasmids inserted into its genome in close proximity to each other in the same locus, but not in tandem. HTA5 was isolated after transformation on a slant containing arginine and hygromycin B. We found that the conidial stock of HTA5 was heterogeneous: 30% of the conidia grew on plates containing the antibiotic, whereas 70% of the conidia did not grow. A simple genetic experiment was performed to determine if this difference was just phenotypical or also genotypical as shown in Figure 2. It could be deduced that all the conidia of the HTA5.7 stock culture had an identical genetic constitution, and that the inactivation of the *hph* gene in this stock was not due to a mutation. This inactivation could be reversed if the mycelia were grown on media containing hygromycin (see Fig. 2B). The hygromycin resistance of conidial stocks grown on this medium was about 30%. We called this phenomenon of reversible inactivation "silencing."

METHYLATION OF THE *HPH* GENE

Since it is known that in other organisms methylation can regulate gene expression, we looked at methylation of the *hph* gene. *N. crassa* has a low but measurable level of 5-methylcytosine (m^5C). Measured by isotope dilution gas chromatography-mass spectrometry, the level of m^5C is 1.44% of total cytosines for the DNA of conidia. During development from conidia to mycelium, this level decreases to 0.96% for the DNA of the mycelium during the early and middle-exponential phase and returns to 1.6% at the stationary phase (Russell et al. 1987). Although the percentage of methylcytosines seems low, one has to consider that *Neurospora* has the potential to code for more than 10,000 genes. Therefore, 1% of 5-methylcytosines could represent the methylation of all the cytosines of 100 different genes, and the number of genes could be higher if some are only partially methylated. Thus, the possible number of methylated genes is sufficient to play an important developmental role. Unfortunately, very little is known about the distribution of 5-methylcytosines in the genome of *Neurospora* and their potential role in the life cycle of this fungus.

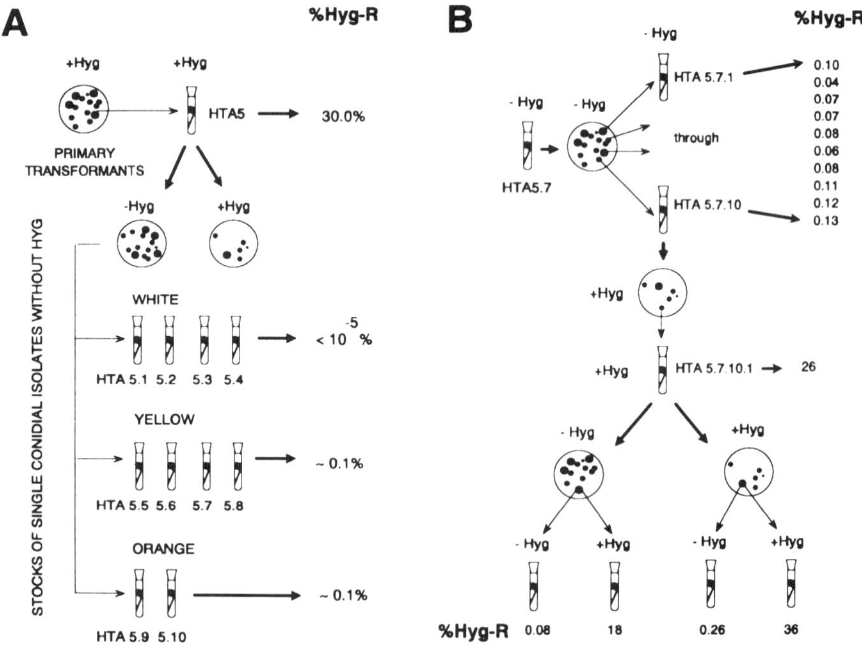

Figure 2 (See facing page for legend.)

Figure 2 Pedigree analysis of the primary transformant HTA5 showing an example of gene silencing. Open circles indicate culture plates containing colonies, and tubes represent conidial stocks. +Hyg indicates that the medium contained hygromycin B (200 μg/ml). %Hyg-R indicates the percentage of conidia resistant to hygromycin B in a particular conidial stock. All the media contained arginine. (*A*) The primary transformant HTA5 was plated with and without hygromycin to determine the %Hyg-R. Ten conidial isolates taken randomly from the plates without the antibiotic were inoculated onto slants without hygromycin. Four slants contained *Neurospora* cultures with white mycelium and conidia, and these cultures had no conidia resistant to hygromycin (%Hyg-R $<10^{-5}$%). Four slants contained cultures that had yellow conidia and mycelium and their percentage of resistance was around 0.1%. Two slants had cultures with orange mycelium and conidia and the cultures had a percentage of resistance of 0.1%. The fact that there were pure white and yellow cultures indicates that the original heterokaryon had segregrated into cultures containing only one type of the original two nuclei. Clearly, all the inserts had been integrated into the nucleus containing the *al*-1 mutation which confers the yellow color, because none of the four pure white cultures containing the *al*-2 nucleus was resistant to hygromycin. Southern blot analysis showed that the white stock cultures had no plasmid inserted, whereas the other stock cultures had the same eight plasmids inserted as did the primary transformant HTA5. This indicated that the large difference in percentage of resistant conidia between HTA5 (30% Hyg-R) and its derivatives such as HTA5.7 (0.1% Hyg-R) was not just due to some instability of plasmid integration. It could, however, be due to some mutation that could not be resolved by Southern blot analysis. (*B*) To test if all the conidia of HTA5.7 had the same genetic constitution, ten conidial isolates taken randomly from the plates without the antibiotic were inoculated onto slants without hygromycin. The %Hyg-R of each of these ten cultures was around 0.1%, indicating that the *hph* gene was present in each nucleus of the HTA5.7 conidial stock. If a culture was obtained from HTA5.7.10 (one of the HTA5.7 derivatives just described) by inoculating a slant containing hygromycin with a conidial isolate from a plate containing the antibiotic, then the new culture (HTA5.7.10.1) had a hygromycin resistance of 26%. This level of resistance is very similar to the one of the great-grandparent HTA5 (%Hyg-R = 30%), indicating that the silencing is reversible. Using HTA5.7.10.1 conidia, it is possible to repeat the cycle of inactivation and reactivation of the *hph* gene, as shown in the figure. The level of hygromycin resistance in HTA5 and its derivatives depends only on the presence or absence of hygromycin in the medium on which the conidial stock is made. (Reprinted, with permission, from Pandit and Russo. 1992.)

In addition to the changes in m^{5}C level during mycelium growth, there is some evidence for changes of methylation of the ribosomal genes during development. In *Neurospora* the 17S, 5.8S, and 25S ribosomal genes are clustered (in this order), and there are 185–200 copies of this rDNA repeat unit arranged in tandem array. Each rDNA repeat is 9 kb long and contains, aside from the three genes mentioned above, a 4-kb-long stretch of a nontranscribed spacer region (Russell et al. 1984). The methylation state of these rDNA repeats was analyzed by Southern blot, digesting the genomic DNA of *Neurospora* with the two isoschizomers *Hpa*II and *Msp*I. Both enzymes recognize and cut the sequence 5′-CCGG if unmethylated, but *Hpa*II does not cut if any of the two cytosines are methylated, and *Msp*I does not cut if the central C is methylated (McClelland et al. 1994). With this technique it was possible to show that the methylated cytosines are not randomly distributed throughout all the CCGG sequences within the rDNA repeat unit, but are clustered specifically within the nontranscribed spacer region of the repeat unit (Russell et al. 1985). Furthermore, the results showed that one or more specific CCGG sites may be methylated or demethylated in correlation with a particular developmental stage. However, no other gene of *Neurospora* has been shown to be epigenetically regulated by methylation.

In contrast to the low level of methylation of the endogenous DNA, *Neurospora* can heavily methylate transforming DNA when it is present in multiple copies on the genome (Bull and Wootton 1984). Heavy methylation has been shown recently also for RIPped DNA (Selker et al. 1993a; see also Hagemann and Selker, this volume). A methylation analysis of representative *hph* transformants, using Southern blot analysis coupled with digestion by the isoschizomers *Hpa*II and *Msp*I, revealed a new peculiarity of methylation in *Neurospora*. Transformants containing multiple copies have indeed many 5-methylcytosines in the transforming integrated plasmid DNA, but transformants with a single integrated copy of the plasmid show no methylation of the integrated plasmid DNA (Pandit and Russo 1992). This new finding prompted us to ask if silencing, which is present only in transformants with multiple copies, is due to cytosine methylation.

IS SILENCING DUE TO METHYLATION?

To test whether methylation is necessary for silencing, one has to perturb methylation of the silenced gene. Theoretically, this can be done in two ways, chemically or genetically. It has been reported that there are mutants defective in methylation (*dim*) in *Neurospora* (Foss et al. 1993).

A possible genetic experiment to show the role of methylation in the silencing phenomenon would be to transform these *dim* strains with the plasmid shown in Figure 1, to isolate transformants with multiple inserts, and then to determine what percentage of these transformants is silenced. This type of experiment has not been done yet. Unfortunately, the classic genetic way to cross the *dim*⁻ strain with the HTA5.7 strain to obtain a double mutant is not possible because RIPping would change the sequence of the many *hph* genes of strain HTA5.7. The alternative chemical way to reduce or eliminate methylation from the silenced gene is to grow the mycelium of HTA5.7 in the presence of 5-azacytidine, an analog of cytidine that inhibits the enzyme methyltransferase. To perform this experiment, we made three consecutive conidial stocks on medium with 5-azacytidine, each stock being initiated by inoculation of 100 conidia from the previous stock. The conidial stocks obtained in this way contained a much higher percentage of Hyg^R than the original HTA5.7 strain, and this percentage increased, reaching more than 50% for the third stock of the series (Pandit and Russo 1992). These results seemed to support the hypothesis that methylation is necessary for silencing. However, 5-azacytidine is also likely to have a side effect, since it induces mitotic recombination in *Drosophila*, an organism that does not have any measurable 5-methylcytosine (Pontecorvo et al. 1992). Experiments done with the strain HTA5.7 confirmed that in *Neurospora* this drug also has the capacity to induce deletions of the multiple copies of the plasmid (Fig. 3A). Such deletions can happen spontaneously and are selected out by growth on hygromycin B (the SRF strains in Fig. 3B). There is a direct correlation between the level of deletion and the level of Hyg^R for the spontaneous resistant strains (Fig. 3B). Furthermore, in the strains SRF11 and SRF15 that have high Hyg^R levels (90% and 91%), the plasmid DNA left after the deletion is practically unmethylated in the *Hpa*II site (Fig. 4). Taken together, these results could be explained in the following way:

1. Repeated *hph* genes are methylated in *N. crassa*, as shown by Pandit and Russo (1992).
2. In HTA5.7, the eight plasmids that are integrated in close proximity to each other on the same chromosome are silenced because perhaps a protein binding specifically methylated DNA creates a chromatin structure that inhibits the transcription of any *hph* gene in this chromatin. In *Neurospora* no such protein has been found yet, but it is well documented in mammalian cells (Tate and Bird 1993).
3. Growth on 5-azacytidine induces deletion of the inserted plasmids due to DNA demethylation (Pandit and Russo 1992). It is known that

the DNA of *Neurospora* mycelia grown on 5-azacytidine is demethylated (Bull and Wootton 1984). The deletions can therefore be explained if one assumes that cytosine methylation inhibits homologous recombination in *Neurospora*. Alternatively, one can also assume that 5-azacytidine induces high mitotic recombination in *Neurospora*, as it does in the meiotic recombination of *Drosophila*.

4. Deletions can happen spontaneously with a low frequency.
5. The inserted plasmid DNA left after deletion has no stretch of DNA that is duplicated.
6. Half of the transformants with multiple methylated copies are not silenced, probably because there is one *hph* gene that is not integrated in the same locus as the others, and therefore is not silenced.

It is difficult to decide at this stage if any of these assumptions are correct, because more experimental work is necessary.

CAN THE *HPH* GENE BE A MODEL FOR OTHER TRANSGENES IN *N. CRASSA*?

The main novelty in the experiments presented above is that a single insert of a foreign gene is not methylated and is not inactivated, whereas multiple copies are methylated and half of the transformants are silenced. That could be a general phenomenon for any gene in *Neurospora* if the following two statements are true:

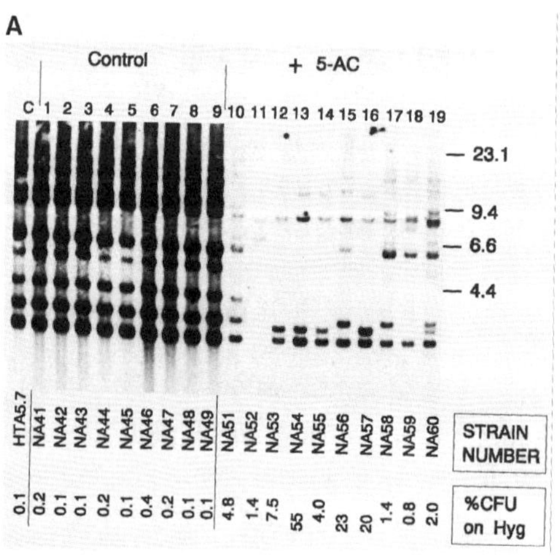

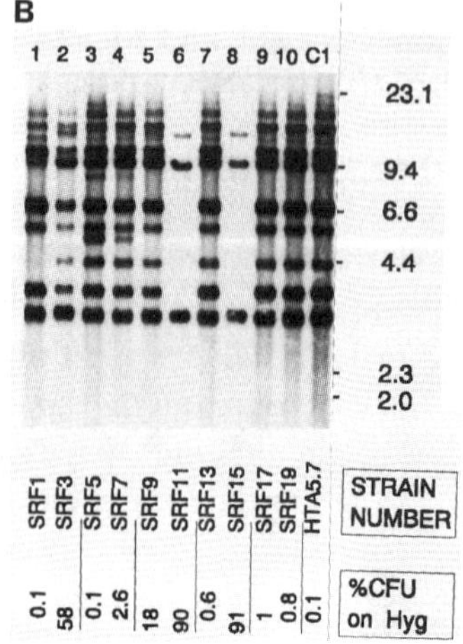

Figure 3 (*See facing page for legend.*)

1. Multiple copies of any transforming DNA, when integrated into the same locus, are methylated independently of their DNA sequence.
2. Single copy of any transgene is not methylated independently of its DNA sequence as long as there is no major duplication.

It is not possible to specify further what a major duplication is, because there are no studies that determine the minimum length and the percentage of identity of duplication DNA that induces methylation in this fungus.

Figure 3 Southern blot analysis of the plasmid integration pattern in the constitutive hygromycin-resistant strains obtained from HTA5.7. (*A*) The resistant strains were obtained by culturing HTA5.7 on 5-azacytidine (5AC), an analog of cytidine that inhibits the enzyme methyltransferase. About 100 conidia of HTA5.7 were inoculated onto a slant containing minimal medium with arginine and 25 μg/ml 5AC, but no hygromycin. About 100 conidia were taken from the stock obtained in this way and were inoculated onto another slant containing the same medium. From this stock a third slant was inoculated in the same way. Conidia from the third stock were plated on minimal medium plus arginine, and ten randomly picked conidial isolates were used to inoculate ten slants with minimal medium plus arginine. The conidial stocks from these slants have been called NA51–NA60. Mycelia were grown from these conidial stocks in liquid medium containing minimal medium plus arginine, and the DNA was isolated and digested with *Eco*RV, which cuts the plasmid pCSN44 only once. The DNA probe for the Southern blot was the whole plasmid pCSN44. The strains NA41–NA49 were obtained in the same way as described above with one exception: The slants they were grown on did not contain any 5AC. The ten strains obtained after 5AC (NA51–NA60) all had higher %Hyg-R values (called %CFU on Hyg in the figure) than the parental strain HTA5.7 and had fewer bands in the Southern blot, indicating that they had deletions of the integrated plasmids. (*B*) The strains SRF3, SRF7, SRF11, SRF15, and SRF19 were obtained in the same way as the 5AC strains NA51–NA60 with two differences: Five consecutive stocks were obtained instead of three, and the slants contained hygromycin (200 μg/ml) instead of 5AC. The other strains, SRF1, SRF5, SRF9, SRF13, and SRF17, were obtained in a similar way as the previous SRF strains except that hygromycin was present only in the slants of the first, third, and fifth stocks. Two strains which are highly resistant to hygromycin, SRF11 and SRF15, both have much less plasmid DNA integrated into the genome due to deletions. We assume that in these strains a spontaneous recombination event occurred, creating a non-silenced *hph* gene that was selected by growing five consecutive stocks on a selective medium with hygromycin. The methylation pattern of these two strains is shown in Fig. 4 (N.N. Pandit and V.E.A. Russo, unpubl.).

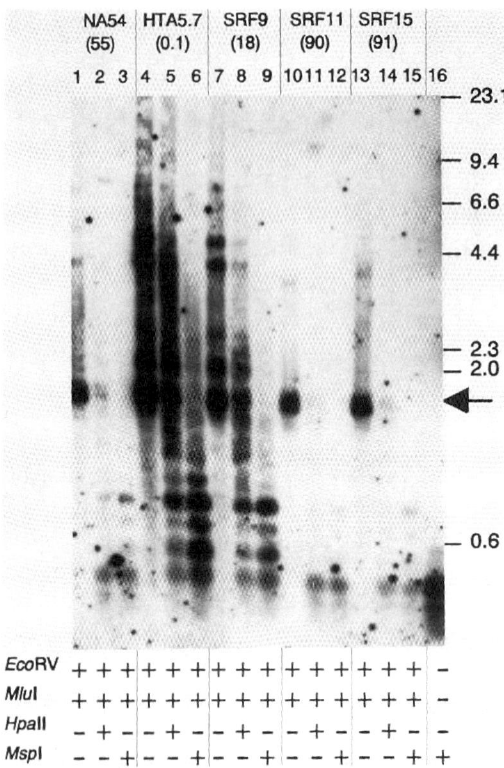

Figure 4 Methylation pattern of the *hph* gene in the transformant HTA5.7 and some of its derivatives with high resistance to hygromycin. Strains NA54, SRF9, SRF11, and SRF15 were described in Fig. 3. DNA was digested with the enzymes listed at the bottom of the figure. *Eco*RV and *Mlu*I have only one site in the plasmid pCSN44 and their restriction sites flank the *hph* gene (see Fig. 1). The DNA probe used for the Southern blot is the stretch of DNA containing the *hph* gene between the sites of *Eco*RV and *Mlu*I. In lane 16, the DNA of the plasmid pCSN44 is loaded as a negative control of *Hpa*II methylation. The results indicate that the HTA5.7 strain has many methylated *Hpa*II sites, whereas the resistant strains have a methylation level that is inversely proportional to the resistant level (N.N. Pandit and V.E.A. Russo, unpubl.).

Are Multiple Copies Always a Signal for Methylation in *N. crassa?*

The first paper to show that in *Neurospora* multiple copies are heavily methylated is the paper of Bull and Wootton (1984). A *N. crassa* strain containing a long deletion of the *am* gene was transformed with a plasmid containing the *am* gene. This plasmid had no homology with the genome of the recipient strain. It was shown that all the nine different transformants analyzed had multiple integrated copies, from 4 to 25, depending on the transformant. The multiple copies of any transformant were integrated into the same locus. No information was given as to whether the integration locus of each transformant was the same or a different one. Two of these nine transformants were further analyzed for methylation of the *am* gene using the Southern blot technique. The genomes of the transformants were cut with the isoschizomer pair *Hpa*II/*Msp*I and *Sau3*A/*Mbo*I, and the probe was the *am* gene. In each isoschizomer pair there is one enzyme that is sensitive to cytosine methylation and another that is not. The specificity and sensitivity of *Hpa*II and *Msp*I were explained above in the section Methylation of the *HPH* Gene. *Sau3*A cuts the restriction site GATC if the cytosine is not methylated, whereas *Mbo*I cuts the same sequence even if the cytosine is methylated. Bull and Wootton (1984) found that the *am* gene was heavily methylated at both of these restriction sites. This is in contrast to the low methylation of cytosine in the genome of *Neurospora*. On the basis of the sequence of the *am* gene, the C in the GATC sites was methylated regardless of the base following it. Bull and Wootton argued, therefore, that multiple copies of DNA induce high methylation in *Neurospora* and, based on the Southern blots using the *Sau3*A/*Mbo*I restriction enzymes, this methylation is not only in the CpG configuration as in mammalian cells, but in any cytosine. A similar result, using one transformant of the *arg* gene, was obtained by Orbach et al. (1988).

Romano and Macino (1992) transformed *Neurospora* with different portions of both the carotenogenic *al*-1 and *al*-3 genes from *Neurospora*. The copy number of the exogenous integrated sequence ranged from 2 to 20, and from each of the *al*-1 and *al*-3 transformants, two were further analyzed for methylation. All four strains tested were heavily methylated at the GATC sites in the ectopic *al*-1 (or *al*-3) genes but not in the endogenous albino genes. The transformants had a peculiarity—the carotenoid production was abnormally low. This phenomenon is similar to silencing and was called "quelling." The main difference between silencing and quelling is that silencing is stable over at least 100 nuclear divisions, whereas quelling is detectably unstable after 25 nuclear divisions. The other difference is that in silencing, all the *hph* genes are in the same locus, whereas in quelling, the exogenous genes are together but in a different locus from the endogenous one, and the endogenous genes are not methylated. It is not clear yet if the difference of stability between silencing and quelling is due to the facts just mentioned or to the different origin of the genes: *al*-1 and *al*-3 are *Neurospora* genes, whereas *hph* is of bacterial origin.

Pandit and Russo in their 1992 paper, already discussed in this chapter, have found that four different transformants containing the multiple copies of the *hph* gene were methylated, whereas three transformants with single copy insert were not methylated.

Chang and Staben (1994) show one Southern blot with one digestion with the restriction enzymes *Eco*RI (which does not cut GAATTmC) and *Eco*RV (insensitive to methylation). In the duplicated region containing the *hph* gene and the **a** mating-type region, there are two sites for *Eco*RI and three for *Eco*RV. The Southern blot shows several bands, including a 8454 band, indicating an incomplete digestion of the duplicated region. Chang and Staben do not comment on these results, perhaps because they are not interested in methylation. In the paper of Mautino et al. (1996) there is the statement that the *Neurospora eth*-1 gene did not become methylated in multicopy transformants, but no data are shown. There are some unpublished reports that certain transgenic constructs can escape methylation (see Hagemann and Selker, this volume), but these have not been confirmed. Our working hypothesis remains, therefore, that any multiple-copy transgenic DNA in *Neurospora* is likely to be methylated. The genes tested so far are *am*, *arg*, *hph*, *al*-1, and *al*-3. Therefore, it seems that the methyltransferases recognize multiple copies independently from the DNA sequence.

Is There a "Portable Signal" for Cytosine Methylation in *N. crassa?*

The first paper that introduced the concept of a "portable signal" for C methylation in *N. crassa* is by Selker et al. (1987). In this paper the ζ–η region of *Neurospora* (see Hagemann and Selker, this volume) was shown to be remethylated when introduced by transformation into *Neurospora*, after having been demethylated through growth in *Escherichia coli*, even when the transformant had only one exogenous copy of the ζ–η region. A fact which complicates the interpretation of these data is that the ζ region and the η region, both about 800 bases long, have 83% identity with each other (Grayburn and Selker 1989). We do not know how long duplicated DNA must be and how much identity is necessary in order to be a signal for the hypothetical methyltransferase in *Neurospora*. It may be that a duplication of 800 bases with 83% identity is enough.

This possibility was recognized by Selker et al. (1993b), who tried to dissect the signal for DNA methylation in the ζ–η region of *Neurospora*. They cloned several deletions of this region, transformed *Neurospora* with the cloned constructs, and isolated transformants with only single inserts. Southern blot analysis showed that all constructs had been methylated. This was true also for the constructs that had only either the ζ region or the η region. These results were interpreted as evidence that the ζ–η region of *Neurospora* has a portable signal and that it is not the duplication per se that is the signal. Curiously, the two plasmids used in those experiments, pUC12 and pBR322, were methylated. As the authors noted, that was peculiar because in a previous paper (Selker et al. 1987) the vector pUC8 was not methylated. It is possible that the vector could itself carry a duplicated sequence that induces methylation of the vector and methylation spreads into the bits of the ζ–η region cloned into it, which could be tested with appropriate controls. Thus, we think that the evidence for a portable signal of methylation in *N. crassa* is not strong.

CONCLUSIONS

Multiple copies of the *hph* gene introduced in *N. crassa* by transformation are methylated, and in about 50% of the transformants the gene is silenced. It has not yet been possible to prove that there is a cause and effect relationship between these two phenomena. We hypothesize that multiple copies of the *hph* genes that are integrated in close proximity to each other on the same chromosome are silenced because of a methylation-directed chromatin structure. It is not known why only half of the transformants with multiple copies are silenced. One possiblity is that in the transformants which are not silenced there is one *hph* gene integrated in a locus different from the locus in which the bulk of the plasmids are integrated, resulting in the bulk of *hph* genes being silenced but not the "solitary" *hph* gene. Silencing seems to be a widespread phenomenon in plants (see Matzke and Matzke, Meyer, both this volume), and we think that all duplicated DNAs analyzed so far are methylated in *N. crassa*, independent of the DNA sequence.

ACKNOWLEDGMENTS

I thank all the coworkers in my laboratory for their excellent results, in particular Niketan Pandit and Paola Crucitti. Many thanks also to Angie Hoffman for reading and criticizing the manuscript.

REFERENCES

Beadle, G.W. and E.L. Tatum. 1941. Genetic control of biochemical reactions in *Neurospora*. *Proc. Natl. Acad. Sci.* **27:** 499–506.

Bull, J.H. and J.C. Wootton. 1984. Heavy methylated amplified DNA in transformants of *Neurospora crassa*. *Nature* **310:** 701–704.

Chang, S. and C. Staben. 1994. Directed replacement of mt A by mt a-1 effects a mating type switch in *Neurospora crassa*. *Genetics* **138:** 75–81.

Foss, H.M., C.J. Roberts, K.M. Claeys, and E.U. Selker. 1993. Abnormal chromosome behavior in *Neurospora* mutants defective in DNA methylation. *Science* **262:** 1737–1741.

Grayburn, W.S. and E.U. Selker. 1989. A natural case of RIP: Degeneration of the DNA sequence in an ancestral tandem duplication. *Mol. Cell. Biol.* **9:** 4416–4421.

Mautino, M.R., J.L. Barra, and A.L. Rosa. 1996. eth-1, the *Neurospora crassa* locus encoding S-adenosylmethionine synthetase: Molecular cloning, sequence analysis and in vivo overexpression. *Genetics* **142:** 789–800.

McClelland, M., M. Nelson, and E. Raschke. 1994. Effect of site-specific methylation on DNA modification methyltransferases and restriction endonucleases. *Nucleic Acids Res.* **22:** 3640–3659.

Orbach, M.J., W.P. Schneider, and C. Yanofsky. 1988. Cloning of methylated transforming DNA from *Neurospora crassa* in *Escherichia coli*. *Mol. Cell. Biol.* **8:** 2211–2213.

Pandit, N.N. and V.E.A. Russo. 1992. Reversible inactivation of a foreign gene, *hph*, during the asexual cycle in *Neurospora crassa* transformants. *Mol. Gen. Genet.* **234**: 412–422.

Pontecorvo, G., A. Avitabile, G. Esposito, G. Migliaccio, and M. Carfagna. 1992. Induced crossing-over in *Drosophila melanogaster* germ cells of DNA repair-proficient and repair-deficient (mei-9^{L1}) males following larval feeding with 5-azacytidine and mitomycin C. *Mutat. Res.* **266:** 93–98.

Romano, N. and G. Macino. 1992. Quelling: Transient inactivation of gene expression in *Neurospora crassa* by transformation with homologous sequences. *Mol. Microbiol.* **6:** 3343–3353.

Russell, P.J., K.D. Rodland, E.M. Rachlin, and J.A. McCloskey. 1987. Differential DNA methylation during the vegetative life cycle of *Neurospora crassa*. *J. Bacteriol.* **169:** 2902–2905.

Russell, P.J., K.R. Rodland, J.E. Cutler, E.M. Rachlin, and J.A. McCloskey. 1985. DNA methylation in *Neurospora*: Chromatographic and isoschizomer evidence for changes during development. *UCLA Symp. Mol. Cell. Biol. New Ser.* **34:** 321–332.

Russell, P.J., S. Wagner, K.D. Rodland, R.L. Feinbaum, J.P. Russell, M.S. Bret-Harte, S.J. Free, and R.L. Metzenberg. 1984. Organization of the ribosomal ribonucleic acid genes in various wild-type strains and wild-collected strains of *Neurospora*. *Mol. Gen. Genet.* **196:** 275–282.

Russo, V.E.A. and N.N. Pandit. 1992. Development in *Neurospora crassa*. In *Development—The molecular genetic approach* (ed. V.E.A. Russo et al.), pp. 88–102. Springer-Verlag, Berlin.

Selker, E.U., D.Y. Fritz, and M.J. Singer. 1993a. Dense non-symmetrical DNA methylation resulting from repeat-induced point mutation (RIP) in *Neurospora*. *Science* **262:** 1724–1728.

Selker, E.U., B.C. Jensen, and G.A. Richardson. 1987. A portable signal causing faithful DNA methylation de novo in *Neurospora crassa*. *Science* **238:** 48–53.

Selker, E.U., G.A. Richardson, P.W. Garrett-Engele, M.J. Singer, and V. Miao. 1993b. Dissection of the signal for DNA methylation in the ζ-η region of *Neurospora*. *Cold Spring Harbor Symp. Quant. Biol.* **58:** 323–329.

Tate, P.H. and A.P. Bird. 1993. Effects of DNA methylation on DNA-binding proteins and gene expression. *Curr. Opin. Genet. Dev.* **3:** 226–231.

Vollmer, S.J. and C. Yanofsky. 1986. Efficient cloning of genes of *Neurospora crassa*. *Proc. Natl. Acad. Sci.* **83:** 4869–4873.

Prokaryotic Model of Epigenetic Inactivation: Chromosomal Silencing in *Bacillus subtilis* Fusion Products

Valérie Grandjean,[1] Françoise Le Hégarat, and Luisa Hirschbein
Institut de Génétique et Microbiologie
CNRS URA 1354, Centre d'Orsay
91405 Orsay CEDEX, France

The study of "DNA silencing" in bacteria by an imprinting-like process is a new but expanding field (Hotchkiss and Gabor 1985; Casadesús and Maldonado 1990; Gratia 1994). To this day, only a few examples of control of gene expression through epigenetic inactivation have been reported in bacteria, unlike eukaryotic systems, where the phenomenon has been reported since 1930 (Muller 1930; Lyon 1993). Silencing involving large DNA fragments has been observed in *Escherichia coli*. For instance, the F plasmid encodes two proteins, SopA and SopB, and a *cis*-acting *sopC* element, which belongs to the partition system. This system is necessary for plasmid inheritance at cell division (Hiraga 1992). Recently, an additional function was demonstrated for the SopB protein, involved in DNA silencing. The SopB product acting at the *sopC* element in vivo promotes the inactivation of genes located within a 10-kb region adjacent to *sopC*. Silencing in this case is associated with the propagation of a nucleoprotein structure that commences at the *sopC* region. This structure restrains the accessibility of the intracellular DNA to proteins such as DNA gyrase or DNA methylase (Lynch and Wang 1995).

Another example is the *E. coli bgl* operon responsible for the utilization of β-glucosides. This operon is normally repressed in wild-type cells. Its silencing might be associated with promoter inactivation through a specific structural chromatin domain. It was demonstrated that promoter inactivation in this case involved the action of both upstream and downstream chromosomal elements (Schnetz 1995). Bacteria can also display heritable genetic inactivation affecting an entire chromo-

[1]Present address: Université de Nice, centre de Biochimie, Parc Valrose, INSERM U273, 06108 Nice, France.

Epigenetic Mechanisms of Gene Regulation

some. This has been found in heterozygous lines of gram-positive bacteria derived from fusion products (Schaeffer et al. 1976; Fleisher and Vary 1985). In this chapter, we recount how the concept of chromosomal inactivation evolved. We describe the genetic and physical studies that have been applied to define the chromosomal status of *B. subtilis* fusion products: recombination analysis, study of ploidy, and study of inactivation stability. A model that accounts for localized chromosome inactivation in the recombinant fusion products is considered.

THE DISCOVERY OF NONCOMPLEMENTING DIPLOIDS

After fusion of *B. subtilis* protoplasts from two parental strains bearing different polyauxotrophic markers using polyethylene glycol, and a period of growth in nonselective medium to promote cell-wall regeneration, diploid bacteria arose which had distinctive phenotypes and stability. The analysis of single colonies from the regeneration medium showed mainly two classes of diploids called recombinant diploids and noncomplementing diploids (Ncds), respectively (Hotchkiss and Gabor 1985; Schaeffer and Hirschbein 1985).

Recombinant Diploids

For the sake of simplicity, we call the unsupplemented glucose medium MO, the parental phenotypes A and B, and their respective properly supplemented media MA and MB. The recombinant diploid class will grow either in MO, MA, MB and M(A+B) (prototroph phenotype) or only on M(A+B) and never on MO, MA, MB (auxotroph phenotype).

Recombinant diploids that displayed an auxotrophic or prototrophic phenotype represented 0.2–1% of the regenerated colonies. Since these clones were stable, they were initially considered haploid recombinants (Schaeffer et al. 1976). However, when the DNA from these stable recombinants was analyzed for genetic linkage by transformation experiments, biosynthetic activities from the nutritionally deficient parental alleles were recovered (Lévi-Meyrueis and Sanchez-Rivas 1984). DNA extracted from recombinant diploids that displayed a prototroph phenotype contained all the genes marked by mutations in the parental strains (Lévi-Meyrueis and Sanchez-Rivas 1984). Hence, it was suggested that the genetic structure of these primary stable recombinants might be diploid, with one chromosome containing the expressed genes, and the other, the inactive genes (Schaeffer and Hirschbein 1985).

Unstable Noncomplementing Diploids

The new notion of diploid bacteria with either a partial or a complete chromosomal inactivation was supported by the discovery by Hotchkiss and Gabor (1980) of the second type of clone produced by protoplast fusion. Those represented from 4% to 10% of the regenerated colonies. They were initially named unstable Ncds. They showed an unstable segregation pattern (biparental traits). Each particular clone displayed the phenotype of one parent, either A or B. However, during growth in selective medium, i.e., MA for clones that had the A parental phenotype or MB for clones that displayed the B parental phenotype, unstable Ncds segregated progeny displaying both the opposite parental phenotype and cells with recombinant phenotypes at a frequency of 10^{-2} to 10^{-4}. The latter, because of their often belated appearance, were named secondary recombinants.

The ability of unstable Ncds to segregate cells bearing the previously unexpressed phenotype was found to be clonal and independent of the growth medium. It could be observed during at least 70–100 cell divisions. These observations suggested that unstable Ncd fusion products are heterodiploids and that both genomes replicate in unstable Ncd cells but are never expressed at the same time (Gabor and Hotchkiss 1983; Hotchkiss and Gabor 1985).

Several results support this conclusion.

1. Artifactual persistence of the biparental traits due to cross-feeding between pure parental strains, suggested by Hauser and Karamata (1992), was excluded by reconstruction experiments. Equal amounts of parental cells A and B were mixed and grown on each of the two selective parental media, MA or MB. In these experiments, the number of counterselected bacteria remained constant or decreased in both selective liquid media. This result is expected in the absence of cross-feeding between the parental cell of the mixture. In contrast, the unstable Ncd progeny in selective medium for the growth of only one parent exhibited, throughout successive cell divisions, a constant increase in the number of cells displaying the phenotype of the counterselected parent.

   Cell adhesion, as a possible artifact that could explain the persistence of the biparental trait (Hauser and Karamata 1992, 1994), was also eliminated by reconstruction experiments. Typically, the two parental strains, one A and the other B, differently marked, were mixed in a 1:1 ratio, treated with polyethylene glycol, and plated onto nonselective media. Well-separated colonies were then picked and placed on media that contained either the A or the B supplement. Of the colonies, 98–99% did not display the biparental traits of the unstable Ncd colonies (Le Dérout et al. 1992). The colonies picked from the mixed master plate grew only in one of the supplemented media, either MA or MB. The remaining 1% of the colonies that were able to grow in MA and MB media and not in MO medium were further analyzed. In colonies grown in MA medium, the cells were always parental, of phenotype A as shown by their inability to grown in medium lacking the A supplements. Equivalent results were obtained by the same procedure for the colonies from the MB medium. On the contrary, the progeny of a single unstable Ncd colony will continue to manifest the biparental traits for 2–4 cycles of subcloning and growth onto nonselective media or in supplemented medium for the growth of the parental strains of identical unstable Ncd phenotypes.

Segregation and reactivation of the previously unexpressed chromosome was the only possible explanation for the biparental traits of unstable Ncds (Grandjean 1994).

2. The silencing of one parental chromosome led to the suppression of dominant sensitivity or resistance markers. This finding was supported by two observations. First, unstable Ncd growth was insensitive to addition of rifampicin when the recessive allele (rfm^R) was located on the expressed chromosome and the dominant allele (rfm^S) on the silent chromosome (Guillen et al. 1982; Lévi-Meyrueis and Sanchez-Rivas 1984). This drug inhibits the activity of the RNA polymerase by association with its β subunit (Boor et al. 1995). Second, unstable Ncds with a chloramphenicol resistance gene (chloramphenicol acetyltransferase, *cat*) on the silent chromosome displayed a Cat^s phenotype (Lopez et al. 1986; Le Dérout et al. 1992).
3. In transformation experiments, with donor DNA purified from unstable Ncds, the efficiency of silent markers was similar to that of the active markers (Bohin et al. 1982; Guillen et al. 1982).
4. Unstable Ncds with persisting biparental traits for at least 60–100 generations displayed a heterogeneous progeny within single colonies. Some of these produced reciprocal recombinants for 8–11 genes. Others accumulated within a single colony several types of recombinants (up to 10 types) including prototrophs. These observations were drawn from the analysis of at least 98 clones representing 6000 colonies tested (Hotchkiss and Gabor 1980; Gabor and Hotchkiss 1983) and were further confirmed by Ftouhi and Guillen (1990). Altogether, these results indicate that unstable Ncd cells bear-

ing the biparental traits can maintain two different replicons for many generations and undergo multiple crossing-overs before segregation occurs.

Stable Noncomplementing Diploids

Ncds that segregate cells of the unexpressed parental phenotype at a frequency of less than 10^{-7} were subsequently described and named stable Ncds (Guillen et al. 1985). Stable Ncds are selected among the 4–10% of unstable Ncds always present in the regeneration medium. Detection of stable Ncds is thus dependent on the knowledge of the previous state of diploidy and of which genes of the parental chromosomes are expressed or silent. Stable Ncds can be isolated either from the unstable Ncd clones arising on the regeneration medium or as segregants from unstable Ncds growing in medium selective for the growth of one parent of identical phenotype (Lévi-Meyrueis and Sanchez-Rivas 1984; Le Dérout et al. 1992). Since stable Ncds cannot be distinguished from haploid cells by their phenotype, the presence of markers from the stabilized silent chromosome had to be detected using several experimental approaches.

1. PEG-induced self-fusion led to the destabilization of both chromosomes and the reactivation of the previously unexpressed markers. This process resulted in the appearance of new phenotypes, unstable Ncds and recombinant diploids, not present in the original fusion products (Guillen et al. 1985).
2. Upon transforming of *B. subtilis* cells with donor DNA purified from stable Ncds, the unexpressed alleles displayed a low but significant transforming efficiency (Guillen et al. 1985; Sanchez-Rivas et al. 1988).
3. Drug-resistance genes cloned from the silent chromosome into a shuttle vector regained their genetic activity when they were transferred into an unrelated species such as *E. coli* (Lopez et al. 1986; Thaler et al. 1990).

SILENCING IS A TWO-STEP PROCESS

Comparative studies on the transforming activity of both purified DNA and nonpurified nucleoids prepared from unstable Ncd and stable Ncd clones indicated that inactive chromosomes are assembled in a structure which prevents gene expression. To gain insight into the basis of chromosomal suppression, nucleoids from single unstable Ncd colonies were prepared by gentle lysis at low ionic strength (Charret et al. 1980; Hirschbein and Guillen 1982). They were then used as donor DNA for transformation. The yield of transformants was 10–100 times higher for markers expressed in vivo than for those located in the silent chromosome. RNase treatment had no effect on the transforming activity of the silent markers. In contrast, proteinase K treatment increased 20- to 50-fold the transforming activity of the silent markers. The stimulating effect of proteinase K was not observed for the markers expressed in vivo. Purified DNA from unstable Ncd clones transformed the silent as well as the expressed markers with the same efficiency (Hotchkiss and Gabor 1980; Bohin et al. 1982). On the contrary, the transforming activity of the silent markers from stable Ncds remained low, even after proteinase K digestion and further DNA purification. The observation that DNA from unstable Ncds and stable Ncds had unequal transforming activities after purification and proteinase K treatment suggests that silent markers in the unstable Ncds are blocked by proteins. It also suggests that chromosomal inactivation in stable Ncds involves a two-stage process, the first step being reversible and associated with DNA tertiary structure. The second step would be, for the most part, irreversible and possibly linked to a modification of the DNA itself (Guillen et al. 1985).

Inactivation Requires the Spo0A Product

The process of chromosomal inactivation was shown to be influenced by the *B. subtilis* Spo0A gene product. The *spo0A* gene encodes a transcription factor that is expressed in the vegetative phase. *Spo0A* mutations are recessive to the wild-type allele and highly pleiotropic (Trowsdale et al. 1978; Ferrari et al. 1985). The Spo0A gene product is essential for the initiation of sporulation (Burbulys et al. 1991). It also has an effect on chromosome condensation (Setlow et al. 1991) and on vegetative growth (Kudoh et al. 1985).

Segregated cells of both parental phenotypes were detected by pedigree analysis of unstable Ncds carrying the *spo0A* mutant allele. When the *spo0A* mutation was carried on the expressed chromosome, and the wild-type allele on the silent chromosome, Ncds remained unstable for an additional 60 generations relative to cultures where the wild-type *spo0A* gene was expressed. Stable Ncds did not arise in the absence of the Spo0A gene product (Lévi-Meyrueis and Sanchez-Rivas 1984; Le Dérout et al. 1992). Conversely, in stable Ncds from spo$^+$ cultures, disruption of the *spo0A* gene (via insertion of the selectable marker erythromycin) promoted partial reactivation of the previously silent genes (see below). These experiments demonstrated that the Spo0A gene product affects the extent of gene inactivation. From these studies, it was concluded that the *spo0A* gene is required for the maintenance of the diploid state as well as for the transition between reversible and irreversible chromosomal silencing (Thaler et al. 1990; Le Dérout et al. 1992).

HYBRID CHROMOSOMES CONTAINING LARGE FRAGMENTS OF INACTIVATED DNA ARE MAINTAINED THROUGH REPLICATION

Single unstable Ncd clones gave rise to progenies that displayed reciprocal recombinant types, but one member of the reciprocal pairs was always present in at least a 100-fold excess (Sanchez-Rivas et al. 1982; Ftouhi and Guillen 1990). Transformation experiments demonstrated that all nonreciprocal recombinant clones appeared to be diploids (Sanchez-Rivas et al. 1982). Two possible genetic structures for the diploid recombinants were therefore proposed: Either inactivation in recombinants is restricted to a mere chromosomal segment, or it affects a complete recombinant chromosome (Lévi-Meyrueis and Sanchez-Rivas 1984). These predictions were tested using a system able to distinguish each parental DNA in the progeny of unstable Ncds (Itaya and Tanaka 1991; Grandjean 1994).

Experiments were carried out with heterodiploids resulting from the fusion of two parental *B. subtilis* strains which carried several auxotrophic markers and drug-resistance genes and which had different *Not*I restriction fragment length polymorphism (RFLP) patterns (Fig. 1). From this analysis, several important consequences emerged regarding the fate of irreversibly unexpressed DNA. (1) Segregation and inactivation vary independently. Contrary to expectation, a clonal population of stable Ncds is not homogeneous. The population is composed mostly of haploid cells but also includes a small proportion of stable Ncds representing less than 10% of the population. Moreover, haploid cells with a phenotype corresponding to the inactive chromosome were never observed. Such results suggested that upon segregation, irreversible silencing prevented the formation of viable cells. (2) Inactive genes from the unexpressed chromosome can be incorporated into the active chromosome through recombination. (3) The recombinant cells may or may not retain the silenced status of the donor silent chromosome in the expressed genome.

Recombinant bacteria were isolated from an unstable Ncd progeny population, and their *Not*I DNA pattern was determined by pulse-field gel electrophoresis (Grandjean 1994). Figure 2 illustrates the mosaic

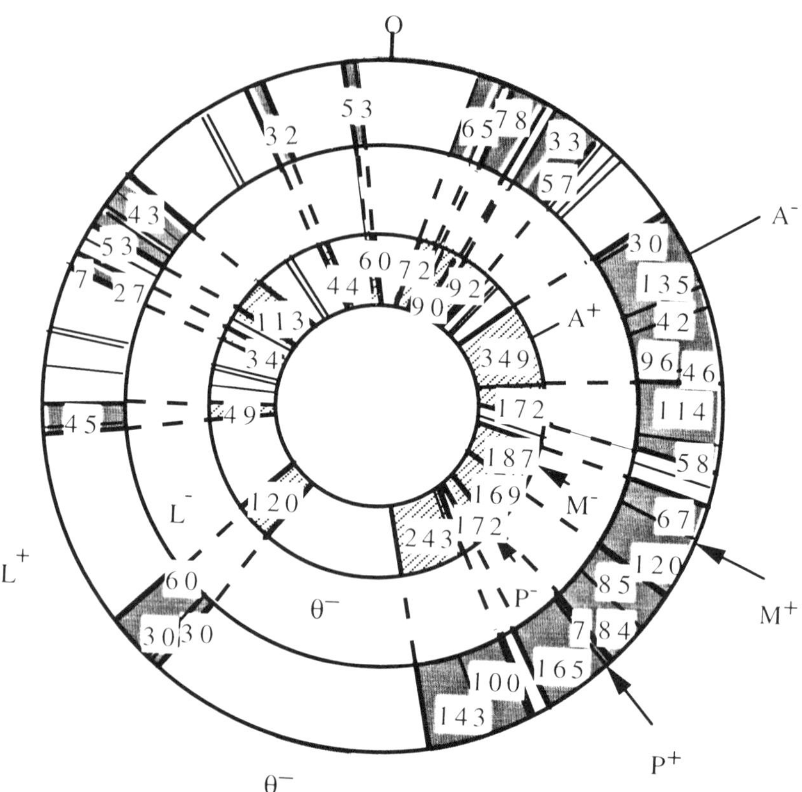

Figure 1 Comparison of the *Not*I fragments between restriction maps of S15 and BEST4087 *B. subtilis* strains. The *Not*I fragments that are identical in both strains are represented in white. Those represented in black or hatched correspond to specific *Not*I fragments for S15 and BEST4087, respectively. The number indicates *Not*I fragment size in kilobases. The phenotypes of S15 and BEST4087 parental strains are: $\theta U^- P^-$ and $M^- L^- P^- \theta^- Cat^R$. (θ) tryptophan; (U) uracil; (P) proline; (M) methionine; (L) leucine; (Cat) chloramphenicol.

structure of one of these recombinant cells (R1) after purification in the appropriate supplemented media.

The DNA from R1 exhibited a pattern that corresponded to a mosaic structure composed of *Not*I fragments characteristic of both parental chromosomes. At least six crossover events that involved fragments of DNA larger than 100 kb are required for the observed *Not*I profile of the R1 recombinant. This result confirmed the genetic interpretations (Hotchkiss and Gabor 1980; Gabor and Hotchkiss 1983; Ftouhi and Guillen 1990) on the origin of the recombinant cells within a single unstable

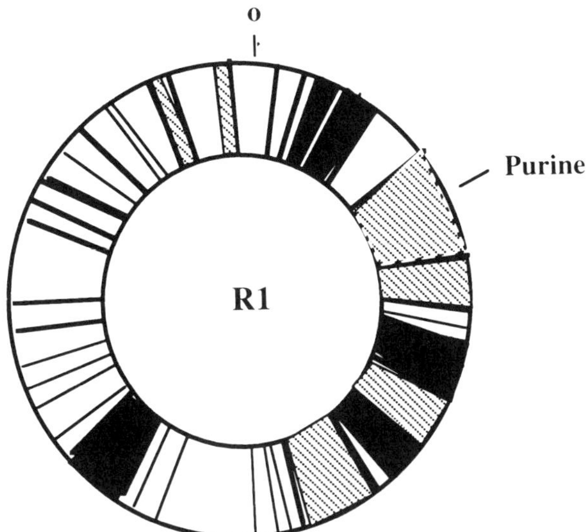

*Figure 2 Not*I restriction map of R1 clone. Filled and hatched regions correspond to S15 and BEST4087 chromosomes, respectively. The recombinant R1 derive from the progeny of unstable Ncds of phenotype identical to the BEST4087 parent indicated in Fig. 1. The R1 phenotype is $Trp^{-}U^{-}Cat^{R}$. 90% of the recombinant cells are auxotroph for purine marker, and the remaining 10% are prototroph.

Ncd colony. Hauser and Karamata (1992, 1994) proposed that the appearance of multiple recombinant types in the unstable Ncd progenies could be explained by the lysis of haploid cells in the mixture and subsequent transformation. These cells, however, are not in a competent state. In addition, in the classic transformation experiments, donor DNA molecules bound to the surface of competent cells undergo double-strand cleavage to produce fragments with an average size of 8 kb to 9 kb (Dubnau and Cirigliano 1972). Thus, this alternative hypothesis for the presence of recombinant cells in the unstable Ncd progenies seems highly unlikely.

Phenotypic analysis of the recombinants indicated that integration of the inactive marker in the expressed chromosome was followed by reexpression in only 10% of the recombinant cells. In the remaining 90%, inactivation, which affected at least a 13-kb segment of the silent chromosome encompassing the pur^{+} operon, was maintained in the active chromosome as illustrated in Figure 2. The inactivated state persisted through cell divisions, between 20 and 40 generations. Silencing was spontaneously reversed with a frequency of 10^{-7}. Pur^+ independent transformants were recovered from the R1 DNA donors.

When the wild-type *Spo0A* gene of the R1 recombinant was disrupted by the insertion of the *cat* gene, only 5–20% of the resulting *Spo0A*$^-$ *Cat*^{R} transformants reexpressed the Pur^+ operon, suggesting that some silencing could be maintained in the absence of the Spo0A gene product.

Silent and reactivated chromosomal regions exhibited identical restriction sites. All attempts to associate the maintenance of the silencing with DNA modification by methylation have been unsuccessful (Grandjean 1994). Thus, the inactivation and reactivation processes do not appear to involve major DNA sequence rearrangements.

All these observations suggest that the mechanism capable of establishing large inactive domains in the chromosome of *B. subtilis* fusion products could be determined by a modification at the level of bacterial chromatin, a general feature of gene inactivation in eukaryotes (Shaffer et al. 1993). A general picture on the fate of unstable Ncd cells that in part reevaluates a preexisting hypothesis on the silencing in diploid recombinant chromosomes, and integrates new observations of clonal heterogeneity, could now be proposed. Both parental chromosomes from unstable Ncds are able to undergo within the clonal population several rounds of recombination, producing recombinant diploids. Upon recombination, the repressed state of the donor silent DNA is either retained in the active chromosome (recombinant diploid I cells) or reexpressed (recombinant diploid II cells).

Since recombinant diploids I and recombinant diploids II are unstable Ncds, their progeny contain haploid recombinants of reciprocal phenotypes. When the inactivation is stabilized, the segregation process is not abolished. Consequently, the recombinant diploid I progeny will show haploid cells harboring a silent domain in the chromosome, whereas in recombinant diploid II progeny, the haploid chromosome is entirely expressed.

A MODEL FOR SILENCING IN FUSION PRODUCTS AND DERIVATIVES

Most prokaryotic cells are considered to possess all their necessary genetic information on a single chromosome. Thus, the partial expression of a diploid genome without DNA rearrangement or mutation was not expected in the prokaryotic systems.

Since DNA sequences do not seem to be modified in the establishment of chromosomal silencing, what other mechanism could be responsible for inactivation? Active and inactive chromosomes coexist in unstable Ncds in the same cellular milieu. Sucrose gradient analysis of nucleoid preparations from unstable Ncds gave a bimodal distribution (Charret et al. 1980; Hirschbein and Guillen 1982). Two peaks were observed for unstable Ncds. Haploid parental cells presented a single peak which behaves like the slow-sedimenting nucleoid, as did their parental cell mixtures. DNA in the two peaks was used as the donor in transformation experiments. The slow-sedimenting nucleoid fraction had a high transformation activity for alleles carried by the chromosome that was expressed in vivo. On the contrary, the rapidly sedimenting nucleoids were devoid of transformation activity both for the in vivo expressed and unexpressed markers (Guillen et al. 1982). It was also shown that proteins associated with silent nucleoids in the unstable Ncd masked the transforming activity of the in vivo unexpressed markers (Bohin et al. 1982) and that inactivation required cell-wall synthesis (Dancer 1981; Karmazyn-Campelli et al. 1985). These effects strongly suggest that chromosomal constraints play a role in partial or total chromosomal inactivation.

It is tempting to speculate that the earliest step in chromosomal silencing is achieved through the labeling of one of the chromosomes by an epigenetic mark. Hotchkiss and Gabor (1985) proposed a model in which the two chromosomes of a diploid cell compete for attachment to specific membrane sites during cell-wall regeneration. One possibility is that abnormal DNA-membrane complexes determine which of the parental chromosomes are inactivated. These abnormal structures would have to be capable of replication and propagation.

An alternative model has been advanced based on the fact that the formation of complementing diploids (diploids in which both chromosomes are metabolically active) is rarely observed. According to this second model, bacteria have functions that are essential in maintaining the haploid life style. In the heterodiploid state, the product(s) of these genes bind on one of the chromosomes at random, thus marking the chromosome. This in turn allows the attachment on the chromosome of additional proteins, resulting in the propagation and stabilization of chromosomal inactivation.

Besides the involvement of the product of the *spo0A* gene in the transition between the reversible and irreversible step of chromosomal inactivation, specific protein factors that prove chromosome suppression in the heterodiploids and derivatives are yet to be identified. Proteins which organized the nucleoids through supercoiled constraints of the DNA are good candidates (Le Hégarat et al. 1993; Arnold-Schulz-Gahmen et al. 1994).

A putative H-NS protein equivalent, in gram-positive bacteria, may act as a global repressor in chromosomal silencing (Shore 1994). First identified as a major component of the *E. coli* nucleoid (Cukier-Kahn et al. 1972; Pon et al. 1988), H-NS affects the expression of many apparently unrelated genes. In the majority of them, the protein represses the expression at the level of transcription, either by binding directly to DNA (Yamada et al. 1991; Zuber et al. 1994) or by stabilizing a nucleoprotein complex (Lucht et al. 1994). However, given a central role in silencing of a change in chromatin-like structure, the selection of mutations that modify the extent of inactivation as well as the characterization of the affected genes should be informative.

CONCLUSION AND PROSPECT

The majority of products generated by the fusion of *B. subtilis* protoplasts are diploid cells bearing an inactive parental chromosome that undergoes many rounds of replication before segregating. Ncds derived from fused protoplasts are not restricted to *B. subtilis* species. Ncd cells were also detected among colonies regenerated from fusion crosses of protoplasts of *Bacillus stearothermophilus* (Chen et al. 1986), *Bacillus megaterium* (Fleicher and Vary 1985), and *Streptomyces* (Hopwood and Wright 1981).

Like the epigenetic controls observed in eukaryotes, gene silencing in prokaryotes has only been observed at the chromosomal level. All the published work analyzes the differential expression of genes between homologous chromosomal regions in diploid bacterial strains generated by protoplast fusion. It must be emphasized, however, that, because of their intrinsic instability, it is difficult to say whether duplicated sequences in merodiploid strains generated by additional autonomous plasmid copies could also be affected. Plasmid sequences recombine with the chromosome with high frequencies. In addition, extensive loss and release in the medium of the plasmid DNA was observed after protoplast fusion (Novick et al. 1980; Cofre and Sanchez-Rivas 1983). It is therefore only in strains generated by the fusion technique which maintain two distinct chromosomes that the concept of chromosomal inactivation could be established.

Some questions relating the physical structure and the fate of the two chromosomes in *B. subtilis* Ncd cells within a clonal population have been resolved (Grandjean 1994). First, protoplast fusion gives rise to diploid cells which behave as a heterokaryon harboring an entire inactive chromosome. Second, different degrees of chromosomal suppression of gene expression do not interfere with the segregation process. Consequently, irreversible inactivation does not produce a homogeneous population in a diploid clone, as was previously supposed (Guillen et al. 1985; Schaeffer and Hirschbein 1985; Sanchez-Rivas et al. 1988). Two separate phenomena determine the composition of the progeny of noncomplementing diploids: (1) the level of viability of segregants containing the unexpressed chromosome, (2) the progressive change of silent DNA from reversible to irreversible inactivation.

The biochemical mechanism responsible for the initiation and propagation of the silencing is not yet known. The organization of the bacterial chromosome is not clearly established (Charret et al. 1980; Petitjohn 1988), and a direct relationship between a chromatin-like structure and a defined regulatory state has yet to be demonstrated.

There is evidence in *E. coli* that silencing of a large chromosomal domain is correlated with protein-protein and DNA-protein interactions, which stabilize a particular chromatin structure at the silent loci. The DNA within such nucleoprotein complexes is refractory to in vivo methylation and to topological modifications (Lynch and Wang 1995).

These results support the hypothesis that epigenetic inactivation in bacteria might be promoted through local changes of chromatin packing. Models of this type are easy to test in recombinants derived from unstable Ncds in which a portion of the nonexpressed chromosome continues to be silent upon integration into an active chromosome after recombination. Contrary to the Ncds, these types of diploid recombinants gave rise to a homogeneous clonal population of haploid cells in which the silent domain persisted through many cell divisions. Thus, reactivation mutants could be readily selected from this type of recombinant.

The epigenetic-like inactivation phenomena described here seem to occur also in gram-negative species. A double mutant of *E. coli K12* resistant to a phage and to a particular chemical that manifests an "abnormal" behavior in the descendants has been described recently (Gratia 1994). The pedigree analysis of the initial mutant strain strongly suggested that they are Ncds displaying an alternated chromosome expression. Experiments on fusion products in gram-positive bacteria seem to indicate that gene silencing mediated by an epigenetic mode of regulation may have appeared and evolved in prokaryotes before it became widespread among eukaryotic organisms.

ACKNOWLEDGMENTS

This work was supported by grants from the Centre National de la Recherche Scientifique URA-1354 and ARC-6794 (L.H. and F.LH) and

INSERM 900 106 (L.H.). V.G. acknowledges the receipt of an ARC fellowship.

REFERENCES

Arnold-Schulz-Gahmen, B., V. Salti-Montesanto, J. Nguyen, L. Hirschbein, and F. Le Hégarat. 1994. The *Bacillus subtilis* nucleoid-associated protein HPB12 strongly compacts DNA. *J. Bacteriol.* **176:** 50–60.

Bohin, J.P., K. Ben Khalifa, N. Guillen, P. Schaeffer, and L. Hirschbein. 1982. Phenotypic expression *in vivo* and transforming activity *in vitro*: Two related functions of folded bacterial chromosome. *Mol. Gen. Genet.* **185:** 65–68.

Boor, K.J., M.L. Duncan, and C.W. Price. 1995. Genetic and transcriptional organization of the region encoding the beta subunit of *Bacillus subtilis* RNA polymerase. *J. Biol. Chem.* **270:** 20329–20336.

Burbulys, D., K.A. Trach, and J.A. Hoch. 1991. The initiation of sporulation in *Bacillus subtilis* is controlled by a multicomponent phosphorelay. *Cell* **64:** 545–552.

Casadesús, J. and R. Maldonado. 1990. Genomic imprinting in microorganisms. *Microbiol. Sem.* **6:** 1–10.

Charret, R., N. Guillen, A.M. Fleury-Guérout, F. Le Hégarat, and L. Hirschbein. 1980. Organization of the isolated *Bacillus subtilis* nucleoid. *Biol. Cell.* **34:** 105–114.

Chen, Z., S.F. Wojcik, and N.E. Welker. 1986. Genetic analysis of *Bacillus staerothermophilus* by protoplast fusion. *J. Bacteriol.* **165:** 994–1001.

Cofre, G. and C. Sanchez-Rivas. 1983. Curing of plasmids during a protoplast-cell wall regeneration cycle in *Bacillus subtilis* strains. *FEMS Microbiol. Lett.* **19:** 11–15.

Cukier-Kahn, R., M. Jacquet, and F. Gros. 1972. Two heat resistant, low molecular weight proteins from *Escherichia coli* that stimulate DNA-directed RNA synthesis. *Proc. Natl. Acad. Sci.* **69:** 3643–3647.

Dancer, B.N. 1981. Control of sporulation in fused protoplasts of *Bacillus subtilis* 168. *J. Gen. Microbiol.***126:** 29–36.

Dubnau, D. and C. Cirigliano. 1972. Fate of transforming DNA following uptake by competent *Bacillus subtilis*: Size and distribution of the integrated donor segments. *J. Bacteriol.* **111:** 488–494.

Ferrari, F., K. Trach, D. LeCoq, J. Spence, E. Ferrari, and J.A. Hoch. 1985. Characterization of the *spo0A* locus and its deduced product. *Proc. Natl. Acad. Sci.* **82:** 2647–2651.

Fleisher, E.R. and P.S. Vary. 1985. Genetic analysis of fusion recombinants and presence of noncomplementing diploids in *Bacillus megaterium*. *J. Gen. Microbiol.* **131:** 919–926.

Ftouhi, N. and N. Guillen. 1990. Genetic analysis of fusion recombinants in *Bacillus subtilis*: Function of the recE gene. *Genetics* **55:** 699–707.

Gabor, M. and R.D. Hotchkiss. 1983. Reciprocal and nonreciprocal recombination in diploid clones from *Bacillus subtilis* protoplast fusion: Association with the replication origin and terminus. *Proc. Natl. Acad. Sci.* **80:** 1426–1430.

Grandjean, V. 1994. "Plasticite du genome bacterienne: Inactivation chromosomique chez *Bacillus subtilis*." Ph.D. thesis, University of Orsay, France.

Gratia, J.P. 1994. Ufr/s variation in *Escherichia coli* K12: A reversible double-mutation or alternate chromosome expression in non-complementing diploids? *Res. Microbiol.* **145:** 309–325.

Guillen, N., M. Amar, and L. Hirschbein. 1985. Stabilized non-complementing diploids (Ncd) from fused protoplast products of *Bacillus subtilis*. *EMBO J.* **4:** 1333–1338.

Guillen, N., M. Gabor, R.D. Hotchkiss, and L. Hirschbein. 1982. Isolation and characterization of the nucleoid of non-complementing diploids from protoplast fusion in *Bacillus subtilis*. *Mol. Gen. Genet.* **185:** 69–74.

Hauser, P. and D. Karamata. 1992. Ploidy of *Bacillus subtilis* exfusants: The haploid nature of cells forming colonies with biparental or prototrophic phenotypes. *J. Gen. Microbiol.* **138:** 1077–1088.

———. 1994. Ploidy of *Bacillus subtilis* exfusants: A controversy. II. No experimental support for diploidy and chromosomal silencing. *Microbiology* **140**: 2–3.

Hiraga, S. 1992. Chromosome and plasmid partition in *Escherichia coli*. *Annu. Rev. Biochem.* **61:** 283–306.

Hirschbein, L. and N. Guillen. 1982. Characterization, assay, and use of isolated bacterial nucleoids. *Methods Biochem. Anal.* **28:** 297–328.

Hopwood, D.A. 1981. Genetic studies with bacterial protoplasts. *Annu. Rev. Microbiol.* **35:** 237–272.

Hotchkiss, R.D. and M.H. Gabor. 1980. Biparental products of bacterial protoplast fusion showing unequal parental chromosome expression. *Proc. Natl. Acad. Sci.* **77:** 3553–3557.

———. 1985. Protoplast fusion in *Bacillus* and its consequence. In *The molecular biology of the bacilli.* (ed. D.A. Dubnau, pp. 109–149. Academic Press, New York.

Itaya, M. and T. Tanaka. 1991. Complete physical map of the *Bacillus subtilis* 168 chromosome constructed by a gene-directed mutagenesis method. *J. Mol. Biol.* **220:** 631–648.

Karmazyn-Campelli, C., B. Savelli, H.J. Rogers, and P. Schaeffer. 1985. Inverse relationship between formation of phenotypically recombinant clones and cell-wall regeneration after fusion of *Bacillus subtilis* protoplasts. *J. Gen. Microbiol.* **131:** 1635–1638.

Kudoh, J., T. Ikeuchi, and K. Kurahashi. 1985. Nucleotide sequences of the sporulation gene *spo0A* and its mutant genes of *Bacillus subtilis*. *Proc. Natl. Acad. Sci.* **82:** 2665–2668.

Le Dérout, J., D.S. Thaler, N. Guillen, and L. Hirschbein. 1992. The *spo0A* gene is implicated in the maintenance of non-complementing diploids in *Bacillus subtilis*. *Mol. Microbiol.* **6:** 1495–1505.

Le Hégarat, F., V. Salti-Montesanto, Y. Hauck, and L. Hirschbein. 1993. Purification and characterization of the HU-like protein HPB9 from the *Bacillus subtilis* nucleoid. *Biochim. Biophys. Acta.* **1172:** 101–107.

Lévi-Meyrueis, C. and C. Sanchez-Rivas. 1984. Complementing and genetic inactivation: Two alternative mechanisms leading to prototrophy in diploid bacterial clones. *Mol. Gen. Genet.* **196:** 488–493.

Lopez, F., N. Guillen, and L. Hirschbein. 1986. Further characterization of the inactive chromosome from *Bacillus subtilis* stabilized noncomplementing diploids. In: Bacillus *molecular genetics and biotechnology applications.* (ed. A.T. Ganesan and J.A. Hoch), pp. 73–86. Academic Press, New York.

Lucht, J.M., P. Dersch, B. Kempf, and E. Bremer. 1994. Interactions of the nucleoid-associated DNA-associated DNA-binding protein H-NS with the regulatory region of the osmotically controlled proU operon of *Escherichia coli*. *J. Biol. Chem.* **269:** 6578–6586.

Lynch, A.S. and J.C. Wang. 1995. SopB protein-mediated silencing of genes linked to the *sopC* locus of *Escherichia coli* F plasmid. *Proc. Natl. Acad. Sci.* **92:** 1896–1900.

Lyon, M.F. 1993. Epigenetic inheritance in mammals. *Trends Genet.* **9:** 123–128.

Muller, H.J. 1930. Types of visible variations induced by X-rays in *Drosophila. J. Genet.* **22:** 299–334.

Novick, R., C. Sanchez-Rivas, A. Gruss, and I. Edelman. 1980. Involvement of the cell envelope in plasmid maintenance: Plasmid curing during the regeneration of protoplasts. *Plasmid* **3:** 348–358.

Petitjohn, D.E. 1988. Histone-like proteins and bacterial chromosome structure. *J. Biol. Chem.* **263:** 12793–12796.

Pon, C.L., R.A. Calogero, and O. Gualerzic. 1988. Identification, cloning, nucleotide sequence and chromosomal map location of H-NS, the structural gene for *Escherichia coli* DNA-binding protein H-NS. *Mol. Gen. Genet.* **212:** 199–202.

Sanchez-Rivas, C., C. Karmazyn-Campelli, and C. Lévi-Meyrueis. 1988. Further studies on recombination in diploid clones from *Bacillus subtilis* protoplast fusion. *Mol. Gen. Genet.* **214:** 321–324.

Sanchez-Rivas, C., C. Lévi-Meyrueis, F. Lazards-Monier, and P. Schaeffer. 1982. Diploid state of phenotypically recombinant progeny arising after protoplast fusion in *Bacillus subtilis. Mol. Gen. Genet.* **188:** 272–278.

Schaeffer, P. and L. Hirschbein. 1985. *Bacillus subtilis* protoplast fusion, nucleoids and chromosome inactivation. In *Fundamental and applied aspects of spores.* (ed. W. Barker, et al.), pp. 77–88. Academic Press, New York.

Schaeffer, P., B. Cami, and R.D. Hotchkiss. 1976. Fusion of bacterial protoplasts. *Proc. Natl. Acad. Sci.* **73:** 2151–2155.

Schnetz, K. 1995. Silencing of *Escherichia coli bgl* promoter by flanking sequence elements. *EMBO J.* **14:** 2545–2550.

Setlow, B., N. Magill, P. Febbroriello, L. Nakhimovsky, D. Koppel, and P. Setlow. 1991. Condensation of the forespore nucleoid early in sporulation of bacillus species. *J. Bacteriol.* **173:** 6270–6278.

Shaffer, C.D., L.L. Wallrath, and S.C.R. Elgin. 1993. Regulating genes by packaging domains: Bits of heterochromatin in euchromatin? *Trends Genet.* **9:** 35–37.

Shore, D. 1994. RAP1: A protean regulator in yeast. *Trends Genet.* **10:** 408–412.

Thaler, D.S., J.R. Roth, and L. Hirschbein. 1990. Imprinting as a mechanism for the control of gene expression. In *The bacterial chromosome* (ed. K. Drlica and M. Riley), pp. 445–456. American Society for Microbiology, Washington, D.C.

Trowsdale, J., S.M. Chen, and J.A. Hoch. 1978. Evidence that *spo0A* mutations are recessive in *spo0A*$^-$/*spo0A*$^+$ merodiploid strains of *Bacillus subtilis. J. Bacteriol.* **135:** 99–133.

Yamada, H., T. Yoshida, K. Tanaka, C. Sasakawa, and C. Mizuno. 1991. Molecular analysis of the *E. coli hns* gene encoding a DNA-binding protein, which preferentially recognizes curved DNA sequences. *Mol. Gen. Genet.* **230:** 332–336.

Zuber, F., D. Kottarz, S. Rimsky, and H. Buc. 1994. Upstream synthetic curved DNA sequences and promoter expression. *Mol. Microbiol.* **12:** 231–240.

Stable Epigenetic States in Differentiated Plant Cells: Implications for Somaclonal Variation and Gene Silencing in Transgenic Plants

Marjori A. Matzke and Antonius J.M. Matzke
Institute of Molecular Biology
Austrian Academy of Sciences
A-5020 Salzburg, Austria

Specialized cells that retain the ability to develop into a complete adult organism are considered totipotent. The totipotency of many plant cell types enables the regeneration of entire fertile plants from cultured somatic or gametic cells of a number of species (see Meins, this volume). However, the ready availability of virtually unlimited numbers of asexually propagated plants has in turn unveiled an unexpected yet ubiquitous phenomenon, which can be described as follows: Because epigenetic modifications of genomes (in the form of varying patterns of methylation or chromatin structure) and not nucleotide changes or structural alterations of genes are believed to underlie cell diffentiation, genetic uniformity is expected in a population of plants regenerated from any given explant source; i.e., regenerants obtained from somatic cells of a single plant should be clones and have a similar appearance. Instead of a monotonous collection of identical plants, however, an astonishing spectrum of phenotypic variability is often observed in sibling regenerants, particularly if an undifferentiated callus phase precedes the regeneration process. Although initially this variation was often slighted or considered artifactual (Skirvin et al. 1993), it was legitimized in 1981 when Larkin and Scowcroft consolidated the scattered observations in a review and proposed the term "somaclonal variation" to refer to the phenotypic variability observed among plants regenerated from cell or tissue culture (Larkin and Scowcroft 1981).

In the intervening years, somaclonal variation has had an impact on both applied and basic plant science. On the positive side, it has captured the interest of breeders, who have recognized its potential as a new source of variability for crop improvement (Larkin and Scowcroft 1981; Evans 1989; Skirvin et al. 1993). A negative side has also emerged, however, as horticulturalists have learned that the asexual propagation of commercial cultivars does not guarantee phenotypic uniformity (Skirvin et al. 1994). For basic researchers, the lure of somaclonal variation has been expressed best by Karp and Bright (1985; also quoted by Peschke and Phillips [1992]): "Resolution of this problem may turn out to make a greater contribution to plant science than direct exploitation of the phenotypic variation itself."

THE ORIGINS OF SOMACLONAL VARIATION

What is the source of the phenotypic variability observed in regenerated plants when, according to the rules of asexual reproduction, "constancy should prevail" (Karp and Bright 1985)? Despite numerous studies, the molecular basis of somaclonal variation is poorly understood. Fundamental questions, such as the relative roles of tissue-culture-induced versus preexisting variability and genetic versus epigenetic changes, remain unresolved. To assess the contributions to somaclonal variation of each of these alternatives, however, they must be clearly distinguished, which is not always a straightforward matter. Other peculiar aspects of somaclonal variation, such as the high frequencies of variants obtained—which routinely exceeds that achieved using conventional mutagens—and the appearance of homozygous changes, also still require an explanation. These considerations are addressed in this chapter.

Although specialized reviews have acknowledged the possible contributions of preexisting variability and epigenetic differences (Larkin and Scowcroft 1981; Karp and Bright 1985; Lee and Phillips 1988; Peschke and Phillips 1992; Skirvin et al. 1994), most textbook accounts of somaclonal variation attribute the bulk of the variation to tissue-culture-induced genetic (i.e., heritable) changes (see, e.g., Kuckuck et al. 1991), and this bias is also evident in many articles (see, e.g., Phillips et al. 1994). This skewed emphasis is based on two debatable assumptions. The first is that epigenetic variation is unstable and nonheritable through sexual cycles (Meins 1983; Skirvin et al. 1994). It is increasingly apparent, however, that somatically acquired epigenetic modifications in plants can be mitotically stable and meiotically heritable (see Meins; Jorgensen et al.; Chandler et al.; Kermicle; Vaucheret et al.; Martienssen; all this volume). Hence, it is not always possible to distinguish between a true genetic mutation (i.e., change in nucleotide sequence or chromosome rearrangement or deletion) and a meiotically stable epigenetic modification.

The second questionable assumption is that unequivocal distinctions can indeed be made between tissue-culture-induced and preexisting genetic or epigenetic variability. This is problematic unless the exact epigenetic and genetic makeup of individual cells is known prior to the culture period. A somatic cell or small group of cells having a unique genetic constitution or epigenetic modification may reproduce preferentially in culture; any plant eventually regenerated from this cell(s) would then probably be scored as a variant acquired during the culture period, although in reality it comprised a miniscule, possibly even undetectable, proportion of the cells in the original explant.

As recognized by Ball (1990), one impediment to research on somaclonal variation has been the lack of an experimental system in which molecular changes of a well-characterized gene with an easily assayable phenotype could be monitored in regenerants. He suggested using defined transgenes encoding antibiotic resistance markers as substrates to study somaclonal variation in plants regenerated from single cells derived from transformants. As described below, this approach has led us to conclude that a substantial proportion of somaclonal variation might be due to diverse, preexisting epigenetic states that are mitotically and meiotically stable in plants regenerated from individual somatic cells.

EXPERIMENTAL SYSTEM

Our work on somaclonal variation has involved double transformation experiments in which two transgene constructs containing different antibiotic selection markers have been introduced into tobacco leaf cells by using sequential transformation steps. Eight transgenic tobacco lines that were homozygous at different single loci for a transgene construct encoding resistance to hygromycin (Hyg^R) were produced using leaf disk transformation and *Agrobacterium* Ti-plasmid-derived vectors. Leaf pieces of these H lines were then retransformed with a second construct that encoded resistance to kanamycin (Kan^R). Only kanamycin selection was applied in the second transformation step. This ensured that plants were regenerated from single Kan^R cells, while simultaneously allowing these cells to have a hygromycin-sensitive (Hyg^S) phenotype. We were thus screening for somaclonal variants at the *H* transgene locus (Fig. 1; note that in this experimental system, the introduction of a K transgene construct did not directly influence the epigenetics of an *H* transgene locus [Neuhuber et al. 1994], but simply guaranteed regeneration of plants from single cells so that the extent of preexisting epigenetic variability in individual cells of the H lines could be analyzed).

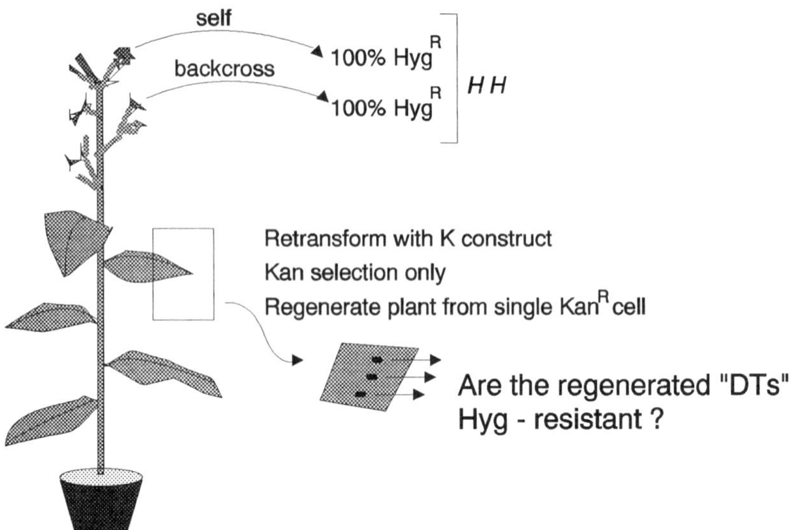

Figure 1 Leaf disk transformation system to study somaclonal variation at *H* transgene loci, which encode Hyg^R, in homozygous (*HH*) tobacco lines (named in Table 1). All eight transgenic tobacco lines that were tested originally transmitted the Hyg^R marker to 100% of selfed and backcross progeny. Two complicating aspects must be noted, however. First, Hyg^R was unstable in progeny of lines that appeared to be epigenetic mosaics (H2P, H7, H2A, and H11), meaning that seedlings that were initially Hyg^R eventually died after several months on hygromycin-containing medium, presumably because inactive (or weakly active) $H^\bullet$ epialleles became predominant (see text). Second, line H59, after several generations of homozygosity, behaved as if it had an epigenotype of $HH^\bullet$ in practically all cells: Selfings produced progeny that were mostly weakly resistant to Hyg and backcrosses produced progeny that segregated 1:1 for Hyg^R:Hyg^S. The Hyg^R weak phenotype in selfed progeny can be attributed both to the partial dominance of the $H^\bullet$ epiallele over *H* and the weakening of the *H* allele in the homozygous state (Neuhuber et al. 1994). (DT) Double transformants containing both an *H* and a *K* locus.

Regenerated double transformants (DTs), which contained both *K* and *H* transgene complexes (as verified by Southern blot analysis), were grown to maturity. Seeds obtained through self-fertilization and backcrosses to untransformed tobacco were sown on medium containing either kanamycin or hygromycin, or a combination of both antibiotics. With respect to sexual transmission of hygromycin resistance, some of the DTs behaved like the homozygous parent; i.e., they produced 100%

Hyg^R progeny in selfings and backcrosses (Table 1; lines H8, H9, H83). A number of the DTs, however, did not segregate the Hyg^R marker like the parental explant source. Several variations were observed: Many DTs from two lines (H2P, H7) did not produce any Hyg^R progeny at all when selfed or backcrossed. One line (H59) produced DTs whose backcross progeny segregated primarily in a 1:1 ratio of Hyg^R:Hyg^S. Another line (H2A) produced a heterogeneous collection of DTs, in which the strength of Hyg^R in progeny could be quite variable. Finally, one line, H11, spawned DTs that clearly behaved as if they were homozygous (*HH*) or hemizygous (*H*-) for the *H* locus, or lacked it altogether ("--") (Table 1) (Neuhuber et al. 1994). DTs of the H11 line proved most informative upon further analysis.

A possible explanation for the results obtained with DTs of line H11 is that the Hyg^S seedlings did not inherit an *H* locus. DNA blot analysis demonstrated that this was not the case (Neuhuber et al. 1994). Therefore, hygromycin sensitivity was due to gene silencing and not to physical loss or disruption of the *H* locus. Another possible explanation for these results is as follows: Perhaps the *H* transgene complex in H11 homozygotes existed as an epigenetic mosaic prior to tissue culture and retransformation. In individual cells, one or the other of the two *H* transgene alleles might have been inactivated, so that regenerants from single cells behaved as homozygous resistant, hemizygous, or homozygous sensitive according to their "epiallelic" constitution at the transgene locus. These "epigenotypes," *HH*, $HH^\bullet$, and $H^\bullet H^\bullet$ ($H^\bullet$ representing a silent "epiallele"), would thus correspond, respectively, to the apparent genotypes of the DTs; i.e., *HH*, *H*- and "--." Any plant regenerated from a single cell of line H11 maintained its specific epigenotype, which was also meiotically heritable through several selfed and backcross generations (Neuhuber et al. 1994).

Because cytosine methylation is an epigenetic modification often associated with gene silencing, we examined methylation in the promoter driving expression of the Hyg^R marker. The *H* transgene complex in the parental H11 line was partially methylated and became even more methylated in DTs in which the Hyg^R marker gene activity was decreased (Neuhuber et al. 1994). Furthermore, other lines that produced DTs with impaired Hyg^R were also found to have partially methylated transgene promoters (Neuhuber et al. 1994). These results are consistent with the *H* transgene complex being present in the original line as an epigenetic mosaic resulting in partial methylation.

Two lines of evidence indicated that different epigenotypes preexisted in leaves of H11 plants and were not induced during the leaf disk culture

Table 1 Somoclonal variation at independent *H* loci, which confer resistance to hygromycin, in different homozygous transgenic tobacco lines

Homozygous transgenic line	No. of regenerants tested for transmission of Hyg^R	Epigenotypes recovered in regenerants	Segregation of Hyg^R in sexual progeny	
			selfed	backcross
H8	7	all *HH*	100% Hyg^R	n.d.
H9	5	all *HH*	100% Hyg^R	100% Hyg^R
H83	19	all *HH*	100% Hyg^R	100% Hyg^R
H2P	12	10/12 $H^{\bullet}H^{\bullet}$	100% Hyg^S	100% Hyg^S
		2/12 *HH*	100% Hyg^R	n.d.
H7	8	6/8 $H^{\bullet}H^{\bullet}$	100% Hyg^S	100% Hyg^S
		1/8 *HH*	100% Hyg^R	n.d.
		1/8 *HH*◐	variable	n.d.

H59	14	all *HH*●	100% Hyg^R_W or Hyg^S	50% Hyg^R; 50% Hyg^R_W or Hyg^S
H2A	12	7/12 *HH*	100%Hyg^R	n.d.
		2/12 *HH*●	100% Hyg^R_W or Hyg^S	n.d.
		1/12 *H*◑*H*◑	variable	
		2/12 *HH*◑	variable	n.d.
H11	13	2/13 *HH*	100% Hyg^R	100% Hyg^R
		4/13 *HH*●	100% HYG^R_W	50% Hyg^R; 50% Hyg^R_W or Hyg^S
		1/13 *H*●*H*●	100% Hyg^S	100% Hyg^S
		3/13 *H*◑*H*◑	variable	n.d.

In each case, homozygous H transgenic lines were retransformed using a K construct, which encodes resistance to kanamycin, and individual plants were regenerated from single cells on the basis of their kanamycin-resistant phenotype (see Neuhuber et al. 1994 and text for details). *H* locus epigenotypes identified in these regenerants are tabulated. (*H*) Active allele conferring moderate to strong resistance to hygromycin (Hyg^R); (*H*●) inactive allele conferring no Hyg^R (i.e., Hyg-sensitive = Hyg^S) or extremely weak Hyg^R (= Hyg^R_W); (*H*◑) intermediate Hyg^R, usually showing considerable variability. (n.d.) Not determined.

and regeneration period. First, even though line H11 produced Hyg^R progeny in selfings and backcrosses, prolonged incubation of these seedlings on hygromycin-containing medium revealed that this resistance was not stable: Beginning approximately 6–8 weeks after germination, H11 seedlings began to turn yellow and eventually died. Second, when sterilized leaf pieces of H11 plantlets used for retransformation with the K construct were placed on shoot-inducing medium containing hygromycin, large portions turned brown even though green shoots also developed. This demonstrated that both Hyg^R and Hyg^S regions existed in the same leaf at this stage (Neuhuber et al. 1994). Together, these observations suggested that a mosaic of epigenotypes developed over time in leaves of line H11; when $HH^{\bullet}$ and $H^{\bullet}H^{\bullet}$ epigenotypes became predominant, seedlings or leaf pieces succumbed to hygromycin. It is important to recall that neither of these two epigenotypes was recovered in the sexual progeny of the H11 parental plant (Fig. 1).

IMPLICATIONS FOR SOMACLONAL VARIATION

Importance of Preexisting Epigenetic Variability and Its Stable Inheritance in Regenerants

Several authors have analyzed what were regarded as tissue-culture-induced changes in DNA methylation patterns in regenerated plants; both increases (Brown 1989) and decreases in methylation have been observed (Kaeppler and Phillips 1993). However, the design of these previous studies, which used gene probes that had no obvious relationship to any observable phenotypic variation in regenerated plants, did not permit a correlation between the methylation "changes" and phenotypic alterations, nor did they exclude the possibility that the methylation changes actually reflected preexisting epigenetic variability in the explant source. In contrast, our data demonstrate directly that preexisting epigenetic differences at a transgene locus encoding resistance to hygromycin can be stably maintained in regenerated plants and can contribute substantially to variation in the Hyg^R phenotype in sibling regenerants. Moreover, a differentiated tissue can conceivably comprise varying proportions of different epialleles, as we have found for leaves (Fig. 2). The exact nature of these transgene epialleles with respect to degree of methylation or heterochromatinization, and their relative proportions in a particular tissue, probably reflect the methylation state/chromatin configuration of flanking plant DNA. In other words, a transgene can be an indicator of the epigenetic state of the context plant DNA in individual cells of a particular tissue at a specific stage of development.

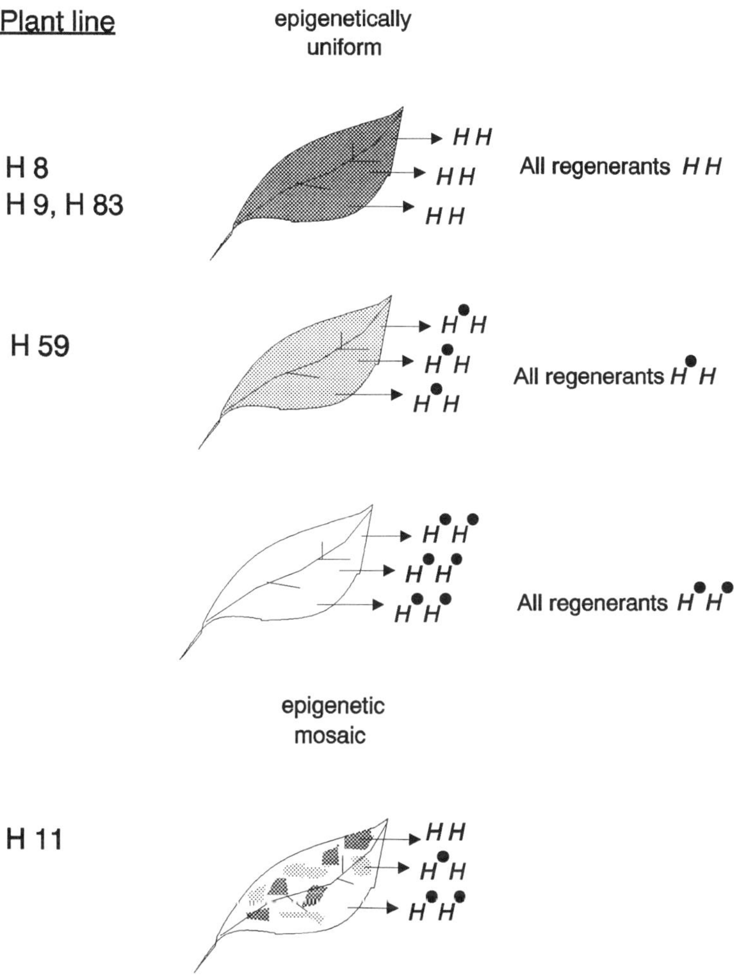

Figure 2 Schematic representation of different possible epigenotypes in leaves of homozygous *HH* plants used for transformation and regeneration. Both epigenetic uniformity and mosaicism were encountered in eight different transgenic tobacco lines, which produced 100% Hyg^R progeny in selfings and backcrosses (exception: H59; Fig. 1 legend). The different epigenotypes could be maintained in plants regenerated from single leaf cells. Examples of specific plant lines exhibiting the indicated epigenetic patterns are shown. So far, no plant line has been recovered that displays complete inactivation of both *H* alleles in all leaf cells (third from top), although this remains a hypothetical possibility. Note that it is not yet known whether the epigenetic variation of H11 is indeed randomly distributed (as shown) or shows a more structured pattern. Lines H2P, H7, and H2A were also epigenetic mosaics, but these did not clearly produce all three possible epigenotypes as did H11.

Because epigenetic changes accompany developmental transitions, endogenous genes also undergo changes in their epigenetic state, and it is reasonable to assume that these states—like those of transgenes—would also be stably maintained in regenerated plants. It is likely, however, that this only applies to a subset of genes: The regeneration of entire plants is not compatible with the impaired expression or permanent inactivation of genes whose products are essential for the implementation of development. Therefore, plants must be regenerated either from cells in which such genes have not been silenced at all or in which their expression can be reset during tissue culture. Genes that affect traits having adaptive value for sessile plants in an unpredictable environment might be particularly apt to exhibit flexible expression patterns based on epigenetic variability. This would be consistent with the types of quantitative changes observed in regenerated plants; e.g., differences in plant height, leaf length and width, fruit production, and days to flowering (Larkin and Scowcroft 1981).

High Frequency of Variation

A distinctive feature of somaclonal variation that requires an explanation is the high frequency of variants that are routinely obtained. Although variation rates as high as 100% have been reported, a more conservative general estimate is 1–3%, which in any case is much greater than the rate of spontaneous mutation at a given locus (10^{-5} to 10^{-6}; Skirvin et al. 1994) and higher than the rate achieved using conventional mutagens such as irradiation or chemicals (Larkin et al. 1989). Our data demonstrate that the frequency of variants obtained would depend on the relative proportions of different epigenotypes present in a given explant source. For line H11, comparable percentages of each epigenotype, *HH*, $HH^{\bullet}$, $H^{\bullet}H^{\bullet}$, were regenerated from leaves (Fig. 2). Because the latter two epigenotypes were recovered only in regenerated plants and not transmitted sexually by the parent H11 plant, approximately 60% of the regenerants would be considered variants or "mutants" at the *H11* locus. This value could even increase to 100% if the predominant epigenotype were $H^{\bullet}H^{\bullet}$. In principle, then, different degrees of epigenetic mosaicism in explant sources could account for the extraordinarily high rates of somaclonal variation.

The stage of development at which an explant source is used for cell culture and plant regeneration influences the types and extent of somaclonal variation (Skirvin et al. 1994). These observations are also compatible with preexisting epigenetic differences as a source of variability. A genome at an early stage of development probably has not yet acquired an extensive and/or fixed epigenetic program. With a smaller pool of epigenetic variability, fewer variants would be expected in a population of regenerants. In contrast, a tissue such as a leaf comprises several highly differentiated cell types—such as epidermal and mesophyll cells, trichomes and stomates—that express different genes. With such a rich reservoir of epigenetic variability, a higher frequency and a wider range of variant phenotypes should appear. The recovery of recurring phenotypic changes in somaclones derived from a particular tissue (Skirvin et al. 1994) might thus be due to regeneration from a specific set of epigenotypes that develop in a given tissue. With our H11 line, we reproducibly regenerated the three possible *HH* epigenotypes from leaves of plantlets that were initially Hyg^R. This suggests that the generation of epigenetic mosaicism at the *H11* locus might be under developmental control, although other tissues, such as roots, would need to be examined for different levels of epigenetic mosaicism to verify this possibility.

ORIGIN OF HOMOZYGOUS CHANGES

Non-segregating, non-parental variants that represent homozygous changes have been frequently observed in populations of regenerated plants (Larkin et al. 1984; Oono 1985; Kaeppler and Phillips 1993). Possible explanations for these include cycles of point mutation followed by monosomy/disomy or trisomy/disomy, and gene conversion (Larkin et al. 1984). Our data, in which $H^{\bullet}H^{\bullet}$ plants were regenerated from an *HH* parent, provide a novel explanation for the source of homozygous changes in somaclonal variants; one must only recognize the possibility of epigenetic mosaicism at a given locus and the stable maintenance of any possible epigenotype in regenerated plants. The frequency with which homozygous variants are recovered would depend on the fraction of explant cells containing two identical epialleles that differed from those that are sexually transmitted by the explant donor. In our experiments, this could range from 0 (lines H9, H83) to 83% (line H2P); for line H11, approximately 30% of the regenerants were $H^{\bullet}H^{\bullet}$.

Previous studies have reported homozygous epigenetic changes in regenerated plants (Oono 1985; Kaeppler and Phillips 1993). Kaeppler and Phillips (1993) reported that 15% of the methylation differences they observed in regenerated maize plants were homozygous changes. Even though these differences were attributed to tissue-culture-induced alterations in DNA methylation, there is no reason not to implicate preexisting epigenetic variability in the explant source.

IMPLICATIONS FOR GENE SILENCING IN TRANSGENIC PLANTS

Leaf-disk transformation is used frequently to obtain transgenic plants for gene expression studies. An initally surprising, but now common, observation is the coordinate silencing of a transgene and homologous endogenous gene (endogene) in transgenic plants, a phenomenon termed cosuppression (see Meins; Jorgensen et al., both this volume). The study of cosuppression has been complicated by unexplained differences in the degree of silencing in sibling regenerants transformed with the same transgene construct. To understand both the mechanism and the quantitative variability of cosuppression, considerable effort has been devoted to analyzing the transgene locus with respect to copy number and arrangement of the transgene construct, degree of methylation, and the sequence of flanking plant DNA. However, a forgotten partner in these two component silencing systems is the homologous endogene. Current thinking is dominated by the assumption that an endogene is uniformly modified and/or expressed in the leaf cells at the time of transformation and regeneration. As we have found, however, a gene can exist in different epigenetic states in leaf cells, and these states can be maintained in regenerated plants. Therefore, variability in silencing must be investigated from the perspectives of both the introduced transgene and the homologous endogene. Conceivably, a methylated or heterochromatinized endogene could "imprint" an incoming transgene with a comparable degree of methylation or heterochromatinization before it integrates into the plant genome. Evidence for crosstalk between an incoming transgene and homologous endogene has been reported by Yibrah and coworkers, who observed structural similarities in independent transgene inserts exposed to the same endogenous locus (Yibrah et al. 1993).

PREEXISTING GENETIC VARIABILITY

Until now, we have discussed the role of preexisting epigenetic variability in producing somaclonal variants. However, preexisting genetic variability in plant cells could also contribute to somaclonal variation. Genetic differences observed in somaclonal variants include changes in ploidy level, chromosome rearrangements, and point mutations of single genes (Lee and Phillips 1988). Because plants can generally tolerate somatic mutations and ploidy changes, all of the genetic differences observed in somaclonal variants can in principle be present in cells of differentiated plant tissues and need not be induced during culture (Marcotrigiano 1990). The extent to which such genetic differences arise spontaneously in plant cells or are produced as part of a developmental program remains an open question. Genetic homogeneity in cells of plants and animals is usually taken for granted; however, the arguments for the constancy of genomes during growth and development are actually quite weak (Spradling 1993). The possibility of programmed genetic changes playing a greater role in differentiation than presently suspected is another area that is largely unexplored but can be approached using transgenes as reporters for somatic genetic and epigenetic alterations in plants regenerated from single cells.

HIGH FREQUENCY OF SOMACLONAL VARIANTS: INVOLVEMENT OF A RIP-LIKE PROCESS?

We have proposed that the high frequency of somaclonal variation can be mainly accounted for by pervasive, preexisting epigenetic variability in differentiated tissues that are used as explant sources, and stable maintenance of a given epigenotype in regenerated plants. However, there is still a role for tissue-culture-induced changes, both genetic and epigenetic. Phillips and coworkers (1994) have attempted to account for the unusually high incidence of somaclonal variation by assigning a central mutagenic role to tissue-culture-induced changes in DNA methylation. A distinctive aspect of their hypothesis invokes the phenomenon of repeat-induced point mutation, or RIP (Phillips et al. 1994).

RIP refers specifically to a process that riddles duplicated sequences with C-G to A-T transition mutations in haploid genomes of special dikaryotic cells formed during the sexual cycle of the fungus *Neurospora crassa*. Sequences altered by RIP then become substrates for de novo cytosine methylation (Selker 1990). Because RIP greatly accelerates the mutation rate, Phillips and coworkers have proposed that the elevated frequency of variation observed in regenerated plants might be explained by the unleashing of a RIP-like process during the stress of tissue culture (Phillips et al. 1994). Although there has been some discussion about whether RIP or a related process operates in vertebrates (Kricker et al. 1992) or plants (Matzke et al. 1994), efficient RIP does not appear to act on repeated transgene sequences in plants (Mittelsten Scheid et al. 1994). Moreover, there is no evidence that either mutations in repeated sequences, which are the exclusive target of RIP in *Neurospora*, or C-G to A-T mutations are involved in somaclonal variation in plants.

Leaving bona fide RIP aside, however, it is still worth considering a role for methylation-induced genetic changes in somaclonal variation because 5-methylcytosine can spontaneously deaminate to produce T. Moreover, the C methylation pathway is inherently mutagenic, particularly under conditions in which the methyl donor S-adenosyl methionine (SAM) is limiting. Such an environment permits the accumulation of an intermediate in the C methylation pathway (5,6-dihydrocytosine), which has a 10^4-fold higher rate of deamination than 5-methylcytosine (Selker 1990; Jones et al. 1992; Laird and Jaenisch 1994; Rossignol and Faugeron 1994). In vitro culture might subject plant cells to SAM-deficient conditions, leading to increased frequencies of C-G to A-T transitions via rapid deamination of 5,6-dihydrocytosine. Thus, a possible cost of maintaining the high content of 5-methylcytosine in plant genomes (up to 30% of Cs can be methylated; Pradhan and Adams 1995) during plant cell culture might be an elevated rate of mutation, which then contributes to somaclonal variation in regenerated plants.

CONCLUSIONS

Our data with transgenes suggest that a developmentally acquired epigenetic state at a locus can be stably maintained in plants regenerated from single cells, leading to mitotically and meiotically heritable changes in the phenotype associated with the locus. It is striking how closely our results on the high frequency of variation and the incidence of homozygous changes conform to previous data obtained from non-transgenic systems, and we believe they can shed light on these heretofore puzzling features of somaclonal variation. Add the variability we have observed at one transgene locus to similar behavior that could occur at each endogenous locus that is subjected to epigenetic modifications, and one can envision a virtually unlimited variety of phenotypes appearing in a population of regenerated plants. Clearly, certain aspects of the developmental program must be reset in somatic cells prior to regeneration, or plants—even variant ones—could not develop at all. However, developmentally nonessential genes that might affect quantitative characters could retain diverse epigenetic modifications without inhibiting the capacity to regenerate.

In summary, somaclonal variation is a fascinating phenomenon that can reveal the spectrum of genomic alterations, both genetic and epigenetic, which are acquired at the single-cell level as plant development unfolds. The exploitation of transgenic plants, with which it is possible to correlate molecular and phenotypic changes, to study this phenomenon is just beginning. Tissue-specific promoters, developmentally regulated genes, and different explant sources derived from transgenic plants offer abundant material to investigate the molecular basis of somaclonal variation.

ACKNOWLEDGMENTS

We thank V.A. Iglesias for help in preparing Table 1. Our work has been supported by grants from the Austrian Fonds zur Förderung der wissenschaftlichen Forschung, the Bundesministerium für Wissenschaft und Forschung, and the Austrian National Bank.

REFERENCES

Ball, S.G. 1990. Molecular basis of somaclonal variation. *Biotechnol. Agric. For.* **11:** 134–152.

Brown, P.T.H. 1989. DNA methylation in plants and its role in tissue culture. *Genome* **31:** 717–729.

Evans, D.A. 1989. Somaclonal variation—Genetic basis and breeding applications. *Trends Genet.* **5:** 46–50.

Jones, P.A., W.M. Rideout, J.-C. Shen, C.H. Spruck, and Y.C. Tsai. 1992. Methylation, mutation and cancer. *BioEssays* **14:** 33–36.

Kaeppler, S.M. and R.L. Phillips. 1993. Tissue culture-induced DNA methylation variation in maize. *Proc. Natl. Acad. Sci.* **90:** 8773–8776.

Karp, A. and S.W.J. Bright. 1985. On the causes and origins of somaclonal variation. *Oxf. Surv. Plant Mol. Cell Biol.* **2:** 199–234.

Kricker, M.C., J.W. Drake, and M. Radman. 1992. Duplication-targeted DNA methylation and mutagenesis in the evolution of eukaryotic chromosomes. *Proc. Natl. Acad. Sci.* **89:** 1075–1079.

Kuckuck, H., G. Kobabe, and G. Wenzel. 1991. *Fundamentals of plant breeding.* Springer-Verlag, Berlin.

Laird, P.W. and R. Jaenisch. 1994. DNA methylation and cancer. *Hum. Mol. Genet.* **3:** 1487–1495.

Larkin, P.J. and W.R. Scowcroft. 1981. Somaclonal variation—A novel source of variability from cell cultures for plant improvement. *Theor. Appl. Genet.* **60:** 197–214.

Larkin, P.J., S.A. Ryan, R.I.S. Brettell, and W.R. Scowcroft. 1984. Heritable somaclonal variation in wheat. *Theor. Appl. Genet.* **67:** 443–455.

Larkin, P.J., P.M. Banks, R. Bhati, R.I.S. Brettell, P.A. Davies, S.A. Ryan, W.R. Scowcroft, L.H. Spindler, and G.J. Tanner. 1989. From somatic variation to variant plants: Mechanisms and applications. *Genome* **31:** 705–711.

Lee, M. and R.L. Phillips. 1988. The chromosomal basis of somaclonal variation. *Annu. Rev. Plant Physiol. Plant Mol. Biol.* **39:** 413–437.

Marcotrigiano, M. 1990. Genetic mosaics and chimeras: Implications in biotechnology. *Biotechnol. Agric. For.* **11:** 85–111.

Matzke, M.A., A.J.M. Matzke, and O. Mittelsten Scheid. 1994. Inactivation of repeated genes—DNA-DNA interaction? In *Homologous recombination and gene silencing in plants* (ed. J. Paszkowski), pp. 271–307. Elsevier, Dordrecht, The Netherlands.

Meins, F. 1983. Heritable variation in plant cell culture. *Annu. Rev. Plant Physiol.* **34:** 327–346.

Mittelsten Scheid, O., K. Afsar, and J. Paszkowski. 1994. Gene inactivation in *Arabidopsis thaliana* is not accompanied by an accumulation of repeat-induced point mutations. *Mol. Gen. Genet.* **244:** 325–330.

Neuhuber, F., Y.-D. Park, A.J.M. Matzke, and M.A. Matzke. 1994. Susceptibility of transgene loci to homology-dependent gene silencing. *Mol. Gen. Genet.* **244:** 230–241.

Oono, K. 1985. Putative homozygous mutations in regenerated plants of rice. *Mol. Gen. Genet.* **198:** 377–384.

Peschke, V.M. and R.L. Phillips. 1992. Genetic implications of somaclonal variation in plants. *Adv. Genet.* **30:** 41–75.

Phillips, R.L., S.M. Kaeppler, and P. Olhoft. 1994. Genetic instability of plant tissue cultures: Breakdown of normal controls. *Proc. Natl. Acad. Sci.* **91:** 5222–5226.

Pradhan, S. and R.L.P. Adams. 1995. Distinct CG and CNG DNA methyltransferases in *Pisum sativum. Plant J.* **7:** 471–481.

Rossignol, J.-L. and G. Faugeron. 1994. Gene inactivation triggered by recognition between DNA repeats. *Experientia* **50:** 307–317.

Selker, E.U. 1990. DNA methylation and chromatin structure: A view from below. *Trends Biochem. Sci.* **15:** 103–107.

Skirvin, R.M., K.D. McPheeters, and M. Norton. 1994. Sources and frequency of somaclonal variation. *HortScience* **29:** 1232–1237.

Skirvin, R.M., M. Norton, and K.D. McPheeters. 1993. Somaclonal variation: Has it proved useful for plant improvement? *Acta Hortic.* **336:** 333–340.

Spradling, A.C. 1993. Position effect variegation and genomic instability. *Cold Spring Harbor Symp. Quant. Biol.* **58:** 585–596.

Yibrah, H.S., D.H. Clapham, and S. Von Arnold. 1993. Antisense RNA inhibition of *uid*A gene expression in transgenic plants: Evidence for interaction between first and second transformation events. *Hereditas* **118:** 273–280.

Sense Cosuppression of Flower Color Genes: Metastable Morphology-based Phenotypes and the Prepattern-threshold Hypothesis

Richard A. Jorgensen, Qiudeng Que, James J. English,[1] and Huai-Yu Wang
Environmental Horticulture
University of California
Davis, California 95616-8587

In flowering plants, transformation with a transgene whose coding sequence is homologous to an endogenous plant gene, and whose expression is driven in the sense orientation by a strong promoter, often produces the opposite outcome of that which was intended. Instead of overexpression of the gene product, transcripts from both the endogenous gene and the transgene are found to be "cosuppressed," resulting in a null phenotype. Nuclear run-on experiments have shown that the cosuppressed state is posttranscriptional; experiments that alter transgene dosage suggest that the induction of cosuppression is threshold-dependent; and cytoplasmic RNA viruses that are homologous to a strongly expressed nuclear gene are subject to cosuppression, suggesting that it occurs in the cytoplasm (de Carvalho et al. 1992, 1995; Lindbo et al. 1993; Smith et al. 1994; Mueller et al. 1995; Goodwin et al. 1996). A preferred interpretation of these observations, presented in detail by Smith et al. (1994), is that the production of nuclear transcripts to levels exceeding some threshold concentration somehow activates a sequence-specific turnover process that acts in the cytoplasm. This phenomenon is referred to variously as "sense suppression," "sense cosuppression," or simply "cosuppression." It can be distinguished from promoter-homology-dependent suppression, a gene silencing phenomenon that occurs at the transcriptional level, requires promoter homology, and is thought to be a consequence of ectopic pairing (for review, see Matzke and Matzke 1995; see also Vaucheret et al.; Meyer; both this volume). Sense cosuppression also can be distinguished from suppression caused by antisense transgenes (Jorgensen et al. 1996).

[1]Present address: Sainsbury Laboratory, John Innes Centre, Colney Lane, Norwich NR4 7UH, United Kingdom.

The purpose of this paper is not to address the posttranscriptional mechanism of sense cosuppression (for review, see Dougherty and Parks 1995); rather, it is to discuss how sense cosuppression appears to act as a highly sensitive reporter of subtle transcriptional variations in anthocyanin pigmentation genes, greatly amplifying the phenotypic effects of epigenetic changes in transgene transcription states. Anthocyanin genes are very useful tools for monitoring gene expression in plants because they confer a phenotype that is visible, dispensable, and cell-autonomous. These genes have been widely used to investigate heritable epigenetic states in plants, as discussed elsewhere (Fedoroff; Chandler et al.; Kermicle; Meyer; Martienssen; all this volume). In contrast with the positive relationship that normally exists between pigment intensity and the transcription rate of an anthocyanin gene, sense cosuppression of an anthocyanin gene reverses this relationship at the threshold for cosuppression, resulting in a steep inverse relationship between pigment intensity and transcription rate and extreme sensitivity to epigenetic changes in transcription states that cross the threshold for cosuppression (Jorgensen 1995).

PATTERNS OF *Chs* SENSE COSUPPRESSION IN PETUNIA FLOWERS

The enzyme chalcone synthase catalyzes an essential reaction in anthocyanin biosynthesis. Sense cosuppression of *Chalcone synthase (Chs)* genes in pigmented petunia flowers can produce pure white flowers but, more frequently, produces striking flower color patterns of which there are many types, ranging from simple to complex, ordered to disordered, and stable to metastable (Napoli et al. 1990; Jorgensen 1995; Jorgensen et al. 1996). Intriguingly, the spatial organization of white sectors of *Chs* cosuppression in these flowers is determined principally by corolla morphology and only rarely by clonal cell lineages. Figure 1 presents typical examples of these patterns. The corolla of the petunia flower is composed of five petals which are joined throughout ontogeny; i.e., congenitally fused. This fusion of petals produces a hollow tube extending upward from the base of the flower that surrounds the sexual organs. At the top of the corolla tube the corolla limbs unfold perpendicular to the tube, each limb remaining joined to its neighbor. Cells located at these junctions between petal limbs are the determinants of the most common pattern of *Chs* cosuppression, the junction-based pattern shown in Figure 1A. Each of the five petals of the petunia flower is bisected by a large midvein that extends from the base of corolla tube to the tip of the limb; this midvein is a determinant of another pattern of

Figure 1 Typical sense cosuppression patterns in petunia flowers. (*A*) Petal junction-based pattern. (Reprinted, with permission, from Jorgensen 1993.) (*B*) Midvein-based pattern. (Reprinted, with permission, from Jorgensen and Napoli 1996.) (*C*) Complex, ordered pattern, known as "Cossack Dancer." (Reprinted, with permission, from Napoli et al. 1990.) (*D*) Complex, disordered pattern. (Reprinted, with permission, from Jorgensen 1994.)

Chs cosuppression (Fig. 1B). Many other patterns are determined by multiple morphological parameters, such as the pattern shown in Figure 1C, which is based on both veins and junctions, as well as the outer edge of the corolla. Other patterns can be disordered and irregular but still exhibit some degree of morphological dependence, as in Figure 1D. For a detailed discussion of pattern types and determinants, see Jorgensen et al. (1996).

Fully pigmented sectors in *Chs* cosuppression patterns are separated from white sectors by a broad, multicellular boundary zone in which pigment intensity declines in a steep, continuous gradient. In petunia flowers, only epidermal cells are pigmented. In each flower, sector boundaries in the upper epidermis of the corolla (i.e., the "face" of the flower, seen in Fig. 1) are coincident with the boundaries in the lower epidermis (i.e., the back of the flower, not seen in Fig. 1), indicating that cosuppression patterns are organized three-dimensionally. In sum then, sense cosuppression of *Chs* in petunia flowers appears to be subject to three-dimensional fields of positional information, generated mainly by petal veins, junctions, and edges.

Plant transgenes are often subject to heritable, but reversible, epigenetic events that alter their transcription states (for review, see Jorgensen 1993; Matzke and Matzke 1993). Heritable, reversible epigenetic variations have also been observed in many *Chs* transgenotes. Some epigenetic variants of *Chs* cosuppression phenotypes affect only the size of cosuppression sectors. However, most epigenetic variants in this system produce a qualitatively new pattern based on different morphological determinants. Thus, a plant producing a junction-based pattern of cosuppression may give rise to an epigenetic variant producing a vein-based pattern of cosuppression. In fact, the flowers shown in Figure 1A and B are epigenetic variants of the same transgenic plant. For further examples and discussion, see Jorgensen (1995) and Jorgensen and Napoli (1996).

THE PREPATTERN-THRESHOLD HYPOTHESIS

How might sense cosuppression of *Chs* be responsive to positional information? Considering the apparent threshold dependence of cosuppression and the epigenetic variation in cosuppression patterns, Jorgensen (1995) proposed that (1) fields of morphological information in flowers comprise shallow spatial gradients ("prepatterns") of transcription factors or their effector molecules, each prepattern being determined by a particular feature of morphology (e.g., the junctions between petals, veins, or the edges of petals); (2) such factors subtly influence transgene transcription to produce a shallow gradient of *Chs* transcription rate and nuclear transcript concentration that follows the prepattern; (3) the cosuppression turnover state is triggered wherever the nuclear transcript concentration gradient crosses the threshold concentration; and (4) this produces a steep gradient of transcript turnover rate that becomes visible as a corresponding gradient of anthocyanin pigmentation. We call this the "prepattern-threshold" hypothesis. Consistent with the prepattern-threshold hypothesis, we have found that a promoter is required to produce the junction pattern and that small changes in transgene promoter strength have the predicted effect on the degree of cosuppression observed in a population of transformants; i.e., weaker promoters produce more plants in which white sectors are small, and stronger promoters produce more plants in which white sectors are large (Q. Que et al., in prep.).

The prepattern-threshold hypothesis is an extension of the model of Coen et al. (1988) that patterns of anthocyanins were determined by prepatterns of regulator molecules that control anthocyanin gene expression. Typically, patterns of anthocyanin gene expression are determined at the transcriptional level, often by regulation of the entire anthocyanin pathway. This level of control is responsible for the epidermal-specific expression of the anthocyanin genes in petunia petals, for example. In the case of sense cosuppression, patterns within the corolla are determined by posttranscriptional silencing of one step in the pathway, that encoded by *Chs*. The prepattern-threshold hypothesis simply suggests that the threshold-dependence of sense cosuppression phenotypes results in great sensitivity to transcriptional variations near the threshold.

Given such sensitivity to variations near the threshold, cosuppression phenotypes should be very sensitive reporters of those epigenetic changes in transcription states that cause the threshold to be crossed. It is easy to see how such changes could alter the size of white cosuppression sectors in petunia flowers, but how could a heritable change in transgene transcription state lead to a *qualitative* change in pattern type? One possibility is that, if the transgene has available to it a sufficient variety of transcription factors that vary in concentration gradients across the flower petal (each according to different morphological parameters as was argued above), a change in the epigenetic state of the transgene could cause the transcription complex to favor a different factor than it had favored previously. In this way, different epigenetic states of the transgene could elicit fundamentally different pigmentation patterns.

Could the CaMV 35S promoter of the *Chs* transgene be so flexible? It is known that this promoter possesses a variety of transcription-factor-binding sites (Benfey et al. 1990) and also that several members of a family of transcription factors can be expressed in the same tissue (Miao and Lam 1995). Whether any such factors exist in spatially determined concentration gradients remains to be determined. However, it is clear that fields of morphological information must exist in developing organs such as petals, and it seems reasonable to consider that these might determine or comprise gradients of transcription factors or molecules that modulate transcription factor activity. Assuming such gradients exist, our model simply suggests that the response of the 35S promoter to each factor depends on its epigenetic state. For example, in one epigenetic state of the promoter, it may respond to a vein-based transcription factor, whereas in another state, it may respond to a junction-based factor; thus, a change in epigenetic state would lead to a change in the pattern of transcription, and thus a change in the pattern of cosuppression. One way we are testing this model is by making a series of modifications to the 35S promoter and evaluating whether alteration of transcription-factor-binding sites in the transgene's promoter will alter cosuppression pattern type and epigenetic properties of the transgene.

A ROLE FOR CHROMATIN?

Indirect evidence that, as is assumed in the model, epigenetic states of *Chs* sense cosuppression are chromosomally based derives from analyses of phenotypic variation among plants in a large population of sense *Chs* transformants. We have found that the particular type of pattern (i.e., junction-based, vein-based, or complex) is determined principally by the copy number and organization pattern of the transgene (e.g., single copy vs. dispersed repeats vs. tandem repeats) and by the epigenetic state of the transgene, not by the effect of plant DNA sequences flanking the transgene (Jorgensen et al. 1996). Single-copy transgenes almost always produce the junction pattern. Multiple, dispersed copies of the transgene can determine either the junction pattern or a vein pattern, depending on the epigenetic state of the transgenes. Inverted repeats of the transgene almost always produce complex patterns, although occasional epigenetic variants produce vein- or junction-based patterns.

Interestingly, it has been shown in *Drosophila* that different sequence organization patterns have distinct chromatin properties (Dorer and Henikoff 1994; Henikoff, this volume). Increased levels of DNA methylation have been observed in multicopy transgenes as compared to single-copy transgenes in plants (Matzke et al. 1994) and thus may play a role in repeat-induced variations of cosuppression patterns. Thus, the dependence of cosuppression pattern types on transgene organization patterns suggests that they may be due to chromatin state differences or to changes in DNA methylation patterns and that the epigenetic states defined by different patterns are transcriptional.

It appears then that multicopy transformants affect *Chs* expression in two distinct ways, one being transcriptional and involving transgene-transgene interactions and the other being posttranscriptional cosuppression of *Chs* transgenes and endogenous *Chs* genes. These mechanisms are likely to work in opposition; i.e., the increase in gene dosage in a multicopy transformant would produce an increase in transcription leading to larger cosuppression sectors or completely white flowers, whereas intertransgene interactions would cause a decrease in transcription, possibly below the threshold for cosuppression and/or a change in the morphological pattern of transcription. Thus, a hierarchy of transcriptional and posttranscriptional events seems likely to be responsible for at least some of the diversity and epigenetic variability of cosuppression patterns.

PRESETTING AND *PARACHROMATIN*: RELATED CONCEPTS FROM MCCLINTOCK AND BRINK

Our model for transcriptional determination of the diverse patterns of *Chs* cosuppression in petunia flowers and for chromatin determination of epigenetic states that select among the available prepatterns of regulatory molecules is deeply rooted in observations and concepts presented more than 30 years ago by Barbara McClintock and R. Alexander Brink. McClintock (1967) and Brink (1960) independently discovered that certain forms of maize genes can adopt heritable, multistable states, and each proposed that these states can be set and reset by physiological and developmental events (for review, see Jorgensen 1994; see also Fedoroff; Kermicle; both this volume). Both believed that heritable abnormalities such as they observed could inform us to some extent about normal gene regulatory mechanisms, and each proposed that normal genes too can adopt somatically heritable states that can be set and reset as new regulatory factors arise in physiology and development.

McClintock made the important observation that an epigenetic state could be "preset" at a stage of development much earlier than the stage at which it would be expressed and that this presetting could be efficiently erased still later. It was especially significant that newly arising states of genes were shown by McClintock to differ from preceding states not only in the pattern of expression they permit, but also in their *responsiveness* to further resetting and presetting later in development. She suggested that normal genes are reset to some sort of ground state during or after sexual reproduction and that those genes that can be observed to transmit altered states through the germ line are simply rare, aberrant forms of a gene.

The sequential process of resetting chromosome-based states was discussed by Brink as a "paragenetic" function of chromatin; i.e., one that can receive, record, and mitotically transmit developmental information and that can replicate in phase with the DNA. Thus, the paragenetic component of chromatin, which he named parachromatin, can act as a programmable memory of ontogeny that can be used for directing subsequent patterns of gene expression and can be reset at each sexual generation. This paragenetic function was proposed to complement the genetic function of chromosomes (which is simply to encode genetic information and transmit it unaltered to the next sexual generation). Unfortunately, knowledge available in the 1960s was insufficient to test such a hypothesis directly. Nonetheless, the general concepts provided then are still pertinent today, and new information and technologies exist by which to build and test more specific hypotheses at the molecular level. An increasing body of evidence suggests that the molecular basis for paragenetic states of genes may involve chromatin-stabilized transcription complexes (Bestor et al. 1994). In this light, Brink's proposal can be viewed as simply a higher-order elaboration on the Jacob-Monod model of genetic control. Applying Brink's concept of parachromatin to the *Chs* sense cosuppression system, we can view the paragenetic state of the transgene as being at the center of the prepattern-threshold hypothesis, its role being to discriminate among prepatterns, to direct expression accordingly, and to be reprogrammed by certain prepatterns.

The evidence that epigenetic states can be reprogrammed by developmental prepatterns is based on the observation that nonclonal epigenetic events occur in shoot apical meristems (Jorgensen 1994, 1995; Jorgensen and Napoli 1996). Plant growth and development are based largely on apical meristems; i.e., relatively undifferentiated groups of cells at the apex of a shoot or root. The shoot apical meristem is the ultimate source of all cells in the primary body of the shoot and, in petunia, produces leaf primordia that arise in a spiral fashion on the meristem's flanks. In the axil of each leaf, an axillary meristem is organized from a nonclonally derived group of cells; this becomes a new shoot apical meristem and produces a branch if it is released from dormancy. By observation of epigenetic events arising in shoots, it has been found that most epigenetic events in the *Chs* sense cosuppression system that create a new flower color pattern do not arise in a single cell, but rather they arise in nonclonal fashion as if the event had occurred in all the cells of the meristem. Thus, meristems appear to mediate the imposition of a new epigenetic state onto their component cells. Although these events do not occur in every shoot apical meristem of the plant, their relatively ordered nature approaches the ordered, resettable epigenetic behavior postulated for normal genes by Brink and McClintock.

CONCLUDING REMARKS

When paragenetic transcriptional control mechanisms are combined with a threshold-dependent posttranscriptional response, as seems to be the case in sense cosuppression of *Chs*, one consequence can apparently be a dynamic, multistable, hierarchical gene regulatory system that possesses very great sensitivity to what might otherwise be regarded as inconsequential shifts in transcription rate. It is likely that the more erratic aspects of the *Chs* system derive simply from the fact that it is based on a transgene that has not been subjected to natural selection. It seems reasonable to suppose that during evolution similar, but less erratic, regulatory systems could arise, and so it also seems reasonable to regard sense cosuppression as potentially more than just a response to the overexpression of foreign genes. Although such a response may have been the primitive function of sense cosuppression, it seems likely that during the evolution of multicellularity, differentiation, and development in the plant kingdom, sense cosuppression would have been exploited for more subtle and interesting purposes by coordination with transcription-level controls. It will not be surprising if sense cosuppression is ultimately found to play a significant role in naturally occurring hierarchical systems of gene regulation in higher plants.

REFERENCES

Benfey, P.N., L. Ren, and N.-H. Chua. 1990. Combinatorial and synergistic properties of CaMV 35S enhancer subdomains. *EMBO J* **9:**1685–1696.

Bestor, T.H., V.L. Chandler, and A.P. Feinberg. 1994. Epigenetic effects in eukaryotic gene expression. *Dev. Genet.* **15:** 458–462.

Brink, R.A. 1960. Paramutation and chromosome organization. *Q. Rev. Biol.* **35:** 120–137.

Coen, E.S., J. Almeida, T.P. Robbins, A. Hudson, and R. Carpenter. 1988. Molecular analysis of genes determining spatial patterns in *Antirrhinum majus*. In *Temporal and spatial regulation of plant genes* (ed. D.P.S. Verma and R.B. Goldberg), pp. 63–82. Springer-Verlag, Wien.

de Carvalho Niebel, F., P. Frendo, M. van Montagu, and M. Cornelissen. 1995. Posttranscriptional cosuppression of β-1,3-glucanase genes does not affect accumulation of transgene nuclear mRNA. *Plant Cell* **7:** 347–358.

de Carvalho, F., G. Gheysen, S. Kushnir, M. van Montagu, D. Inze, and C. Castresana. 1992. Suppression of β-1,3-glucanase transgene expression in homozygous plants. *EMBO J.* **11:** 2595–2602.

Dorer, D.R. and S. Henikoff. 1994. Expansions of transgene repeats cause heterochromatin formation and gene silencing in *Drosophila*. *Cell* **77:** 993–1002.

Dougherty, W.G. and T.D. Parks. 1995. Transgenes and gene suppression: Telling us something new? *Curr. Opin. Cell Biol.* **7:** 399–405.

Goodwin, J., K. Chapman, S. Swaney, T.D. Parks, E.A. Wernsman, and W.G. Dougherty.

1996. Genetic and biochemical dissection of transgenic RNA-mediated virus resistance. *Plant Cell* **8:** 95–105.

Jorgensen, R. 1993. The germinal inheritance of epigenetic information in plants. *Philos. Trans. R. Soc. Lond. B* **339:** 173–181.

———. 1994. Developmental significance of epigenetic impositions on the plant genome: A paragenetic function for chromosomes. *Dev. Genet.* **15:** 523–532.

———. 1995. Cosuppression, flower color patterns, and metastable gene expression states. *Science* **268:** 686–691.

Jorgensen, R.A. and C.A. Napoli. 1996. A responsive regulatory system is revealed by sense suppression of pigment genes in *Petunia* flowers. In *Genomes: Proceedings of the 22nd Stadler Genetics Symposium* (ed. J.P. Gustafson and R.B. Flavell), pp. 159–176. Plenum Press, New York.

Jorgensen, R.A., P.D. Cluster, J. English, Q. Que, and C.A. Napoli. 1996. Chalcone synthase cosuppression phenotypes in petunia flowers: Comparison of sense vs. antisense constructs and single copy vs. complex T-DNA sequences. *Plant Mol. Biol.* (in press).

Lindbo, J.A., L. Silva-Rosales, W.M. Proebsting, and W.G. Dougherty. 1993. Induction of a highly specific antiviral state in transgenic plants: Implications for gene regulation and virus resistance. *Plant Cell* **5:** 1749–1759.

Matzke, M.A. and A.J.M. Matzke. 1993. Genomic imprinting in plants: Parental effects and *trans*-inactivation phenomena. *Annu. Rev. Plant Physiol. Plant Mol. Biol.* **44:** 53–76.

———. 1995. How and why do plants inactivate homologous (trans)genes? *Plant Physiol.* **107:** 679–685.

Matzke, A.J.M., F. Neuhuber, Y.-D. Park, P.F. Ambros, and M.A. Matzke. 1994. Homology-dependent gene silencing in transgenic plants: Epistatic silencing loci contain multiple copies of methylated transgenes. *Mol. Gen. Genet.* **244:** 219–229.

McClintock, B. 1967. Regulation of pattern of gene expression by controlling elements in maize. *Carnegie Inst. Wash. Year Book* **65:** 568–578.

Mueller, E., J.E. Gilbert, G. Davenport, G. Brigneti, and D.C. Baulcombe. 1995. Homology-dependent resistance: Transgenic virus resistance in plants related to homology-dependent gene silencing. *Plant J.* **7:** 1001–1013.

Miao, Z.H. and E. Lam. 1995. Construction of a trans-dominant inhibitor for members of the TGA family of transcription factors conserved in higher plants. *Plant J.* **7:** 887–896.

Napoli, C., C. Lemieux, and R. Jorgensen. 1990. Introduction of a chimeric chalcone synthase gene into petunia results in reversible co-suppression of homologous genes *in trans*. *Plant Cell* **2:** 279–289.

Smith, H.A., S.L. Swaney, T.D. Parks, E.A. Wernsman, and W.G. Dougherty. 1994. Transgenic plant virus resistance mediated by untranslatable sense RNAs: Expression, regulation, and fate of nonessential RNAs. *Plant Cell* **6:** 1441–1453.

Analysis of a Tobacco Transgene Locus That Triggers Both Transcriptional and Posttranscriptional Silencing

Hervé Vaucheret, Taline Elmayan, Philippe Mourrain, and Jean-Christophe Palauqui
Laboratoire de Biologie Cellulaire
INRA, F-78026 Versailles Cedex
France

INTRODUCTION

Numerous studies have demonstrated that introduction of a transgene in plants can result in transgenic plants which either express the transgene or not (for recent reviews, see Flavell 1994; Matzke and Matzke 1995). Although not all silencing phenomena have been studied in detail at the molecular level, it is known that gene silencing in plants can occur either at the transcriptional level—i.e., transcription is blocked—or at the posttranscriptional level—i.e., RNA turnover is increased.

Transcriptional Silencing

Transcriptional silencing is defined as *cis*-inactivation when it affects transcription of either single or multiple copies of a transgene inserted at a single locus (Linn et al. 1990; Meyer et al. 1993). It is defined as dominant *trans*-inactivation when a transgene driven by a given promoter inhibits transcription of unlinked transgenes driven by the same promoter (Matzke et al. 1989, 1994; Vaucheret 1993; Park et al. 1996).

Posttranscriptional Silencing

Posttranscriptional silencing events have been described as the result of an increase of the turnover of RNA encoded either by single or multiple copies of a transgene. Although the exact step at which silencing occurs remains unknown, it has been shown that RNAs are produced in the nucleus but do not accumulate in the cytosol (Dehio and Schell 1994; Ingelbrecht et al. 1994; Elmayan and Vaucheret 1996). Posttranscriptional silencing is known as cosuppression when it simultaneously affects endogenous host genes that are homologous to the introduced transgenes

Epigenetic Mechanisms of Gene Regulation

(Napoli et al. 1990; van der Krol et al. 1990; van Blockland et al. 1994; de Carvalho-Niebel et al. 1995).

Up to now, transcriptional and posttranscriptional silencing phenomena have been described separately, each occurring as a consequence of the introduction of a particular transgene. In this review, an atypical silencing locus that triggers both posttranscriptional silencing of homologous host genes and transcriptional silencing of homologous transgenes is described and compared to typical silencing loci triggering cosuppression of the same homologous host genes.

POSTTRANSCRIPTIONAL SILENCING OF TOBACCO NITRITE REDUCTASE GENES

Nitrate Assimilation Pathway in Higher Plants

Nitrate is the major source of nitrogen for higher plants. Nitrate is reduced into ammonium by nitrate reductase (NR) and nitrite reductase (NiR). The incorporation of ammonium onto the carbon skeleton produced by photosynthesis provides reduced nitrogen sources for the synthesis of proteins and nucleic acids. NR is a cytoplasmic enzyme encoded by a single gene per haploid genome in tobacco, whereas NiR is a chloroplastic enzyme encoded in the nucleus by a small multigene family of root- and leaf-specific genes (Kronenberger et al. 1993). NR-deficient tobacco mutants have been selected that are unable to use nitrate as a nitrogen source but can grow with ammonium. Transgenic tobacco plants that overexpress NR activity and exhibit a better assimilation of nitrate have also been obtained (Dorlhac de Borne et al. 1994). During the attempt to study the physiological effect of the deregulation of NiR, transgenic tobacco plants that overexpress NiR activity, and NiR-deficient transgenic tobacco plants, were produced using sense and antisense transgenes, respectively. Unexpectedly, silencing of NiR was observed with both types of constructs.

Introduction of a Sense Transgene Can Trigger NiR Silencing (Cosuppression)

In an attempt to increase NiR activity in transgenic plants, the complete tobacco *Nii1* gene with its own introns, promoter, and termination signals was cloned downstream from a tandemly duplicated enhancer of the cauliflower mosaic virus (CaMV) 35S promoter (Fig. 1) and introduced into tobacco by *Agrobacterium*-mediated transformation. As expected, most of the resulting transformants had elevated NiR mRNA levels.

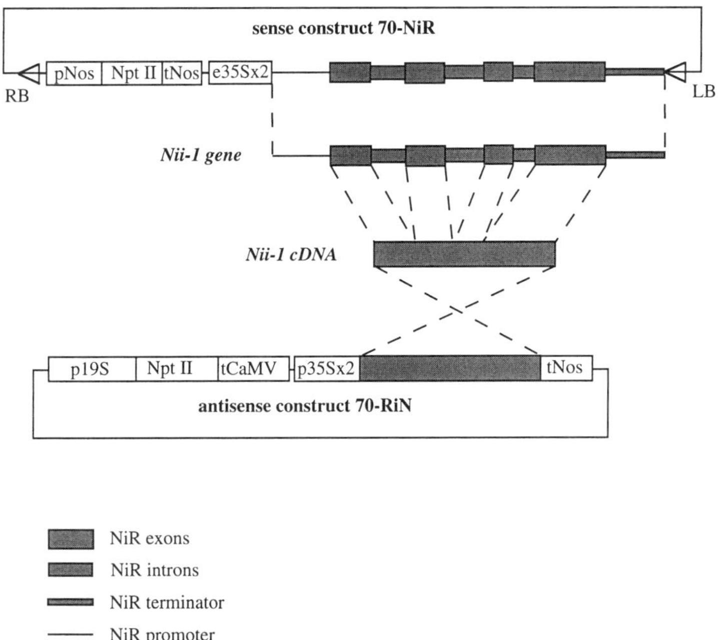

Figure 1 Physical map of sense 70-NiR and antisense 70-RiN transgenes. Sense 70-NiR construct. The complete tobacco *Nii1* gene with its own exons, introns, promoter, and termination signals was cloned downstream from a tandemly duplicated enhancer from the 35S promoter (e35Sx2) and introduced into the binary vector pBIN19 carrying a selectable *Npt II* marker gene under the control of the promoter of the nopaline synthase gene (pNos). Left (*LB*) and right (*RB*) borders of the T-DNA are indicated. Introduction of the T-DNA construct into tobacco plants was achieved by *Agrobacterium*-mediated transformation. Antisense p70-RiN construct. The tobacco *Nii1* cDNA cloned in the antisense orientation downstream from the 35S promoter with a tandemly duplicated enhancer (p35Sx2) was introduced into a Bluescript plasmid carrying the *Npt II* gene cloned downstream from the 19S promoter. Introduction of the plasmid DNA construct into tobacco plants was achieved by electroporation of protoplasts.

However, one of these transformants, plant 461-2, which strongly overexpressed NiR mRNA and carried the transgene inserted at a single locus, gave rise to homozygous progeny that were NiR-deficient, i.e., neither host gene nor transgene expression was observed. Conversely, both host and transgene expression were observed in hemizygous progeny, and host gene expression was observed in wild-type progeny that did not inherit the transgene locus, indicating that silencing is reversible (Vaucheret et al. 1995).

Silencing Occurs at Each Generation during the Early Development of Homozygous Progeny of Plant 461-2

In wild-type plants, the level of NiR mRNA increases progressively during the first 5 weeks of development. In hemizygous progeny derived from plant 461-2, the initial level of NiR mRNA is higher than in wild type, due to the presence of an enhancer in the transgene, and the same increase is observed, due to the conservation of the regulatory signals of the NiR gene in the transgene. In homozygous progeny derived from plant 461-2, the initial level of NiR mRNA is higher than in hemizygous plants, but it starts to decrease after 2 weeks and is undetectable after one month, leading to deficiency, and the phenomenon occurs at each generation (Vaucheret et al. 1995). Taken together, these results suggest that silencing of NiR by cosuppression is triggered cyclically at each generation when host and transgene NiR mRNA accumulate up to a threshold level which is reached only in the homozygous progeny of plant 461-2 (see other chapters in this volume about cosuppression).

Introduction of an Antisense Transgene Can Trigger NiR Silencing

In an attempt to decrease NiR activity in transgenic plants, a plasmid carrying the bacterial neomycin phosphotransferase gene (Npt) cloned downstream from the CaMV 19S promoter, and the tobacco *Nii1* cDNA cloned in the antisense orientation downstream from the CaMV 35S promoter with a tandemly duplicated enhancer (Fig. 1), was introduced into tobacco protoplasts by electroporation. Plants were regenerated from kanamycin-resistant cells and screened for a partial or a total deficiency in NiR activity. One plant with no NiR activity, plant 271, was identified by this screen (Vaucheret et al. 1992). Molecular and genetic analysis revealed that NiR deficiency in plant 271 behaves as a monogenic dominant trait; i.e., both hemizygous and homozygous progeny were NiR-deficient, thus indicating that this plant carries a single transgene locus. NiR enzymatic activity and NiR mRNA accumulation recovered uniformly to wild-type levels in wild-type progeny that did not inherit the transgene locus, indicating that silencing is reversible (Vaucheret et al. 1992; Park et al. 1996).

Silencing Occurs at Each Generation during Early Development of Hemizygous and Homozygous Progeny of Plant 271

Progeny that had inherited the transgene locus 271 recovered partial NiR activity after 2 weeks of growth. Subsequently, NiR activity decreased rapidly to reach undetectable levels after one month's growth, and the phenomenon occurred at each generation. Northern blot analysis showed that neither NiR (sense) nor RiN (antisense) mRNAs accumulated in this plant. Run-on transcription of sense NiR RNAs was detected in the nucleus, suggesting that NiR silencing occurred at the posttranscriptional level (Park et al. 1996). However, run-on transcription of antisense RiN RNAs could not be detected in the nucleus (T. Elmayan and H. Vaucheret, unpubl.). Taken together, these results indicate that the transgene locus of plant 271 is responsible for a posttranscriptional silencing phenomenon that occurs cyclically at each generation during the early development of the plants, and that closely resembles sense silencing (i.e., cosuppression of homologous host genes and transgenes) in the transgenic plant 461-2 (Vaucheret et al. 1995).

TRANSCRIPTIONAL SILENCING OF 19S AND 35S PROMOTERS

Plant 271 Carries Silent 19S and 35S Transgenes

The plasmid containing the antisense RiN construct used to generate primary transformant 271 carries an Npt gene driven by the CaMV 19S promoter. This gene normally confers resistance to kanamycin. However, after regeneration from callus, no expression could be detected from this marker gene in plant 271, even though the callus was selected on kanamycin. Progeny derived by six generations of self-pollinations or backcrosses were all kanamycin-sensitive. Run-on transcription could not be detected from the 19S-Npt transgene in these plants (Y.-D. Park and A.J.M. Matzke, unpubl.), and the transgene locus was extensively methylated (Vaucheret 1993). Therefore, the absence of transcription of the 19S-Npt and 35S-RiN transgenes suggests that the transcription from both 19S and 35S promoters is impeded at this locus.

The Transgene Locus of Plant 271 Silences Unlinked 19S and 35S Transgenes

Unexpectedly, introduction of additional transgenes under the control of the 19S promoter or the 35S promoter into plant 271 by sexual crossing or transformation led to their silencing, irrespective of the nature of the coding sequence being expressed and of their position of integration within the genome (Vaucheret 1993). Silencing occurs during the early development of hybrid plants in the case of sexual crosses or at some stage of the regeneration of transformed calli in the case of direct transformation (Vaucheret 1994). Deletion analysis revealed that transgenes driven by a fragment of the 35S promoter as short as 90 bp were also subject to silencing in plants carrying the transgene locus of plant 271 (Vaucheret 1993). The promoter sequences of the inactivated 35S transgenes become methylated de novo and their transcription is blocked (Elmayan and Vaucheret 1996; Park et al. 1996). Elimination of the so-called "271 silencing locus" by segregation leads to the reactivation of the target 19S or 35S transgenes. However, as opposed to the immediate reactivation of host NiR genes, reexpression of 19S and 35S transgenes in the absence of the 271 silencing locus occurred only after a lag period of up to several generations. The persistence of a silent state correlates with the maintenance of cytosine methylation in symmetrical CG and CNG sites (Vaucheret 1994; Park et al. 1996).

The 271 Silencing Locus Behaves Like the H2 Silencing Locus

Transcriptional silencing of unlinked 19S and 35S transgenes in the presence of the 271 silencing locus resembles silencing of unlinked nopaline synthase (Nos) transgenes in the presence of the H2 silencing locus, another tobacco silencing locus that carries multiple copies of the *Agrobacterium tumefaciens* Nos promoter (Matzke et al. 1989, 1993). In each case, the silencing locus is able to trigger heritable methylation and inactivation of unlinked target transgenes driven by homologous promoters (Fig. 2). One model to explain silencing *in trans* involves the pairing of promoter-homologous target transgenes with the silencing locus and the transfer of methylation between them (Matzke et al. 1994). The 271 silencing locus is completely methylated and carries transcriptionally silent 19S-Npt and 35S-RiN transgenes. The H2 silencing locus is completely methylated in the Nos promoter, and it is able to silence the expression of transgenes driven by the Nos promoter. The H2 silencing locus also carries an active 35S-Hph transgene which is only partially methylated. However, it fails to silence the expression of transgenes driven by the 35S promoter. Perhaps silencing of strong promoters like 35S requires a denser pattern of DNA methylation, such as that found in the 271 silencing locus.

The H2 Silencing Locus Cannot Acquire the Properties of the 271 Silencing Locus

To test whether an increase in the methylation state of the H2 silencing locus would allow it to silence the expression of unlinked transgenes driven by the 35S promoter, plants homozygous for the H2 silencing

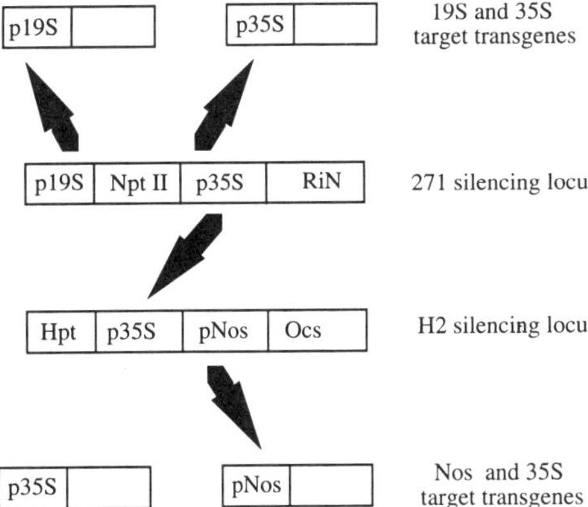

Figure 2 Silencing properties of the 271 and H2 silencing loci. The transgene locus of plant 271 carries multiple copies of a plasmid carrying the p19S-Npt and p35S-RiN transgenes. The transgene locus of plant H2 carries multiple copies of a T-DNA carrying the pNos-Ocs and p35S-Hpt transgenes. The 271 silencing locus is responsible for methylation and transcriptional *trans*-silencing of target transgenes driven by the 19S or 35S promoters, including the 35S-Hpt transgene carried on the transgene locus of plant H2. The H2 silencing locus is responsible for methylation and transcriptional *trans*-silencing of target transgenes driven by the Nos promoter, but not of target transgenes driven by the 35S promoter. When the H2 silencing locus is brought into the presence of the 271 silencing locus and then separated by segregation, it retains dense methylation in the 35S promoter. However, it still fails to silence target transgenes driven by the 35S promoter.

locus were crossed with plants homozygous for the 271 silencing locus (Park et al. 1996). The 35S-Hph transgene carried by the H2 silencing locus became transcriptionally inactive and strongly methylated. This hybrid plant was backcrossed to a wild type, and plants that had lost the 271 silencing locus by segregation, but retained the H2 silencing locus, were analyzed. Some of these plants still carried a silent and highly methylated 35S-Hph transgene. These plants were crossed to plants carrying unlinked target 35S transgenes. However, the hypermethylated H2 silencing locus failed to silence these target transgenes, suggesting that the silencing properties of a transgene locus cannot be explained solely by its methylation state.

Both H2 and 271 Silencing Loci Are Localized at Telomeres

The location of the silencing locus within the nucleus might contribute to its ability to search the entire genome for homology. In situ hybridization was performed to determine the chromosomal locations of these two silencing loci. The 271 silencing locus was localized at a telomere, whereas the H2 silencing locus was localized close to a telomere, suggesting that both loci occupy chromosomal sites that might be favorable for interactions that lead to silencing (Park et al. 1996). Therefore, some other parameter(s) might contribute to the *trans*-silencing properties of the 271 silencing locus.

Silencing Properties of the 271 Locus Are Altered by DNA Rearrangement

The structural arrangement and/or the number of copies are supposed to play a role in silencing (for review, see Flavell 1994; Matzke and Matzke 1995). Indeed, the H2 silencing locus contains four closely linked T-DNA copies (Matzke et al. 1993), whereas six or seven plasmid copies are present in the 271 silencing locus (Vaucheret 1993). Conversely, transgenic plants carrying a lower number of copies of the same constructs inserted at different locations do not show silencing. A spontaneous derivative allele of the 271 silencing locus has been isolated (H. Vaucheret, unpubl.). This allele lacks some of the multiple copies of the plasmid present at the locus of the original allele. It has retained dense methylation and *cis*-inactivation of the 19S-Npt transgene but has lost all the *trans*-inactivation properties associated with the original allele, suggesting that the structural arrangement and/or the number of copies is essential for the triggering of silencing *in trans* in combination with methylation and chromosomal location.

CONCLUSIONS

The analysis of various silencing phenomena in plants has clearly demonstrated that transcriptional and posttranscriptional silencing differ in their heritability and in their requirement for homology between silencing loci and target transgenes (Matzke and Matzke 1994). Transcriptional silencing requires homology between promoter sequences and leads to heritable methylation and inactivation. Conversely, posttranscriptional silencing requires homology between coding sequences and leads to progressive inactivation during development. In addition, posttranscriptional silencing is not heritable. The analysis of plant 271 shows that these two distinct silencing phenomena can be triggered by a single transgene locus.

Transcriptional Silencing

The 271 silencing locus carries multiple copies of two transgenes driven by the 19S and 35S promoters, respectively, inserted in a particular structure which is probably responsible for methylation and transcriptional *cis*-inactivation of the two transgenes. In addition, the location of this silencing locus at a telomere may be responsible for its ability to scan the entire genome for promoter-homologous partners to inactivate *in trans*. Dominant (epistatic) transcriptional silencing is proposed to occur through the transfer of the methylated and inactive states of the silencing locus to the promoter of unlinked homologous target transgenes during transient somatic pairing (Matzke et al. 1994). This phenomenon resembles dominant position effect variegation in *Drosophila* (see the corresponding chapters in this volume) and paramutation in plants (see the corresponding chapters in this volume). However, paramutation takes place between alleles, whereas transgene interactions involving the 271 silencing locus or the H2 silencing locus can affect nonallelic but promoter-homologous transgenes, irrespective of their location within the genome. It remains to be determined whether this new phenomenon represents a different type of silencing mechanism or a paramutation-like mechanism with an additional ectopic *trans*-effect due to the telomeric localization of the silencing loci in question.

Posttranscriptional Silencing

Transcriptional *trans*-inactivation of promoter-homologous partners is not the only characteristic of the 271 silencing locus. This locus is also able to inactivate host NiR genes by a posttranscriptional mechanism; i.e., transcription of sense NiR RNAs is initiated in the nucleus, but RNAs do not accumulate in the cytosol. Since no transcription of antisense RiN RNAs can be detected at the 271 silencing locus, the actual mechanism of silencing remains to be determined. Three possibilities can be evoked:

1. A small amount of antisense RNA might go undetected by run-on experiments. These might form transient unstable hybrids with host transcripts in the manner suggested by Dougherty and Parks (1995). This might account for the rapid degradation of host gene transcripts in these transgenic plants.
2. The mode of integration of the plasmid DNA into the genome might have placed one copy of the NiR sequence in sense orientation under the control of an adjacent plant promoter. This could allow transcrip-

tion of host and transgene sense NiR RNAs above the threshold level which is responsible for the triggering of cosuppression (see the other chapters in this volume about cosuppression), thus explaining why silencing in plant 271 resembles silencing in plant 461-2, which carries a sense NiR transgene.

3. The 271 silencing locus might interact with host NiR genes in the same manner as it interacts with promoter-homologous partners, thus resulting in the transfer of its methylated and inactive states to the homologous region of host NiR genes; i.e., their coding sequence. Run-on experiments indicate that methylation of the coding sequence does not impede the initiation of transcription (Ingelbrecht et al. 1994). However, coding sequence methylation might lead to premature termination as suggested in the process of methylation induced premeiotically (MIP) in the fungus *Ascobolus immersus* (Barry et al. 1993). Alternatively, structural changes in chromatin structure, associated with DNA methylation, might also modify the nuclear architecture and impede RNA maturation and/or transport, thus leading to the formation of unproductive RNA (for review, see Flavell 1994).

In conclusion, transcriptional and posttranscriptional silencing phenomena are two unpredictable processes that can arise separately or together as the consequence of the introduction of a transgene in a plant genome. Since the reliable expression of transgenes in transgenic plants is essential for biotechnological applications, it is important to rapidly discard from breeding programs those plants that are susceptible to silencing. Although it is now possible to apply some diagnostic tests to search for silencing susceptibility (e.g., the presence of multiple and/or methylated copies of the transgene), a large amount of work is still required to define precisely the various physiological, developmental, environmental, genetic, and molecular factors that control silencing in plants.

ACKNOWLEDGMENTS

We thank all the members of the European Network on Gene Silencing for fruitful discussion, and especially Marjori and Antonius Matzke for excellent collaboration. Thanks are also due to Ian Small and Rob Martienssen for correcting the manuscript.

REFERENCES

Barry, C., G. Faugeron, and J.-L. Rossignol. 1993. Methylation induced premeiotically in *Ascobolus*: Coextension with DNA repeat lengths and effect on transcript elongation. *Proc. Natl. Acad. Sci.* **90:** 4557–4561.

de Carvalho-Niebel, F., P. Frendo, M. Van Montagu, and M. Cornelissen. 1995. Posttranscriptional cosuppression of beta-1,3-glucanase genes does not affect accumulation of transgene nuclear mRNA. *Plant Cell* **7:** 347–358.

Dehio, C. and J. Schell. 1994. Identification of plant genetic loci involved in a posttranscriptional mechanism for meiotically reversible transgene silencing. *Proc. Natl. Acad. Sci.* **91:** 5538–5542.

Dorlhac de Borne, F., M. Vincentz, Y. Chupeau, and H. Vaucheret. 1994 Co-suppression of nitrate reductase host genes and transgenes in transgenic tobacco plants. *Mol. Gen. Genet.* **243:** 613–621.

Dougherty, W.G. and T.D. Parks. 1995. Transgenes and gene suppression: Telling us something new? *Curr. Opin. Cell Biol.* **7:** 399–405.

Elmayan, T. and H. Vaucheret. 1996. Expression of single copies of a strongly-expressed 35S-transgene can be silenced post-transcriptionally. *Plant J.* **9:** 787–797.

Flavell, R.B. 1994. Inactivation of gene expression in plants as a consequence of specific sequence duplication. *Proc. Natl. Acad. Sci.* **91:** 3490–3496.

Ingelbrecht, I., H. Van Houdt, M. Van Montagu, and A. Depicker. 1994. Posttranscriptional silencing of reporter transgenes in tobacco correlates with DNA methylation. *Proc. Natl. Acad. Sci.* **91:** 10502–10506.

Jorgensen, R. 1992. Silencing of plant genes by homologous transgenes. *Agric. Biotechnol. News* **4:** 265N–273N.

Kronenberger, J., A. Lépingle, M. Caboche, and H. Vaucheret. 1993. Cloning and expression of distinct nitrite reductases in tobacco leaves and roots. *Mol. Gen. Genet.* **236:** 203–208.

Linn, F., I. Heidmann, H. Saedler, and P. Meyer. 1990. Epigenetic changes in the expression of the maize A1 gene in *Petunia hybrida*: Role of the numbers of integrated gene copies and state of methylation. *Mol. Gen. Genet.* **222:** 329–336.

Matzke, A.J.M. and M.A. Matzke. 1994. *Trans*-inactivation of homologous sequences in *Nicotiana tabacum*. *Curr. Top. Microbiol. Immunol.* **197:** 1–14

Matzke, A.J.M., F. Neuhuber, Y.D. Park, P.F. Ambros, and M.A. Matzke. 1994. Homology-dependent gene silencing in transgenic plants: Epistatic silencing loci contain multiple copies of methylated transgenes. *Mol. Gen. Genet.* **244:** 219–229.

Matzke, M.A. and A.J.M. Matzke. 1995. How and why do plants inactivate homologous (*trans*)genes? *Plant. Physiol.* **107:** 679–685.

Matzke, M.A., F. Neuhuber, and A.J.M. Matzke. 1993. A variety of epistatic interactions can occur between partially homologous transgene loci brought together by sexual crossing. *Mol. Gen. Genet.* **236:** 379–386.

Matzke, M.A., M. Primig, J. Trnovsky, and A.J.M. Matzke. 1989. Reversible methylation and inactivation of marker genes in sequentially transformed tobacco plants. *EMBO J.* **8:** 643–649.

Meyer, P., I. Heidmann, and I. Niedenhof. 1993. Differences in DNA-methylation are associated with a paramutation phenomenon in transgenic petunia. *Plant J.* **4:** 89– 100.

Napoli, C., C. Lemieux, and R. Jorgensen. 1990. Introduction of a chimeric chalcone synthase gene into petunia results in reversible cosuppression of homologous genes *in trans*. *Plant Cell* **2:** 279–289.

Park, Y.-D., I. Papp, E.A. Moscone, V.A. Iglesias, H. Vaucheret, M.A. Matzke, and A.J.M. Matzke. 1996. Gene silencing mediated by promoter homology occurs at the level of transcription and results in meiotically heritable alterations in methylation and gene activity. *Plant J.* 183–194.

van Blokland, R., N. Vandergeest, J.N.M. Mol, and J.M. Kooter. 1994. Transgene-mediated suppression of chalcone synthase expression in *Petunia hybrida* results from an increase in RNA turnover. *Plant J.* **6:** 861–877.

van der Krol, A.R., L.A. Mur, M. Beld, J.N.M. Mol, and A.R. Stuitje. 1990. Flavonoid genes in petunia: Addition of a limited number of gene copies may lead to a suppression of gene expression. *Plant Cell* **2:** 291–299.

Vaucheret, H. 1993. Identification of a general silencer for 19S and 35S promoters in a transgenic tobacco plant: 90 bp of homology in the promoter sequences are sufficient for *trans*-inactivation. *C.R. Acad. Sci.* **316:** 1471–1483.

———. 1994. Promoter-dependent *trans*-inactivation in transgenic tobacco plants: Kinetic aspects of gene silencing and gene reactivation. *C.R. Acad. Sci.* **317:** 310–323.

Vaucheret, H., J.-C. Palauqui, T. Elmayan, and B. Moffatt. 1995. Molecular and genetic analysis of nitrite reductase cosuppression in transgenic tobacco plants. *Mol. Gen. Genet.* **248:** 311–317.

Vaucheret, H., J. Kronenberger, A. Lepingle, F. Vilaine, J.-P. Boutin, and M. Caboche. 1992. Inhibition of tobacco nitrite reductase activity by expression of antisense RNA. *Plant J.* **2:** 559–569.

Epigenetic Modifications and Gene Silencing in Plants

Frederick Meins, Jr.
The Friedrich Miescher Institute
CH-4002 Basel, Switzerland

During the development of plants, meristems and organ primordia become progressively committed to form specific and definite structures. This process, called determination, results from stable changes in phenotype that persist in the absence of the agent that originally induced the change. As a consequence, parts of an organism can "remember" their past, and this permits new formation of structures. Nevertheless, plants also show remarkable developmental plasticity. The shape and size of organs are strongly influenced by the environment, and plants have a pronounced capacity for regeneration. The developmental plasticity of determined structures illustrates the important point that stable changes are not necessarily irreversible. Determination is a relative process: Developmental states stable in one environment are not necessarily stable in some other environment (Meins and Wenzler 1986).

Little is known about the molecular mechanisms underlying stable reversible changes in plants. Recent studies show that multiple copies of transgenes introduced into plant cells can interact with one another or with homologous host genes *in trans*, resulting in the inactivation of expression of both genes. This phenomenon, called *silencing* (Jorgensen 1992) or *cosuppression* (Napoli et al. 1990), is of particular interest because it provides well-defined experimental systems for investigating the molecular basis for stable reversible changes. The aim of this chapter is to call attention to the similarities between stable changes in plant development and gene silencing and to propose general biochemical switch models that can account for many features of both forms of variation.

THE CONCEPT OF EPIGENETIC CHANGE

Stable Changes in Meristematic Activity

Stable, reversible changes are well documented in plant development. The plant life cycle consists of morphologically distinct haploid gametophytic and diploid sporophytic generations that are stable, self-perpetuating stages (Bell 1992). The sporophyte of higher plants is

Epigenetic Mechanisms of Gene Regulation

formed from root and shoot meristems established during early embryonic development. These meristems exhibit different developmental potencies that persist in populations of dividing cells (Ball 1952; Torrey 1954). The shoot meristem goes through successive stable changes in activity during the course of postembryonic development (Poethig 1990). Most higher plants exhibit a variable juvenile phase in which the shoot apex is incapable of forming reproductive structures. The activity of the shoot meristem then undergoes gradual, stable changes to generate the adult phase capable of reproductive growth. In the case of woody perennial species such as English ivy (*Hedera helix* L.) (Hackett et al. 1987), the morphologically distinct juvenile and adult phases persist in cuttings and phase-specific characters are maintained in cultured tissues, indicating that stability does not depend on the organized structure of the meristem.

The morphological transformation of the adult shoot meristem to an inflorescence meristem involves at least two successive stable changes. First, a state of competence for flowering is established that shows variable stability and depends on a variety of developmental and environmental factors (Bernier et al. 1981). In some cases, the flowering state can persist under noninductive conditions for the lifetime of the plant. At the other extreme, partial or complete reversion to the vegetative state can occur under noninductive conditions even after several whorls of the flower have formed (Krishnamoorthy and Nanda 1968; Battey and Lyndon 1984). In a second major stable change, the floral meristem is established and forms inflorescences and floral organs. These processes are regulated in a hierarchical fashion by homeotic genes encoding transcription factors (Weigel and Meyerowitz 1994; Mandel and Yanofsky 1995; Weigel and Nilsson 1995).

Stabilization at the Cellular and Supracellular Levels

Many specialized plant cell types have been shown to be totipotent (Narayanaswamy 1977). This raises the fundamental problem of how structures consisting of genetically equivalent cells can exhibit different stable states. One solution, which is favored by many botanists, is that stability is a supracellular phenomenon, i.e., it occurs primarily at the tissue and organ level (Henshaw et al. 1982). Detailed studies of cell lineage support this view: In general, there is no strict cell lineage in plants, and developmental compartments are not clonal (Poethig 1987).

Several mechanisms have been proposed for determination at the supracellular level. One possibility is that the spatial distribution of cells results from self-perpetuating patterns of division. For example, in the water fern *Azolla*, the root is formed from a single apical cell through a fixed cell lineage and a regular pattern of divisions (Gunning 1978). Depending on the division planes, two self-perpetuating enantiomorphic forms of roots develop. Another class of mechanisms first proposed by A.M. Turing (1952) is that regulatory interactions among cells can lead to stable patterns of cellular activity. This could involve diffusible signals such as the plant hormones auxin and cytokinin, which are known to induce stable patterns of differentiation when applied to plants (Sachs 1981). Most plant cells are connected by plasmodesmata, which provide regulated channels of cytoplasmic contact between cells (Lucas and Wolf 1993). Recent findings have shown that the product of the *Knotted* gene in maize, which is a transcription factor, can move from cell to cell via plasmodesmata, raising the intriguing possibility that transcription factors could play a role in supracellular determination (Lucas et al. 1995).

Stability could also result from mitotically transmissible cell determination. Organisms arise from a germ by *epigenesis*; i.e., the progressive formation of new structures, resulting from selective gene expression rather than from the sorting out of genetic determinants (Wardlaw 1970). Nevertheless, there are many well-documented examples of mitotic transmission of determination in animals (Stent 1985). This led Nanney (1958) to propose that there are two systems of inheritance: a genetic system concerned with the transmission of developmental potentialities between sexual generations of organisms, and an epigenetic (i.e., developmental) system concerned with the somatic transmission of patterns of gene expression. On the basis of similar considerations, Harris (1964) introduced the term *epigenetic change* to denote heritable cellular alterations that do not result from permanent genetic modifications and differ from classic mutations. This concept was later extended to variation in plant development by Braun and Meins (1970).

Criteria for Epigenetic Changes

Finding a workable definition for epigenetic changes acceptable to developmental biologists, geneticists, and molecular biologists is a challenging problem. Meins (1983) proposed an operational definition to distinguish between epigenetic changes and rare, random mutations in plants. Epigenetic changes are directed; i.e., they occur regularly in response to specific inducers. Under inductive conditions the rate of epigenetic variation is high, greater than 10^{-3} events per cell generation. In contrast, well-characterized gametic mutations in plants usually exhibit spontaneous rates of less than 10^{-5}. The variant phenotype, although mitotically transmissible, is potentially reversible, and the reversal process is directed and occurs at high rates. The range of phenotypes generated by epigenetic changes is limited by the genetic potential of the cell. Finally, by definition, epigenetic changes are not transmitted meiotically.

The specific test for epigenetic changes in totipotent plant cells has two steps. First, cells showing the original and variant phenotypes are cloned. If these states are inherited at the cellular level, then they should persist in cloned lines. Second, reversibility of the cell-heritable change is tested by regenerating complete plants from the clones and assaying tissues in the plants and their meiotic progeny for the variant trait. If the trait is regularly lost during development prior to meiosis, then one may conclude that the variant phenotype is likely to be the result of an epigenetic change.

Although reversion of a variant phenotype at high rates during development provides good evidence for epigenetic changes, many stable, reversible changes have now been identified that are meiotically transmitted. Examples include paramutation (Brink 1973; Hagemann and Berg 1977; Patterson and Chandler 1995) and presetting of transposable element activity (Fedoroff et al. 1989) in plants and genomic imprinting in animals (Latham et al. 1995). Holliday (1987) coined the term *epimutation* to denote epigenetic variation that is inherited through the germ line and results from modifications in the pattern of DNA methylation. In the discussion that follows, *epigenetic modification* is applied rather broadly for stable developmental phenomena. As recommended by Jorgensen (1993), epimutation is used without specifying an underlying mechanism for epigenetic changes that are transmitted meiotically.

EPIGENETIC CHANGES IN GROWTH FACTOR REQUIREMENT

There are few well-documented examples of epigenetic changes in the normal development of plants; e.g., apical-cell determination in the simple marine alga *Sphacelaria cirrosa* (Ducreux 1983), the determination of secretory-cell idioblasts in *Ricinus* (Bloch 1948), and certain forms of color variation in cultured carrot and red-beet cells (Dougall et al. 1980; Girod and Zryd 1987).

Compelling evidence for epigenetic changes comes from studies of cell-heritable variation in the requirement of cells for growth factors. Cultured plant tissues sometimes lose their exogenous requirement for specific growth factors and thereafter can be serially subcultured in the absence of the added factor. This phenomenon, called *habituation*, has been the subject of several recent reviews (Meins 1989a, 1994; Jackson and Lyndon 1990; Syono and Fujita 1994). Cultured explants of tobacco pith parenchyma require exogenous supplies of two growth hormones, cytokinin and auxin, for sustained growth on an otherwise complete medium. Cultures of pith tissue can give rise to variants that show either a cytokinin-habituated phenotype (cytokinin-independent, auxin-dependent, A^-C^+) or a fully habituated phenotype (cytokinin- and auxin-independent, A^+C^+). Cloning experiments have shown that these forms of habituation result from gradual, progressive alterations in cell heredity (Lutz 1966; Tandeau de Marsac and Jouanneau 1972; Binns and Meins 1973; Meins and Binns 1977).

The Nature of the Cell-heritable Change

Cytokinin habituation can be induced in primary explants of tobacco pith by heat treatment or by briefly incubating the tissues with trace amounts of cytokinins (Meins and Lutz 1980; Meins et al. 1980b). Under these conditions, some of the explants form dense, white clusters of rapidly dividing cells which are clearly discernible from the surrounding translucent tissue. Using these clusters as a marker, it was possible to estimate that the minimum rate of habituation, more than 5×10^{-3} per cell generation, is 100 to 1000 times faster than the rate of somatic mutation in *Nicotiana* species (Meins and Lutz 1980). Unlike classic mutations, the rate of cytokinin habituation is strongly dependent on physiological and developmental factors; e.g., the season when plants are sampled (Meins and Lutz 1980), the position of the pith tissue in the plant (Meins et al. 1980b; Turgeon 1982; Jackson and Lyndon 1988), and cell type (Meins and Lutz 1979; Meins et al. 1980a).

Normal, fertile plants can be regularly regenerated from clones of cytokinin-habituated pith cells (Binns and Meins 1973; Carlson 1979). Pith tissues from these regenerates, like untreated seed-grown plants, once again exhibit a nonhabituated phenotype, indicating that cytokinin habituation is a potentially reversible process. Finally, reversion to the nonhabituated state, like habituation, occurs at very high rates, about 4×10^{-3} per cell generation, during the regeneration process (Meins and Binns 1982). This observation, and the fact that habituation is extremely rapid and depends on both the physiological and developmental state of pith tissue, provide strong evidence that cytokinin habituation results from epigenetic changes rather than from rare, random mutations.

Epigenetic Variation in Competence for Cytokinin Habituation

Cells from different parts of the tobacco plant vary in cytokinin requirement and competence to habituate (Meins and Lutz 1979; Meins et al. 1980a). Primary pith tissue consists of *inducible* C^- cells that habituate rapidly under inductive conditions and *noninducible* C^- cells that do not habituate under these conditions. Cortex tissues of the stem consist of *constitutive* C^+ cells, and explants of leaf lamella consist of noninducible C^- cells.

The cell-cloning and plant-regeneration tests summarized in Figure 1 show that these phenotypes result from epigenetic changes. The great majority of clones (~90–95%) isolated from pith, leaf, and cortex tissues

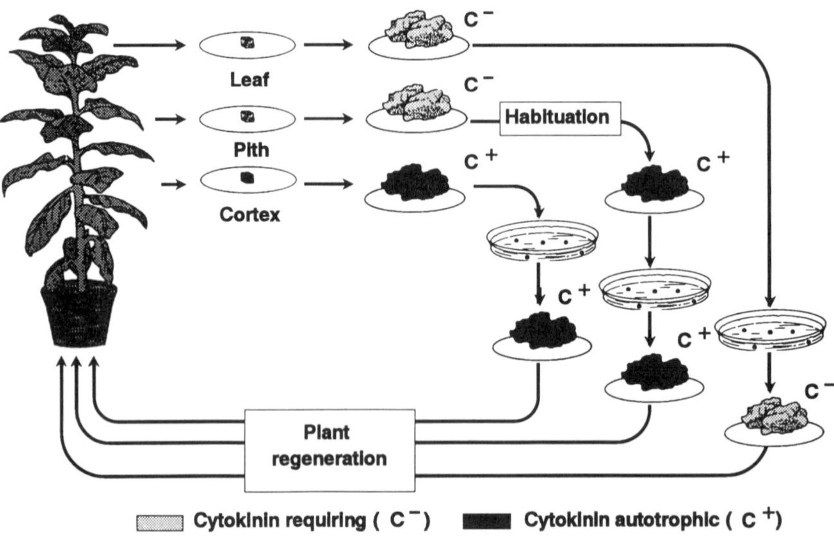

Figure 1 The plant-regeneration test used to demonstrate epigenetic changes in cytokinin requirement. Cultured tissue lines were started from leaf, stem cortex, and pith of a tobacco plant. The leaf and cortex lines showed stable C^- and C^+ phenotypes, respectively. Pith tissues exhibited habituation; i.e., the initially C^- tissues rapidly converted to the C^+ phenotype when subcultured on cytokinin-containing medium. Clones were obtained by plating single cells from the tissue lines in soft agar. Almost all of the clones expressed the cytokinin phenotype of the parent tissue lines. Complete, fertile tobacco plants were regenerated from the clones by incubating tissues on a shoot-inducing medium. Independent of the clonal origin of the regenerated plants, leaf, cortex, and pith tissues expressed the cytokinin phenotype of comparable tissues from seed-grown plants. These experiments show that the leaf-, cortex-, and pith-specific cytokinin phenotypes are cell heritable, but potentially reversible.

retain their tissue-specific phenotype, indicating that these states of cytokinin requirement are inherited by individual cells (Meins and Lutz 1979; Meins et al. 1980a). Regardless of the clone used, the pith, cortex, and leaf tissues of the regenerated plants exhibit the phenotype expected for comparable tissues from untreated, seed-grown plants (Meins and Foster 1985, 1986b). Thus, at some point during the plant regeneration process, different epigenetic states of cytokinin requirement are "reset" to give tissue types that express the "correct" phenotype.

Pseudodirected Variation

Cytokinin habituation of pith cells results from epigenetic changes in the expression of preexistent genetic potentialities. The C^+ phenotype can also result from genetic changes. Dominant genetic mutations at the *Habituated leaf* loci *Hl-1* and *Hl-2* have been found in which leaf and pith tissues show the constitutive C^+ phenotype of cortex tissues (Meins et al. 1983; Meins and Foster 1986a). Recent studies show that variation in cytokinin requirement can also occur in a "gray zone" between epigenetic change and classic genetic mutation (Meins and Seldran 1994).

The first hints came from experiments showing that an extremely high incidence of C^+ variants are in populations of C^- leaf cells cultured on suboptimal concentrations of cytokinins. To test the hypothesis that these variants arose by a directed process, a method was devised to measure rates of variation in growth factor requirement indirectly. Application of this method to numerous cloned lines of C^- cells gave the surprising result that both the forward and backward rates of variation were on the order of 10^{-2} per cell generation. Evidently, cultured cells are oscillating between the C^- and C^+ states at very high rates. This form of variation, called *pseudodirected variation*, occurs so rapidly that the distinction between random and induced events is blurred. Seemingly directed changes induced by cytokinins at the tissue level result from selection for the different cell types acting on an interconverting population of the two cell types.

The C^+ phenotype resulting from pseudodirected variation persists in regenerated plants and, when the plants are self-fertilized, is inherited at the *Hl-2* locus as a dominant Mendelian character (Meins and Seldran 1994). These findings, taken together with cell cloning and breeding studies with lines derived from homozygous *Hl-2*/*Hl-2* plants, provide strong evidence that, in contrast to habituation, pseudodirected variation can be meiotically transmitted (F. Meins, in prep.). Unlike most classic mutations, however, the meiotically transmissible C^+ state undergoes rapid reversion in successive sexual generations at frequencies up to 100%. Thus, it appears that pseudodirected variation has as its basis some form of epimutation.

EPIGENETIC CHANGES IN NEOPLASTIC DEVELOPMENT

Crown Gall and Hairy Root

Plants, like many animals, can form tumors capable of abnormal, autonomous growth. Epigenetic modifications regulate the development of plant tumors. In crown gall disease, tumors result when a portion of the Ti plasmid carried by *Agrobacterium tumefaciens*, the T-DNA, is integrated into the nuclear genome of the host cell (for review, see Binns and Thomashow 1988). Several genes in the T-DNA encode enzymes involved in the metabolism of auxins and cytokinins. Of particular importance are the loci *tms1* and *tms2*, which encode enzymes for indole-3-acetic acid synthesis, and the *ipt* locus encoding isopentenyl transferase, a key enzyme in cytokinin biosynthesis. Expression of these genes is required for the neoplastic character of cultured tumor cells, i.e., their capacity for autonomous growth and hormone autotrophic phenotype.

Although the neoplastic character of crown gall cells is stable and mitotically transmitted, it can also be lost (Braun and Wood 1976; Turgeon et al. 1976; Yang et al. 1980). Transformation of tobacco with certain strains of *A. tumefaciens* results in the formation of teratomas. These are highly organized tumors consisting of a chaotic mass of abnormal leaf- and stem-like structures. Cloning experiments have established that individual teratoma cells are multipotential and give rise to the tissue types and organs characteristic of the teratoma from which they were derived. Tumor suppression occurs when cloned teratoma tissues are grafted onto the cut-stem tip of tobacco plants from which the axillary buds have been removed. Under these conditions, the tissues develop shoots that become progressively more normal in function, gross morphology, and histological detail. These suppressed-tumor shoots still contain T-DNA (Yang et al. 1980), do not form roots, and are immune to *A. tumefaciens* transformation. Explants and even individual epidermal hair cells and leaf mesophyll cells regularly revert to the neoplastic phenotype when grown in culture (Binns et al. 1981; Huff and Turgeon 1981). Therefore, crown gall tumor cells have the capacity to differentiate, participate in organogenesis, and exhibit normal growth control while covertly retaining their neoplastic character.

Tumor reversal, which is a permanent loss of the neoplastic phenotype, occurs when suppressed-tumor shoots flower and set seed. Plant raised from the seeds are perfectly normal and have lost either all of their T-DNA or those parts of the T-DNA required for growth autonomy. These experiments argue strongly that the expression of the neoplastic state in crown gall depends on both genetic competence resulting from the genes introduced into host cells during tumor inception and the appropriate epigenetic state of the competent cells.

Similar conclusions may be drawn from studies of hairy-root disease caused by *Agrobacterium rhizogenes* Ri-plasmids (for review, see White and Sinkar 1987). Hairy-root tumors are highly organized root teratomas that result from the expression of T-DNA genes integrated into the host genome during tumor inception. When cultured, root teratomas of many plant species can be induced to form complete plants showing the hairy-root syndrome: reduced apical dominance of roots and stems; wrinkled, elongated leaves; high rates of root growth in culture; and a loss of root geotropism.

Hairy-root plants are analogous to suppressed-tumor shoots in crown gall. The important difference is that both the aberrant phenotype and T-DNA sequences are transmitted through seed. Hairy-root plants also can undergo stable changes in state. Tobacco plants occasionally give rise to lateral shoots showing either an exaggerated hairy-root phenotype or a revertant, normal phenotype. The revertant phenotype can be transmitted both in vegetative cuttings and meiotically; reversion results from the persistent inactivation of expression of T-DNA *rol* genes required for the hairy-root phenotype (Sinkar et al. 1988).

Genetic Hybrid Tumors

Hybrids formed between certain species in the genus *Nicotiana* are tumor-prone: They spontaneously form teratomatous outgrowths, most commonly on regions of the stem subject to mechanical stress (Bayer 1983; Ichikawa and Syono 1991). This phenotype depends on one or a small number of genes controlling tumor initiation contributed by one parent and more numerous genes influencing degree of tumor expression contributed by the other parent (Ahuja 1968; Smith 1988). In the case of *N. glauca* x *langsdorffii* hybrids (GGLL), explants cultured from the spontaneous teratomas show a fully habituated A^+C^+ phenotype, whereas normal-appearing tissues require an initial treatment with auxin to trigger habituation (Schaeffer and Smith 1963; Cheng and Smith 1973). Cultured tissues of both types have been shown to be tumorigenic when grafted onto the *N. glauca* parent, indicating that teratomatous as well as normal-appearing tissues from tumor-prone plants have the potential for neoplastic growth (White 1944; Kehr and Smith 1954).

The *N. glauca* x *langsdorffii* hybrid can be generated parasexually by fusing leaf protoplasts of the two parental species (Carlson et al. 1972; Smith et al. 1976). Cloned lines derived from the fusion products exhibit the teratomatous phenotype but can be induced to form shoots, which, when grafted onto *N. glauca* plants, develop normally, eventually flower, and set seed. Plants raised from the seed express the tumor-prone phenotype. Taken together, these findings show that cells of tumor-prone hybrids are totipotent and that tumor initiation and reversion result from epigenetic changes.

Possible Mechanisms for Epigenetic Changes in Growth Autonomy

The studies briefly summarized above have led to the concept that the conversion of normal cells to the fully autonomous state involves two key steps (Meins 1994). First, a competent or tumor-prone state is established by mutation, by plasmid-mediated transformation, by interspecific hybridization, or by a combination of these genetic events. Second, cells competent for neoplastic development must be triggered to express these potentialities in a stable fashion. These changes are potentially reversible and depend on epigenetic mechanisms that appear to regulate hormone autotrophic growth in normal development. Each tumor disease shows a distinctive pattern of "normal" and neoplastic phases ranging from fully autonomous, unorganized tumors to normal-appearing entire plants showing a tumor-prone phenotype.

The molecular basis for the alternation in normal and tumorous phases is not known. One possible mechanism discussed in detail below is that growth factors can activate a positive-feedback loop leading to persistent growth autonomy. Tobacco cells transformed with an *ipt*-deficient Ti-plasmid form fertile, normal-appearing plants (Binns 1983). Treatment with cytokinins, which induces cytokinin habituation of normal pith, induces the neoplastic phenotype in pith tissues of these *ipt*⁻ plants. Similar induction of persistent autonomous growth by hormones has been reported for GGLL hybrids (Cheng and Smith 1973).

Hypermethylation of cytosines in DNA appears to be important in the reversion of plant tumor cells. Crown gall tumors of tobacco, which grow as unorganized tissues in culture, can give rise to teratomatous variants capable of regenerating complete, normal-appearing plants (Amasino et al. 1984; John and Amasino 1989). Although the parent line, teratomatous variants, and regenerated plants carry at least one copy of T-DNA, they show persistent differences in T-DNA gene expression. Several T-DNA genes, including *tms1* and *tms2* but not *ipt*, are expressed in the parent line; only *ipt* is expressed in the teratomatous variants; and no T-DNA genes are expressed in the regenerated plants and their progeny. The important point is that the inactivated T-DNA genes show hypermethylation. When tissues derived from normal-appearing regenerates are treated with the methylation inhibitor 5-azacytidine, the genes are demethylated, T-DNA transcripts are detectable, and expression of the neoplastic phenotype is restored.

Although genes required for the autonomous growth of GGLL hybrids have not been identified, there is evidence that one of the genetic factors has an *ipt*-like function and is subject to hypermethylation. A nontumorous mutation of GGLL has been obtained by X-ray irradiation that appears to be inherited as a monogenic, dominant trait (Izard 1957; Smith and Stevenson 1961; Durante et al. 1982). The neoplastic phenotype of this mutant could be restored by transforming cultured cells with the Ti plasmid *ipt* gene (Feng et al. 1990) or by treating cells with 5-azacytidine, which also decreased the methylcytosine content of the bulk DNA (Durante et al. 1989).

GENE SILENCING

Silencing is a general phenomenon reported for several plant species transformed with a variety of different chimeric genes. For detailed information the reader is referred to the contributions of Jorgensen et al.; Matzke and Matzke; Meyer; and Vaucheret et al. in this volume and a recent collection of reviews (Meyer 1995a). The basic observation is that transgenes introduced into plant cells can interact with one another or with homologous host genes *in trans*, resulting in a marked inhibition of expression of the genes involved. This inhibition reflects a variable but substantial decrease in steady-state levels of mRNA. Silencing appears to be an epigenetic modification: It does not result from permanent changes in genetic constitution. Southern-blot analyses of genomic DNA from silent plants have shown that the transgene is still present and has not undergone conspicuous rearrangement or deletions. Moreover, silent genes can undergo spontaneous as well as developmentally regulated reactivation, by a process called *resetting* (Meins and Kunz 1995).

Stability

The triggering of silencing is usually a stochastic event. Only some plants in a population derived from the same transformation event show the silent phenotype, and variable patterns of silencing are sometimes observed in the same plant. Once established, the silent state shows some degree of stability. It often persists for long periods of time during vegetative growth of the transgenic plant.

Plants showing variable patterns of silencing provide insight into how the silent state is maintained in vegetatively growing plants. The diagnostic feature of mitotically transmitted stability is that it generates clonal patterns of variegation, which are well characterized (Poethig 1987). In contrast, patterns resulting from supracellular mechanisms such as cell-cell interactions are not generally clonal. Detailed analyses of flower-color patterns suggest that the silencing of *chalcone synthase* (*CHS*) genes in petunia is not clonal (Jorgensen 1995). Similar conclusions have been drawn for the distribution of silent leaves on *N. sylvestris* plants transformed with a *chitinase* (*CHN*) gene (Hart et al. 1992). Clonal patterns of silencing can also be found, for example in the case of a bacterial β-*glucuronidase* gene transformant of tobacco (Schmülling and Röhrig 1995). Thus, depending on the experimental system, it appears that both cell-cell interactions and mitotically transmitted stability can have a role in maintaining the silent state.

Heritability of the silent state is variable. Silencing dependent on the interaction of similar transcribed sequences and sensitive to transgene dose is not generally inherited in sexual crosses (see, e.g., Dehio and Schell 1994; Dorlhac de Borne et al. 1994; Vaucheret et al. 1995; de Carvalho et al. 1992; Hart et al. 1992; Meins and Kunz 1994). In other cases, the silent state shows some degree of transmission in sexual crosses and appears to be inherited as a dominant trait (Jorgensen 1993). The pattern of inheritance is often complex and consistent with the hypothesis that the silent phenotype is somewhat unstable in the hemizygous state (Meyer et al. 1993). There is also evidence that silent and active alleles at the transgene locus can interact, resulting in a heritable change in state reminiscent of paramutation (Matzke and Matzke 1993; Meyer et al. 1993; Matzke et al. 1994).

Developmental and Environmental Effects

Forms of silencing that are not transmitted meiotically are developmentally regulated (Meins and Kunz 1995). For a variety of transgenes regulated by the cauliflower mosaic virus 35S RNA promoter (*P35S*), plants initially show a high expressing phenotype before the onset of silencing (Smith et al. 1990; de Carvalho et al. 1992; Hart et al. 1992; Seymour et al. 1993; Dorlhac de Borne et al. 1994; Vaucheret et al. 1995; Dehio and Schell 1994). Several different transgenes regulated by *P35S* in tobacco, *Arabidopsis*, and *N. sylvestris* are silenced at roughly the same time during seedling development (de Carvalho et al. 1992; Dehio and Schell 1994; Dorlhac de Borne et al. 1994; Vaucheret et al. 1995). This suggests there is a critical period when silencing is triggered. Stability and incidence of silencing in sibling plants are also influenced by genetic factors of the host, transgene copy number, and the site of transgene integration (de Carvalho et al. 1992; Hart et al. 1992; Dorlhac de Borne et al. 1994; Vaucheret et al. 1995; Palauqui and Vaucheret 1995). In the few cases studied in detail, e.g., *P35S-CHN* in *N. sylvestris* and *P35S-rolB* in *Arabidopsis*, resetting of the silent state regularly occurs during seed development (Dehio and Schell 1994; Meins and Kunz 1994).

In addition to transgene and host factors, environment can have a strong influence on silencing. The incidence of silencing can vary dramatically depending on whether plants are raised in culture vessels under laboratory conditions, in a greenhouse, or in the field (Hart et al. 1992; Meyer et al. 1992; Dorlhac de Borne et al. 1994; Meins and Kunz 1994; Palauqui and Vaucheret 1995). There are also reports that silencing can be promoted by high light intensity (van der Krol et al. 1990) or even by transplantation of seedlings (Brandle et al. 1995).

Transcriptional *versus* Posttranscriptional Inhibition of Gene Activity

Silencing can act both transcriptionally and posttranscriptionally. Methylation of cytosines in the promoter region is known to inhibit transcription of plant genes and has been proposed as a mechanism for silencing (Matzke et al. 1989; Linn et al. 1990). In the case of *P35S-dihydroflavanol-4-reductase* (*DFR*) silencing in *DFR*-deficient petunia, the degree and incidence of silencing are roughly correlated with methylation of a particular site in *P35S*, and resetting is accompanied by decreased methylation at this site (Meyer et al. 1993). Runoff transcription studies confirmed that silencing of this transgene is at the level of transcription. On the other hand, runoff transcription studies have shown that silencing of several other transgenes in *Arabidopsis* (Dehio and Schell 1994), tobacco (de Carvalho et al. 1992; Ingelbrecht et al. 1994), and petunia (van Blokland et al. 1994) is at the posttranscriptional level. It is of interest that methylation of the transgene and posttranscriptional silencing are not mutually exclusive. In *P35S-neomycin transferase II* transformants of tobacco, increased methylation of the transgene is associated with silencing even though transcription of the transgene is not reduced (Ingelbrecht et al. 1994).

Transgenic tobacco expressing untranslatable transcripts of tobacco etch virus, potato virus Y, and potato virus X, which replicate in the cytoplasm, show resistance to infection with these viruses (Lindbo and Dougherty 1992; Smith et al. 1994; Mueller et al. 1995). Resistance is viral-gene sequence-specific and is associated with decreased steady-state levels of the transcripts. As judged from runoff transcription measurements, decreases in transcript accumulation are regulated by a posttranscriptional mechanism (Smith et al. 1994; Mueller et al. 1995). These intriguing findings suggest that posttranscriptional silencing might involve events occurring in the cytoplasm.

Possible Mechanisms

Silencing poses fundamental and challenging problems. How are genes with the same or similar sequences recognized by cells? What is the nature of the *trans*-acting signal or signals? How is the accumulation of specific transcripts inhibited? What stabilizes the silent state in a potentially reversible fashion? How are silencing and resetting developmentally regulated and influenced by the environment?

Several mechanisms for silencing have been proposed, including DNA methylation, production of antisense transcripts, ectopic gene pairing, chromatin modification, threshold effects, and biochemical switches (for reviews, see Matzke and Matzke 1995; Meyer 1995b). It should be emphasized that these mechanisms are not necessarily mutually exclusive and, at present, none can fully account for silencing. For example, DNA methylation provides the stability and potential heritability of the inactive state and could account for the inhibition of transcription. It does not, however, provide a solution to the more general problem of DNA sequence recognition and the nature of the *trans*-acting signal. Simple models based on production of antisense transcripts provide a plausible explanation for sequence specificity and posttranscriptional inactivation *in trans* but do not explain why the silent state is stable.

BIOCHEMICAL SWITCH MODELS FOR EPIGENETIC MODIFICATIONS

Biochemical switch models provide an attractive "systems" explanation for the stable, reversible behavior of epigenetic modifications without regard to a specific molecular mechanism. The underlying theory has been described in detail previously (Monod and Jacob 1961; Rosen 1972; Meins 1989b; Jablonka et al. 1992). In brief, certain systems of reactions and regulatory networks can exist in alternative, stable steady states. To illustrate this point, consider a simple system in which some metabolite induces its own biosynthesis in an autocatalytic fashion and is subject to degradation by a first-order process. Thus, the rates of metabolite synthesis (dC_S/dt), degradation (dC_D/dt), and net accumulation (dC_{net}/dt) are given by the expressions

$$dC_S/dt = F(C)$$

$$dC_D/dt = kC$$

$$dC_{net}/dt = dC_S/dt - dC_D/dt$$

where C is the concentration of the metabolite, F(C) is the rate of synthesis, and k is the degradation constant. Assuming that biosynthesis is subject to positive autoregulation, F(C) will be some type of sigmoid function of concentration.

Figure 2 shows a graphical representation of the dependence of degradation and synthesis on concentration. The points of intersection of the synthesis and degradation curves correspond to steady states in which $dC_{net}/dt = 0$. Where and how often the curves intersect depends on the kinetic constant. For examples, by varying the slope k of the degradation curve, three different types of behavior can be obtained: For large slopes (A), the curves intercept at one point corresponding to a stable "off" state in which the steady-state concentration is low. For small slopes (B), the curves intersect at one point corresponding to a stable "on" state in which the steady-state concentration is high. Finally, for intermediate slopes (C), the curves intersect three times. These intersections specify stable high and low states separated by an unstable intermediate state. Thus, given the proper kinetic parameters, this system can function as a reversible biochemical switch. Because the stability and variety of states depend on the kinetic parameters and, hence, ultimately on the genetic constitution of the cells, genetically identical cells or groups of cells can exhibit different stable states depending on their history.

Applications to Epigenetic Changes in Cytokinin Requirement

A biochemical switch model has been proposed for epigenetic changes in the cytokinin habituation of tobacco pith cells (Meins and Binns 1978; Meins 1989b). According to this model, the C^+ state is maintained by a

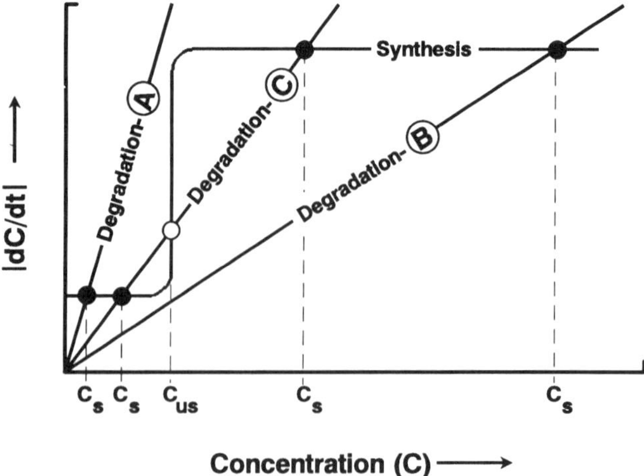

Figure 2 Graphical representation of a regulatory pathway showing biochemical switch behavior. The absolute values of metabolite synthesis and degradation rates ($|dC/dt|$) are plotted versus metabolite concentration (C). The system is at steady states where the synthesis and degradation curves intersect, i.e., when $dC_S/dt = dC_D/dt$ and $dC_{net}/dt = 0$. The concentrations corresponding to the stable (*filled circles*) and unstable (*open circles*) steady states are C_s and C_{us}.

positive feedback loop in which cytokinins, or physiologically related cell-division-promoting substances, induce their own biosynthesis and are degraded by a first-order process. It is further assumed that cells exhibit the C^+ phenotype because they have higher steady-state levels of cytokinins than do C^- cells.

Several predictions of this model have been experimentally verified. First, treating inducible C^- pith cells with the synthetic cytokinin kinetin induces cytokinin habituation. The low concentrations required—1000-fold less than is required for cell division—and sharp concentration threshold are expected for positive autoregulation (Meins and Lutz 1980). Second, C^+ pith cells shift back to the inducible C^- state under conditions that block cytokinin accumulation. As judged by cytokinin requirement and capacity to cross-feed neighboring cells, reducing the incubation temperature from 25°C to 16°C or reducing the levels of auxin in the culture medium inhibits cytokinin accumulation (Binns 1979; Binns and Meins 1979). When C^+ pith tissues are incubated under these nonpermissive conditions and then shifted to permissive conditions, the tissues exhibit an inducible C^- phenotype and revert to a C^+ phenotype when treated with cytokinin. Third, treatment of tobacco cells with synthetic cytokinins can induce the accumulation of naturally occurring cytokinins (Hansen et al. 1985; Vankova et al. 1987; Zhang et al. 1995).

The weakest point in the model is the relationship between cytokinin content and the capacity of cells for cytokinin-autotrophic growth. Although the literature is conflicting, the data suggest that C^+ cells have higher steady-state levels of zeatin riboside or functionally related cytokinins than do C^- cells (Hansen et al. 1987; Meins 1994). The available evidence strongly supports some type of positive autoregulation involving cytokinins, but it cannot be ruled out that cytokinins other than the ones measured, or an entirely different class of cell-division-promoting substances, are the key components of the feedback loop.

Applications to Gene Silencing

Biochemical switch models also provide a plausible explanation for stable reversible switching between high- and low-expressing states in gene silencing. This is illustrated by the heuristic model shown in Figure 3A for transcriptional forms of silencing that depend on the interaction of transcribed sequences *in trans* (Meins and Kunz 1995). The key assumptions of the model are: (1) Expression of a host gene (X) with its own promoter and a homologous transgene (X′) regulated by the 35S promoter is regulated by a diffusible activator. (2) The activator interacts with the transcribed regions of both genes and increases the production of complete X and X′ transcripts. (3) Induction of activator production depends on the combined concentration of X and X′ transcripts. Because the activator increases transcript concentration and the transcripts, in turn, induce the activator, a positive autoregulatory loop is generated. If the concentration of transcribed sequences X and X′ is limiting, then the synthesis rate will be a sigmoid function of RNA concentration. (4) RNA degradation is a first-order process.

The behavior of the system can be described by the synthesis and degradation curves shown in Figure 3B. To see how this system can explain silencing and resetting, imagine that X and X′ encode an enzyme essential for red pigmentation of flowers. If the gene is expressed in the untransformed plant, then flowers will be pink in color, corresponding to the stable, high-expressing state where the synthesis curve and degradation curve C intersect at position 1 (Fig. 3B). After transformation with X′, the dose of genes, and hence rate of transcription, are increased. Thus, in the transformants, the stable, high-expressing state for the same degradation rate C will be higher (position 2) and the flowers will be darker red in color.

Silencing in this system (i.e., white-colored flowers) can be triggered by perturbing the rates of either degradation or synthesis. Suppose, for example, that as a result of environmental influences or developmental regulation, there is a momentary increase in degradation rate (curve A) such that the RNA synthesis and degradation curves now intersect at just one point (position 3). Here the steady-state level of RNA is very low and the flowers will be white in color. After the perturbation, when the rate of degradation is restored to its original level (curve C), the system will shift from position 3 to position 4, which is also a low RNA state, and the flowers will remain white. The flowers remain in the silent state because both positions 3 and 4 correspond to RNA concentrations below

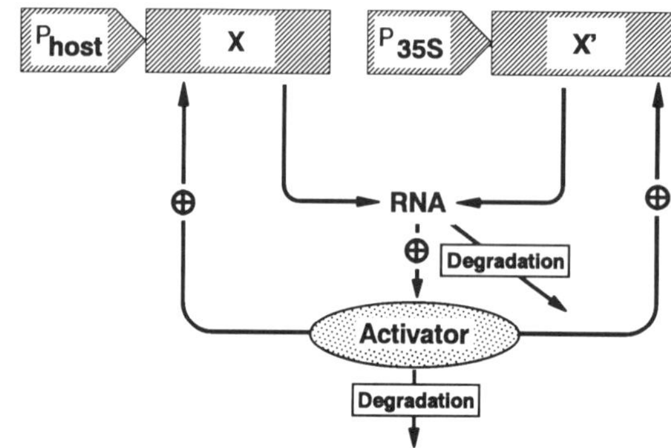

B.

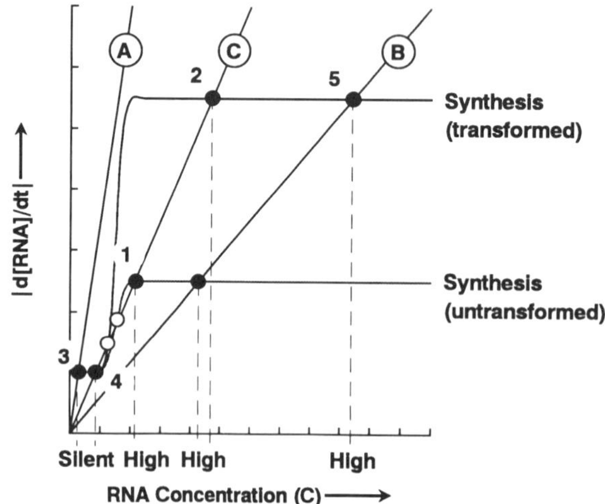

Figure 3 (*See facing page for legend.*)

the threshold required to return to the high-expressing state at position 2.

By a similar process, a momentary decrease in degradation rate (curve B) could trigger the stable resetting of silent cells to the high-expressing state. Curve B intercepts the synthesis curve once, at position 5, and there is no threshold for switching. Thus, the system will move from a stable silent state (position 4) to a stable high-expressing state (position 5) and the flowers will change in color from white to red. When the rate of degradation is restored to its original level (curve C), the system will shift from position 5 back to position 2, which is also a high-expressing state, and the flowers will remain red in color.

Related, somewhat more elaborate versions of this model have been proposed for multiple stable states and for posttranscriptional silencing (Meins and Kunz 1995). These models can account for several important features of silencing, including sequence recognition and specificity, which depend on the nature of the activator; *trans*-activation of multiple target genes, which results from interactions with the diffusible activator; and, alternative, stable high and silent states of target-gene expression, which result from the kinetic structure of the regulatory system. The models also provide a plausible explanation for developmental regulation and environmental effects often reported for silencing. The number of steady states, as well as their stability, depends on the rates of regulator synthesis and degradation. This in turn depends on transcription of the target genes and the stability of the transcripts. Thus, developmental regulation of the target genes or even momentary perturbation of their expression could result in stable, but potentially reversible, switching between the high-expressing and silent states.

Figure 3 Transcriptional activator (*A*) for silencing resulting from the interaction of the transcribed sequence (X) of a host gene regulated by its own promoter (P_{host}) with the transcribed sequence (X′) of a homologous transgene regulated by the 35S RNA promoter (P_{35S}). It is assumed that production of the diffusible activator depends on the combined concentration of X and X′ transcripts (RNA). Pathways for the first-order degradation as well as positive (+) and negative (–) feedback circuits are indicated. In the graphical representation of the regulatory pathway (*B*), the absolute rates and steady states are labeled as shown in Fig. 2. Stable states discussed in the text are numbered. It is assumed that in transgenic plants with a potential for silencing, the combined transcription of the host genes and transgenes results in a synthesis curve which intersects with degradation curve C to give alternative silent and high-expressing states. Perturbations of the system by increasing (curve A) or decreasing (curve B) degradation can trigger stable switching between the silent and high-expressing states. For illustration, it is assumed that states 3 and 4 give a white flower color, state 1 gives a pink flower color, and states 2 and 5 give a red flower color.

CONCLUDING REMARKS

Epigenetic modifications are the basis for the progressive, but plastic, nature of plant development. Although determination associated with organ formation is likely to be primarily at the supracellular level, there is also strong evidence for epigenetic changes affecting growth factor requirement and expression of neoplastic phenotypes. Cell-heritable changes in cytokinin requirement are of particular interest because cytokinins act in concert with auxins as primary morphogens and have a key function in tumorigenesis (Skoog and Miller 1957; Braun 1978). To account for the plasticity of plant development, Sachs (1988) proposed that mature structures are formed by epigenetic selection. According to this hypothesis, internal and environmental factors act to select a suitable developmental event from a repertoire of alternatives. Pseudodirected variation provides an attractive paradigm for epigenetic selection. Hormones acting on a population of rapidly switching cells could generate regions differing in hormone responsiveness and, hence, different developmental potentialities. This raises the intriguing possibility that gradients in hormone concentration might act by selection rather than by induction to give predictable patterns at the tissue and organ level.

At present, it is not clear whether gene silencing is an untoward side effect of genetic transformation of plants or reflects more fundamental mechanisms important in regulating development. Some forms of silencing show great similarity to epigenetic modifications encountered in normal and neoplastic development. There is also evidence that silencing of genes for MADS-box transcription factors important in floral determination results in homeotic transformations (Angenent et al. 1993, 1994) and that silencing in *P35S-CHS* transformants of petunia generates patterns of flower pigmentation described for certain spontaneous mutants (Mol et al. 1994).

The heuristic models presented here emphasize the belief that silencing and other types of epigenetic modification might have similar underlying mechanisms. Biochemical switch models provide a plausible explanation for stable, potentially reversible changes at the supragenetic level in response to diverse developmental and environmental signals. They are sufficiently flexible to account for states stable at the supracellular level that are not clonally transmitted, for mitotically transmitted epigenetic changes, and for meiotically transmitted epimutations.

There are several regulatory systems and metabolic pathways known in plants that might be candidates for switching mechanisms. For example, autocatalytic production of hormones has been reported for auxins, ethylene, and gibberellins (for review, see Meins and Wenzler 1986).

The photoinduced state can be transmitted in *Perilla* by serial grafting of induced leaves to uninduced plants, suggesting that autocatalytic production of a graft-transmissible, diffusible factor is involved (Zeevaart 1958). Positive autoregulation has been implicated in the stable, transcriptional regulation of flowering by MADS-box transcription factors in *Antirrhinum* (Schwarz-Sommer et al. 1992a,b; Halfter et al. 1994), in the stable regulation of *Spm* transposable-element activity in maize (Fedoroff et al. 1989), and in the differentiation of vascular tissue (Sachs 1981).

The specific models proposed for gene silencing are not totally unrealistic. For example, the transcribed regions of plant genes are known to contain elements that affect steady-state levels of the transcripts (Callis et al. 1987; An et al. 1989; Dean et al. 1989; Douglas et al. 1991). Although these models cannot account for certain features of silencing, e.g., dependence on gene copy number and silencing due to promoter-promoter interactions, they lead to a number of predictions amenable to experimental verification that might provide insight into silencing and the more general problem of epigenetic modification.

ACKNOWLEDGMENTS

Many fruitful discussions with other participants in the EC Network for Gene Silencing in Transgenic Plants and the original work described here were supported in part by grant No. 93.03992 from the Swiss Federal Office for Science and Education.

REFERENCES

Ahuja, M.R. 1968. An hypothesis and evidence concerning the genetic components controlling tumor formation in *Nicotiana*. *Mol. Gen. Genet.* **103:** 176–184.

Amasino, R.M., A.L.G. Powell, and M.P. Gordon. 1984. Changes in T-DNA methylation and expression are associated with phenotypic variation and plant regeneration in a crown gall tumor line. *Mol. Gen. Genet.* **197:** 437–446.

An, G., A. Mitra, H.K. Choi, M.A. Costa, K. An, R.W. Thornburg, and C.A. Ryan. 1989. Functional analysis of the 3′ control region of the potato wound-inducible protease inhibitor II gene. *Plant Cell* **1:** 115–122.

Angenent, G.C., J. Franken, M. Busscher, L. Colombo, and A.J. van Tunen. 1993. Petal and stamen formation in petunia is regulated by the homeotic gene *fbp1*. *Plant J.* **4:** 101–112.

Angenent, G.C., J. Franken, M. Busscher, D. Weiss, and A.J. van Tunen. 1994. Co-suppression of the petunia homeotic gene *fbp2* affects the identity of the generative meristem. *Plant J.* **5:** 33–44.

Ball, E. 1952. Morphogenesis of shoots after isolation of the shoot apex of *Lupinus albus*. *Am. J. Bot.* **39:** 167–191.

Battey, N.H. and R.F. Lyndon. 1984. Changes in apical growth and phyllotaxis on flowering and reversion in *Impatiens balsamina* L. *Ann. Bot.* **54:** 553–567.

Bayer, M.H. 1983. Genetic tumors: Physiological aspects of tumor formation in interspecific hybrids. In *Molecular biology of plant tumors* (ed. G. Kahl and J.S. Schell), pp. 33–67. Academic Press, New York.

Bell, P.R. 1992. Apospory and apogamy: Implications for understanding the plant life cycle. *Int. J. Plant Sci.* **153:** S123–S136.

Bernier, G., J.-M. Kinet, and R.M. Sachs. 1981. *The physiology of flowering.* CRC Press, Boca Raton, Florida.

Binns, A.N. 1979. "Habituation of tobacco pith cells for factors promoting cell division." Ph. D. thesis. Princeton University, Princeton, New Jersey.

———. 1983. Host and T-DNA determinants of cytokinin autonomy in tobacco cells transformed by *Agrobacterium tumefaciens. Planta* **158:** 272–279.

Binns, A. and F. Meins, Jr. 1973. Habituation of tobacco pith cells for factors promoting cell division is heritable and potentially reversible. *Proc. Natl. Acad. Sci.* **70:** 2660–2662.

———. 1979. Cold sensitive expression of cytokinin habituation by tobacco pith cells in culture. *Planta* **145:** 365–369.

Binns, A.N. and M.F. Thomashow. 1988. Cell biology of *Agrobacterium* infection and transformation of plants. *Annu. Rev. Microbiol.* **42:** 575–606.

Binns, A.N., H.N. Wood, and A.C. Braun. 1981. Suppression of the tumorous state in crown-gall teratomas of tobacco: A clonal analysis. *Differentiation* **19:** 97–102.

Bloch, R. 1948. The development of the secretory cells of *Ricinus* and the problem of cellular differentiation. *Growth* **12:** 271–284.

Brandle, J.E., S.G. McHugh, L. James, H. Labbé, and B.L. Miki. 1995. Instability of transgene expression in field grown tobacco carrying the *csr1-1* gene for sulfonylurea herbicide resistance. *Bio/Technology* **13:** 994–998.

Braun, A.C. 1978. Plant tumors. *Biochim. Biophys. ACTA* **516:** 167–191.

Braun, A.C. and F. Meins, Jr. 1970. The regulation of the expression of cellular phenotypes in crown-gall teratoma tissue of tobacco. *Symp. Int. Soc. Cell Biol.* **9:** 193–204.

Braun, A.C. and H.N. Wood. 1976. Suppression of the neoplastic state with the acquisition of specialized function in cells, tissues and organs of crown gall teratomas of tobacco. *Proc. Natl. Acad. Sci.* **73:** 496–500.

Brink, R.A. 1973. Paramutation. *Annu. Rev. Genet.* **7:** 129–152.

Callis, J., M. Fromm, and V. Walbot. 1987. Introns increase gene expression in cultured maize cells. *Genes Dev.* **1:** 1183–1200.

Carlson, P.S. 1979. Peptides of normal and variant cells of tobacco. *Dev. Genet.* **1:** 3–12.

Carlson, P.S., H.H. Smith, and R.D. Dearing. 1972. Parasexual interspecific plant hybridization. *Proc. Natl. Acad. Sci.* **69:** 2292–2294.

Cheng, T.-Y. and H.H. Smith. 1973. The influence of genomes on the autonomous growth of pith cultures of *Nicotiana glauca-langsdorffii* hybrids. *Planta* **113:** 29–34.

Dean, C., C.M. Favreau, D. Bond-Nutter, J. Bedbrook, and P. Dunsmuir. 1989. Sequences downstream of translation start regulate quantitative expression of two petunia *rbcS* genes. *Plant Cell* **1:** 201–208.

de Carvalho, F., G. Gheysen, S. Kushnir, M. van Montagu, D. Inzé, and C. Castresana. 1992. Suppression of β-1,3-glucanase transgene expression in homozygous plants. *EMBO J.* **11:** 2595–2602.

Dehio, C. and J. Schell. 1994. Identification of plant genetic loci involved in a post-transcriptional mechanism for meiotically reversible transgene silencing. *Proc. Natl. Acad. Sci.* **91:** 5538–5582.

Dorlhac de Borne, F., M. Vincentz, Y. Chupeau, and H. Vaucheret. 1994. Co-suppression of nitrate reductase host genes and transgenes in transgenic tobacco plants. *Mol. Gen. Genet.* **243:** 613–621.

Dougall, D.K., J.M. Johnson, and G.H. Whitten. 1980. A clonal analysis of anthocyanin accumulation by cell cultures of wild carrot. *Planta* **149:** 292–297.

Douglas, C.J., K.D. Hauffe, M.-E. Ites-Morales, M. Ellard, U. Paszkowski, K. Hahlbrock, and J.L. Dangl. 1991. Exonic sequences are required for elicitor and light activation of a plant defense gene, but promoter sequences are sufficient for tissue specific expression. *EMBO J.* **10:** 1767–1775.

Ducreux, G. 1983. Isolement expérimental des cellules terminales de l'apex de *Sphacelaria cirrosa* (Roth) C. Agardh (Sphacelariales, Phéophycées) et analyse comparée de leurs potentialités morphogénétiques. *Phycologia* **22:** 415–429.

Durante, M., E. Cecchini, L. Natali, L. Citti, C. Geri, R. Parenti, and V.N. Ronchi. 1989. 5-Azacytidine induced tumorous transformation and DNA hypomethylation in *Nicotiana* tissue cultures. *Dev. Genet.* **10:** 298–303.

Durante, M., C. Geri, M. Buiatti, S. Baroncelli, R. Parenti, V.N. Ronchi, G. Martini, F.C. Grenci, J. Grisvard, and E. Guillé. 1982. DNA heterogeneity and genetic control of tumorigenesis in *Nicotiana* tumorous and nontumorous genetypes. *Dev. Genet.* **3:** 25–39.

Fedoroff, N., P. Masson, and J.-A. Banks. 1989. Mutations, epimutations, and the developmental programming of the maize *suppressor-mutator* transposable element. *BioEssays* **10:** 139–144.

Feng, X.-H., S.K. Dube, P.J. Bottino, and S.-D. Kung. 1990. Restoration of shooty morphology of a nontumorous mutant of *Nicotiana glauca* X *N. langsdorffii* by cytokinin and the isopentenyltransferase gene. *Plant Mol. Biol.* **15:** 407–420.

Girod, P.-A. and J.-P. Zryd. 1987. Clonal variability and light induction of betalain synthesis in red beet cell cultures. *Plant Cell Rep.* **6:** 27–30.

Gunning, B.E.S. 1978. Formative and proliferative cell divisions, cell differentiation and developmental changes in the meristem of *Azolla* roots. *Planta* **143:** 121.

Hackett, W.P., R.E. Cordero, and C. Srinivasan. 1987. Apical meristem characteristics and activity in relation to juvenility in *Hedera*. In *Manipulation of flowering* (ed. J.G. Atherton), pp. 93–99. Butterworth, London.

Hagemann, R. and W. Berg. 1977. Vergleichende Analyse der Paramutationssysteme bei höheren Pflanzen. *Biol. Zentralblatt* **96:** 257–301.

Halfter, U., N. Ali, J. Stockhaus, L. Ren, and N.-H. Chua. 1994. Ectopic expression of a single homeotic gene, the *Petunia* gene *green petal*, is sufficient to convert sepals to petaloid organs. *EMBO J.* **13:** 1443–1449.

Hansen, C.E., F. Meins, Jr., and R. Aebi. 1987. Hormonal regulation of zeatin-riboside accumulation by cultured tobacco cells. *Planta* **172:** 520–525.

Hansen, C.E., F. Meins, Jr., and A. Milani. 1985. Clonal and physiological variation in the cytokinin content of tobacco-cell lines differing in cytokinin requirement and capacity for neoplastic growth. *Differentiation* **29:** 1–6.

Harris, M. 1964. *Cell culture and somatic variation.* Holt, Rinehart and Winston, New York.

Hart, C.M., B. Fischer, J.-M. Neuhaus, and F. Meins, Jr. 1992. Regulated inactivation of homologous gene expression in transgenic *Nicotiana sylvestris* plants containing a defense-related tobacco chitinase gene. *Mol. Gen. Genet.* **235:** 179–188.

Henshaw, G.G., J.F. O'Hara, and K.J. Webb. 1982. Morphogenetic studies in plant tissue cultures. *Brit. Soc. Cell Biol. Symp.* **4:** 231–251.

Holliday, R. 1987. The inheritance of epigenetic defects. *Science* **238:** 163–170.

Huff, M. and R. Turgeon. 1981. Neoplastic potential of trichomes isolated from tobacco crown gall teratomas. *Differentiation* **19:** 93–96.

Ichikawa, T. and K. Syono. 1991. Tobacco genetic tumors. *Plant Cell Physiol.* **32:** 1123–1128.

Ingelbrecht, I., H. van Houdt, M. van Montagu, and A. Depicker. 1994. Posttranscriptional silencing of reporter transgenes in tobacco correlates with DNA methylation. *Proc. Natl. Acad. Sci.* **91:** 10502–10506.

Izard, C. 1957. Obtention et fixation de lignées tumorales et non tumorales a partir de mutations expérimentales de l'hybride *N. glauca* X *N. langsdorffii*. *C. R. Acad. Agric.* **43:** 325–327.

Jablonka, E., M. Lachmann, and M.J. Lamb. 1992. Evidence, mechanisms and models for the inheritance of acquired characters. *J. Theor. Biol.* **158:** 245–268.

Jackson, J.A. and R.F. Lyndon. 1988. Cytokinin habituation in juvenile and flowering tobacco. *Plant Physiol.* **132:** 575–579.

———. 1990. Habituation: Cultural curiosity or developmental determinant? *Physiol. Plant.* **79:** 579–583.

John, M.C. and R.M. Amasino. 1989. Extensive changes in DNA methylation patterns accompany activation of a silent T-DNA ipt gene in *Agrobacterium tumefaciens*-transformed plant cells. *Mol. Cell Biol.* **9:** 4298–4303.

Jorgensen, R. 1992. Silencing of plant genes by homologous transgenes. *Agric. Biotechnol. News* **4:** 265N–273N.

———. 1993. The germinal inheritance of epigenetic information in plants. *Philos. Trans. R. Soc. Lond. B Biol. Sci.* **339:** 173–181.

———. 1995. Cosuppression, flower color patterns, and metastable gene expression states. *Science* **268:** 686–691.

Kehr, A.E. and H.H. Smith. 1954. Genetic tumors in *Nicotiana* hybrids. *Brookhaven Symp. Biol.* **6:** 55–76.

Krishnamoorthy, H.N. and K.K. Nanda. 1968. Floral bud reversion in *Impatiens balsamina* under non-inductive photoperiods. *Planta* **80:** 43–51.

Latham, K.E., J. McGrath, and D. Solter. 1995. Mechanistic and developmental aspects of genetic imprinting in mammals. *Int. Rev. Cytol.* **160:** 53–98.

Lindbo, J.A. and W.G. Dougherty. 1992. Untranslatable transcripts of the tobacco etch virus coat protein gene can interfere with tobacco etch virus replication in transgenic plants and protoplasts. *Virology* **189:** 725–733.

Linn, F., I. Heidmann, H. Saedler, and P. Meyer. 1990. Epigenetic changes in the expression of the maize *A1* gene in *Petunia hybrida*: Role of numbers of integrated gene copies and state of methylation. *Mol. Gen. Genet.* **222:** 329–336.

Lucas, W.J. and S. Wolf. 1993. Plasmodesmata: The intercellular organelles of green plants. *Trends Cell Biol.* **3:** 308–315.

Lucas, W.J., S. Bouché-Pillon, D.P. Jackson, L. Nguyen, L. Baker, B. Ding, and S. Hake. 1995. Selective trafficking of KNOTTED1 homeodomain protein and its mRNA through plasmodesmata. *Science* **270:** 1980–1983.

Lutz, A. 1966. Obtention de plantes de tabac a partir de cultures unicellulaires provenant d'une souche anergée. *C. R. Acad. Sci.* **262:** 1856–1858.

Mandel, M.A. and M.F. Yanofsky. 1995. A gene triggering flower formation in *Arabidopsis. Nature* **377:** 522–524.

Matzke, M. and A.J.M. Matzke. 1993. Genomic imprinting in plants: Parental effects and *trans*-inactivation phenomena. *Annu. Rev. Plant Physiol. Plant Mol. Biol.* **44:** 53–76.

———. 1995. How and why do plants inactivate homologous (trans)genes? *Plant Physiol.* **107:** 679–685.

Matzke, M.A., M. Primig, J. Trnovsky, and A.J.M. Matzke. 1989. Reversible methylation and inactivation of marker genes in sequentially transformed tobacco plants. *EMBO J.* **8:** 643–649.

Matzke, A.J.M., F. Neuhuber, Y.-D. Park, P.F. Ambros, and M.A. Matzke. 1994. Homology-dependent gene silencing in transgenic plants: Epistatic silencing loci contain multiple copies of methylated transgenes. *Mol. Gen. Genet.* **244:** 219–229.

Meins, F. Jr. 1983. Heritable variation in plant cell culture. *Annu. Rev. Plant Physiol.* **34:** 327–346.

———. 1989a. Habituation: Heritable variation in the requirement of cultured plant cells for hormones. *Annu. Rev. Genet.* **23:** 395–408.

———. 1989b. A biochemical switch model for cell-heritable variation in cytokinin requirement. In *Molecular basis of plant development* (ed. R. Goldberg), pp. 13–24. A.R. Liss, New York.

———. 1994. Habituation of cultured cells for cytokinins. In *Cytokinins: Chemistry, activity, and function* (ed. D.W.S. Mok and M.C. Mok), pp. 269–287. CRC Press, Boca Raton, Florida.

Meins, F., Jr. and A.N. Binns. 1977. Epigenetic variation of cultured somatic cells: Evidence for gradual changes in the requirement for factors promoting cell division. *Proc. Natl. Acad. Sci.* **74:** 2928–2932.

———. 1978. Epigenetic clonal variation in the requirement of plant cells for cytokinins. In *The clonal basis for development* (ed. S. Subtelny and I.M. Sussex), pp. 185–201. Academic Press, New York.

———. 1982. Rapid reversion of cell-division-factor habituated cells in culture. *Differentiation* **23:** 10–12.

Meins, F., Jr. and R. Foster. 1985. Reversible, cell-heritable changes during the development of tobacco pith tissues. *Dev. Biol.* **108:** 1–5.

———. 1986a. A cytokinin mutant derived from cultured tobacco cells. *Dev. Genet.* **7:** 159–165.

———. 1986b. Transdetermination of plant cells. *Differentiation* **30:** 188–189.

Meins, F., Jr. and C. Kunz. 1994. Silencing of chitinase expression in transgenic plants: An autoregulatory model. In *Homologous recombination and gene silencing in plants* (ed. J. Paszkowski), pp. 335–348. Kluwer Academic, Dordrecht, The Netherlands.

———. 1995. Gene silencing in transgenic plants: A heuristic autoregulation model. *Curr. Top. Microbiol. Immunol.* **197:** 105–120.

Meins, F., Jr. and J. Lutz. 1979. Tissue specific variation in the cytokinin habituation of cultured tobacco cells. *Differentiation* **15:** 1–6.

———. 1980. The induction of cytokinin habituation in primary pith explants of tobacco. *Planta* **149:** 402–407.

Meins, F., Jr. and M. Seldran. 1994. Pseudodirected variation in the requirement of cultured plant cells for cell-division factors. *Development* **120:** 1163–1168.

Meins, F., Jr. and H. Wenzler. 1986. Stability of the determined state. *Symp. Soc. Expl.*

Biol. **40:** 155–170.

Meins, F., Jr., R. Foster, and J.D. Lutz. 1983. Evidence for a Mendelian factor controlling the cytokinin requirement of cultured tobacco cells. *Dev. Genet.* **4:** 129–141.

Meins, F., Jr., J. Lutz, and A.N. Binns. 1980a. Variation in the competence of tobacco pith cells for cytokinin habituation in culture. *Differentiation* **16:** 71–75.

Meins, F., Jr., J. Lutz, and R. Foster. 1980b. Factors influencing the incidence of habituation for cytokinin of tobacco pith tissue in culture. *Planta* **150:** 264–268.

Meyer, P. 1995a. *Gene silencing in higher plants and related phenomena in other eukaryotes.* Springer-Verlag, Heidelberg.

———. 1995b. Understanding and controlling transgene expression. *Trends Biotechnol.* **13:** 332–337.

Meyer, P., I. Heidmann, and I. Neidenhof. 1993. Differences in DNA-methylation are associated with a paramutation phenomenon in transgenic petunia. *Plant J.* **4:** 89–100.

Meyer, P., F. Linn, I. Heidmann, H. Meyer, I. Niedenhof, and H. Saedler. 1992. Endogenous and environmental factors influence 35S promoter methylation of a maize A1 gene construct in transgenic petunia and its colour phenotype. *Mol. Gen. Genet.* **231:** 345–352.

Mol, J.N.M., R. van Blokland, P. de Lange, M. Stam, and J.M. Kooter. 1994. Post-transcriptional inhibition of gene expression: Sense and antisense genes. In *Gene inactivation and homologous recombination in plants* (ed. J. Paszkowski), pp. 309–334. Kluwer Academic, Dordrecht, The Netherlands.

Monod, J. and F. Jacob. 1961. General conclusions: Telenomic mechanisms in cellular metabolism, growth and differentiation. *Cold Spring Harbor Symp. Quant. Biol.* **26:** 389–401.

Mueller, E., J. Gilbert, G. Davenport, G. Brigneti, and D.C. Baulcombe. 1995. Homology-dependent resistance: Transgenic virus resistance in plants related to homology-dependent gene silencing. *Plant J.* **7:** 1001–1013.

Nanney, D.L. 1958. Epigenetic control systems. *Proc. Natl. Acad. Sci.* **44:** 712–717.

Napoli, C., C. Lemieux, and R. Jorgensen. 1990. Introduction of a chimeric chalcone synthase gene into petunia results in reversible co-suppression of homologous genes *in trans*. *Plant Cell* **2:** 279–289.

Narayanaswamy, S. 1977. Regeneration of plants from tissue cultures. In *Plant cell, tissue and organ culture* (ed. J. Reinert and Y.P.S. Bajaj), pp. 179–248. Springer-Verlag, Heidelberg.

Palauqui, J.-C. and H. Vaucheret. 1995. Field trial analysis of nitrate reductase cosuppression: A comparative study of 38 combinations of transgene loci. *Plant Mol. Biol.* **29:** 149–159.

Patterson, G.I. and V.L. Chandler. 1995. Paramutation in maize and related allelic interactions. *Curr. Top. Microbiol. Immunol.* **197:** 121–141.

Poethig, R.S. 1987. Clonal analysis of cell lineage patterns in plant development. *Am. J. Bot.* **74:** 581–594.

———. 1990. Phase change and the regulation of shoot morphogenesis in plants. *Science* **250:** 923–930.

Rosen, R. 1972. Mechanics of epigenetic control. In *Foundations of mathematical biology* (ed. R. Rosen), pp. 79–140. Academic Press, New York.

Sachs, T. 1981. The control of patterned differentiation of vascular tissues. *Adv. Bot. Res.* **9:** 151–262.

———. 1988. Epigenetic selection: An alternative mechanism of pattern formation. *J.*

Theor. Biol. **134:** 547–559.

Schaeffer, G.W. and H.H. Smith. 1963. Auxin-kinetin interaction in tissue cultures of Nicotiana species and tumor-conditioned hybrids. *Plant Physiol.* **38:** 291–297.

Schmülling, T. and H. Röhrig. 1995. Gene silencing in transgenic tobacco hybrids: Frequency of the event and visualization of somatic inactivation pattern. *Mol. Gen. Genet.* **249:** 375–390.

Schwarz-Sommer, Z., H. Saedler, and H. Sommer. 1992a. Homeotic genes in the genetic control of flower morphogenesis in *Antirrhinum majus*. In *Development: The molecular genetic approach* (ed. V.E.A. Russo et al.), pp. 242–256. Springer-Verlag, Berlin.

Schwarz-Sommer, Z., I. Hue, P. Huijser, P.J. Flor, R. Hansen, F. Tetens, W.E. Loennig, H. Saedler, and H. Sommer. 1992b. Characterization of the Antirrhinum floral homeotic MADS-box gene deficiens. Evidence for DNA binding and autoregulation of its persistent expression throughout flower development. *EMBO J.* **11:** 251–264.

Seymour, G.B., R.G. Fray, P. Hill, and G.A. Tucker. 1993. Down regulation of two non-homologous endogenous tomato genes with a single chimeric sense gene construct. *Plant Mol. Biol.* **23:** 1–9.

Sinkar, V.P., F.F. White, I.J. Furner, M. Abrahamsen, F. Pythoud, and M.P. Gordon. 1988. Reversion of aberrant plants transformed with *Agrobacterium rhizogenes* is associated with the transcriptional inactivation of the T_L-DNA genes. *Plant Physiol.* **86:** 584–590.

Skoog, F. and C.O. Miller. 1957. Chemical regulation of growth and organ formation in plant tissues cultured *in vitro*. *Symp. Soc. Exp. Biol.* **11:** 118–131.

Smith, C.J.S., C.F. Watson, C.R. Bird, J. Ray, W. Schuch, and D. Grierson. 1990. Expression of a a truncated tomato polygalacturonase gene inhibits expression of the endogenous gene in transgenic plants. *Mol. Gen. Genet.* **224:** 477–481.

Smith, H.A., S.L. Swaney, T.D. Parks, E.A. Wernsman, and W.G. Dougherty. 1994. Transgenic plant virus resistance mediated by untranslatable sense RNAs: Expression, regulation, and fate of nonessential RNAs. *Plant Cell* **6:** 1441–1453.

Smith, H.H. 1988. The inheritance of genetic tumors in *Nicotiana* hybrids. *J. Hered.* **79:** 277–283.

Smith, H.H. and H.Q. Stevenson. 1961. Genetic control and radiation effects in *Nicotiana* tumors. *Z. Vererbungsl.* **92:** 100–118.

Smith, H.H., K.M. Kao, and N.C. Combatti. 1976. Interspecific hybridization by protoplast fusion in *Nicotiana*. *J. Hered.* **67:** 123–128.

Stent, G.S. 1985. The role of cell lineage in development. *Philos. Trans. R. Soc. Lond. B Biol. Sci.* **312:** 3–19.

Syono, K. and T. Fujita. 1994. Habituation as a tumorous state that is interchangeable with a normal state in plant cells. *Int. Rev. Cytol.* **152.** 265–299.

Tandeau de Marsac, N. and J.-P. Jouanneau. 1972. Variation de l'exigence en cytokinine de lignées clonales de cellules de tabac. *Physiol. Veg.* **10:** 369–380.

Torrey, J.G. 1954. The role of vitamines and micronutrient elements in the nutrition of the apical meristem of pea roots. *Physiol. Plant.* **29:** 265–275.

Turgeon, R. 1982. Cytokinesis, cell expansion, and the potential for cytokinin-autonomous growth in tobacco pith. *Plant Physiol.* **70:** 1071–1074.

Turgeon, R., H.N. Wood, and A.C. Braun. 1976. Studies on the recovery of crown gall tumor cells. *Proc. Natl. Acad. Sci.* **73:** 3562–3564.

Turing, A.M. 1952. The chemical basis of morphogenesis. *Philos. Trans. R. Soc. Lond. B Biol. Sci.* **237:** 5–72.

van Blokland, R., N. van der Geest, J.N.M. Mol, and J.M. Kooter. 1994. Transgene-mediated suppression of chalcone synthase expression in *Petunia hybrida* results from an increase in RNA turnover. *Plant J.* **6:** 861–877.

van der Krol, A.R., L.A. Mur, M. Beld, J.N.M. Mol, and A.R. Stuitje. 1990. Flavonoid genes in petunia: Addition of a limited number of gene copies may lead to a suppression of gene expression. *Plant Cell* **2:** 291–299.

Vankova, R., M. Kaminek, J. Eder, and T. Vanek. 1987. Dynamics of production of *trans*-zeatin and *trans*-zeatin riboside by immobilized cytokinin-autonomous and cytokinin-dependent tobacco cells. *J. Plant Growth Regul.* **6:** 147–157.

Vaucheret, H., J.-C. Palauqui, T. Elmayan, and B. Moffatt. 1995. Molecular and genetic analysis of nitrite reductase co-suppression in transgenic tobacco plants. *Mol. Gen. Genet.* **248:** 311–317.

Wardlaw, C.W. 1970. *Cellular differentiation in plants and other essays.* Manchester University Press, Manchester.

Weigel, D. and E.M. Meyerowitz. 1994. The ABCs of floral homeotic genes. *Cell* **78:** 203–209.

Weigel, D. and O. Nilsson. 1995. A developmental switch sufficient for flower initiation in diverse plants. *Nature* **377:** 495.

White, F.F. and V.P. Sinkar. 1987. Molecular analysis of root induction by *Agrobacterium rhizogenes.* In *Plant DNA infectious agents* (ed. T. Hohn and J. Schell), pp. 149–177. Springer-Verlag, Vienna.

White, P.R. 1944. Transplantation of plant tumors of genetic origin. *Cancer Res.* **4:** 791–794.

Yang, F.-M., A. Montoya, D.J. Merlo, M.H. Drummond, M.-D. Chilton, E.W. Nester, and M.P. Gordon. 1980. Foreign DNA sequences in crown gall teratomas and their fate during the loss of the tumorous trait. *Mol. Gen. Genet.* **177:** 707–714.

Zeevaart, J.A.D. 1958. Flower formation as studied by grafting. *Meded. Landbouwhoeschool Wageningen* **58:** no. 3.

Zhang, R., X. Zhang, J. Wang, D.S. Letham, S.A. McKinney, and T.J.V. Higgins. 1995. The effect of auxin on cytokinin levels and metabolism in transgenic tobacco tissue expressing an *ipt* gene. *Planta* **196:** 84–94.

Transcriptional Silencing in the Fission Yeast: A Manifestation of Higher Order Chromosome Structure and Functions

Robin C. Allshire
MRC Human Genetics Unit
Western General Hospital
Edinburgh EH4 2XU,
Scotland, United Kingdom

Centromeres and telomeres perform a variety of important functions during the cell cycle. The *cis*-acting centromeric DNA nucleates the assembly of a large protein complex, the kinetochore, which performs several functions crucial for the faithful transmission of chromosomes to daughter nuclei. Sister chromatids remain tethered at the kinetochore until the transition from metaphase to anaphase. During prometaphase, sister kinetochores must capture microtubules emanating from opposite spindle poles, and only when all sister kinetochores on all chromosomes have achieved bilateral attachment can anaphase be triggered by dissolution of the link between sister kinetochores (Gorbsky 1995). Failure of centromere/kinetochore function leads to aberrant segregation of chromosomes due to nondisjunction or premature separation of sister chromatids, resulting in chromosome loss and gain. Since telomeres are required to protect the ends of linear chromosomes against replicative diminution, exonucleolytic degradation, and end-to-end fusion, loss of telomere function also results in the destabilization of chromosomes (Sandell and Zakian 1993).

In many organisms, both centromeres and telomeres are associated with large protein complexes that are frequently found to contact the nuclear periphery during interphase (Karpen 1994; Gilson et al. 1995). In addition, these chromosomal regions tend to be heterochromatic, transcriptionally silent, suppressed for recombination, and late-replicating (Karpen 1994; Palladino and Gasser 1994; Gilson et al. 1995). When reporter genes are placed within or adjacent to these regions, they regularly become subject to a form of epigenetic regulation. Although all cells in a population are genetically identical, some lineages express the gene, whereas in others it is repressed or silent. This frequently leads to a variegated phenotype such as mottled red/white eye color in *Drosophila*

 0-87969-490-4/96 $5 + .00

or red/white sectoring of colony color in yeast (Karpen 1994; Palladino and Gasser 1994; Gilson et al. 1995; Reuter and Spierer 1995).

The phenomenon of silencing at these chromosomal regions may be a manifestation of their specialized structures and the assembly of large protein complexes required to perform their functions. The variegated phenotypes remain mysterious but may reflect different states of assembly or "molecular breathing" of these structures. Here these phenomena are discussed with respect to centromere and telomere function and the silent mating-type loci in the fission yeast, *Schizosaccharomyces pombe.*

CENTROMERES AND TELOMERES OF *S. POMBE*

The fission yeast is an attractive unicellular eukaryote in which to study chromosome biology and chromosome-related phenomena (Allshire 1995). It is genetically tractable and has a good genetic map coupled to a detailed physical map of the three chromosomes. There are complete sets of overlapping cosmids, P1 phage, and YAC clones covering the whole 14-Mbp genome (Hoheisel et al. 1993; Mizukami et al. 1993). Projects are under way to complete the sequence of the whole genome within the next 5 years.

As in the budding yeast, *Saccharomyces cerevisiae*, three functionally important chromosomal elements have been identified and characterized in fission yeast: centromeres, telomeres, and replication origins (Allshire 1995). Mitotically stable artificial linear minichromosomes can be created by combining all three elements (Takahashi et al. 1992; Baum et al. 1994). Unlike the minute centromeres of *S. cerevisiae*, where only 125 bp of DNA is required to form a completely functional centromere, the centromere regions of *S. pombe* are large and complex. Large repetitive structures that surround a central region in a symmetrical manner are found at all three centromeres (Fig. 1). At least 15 kb (central domain plus a flanking repeat) of this fission yeast centromeric DNA is required to provide reasonable mitotic stability on minichromosomes (Baum et al. 1994). Remarkably, this overall arrangement at fission yeast centromeres bears considerable resemblance to organization of the centromeric heterochromatin of *Drosophila* recently shown to be required for centromere function (Murphy and Karpen 1995). The sequences residing at *S. pombe* telomeres have been cloned and sequenced and, as in many organisms, they have approximately 300 bp of a degenerate G strand biased repeat with the modal sequence TTACAGG running 5′to 3′ toward the chromosome end (Sugawara 1989; Allshire 1995). Up to 20 kb of telomere-associated sequences (TAS) lie internal to this terminal

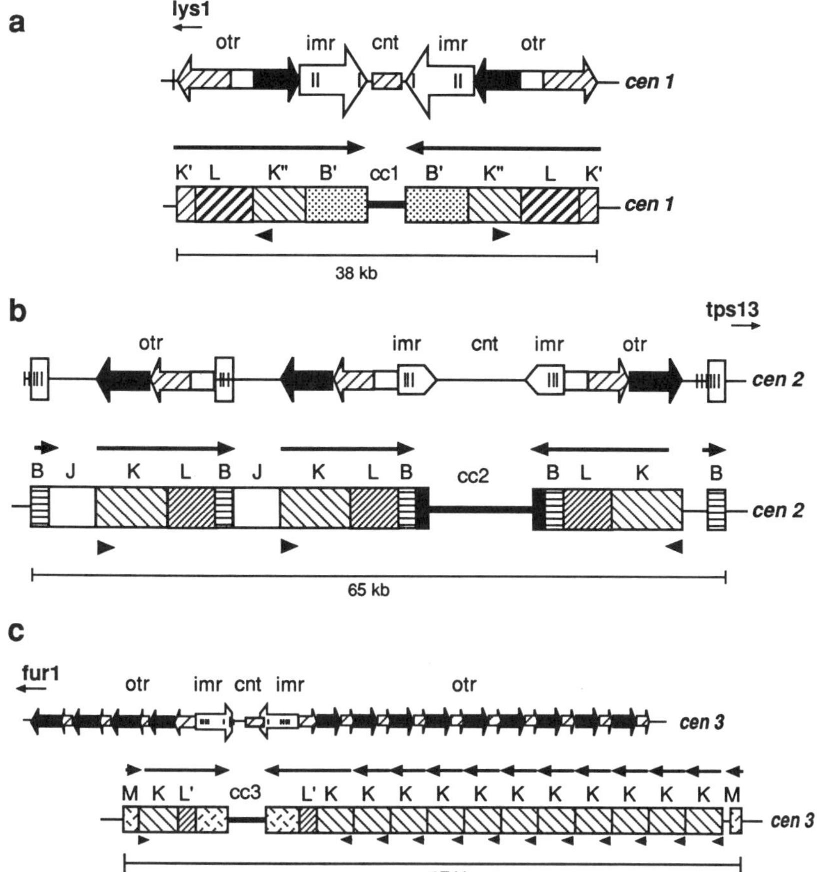

Figure 1 Organization of DNA elements at fission yeast centromeres. The arrangements of repeated sequences at each centromere are approximately aligned using both systems of nomenclature (Takahashi et al. 1992; Steiner and Clarke 1994). The otr repeat at each centromere is composed of the dg (*cross-hatched arrow*) and dh (*filled arrows*) repeats. A large inverted repeat surrounds the central region of each centromere. (Modified from Takahashi et al. 1992; Steiner and Clarke 1994.)

repeat at both ends of chromosomes 1 and 2; the terminal repeat directly abuts terminal arrays of rDNA units on both ends of chromosome 3 (Sugawara 1989).

The coordination of cell-cycle events and their regulation in *S. pombe* are the subject of intensive studies (Carr 1995; Woolard and Nurse 1995; Yanagida 1995). Any region of a chromosome can now be visualized at any stage of the cell cycle using fluorescent in situ hybridization (FISH) with labeled DNA probes (Uzawa and Yanagida 1992; Funabiki et al. 1993). The FISH technique allows centromeres and telomeres to be localized with respect to the spindle pole bodies (SPBs), microtubules, and other proteinaceous structures throughout mitotic and meiotic cell cycles (Uzawa and Yanagida 1992; Funabiki et al. 1993; Chikashige et al. 1994; Ekwall et al. 1995). Using this approach, the dynamic movement of centromeres and telomeres can be tracked. In mitotic cycles centromeres can be seen to cluster together on the nuclear periphery adjacent to the SPB in interphase; congress in the center of the early, short, mitotic spindles; move toward the spindle poles in early anaphase; and move apart with the adjacent SPBs in mid to late anaphase (Fig. 2). Telomeres also aggregate in clusters at the nuclear periphery in interphase but detach from the periphery upon entry into mitosis. During chromosome segregation, they appear to be the last regions of the sister chromatids to separate from each other (Funabiki et al. 1993). In preparation for meiosis, telomeres in fission yeast take on a new role (Chikashige et al. 1994). Just prior to karyogamy (fusion of nuclei to form a zygote), the telomeres move from sites elsewhere on the nuclear periphery to join the centromeres clustered at the SPB. Once both nuclei have fused, the centromeres are released from the SPB, and there follows a series of telomere/SPB-led rapid traverses of the cell lasting for approximately 15 minutes. Because of the shape of the stained chromatin in these cells, this has been termed "horsetail" movement. The purpose of these movements may be to allow pairing of parental homologous chromosomes and to initiate recombination.

SILENCING AND SWITCHING AT THE MATING-TYPE LOCI

Situated in the middle of the right arm of chromosome 2 several hundred kilobases from the centromere or telomeres are the mating-type loci (Fig. 3). The *mat1* locus is transcriptionally active (encoding the P-plus or M-minus genes) and accepts genetic information donated by gene conversion from either of the silent loci, *mat2* and *mat3*, lying approximately 15 and 30 kb away (Klar 1992a,b). Although the three *mat* loci are physically equidistant from each other, the genetic distance between *mat2* and *mat3* is minuscule (<0.001 cM), whereas recombination occurs at a normal rate in the *mat1-mat2* interval (1 cM) (Egel 1984b). This suppression of recombination and repression of transcription in the *mat2-mat3* interval is thought to reflect an unusual chromatin organization in this region.

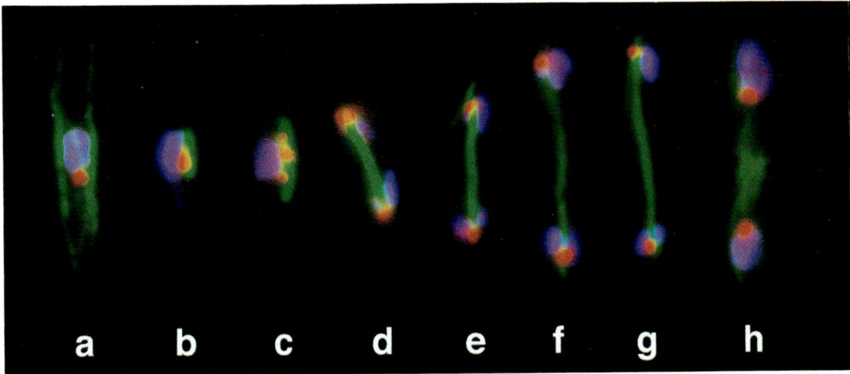

Figure 2 Localization of fission yeast centromeres through the cell cycle. Wild-type haploid cells stained with DAPI (which detects chromosomal DNA; *blue*) and with the anti-tubulin antibody TAT1 (which detects cytoplasmic and spindle microtubules; *green*) and hybridized with fluorescently labeled centromeric DNA probe (*red*) (see Ekwall et al. 1995). The microtuble staining allows cells in different stages of the cell cycle to be identified. (*a*) Interphase, (*b*) metaphase, (*c*) early anaphase, (*d,e*) mid anaphase, (*f,g*) late anaphase/telophase, (*h*) cytokinesis.

Switching of mating type requires the generation of a double-strand break (DSB) at a specific site on the right side of the *mat1* locus (Beach 1983; Beach and Klar 1984; Egel et al. 1984). This break stimulates conversion of *mat1* by interaction with the donor loci, *mat2* and *mat3*, resulting in the replacement of P with M or M with P information (Klar 1992a,b). The pattern of switching has been well characterized and conforms to a set of rules. The main observations are that, in general, only one granddaughter out of four derived from a cell of a particular mating type is switched and that the sister of a switched cell produces one switched and one unswitched daughter, recurrently (Miyata and Miyata 1981; Egel 1984a; Egel and Eie 1987). As more details of the mating-type loci have accumulated, it has become clear that some form of imprinting must be invoked to explain the pattern of switching in pedigrees (Klar 1987, 1990, 1992a,b). The DSB has itself been demonstrated to be an imprinting event, since its presence can direct chromosome-specific gene conversion events in the following generation (Klar and Bonaduce 1993). However, a mechanism of strand marking must also exist to distinguish the sister chromatid that develops the DSB from the one that does not (Klar 1992b; Klar and Bonaduce 1993). The molecular details of this imprinting mechanism remain to be elucidated. Because 80% of P

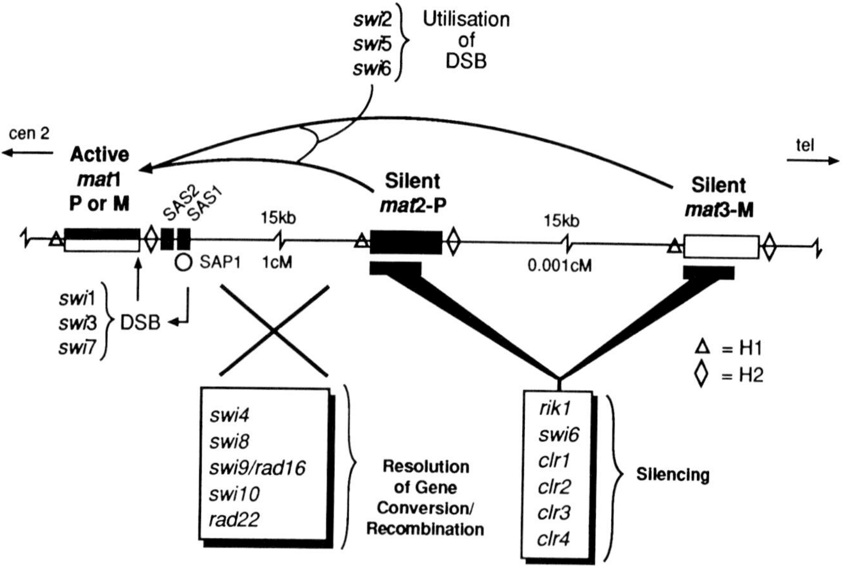

Figure 3 Schematic representation of the silent mating-type loci and the components involved in silencing and switching. H1 and H2 are shared blocks of homology flanking *mat1*, *mat2*, and *mat3*, which mediated the switching. DSB is the site of the double-strand break within *mat1*. See text for explanation.

cells switch to M cells and vice versa, there must also be some mechanism governing the choice of donor available. It has been postulated that the same specialized structure responsible for maintaining *mat2* and *mat3* in a repressed state is also required to fold this chromosomal region in distinctly different ways in P and M cells. Thus, when P is at *mat1*, it can recombine more easily with *mat3*-M, and when M is at *mat1* it contacts *mat2*-P more readily than *mat3*-M (Fig. 4). Several observations suggest that the unusual chromatin structure in this region promotes *mat1*-M interactions with *mat2*-P, since in the presence of certain mutations (*swi2* and *swi6*; see below) the default appears to be *mat1*-*mat3* interactions so that M cells predominate (Egel et al. 1984; Gutz and Schmidt 1985; Thon and Klar 1993).

The DSB at *mat1*, which initiates mating-type switching, is known to persist in about 25% of chromosomes in an actively growing culture throughout most of the cell cycle (Beach 1983). This break must be protected in some manner so as not to be recognized by the surveillance systems that normally sense DNA damage and elicit a cell-cycle arrest (Hartwell and Kastan 1994; Carr 1995; Carr and Hoekstra 1995). Therefore, this DSB has similarities with telomeres, which are in essence one

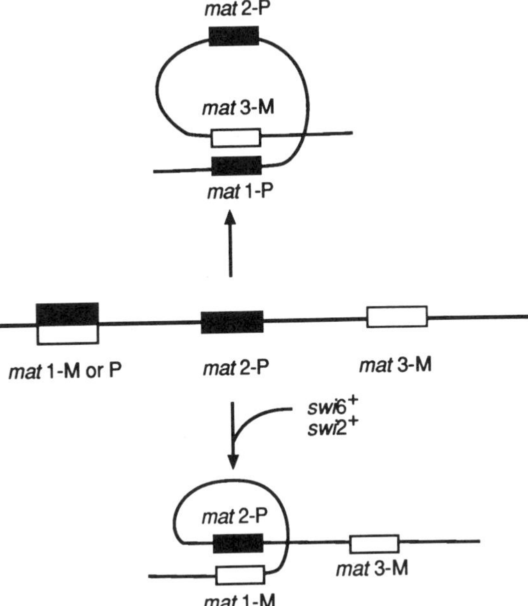

Figure 4 Formation of alternative intrachromosomal folded structures in the mating-type region. In wild-type h^{90} homothallic strains, P or M information occupies *mat1*. P cells predominantly switch to M cells and vice versa, suggesting the formation of alternative structures in P and M cells. *swi6* and *swi2* mutants mate predominantly as M cells, suggesting that M information is present at *mat1* in most cells. Thus, the *swi6* and *swi2* gene products must promote interactions with *mat2*-P in M cells. See text for further details. (Modified from Thon and Klar 1993.)

end of a DSB and are also camouflaged in some manner so as to remain undetected by the cell (Sandell and Zakian 1993). The site of the break in *mat1* has been mapped precisely on one strand. Interestingly, it is apparently blocked on the opposite strand, suggesting some factor is bound tightly at the break (Nielsen and Egel 1989).

A number of switching mutants (*rad22* and *swi1–swi10*; unrelated to the *S. cerevisiae SWI* genes that regulate *HO* endonuclease expression) have been isolated that influence the frequency and ability to switch (Egel et al. 1984; Gutz and Schmidt 1985; Schmidt et al. 1987, 1989; Klar 1992b). These define three categories of defects in switching: (1) DSB formation and stabilization, (2) DSB utilization, and (3) resolution of the gene conversion event (see Table 1). Given the complexity of gene conversion events, it is not surprising that mutations in genes encoding components of the excision (Rodel et al. 1992; Carr et al. 1994), mis-

Table 1 Factors involved in mating-type switching and silencing

Locus	Defect	Function	Similarities	Reference
rad22	DSB repair, UV sensitive, recombination resolution	recombination repair	*S. cerevisiae RAD52*	Osterman et al. (1993)
swi1	DSB formation		n.a.	Egel et al. (1984)
swi3	DSB formation		n.a.	Egel et al. (1984)
swi7	DSB formation	DNA synthesis	DNA polymerase α	Singh and Klar (1993)
swi2	switching, utilization of DSB		n.a.	Egel et al. (1984)
swi5	switching, UV sensitive, utilization of DSB		n.a.	Egel et al. (1984)
swi6	switching, utilization of DSB, *mat2/3* silencing, *cen* silencing and segregation	formation of higher-order chromatin structure	heterochromatin proteins in *Drosophila* (HP1, Pc) and mammals (m31, m32, hsα)	Lorentz et al. (1992, 1994); Ekwall et al. (1995)
swi4	recombination resolution	mismatch repair	Duc-1, Rep-3, MutS	Fleck et al. (1992)
swi8	recombination resolution		n.a.	Fleck et al. (1994)
swi9	recombination resolution	excision repair	excision repair component	Carr et al. (1994)
(*rad16*)	UV sensitive		*S. cerevisiae RAD1*	

swi10	recombination resolution, UV sensitive	excision repair	excision repair component human ERCC1, *S. cerevisiae RAD10*	Rodel et al. (1992)
clr1	*mat2/3* silencing, *cen* silencing		3 zinc fingers	Thon and Klar (1992); G. Thon (pers. comm.)
clr2	*mat2/3* silencing, *cen* silencing		n.a.	Ekwall and Ruusala (1994); Thon et al. (1994)
clr3	*mat2/3* silencing, *cen* silencing		n.a.	Ekwall and Ruusala (1994); Thon et al. (1994)
clr4	*mat2/3* silencing, *cen* silencing and segregation		n.a.	Ekwall and Ruusala (1994); Thon et al. (1994)
rik1	*mat2/3* silencing, *cen* silencing and segregation		no significant similarities in database	Egel et al. (1989)

match (Fleck et al. 1992, 1994), and recombination (Ostermann et al. 1993) repair processes influence the resolution of switching events within the mating-type region. The *swi4*, *swi8 swi9*(*rad16*), *swi10*, and *rad22* mutations frequently lead to large-scale rearrangements of the mating-type loci, resulting in heterothallic derivatives (Beach and Klar 1984; Egel et al. 1984; Gutz and Schmidt 1985; Fleck et al. 1992, 1994; Klar 1992b; Ostermann et al. 1993). Two *cis*-acting sequences, SAS1 and SAS2 (switch-activating site), lying distal to the *mat1* locus, are required for efficient DSB formation and switching. The SAP1 protein (switch-activating protein) binds at SAS1, and mutations in SAS1 that disrupt binding of SAP1 reduce switching frequency (Arcangioli and Klar 1991). The *sap1*$^+$ gene (*note:* in *S. pombe* the allele name is lowercase italics with $^+$ indicating wild type) encodes a novel protein and is essential for viability, suggesting that it is required for other cellular processes apart from generating a DSB at *mat1* (Arcangioli et al. 1994). A 263-bp deletion of sequences distal to *mat1* removes both the SAS1 and SAS2 sites, generating the *smt-0* mutation which abolishes mating-type switching (Styrkarsdottir et al. 1993).

Six mutations have been identified which release recombinational suppression and alleviate transcriptional silencing of the *mat2-mat3* region (*clr1*, *clr2*, *clr3*, *clr4*, *rik1*, and *swi6*, see Table 1) (Egel et al. 1989; Klar and Bonaduce 1991; Lorentz et al. 1992; Thon and Klar 1992; Ekwall and Ruusala 1994; Thon et al. 1994). Mutations at *swi6* are unusual in that they affect the ability to switch mating type in addition to transcriptional repression and recombination suppression in the *mat2-mat3* interval (Klar and Bonaduce 1991; Lorentz et al. 1992). It is thought that *swi6*$^+$ is primarily required to promote utilization of the donor locus, *mat2*-P, for gene conversion, since *swi6* mutants still form a DSB at *mat1* with a normal frequency (Egel et al. 1984; Thon and Klar 1993). Several of these genes have been cloned and sequenced, some of which are discussed further below (Table 1). Four *cis*-acting DNA elements adjacent to *mat2* are required to maintain the silent state on episomal plasmids (Ekwall et al. 1991). Two of these elements act in concert with the *clr1*, *clr2*, *clr3*, *clr4*, and *swi6* gene products to silence *mat2* (Thon et al. 1994).

The *mat2-mat3* region has properties that are in essence similar to regions of heterochromatin originally described in larger eukaryotes: It is transcriptionally inert and recombination is suppressed. In addition, although it has not been rigorously tested, this chromosomal locus might be associated with the nuclear periphery in wild-type cells (Ekwall et al. 1995).

CENTROMERIC SILENCING IN *S. POMBE*

To ascertain if, as in other eukaryotes, a heterochromatin-like structure exists at *S. pombe* centromeres and telomeres, reporter genes have been placed within or adjacent to these regions (Allshire et al. 1994, 1995). When the *ura4*+ gene is placed at any of 11 sites within *cen1*, these strains have the unusual ability to form colonies on both selective (–uracil) and counterselective (+5-fluoro-orotic acid) plates. Two populations must exist in these strains: those where the *ura4*+ gene is expressed, and those where it is repressed. Thus, these strains variegate between the expressed and repressed states. Northern analyses indicated that transcription at some of these sites is more repressed than at others (Allshire et al. 1995). The pattern of this repression is symmetrically related to the inverted organization of DNA elements across the centromere (Fig. 5). At insertion sites (sites 1 and 2) lying just 900 bp and 1200

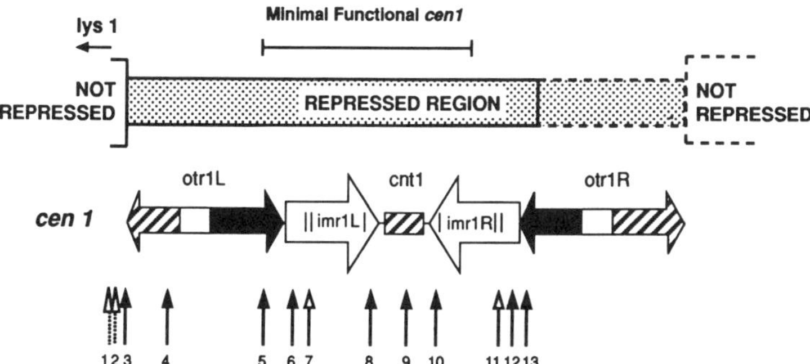

Figure 5 Transcriptional repression across *cen1*. The minimal region of *cen1* that provides centromere function is indicated (Baum et al. 1994). The *ura4*+ gene has been inserted at 13 sites across *cen1* (Allshire et al. 1995). Sites 1 and 2 reside oustide the DNA elements which demarcate *cen1*; no repression is observed at these sites since no colonies can be formed on counterselective plates (*hatched open arrows*). At all other sites, the *ura4*+ gene has been found to be repressed or repressible. Although sites 7 and 11 display a relatively high level of expression under nonselective conditions, the *ura4*+ gene is clearly repressible since colonies are formed on counterselective plates. The *ura4*+ gene has only been inserted at one site in the right hand-side otr; however, given that the structure is symmetrical, it is likely that transcription is repressed throughout the entire centromeric region (repressed/nonrepressed structure is therefore only indicated by a dashed outline on the right-hand side). The *clr4* and *rik1* mutations completely abolish repression at insertion sites 12 and 13 (implies site 6 also). *clr1*, *2*, *3*, *4*, *rik1*, or *swi6* mutations have minimal effects on repression at the central site 9; no other sites have been tested (see Allshire et al. 1995).

bp outside the defined centromeric region, no repression can be detected. The entire centromere therefore appears to be enclosed in an inaccessible, repressive, heterochromatin-like structure, which is refractory to transcription (Allshire et al. 1995).

Insertion of the *ade6*$^+$ reporter gene into the middle of centromeres 1 and 3 (TM1/site9 and TM3) resulted in colonies that display a variegated phenotype in that red/white-sectored colonies are generated (Allshire et al. 1994). Colony color ranges from white ("on" in many cells) to pink (suggesting intermediate levels of expression/repression) to red ("off" or repressed in most cells). Because these colonies contain at least 2×10^7 cells, the expressed/repressed state is stable for up to 25 generations. This suggests that the chromatin structure at individual centromeres is dynamic, but a particular state can be fixed and inherited through several mitotic cell cycles in a given lineage. It is not known if these different "open" and "closed" states influence centromere function within a particular lineage; however, the ability to assemble a functional centromere can be regulated epigenetically (Steiner and Clarke 1994). Different levels of repression at different regions within the centromere are also apparent with the *ade6*$^+$ marker. When the *ade6*$^+$ gene is placed within the outer repeats of *cen1* (site 13), most colonies are red in color on indicator plates, and only rarely is a white sector detected (G. Cranston and R.C. Allshire, unpubl.). These differences in repression and repressibility at different sites presumably reflect variation in chromatin organization and accessibility across the centromere. It seems reasonable to assume that this repression results from kinetochore assembly on this centromeric DNA, thus excluding transcription factors from the underlying marker gene.

TELOMERIC SILENCING IN *S. POMBE*

In the budding yeast, *S. cerevisiae*, the phenomenon of telomeric silencing has been termed telomere position effect (TPE) and has been examined in depth (see Holmes et al., this volume). Briefly, the *URA3*, *ADE2*, *TRP1*, and *HIS3* genes (*note:* in *S. cerevisiae* the wild-type allele is written in uppercase italics and mutant alleles are witten in lowercase) can all be transcriptionally silenced when placed adjacent to a telomere in *S. cerevisiae* (Gottschling et al. 1990). Many factors are known to be required to affect this silencing. For example, mutations in the *RAP1*, *SIR2*, *SIR3*, and *SIR4* genes and the genes encoding histone H3 and H4 alleviate the silent state allowing expression of the telomeric marker

(Aparicio et al. 1991; Kyrion et al. 1993; Renauld et al. 1993). These genes, in addition to *SIR1*, are also required to maintain transcriptional repression at the *S. cerevisiae* silent mating-type loci *HML* and *HMR* (see Holmes et al., this volume). There is good evidence that these proteins are assembled into a complex at telomeres which frequently forms aggregates at the nuclear periphery (Wright et al. 1992; Palladino et al. 1993; Hecht et al. 1995). The formation of this telomere complex, in association with other proteins, probably performs a number of critical functions, such as protecting the telomere from exonucleolytic attack and end-to-end fusion, facilitating the engagement of the telomere replication machinery (telomerase), and disguising this natural end of a DNA molecule so that it is undetected by DNA damage checkpoints. Defects in these components can disrupt the formation of the telomere complex and the association of telomeres with the nuclear periphery, alter terminal repeat tract length, and cause increased rates of chromosome loss (Conrad et al. 1990; Kyrion et al. 1993; Palladino et al. 1993; Hecht et al. 1995). The phenomenon of TPE is probably a consequence of the assembly of this large complex adjacent to reporter genes.

When the *ura4*$^+$ and *ade6*$^+$ genes are placed adjacent to newly formed telomeres in *S. pombe*, a similar transcription repression effect on expression is seen. Strains with a telomeric *ura4*$^+$ gene grow on both selective and counterselective plates, and a telomeric *ade6*$^+$ marker results in red/pink/white-sectored colonies (Nimmo et al. 1994). This is consistent with a telomeric marker being capable of adopting different states, resulting in variegated phenotypes. There is also some evidence which suggests that fission yeast telomere is enclosed in a nonnucleosomal, protected structure (Chikashige et al. 1989).

FACTORS AFFECTING TRANSCRIPTIONAL REPRESSION AND CENTROMERE FUNCTION IN *S. POMBE*

The paradigm set by the studies linking TPE with silencing at *HML* and *HMR* in *S. cerevisiae* stimulated investigation of whether factors affecting centromeric and mating-type silencing in *S. pombe* were similarly shared (Allshire et al. 1995). The *clr1*, *clr2*, *clr3*, *clr4*, *rik1*, and *swi6* mutants were shown to affect silencing of the *ura4*$^+$ gene at the three centromeric sites tested (sites 9, 12, and 13; Fig. 5). However, only *clr4-S5* and *rik1-304* mutants result in complete derepression within the centromere and only at specific sites (flanking sites 12 and 13 but not the central site 9; Fig. 5). The *swi6-115 and swi6*Δ mutants had a somewhat intermediate effect; little or no derepression could be detected in the central region, but the level of marker gene expression was clearly increased in the outer flanking repeats.

If the assembly of a fully functional kinetochore complex results in centromeric silencing, then some mutations that interfere with kinetochore function should missegregate chromosomes and alleviate transcriptional repression within the centromere. Conforming with this notion, both nonessential minichromosomes and endogenous chromosomes are lost at highly elevated rates (100x) in *clr4-S5*, *rik1-304*, *rik1*Δ, *swi6-115*, and *swi6*Δ mutants, all of which have intermediate to strong effects on centromeric silencing. No significant increase in the rate of chromosome loss is detectable in *clr1*, *clr2*, and *clr3* point mutants (null alleles have not been created/tested), which affect centromeric silencing to a lesser extent. The possibility that *clr4*, *rik1*, and *swi6* mutations act through telomeres with respect to chromosome stability was discounted by the fact that centromere activity of a compromised circular minichromosome was completely obliterated in these backgrounds (Allshire et al. 1995). These six mutations have only a weak effect on transcriptional repression adjacent to telomeres, since no increase in expression of a telomeric *ura4*$^+$ gene can be detected by Northern analyses (Allshire et al. 1995).

FUNCTIONS OF THE *clr4*, *rik1*, AND *swi6* GENES

The *swi6*$^+$ gene has recently been cloned by complementation of the switching defect with genomic libraries (Lorentz et al. 1994). Sequence analyses demonstrated that it encoded a predicted protein of 37 kD. Intriguingly, database searches demonstrated that the Swi6 protein shares homology with the chromo-domain motif found in several higher eukaryotic proteins known to be located at heterochromatic regions or involved in heterochromatin formation (Messmer et al. 1992; Powers and Eissenberg 1993; Saunders et al. 1993; Wreggett et al. 1994; Kellum et al. 1995). The most extensively analyzed chromodomain proteins, HP1 and Pc, are found in *Drosophila*. Heterochromatin protein 1 (HP1) is encoded by the *Su(var)205* gene, mutations in which act to suppress position-effect variegation adjacent to heterochromatin (Eissenberg et al. 1992; see Henikoff, this volume). Embryos lacking HP1 display a high frequency of anomalous chromosome segregation events (Kellum and Alberts 1995). Polycomb (Pc) is required for the coordinate shutdown of homeotic genes during the latter stages of *Drosophila* development and is associated in a multisubunit complex with these loci on *Drosophila* chromosomes (Paro 1993; Bienz and Muller 1995; see Paro and Harte, this volume).

Further analyses of all chromodomain-containing proteins in the databases have revealed the presence of a subclass that contain two types of chromodomains, the "classic" chromodomain, and the chromo shadow domain (Aasland and Stewart 1995). Both HP1 and Swi6 proteins belong to this subclass containing an amino-terminal chromodomain and a carboxy-terminally located chromo shadow domain. With the identification of this new domain, it is apparent that the chromo shadow domain is responsible for targeting HP1 to centromeric heterochromatin in *Drosophila* (Powers and Eissenberg 1993; Platero et al. 1995). The amino-terminal chromodomain may be involved in recruiting other proteins to that site. These roles may be reversed in other chromodomain proteins. Together, these two domains appear to act to promote the assembly of macromolecular complexes generating transcriptionally silent regions of chromatin.

Immunostaining combined with FISH has shown that the Swi6 protein is located at the three known sites of transcriptional silencing (centromeres, telomeres, and the silent mating-type loci) in the fission yeast throughout the cell cycle (Ekwall et al. 1995). This localization is disrupted in cells bearing the *clr4-S5*, *rik1-304*, or *swi6-115* mutations (K. Ekwall and R.C. Allshire, in prep.). The *swi6-115* allele carries a substitution of basic arginine for a highly conserved hydrophobic residue (tryptophan) at amino acid 269 within the chromo shadow domain. FISH analyses, however, indicate that the associations of centromeres and telomeres with the nuclear periphery are still maintained in *clr4*, *rik1*, and *swi6* mutants (K. Ekwall et al., unpubl.). These three mutations have a specific effect on centromere function; analyses of cells by FISH shows that centromeres are delayed in their migration to the poles of the spindle so that they lag on late-anaphase spindles in a high proportion of cells (Ekwall et al. 1995; K. Ekwall and R.C. Allshire, in prep.). This indicates that the Swi6 protein is a key component at fission yeast centromeres, and its recruitment to the centromere is dependent on Clr4p and Rik1p. All three mutants affect transcriptional silencing within the centromere and cause high rates of chromosome loss and a high incidence of centromere lagging in anaphase. The *rik1*$^+$ gene has also been cloned and sequenced; however, its sequence is uninformative with respect to its function, and so far it has not been localized (B. Borgstrøm and R. Egel, pers. comm.). The *clr1*$^+$ gene encodes a protein containing three zinc fingers (G. Thon and A.J.S. Klar, pers. comm.).

As noted above, in *S. cerevisiae*, factors required for maintaining silencing at *HML* and *HMR* have a shared role in silencing at telomeres. It was therefore surprising that, in fission yeast, none of the *mat2-mat3* silencing factors tested (*clr1*, *clr2*, *clr3*, *clr4*, *rik1*, or *swi6*) has a very significant effect on transcriptional repression of marker genes placed adjacent to telomeres in fission yeast. The *HML* and *HMR* loci, however, lie in relatively close proximity to telomeres in *S. cerevisiae* and, indeed, this location imposes an additional layer of silencing on *HMR* (Thompson et al. 1994). Thus, it is possible that the mechanism of *HML* and *HMR* silencing has integrated factors from this locality to tighten *HM* regulation.

Of the six genes known to be required for silencing at *mat2-mat3* in *S. pombe*, three have been completely sequenced (see Table 1). No similarity exists between any of these and the factors known to affect *HM* or telomeric silencing in *S. cerevisiae*. It appears that in *S. pombe*, silencing of *mat2-mat3* may have recruited factors involved in centromere structure and function. Screens to identify factors involved in mediating silencing at fission yeast telomeres are under way. It will be intriguing to see if these telomeric factors share homology with the known factors affecting TPE in *S. cerevisiae*. This could pave the way to the identification of conserved telomeric silencing proteins in eukaryotes in general.

The connection between centromere structure and function and the silent mating-type region has recently been further cemented by the astonishing finding that 4.3 kb of the region lying between *mat2* and *mat3* is 96% identical to the K or otr (dgII and dhIIa) repeats of centromere 2 (S. Grewel and A. Klar, pers. comm.) To maintain such a high degree of homology strongly suggests that these regions of the genome must frequently interact. FISH analysis indicates that centromeres, telomeres, and the mating-typing loci tend to share the same nuclear peripheral compartment (Funabiki et al. 1993; Ekwall et al. 1995). This location may provide the opportunity, for whatever unknown reason, for these regions to interact.

VARIEGATION IN CENTROMERE FUNCTION

The sequences required to provide reasonable mitotic centromere function in *S. pombe* have been elucidated by a series of experiments creating constructs with minimal amounts of centromeric DNA (Fig. 5; see Fig. 1 diagram and explanation of DNA element nomenclature) (Takahashi et al. 1992; Baum et al. 1994). It has become clear that a central domain, flanking B/imr repeats and some K/otr(dg) repeat, allows fairly stable propagation (4-12% loss/division). The symmetrical arrangement of sequences surrounding a central region is not an absolute requirement, since constructs with only part of a central domain, a B/imr repeat and multiple (2-3) flanking K"/dg repeats are also functional (0.4–2.3%) (Baum et al. 1994). The construction of minichromosome vectors with the *ade6-704/sup3e* marker system and containing minimal centromere DNA from *cen3* revealed a curious phenomenon (Steiner and Clarke 1994). Two classes of transformants could be identified containing intact, unrearranged construct: those that segregated it relatively stably (1–2% loss per division) and those where it was extremely unstable (92–99% loss per division). Stability was indicated by predominantly white colony color on nonselective indicator plates, whereas pure red colonies implied extreme instability. One credible explanation for this phenomenon is that these minimal centromere constructs are only slowly assembled into a higher-order chromatin structure necessary for kinetochore function. Therefore, the same construct in different transformants may form a functional centromere or segregate as an unstable element. This time-dependent assembly of a structure necessary for kinetochore function results in an epigenetic effect with respect to the function of these minimal centromeres and plasmid stability.

Since these constructs lack significant amounts of B/imr inverted repeat surrounding the central domain and, in addition, only contain a single copy of the K"/dg outer repeat, it is worth noting that *clr4*$^+$, *rik1*$^+$, and *swi6*$^+$ mediate their effects on silencing primarily through the otr and B/imr regions of the centromere and not through the central region (see Fig. 5) (Allshire et al. 1995). Mutations in *clr4*, *rik1*, and *swi6* completely obliterate stability of a normally stable minichromosome containing symmetrically arranged outer repeats (Allshire et al. 1995). However, preliminary data suggest that a low frequency of stable (white) colonies may exist in these populations (R. Allshire, unpubl.). This suggests that the epigenetic effect of having minimal centromeric sequences on centromere function observed by Steiner and Clarke (1994) may be mimicked by minichromosomes bearing fully functional centromeres in the absence of *clr4*$^+$, *rik1*$^+$, or *swi6*$^+$. This implies that *clr4*$^+$, *rik1*$^+$, and *swi6*$^+$ facilitate the formation of a fully functional centromere by mediating the formation of a higher-order chromatin structure.

INTRACHROMOSOMAL FOLDING AND THE FORMATION OF A FUNCTIONAL CENTROMERE

The *swi6*$^+$ gene not only is required for silencing within the centromere and the mating-type loci, but also facilitates switching of mating type.

Analyses of the configuration of the mating-type cassettes in homothallic h^{90} cells in a *swi6-115* background revealed that most cells in the population had M information at *mat1* (Thon and Klar 1993). Thus, *swi6*$^+$ not only ensures transcriptional repression at *mat2* and *mat3*, but is also required to allow M information at *mat1* to contact *mat2*-P and mediate a switch of mating-type information. A clever experiment demonstrated that switching is dependent on the relative chromosomal positioning of *mat2* and *mat3* with respect to *mat1* (Thon and Klar 1993). A strain was constructed where the silent M information was engineered into the *mat2* locus (normally at *mat3*) and the silent P information was placed at *mat3* (normally at *mat2*). In this h^{09} configuration in a wild-type background, only 18% of cells switch, since now *mat1*-P only finds P information at *mat3* and, likewise, *mat1*-M can only see M information at *mat2*. In the presence of the *swi6-115* mutation, most h^{09} cells have P information at *mat1*. Therefore, since *swi6* mutations hinder the ability of *mat2*-P to interact with *mat1*-M in a h^{90} strain and *mat2*-M with *mat1*-P in h^{09} strains, it must be required to mediate a specific intrachromosomal folding event facilitating the interaction of *mat2* with *mat1* (Fig. 4).

The following observations must be taken into account by any model speculating on the role of *clr4*$^+$, *rik1*$^+$, and *swi6*$^+$ at centromeres:

1. *clr4*$^+$ and *rik1*$^+$ are required for Swi6p localization.
2. Mutations of *swi6* only alleviate and, unlike *clr4* and *rik1* mutations, do not completely derepress transcriptional repression at two sites tested within the centromere.
3. Mutations at *clr4*, *rik1*, and *swi6* cause a similar elevation in the rate of chromosome loss.
4. *swi6*$^+$ apparently mediates an intrachromosomal folding event to enable efficient switching of mating type.
5. Minimal centromere constructs inefficiently nucleate kinetochore assembly.

Since all three *S. pombe* centromeres are symmetrically organized structures (Fig. 1), it is possible that as at the mating-type locus, *swi6*$^+$ is required to facilitate the formation of an intrachromosomal folded structure within the centromere, perhaps allowing the flanking long inverted repeats to make contact in a hairpin-like or more complex structure as depicted in the cartoon in Figure 6. In this model, Clr4p and Rik1p (presumed to be structural components) would be in closer contact with the underlying chromatin and would also contact (directly or indirectly) Swi6p. Loss of Swi6p would lead to inefficient folding which, if necessary for kinetochore assembly, would lead to weakened kinetochore activity (centromere-microtubule attachment). The products of *clr4*$^+$ and *rik1*$^+$ might be more involved in the formation of the underlying chromatin and would therefore have a greater effect on silencing but also would lead to weakened kinetochores due to the lack of recruitment of Swi6p to the centromere and inefficient formation of the necessary folded structure (Fig. 6).

The complete characterization of the *clr4*$^+$ and *rik1*$^+$ gene products, and the development of reagents allowing their detection in wild-type and mutant strains of fission yeast, should aid in the elucidation of their role in centromere function. Until such time, the above should only be regarded as a framework requiring much more detailed examination. Genetic analyses indicate that many other factors are also involved in silencing and centromere function. Six additional loci have been identified with similar phenotypes as *clr4*, *rik1*, and *swi6* mutants (*csp7–12*, *c*entromere *s*uppressor of *p*osition effect; K. Ekwall and R.C. Allshire, unpubl.).

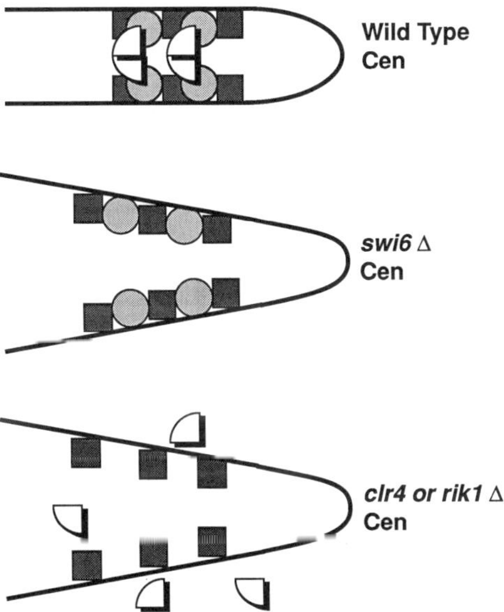

Figure 6 Cartoon model for role of *clr4*$^+$, *rik1*$^+$, and *swi6*$^+$ gene products in centromere structure and function. Clr4p and Rik1p are represented by either the circles or squares. Swi6p is represented by the quarter circle. *clr4* or *rik1* mutations cause delocalization of Swi6p and alter the higher-order structure, allowing increased access for transcription. All three proteins are required to assemble a complex that facilitates the assembly of completely functional kinetochore. See text for details.

CONCLUSIONS

The phenomenon of silencing in the mating-type region of *S. pombe* has been known for some time. Recent studies have revealed that transcriptional silencing also occurs within centromeres and adjacent to telomeres, where large protein complexes are formed. It is likely that silencing results from the assembly of these complexes excluding transcription factors from nearby or embedded genes. These silencing phenomena can now be utilized to genetically dissect components of the kinetochore and telomere complexes. Initial studies have indicated that several factors required for maintaining silencing at the mating-type loci are also required for centromere function. These silencing factors provide an entry point for identifying components required for kinetochore assembly in fission yeast.

ACKNOWLEDGMENTS

Jean-Paul Javerzat, Karl Ekwall, Elaine Nimmo, and Gwen Cranston are thanked for their large contribution to much of the centromere and telomere work described. I also thank Karl Ekwall, Elaine Nimmo, Phillip Lord, Toru Nakamura, and Genevieve Thon for critical reading of the manuscript, and Amar Klar for communicating unpublished results. I am also grateful to Norman Davidson, Sandy Bruce, and Douglas Stewart for the artwork. This work was supported by the Medical Research Council of Great Britain with additional funding from the Wellcome Trust, European Union, European Molecular Biology Organization, and Human Frontiers in Science Program.

REFERENCES

Aasland, R. and A.F. Stewart. 1995. The chromo shadow domain, a second chromo domain in heterochromatin-binding protein 1, HP1. *Nucleic Acids Res.* **23:** 3168–3173.

Allshire, R.C. 1995. Elements of chromosome structure and function in fission yeast. *Semin. Cell Biol.* **6:** 55–64.

Allshire, R.C., J.P. Javerzat, N.J. Redhead, and G. Cranston. 1994. Position effect variegation at fission yeast centromeres. *Cell* **76:** 157–169.

Allshire, R.C., E.R. Nimmo, K. Ekwall, J.P. Javerzat, and G. Cranston. 1995. Mutations derepressing silent domains within fission yeast centromeres disrupt chromosome segregation. *Genes Dev.* **9:** 218–233.

Aparicio, O.M., B.L. Billington, and D.E. Gottschling. 1991. Modifiers of position effect are shared between telomeric and silent mating-type loci in *Saccharomyces cerevisiae. Cell* **66:** 1279–1287.

Arcangioli, B. and A.J.S. Klar. 1991. A novel switch-activating site (SAS1) and its cognate binding-factor (SAP1) required for efficient mat1 switching in *Schizosac-*

charomyces pombe. EMBO J. **10:** 3025–3032.

Arcangioli, B., T.D. Copeland, and A.J.S. Klar. 1994. Sap1, a protein that binds to sequences required for mating-type switching, is essential for viability in *Schizosaccharomyces pombe. Mol. Cell. Biol.* **14:** 2058–2065.

Baum, M., V.K. Ngan, and L. Clarke. 1994. The centromeric K-type repeat and the central core are together sufficient to establish a functional *Schizosaccharomyces pombe* centromere. *Mol. Biol. Cell* **5:** 747–761.

Beach, D.H. 1983. Cell type switching by DNA transposition in fission yeast. *Nature* **305:** 682–688.

Beach, D.H. and A.J.S. Klar. 1984. Rearrangements of the transposable mating-type cassettes of fission yeast. *EMBO J.* **3:** 603–610.

Bienz, M. and J. Muller. 1995. Transcriptional silencing of homeotic genes in *Drosophila. BioEssays* **17:** 775–784.

Carr, A.M. 1995. DNA-structure checkpoints in fission yeast. *Semin. Cell Biol.* **6:** 65–72.

Carr, A.M. and M.F. Hoekstra. 1995. The cellular responses to DNA-damage. *Trends Cell Biol.* **5:** 32–40.

Carr, A.M., H. Schmidt, S. Kirchhoff, W.J. Muriel, K.S. Sheldrick, D.J. Griffiths, C.N. Basmacioglu, S. Subramani, M. Clegg, A. Nasim, and A.R. Lehmann. 1994. The rad16 gene of *Schizosaccharomyces pombe:* A homolog of rad1 gene of *Saccharomyces cerevisiae. Mol. Cell. Biol.* **14:** 2029–2040.

Chikashige, Y., D.Q. Ding, H. Funabiki, T. Haraguchi, S. Mashiko, M. Yanagida, and Y. Hiraoka. 1994. Telomere-led premeiotic chromosome movement in fission yeast. *Science* **264:** 270–273.

Chikashige, Y., N. Kinoshita, Y. Nakaseko, T. Matsumoto, S. Murakami, O. Niwa, and M. Yanagida. 1989. Composite motifs and repeat symmetry in *S. pombe* cemtromeres: Direct analyses by integration of Not1 restriction sites. *Cell* **57:** 739–751.

Conrad, M.N., J.H. Wright, A.J. Wolf, and V.A. Zakian. 1990. RAP1 protein interacts with yeast telomeres in vivo: Overproduction alters telomere structure and decreases chromosome stability. *Cell* **63:** 739–750.

Egel, R. 1984a. The pedigree pattern of mating-type switching in *Schizosaccharomyces pombe. Curr. Genet.* **8:** 205–210.

——— 1984b. Two tightly linked silent cassettes in the mating-type region of *Schizosaccharomyces pombe. Curr. Genet.* **8:** 199–203.

Egel, R. and B. Eie. 1987. Cell lineage asymmetry in *Schizosaccharomyces pombe:* Unilateral transmission of a high-frequency state for mating-type switching in diploid pedigrees. *Curr. Genet.* **12:** 429–433.

Egel, R., D.H. Beach, and A.J.S. Klar. 1984. Genes required for initiation and resolution steps of mating-type switching in fission yeast. *Proc. Natl. Acad. Sci.* **81:** 3481–3485

Egel, R., M. Willer, and O. Nielsen. 1989. Unblocking of meiotic crossing-over between the silent mating-type cassettes of fission yeast, conditioned by the recessive, pleiotropic mutant rik1. *Curr. Genet.* **15:** 407–410.

Eissenberg, J.C., G.D. Morris, G. Reuter, and T. Hartnett. 1992. The heterochromatin-associated protein HP-1 is an essential protein in *Drosophila* with dosage-dependent effects on position-effect variegation. *Genetics* **131:** 345–352.

Ekwall, K. and T. Ruusala. 1994. Mutations in rik1, clr2, clr3 and clr4 genes asymmetrically derepress the silent mating-type loci in fission yeast. *Genetics* **136:** 53–64.

Ekwall, K., O. Nielsen, and T. Ruusala. 1991. Repression of a mating type cassette in the fission yeast by four elements. *Yeast* **7:** 745–755.

Ekwall, K., J. Javerzat, A. Lorentz, H. Schmidt, G. Cranston, and R.C. Allshire. 1995. The chromodomain protein Swi6: A key component at fission yeast centromeres. *Science* **269:** 1429–1431.

Fleck, O., H. Michael, and L. Heim. 1992. The swi4+ gene of *Schizosaccharomyces pombe* encodes a homolog of mismatch repair enzymes. *Nucleic Acids Res.* **20:** 2271–2278.

Fleck, O., C. Rudolph, A. Albrecht, A. Lorentz, P. Schar, and H. Schmidt. 1994. The mutator gene swi8 effects specific mutations in the mating-type region of *Schizosaccharomyces pombe. Genetics* **138:** 621–632.

Funabiki, H., I. Hagan, S. Uzawa, and M. Yanagida. 1993. Cell cycle-dependent specific positioning and clustering of centromeres and telomeres in fission yeast. *J. Cell Biol.* **121:** 961–976.

Gilson, E., T. Laroche, and S.M. Gasser. 1995. Telomeres and the functional architecture of the nucleus. *Trends Cell Biol.* **3:** 128–134.

Gorbsky, G.J. 1995. Kinetochores, microtubules and the metaphase checkpoint. *Trends Cell Biol.* **5:** 143–148.

Gottschling, D.E., O.M. Aparicio, B.L. Billington, and V.A. Zakian. 1990. Position effect at *S. cerevisiae* telomeres: Reversible repression of Pol II transcription. *Cell* **63:** 751–762.

Gutz, H. and H. Schmidt. 1985. Switching genes in *Schizosaccharomyces pombe. Curr. Genet.* **9:** 325–331.

Hartwell, L.H. and M.B. Kastan. 1994. Cell-cycle control and cancer. *Science* **266:** 1821–1828.

Hecht, A., T. Laroche, S. Strahlbolsinger, S.M. Gasser, and M. Grunstein. 1995. Histone H3 and H4 N-termini interact with sir3 and sir4 proteins: A molecular model for the formation of heterochromatin in yeast. *Cell* **80:** 583–592.

Hoheisel, J.D., E. Maier, R. Mott, L. McCarthy, A.V. Grigoriev, L.C. Schalkwyk, D. Nizetic, F. Francis, and H. Lehrach. 1993. High-resolution cosmid and P1 maps spanning the 14-Mb genome of the fission yeast *S. pombe. Cell* **73:** 109–120.

Karpen, G.H. 1994. Position-effect variegation and the new biology of heterochromatin. *Curr. Opin. Genet. Dev.* **4:** 281–291.

Kellum, R. and B.M. Alberts. 1995. Heterochromatin protein 1 is required for correct chromosome segregation in *Drosophila* embryos. *J. Cell Sci.* **108:** 1419–1431.

Kellum, R., J.W. Raff, and B.M. Alberts. 1995. Heterochromatin protein-1 distribution during development and during the cell-cycle in *Drosophila* embryos. *J. Cell Sci.* **108:** 1407–1418.

Klar, A.J.S. 1987. Differentiated parental DNA strands confer developmental asymmetry on daughter cells in fission yeast. *Nature* **326:** 466–470.

———. 1990. The developmental fate of fission yeast cells is determined by the pattern of inheritance of parental and grandparental DNA strands. *EMBO J.* **9:** 1407–1415.

———. 1992a. Developmental choices in mating-type interconversion in fission yeast. *Trends Genet.* **8:** 208–213.

———. 1992b. Molecular genetics of fission yeast cell type: Mating type and mating-type interconversion. In *The molecular and cellular biology of the yeast* Saccharomyces cerevisiae: *Gene expression* (ed. E.W. Jones et al.), vol. 2, pp. 745–777. Cold Spring Harbor Laboratory Press, Cold Spring Harbor, New York.

Klar, A.J.S. and M.J. Bonaduce. 1991. Swi6, a gene required for mating-type switching, prohibits meiotic recombination in the mat2-mat3 cold spot of fission yeast. *Genetics*

129: 1033–1042.

———. 1993. The mechanism of fission yeast mating-type interconversion: Evidence for 2 types of epigenetically inherited chromosomal imprinted events. *Cold Spring Harbor Symp. Quant. Biol.* **58:** 457–465.

Kyrion, G., K. Liu, C. Liu, and A.J. Lustig. 1993. RAP1 and telomere structure regulate telomere position effects in *Saccharomyces cerevisiae. Genes Dev.* **7:** 1146–1159.

Lorentz, A., L. Heim, and H. Schmidt. 1992. The switching gene swi6 affects recombination and gene-expression in the mating-type region of *Schizosaccharomyces pombe. Mol. Gen. Genet.* **233:** 436–442.

Lorentz, A., K. Ostermann, O. Fleck, and H. Schmidt. 1994. Switching gene swi6, involved in repression of silent mating-type loci in fission yeast, encodes a homolog of chromatin-associated proteins from *Drosophila* and mammals. *Gene* **143:** 139–143.

Messmer, S., A. Franke, and R. Paro. 1992. Analysis of the functional role of the *Polycomb* chromo domain in *Drosophila melanogaster. Genes Dev.* **6:** 1241–1254.

Miyata, H. and M. Miyata. 1981. Mode of conjugation in homothallic cells of *Schizosaccharomyces pombe. J. Gen. Appl. Microbiol.* **27:** 365–371.

Mizukami, T., W.I. Chang, I. Garkavtsev, N. Kaplan, D. Lombardi, T. Matsumoto, O. Niwa, A. Kounosu, M. Yanagida, T.G. Marr, and D. Beach. 1993. A 13 kb resolution cosmid map of the 14 Mb fission yeast genome by nonrandom sequence-tagged site mapping. *Cell* **73:** 121–132.

Murphy, T.D. and G.H. Karpen. 1995. Localization of centromere function in a *Drosophila* minichromosome. *Cell* **82:** 599–610.

Nielsen, O. and R. Egel. 1989. Mapping the double-strand breaks at the mating-type locus in fission yeast by genomic sequencing. *EMBO J.* **8:** 269–276.

Nimmo, E.R., G. Cranston, and R.C. Allshire. 1994. Telomere-associated chromosome breakage in fission yeast results in variegated expression of adjacent genes. *EMBO J.* **13:** 3801–3811.

Ostermann, K., A. Lorentz, and H. Schmidt. 1993. The fission yeast rad22 gene, having a function in mating-type switching and repair of DNA damages, encodes a protein homolog to rad52 of *Saccharomyces cerevisiae. Nucleic Acids Res.* **21:** 5940–5944.

Palladino, F. and S.M. Gasser. 1994. Telomere maintenance and gene repression: A common end. *Curr. Opin. Cell Biol.* **6:** 373–379.

Palladino, F., T. Laroche, E. Gilson, A. Axelrod, L. Pillus, and S.M. Gasser. 1993. SIR3 and SIR4 proteins are required for the positioning and integrity of yeast telomeres. *Cell* **75:** 543–555.

Paro, R. 1993. Mechanisms of heritable gene repression during development of *Drosophila. Curr. Opin. Cell Biol.* **5:** 999–1005.

Platero, J.S., T. Hartnett, and J.C. Eissenberg. 1995. Functional analysis of the chromo domain of HP1. *EMBO J.* **14:** 3977–3986.

Powers, J.A. and J.C. Eissenberg. 1993. Overlapping domains of the heterochromatin-associated protein HP1 mediate nuclear localization and heterochromatin binding. *J. Cell Biol.* **120:** 291–299.

Renauld, H., O.M. Aparicio, P.D. Zierath, B.L. Billington, S.K. Chhablani, and D.E. Gottschling. 1993. Silent domains are assembled continuously from the telomere and are defined by promoter distance and strength, and by SIR3 dosage. *Genes Dev.* **7:** 1133–1145.

Reuter, G. and P. Spierer. 1995. Position effect variegation and chromatin proteins. *BioEssays* **14:** 605–612.

Rodel, C., S. Kirchhoff, and H. Schmidt. 1992. The protein-sequence and some intron positions are conserved between the switching gene swi10 of *Schizosaccharomyces pombe* and the human excision repair gene ERCC1. *Nucleic Acids Res.* **20:** 6347–6353.

Sandell, L.L. and V.A. Zakian. 1993. Loss of a yeast telomere: Arrest, recovery, and chromosome loss. *Cell* **75:** 729–739.

Saunders, W.S., C. Chue, M. Goebl, C. Craig, R.F. Clark, J.A. Powers, J.C. Eissenberg, S.C. Elgin, N.F. Rothfield, and W.C. Earnshaw. 1993. Molecular cloning of a human homologue of *Drosophila* heterochromatin protein HP1 using anti-centromere autoantibodies with anti-chromo specificity. *J. Cell Sci.* **104:** 573–582.

Schmidt, H., P. Kapitza, and H. Gutz. 1987. Switching genes in *Schizosaccharomyces pombe*—Their influence on cell viability and recombination. *Curr. Genet.* **11:** 303–308.

Schmidt, H., P. Kapitzafecke, E.R. Stephen, and H. Gutz. 1989. Some of the swi genes of *Schizosaccharomyces pombe* also have a function in the repair of radiation damage. *Curr. Genet.* **16:** 89–94.

Singh, J. and A.J.S. Klar. 1993. DNA polymerase-α is essential for mating type switching in fission yeast. *Nature* **361:** 271–273.

Steiner, N.C. and L. Clarke. 1994. A novel epigenetic effect can alter centromere function in fission yeast. *Cell* **79:** 865–874.

Styrkarsdottir, U., R. Egel, and O. Nielsen. 1993. The smt-0 mutation which abolishes mating-type switching in fission yeast is a deletion. *Curr. Genet.* **23:** 184–186.

Sugawara, N.F. 1989. "DNA sequences at the telomeres of the fission yeast *S. pombe.*" Ph.D. thesis, Harvard University, Cambridge, Massachusetts.

Takahashi, K., S. Murakami, Y. Chikashige, H. Funabiki, O. Niwa, and M. Yanagida. 1992. A low copy number central sequence with strict symmetry and unusual chromatin structure in fission yeast centromere. *Mol. Biol. Cell* **3:** 819–835.

Thompson, J.S., L.M. Johnson, and M. Grunstein. 1994. Specific repression of the yeast silent mating locus HMR by an adjacent telomere. *Mol. Cell Biol.* **14:** 446–455.

Thon, G. and A.J.S. Klar. 1992. The clr1 locus regulates the expression of the cryptic mating-type loci of fission yeast. *Genetics* **131:** 287–296.

———. 1993. Directionality of fission yeast mating-type interconversion is controlled by the location of the donor loci. *Genetics* **134:** 1045–1054.

Thon, G., A. Cohen, and A.J. Klar. 1994. Three additional linkage groups that repress transcription and meiotic recombination in the mating-type region of *Schizosaccharomyces pombe. Genetics* **138:** 29–38.

Uzawa, S. and M. Yanagida. 1992. Visualization of centromeric and nucleolar DNA in fission yeast by fluorescence in situ hybridization. *J. Cell Sci.* **101:** 267–275.

Woollard, A. and P. Nurse. 1995. G1 regulation and checkpoints operating around START in fission yeast. *BioEssays* **17:** 481–490.

Wreggett, K.A., F. Hill, P.S. James, A. Hutchings, G.W. Butcher, and P.B. Singh. 1994. A mammalian homologue of *Drosophila* heterochromatin protein 1 (HP1) is a component of constitutive heterochromatin. *Cytogenet. Cell Genet.* **66:** 99–103.

Wright, J.H., D.E. Gottschling, and V.A. Zakian. 1992. *Saccharomyces* telomeres assume a non-nucleosomal chromatin structure. *Genes Dev.* **6:** 197–210.

Yanagida, M. 1995. Frontier question about sister chromatid separation in anaphase. *BioEssays* **17:** 519–526.

Transcriptional Silencing of the Yeast Mating-type Genes

Scott G. Holmes, Miriam Braunstein, and James R. Broach
Department of Molecular Biology
Princeton University
Princeton, New Jersey 08544

Differentiated cells rely on stable patterns of gene expression to maintain their committed state. Stability of gene expression in committed cell lineages depends on the interaction of a complex set of activating and repressing complexes with the DNA template. In each new cell division, these interactions must be precisely re-formed. One mechanism to ensure the stable inheritance of gene expression is to have the expression state of a gene play a role in its propagation. In this strategy, an activating or repressing complex acts as a template for its own re-formation through multiple cell divisions. Depending on the nature and regulation of the initial establishment of the expression state, this could lead to identical loci within the same cell having distinct, stable expression states.

Self-templating regulation has been proposed to explain the epigenetic pattern of inheritance of repressed transcription in organisms that exhibit heterochromatin-mediated position effects. Position effect refers to a phenomenon in which the expression state of a gene is influenced by its position on a chromosome. First described in *Drosophila* from adventitious juxtaposition of actively transcribed genes in euchromatic regions with a repressive heterochromatic domain (Henikoff 1990; Wilson et al. 1990), position effect resulting from heterochromatin also functions in many organisms to control regional gene expression in a regulated manner. A striking example is provided by dosage compensation in mammals. Female mammals have two copies of the X chromosome, whereas males have one. To prevent deleterious gene dosage effects, one X chromosome is inactivated in females. This is achieved by maintaining one in a largely inactive, heterochromatic state. Heterochromatin formation emanates from a *cis*-acting sequence, the X-inactivation center (XIC), and may depend on the function of the XIC-linked Xist gene. After a single X chromosome is randomly inactivated early in development, the relative expression states of the two X chromosomes are stable through the life of the organism (Migeon 1994; Rastan 1994; Riggs and

Epigenetic Mechanisms of Gene Regulation

Pfeiffer 1992). The mechanistic basis of this repression and its self-templating capacity are unresolved. Studies from other experimental systems, such as the one described in this chapter, may help answer many of the unresolved questions.

SILENCING IN *SACCHAROMYCES CEREVISIAE*

The developmental program of the yeast *S. cerevisiae* is also regulated at the level of gene expression by a position-effect mechanism (Herskowitz et al. 1992; Laurenson and Rine 1992). Haploid yeast exist in one of two distinct mating types, known as **a** and α. Mating type is determined by genes present at the expressed *MAT* locus. Similar or identical genes are also present at the *HML* and *HMR* loci located at either end of the same chromosome (see Fig. 1). Despite the fact that *HML* and *HMR* contain all the information necessary for expression of their **a** and α genes, they are not transcribed. Mating-type genes present at *HML* and *HMR* are normally expressed only after they are transposed to *MAT* following mating-type switching. This repression of the *HML* and *HMR* loci is known as silencing. Defects in silencing cause expression of the mating-type genes at *HML* and *HMR*, disrupting the developmental program of the cell and leading to a non-mating phenotype due to expression of both **a** and α in-

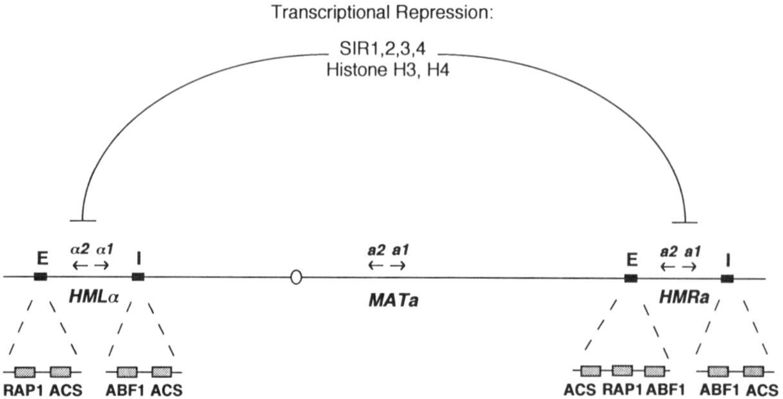

Figure 1 Diagram of the mating-type loci on chromosome III in *S. cerevisiae*. The active (*MAT*) and two silenced (*HML* and *HMR*) mating-type loci share extensive sequence identity as well as allele-specific domains. The silenced loci are bracketed by E and I silencers that are required *in cis* for transcriptional repression. Silencers comprise DNA-binding sites for Rap1p, Abf1p, and ORC, which binds ACS elements. Silencing also depends on several *trans*-acting factors, including the products of the *SIR* genes and histone H3 and H4.

formation in the same cell. In this review, we examine the mechanistic basis of this example of position-effect regulation.

COMPONENTS OF SILENCING

Silencers

Silencing depends on the concerted action of both *cis*- and *trans*-acting elements. Each of the silent mating-type loci is flanked by DNA sequences, known as the E and I silencers, that mediate repression (Abraham et al. 1984; Feldman et al. 1984). The silencers consist of different combinations of three defined elements, the binding sites for Rap1p, Abf1p, and ORC; the latter a complex of proteins involved in the initiation of DNA replication that binds yeast ACS (*A*RS *c*onsensus *s*equence) elements (Shore et al. 1987; Buchman et al. 1988a,b; Diffley and Stillman 1988; Bell et al. 1993; Foss et al. 1993; Micklem et al. 1993.). At *HML* the silencers have redundant function; either *HML*-E or *HML*-I is sufficient to repress *HML*α (Mahoney and Broach 1989). However, whereas *HMR*-E is sufficient for repression of *HMR***a**, *HMR*-I cannot mediate repression on its own (Brand et al. 1985). Silencer sequences are both necessary and sufficient to nucleate formation of the repressive structure; removal of the silencer sequences leads to full expression of *HML* and *HMR* (Brand et al. 1985; Mahoney and Broach 1989), whereas introduction of silencer sequences next to the *MAT* locus represses *MAT* transcription (Shei and Broach 1995). Silencers have similar functions but different potencies; a weaker silencer can be improved simply by inserting more copies of the silencer sequences. Finally, silencers are orientation-dependent (Shei and Broach 1995).

Silencer-binding Proteins

Silencer-binding factors play an essential role in mediating repression. ORC is a six-subunit protein complex that binds to yeast ACS elements and is required for initiation of DNA replication (Bell et al. 1993; Fox et al. 1995). In addition, temperature-sensitive alleles of two of the genes encoding ORC subunits, *ORC2* and *ORC5*, cause a defect in silencing (Foss et al. 1993; Fox et al. 1995), documenting the role of ORC in silencing. Rap1p is an essential protein that binds to the promoters of many genes, and in these contexts, Rap1p generally acts as an activator of transcription (Shore 1994). However, Rap1p also plays a critical role in transcriptional silencing: Certain alleles of *RAP1* allow cell growth but exhibit defects in silencing (Kurtz and Shore 1991; Liu et al. 1994).

ABF1 shows limited sequence similarity to *RAP1* (Diffley and Stillman 1989a) and also is involved in the positive regulation of a variety of genes. However, as with *RAP1*, specific alleles of *ABF1* diminish silencing (Loo et al. 1995).

Other *trans*-Acting Factors

In addition to the silencer-binding factors, several *trans*-acting factors affect repression of *HML* and *HMR*. These include histone H3 (Thompson et al. 1994), histone H4 (Kayne et al. 1988), and the products of the four *SIR* (*s*ilent *i*nformation *r*egulator) genes (Haber and George 1979; Klar et al. 1979; Rine et al. 1979; Rine and Herskowitz 1987). Deletion of the histone H4 amino terminus, or specific point mutations in the same region, lead to full expression of the silent mating-type loci (Kayne et al. 1988; Johnson et al. 1992; Megee et al. 1990; Park and Szostak 1990). Mutations in the histone H3 amino terminus also cause silencing defects, but are only revealed in backgrounds that are already compromised in repression (Thompson et al. 1994). The *SIR* proteins are encoded by nonessential genes whose products do not bind DNA themselves, but do interact with silencer-binding factors and histones (Moretti et al. 1994; Hecht et al. 1995). Null mutations in the *SIR2*, *SIR3*, or *SIR4* genes lead to complete expression of *HML* and *HMR*.

YEAST SILENCING AND HETEROCHROMATIN

Several features of yeast silencing suggest that repression is achieved by formation of a repressive chromatin structure which is the equivalent of metazoan heterochromatin. Heterochromatin is defined cytologically in flies and mammalian cells as regions of chromosomes that remain condensed through interphase (Heitz 1928; John 1988). Due to the small size and difficulty in visualizing yeast chromosomes, cytological criteria cannot be used to address whether heterochromatin is present in yeast. However, silenced domains in yeast and heterochromatic domains in metazoans share a number of genetic and biochemical features, which indicate that silencing in yeast results from formation of heterochromatin. First, silencing in yeast and heterochromatin in metazoans exert a nonspecific repression of gene transcription. A variety of yeast genes introduced at *HML* and *HMR* become subject to repression that depends on the silencers and *SIR* proteins (Brand et al. 1985; Schnell and Rine 1986; Mahoney and Broach 1989). Second, as with heterochromatin in *Drosophila* and the inactive X chromosome in mammals, *HML* and *HMR* are replicated late in S phase (McCarroll and Fangman 1988; Reynolds et al. 1989; Ferguson and Fangman 1992). Third, silenced domains in yeast are packaged in chromatin containing a specific form of histone H4 (see below), which is found exclusively in heterochromatin in *Drosophila* and mammals (Braunstein et al. 1993, 1996). Fourth, position-effect variegation in *Drosophila* and silencing in yeast each show acute sensitivity both to the gene dosage of components or regulators of silencing (Marshall et al. 1987; Stone et al. 1991; Renauld et al. 1993; Sussel et al. 1993) and to the amount of heterochromatin present in the cell (Buck and Shore 1995; Wiley and Zakian 1995). Finally, both yeast silencing and metazoan heterochromatin promote epigenetic inheritance, in which the expression of a gene varies between two states, but each state exhibits relatively stable mitotic inheritance (Pillus and Rine 1989; Gottschling et al. 1990; Mahoney et al. 1991; Sussel et al. 1993). Thus, silencing appears to be the yeast equivalent of metazoan heterochromatin.

EPIGENETIC INHERITANCE OF SILENCING

Transcriptional repression of the yeast mating-type cassettes is extremely efficient. In a wild-type background, less than 1 cell in 10^6 is derepressed for either of the silent cassettes (J. Strathern, pers. comm.). This exceptionally high efficiency of repression precludes assessing directly whether the repression mechanism is self-templating. However, the underlying epigenetic mechanism aiding the stability of repression at *HML* and *HMR* can be readily observed in mutant backgrounds that weaken, but do not eliminate, silencing. These observations, summarized below, document that a significant component contributing to efficient repression is the ability of the repressed state to promote its own persistence.

Sir1p

Epigenetic inheritance of the repressed state in yeast was first described by Pillus and Rine (1989) in strains deleted for the *SIR1* gene. Cells deleted for *SIR1* exhibit a decrease in the efficiency of silencing, when measured on a population basis by mating efficiency. However, by examining the expression state of individual cells using sensitivity to mating pheromone as an assay, Pillus and Rine found that, rather than showing a uniform decrease in mating efficiency, cells lacking Sir1p consisted of distinct populations of fully repressed and fully derepressed cells. By following the pedigree of individual cells to examine the expression state in succeeding generations, Pillus and Rine found that these cells exhibited a stable inheritance of their expression state; repressed cells gave rise to repressed cells at a far higher rate than would be predicted from their ratio in the population, whereas derepressed cells gave rise to derepressed cells, with rare conversions to the repressed state (see Fig. 2). This observation prompted the proposal that the repressed state was self-templating; once established, the repressive structure could enhance the probability that it would be inherited. Furthermore, the low frequency of conversion from derepressed cells to repressed cells in *sir1*⁻ strains is consistent with the hypothesis that Sir1p functions in the establishment of silencing but not in the maintenance of repression (Pillus and Rine 1989; Pillus 1992).

Telomeric Silencing

An epigenetic pattern of inheritance of silenced chromatin in yeast has been observed in several other contexts. Yeast telomeres exert a revers-

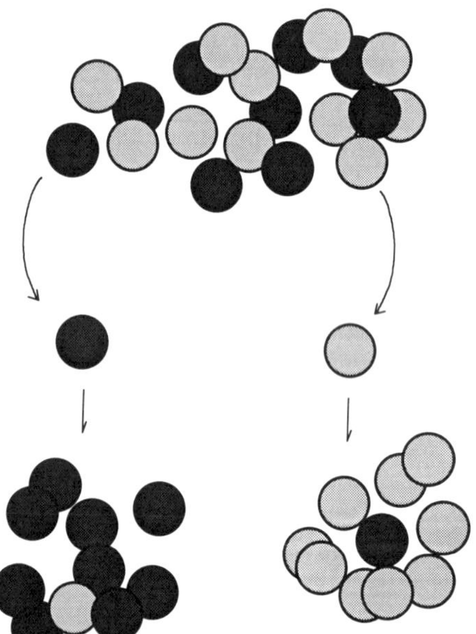

Figure 2 Epigenetic inheritance of a transcription state. Deletions of the *SIR1* gene or specific mutations in the *HML*-E or *HMR*-E silencer yield a mixed population of completely repressed (*dark circles*) or completely derepressed (*light circles*) cells. These cells exhibit a stable inheritance of their expression state such that repressed cells give rise to repressed cells, and derepressed cells give rise to derepressed cells, at a far higher rate than would be predicted from their ratio in the population.

ible transcriptional repression on yeast genes placed near them (Gottschling et al. 1990). Each yeast telomere contains approximately 20 Rap1p-binding sites (Wang and Zakian 1990; Gilson et al. 1993), which mediate telomere position effect (Liu et al. 1994; Stavenhagen and Zakian 1994). Repression at telomeres depends on almost all the *trans*-acting factors that control repression of the silent mating-type genes (the sole exception is Sir1p) (Aparicio et al. 1991), suggesting by analogy that telomere position effect is also likely to involve the formation of heterochromatin. This hypothesis is fortified by the observation that heterochromatin is formed at the telomeres of other eukaryotes. Genes placed at yeast telomeres exhibit variegation in their expression, producing mixed populations of repressed and derepressed cells. Similar to the effect seen at *HML* in *sir1*$^-$ backgrounds, these distinct expression states are mitotically stable, persisting for several generations before switching states (Gottschling et al. 1990).

RAP1 Alleles

Specific alleles of the *RAP1* gene (*rap1*^{s} alleles) also produce an epigenetic pattern of inheritance, as assayed by the expression of a reporter gene placed at *HMR* and visualized by a colony-sectoring assay. This phenotype is evident only in the background of a partially disabled silencer (Sussel et al. 1993) and may be indirect. That is, the *RAP1* alleles that elicit an epigenetic response also increase the efficiency of silencing at telomeres. This effect on telomeres appears to diminish efficiency of silencing at *HMR* through some form of competition (Buck and Shore 1995). Therefore, the heritability of expression at *HML* in a *rap1*^{s} background may be the indirect consequence of the epigenetic effect exhibited by telomeres.

Silencer Mutations

Specific mutations in the silencer sequences also reveal an epigenetic mechanism (Mahoney et al. 1991; Sussel et al. 1993). Partial deletions of the *HML*-E silencer—those deleted for either the ACS element, the Rap1p site, or a third, less well defined region—decrease the efficiency of silencing, as measured by the overall mating efficiency of populations of cells (Mahoney and Broach 1989; Mahoney et al. 1991). However, as with *sir1* mutants, a population of cells carrying any of the partial deletions of *HML*-E are a mix of repressed and derepressed cells whose expression states are relatively mitotically stable. Thus, mutations of the silencer appear to yield two genetically identical but phenotypically distinct populations that differ only in the expression states of *HML*.

Further studies in this system eliminated a competing hypothesis to explain the epigenetic nature of the *sir1* and silencer mutants; namely, that the distinct states simply represent the division between cells expressing *HML*α information above or below some threshold level. A strain designed to mimic this scenario—i.e., one that expressed a constant low level of α2 message—does yield a mixed population of repressed and derepressed cells. However, in this strain a single cell wavers between repressed and derepressed states and, moreover, such repressed cells are no more likely to give rise to repressed cells than to derepressed cells. This indicated that the expression state at *HML* in a *sir1* or silencer deletion background is not continuously variable. Rather, variation in repression states is binary, fully repressed or fully derepressed (Mahoney et al. 1991).

Other Epigenetic Silencing Mutants

To date, an epigenetic pattern of inheritance of the repressed state at the silent mating-type loci has only been observed in strains bearing mutations in elements that function in the nucleation of the silenced state. This includes the silencer sequences themselves, the silencer-binding protein Rap1p, and Sir1p (Sir1p can mediate the formation of limited silencing on its own when fused to a Gal4p DNA-binding domain and recruited to an *HMR* locus in which the E silencer has been replaced with a *GAL4* UAS sequence [Chien et al. 1993]). Thus, we might anticipate that some mutations in other silencer associated proteins—Abf1p or the ORC components—would exhibit epigenetic patterns of expression. The fact that none has been identified may reflect the limited number of alleles examined. In contrast, as described below, other components of the silencing apparatus—the other Sir proteins and the histones—act in a role distinct from that of the silencer. Accordingly, we predict that mutations in these other genes would be less likely to yield an epigenetic phenotype. These considerations have been incorporated into the model of silencing described below.

STAGES OF SILENCING

Silencing can formally be divided into three discrete stages: establishment, maintenance, and inheritance (Holmes and Broach 1996). Establishment refers to the formation of a silenced domain over a region that was formerly not silenced. This would occur, for instance, following integration of a silencer at an ectopic site. Establishment also occurs at normally silenced genes, either after silencing has been conditionally disrupted by mutation and the cell then returned to the permissive conditions, or in the transition from a derepressed to a repressed state in strains carrying a mutation exhibiting epigenetic inheritance. However, since these two conditions may not deactivate silencing in the same way, the mechanism of establishment may not be the same in both cases. Maintenance can refer to several processes, but in this review we restrict the definition to persistence of the silenced domain in the absence of cell cycle progression. We refer to inheritance as the process by which the silenced domain is perpetuated mitotically from one generation to the next. Each of these stages is examined in detail in the following.

Establishment

Silencing is initiated by association of a subset of the silencer-binding proteins—Rap1p, Abf1p, and ORC—with a nucleation site, which can be either a silencer or a telomere sequence. The primary function of these proteins is to recruit a *SIR* protein complex (Moretti et al. 1994; Cockell et al. 1995). However, recruitment of a Sir protein complex may not be sufficient to promote silencing; tethering Sir2p, Sir3p, or Sir4p to an *HMR* locus that lacks the *HMR*-E silencer by fusing these proteins to a Gal4p-DNA-binding domain and placing a *GAL4* UAS at *HMR*-E does not establish silencing (Chien et al. 1993). However, tethering Rap1p (Buck and Shore 1995) or Sir1p (Chien et al. 1993) in a similar manner does result in a limited amount of silencing. Although the Sir1p does not appear to bind DNA, it may exert its effects through the ACS element by interacting with a component of the ORC complex (Chien et al. 1993; Triolo and Sternglanz 1996). Thus, since certain silencer-binding proteins can promote silencing when tethered to the DNA, but individual *SIR* proteins do not, either multiple *SIR* proteins have to be recruited in order to establish silencing or silencer-binding proteins have critical functions in addition to recruitment.

One proposed additional function of the silencer-binding proteins is to localize *HML* and *HMR* to the nuclear periphery. Sir3p, Sir4p, and Rap1p are normally restricted to the nuclear periphery, a property that is dependent on the integrity of all three proteins. This provides additional evidence that these proteins function as a complex in yeast (Palladino et al. 1993; Cockell et al. 1995) and is consistent with the observation that Sir4p has regions of similarity to nuclear lamins (Diffley and Stillman

1989b). The inactive X chromosome in mammals also localizes to the nuclear periphery, suggesting that this interaction is important. However, the functional significance of this localization is not clear. In yeast, the peripheral localization of Rap1 protein does not strictly correlate with silencing. Mutations in the *RLF6* gene abolish Rap1p localization without affecting silencing (Konkel et al. 1995), and silencing can be abolished in other backgrounds without affecting Rap1p localization (Cockell et al. 1995). The localization of DNA sequences in these studies has been inferred from the position of the Rap1 protein; determining the position of telomere and silent mating-type locus sequences will be necessary to establish the role of subnuclear localization in the repression mechanism.

Following recruitment to the silencers, the *SIR* proteins mediate a change in the local chromatin structure that spreads outward from the silencer sequences. The observation that silencing has only two states, all or nothing, suggests that this may be a cooperative process (Mahoney et al. 1991). The precise composition and organization of silenced chromatin are not known, but it differs structurally and functionally from bulk chromatin. Silencing does not involve a change in the number or position of nucleosomes (Nasmyth 1982), but it is associated with a specific posttranslational modification of histone H4. Histones are subject to a reversible acetylation of lysine residues present along their amino-terminal tails, and the acetylation state of histones can affect the accessibility of transcription factors to promoters bound by nucleosomes containing these histones in vitro (Lee et al. 1993). In yeast, as well as in *Drosophila* and mammals, silenced chromatin contains primarily deacetylated histone H3 and H4 (Braunstein et al. 1993, 1996). Genetic evidence indicates that the specific pattern of histone H4 acetylation found on silenced chromatin in yeast is important for repression, suggesting that a decrease in acetylation is not merely an indirect effect of heterochromatin formation (Megee et al. 1990; Park and Szostak 1990; Johnson et al. 1992; Braunstein et al. 1996).

Several functional interactions are affected by the establishment of silencing. Besides the well-characterized effects on repressing transcription, silenced chromatin in yeast also restricts the access of a number of proteins to the silent mating-type loci. These include DNA repair factors (Terleth et al. 1989), endonucleases (Loo and Rine 1994), and ectopically expressed DNA methylases (Gottschling 1992; Singh and Klar 1992). However, the silent mating-type loci are readily accessible to DNase I, micrococcal nuclease (Nasmyth 1982), and Flp1p, a site-specific recombinase (Holmes and Broach 1996), and are substrates for the mating-type switching event that involves a gene conversion event between *HML*/*HMR* and *MAT*. These observations suggest a selective, not exclusive, resistance to protein interactions. Several experiments indicate that these alterations in chromatin are mediated by an effect of the *SIR* proteins on nucleosome structure. First, mutations in histone H4 that derepress silencing can be suppressed by specific mutations in *SIR3* (Johnson et al. 1990). Second, Sir3p and Sir4p physically interact with histones H3 and H4 in vitro (Hecht et al. 1995). Finally, overexpression of Sir2p in yeast leads to a global deacetylation of histone proteins (Braunstein et al. 1993).

Following a transient temperature shift of a strain carrying a temperature-sensitive allele of *SIR3,* reestablishment of silencing requires progression through the cell cycle (Miller and Nasmyth 1984). This result is consistent with the proposal that transit through S phase is required to establish silencing. Each silencer contains an ARS element, which serves as an origin of DNA replication on plasmids in yeast. These two observations initially led to the conjecture that initiation of DNA replication and the establishment of silencing were linked (Miller and Nasmyth 1984). However, several lines of evidence suggest that the establishment of silencing does not depend on initiation events occurring at the silencers. Although the *HMR*-E ARS does function as an origin at its chromosomal location (Rivier and Rine 1992), the *HML* ARS elements do not have measurable origin activity (Dubey et al. 1991), and sequences required for the *HML*-E silencer to function as an origin on plasmids are not required for silencing (Mahoney et al. 1991). In addition, alleles of the *ORC5* gene, a component of the initiation complex that binds ACS elements, have been isolated that are defective for silencing but not for DNA replication (Fox et al. 1995). The requirement for S-phase transit is most likely to allow remodeling of chromatin following duplication of the DNA to establish a nucleoprotein complex or DNA structure. This alteration in structure could be instigated directly by the action of *SIR* proteins, or it could be a passive event, in which the *SIR* proteins prevent access of the silent regions to an agent that induces an active chromatin conformation. For instance, the silencing complex could protect the DNA from the action of histone acetylases; S-phase transit would be required after derepression of silencing for the deposition of new, hypoacetylated histones.

The epigenetic pattern of inheritance resulting from specific mutations of certain components of the silencing apparatus has been taken as evidence that the exclusive role of these components is in establishment of silencing (Pillus and Rine 1989; Sussel et al. 1993). This is likely not the case. In each observable case of epigenetic inheritance of repression of the silent cassettes, the overall efficiency of silencing is reduced as a result of some specific mutation. This decrease in efficiency reflects a change in the equilibrium established by interconversion from the repressed to the derepressed state and vice versa. This change could result from a decrease in the rate of derepressed cells giving rise to repressed cells (a defect in establishment) or an increase in the rate of repressed cells converting to derepressed cells (a defect in inheritance) or both. The decrease in silencing observed in cells deleted for the *SIR1* gene is likely due to a combination of these defects. The rates of interconversion have been measured directly for *sir1* strains. A derepressed *sir1* cell has only a 0.004 chance of giving rise to a repressed cell in the next generation, whereas a repressed cell has a 0.02 to 0.1 chance of giving rise to a derepressed cell (Pillus and Rine 1989). These numbers can be compared with those for wild-type cells. The rate of decay of silencing in wild-type cells is less than 1 in 10^6, as measured by the percentage of α-mating cells in an *HMLα mata⁻ HMRa* strain (J. Strathern, pers. comm.). Thus, the presence of Sir1p decreases the rate of loss of silencing from 1 in 50 to 1 in 10^6, clearly demonstrating its role in inheritance of silencing. Because silencing is so efficient, the rate of establishment has not been measured in wild-type cells. However, establishment experiments using temperature-sensitive alleles of *SIR3* suggest that the rate is no smaller than 0.5 (Miller and Nasmyth 1984). Thus, Sir1p increases the rate of establishment by 100-fold. Accordingly, to the extent that reestablishment in this mutant background reflects the wild-type situation, these observations indicate that Sir1p also plays a role in establishment. Therefore, Sir1p (as well as other components whose mutation yields epigenetic inheritance) must be important for the inheritance of the repressed state as well as for its establishment.

Maintenance

Once established, the repressive structure must be maintained for the life of the cell. Although all the various elements that are essential for silencing have not been systematically tested for their functions at different phases of silencing, the silencers and the *SIR* proteins have distinct roles in the maintenance of silencing in G_1. Studies with temperature-sensitive alleles of *SIR3* have shown that silencing is rapidly lost in G_1-arrested cells if the culture is shifted to the nonpermissive temperature (Miller and Nasmyth 1984; Holmes and Broach 1996). In contrast, deletion of the *HMLα* silencer sequences in cells arrested at G_1 phase has no effect on repression of *HML*α (Holmes and Broach 1996). Therefore, the role of the silencers can be distinguished from that of Sir3p at this stage. Sir3p, in conjunction with other *SIR* proteins, may be an integral part of the repressive structure formed over the silent mating-type loci. This would account for its continuous requirement for maintenance of silencing.

The properties of the silenced chromatin are not static during the cell cycle. Aparicio and Gottschling (1994) examined the nature of silencing by characterizing the ability of transcription factors to gain access to a gene placed near telomere sequences. The repressed structure at telomeres exhibits varying sensitivity to challenge by competition with promoter-binding factors as a function of the cell cycle. Immediately following S phase, telomeric genes are repressed for basal transcription but can be activated when high levels of a specific inducer of the test gene are present. However, cells in G_1 phase of the cell cycle are repressed for both basal and activated transcription, suggesting a more fully repressive configuration of telomeric chromatin. Despite these observations that the accessibility of silenced domains can change during the cell cycle, repression at the silent cassettes is fully maintained. No transcription from the silent cassettes can be detected at any stage during the cell cycle. Thus, the structure of heterochromatin may fluctuate during the cell cycle but maintain its repressive capacity throughout.

Inheritance

In various contexts, heterochromatin can promote its own inheritance. As described above, an epigenetic mechanism contributes to the stable propagation of silencing to successive generations. Specific aspects of the silencing mechanism in yeast may be incorporated into general models of how inheritance of a repressed state is achieved.

A simple model for the inheritance of an expression state is that, once established, the nucleoprotein complexes determining gene activity or repression are capable of reproducing these structures following DNA replication (see Fig. 3). In this model, the repressive complex that is formed over the silent mating-type loci remains associated with the DNA during replication. Elements of the complex remain associated with each strand of the duplex as it separates and, after completion of DNA replication, the partial complex present on each new DNA duplex is sufficient to seed formation of a complete repressive structure. There are several candidates for the components that are segregated and serve to crystallize formation of new repressive chromatin. First, silencing is associated with

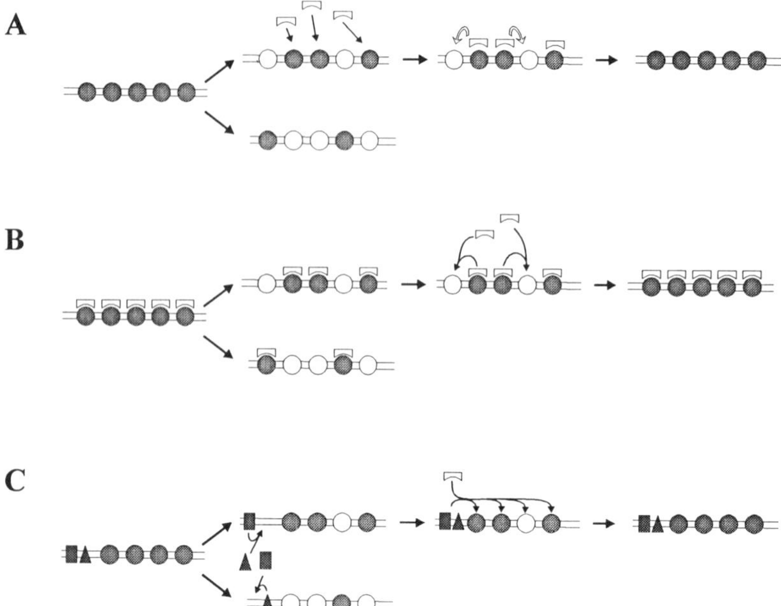

Figure 3 Mechanisms of epigenetic inheritance at the silent mating-type loci. In each model, some feature of the nucleoprotein complex determining repression is segregated with the old DNA strands to nucleate re-formation of the repressive structure on the nascent duplexes. The models are not mutually exclusive. (*A*) Modified nucleosomes are randomly segregated, then specifically attract a complex of Sir proteins, which modify new nucleosomes. (*B*) Sir proteins bound to nucleosomes over the silent mating-type loci remain associated with the nucleosomes as they are distributed to the new duplexes. New nucleosomes are modified by these Sir proteins, then attract more Sir complexes, which associate with new nucleosomes, or modify new nucleosomes themselves. (*C*) Silencer-binding proteins distribute to each new DNA duplex and direct re-formation of the repressive structure by attracting a complex of Sir proteins.

hypoacetylated forms of histones H3 and H4 (Braunstein et al. 1993, 1996). Distribution of nucleosomes containing these specifically modified histones, and subsequent specific recognition by components of the silencing machinery, could lead to a rapid re-formation of the structure. Sir3p and Sir4p bind to unacetylated histone H3 and H4 in vitro (Hecht et al. 1995). Whether these *SIR* proteins show a diminished ability to bind acetylated histones has not been shown, but this could potentially provide specificity for the interaction. Alternatively, the *SIR* proteins, particularly Sir3p and Sir4p, may remain bound to the nucleosomes, segregate to the newly formed DNA, and reseed formation of the complex. Finally, proteins bound to the silencer sequences may be distributed to each new DNA duplex and initiate re-formation of the repressive structure in each cell division.

Silencer sequences play a critical role in the inheritance of repression. The role of the silencers in postestablishment events has been examined using a system in which the silencer sequences could be deleted by an inducible, in vivo recombination system (Holmes and Broach 1996). Using this system, we found that silencers are not continuously required for maintenance of the repressed state; repression is unaffected by deletion of the silencer in cells blocked in G_1. In contrast, repression in the absence of the silencers is erased in cycling cells within a single generation. This indicates that a structure that is sufficient to maintain repression is not sufficient to promote its own inheritance. The role of silencers could be to promote the inheritance of the repressed state following DNA replication, or to prevent disassembly of the repressed state at other phases of the cell cycle, or to perform both functions. Proteins may remain bound to the silencers through the cell cycle. Following DNA replication, proteins that remain bound could help the re-formation of heterochromatin. Since the silencers are dispensable for maintenance of repression in G_1 but are essential for the initial establishment of silencing, establishment and inheritance functions may be coincident. This suggests that inheritance or self-templating may require an establishment event each cell cycle. This would be consistent with the requirement for the silencers in each cell division cycle. This establishment event would not be equivalent to a de novo formation of silenced chromatin from a derepressed cell; if that were true, then derepressed cells and repressed cells would be expected to have an equal probability of being silenced in the next generation. Since silenced cells are far more likely to pass on the repressed state in the context of a partially active silencer, inheritance must be far more efficient with a silenced template. This suggests that features of the chromatin besides the silencer may be distributed and initiate heterochromatin re-formation.

A series of physical interactions link silencer-binding proteins, the *SIR* proteins, and the nucleosome. Combined with the cooperative nature of establishment of silencing, these observations suggest that a feedback loop of mutual attractions may function efficiently to reestablish silencing each division cycle. The persistence of silencer-binding proteins could attract a *SIR* protein complex, whereas deacetylated histones bound by a *SIR* protein complex could predispose recruitment of silencer-binding factors. At no time in the cell cycle are *HML* and *HMR* expressed, suggesting the re-formation of silencing is rapid. This mass-action model also predicts the dosage sensitivity exhibited by silencing factors (Locke et al. 1988).

Heterochromatin in yeast and in other organisms replicates late in S phase. This observation has given rise to a distinct class of models for the persistence of the structure in which maintaining the distinct timing of replication is crucial to maintaining repression. A competition may take place between the silencing machinery and transcription factors attempting to gain access to the silenced gene's promoters. By the end of S phase, the concentration of transcription factors may be low, due to titration by the promoters of previously replicated genes, shifting the equilibrium to formation of the repressive structure. In this model, the silencing machinery would have the primary task of ensuring that the encompassed regions become, and remain, late-replicating. Consistent with this proposal, high expression of a transcription factor at the end of S phase can activate genes placed at yeast telomeres, suggesting a competition does take place (Aparicio and Gottschling 1994). Alternatively, there may be qualitative differences in the chromatin formed late in S phase, or it may be localized to a novel site in the nucleus that promotes repression.

The study of silencing in yeast has allowed identification of many components essential for the process and a clear description of an underlying epigenetic pattern of inheritance. Progress has been made toward determining how the various factors interact to form and propagate the repressed structure. However, substantial detail remains to be resolved, and efforts to do so should yield insights into the structure and function of heterochromatin-based repression in yeast as well as in larger eukaryotes.

CONCLUSIONS

Yeast utilizes a position-effect mechanism known as silencing to regulate expression of genes required for determining cell type. Silencing involves the formation of a repressive chromatin structure that by several criteria is equivalent to metazoan heterochromatin. Similar to the pattern seen in other organisms, heterochromatic regions of the yeast genome exhibit a mitotically stable inheritance that suggests a self-templating mechanism. In yeast, a structure that is sufficient to repress transcription is not sufficient to promote its own inheritance; the sequences required for nucleation of the structure are also required. The inheritance of silencing could involve a rapid disassembly and reassembly mechanism in which some aspect of the silencing complex remains associated with the old DNA strands after DNA replication and is sufficient to seed reformation of the structure on the new DNA duplexes in a cooperative assembly process.

ACKNOWLEDGMENTS

We thank members of our laboratory and other colleagues for helpful discussions. Work in the author's laboratory is supported by National Institutes of Health grant GM-48540 to J.R.B. and a National Institutes of Health postdoctoral fellowship to S.G.H.

REFERENCES

Abraham, J., K.A. Nasmyth, J.N. Strathern, A.J.S. Klar, and J.B. Hicks. 1984. Regulation of mating-type information in yeast: Negative control requiring sequences both 5′ and 3′ to the regulated region. *J. Mol. Biol.* **176:** 307–331.

Aparicio, O.M. and D.E. Gottschling. 1994. Overcoming telomeric silencing: A *trans*-activator competes to establish gene expression in a cell cycle-dependent way. *Genes Dev.* **8:** 1133–1146.

Aparicio, O.M., B.L. Billington, and D.E. Gottschling. 1991. Modifiers of position effect are shared between telomeric and silent mating-type loci in *S. cerevisiae.* **Cell 66:** 1279–1287.

Bell, S.P., R. Kobayashi, and B. Stillman. 1993. Yeast origin recognition complex functions in transcription silencing and DNA replication. *Science* **262:** 1844–1849.

Brand, A., L. Breeden, J. Abraham, R. Sternglanz, and K.A. Nasmyth. 1985. Characterization of a silencer in yeast: A DNA sequence with properties opposite to those of a transcriptional enhancer. *Cell* **41:** 41–48.

Braunstein, M., C.D. Allis, B.M. Turner, and J.R. Broach. 1996. Transcriptional silencing in yeast requires a heterochromatin histone acetylation pattern. *Mol. Cell. Biol.* (in press).

Braunstein, M., A.B. Rose, S.G. Holmes, C.D. Allis, and J.R. Broach. 1993. Transcriptional silencing in yeast is associated with reduced nucleosome acetylation. *Genes Dev.* **7:** 592–604.

Buchman, A.R., N.F. Lue, and R.D. Kornberg. 1988a. Connections between transcriptional activators, silencers, and telomeres as revealed by functional analysis of a yeast DNA-binding protein. *Mol. Cell. Biol.* **8:** 5086–5099.

Buchman, A.R., W.J. Kimmerly, J. Rine, and R.D. Kornberg. 1988b. Two DNA-binding factors recognize specific sequences at silencers, upstream activating sequences, autonomously replicating sequences, and telomeres in *Saccharomyces cerevisiae. Mol. Cell. Biol* **8:** 210–225.

Buck, S.W. and D. Shore. 1995. Action of a RAP1 carboxy-terminal silencing domain reveals an underlying competition between HMR and telomeres in yeast. *Genes Dev.* **9:** 370–384.

Chien, C.T., S. Buck, R. Sternglanz, and D. Shore. 1993. Targeting of SIR1 protein establishes transcriptional silencing at HM loci and telomeres in yeast. *Cell* **75:** 531–541.

Cockell, M., F. Palladino, T. Laroche, G. Kyrion, C. Liu, A.J. Lustig, and S.M. Gasser.

1995. The carboxy termini of Sir4 and Rap1 affect Sir3 localization: Evidence for a multicomponent complex required for yeast telomeric silencing. *J. Cell Biol.* **129:** 909–924.

Diffley, J.F. and B. Stillman. 1988. Purification of a yeast protein that binds to origins of DNA replication and a transcriptional silencer. *Proc. Natl. Acad. Sci.* **85:** 2120–2124.

———. 1989a. Similarity between the transcriptional silencer binding proteins ABF1 and RAP1. *Science* **246:** 1034–1038.

———. 1989b. Transcriptional silencing and lamins. *Nature* **342:** 24.

Dubey, D.D., L.R. Davis, S.A. Greenfeder, L.Y. Ong, J.G. Zhu, J.R. Broach, C.S. Newlon, and J.A. Huberman. 1991. Evidence suggesting that the ARS elements associated with silencers of the yeast mating-type locus HML do not function as chromosomal DNA replication origins. *Mol. Cell. Biol.* **11:** 5346–5355.

Feldman, J.B., J.B. Hicks, and J.R. Broach. 1984. Identification of the sites required for repression of a silent mating type locus in yeast. *J. Mol. Biol.* **178:** 815–834.

Ferguson, B.M. and W.L. Fangman. 1992. A position effect on the time of replication origin activation in yeast. *Cell* **68:** 333–339.

Foss, M., F.J. McNally, P. Laurenson, and J. Rine. 1993. Origin recognition complex (ORC) in transcriptional silencing and DNA replication in *S. cerevisiae*. *Science* **262:** 1838–1844.

Fox, C.A., S. Loo, A. Dillin, and J. Rine. 1995. The origin recognition complex has essential functions in transcriptional silencing and chromosomal replication. *Genes Dev.* **9:** 911–924.

Gilson, E., M. Roberge, R. Giraldo, D. Rhodes, and S. Gasser. 1993. Distortion of the DNA double helix by RAP1 at silencers and multiple telomeric binding sites. *J. Mol. Biol.* **231:** 293–310.

Gottschling, D.E. 1992. Telomere-proximal DNA in *Saccharomyces cerevisiae* is refractory to methyltransferase activity in vivo. *Proc. Natl. Acad. Sci.* **89:** 4062–4065.

Gottschling, D.E., O.M. Aparicio, B.L. Billington, and V.A. Zakian. 1990. Position effect at *S. cerevisiae* telomeres: Reversible repression of pol II transcription. *Cell* **63:** 751–762.

Haber, J.E. and J.R. George. 1979. A mutation that permits the expression of normally silent copies of mating-type information in *Saccharomyces cerevisiae*. *Genetics* **93:** 13–35.

Hecht, A., T. Laroche, S. Strahl-Bolsinger, S.M. Gasser, and M. Grunstein. 1995. Histone H3 and H4 N-termini interact with SIR3 and SIR4 proteins: A molecular model for the formation of heterochromatin in yeast. *Cell* **80:** 583–592.

Heitz, E. 1928. Das heterochromatin der moose. *Jahrb. Wiss. Bot.* **69:** 726–818.

Henikoff, S. 1990. Position-effect variegation after 60 years. *Trends Genet.* **6:** 422–426.

Herskowitz, I., J. Rine, and J. Strathern. 1992. Mating-type determination and mating type interconversion in *Saccharomyces cerevisiae*. In *The molecular and cellular biology of the yeast* Saccharomyces: *Gene Expression* (ed. E.W. Jones et al.), vol. 2, pp. 583–656. Cold Spring Harbor Laboratory Press, Cold Spring Harbor, New York.

Holmes, S. and J.R. Broach. 1996. Silencers are required for inheritance of the repressed state in yeast. *Genes Dev.* **10:** 1021–1032.

John, B. 1988. The biology of heterochromatin. In *Heterochromatin: Molecular and structural aspects* (ed. R.S. Verma), pp. 1–147. Cambridge University Press, England.

Johnson, L.M., G. Fisher-Adams, and M. Grunstein. 1992. Identification of a non-basic domain in the histone H4 N-terminus required for repression of the yeast silent mating loci. *EMBO J.* **11:** 2201–2209.

Johnson, L.M., P.S. Kayne, E.S. Kahn, and M. Grunstein. 1990. Genetic evidence for an interaction between *SIR3* and histone H4 in the repression of the silent mating loci in *Saccharomyces cerevisiae. Proc. Natl. Acad. Sci.* **87:** 6286–6290.

Kayne, P.S., U.-J. Kim, M. Han, J.R. Mullen, F. Yoshizaki, and M. Grunstein. 1988. Extremely conserved histone H4 N terminus is dispensable for growth but essential for repressing the silent mating loci in yeast. *Cell* **55:** 27–39.

Klar, A.J.S., S. Fogel, and K. MacLeod. 1979. MAR1—A regulator of *HM*a and *HM*α loci in *Saccharomyces cerevisiae. Genetics* **93:** 37–50.

Konkel, L.M., S. Enomoto, E.M. Chamberlain, P. McCune-Zierath, S.J. Iyadurai, and J. Berman. 1995. A class of single-stranded telomeric DNA-binding proteins required for Rap1p localization in yeast nuclei. *Proc. Natl. Acad. Sci.* **92:** 5558–5562.

Kurtz, S. and D. Shore. 1991. RAP1 protein activates and silences transcription of mating-type genes in yeast. *Genes Dev.* **5:** 616–628.

Laurenson, P. and J. Rine. 1992. Silencers, silencing, and heritable transcriptional states. *Microbiol. Rev.* **56:** 543–560.

Lee, D.Y., J.J. Hayes, D. Pruss, and A.P. Wolffe. 1993. A positive role for histone acetylation in transcription factor access to nucleosomal DNA. *Cell* **72:** 73–84.

Liu, C., X. Mao, and A.J. Lustig. 1994. Mutational analysis defines a C-terminal tail domain of RAP1 essential for telomeric silencing in *Saccharomyces cerevisiae. Genetics* **138:** 1025–1040.

Locke, J., M.A. Kotarski, and K.D. Tartof. 1988. Dosage-dependent modifiers of position effect variegation in *Drosophila* and a mass action model that explains their effects. *Genetics* **120:** 181–198.

Loo, S. and J. Rine. 1994. Silencers and domains of generalized repression. *Science* **264:** 1768–1771.

Loo, S., P. Laurenson, M. Foss, A. Dillin, and J. Rine. 1995. Roles of *ABF1*, *NPL3*, and *YCL54* in silencing in *Saccharomyces cerevisiae. Genetics* **141:** 889–902.

Mahoney, D.J. and J.R. Broach. 1989. The *HML* mating-type cassette of *Saccharomyces cerevisiae* is regulated by two separate but functionally equivalent silencers. *Mol. Cell. Biol.* **9:** 4621–4630.

Mahoney, D.J., R. Marquardt, G.J. Shei, A.B. Rose, and J.R. Broach. 1991. Mutations in the *HML* E silencer of *Saccharomyces cerevisiae* yield metastable inheritance of transcriptional repression. *Genes Dev.* **5:** 605–615.

Marshall, M., S. Mahoney, A. Rose, J.B. Hicks, and J.R. Broach. 1987. Functional domains of *SIR4*, a gene required for position effect regulation in *Saccharomyces cerevisiae. Mol. Cell. Biol.* **7:** 4441–4452.

McCarroll, R.M. and W.L. Fangman. 1988. Time of replication of yeast centromeres and telomeres. *Cell* **54:** 505–513.

Megee, P.C., B.A. Morgan, B.A. Mittman, and M.M. Smith. 1990. Genetic analysis of histone H4: Essential role of lysines subject to reversible acetylation. *Science* **247:** 841–845.

Micklem, G., A. Rowley, J. Harwood, K. Nasmyth, and J.F. Diffley. 1993. Yeast origin recognition complex is involved in DNA replication and transcriptional silencing. *Nature* **366:** 87–89.

Migeon, B. 1994. X-chromosome inactivation: Molecular mechanisms and genetic consequences. *Trends Genet.* **10:** 230–235.

Miller, A. and K. Nasmyth. 1984. Role of DNA replication in the repression of silent mating type loci in yeast. *Nature* **312:** 247–251.

Moretti, P., K. Freeman, L. Coodly, and D. Shore. 1994. Evidence that a complex of SIR proteins interacts with the silencer and telomere-binding protein *RAP1. Genes Dev.* **8:** 2257–2269.

Nasmyth, K.A. 1982. The regulation of yeast mating-type chromatin structure by *SIR*: An action at a distance affecting both transcription and transposition. *Cell* **30:** 567–578.

Palladino, F., T. Laroche, E. Gilson, A. Axelrod, L. Pillus, and S.M. Gasser. 1993. SIR3 and SIR4 proteins are required for the positioning and integrity of yeast telomeres. *Cell* **75:** 543–555.

Park, E.-C. and J.W. Szostak. 1990. Point mutations in the yeast histone H4 gene prevent silencing of the silent mating type locus *HML. Mol. Cell. Biol.* **10:** 4932–4934.

Pillus, L. 1992. An acquired state: Epigenetic mechanisms in transcription. *Curr. Opin. Cell Biol.* **4:** 453–458.

Pillus, L. and J. Rine. 1989. Epigenetic inheritance of transcription states in *S. cerevisiae. Cell* **59:** 637–647.

Rastan, S. 1994. X chromosome inactivation and the Xist gene. *Curr. Opin. Genet. Dev.* **4:** 292–297.

Renauld, H., O.M. Aparicio, P.D. Zierath, B.L. Billington, S.K. Chablani, and D.E. Gottschling. 1993. Silent domains are assembled continuously from the telomere and are defined by promoter distance and strength, and by SIR3 dosage. *Genes Dev.* **7:** 1133–1145.

Reynolds, A.E., R.M. McCarroll, C.S. Newlon, and W.L. Fangman. 1989. Time of replication of *ARS* elements along yeast chromosome III. *Mol. Cell. Biol.* **9:** 4488–4494.

Riggs, A.D. and G.P. Pfeiffer. 1992. X-chromosome inactivation and cell memory. *Trends Genet.* **8:** 169–174.

Rine, J. and I. Herskowitz. 1987. Four genes responsible for a position effect on expression from *HML* and *HMR* in *Saccharomyces cerevisiae. Genetics* **116:** 9–22.

Rine, J.D., J.N. Strathern, J.B. Hicks, and I. Herskowitz. 1979. A suppressor of mating-type locus mutation in *Saccharomyces cerevisiae*: Evidence for and identification of cryptic mating type loci. *Genetics* **93:** 877–901.

Rivier, D.H. and J. Rine. 1992. An origin of DNA replication and a transcription silencer require a common element. *Science* **256:** 659–663.

Schnell, R. and J. Rine. 1986. A position effect on the expression of a tRNA gene mediated by the *SIR* gene of *Saccharomyces cerevisiae. Mol. Cell. Biol.* **6:** 494–501.

Shei, G.J. and J.R. Broach. 1995. Yeast silencers can act as orientation-dependent gene inactivation centers that respond to environmental signals. *Mol. Cell. Biol.* **15:** 3496–3506.

Shore, D. 1994. RAP1: A protein regulator in yeast. *Trends Genet.* **10:** 408–412.

Shore, D., D.J. Stillman, A.H. Brand, and K.A. Nasmyth. 1987. Identification of silencer binding proteins from yeast: Possible roles in *SIR* control and DNA replication. *EMBO J.* **6:** 461–467.

Singh, J. and A.J.S. Klar. 1992. Active genes in budding yeast display enhanced in vivo accessibility to foreign DNA methylases: A novel in vivo probe for chromatin structure of yeast. *Genes Dev.* **6:** 186–196.

Stavenhagen, J.B. and V.A. Zakian. 1994. Internal tracts of telomeric DNA act as silencers in *Saccharomyces cerevisiae. Genes Dev.* **8:** 1411–1422.

Stone, E.M., M.J. Swanson, A.M. Romeo, J.B. Hicks, and R. Sternglanz. 1991. The SIR1 gene of *Saccharomyces cerevisiae* and its role as an extragenic suppressor of several mating-defective mutants. *Mol. Cell. Biol.* **11:** 2253–2262.

Sussel, L., D. Vannier, and D. Shore. 1993. Epigenetic switching of transcriptional states: *cis-* and *trans*-acting factors affecting establishment of silencing at the *HMR* locus in *Saccharomyces cerevisiae*. *Mol. Cell. Biol.* **13:** 3919–3928.

Terleth, C., C.A. van Sluis, and P. van de Putte. 1989. Differential repair of UV damage in *Saccharomyces cerevisiae*. *Nucleic Acids Res.* **12:** 4433–4439.

Thompson, J.S., X. Ling, and M. Grunstein. 1994. Histone H3 amino terminus is required for telomeric and silent mating locus repression in yeast. *Nature* **369:** 245–247.

Triolo, T. and R. Sternglanz. 1996. Role of interactions between the origin recognition complex and SIR1 in transcriptional silencing. *Nature* **381:** 251–253.

Wang, S.-S. and V.A. Zakian. 1990. Sequencing of *Saccharomyces* telomeres cloned using T4 DNA polymerase reveals two domains. *Mol. Cell. Biol.* **10:** 4415–4419.

Wiley, E.A. and V.A. Zakian. 1995. Extra telomeres, but not internal tracts of telomeric DNA, reduce transcriptional repression at *Saccharomyces* telomeres. *Genetics* **139:** 67–79.

Wilson, C., H.J. Bellen., and W.J. Gehring. 1990. Position effects on eukaryotic gene expression. *Annu. Rev. Cell Biol.* **6:** 679–714.

Stable Chromatin States Regulating Homeotic Genes in *Drosophila*

Vincenzo Pirrotta
Department of Zoology
University of Geneva
CH1211 Geneva, Switzerland

During the development of the embryo, many of the important developmental decisions that are made by individual cells depend on molecular cues that are transiently present but that vanish as development proceeds. Yet, the commitment of a cell to a developmental pathway is maintained in its cellular progeny. Many kinds of molecular mechanisms can be imagined to preserve such a state of determination, based, for example, on self-maintaining intracellular or intercellular feedback loops, but evidence has been accumulating for a type of mechanism that relies on the establishment of self-perpetuating alternative chromatin states that determine the activity of broad chromatin domains. In *Drosophila*, this type of mechanism involves the products of the Polycomb group of genes (PcG) and is best known for its role in maintaining patterns of homeotic gene expression. The effects of the PcG proteins in establishing a cellularly inherited but variable repressed state are reminiscent of position-effect variegation in *Drosophila*, silencing in yeast, and paramutation in plants. The study of the mechanisms regulating the expression of homeotic genes has provided much of the current knowledge of PcG proteins and their action.

THE REPRESSED STATE OF THE *Ubx* GENE

Drosophila possesses a set of homeotic genes, each of which is expressed in a particular segmental domain along the antero-posterior axis of the embryo and repressed in other segments. The expression of each homeotic gene confers a positional identity to the cells of a particular parasegment and determines their subsequent developmental fate. The mechanisms controlling the expression of the different homeotic genes are probably very similar, but the best-known is that of the *Ultrabithorax* gene (*Ubx*), required to specify the development of the posterior thorax and anterior abdomen. Correspondingly, the gene is active very weakly in parasegment 5 (PS5), strongly in PS6, and at intermediate levels in

PS7–13 but is repressed in more anterior segments (Fig. 1). The parasegments, body units consisting of the posterior part of a segment and the anterior part of the following segment, are the repetitive units in which the expression pattern of homeotic genes is organized. The parasegmental activity of *Ubx* is determined by a set of enhancer modules scattered over a region spanning from 35 kb upstream to 45 kb downstream from the promoter (Simon et al. 1990; Müller and Bienz 1991; Qian et al. 1991; Pirrotta et al. 1995). Each module consists of 400–500 bp and contains a cluster of binding sites for activators which are segmentation pair-rule genes such as *fushi tarazu* (*ftz*) or *even skipped* (*eve*) (Qian et al. 1993). The pattern of expression each module generates depends on the particular combination of activators present in that module. Expression in the anterior half of the embryo is repressed by *hunchback* (*hb*) and in the posterior-most region by *tailless* (*tll*), two segmentation gap genes ac-

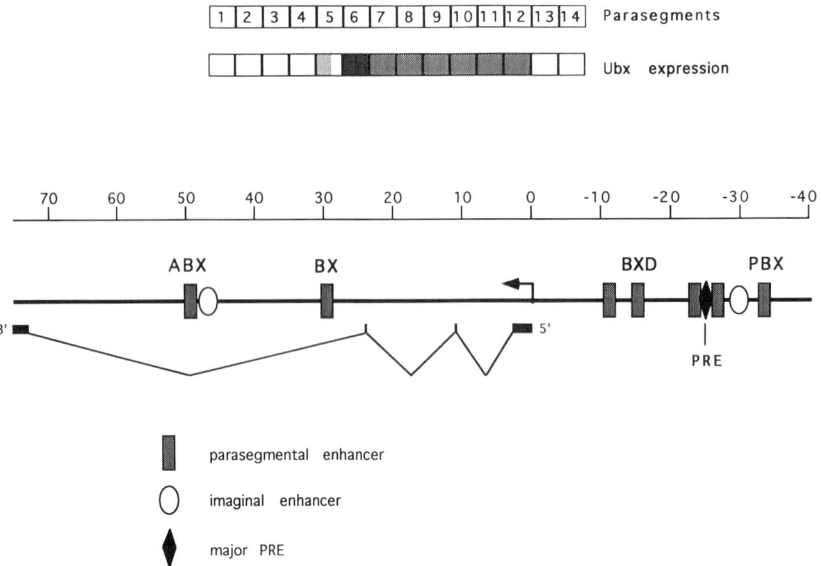

Figure 1 Ubx expression domain and regulatory region. A schematic representation of the embryonic segments and the *Ubx* domain of expression is shown above. Expression levels are indicated by the darkness of the shading. Below is a map of the *Ubx* gene with the transcription start site indicated by the bent arrow. Seven parasegmental enhancer modules are represented by dark rectangular boxes, four of which correspond to the genetically defined *abx*, *bx*, *bxd*, and *pbx* regulatory regions. Two clusters of imaginal enhancers are represented by open ovals. The site of the principal PcG response element (PRE) is indicated by a black diamond. The scale is in kilobases.

tivated in their respective anterior and posterior domains by maternal positional cues (Qian et al. 1991; Zhang et al. 1991). These repressors bind to *Ubx* enhancers at sites that are immediately adjacent to, or overlap with, the binding sites of the activators. In vivo and in vitro experiments indicate that they act by competitive binding and blocking the access of the activators.

Experiments with reporter gene constructs driven by isolated enhancer modules show that the repressing activity fades away shortly after gastrulation, when both *hb* and *tll* stop being expressed in their anterior and posterior domains. As a result, the individual enhancers are derepressed and begin to function more anteriorly and posteriorly (Simon et al. 1990; Müller and Bienz 1991; Qian et al. 1991; Pirrotta et al. 1995). In the embryo, however, expression of the *Ubx* gene in these regions continues to be repressed by the action of the PcG products (Simon et al. 1992) mediated by specific target regions in the *Ubx* gene called the PcG response elements (PREs) (Chan et al. 1994). The function of most parasegmental enhancers cannot continue beyond the germ-band-extension stage because the segmentation gene products that activate them disappear at that point. Two of these enhancers, however, respond to activators present in the nervous system and are responsible for the strong expression of the *Ubx* gene in the central nervous system at later stages. Segmental expression of the *Ubx* gene is confined to the correct embryonic domain specified initially by the repressive action of the HB and TLL proteins, even though these repressors are no longer present. The continued repression in the anterior and posterior regions of the embryo depends on the PcG products, indicating that they maintain the repressed state initiated in the early embryo throughout the rest of embryonic development even in rapidly proliferating tissues such as the nervous system.

In *Drosophila*, adult structures are distinct from larval structures. They develop from the imaginal discs, small clusters of cells set aside in the early embryo, which proliferate in the course of larval growth but remain undifferentiated until the larva initiates metamorphosis. The homeotic genes are required to specify the segmental identity of the adult structures and must therefore preserve the correct segmental restrictions specified in the early embryo. However, expression of *Ubx* in imaginal discs is not a continuation of embryonic expression but is directed by a different set of enhancers (Christen and Bienz 1994; Pirrotta et al. 1995). These are grouped in two regions, one about 30 kb upstream of the promoter and one 45 kb downstream. When these isolated enhancers are tested in a reporter gene construct, they are active in imaginal discs without regard to their segmental position (Fig. 2), showing that, by themselves, they lack positional information to determine the correct segments in which they should function and those in which they should be repressed. In the context of the endogenous gene, however, the repressive state installed in the early embryo by the PcG proteins prevents expression in the anterior imaginal discs throughout the proliferation of the disc cells.

THE PCG RESPONSE ELEMENT

These results imply that the *Ubx* gene contains target sites for PcG products that are not closely associated with the individual enhancers but can

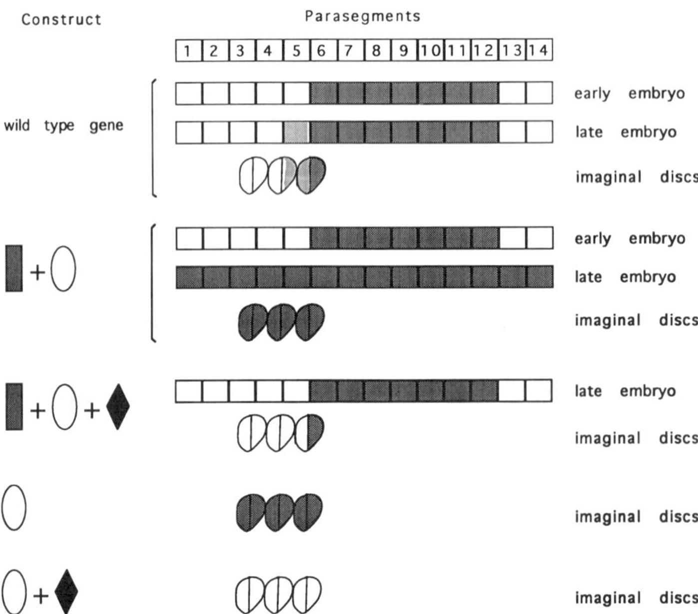

Figure 2 PRE maintenance of expression domain. The expression pattern of reporter constructs controlled by different combinations of parasegmental enhancers (*shaded rectangular boxes*), imaginal disc enhancers (*white ovals*), and PRE (*black diamond*) is shown schematically. Note that half of each imaginal disc derives from one parasegment and half from another. The PRE represses the reporter gene in the parasegments in which it was not active in the early embryo. Since imaginal enhancers are not active at that stage, when alone, they are repressed by the PRE in all parasegments, but when together with a parasegmental enhancer, they remain active in the corresponding parasegmental domain (Poux et al. 1996).

silence their activity despite the fact that they are scattered over more than 80 kb, maintaining the repressed state through many cell divisions after the original events that established it. When the *Ubx* regulatory region was scanned for elements able to confer this kind of repression, one site was found to have the required properties (Chan et al. 1994). This site, included together with one or more of the *Ubx* embryonic enhancers in a reporter gene construct, is able to maintain them repressed in the anterior half of the embryo at later developmental stages. This continued repression is dependent on the PcG genes, indicating that the fragment contains a functional PRE. In the presence of the PRE, imaginal disc enhancers are repressed everywhere, but if they are in the same construct with an embryonic enhancer, they are repressed only in the anterior segments (Fig. 2), indicating that the PRE receives the cues to initiate repression from the early enhancers. The PRE affects not only the reporter gene construct but also other genes, for example the *white* gene, included in the transposon construct used to generate transgenic flies. As a result, the *white* gene, which normally produces eye pigmentation, is totally or partially repressed, resulting in white or, more frequently, variegated eyes. The variegation is often in the form of large clonal patches or streaks (Fig. 3), indicating that the repressed state is established in some but not in other cells in the primordia of the eye and suggesting that the establishment occurs at an early stage, probably shortly after blastoderm, when the precursors of the eye primordium are limited to a few cells.

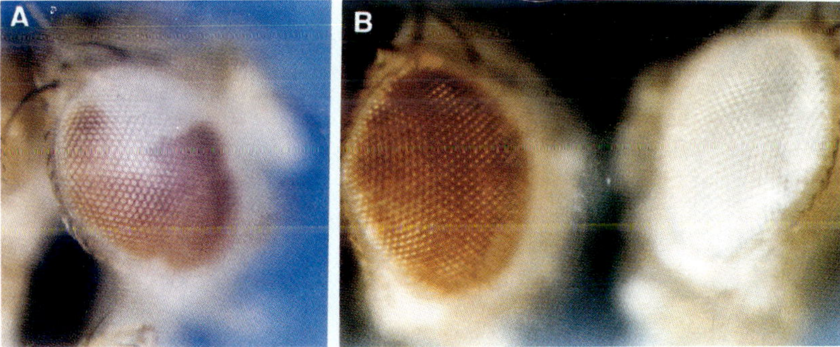

Figure 3 The PRE induces variegation of *white* gene expression. A transposon containing a PRE as well as the *white* gene produces variegated eye pigmentation, often in clonal patches (*A*). Interactions between two copies of such a transposon resulting from homologous chromosome pairing cause a dramatic increase in repression (*B*). A fly with a single copy of the transposon is on the left and a fly homozygous for that transposon insertion is on the right.

PREs with very similar properties have been detected in the regulatory regions of other genes (Fauvarque and Dura 1993; Kassis 1994; Gindhart and Kaufman 1995; Kapoun and Kaufman 1995). Some of these are other homeotic genes such as *Abdominal-B*, *Antennapedia*, or *Sex combs reduced*; others are important patterning genes such as *engrailed*. Particularly interesting is the fact that some PcG genes may be themselves targets of PcG regulation: A PRE has been found in the regulatory region of *polyhomeotic*, a member of the PcG, suggesting that the expression of PcG genes may be autoregulated. These PREs share a number of properties, including the ability to silence a reporter gene and to cause the variegated expression of the *white* gene present in the same transposon construct. Constructs containing a PRE are direct targets of PcG proteins. This has been shown by staining polytene chromosomes with antibodies against different PcG proteins. When this is done with wild-type flies, antibodies against the *Polycomb* protein (Pc) reveal a set of about 100 sites on the euchromatic arms of the polytene chromosomes (Zink and Paro 1989). Other antibodies show that many, although not all, of these sites are also target sites for other members of the PcG such as *polyhomeotic*, *Posterior sex combs* (*Psc*), *Suppressor of zeste-2* (*Su[z]*), or *Polycomblike* (*Pcl*) (DeCamillis et al. 1992; Franke et al. 1992; Rastelli et al. 1993; Lonie et al. 1994). In flies transformed with a transposon containing a PRE, a new binding site for PcG proteins is detected at the chromosomal insertion site of the transposon (Zink et al. 1991; Chan et al. 1994).

PREs from different genes have been localized in regions of several hundred nucleotides. When these regions are further subdivided, weaker activity is found in several subfragments, implying that the PRE is a cluster of individual targets for PcG proteins (Kassis 1994; Gindhart and Kaufman 1995; L. Rastelli et al., unpubl.). The properties of the smaller elements are often different in detail, suggesting that they may contain targets for different subsets of PcG proteins. In some cases, the activity of smaller fragments depends on some PcG mutations but not others, and transposons containing them create new chromosomal binding sites for some but not other PcG proteins, indicating that the PRE is a cluster of elements responding to different subsets of PcG proteins. The fact that the binding sites for different PcG proteins on polytene chromosomes do not always correspond indicates that different PREs may be composed of different elements and do not always involve the same set of PcG proteins. Genetic evidence shows, furthermore, that different subsets of the PcG proteins have been recruited to act in certain situations that do not involve the others. For example, effective repression of the posterior gap genes *knirps* and *giant* requires the participation of the PcG genes *Enhancer of zeste* (*E[z]*), *Psc*, and *pleiohomeotic* (*pho*) but not *Pc* or *extra sex combs* (*esc*) (Pelegri and Lehmann 1994). The silencing of genes placed near *Drosophila* telomeres does not involve the *Suppressor of variegation* (*Su[var]*) genes that act in the centric heterochromatin but requires the products of the PcG genes *Psc* and *Su(z)2* (Wallrath and Elgin 1995). Regulation of the *engrailed* gene involves several members of the PcG but not *esc* (Moazed and O'Farrell 1992). It is likely that these differences involve not only the specific constitution of the different PREs, but also the nature of the sequences surrounding them.

COOPERATIVE INTERACTIONS

Genetic and molecular properties of the PcG products suggest that they act in concert to form a repressive complex. Mutants show that a reduced dosage of members of the group is often insufficient to maintain a complete repression of homeotic gene expression outside of their correct domains of activity. Mutations in different genes of the group show strong synergistic effects, including in many cases lethality of *trans*-heterozygous combinations (Cheng et al. 1994). On the other hand, insufficiency of one PcG gene can in some cases be compensated by extra gene copies of another member. Direct molecular interactions were confirmed by the fact that antibodies against Pc precipitate a large complex containing also other members of the group (Franke et al. 1992). The nature of these interactions is beginning to be understood. The Pc protein contains at least two domains important for function: One is the chromodomain, a motif found also in the heterochromatin protein HP1, the product of the *Su(var)2-3* gene (Messmer et al. 1992). Another domain is at the carboxyl terminus of the Pc protein. When the Pc chromodomain is substituted for the HP1 chromodomain, the hybrid protein is targeted to the Pc-binding sites on the euchromatic arms of polytene chromosomes, as well as the centric heterochromatin where the HP1 protein normally resides (Platero et al. 1995). In addition, the endogenous HP1 protein is now also directed to the Pc sites, as well as to the heterochromatin, while conversely, the endogenous Pc protein is partly redirected to the heterochromatin. These results show that the chromodomain is responsible for interactions between Pc molecules, whereas some other part of the protein, probably the carboxy-terminal region, is responsible for interactions with other PcG proteins.

It is likely that protein-protein interactions of this sort are responsible for the assembly of the silencing complex through the progressive and cooperative recruitment of PcG proteins in a way resembling the formation of heterochromatin by the *Su(var)* gene products (Reuter and Spierer 1992). This cooperative nature is strongly suggested by the variegated expression often seen with transgene constructs containing a PRE, implying that the repressed state tends to be established in an all-or-none fashion. The chromodomain swap experiments also imply that the resulting complex contains many molecules of the Pc protein and presumably of the other PcG proteins, which would explain the sensitivity of silencing to PcG gene dosage, since, by the law of mass action, complex formation would be highly dependent on the concentration of the components (Locke et al. 1988).

The formation of the PcG complex is highly dependent on chromosomal context. The degree of both silencing and variegation depends strongly on the site of insertion of the transposon carrying the construct, suggesting that the silent state involves interactions with sequences outside the transposon. Contributing sequences can also be located on another chromosome brought physically in contact by homologous chromosome pairing. In fact, silencing effects at a given insertion site are often dramatically stronger when the transposon is present on both homologous chromosomes (Fig. 3) and weakened when chromosome pairing is prevented by chromosomal rearrangements, as if the vicinity of another PRE greatly enhanced the formation or stability of the PcG complex (Kassis et al. 1991; Fauvarque and Dura 1993; Chan et al. 1994; Gindhart and Kaufman 1995). A physical interaction between PcG complexes at two different sites is also implied by the "homing" phenomenon: Transposons carrying a PRE frequently insert at chromosomal sites that already contain PREs (Hama et al. 1990; Fauvarque and Dura 1993; Kassis 1994). The formation and the silencing function of the PRE complex are then facilitated by the vicinity of certain sequences, even when these sequences are not able by themselves to form a stable PcG complex. A likely interpretation is that these sequences are weak targets for PcG proteins which contribute to the stability of a complex that initiates at the PRE. In the *Ubx* gene, the PcG silencing can clearly act over a large distance, since it represses different enhancers spread over more than 80 kb. It is unlikely that a single complex formed at the PRE is responsible. In fact, weaker elements that respond to PcG genes, particularly when present in combinations, have been reported at other sites along the *Ubx* regulatory regions (Müller and Bienz 1991; Christen and Bienz 1994; Chiang et al. 1995). In vivo cross-linking experiments show that in a repressed *Ubx* gene, Pc protein is not localized at a single site but is found associated to a greater or lesser extent with all restriction fragments of the gene (Orlando and Paro 1993). This could be interpreted as the extension of an initial complex starting at the PRE and spreading continuously over flanking regions. More likely, however, is the possibility that the spread of the complex is discontinuous and relies on a large number of weaker sites that become stabilized by interactions with the initial complex formed at the PRE.

An interesting consequence of the idea that a silencing complex is stabilized or more repressive when it extends to involve a larger region is that a set of genes could be more stably turned off if they are arranged in a cluster. This has been proposed as a possible explanation for the evolutionary conservation of homeotic gene clusters (*Hox* clusters) (Gaunt and Singh 1990). *Hox* clusters are found in organisms ranging from nematodes to man with a conserved sequential arrangement of the genes within a cluster that reflects their expression domains along the anterior-posterior axis of the embryo. As was first noted by Lewis (1978), for the *Drosophila* bithorax complex, the gene nearest one end of the cluster (the 3′ end) is derepressed most anteriorly and in all segments more posterior. This arrangement means that in the anterior of the embryo only the first gene would be active and the rest would be included in a large contiguous silenced chromatin domain, ensuring that the more posteriorly acting genes would be most reliably repressed in the anterior of the embryo.

ASSEMBLY OF THE PcG COMPLEX

The idea that formation of the PcG complex requires multiple interactions over a larger region is consistent with the fact that none of the PcG proteins so far isolated is able by itself to form stable complexes with DNA in vitro. However, if the Pc protein is supplied with the ability to bind stably to DNA by fusing it with the DNA-binding domain of the yeast GAL4 protein, the resulting hybrid protein can silence in vivo a reporter gene containing GAL4-binding sites (Müller 1995). This silencing is not an intrinsic property of the Pc protein but is dependent on the activity of other members of the PcG. The fact that a stably bound PcG protein constitutes an effective PRE implies that the hybrid protein can recruit other members of the PcG to form a silencing complex. Why then are PcG proteins not provided with a stable DNA-binding domain? It is possible that some of the PcG proteins not yet characterized have such a DNA-binding function. It seems more likely that providing stable binding would defeat the function of PcG silencing: PcG proteins are virtually ubiquitous, but they must silence their target genes only depending on some transient condition, preserving that silent state at later times while leaving the genes active in the appropriate cells. The high degree of cooperative interactions prerequisite for the establishment of stable binding means that the formation of the complex is highly sensitive to initial conditions and has an all-or-none character.

These considerations have suggested a model for the formation of the PcG complex (Pirrotta and Rastelli 1994), according to which assembly begins at the PRE because the clustering of response elements within it permits molecular interactions between the proteins weakly bound at individual target sites, resulting in a relatively stable core (Fig. 4). The long-term stability of this complex depends, however, on being able to recruit additional weaker sites in flanking sequences. A number of such weaker sites have been reported in the *Ubx* gene, each of which, although not able to establish the silenced state alone, appears to have PRE activity when combined with other weak elements. The presence or absence of additional weak contributory sites in sequences flanking the insertion is probably responsible for the variability of PRE silencing in transposons inserted in different chromosomal sites. These would contribute then both to the formation of a stable PcG complex and to its spread to affect a larger region. The spreading of the complex need not be a continuous coating of the chromatin, as has been proposed for the silencing complexes at the yeast telomeres (Hecht et al. 1995), but it could also be saltatory, involving discrete sequences containing weak PcG targets recruited through looping to participate in a larger, stable complex. An attractive feature of this model is that it might provide a mechanism for the persistence of the repressed state through many rounds of cell division. Replication of the DNA would require the unraveling of one loop at a time, leaving a stable core intact at any one

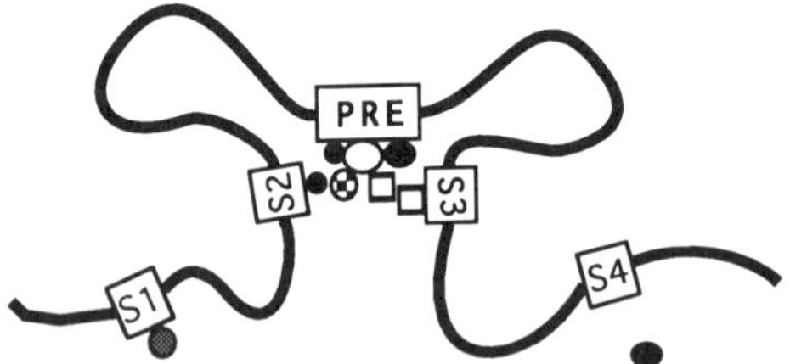

Figure 4 Model of PcG complex assembly. Complex formation initiates at the PRE where several PcG proteins stabilize one another's binding. Secondary sites, labeled S1, S2, etc., are weaker sites that may be able to bind weakly and unstably to subsets of PcG proteins. Such weaker sites interact with a nearby PRE, forming a larger and more stable complex. The complex may spread further by progressively recruiting additional secondary sites.

time. An interesting light is cast on this ability of the complex to reestablish itself by the experiments using a PC-GAL4 hybrid protein (Müller 1995). A reporter gene provided with GAL4-binding sites is silenced by the complex recruited by this protein but remains silent only as long as the PC-GAL4 protein is produced from a construct driven by the heat shock promoter. When the reporter gene includes a weak PRE in addition to the GAL4-binding sites, repression initiated by the PC-GAL4 hybrid can be maintained even when this protein stops being produced. In this case, according to the model, the complex, once recruited by the hybrid protein, has sufficient interactions with the weak PRE that it no longer requires the recruiting protein for its stability.

The initiation of complex formation remains the critical step. One of the PcG genes, *esc*, is known to be specifically required at early stages of development but not at later times (Struhl and Brower 1982). Its product is a protein of the WD repeat family, containing multiple motifs of about 40 amino acids punctuated by tryptophan and aspartate residues (WD repeats) that mediate interactions between proteins (Gutjahr et al. 1995; Sathe and Harte 1995). In yeast, the WD repeat protein TUP1 interacts with DNA-binding proteins such as α2 to recruit a repressing complex that does not itself bind to DNA (Komachi et al. 1994). An attractive possibility is therefore that the ESC protein is necessary to facilitate the recruitment of PcG proteins in the initial stages but is no longer necessary when a stable complex has been achieved. What determines when and where complex formation will initiate? A critical requirement for the correct role of a PcG complex in homeotic regulation is that it must affect only the activity of the gene in the cells in which it was initially repressed and not in the cells in which it was active. Since the PcG genes themselves have a nearly ubiquitous expression, and since they silence different homeotic genes in different domains, it is clear that the positional information must reside in the target gene. One possible mechanism for the decision whether to assemble the silencing complex or not is that the initial repressors, the HB and TLL proteins in the case of the *Ubx* gene, serve as the first recruiting signal (Zhang and Bienz 1992). In this role they would resemble the RAP protein in recruiting silencing complexes in yeast, and they would not be required when a PcG protein is provided with a DNA-binding domain that permits it to bind stably and independently. Although the hypothesis is attractive, a role for HB and TLL in initiating complex formation at the PRE is not clear, since silencing can occur with constructs and in tissues in which these proteins are not involved (Chan et al. 1994; Pirrotta et al. 1995).

An alternative, but less concrete, hypothesis is that the assembly of the complex is dependent on the state of activity of the gene at the time the repressive complex is formed in the early embryo. The assembly of the complex is apparently sensitive to the state of the neighboring chromatin. The ability of a PRE to repress a reporter gene driven by the GAL4 activator is dependent on the level of GAL4 in the cell. High levels of GAL4 appear to act competitively, preventing not only PRE repression, but also the formation of a PcG complex detectable by antibodies on polytene chromosomes (Zink and Paro 1995). This suggests that the formation of the PRE complex may be incompatible with the transcriptionally active state, possibly because the interaction of active enhancer complexes with the promoter is incompatible with the interactions required by the PcG complex. It is important to bear in mind, however, that, given the diversity of PREs and the subsets of PcG proteins acting on them, their functions and the features controlling them may be equally heterogeneous.

TRITHORAX-GROUP PROTEINS

Whereas the PcG genes maintain the repressed state of homeotic genes, genetics has shown that another set of genes is important to maintain a high level of expression of homeotic genes during later development. The function of these genes, the trithorax group (trxG), is not to extend the activity of the embryonic enhancers, which stop functioning after the early embryonic stages, nor is it to activate expression at later stages, since this expression is dependent on other, late-acting enhancers. Many of the trxG genes were identified by mutations that relieve the homeotic derepression phenotypes caused by PcG insufficiency. A likely role for the trxG, therefore, is in counteracting the effects of PcG repression. Only a few of the proteins assigned to the trxG have been molecularly characterized. They are a heterogeneous set, and their mode of action may be equally heterogeneous. Some of the trxG proteins (e.g., ZESTE and GAGA factor) are DNA-binding proteins that appear to enhance the accessibility of activators to chromatin (Biggin and Tjian 1988; Tsukiyama et al. 1994; Tsukiyama and Wu 1995). The BRAHMA protein is related to the yeast SNF2/SWI2 protein, which is part of a large complex that alters nucleosomes and facilitates the binding of activators (Laurent et al. 1993; Coté et al. 1994). The *trx* and *ash1* products are large proteins containing motifs found also in the products of *E(z)*, a PcG protein, and of *Su(var)3-9*, a heterochromatin protein, suggesting that they might interact directly with the silencing complex. Antibody staining of polytene chromosomes has shown that these proteins frequently, although not always, colocalize with PcG proteins (Chinwalla et al. 1995; N. Tripoulas et al., pers. comm.) and, in fact, binding sites for ZESTE, GAGA factor, and TRX protein are closely associated with the *Ubx* PRE (Chang et al. 1995; C. Tatout et al., unpubl.). These results suggest that the role of at least some of the trxG proteins is to prevent the formation or spread of PcG complexes at later stages, to avoid repressing the target genes even in cells in which they should remain functional.

CONCLUSIONS

Silencing involving members of the PcG is not confined to the homeotic genes in *Drosophila*. The binding sites on polytene chromosomes suggest that more than 100 genes may be subject to this type of mechanism, of which a few have been identified. These include segmentation genes such as *engrailed*, *wingless*, *knirps*, and *giant*, and members of the PcG complex such as *polyhomeotic*. In addition, an important role of PcG genes in the development of the central nervous system is implied by the effects of some PcG mutations. Although none of the molecularly characterized PcG proteins is found in the centric heterochromatin, the effects of some mutations suggest that they may contribute to the silencing of genes placed in the vicinity of heterochromatin. In contrast, a direct role in telomeric silencing has been demonstrated for the products of the *Su(z)2* and *Psc* genes, which have prominent binding sites at telomeres. Finally, although its role is not clear, the product of the *E(z)* gene is essential for chromosome integrity in proliferating cells.

The PcG genes have an essential and widespread role in a variety of processes. It would not be surprising, therefore, if both the genes and the mechanisms involved were substantially conserved in evolution. Although there are important similarities with the mechanism of silencing in yeast, none of the yeast proteins so far shown to be involved in silencing has significant homology with *Drosophila* PcG genes. In contrast, homologs for several PcG proteins have been found in mammalian genomes. The homology is not limited to single structural motifs, since a mammalian *Pc* homolog, introduced in *Drosophila*, can substitute for the *Drosophila* gene (Müller et al. 1995). In addition, the *Drosophila Pc*, *Su(z)2*, and *Psc* products, when targeted by the addition of a *lexA* DNA-binding domain, can silence a reporter gene in mammalian cells (Bunker and Kingston 1994). Like similar experiments in *Drosophila*, this implies that the targeted protein can serve as a nucleus to recruit other members of a repressive complex and reveals the conservation of an analogous mechanism in mammals that can recognize the insect proteins.

If a closely related silencing mechanism exists in mammalian cells, what could be its targets? An interesting possibility is that mammalian *Hox* gene clusters, like the *Drosophila* homeotic genes, are regulated by a silencing mechanism. In a speculative vein, it might be supposed that protein complexes bound to chromatin regions serve as the first stage of a chromatin-marking mechanism necessary for the establishment of a cellular memory in mammalian cells. In *Drosophila*, with its smaller genome, short life span, and small number of cell divisions, this first stage would be sufficient to maintain patterns of gene expression established in early development, whereas in vertebrates a second, more stable marking based on methylation might be recruited to ensure the long-term memory of the repressed state. It would be interesting to explore the relationship between PcG silencing complexes and methylation in phenomena like genetic imprinting or X-chromosome inactivation.

ACKNOWLEDGMENT

This work was supported by the Swiss National Science Foundation.

REFERENCES

Biggin, M.D. and R. Tjian. 1988. Transcription factors that activate the Ultrabithorax promoter in developmentally staged extracts. *Cell* **53:** 699–711.

Bunker, C.A. and R.E. Kingston. 1994. Transcriptional repression by *Drosophila* and mammalian Polycomb group proteins in transfected mammalian cells. *Mol. Cell. Biol.* **14:** 1721–1732.

Chan, C.-S., L. Rastelli, and V. Pirrotta. 1994. A *Polycomb* response element in the *Ubx* gene that determines an epigenetically inherited state of repression. *EMBO J.* **13:** 2553–2564.

Chang, Y.-L., B.O. King, M. O'Connor, A. Mazo, and D.-H. Huang. 1995. Functional reconstruction of trans regulation of the *Ultrabithorax* promoter by the products of two antagonistic genes, *trithorax* and *Polycomb*. *Mol. Cell. Biol.* **15:** 6601–6612.

Cheng, N.N., D.A.R. Sinclair, R.B. Campbell, and H.W. Brock. 1994. Interactions of *polyhomeotic* with Polycomb group genes of *Drosophila melanogaster*. *Genetics* **138:** 1151–1162.

Chiang, A., M.B. O'Connor, R. Paro, J. Simon, and W. Bender. 1995. Discrete *Polycomb*-binding sites in each parasegmental domain of the bithorax complex. *Development* **121:** 1681–1689.

Chinwalla, V., E.P. Jane, and P.J. Harte. 1995. The *Drosophila trithorax* protein binds to specific chromosomal sites and is co-localized with *Polycomb* at many sites. *EMBO J.* **14:** 2056–2065.

Christen, B. and M. Bienz. 1994. Imaginal disc silencers from *Ultrabithorax*: Evidence for *Polycomb* response elements. *Mech. Dev.* **48:** 255–266.

Coté, J., J. Quinn, J.L. Workman, and C.L. Peterson. 1994. Stimulation of GAL4 derivative binding to nucleosomal DNA by the yeast SWI/SNF complex. *Science* **265:** 53–60.

DeCamillis, M., N. Cheng, D. Pierre, and H.W. Brock. 1992. The *polyhomeotic* gene of *Drosophila* encodes a chromatin protein that shares polytene chromosome-binding sites with *Polycomb*. *Genes Dev.* **6:** 223–232.

Fauvarque, M.-O. and J.-M. Dura. 1993. *polyhomeotic* regulatory sequences induce developmental regulator-dependent variegation and targeted P-element insertions in *Drosophila*. *Genes Dev.* **7:** 1508–1520.

Franke, A., M. DeCamillis, D. Zink, N. Cheng, H.W. Brock, and R. Paro. 1992. *Polycomb* and *polyhomeotic* are constituents of a multimeric protein complex in chromatin of *Drosophila melanogaster*. *EMBO J.* **11:** 2941–2950.

Gaunt, S.J. and P.B. Singh. 1990. Homeogene expression patterns and chromosomal imprinting. *Trends Genet.* **6:** 208–212.

Gindhart, J.G., Jr. and T.C. Kaufman. 1995. Identification of *Polycomb* and *trithorax* group responsive elements in the regulatory region of the *Drosophila* homeotic gene *Sex combs reduced*. *Genetics* **139:** 797–814.

Gutjahr, T., E. Frei, C. Spicer, S. Baumgartner, R.A.H. White, and M. Noll. 1995. The Polycomb-group gene, *extra sex combs*, encodes a nuclear member of the WD-40 repeat family. *EMBO J.* **14:** 4296–4306.

Hama, C., Z. Ali, and T.B. Kornberg. 1990. Region-specific recombination and expression are directed by portions of the *Drosophila engrailed* promoter. *Genes Dev.* **4:** 1079–1093.

Hecht, A., T. Laroche, S. Strahl-Bolsinger, S.M. Gasser, and M. Grunstein. 1995. Histone H3 and H4 N-termini interact with SIR3 and SIR4 proteins: A molecular model for the formation of heterochromatin in yeast. *Cell* **80:** 583–592.

Kapoun, A.M. and T.C. Kaufman. 1995. Regulatory regions of the homeotic gene *proboscipedia* are sensitive to chromosomal pairing. *Genetics* **140:** 643–658.

Kassis, J.A. 1994. Unusual properties of regulatory DNA from the *Drosophila engrailed* gene: Three "pairing-sensitive" sites within a 1.6 kb region. *Genetics* **136:** 1025–1038.

Kassis, J.A., E.P. VanSickle, and S.M. Sensabaugh. 1991. A fragment of *engrailed* regulatory DNA can mediate transvection of the *white* gene in *Drosophila*. *Genetics* **128:** 751–761.

Komachi, K., M.J. Redd, and A.D. Johnson. 1994. The WD repeats of Tup1 interact with the homeodomain protein alpha 2. *Genes Dev.* **8:** 2857–2867.

Laurent, B.C., I. Treich, and M. Carlson. 1993. The yeast SNF2/SWI2 protein has DNA-stimulated ATPase activity required for transcriptional activation. *Genes Dev.* **7:** 583–591.

Lewis, E.B. 1978. A gene complex controlling segmentation in *Drosophila*. *Nature* **276:** 565–570.

Locke, J., M.A. Kotarski, and K.D. Tartof. 1988. Dosage-dependent modifiers of position effect variegation in *Drosophila* and a mass action model that explains their effect. *Genetics* **120:** 181–198.

Lonie, A., R. D'Andrea, R. Paro, and R. Saint. 1994. Molecular characterization of the *Polycomblike* gene of *Drosophila melanogaster*, a trans-acting negative regulator of homeotic gene expression. *Development* **120:** 2629–2636.

Messmer, S., A. Franke, and R. Paro. 1992. Analysis of the functional role of the *Polycomb* chromo domain in *Drosophila melanogaster*. *Genes Dev.* **6:** 1241–1254.

Moazed, D. and P.H. O'Farrell. 1992. Maintenance of the engrailed expression pattern by Polycomb group genes in *Drosophila*. *Development* **116:** 805–810.

Müller, J. 1995. Transcriptional silencing by the *Polycomb* protein in *Drosophila* embryos. *EMBO J.* **14:** 1209–1220.

Müller, J. and M. Bienz. 1991. Long range repression conferring boundaries of *Ultrabithorax* expression in the *Drosophila* embryo. *EMBO J.* **10:** 3147–3155.

Müller, J., S. Gaunt, and P.A. Lawrence. 1995. Function of the *Polycomb* protein is conserved in mice and flies. *Development* **121:** 2847–2852.

Orlando, V. and R. Paro. 1993. Mapping *Polycomb*-repressed domains in the bithorax complex using *in vivo* formaldehyde cross-linked chromatin. *Cell* **75:** 1187–1198.

Pelegri, F. and R. Lehmann. 1994. A role of Polycomb group genes in the regulation of gap gene expression in *Drosophila*. *Genetics* **136:** 1341–1353.

Pirrotta, V. and L. Rastelli. 1994. *white* gene expression, repressive chromatin domains and homeotic gene regulation in *Drosophila*. *BioEssays* **16:** 549–556.

Pirrotta, V., C.S. Chan, D. McCabe, and S. Qian. 1995. Upstream parasegmental enhancers and the establishment of the expression pattern of the *Ubx* gene. *Genetics* **141:** 1439–1450.

Platero, J.S., T. Hartnett, and J.C. Eissenberg. 1995. Functional analysis of the chromodomain of HP1. *EMBO J.* **14:** 3977–3986.

Poux, S., C. Kostić, and V. Pirrotta. 1996. Hunchback-independent silencing of late Ubx enhancers by a Polycomb group response element. *EMBO J.* (in press).

Qian, S., M. Capovilla, and V. Pirrotta. 1991. The bx region enhancer, a distant cis-control element of the D*rosophila Ubx* gene and its regulation by *hunchback* and other segmentation genes. *EMBO J.* **10:** 1415–1425.

———. 1993. Molecular mechanisms of pattern formation by the BRE enhancer of the *Ubx* gene. *EMBO J.* **12:** 3865–3877.

Rastelli, L., C.S. Chan, and V. Pirrotta. 1993. Related chromosome binding sites for *zeste*, suppressors of *zeste* and Polycomb group proteins in *Drosophila* and their dependence on *Enhancer of zeste* function. *EMBO J.* **12:** 1513–1522.

Reuter, G. and P. Spierer. 1992. Position effect variegation and chromatin proteins. *BioEssays* **14:** 605–612.

Sathe, S.S. and P.J. Harte. 1995. The *extra sex comb* protein contains WD motifs essential for its function as a repressor of the homeotic genes. *Mech. Dev.* **52:** 77–87.

Simon, J., M. Peifer, W. Bender, and M. O'Connor. 1990. Regulatory elements of the bithorax complex that control expression along the anterior-posterior axis. *EMBO J.* **9:** 3945–3956.

Simon, J., A. Chiang, and W. Bender. 1992. Ten different Polycomb group genes are required for spatial control of the *abdA* and *AbdB* homeotic products. *Development* **114:** 493–505.

Struhl, G. and D. Brower. 1982. Early role of the *esc*$^+$ gene product in the determination of segments in *Drosophila*. *Cell* **31:** 285–292.

Tsukiyama, T. and C. Wu. 1995. Purification and properties of an ATP-dependent nucleosome remodeling factor. *Cell* **83:** 1011–1020.

Tsukiyama, T., P.B. Becker, and C. Wu. 1994. ATP-dependent nucleosome disruption at a heat-shock promoter mediated by binding of GAGA transcription factor. *Nature* **367:** 525–532.

Wallrath, L.L. and S.C.R. Elgin. 1995. Position effect variegation in *Drosophila* is associated with an altered chromatin structure. *Genes Dev.* **9:** 1263–1277.

Zhang, C.-C. and M. Bienz. 1992. Segmental determination in *Drosophila* conferred by *hunchback* (*hb*), a repressor of the homeotic gene *Ultrabithorax* (*Ubx*). *Proc. Natl.*

Acad. Sci. **89:** 7511–7515.

Zhang, C.-C., J. Müller, M. Hoch, H. Jäckle, and M. Bienz. 1991. Target sequences for *hunchback* in a control region conferring *Ultrabithorax* expression boundaries. *Development* **113:** 1171–1179.

Zink, B. and R. Paro. 1989. *In vivo* binding pattern of a trans-regulator of homoeotic genes in *Drosophila melanogaster*. *Nature* **337:** 468–471.

———. 1995. *Drosophila* Polycomb-group regulated chromatin inhibits the accessibility of a trans-activator to its target DNA. *EMBO J.* **14:** 5660–5671.

Zink, B., Y. Engström, W.J. Gehring, and R. Paro. 1991. Direct interaction of the Polycomb protein with *Antennapedia* regulatory sequences in polytene chromosomes of *Drosophila melanogaster*. *EMBO J.* **10:** 153–162.

The Role of Polycomb Group and Trithorax Group Chromatin Complexes in the Maintenance of Determined Cell States

Renato Paro
ZMBH
University of Heidelberg
69120 Heidelberg, Germany

Peter J. Harte
Department of Genetics
Case Western Reserve University
Cleveland, Ohio 44106-4955

In a developing organism, determined states of cells must be maintained over many cell generations. In *Drosophila* the genes of the Polycomb group (PcG) and trithorax group (trxG) are part of a cellular memory system that maintains the differential expression patterns of genes encoding developmental regulators involved in defining the determined states of cells. The PcG is responsible for maintaining the inactive expression states of regulatory genes such as the homeotic selector genes. Evidence increasingly suggests that the PcG proteins act by an epigenetic mechanism to keep genes inactive in a stable and heritable manner. The PcG genes share some similarities with another class of *Drosophila* genes, modifiers of position-effect variegation (PEV), some of which encode structural components of heterochromatin (see Henikoff, this volume). These similarities suggest that the permanent silencing of developmental regulatory genes might involve heterochromatin-like repressive complexes.

The products of the trxG genes are required for long-term maintenance of the active transcriptional states of developmental regulatory genes. They appear to counter the formation of repressive PcG-chromatin structures to render the genes accessible to transcription factors. Accumulating evidence suggests that an intricate interplay between the PcG and trxG proteins determines the distinct higher-order chromatin structures responsible for long-term maintenance of stable transcriptional states of developmental regulatory genes.

Epigenetic Mechanisms of Gene Regulation

PcG PROTEINS HERITABLY SILENCE GENE EXPRESSION

Mutations in any of the PcG genes (see Table 1) cause embryonic lethality and exhibit homeotic transformations that result from widespread ectopic expression of homeotic selector genes outside their normal realms of action. However, this derepression only becomes apparent at later embryonic stages. The early spatially restricted patterns of homeotic gene expression, which are controlled by the activities of the early patterning genes (maternal and segmentation genes), are not affected in PcG mutants and are thus normally established without any requirement for PcG activities. This led to the conclusion that the PcG proteins recognize some feature associated with the early inactive states of homeotic genes and "freeze" or "lock in" these states over the rest of development. PcG-mediated repression is not restricted to the silencing of homeotic genes; it has also been found to affect several genes of the maternal and segmentation class (McKeon et al. 1994; Pelegri and Lehmann 1994), suggesting that it is utilized at repeated times during development. PcG-mediated repression probably represents a specialized restrictive process that ensures faithful and heritable silencing of important developmental regulators in appropriate parts of the body, since misexpression of these factors would result in dramatic alterations of the body pattern.

A first clue to the molecular nature of this type of transcriptional regulation came through the cloning of the *Polycomb (Pc)* gene. The Pc protein (390 amino acids) contains a highly conserved 48-residue motif also found in the heterochromatin-associated protein HP1, termed the chromodomain (Paro and Hogness 1991). Interestingly, HP1 is encoded by the *Su(var)205* gene, which was initially identified by a mutation acting as suppressor of PEV (Eissenberg et al. 1990). Thus, HP1 is analogous to Pc in being required for the repressive effects of heterochromatin seen in PEV. Other parallels between PcG-mediated silencing and heterochromatin-mediated silencing have been observed at the genetic level. Mutations in several PcG genes have been reported to act as modifiers of PEV, and mutations in some modifiers of PEV have been found to exhibit PcG-like homeotic phenotypes (Grigliatti 1991; Dorn et al. 1994), suggesting that PcG-mediated silencing and heterochromatin-mediated silencing may share some common components and mechanisms.

Members of the PcG show strong genetic interactions among themselves, as do the modifiers of PEV. This has been interpreted to be an indication of their cooperation in multimeric complexes (Locke et al. 1988). Indeed, several members of the PcG have been found to be part of a large (2.5 MD) multiprotein complex(es) (Franke et al. 1992; H. Strutt and R. Paro, unpubl.). Correspondingly, the individual PcG proteins analyzed to date share many of their target sites on larval polytene chromosomes (DeCamillis et al. 1992; Franke et al. 1992; Martin and Adler 1993; Rastelli et al. 1993; Lonie et al. 1994). These data have led to the proposition that large multimeric complexes composed of many PcG proteins package inactive genes into heterochromatin-like states, which are then epigenetically transmitted to progeny cells.

The chromodomain appears to play an important role in the assembly and/or targeting of the multiprotein PcG complexes. Mutations in the Pc chromodomain abolish the binding of Pc and at least one other PcG protein, Polyhomeotic, to their respective target genes (Messmer et al. 1992; Franke et al. 1995). Similarly, mutations in the HP1 chromodomain also affect its function, causing a typical suppressor-of-PEV phenotype (Platero et al. 1995). HP1 has recently been shown to contain a second more divergent chromodomain (termed chromo shadow domain) near its carboxyl terminus (Aasland and Steward 1995), which is responsible for targeting HP1 specifically to heterochromatin (Powers and Eissenberg 1993). The molecular function of the chromodomain remains unclear. However, domain-swap experiments suggest that it mediates protein-protein interactions. A chimeric HP1 protein containing the Pc chromodomain in the amino-terminal position is targeted to all the euchromatic Pc-binding sites in addition to heterochromatin. Furthermore, in lines containing this chimera, the endogenous Pc protein is also recruited to the centric heterochromatin of polytene chromosomes (Platero et al. 1995). Interestingly, this chimeric protein can rescue the suppressor-of-PEV phenotype exhibited by the HP1 chromodomain mutations. This interchangeability of components between PcG complexes and heterochromatin further suggests that they utilize common basic mechanisms.

The targeting of the PcG multiprotein complexes occurs through *cis*-acting DNA elements termed PcG response elements (PREs) (Simon et al. 1990; Fauvarque and Dura 1993; Chan et al. 1994; Zink and Paro 1995; see also Pirrotta, this volume). In transgenic fly lines, constructs containing isolated PREs can recruit PcG proteins to their integration sites on polytene chromosomes (Chan et al. 1994). PREs act to silence nearby as well as distant reporter genes. However, isolated PREs are apparently only able to nucleate the formation of unstable PcG silencing complexes. Additional DNA elements are necessary to form large silencing complexes that are mitotically stable (Müller 1995; Paro 1995). The nature of these other sequences is unknown. They could be weak PREs

Table 1 PcG and trxG genes whose products have been characterized

Gene	Protein (no. amino acids)	Motifs	Homologs	References[b]
Polycomb group				
Polycomb (Pc)	390	chromodomain	mouse M33	1, 2
			Xenopus Pc	30
polyhomeotic (ph)	1589	C_4 zinc finger	mouse Rae-28	3, 4
Posterior sex combs (Psc)	1603	C_3HC_4 RING finger	bmi and mel18	5–7
Suppressor-2 of zeste (Su[z[2)	1365	C_3HC_4 RING finger	bmi and mel18	5–7
Polycomb-like (Pcl)	857	2 C_4HC_3 PHD fingers	human EST[a]	8
extra sex combs (esc)	425	5 WD motifs	mouse homolog	9–11
Enhancer of zeste (E[z]) alias *polycombeotic (pco)*	760	SET domain	human ESTs[a]	12

trithorax group				
trithorax (trx)	3726 3358	SET domain 4 C_4HC_3 PHD fingers nuclear receptor type DNA-binding motif	human MLL mouse Mll	13–16
absent small or homeotic 1 (ash1)	2144	SET domain C_4HC_3 PHD finger		17
brahma (brm)	1638	ATPase motifs bromodomain	yeast SWI2 human BRG, hbrm	18–20
Trithorax-like (Trl) (GAGA factor)	519	zinc finger POZ/BTB domain		21, 22
ISWI	1027	ATPase motifs	yeast SWI2	23
snf5-related 1 (snr1)	370		yeast SWI5	24
female sterile homeotic (fs[1]h)	1118	bromodomain	human RING3	25, 26

Other PcG genes (Simon 1995) whose products have not yet been characterized include: *Additional sex combs*, *Sex comb on midleg*, *Sex comb extra*, *super sex combs*, *multi sex combs*, *pleiohomeotic*, *Enhancer of Polycomb*. Other trxG genes (Kennison 1993) whose products have not yet been characterized include: *ash2* (Shearn 1989), *devinir*, *homothorax*, *kismet*, *kohtalo*, *moira*, *osa*, *sallimus*, *skuld*, *Suppressor of Polycomb 37D*, *urdur*, *verthandi*.

[a]Indicates highly similar partial cDNA sequences present in the NCBI database of expressed sequence tags (ESTs).

[b]References containing sequences of these PcG and trxG genes and their homologs: (1) Paro and Hogness 1991; (2) Pearce et al. 1992; (3) DeCamillis et al. 1992; (4) Nomura et al. 1994; (5) Brunk et al. 1991; (6) van Lohuizen et al. 1991; (7) Ishida et al. 1993; (8) Lonie et al. 1994; (9) Sathe and Harte 1995a; (10) Simon et al. 1995; (11) Gutjahr et al. 1995; (12) Jones and Gelbart 1993; (13) Sedkov et al. 1994; (14) Stassen et al. 1995; (15) Gu et al. 1992; (16) Ma et al. 1993; (17) D. LaJeunuesse and A. Shearn, unpubl.; (18) Tamkun et al. 1992; (19) Khavari et al. 1993; (20) Muchardt and Yaniv 1993; (21) Soeller et al. 1993; (22) Farkas et al. 1994; (23) Elfring et al. 1994; (24) Dingwall et al. 1995; (25) Haynes et al. 1989; (26) Beck et al. 1992; (27) Simon 1995; (28) Kennison 1993; (29) Shearn 1989; (30) Reijnen et al. 1995.

that interact with specific DNA-binding components of the PcG complex, or more general features of chromosomal topology like Z-DNA, histone modifications, or repeated DNA sequences with which PcG complexes can interact. In particular instances, isolated PREs can induce a local constriction in polytene chromosomes (D. Zink et al., unpubl.). Since no underreplication has been observed, such a constriction may be caused by chromatin compaction, indicating that local concentrations of PcG protein complexes can result in macroscopically visible chromosome condensation.

The molecular architecture of the chromatin fiber regulated by PcG silencing complexes remains an important issue to be resolved. An interesting model is suggested by a related epigenetic phenomenon observed in yeast, where telomeric silencing is accompanied by cooperative spreading of SIR protein complexes from a telomeric nucleation site defined by binding of the RAP1 protein. These complexes compact the chromatin fiber into a higher-order repressive structure by defined interactions with the amino-terminal tails of histones H3 and H4 (Hecht et al. 1995; Palladino and Gasser 1995).

What is the analogous mechanism underlying the targeting and assembly of the first PcG silencing complexes, which depends critically on the prior transcriptional status of a target gene in each cell? Among the PcG proteins, ESC may play a unique role in this process. Unlike all the other PcG proteins, which are required continuously, ESC is present and required for only the first few hours of embryogenesis, the time when PcG silencing is first being implemented (Struhl and Brower 1982; Sathe and Harte 1995a). The presence of multiple WD motifs in ESC, which are known in other proteins to mediate protein-protein interactions, is also consistent with a role in initial complex assembly (see Pirrotta, this volume).

trxG PROTEINS AND MAINTENANCE OF ACTIVE TRANSCRIPTIONAL STATES

The trithorax group (trxG) genes exhibit antagonistic genetic interactions with PcG genes, which has led to the suggestion that trxG proteins may function to counteract or modulate PcG-mediated silencing. Most of the 20 trxG members (see Table 1) were identified in genetic screens for suppressors of the dominant homeotic phenotypes of *Pc* or *Antennapedia* mutations (Kennison and Russell 1987; Kennison and Tamkun 1988; see also Ingham and Whittle 1980; Shearn 1989). As a result, they are likely to represent a group of proteins with diverse functions, since any mutation resulting in reduced levels of homeotic gene expression or impairment of homeotic protein function could, in principle, be eligible for detection in such screens. Genetic evidence also suggests that it is unlikely that all trxG proteins function in concert or as a complex (LaJeunuesse and Shearn 1995). However, recent molecular and biochemical analyses of several trxG proteins suggest that a unifying theme for the role of at least some trxG proteins may lie in their chromatin remodeling activities.

brahma, snr1, and ISWI: trxG Homologs of Yeast SWI/SNF Proteins

The role of *brahma* (*brm*) was reviewed recently (Peterson and Tamkun 1995; Tamkun 1995). The original *brm* mutation was identified as a suppressor of the dominant haploinsufficient *Pc* phenotype. The BRM protein is related to the yeast transcriptional activator SWI2 (also known as SNF2) (Tamkun et al. 1992), sharing a conserved ATPase domain (500 amino acids) and carboxy-terminal bromodomain (77 amino acids), a motif found in a number of other proteins known or suspected to have roles in transcriptional regulation as "adapters" or co-activators (Haynes et al. 1992; Tamkun 1995). Mutations in the ATPase domain of SWI2 block its function. A hybrid SWI2 protein containing the BRM ATPase domain complements yeast SWI2 mutations, indicating functional conservation of this domain (Elfring et al. 1994). However, the entire BRM protein does not complement yeast SWI2 mutations, raising questions about the functional equivalence of BRM and SWI2. Screens for other SWI2 homologs by low-stringency hybridization indicate that BRM is the *Drosophila* protein most closely related to SWI2, but another more distantly related SWI2-like protein, named ISWI, was identified (Elfring et al. 1994). There are several different mammalian proteins that are related to SWI2 (Khavari et al. 1993; Muchardt and Yaniv 1993). The bromodomain of a human BRM homolog, hbrm, is not required for stimulation of transcription in a mammalian transient transfection assay (Muchardt and Yaniv 1993), and deletion of the BRM bromodomain in vivo indicates that it is dispensable for complete rescue of *brm* null mutations (J. Tamkun, pers. comm.). Mutations in SWI2 and other yeast SWI genes exhibit similar phenotypes and encode subunits of an approximately 2-MD multiprotein complex containing at least 10 proteins. Similar SWI-like complexes have been identified and purified from humans (Imbalzano et al. 1994; Kwon et al. 1994) and *Drosophila* (Dingwall et al. 1995; J. Tamkun, pers. comm.), suggesting widespread evolutionary conservation of SWI-like protein complexes and their functions.

A purified human SWI complex containing the BRM homolog BRG1 is not required for basal transcription, but it is required for activation of transcription by transcriptional activators, as well as TBP, the TATA-binding subunit of the basal transcription factor TFIID. Purified SWI complexes have an ATP-dependent nucleosome displacement activity and greatly stimulate the in vitro binding of these transcriptional activators to nucleosomal DNA, but not naked DNA, suggesting that they stimulate transcription by overcoming inhibitory effects of chromatin on the binding of other activators (Imbalzano et al. 1994; Kwon et al. 1994). The yeast SWI complex is associated in stoichiometric quantities with the RNA polymerase II holoenzyme complex, the likely active form of the enzyme in vivo. It is tightly bound to the highly conserved carboxy-terminal domain of the largest pol II subunit (Wilson et al. 1996). This suggests that SWI complexes are present, if not absolutely required, at all active promoters in vivo, at least in yeast. Interestingly, SWI-dependent activation in yeast can be rendered SWI-independent in some cases by multimerizing the activator binding site. This suggests that the differential SWI-dependence of individual yeast genes may simply reflect the effective binding affinities of particular activators within the specific nucleosomal contexts of each promoter. In *Drosophila*, the BRM protein can be detected bound to approximately 50 specific sites on the polytene chromosomes of the larval salivary gland, including several puffed regions that mark transcriptionally active genes (V. Orlando and R. Paro, unpubl.). This is many fewer chromosomal sites than detected with antibodies to *Drosophila* RNA polymerase II subunits (Weeks et al. 1993), raising the possibility that SWI complexes may have a less constitutive and more selective regulatory association with the basal transcriptional machinery in higher eukaryotes. If so, it is not yet known which components of SWI/BRM complexes are responsible for such binding-site specificity. However, the SWI2 homolog BRG1 has been shown to interact directly with specific transcriptional activators, including Rb, and this may serve to target SWI activities to specific genes (Dunaief et al. 1994).

A *Drosophila* homolog of the yeast SWI/SNF5 gene, *snf5-related 1* (*snr1*) was recently isolated in a screen for *Drosophila* homologs of other known subunits of SWI complexes (Dingwall et al. 1995). Mutations in *snr1* are recessive lethals that exhibit strong trx-like phenotypes in double heterozygotes with *brm* or *trx* mutations. *snr1* is expressed in an identical pattern to *brm* throughout development, and both proteins are present in the same high-molecular-weight complex, presumably a *Drosophila* SWI complex (Dingwall et al. 1995).

Trithorax-like/GAGA

Another member of the trxG, *Trithorax-like* (*Trl*), was identified by a P-element-induced mutation that exhibits dominant *trx*-like phenotypes, interacts synergistically with *trx* mutations, and also behaves as an enhancer of PEV (Farkas et al. 1994). *Trl* was found to encode the GAGA factor, whose function was recently reviewed in detail (Granok et al. 1995). GAGA was originally identified as a transcriptional activator that bound specifically to multiple GA/CT-rich target sequences in the *Ultrabithorax* (*Ubx*) gene (Biggin and Tjian 1988) but was also identified independently on the basis of binding similar sequences in the *hsp26*, *hsp70*, and histone H3 and H4 genes (Gilmour et al. 1989). More recently, it was identified as a heterochromatin-associated protein, presumably due to its binding to tandem arrays of a specific satellite DNA sequence (AAGAG and/or AAGAGAG) that contains a GAGA-binding site (Raff et al. 1994). The latter association is seen in *Drosophila melanogaster*, but not *Drosophila virilis*, which does not contain these satellite species. This makes it likely that this heterochromatic binding is "adventitious" and that the enhancing effects of *Trl* mutations on PEV are mediated by its action at its euchromatic binding sites within its many target genes, rather than any effect on heterochromatin. The GAGA protein contains a single zinc finger and a highly conserved motif, the POZ or BTB domain (120 amino acids), which mediates homo- and heterodimerization in vitro (Zollman et al. 1994; Chen et al. 1995) and also inhibits DNA binding by purified GAGA in vitro (Bardwell and Treisman 1994). Some further functional diversity may be dictated by GAGA isoforms generated by alternative splicing (Soeller et al. 1993; Granok et al. 1995).

Most GAGA-binding sites are located within a few hundred base pairs of the promoter, but the large *Ubx* 5′ regulatory region contains multiple potential GAGA-binding sites distributed in the vicinity of both promoter and enhancer elements, consistent with evidence that GAGA acts locally to promote the formation of "open" chromatin configurations necessary for the efficient binding of other activators. GAGA binding promotes the formation of local DNase I hypersensitive sites, which are observed even in the absence of binding of these other activators, consistent with the observed constitutive binding of GAGA to its target genes regardless of their transcriptional status (Tsukiyama et al. 1994). GAGA binding by itself is thus insufficient to activate transcription in the absence of the other activators whose binding it facilitates. GAGA has been shown to act as an antirepressor of histone H1-mediated inhibition of RNA polymerase II transcription (Croston et al. 1991). Purified GAGA has been shown to possess an ATP-dependent nucleosome-displacement activity that facilitates the binding of other activators (Tsukiyama et al. 1994). GAGA itself does not bind ATP, and its ATP-dependent nucleosome-displacement activity has recently been shown to be due to its association with another multisubunit factor, NURF, one component of which is the *Drosophila* BRM-related protein ISWI (Tsukiyama and Wu 1995; Tsukiyama et al. 1995). ISWI (1027 amino acids) has a shorter and less conserved version of the BRM/SWI2 ATPase domain and contains no bromodomain. NURF/ISWI and GAGA are found in a complex that is separate from and smaller than the SWI2 complex (Tsukiyama and Wu 1995; Tsukiyama et al. 1995). This suggests an intriguing mechanistic link between the chromatin-remodeling activities of the two biochemically best-characterized trxG members. Whether NURF/GAGA complexes, like SWI complexes, are directly associated with RNA polymerase II initiation complexes remains to be determined. Genetic analysis of ISWI remains to be done, but ISWI mutants would be expected to have overlapping, if not identical, phenotypes to *Trl*/GAGA mutants, depending on how exclusive the GAGA:ISWI association is.

trithorax

The *trx* gene encodes two very large proteins (Kuzin et al. 1994; Sedkov et al. 1994; Stassen et al. 1995) that contain a novel variant of the nuclear receptor-type DNA-binding motif (Stassen et al. 1995) and two other novel motifs found in other proteins from evolutionarily diverse taxa. One of these, the carboxy-terminal SET domain (150 residues), is also found in ASH1 (another trxG protein) (D. LaJeunuesse and A. Shearn, pers. comm.), in Enhancer of zeste (E[Z], a PcG protein) (Jones and Gelbart 1993), and in SU(VAR)3-9, which is required for the heterochromatin-mediated repression seen in PEV (Tschiersch et al. 1994). The SET domain is present in more than 20 other proteins, including several fungal and plant proteins, indicating its ancient origin (Stassen et al. 1995). The TRX protein also contains four copies of a novel C_4HC_3 zinc finger motif, termed the PHD finger (Aasland et al. 1995; Stassen et al. 1995), which is also present in two copies in Polycomb-like (PCL, another PcG protein) (Stassen et al. 1995) and in one copy in ASH1 (D. LaJeunuesse and A. Shearn, pers. comm.). This motif is also found in at least 50 other proteins, several of which appear to function as transcriptional "adapters" between specific activator proteins and other components of the transcriptional machinery, although it is not known if this motif mediates this activity.

Sequence-specific DNA-binding activity has not been demonstrated for TRX in vitro. However, in vivo, TRX binds specifically to about 75 specific chromosomal sites, including those of known targets of TRX: the Bithorax complex (*BXC*), the Antennapedia complex (*ANTC*), and the *forkhead* and *engrailed* genes (Kuzin et al. 1994; Breen et al. 1995; Chinwalla et al. 1995). This indicates that TRX either binds to specific DNA sequences itself or is associated with other proteins which do so. The PRE identified in the 5′ regulatory region of the *Ubx* gene also contains a TRX response element (TRE) (Chan et al. 1994), which has been delimited to several hundred base pairs by the ability of isolated fragments of *Ubx* located at ectopic chromosomal positions to recruit TRX protein (Chinwalla et al. 1995 and in prep.) and also by its ability to stimulate expression of a reporter gene in a transient transfection assay in conjunction with a TRX expression plasmid (Chang et al. 1995).

INTERACTIONS BETWEEN PcG AND TrxG

Mutations in PcG genes lead to homeotic transformations of segment identity due to the ectopic expression of homeotic genes. In wild-type flies, these homeotic genes are repressed by PcG. Loss-of-function mutations in trxG genes reduce or completely suppress PcG mutant phenotypes. In contrast, increasing trxG gene dosage can sometimes enhance homeotic transformations seen in PcG mutants (Capdevila and García-Bellido 1981; Ingham 1983; Capdevila et al. 1986; Jones and Gelbart 1993). Thus, TrxG and PcG proteins appear to act antagonistically and compete at some level. This raises the possibility that the final determination of segment identity may depend critically on the balance of these competing positive and negative regulatory factors acting on the homeotic genes. However, the suppression of PcG phenotypes by *trx* mutations is not directed selectively at the ectopic expression seen in PcG mutants. *trx* mutations reduce the level of homeotic gene expression in their normal domains to the same extent as they do in the ectopic domains of expression found in PcG mutants (T.R. Breen and P.J. Harte, unpubl.). This suggests that the apparent antagonism between the two groups could be quite indirect and that the role of TRX is not likely to be restricted to countering PcG silencing.

Genetic interactions between mutations in *trx* and PcG genes have fueled speculation about their potentially intimate physical and mechanistic associations. However, there is presently no evidence for direct interactions of any PcG with trxG proteins at the molecular level. It has been suggested that the trxG proteins, as a group, antagonize or counter the proposed "heterochromatinizing" action of PcG silencing to promote long-term maintenance of transcriptionally active states (Paro 1990; Travers 1992; Winston and Carlson 1992). In the cases of GAGA and SWI complexes, "antirepression" of nucleosomal inhibition of transcription has been directly demonstrated. The apparently similar ATP-dependent nucleosome-displacement activities of the GAGA/NURF and SWI complexes suggest that at least some of the other trxG proteins may also act through similar mechanisms to promote alterations in chromatin structure. However, this does not mean that they must act as direct antirepressors of PcG-mediated silencing. Indeed, the activities of GAGA and SWI complexes are likely to exert their effects on PcG silencing indirectly, ultimately through the action of other activators whose binding they facilitate. Recent evidence from in vivo studies in *Drosophila* indicate that strong transcriptional activators such as GAL4, when present at high enough levels, can counter PcG-mediated silencing even after PcG silencing complexes have been assembled (Fig. 1) (Zink and Paro 1995). These results predict that any strong, constitutively active enhancer should render a promoter relatively refractory to silencing activities of a neighboring PRE. In the above case, because the GAL4 activator is exogenous to *Drosophila*, it is most likely that its effect on PcG silencing is indirect; i.e., the binding of an activator to its binding site(s) can perturb preassembled PcG silencing complexes without specifically "interacting" with them. This is an important insight, since it suggests that the antagonistic effects of different trxG proteins could reflect either (1) direct "antirepression" effects on the PcG silencing mechanism, e.g., through direct interactions with PcG proteins to influence the stability or behavior of nascent PcG complexes or (2) indirect effects on the PcG silencing mechanism, either by acting as direct activators or by facilitating or stabilizing the binding of other activators, which could indirectly perturb PcG silencing or prevent its initial implementation. Indeed, the recently demonstrated epigenetic "inheritance" by the *Ubx* imaginal enhancers of the earlier transcriptional status of the embryonic enhancers in each cell probably involves such "indirect" antagonism by any transcriptionally engaged embryonic enhancers (Pirrotta et al. 1995).

Recent studies raise the possibility that TRX may exert its antagonistic effects on PcG silencing by acting more directly on PcG proteins. In double-labeling experiments, the TRX and Pc proteins colocalize at many of their binding sites on polytene chromosomes, suggesting that their proximity may be functionally important (Chinwalla et al. 1994). The 120-kb *Ubx* gene appears to contain only a single TRE, and it is so far inseparable from the single (or at least stronger of two)

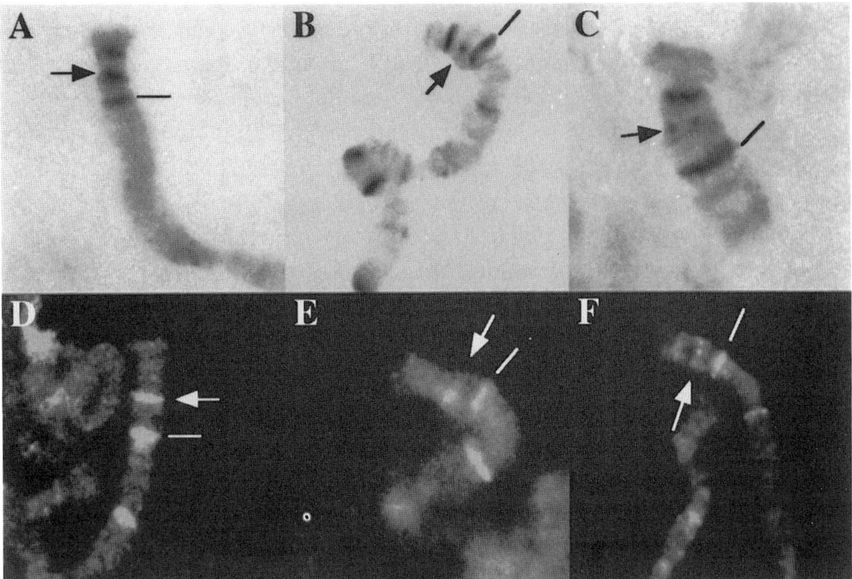

Figure 1 Immunostaining of a larval polytene chromosome carrying a transgene construct containing an isolated PRE to demonstrate competition between the repressor protein Polycomb and the yeast *trans*-activator GAL4. In this construct the PRE (the FAB regulatory element of the bithorax complex) is flanked by GAL4-binding sites ($GAL4_{UAS}$) and a *lacZ* reporter gene (Zink and Paro 1995). To activate the *lacZ* reporter, this transgenic strain was crossed to a GAL4-expressing strain. (*A–C*) Chromosome stained with anti-Pc antibodies. The integration site of the PRE transgene is shown by the arrow (61C4 on the 3rd chromosome). (*A*) In the absence of GAL4, the Pc protein binds to the PREs over the entire width of the chromosome. (*B*,*C*) In the presence of GAL4, only dots of Pc staining are visible at the PRE site. Note that adjacent endogenous sites of Pc staining (*line*) are not affected. (*D–F*) Chromosomes containing the same PRE construct at 61C4 stained with anti-GAL4 antibodies. The GAL4 and Pc proteins exhibit complementary staining patterns. (*D*) High amounts of GAL4 completely remove Pc protein and cover the chromosome over the entire width. (*E*,*F*) Intermediate amounts of GAL4 produce a dotted distribution. Note that the patchy distribution of the two proteins suggests a clonal event. (Adapted, with permission, from Zink and Paro 1995 [copyright Oxford University Press].)

Ubx PRE (Chan et al. 1994; Chang et al. 1995; V. Chinwalla et al., in prep.). Furthermore, TRX and PcG proteins are simultaneously associated with the same TRE/PRE in the same cell, indicating that their genetic antagonism is not likely to reflect competition for binding, although quantitative effects of any such competition cannot be ruled out using the chromosome-binding assay. Most interestingly, in isolation, this PRE/TRE does not exhibit any detectable enhancer activity in vivo when coupled to a *lacZ* reporter previously used to identify and delimit *Ubx* enhancers (Chan et al. 1994). This suggests that TRX does not act as a conventional transcriptional activator. Furthermore, TRX is bound to the transcriptionally repressed *BXC* and *ANTC* loci in larval salivary glands (and also binds to the isolated *Ubx* TRE), suggesting that TRX, like GAGA, is constitutively bound to its target genes at some level, regardless of their transcriptional status. This suggests that binding of TRX, like GAGA, is not sufficient to activate transcription, or at least that PcG binding preempts any such activity. Transient transfection assays in a *Drosophila* cell line using a TRX expression plasmid and a similar *Ubx-lacZ* reporter containing a TRE/PRE element suggest that TRX itself may have some transcriptional activating activity, although the observed activation is quite modest (three- to fourfold). Interestingly, this activation can be directly competed by increasing levels of exogenously provided Pc protein (Chang et al. 1995), raising the possibility that weak activation by TRX reflects direct antirepression of PcG-silencing effects. Further characterization of the relationship between PREs and TREs, and investigation of the possibilities of direct interactions between TRX and PcG proteins, should shed more light on this possibility. This possibility is further hinted at by the presence of a SET domain TRX, ASH1 (trxG proteins) and E(Z) (a PcG protein) and the presence of C_4HC_3 PHD finger motifs in TRX, ASH1, and the PcG protein PCL.

CONCLUSIONS

In a developing embryo, the proteins of the PcG and trxG are for the most part homogeneously distributed in all cells. Yet they control a particular set of genes differently in diverse parts of the body. How is a gene tagged to stay permanently inactivated through multiple cell divisions if no covalent mark is available, like methylation (which seems to play no role in *Drosophila*)? The solution could lie in the self-organizing properties of the chromatin structures generated by the PcG and trxG complexes. At a particular time in development, the patterning system establishes the expression states of genes encoding determinants of cell fates. In the case of homeotic genes, this is done by the transiently expressed transcription factors encoded by the maternal and segmentation genes. This initiation signal is translated into an appropriate chromatin structure controlled by the PcG and trxG. The apparently extended chromatin structures they implement subsequently ensure the mitotic stability of transcriptional states such that either the inactive or the active state can be faithfully maintained over many cell generations. For the PcG-silencing complexes, interactions with multiple DNA sites are necessary to achieve this stability. During passage of the replication machinery, this could allow some silencing complexes to maintain contact with some part of a specific target gene at all times and serve to "template" their reassembly on both daughter strands through cooperative interactions (Fig. 2). The cooperative and self-organizing features of these multiprotein complexes would drive the reassembly of chromatin structures necessary to stably maintain the appropriate active or inactive transcriptional states in the following interphase and S phase.

Evidence is accumulating that this type of epigenetic transcriptional regulation has been conserved during evolution. A number of proteins that share homology with PcG and trxG proteins have been identified in a variety of species, and recent results confirm their functional similarities. Mutations in the murine *bmi-1* gene, a homolog of the PcG gene *Posterior sex combs* (*Psc*), result in homeotic transformations in the mouse embryo (van der Lugt et al. 1994; Alkema et al. 1995). A murine gene, M33 (Pearce et al. 1992), encoding a Pc homolog, can partially rescue the phenotype of *Pc* mutations when transformed into *Drosophila* (Müller et al. 1995). In *Xenopus*, a Pc and bmi-1/PSC homolog were found to physically interact (Rejinen et al. 1995). The PcG protein ESC was predicted to be widely conserved on the basis of its high degree of conservation in a distant *Drosophila* species (Sathe and Harte 1995b), and its recently identified mammalian counterpart is indeed very highly conserved (A. Schumacher et al., pers. comm.). If ESC is uniquely involved in the initial assembly of PcG-silencing complexes, as has been suggested, then it seems likely that this function is also conserved in vertebrates.

In addition to the BRM/SWI homologs discussed above, a TRX homolog, MLL (also known as ALL-1, HRX, Htrx), has been identified in humans (Gu et al. 1992; Tkachuk et al. 1992; Domer et al. 1993) and mice (Ma et al. 1993) and is implicated in the majority of infantile acute lymphocytic and mixed lineage leukemias. Disruption of the mouse *MLL* gene by targeted mutagenesis causes homeotic transformations of the vertebrae in heterozygotes and loss of HOX gene expression, indicating that MLL is a functional equivalent of TRX in mammals (Yu et al. 1995). Such evolutionary conservation of multiple components of the PcG and trxG strongly suggests that in vertebrates the long-term maintenance of the transcriptional states of genes encoding developmentally important cell-fate-determining factors is likely to be implemented

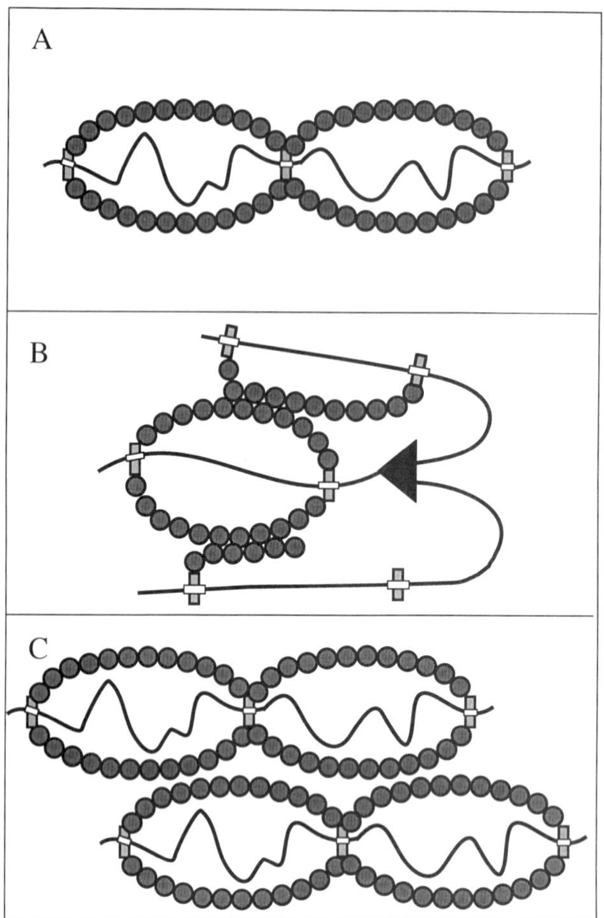

Figure 2 Model for the epigenetic inheritance of silencing complexes through mitoses. (*A*) PcG complexes (*circles*) organize an extended region of the chromatin fiber in a heterochromatin-like fashion and interact (*gray rectangles*) with the DNA at specific *cis*-regulatory elements (PREs, maintenance elements, etc.; *white boxes*). (*B*) During DNA synthesis, functional silencing complexes may be completely disassembled locally and dissociate temporarily from the DNA as the replication machinery (*black triangle*) passes. However, those intact silencing complexes ahead of the replication fork may rapidly "template" their reassembly on both daughter strands by stabilizing weak constitutive associations of PcG proteins with PREs through cooperative interactions with them. (*C*) After completion of mitosis, the self-organizing features of the PcG complexes allow a full reconstitution of the silenced chromosomal domain.

by similar chromatin structures controlled by PcG and trxG factors. It will be interesting to see whether other epigenetic phenomena, like X-chromosome inactivation, genomic imprinting (see corresponding chapters in this volume), or odor receptor gene silencing (Chess et al. 1994), utilize similar mechanisms based on higher-order chromatin structures.

ACKNOWLEDGMENTS

Thanks to John Tamkun, Allen Shearn, Armin Schumacher, and Terry Magnuson for communicating information prior to publication. The work of R. Paro was supported by the Deutsche Forschungsgemeinschaft and the Fonds der Chemischen Industrie, and the work of P.J. Harte was supported by the National Institutes of Health, The American Cancer Society, The American Heart Association, and the Council for Tobacco Research.

REFERENCES

Aasland, R. and F.A. Stewart. 1995. The chromo shadow domain, a second chromo domain in heterochromatin-binding protein, HP1. *Nucleic Acids Res.* **23:** 3168–3173.

Aasland, R., T.J. Gibson, and F.A. Stewart. 1995. The PHD finger: Implications for chromatin-mediated transcriptional regulation. *Trends Biochem. Sci.* **95:** 56–59.

Alkema, M.J., N.M.T. van der Lugt, R.C. Bobeldijk, A. Berns, and M. van Lohuizen. 1995. Transformation of axial skeleton due to overexpression of *bmi-1* in transgenic mice. *Nature* **374:** 724-727.

Bardwell, V.J. and R. Treisman. 1994. The POZ domain: A conserved protein-protein interaction motif. *Genes Dev.* **15:** 1664–1677.

Beck, S., I. Hanson, A. Kelley, D.J.C. Pappin, and J. Trowsdale. 1992. A homologue of the *Drosophila female sterile homeotic* (*fsh*) gene in the class II region of the human MHC. *DNA Seq.* **2:** 203–210.

Biggin, M.D. and R. Tjian. 1988. Transcription factors that activate the *Ultrabithorax* promoter in developmentally staged extracts. *Cell* **53:** 699–711.

Breen, T.R., V. Chinwalla, and P.J. Harte. 1995. *Trithorax* is required to maintain *engrailed* expression in a subset of *engrailed*-expressing cells. *Mech. Dev.* **52:** 89–98.

Brunk, B.P., E.C. Martin, and P.N. Adler. 1991. *Drosophila* genes *Posterior sex combs* and *Suppressor two of zeste* encode proteins with homology to the murine *bmi-1* oncogene. *Nature* **353:** 351–353.

Capdevila, M.P. and A. García-Bellido. 1981. Genes involved in the activation of the bithorax complex of *Drosophila*. *Roux's Arch. Dev. Biol.* **190:** 339–350.

Capdevila, M.P., J. Botas, and A. García-Bellido. 1986. Genetic interactions between the *Polycomb* locus and the Antennapedia and bithorax complexes of *Drosophila*. *Roux's Arch. Dev. Biol.* **195:** 417–432.

Chan, C.S., L. Rastelli, and V. Pirrotta. 1994. A Polycomb response element in the *Ubx* gene that determines an epigenetically inherited state of repression. *EMBO J.* **13:** 2553–2564.

Chang, Y., B.O. King, M. O'Connor, A. Mazo, and D. Huang. 1995. Functional reconstruction of trans regulation of the *Ultrabithorax* promoter by the products of two antagonistic genes, *trithorax* and *Polycomb*. *Mol. Cell. Biol.* **15:** 6601–6612.

Chen, W., S. Zollman, J.L. Couderc, and F.A. Laski. 1995. The BTB domain of *bric à brac* mediates dimerization *in vitro*. *Mol. Cell. Biol.* **15:** 3424–3429.

Chess, A., I. Simon, H. Cedar, and R. Axel. 1994. Allelic inactivation regulates olfactory receptor gene expression. *Cell* **78:** 823–834.

Chinwalla, V., E.P. Jane, and P.J. Harte. 1995. The *Drosophila* trithorax protein binds to specific chromosomal sites and is co-localized with *Polycomb* at many sites. *EMBO J.* **14:** 2056–2065.

Croston, G.E., L.A. Kerrigan, L.M. Lira, D.R. Marshak, and J. Kadonaga. 1991. Sequence-specific antirepression of histone H1-mediated inhibition of basal RNA polymerase II transcription. *Science* **251:** 643–649.

DeCamillis, M., N.S. Cheng, D. Pierre, and H. Brock. 1992. The *polyhomeotic* gene of *Drosophila* encodes a chromatin protein that shares polytene chromosome-binding sites with Polycomb. *Genes Dev.* **6:** 223–232.

Dingwall, A.K., S.J. Beek, C.M. McCallum, J.W. Tamkun, G.V. Kalpana, S.P. Goff, and M.P. Scott. 1995. The *Drosophila* snr1 and brm proteins are related to yeast SWI/SNF proteins and are components of a large protein complex. *Mol. Biol. Cell* **6:** 777–791.

Domer, P.H., S.S. Fakharzadeh, C.S. Chen, J. Jockel, L. Johansen, G.A. Silverman, J.H. Kersey, and S.J. Korsmeyer. 1993. Acute mixed-lineage leukemia t(4;11)(q21;q23) generates an MLL-AF4 fusion product. *Proc. Natl. Acad. Sci.* **90:** 7884–7888.

Dorn, R., V. Krauss, G. Reuter, and H. Saumweber. 1994. The enhancer of position-effect variegation of *Drosophila*, *E(var)3-93D*, codes for a chromatin protein containing a conserved domain common to several transcriptional regulators. *Proc. Natl. Acad. Sci.* **90:** 11376–11380.

Dunaief, J.L., B.E. Strober, S. Guha, P.A. Khavari, K. Alin, J. Luban, M. Begemann, G.R. Crabtree, and S.P. Goff. 1994. The retinoblastoma protein and BRG1 form a complex and cooperate to induce cell cycle arrest. *Cell* **79:** 119–130.

Elfring, L.K., R. Deuring, C.M. McCallum, C.L. Peterson, and J.W. Tamkun. 1994. Identification and characterization of *Drosophila* relatives of the yeast transcriptional activator SNF2/SWI2. *Mol. Cell. Biol.* **14:** 2225–2234.

Eissenberg, J.C., T.C. James, D.M. Foster-Hartnett, T. Hartnett, V. Ngan, and S.C. Elgin. 1990. Mutation in a heterochromatin-specific chromosomal protein is associated with suppression of position-effect variegation in *Drosophila melanogaster*. *Proc. Natl. Acad. Sci.* **87:** 9923–9927.

Farkas, G., J. Gausz, M. Galloni, G. Reuter, H. Gyurkovics, and F. Karch. 1994. The *Trithorax-like* gene encodes the *Drosophila* GAGA factor. *Nature* **371:** 806–808.

Fauvarque, M.O. and J.M. Dura. 1993. *polyhomeotic* regulatory sequences induce developmental regulator-dependent variegation and targeted P element insertions in *Drosophila*. *Genes Dev.* **7:** 1508–1520.

Franke, A., S. Messmer, and R. Paro. 1995. Mapping functional domains of the Polycomb protein of *Drosophila melanogaster*. *Chromosome Res.* **3:** 351–360.

Franke, A., M. DeCamillis, D. Zink, N. Cheng, H.W. Brock, and R. Paro. 1992. Polycomb and polyhomeotic are constituents of a multimeric protein complex in chromatin of *Drosophila melanogaster*. *EMBO J.* **11:** 2941–2950.

Gilmour, D.S., G.H. Thomas, and S.C. Elgin. 1989. *Drosophila* nuclear proteins bind to regions of alternating C and T residues in gene promoters. *Science* **245:** 1487–1490.

Granok, H., B.A. Leibovitch, C.D. Shaffer, and S.C. Elgin. 1995. Chromatin. Ga-ga over GAGA factor. *Curr. Biol.* **5:** 238–241.

Grigliatti, T. 1991. Position-effect variegation—An assay for nonhistone chromosomal proteins and chromatin assembly and modifying factors. In *Functional organization of the nucleus* (ed. B. Hamkalo and S. Elgin), pp. 587–627. Academic Press, San Diego.

Gu, Y., T. Nakamura, H. Alder, R. Prasad, O. Canaani, G. Cimino, C.M. Croce, and E. Canaani. 1992. The t(4;11) chromosome translocation of human acute leukemias fuses the ALL-1 gene, related to *Drosophila* trithorax, to the AF-4 gene. *Cell* **71:** 701–708.

Gutjahr, T., E. Frei, C. Spicer, S. Baumgartner, R.A. White, and M. Noll. 1995. The Polycomb-group gene, extra sex combs, encodes a nuclear member of the WD-40 repeat family. *EMBO J.* **14:** 4296–4306.

Haynes, S.R., B.A. Mozer, N. Bhatia-Dey, and I.B. Dawid. 1989. The *Drosophila fsh* locus, a maternal effect homeotic gene, encodes apparent membrane proteins. *Dev. Biol.* **134:** 246–257.

Hecht, A., T. Laroche, S.S. Bolsinger, S.M. Gasser, and M. Grunstein. 1995. Histone H3 and H4 N-termini interact with the silent information regulators Sir3 and Sir4: A molecular model for the formation of heterochromatin in yeast. *Cell* **80:** 583–592.

Imbalzano, A.Y., H. Kwon, M.R. Green, and R.E. Kingston. 1994. Facilitated binding of TATA-binding protein to nucleosomal DNA. *Nature* **370:** 481–485.

Ingham, P.W. 1983. Differential expression of bithorax complex genes in the absence of the *extra sex combs* and *trithorax* genes. *Nature* **306:** 591–593.

Ingham, P.W. and R. Whittle. 1980. *Trithorax*: A new homeotic mutation of *Drosophila melanogaster* causing transformations of abdominal and thoracic imaginal segments. I. Putative role during embryogenesis. *Mol. Gen. Genet.* **179:** 607–614.

Ishida, A., H. Asano, M. Hasegawa, H. Koseki, T. Ono, M.C. Yoshida, M. Taniguchi, and M. Kanno. 1993. Cloning and chromosome mapping of the human Mel-18 gene which encodes a DNA-binding protein with a new "RING-finger" motif. *Gene* **129:** 249–255.

Jones, R.S. and W.M. Gelbart. 1993. The *Drosophila* polycomb-group gene *Enhancer of zeste* contains a region with sequence similarity to *trithorax*. *Mol. Cell. Biol.* **13:** 6357–6366.

Kennison, J.A. 1993. Transcriptional activation of *Drosophila* homeotic genes from distant regulatory elements. *Trends Genet.* **9:** 75–79.

Kennison, J.A. and M.A. Russell. 1987. Dosage-dependent modifiers of homeotic mutations in *Drosophila melanogaster*. *Genetics* **116:** 75–86.

Kennison, J.A. and J.W. Tamkun. 1988. Dosage-dependent modifiers of *Polycomb* and *Antennapedia* mutations in *Drosophila*. *Proc. Natl. Acad. Sci.* **85:** 8136–8140.

Khavari, P.A., C.L. Peterson, J.W. Tamkun, D.B. Mendel, and G.R. Crabtree. 1993. BRG1 contains a conserved domain of the SWI2/SNF2 family necessary for normal mitotic growth and transcription. *Nature* **366:** 170–174.

Kuzin, B., S. Tillib, Y. Sedkov, L. Mizrochi, and A. Mazo. 1994. The *Drosophila trithorax* gene encodes a chromosomal protein and directly regulates the region-specific homeotic gene *fork head*. *Genes Dev.* **8:** 2478–2490.

Kwon, H., A.N. Imbalzano, P.A. Khavari, R.E. Kingston, and M.R. Green. 1994. Nucleosome disruption and enhancement of activator binding by a human SW1/SNF complex. *Nature* **370:** 477–481.

LaJeunuesse, D. and A. Shearn. 1995. Transregulation of thoracic homeotic selector genes of the Antennapedia and bithorax complexes by the trithorax group genes *absent, small or homeotic discs 1 and 2*. *Mech. Dev.* **53:** 123–140.

Locke, J., M.A. Kotarski, and K.D. Tartof. 1988. Dosage-dependent modifiers of position effect variegation in *Drosophila* and a mass action model that explains their effect. *Genetics* **120:** 181–198.

Lonie, A., R. D'Andrea, R. Paro, and R. Saint. 1994. Molecular characterisation of the *Polycomblike* gene of *Drosophila melanogaster*, a *trans*-acting negative regulator of homeotic gene expression. *Development* **120:** 2629–2636.

Ma, Q., H. Alder, K.K. Nelson, D. Chatterjee, Y. Gu, T. Nakamura, E. Canaani, C.M. Croce, L.D. Siracusa, and A.M. Buchberg. 1993. Analysis of the murine *All-1* gene reveals conserved domains with human ALL-1 and identifies a motif shared with DNA methyltransferases. *Proc. Natl. Acad. Sci.* **90:** 6350–6354.

Martin, E.C. and P.N. Adler. 1993. The Polycomb group gene *Posterior Sex Combs* encodes a chromosomal protein. *Development* **117:** 641–655.

McKeon, J., E. Slade, D.A. Sinclair, N. Cheng, M. Couling, and H.W. Brock. 1994. Mutations in some Polycomb group genes of *Drosophila* interfere with regulation of segmentation genes. *Mol. Gen. Genet.* **244:** 474–483.

Messmer, S., A. Franke, and R. Paro. 1992. Analysis of the functional role of the Polycomb chromo domain in *Drosophila melanogaster*. *Genes Dev.* **6:** 1241–1254.

Muchardt, C. and M. Yaniv. 1993. A human homologue of *S. cerevisiae* SNF2/SWI2 and *Drosophila* brm genes potentiates transcriptional activation by the glucocorticoid receptor. *EMBO J.* **12:** 4279–4290.

Müller, J. 1995. Transcriptional silencing by the Polycomb protein in *Drosophila* embryos. *EMBO J.* **14:** 1209–1220.

Müller, J., S. Gaunt, and P.A. Lawrence. 1995. Function of the Polycomb protein is conserved in mice and flies. *Development* **121:** 2847–2852.

Nomura, M., Y. Takihara, and K. Shimada. 1994. Isolation and characterization of retinoic acid-inducible cDNA clones in F9 cells: One of the early inducible clones encodes a novel protein sharing several highly homologous regions with a *Drosophila polyhomeotic* protein. *Differentiation* **57:** 39–50.

Palladino, F. and S.M. Gasser. 1995. Telomere maintenance and gene repression: A common end? *Curr. Opin. Cell Biol.* **6:** 373–379.

Paro, R. 1990. Imprinting a determined state into the chromatin of *Drosophila*. *Trends Genet.* **6:** 416–421.

———. 1995. Propagating memory of transcriptional states. *Trends Genet.* **11:** 295–297.

Paro, R. and D. S. Hogness. 1991. The Polycomb protein shares a homologous domain with a heterochromatin-associated protein of *Drosophila*. *Proc. Natl. Acad. Sci.* **88:** 263–267.

Pearce, J.J.H., P.B. Singh, and S.J. Gaunt. 1992. The mouse has a *Polycomb*-like chromobox gene. *Development* **114:** 921–929.

Pelegri, F. and R. Lehmann. 1994. A role of Polycomb group genes in the regulation of *gap* gene expression in *Drosophila*. *Genetics* **136:** 1341–1353.

Peterson, C.L. and J.W. Tamkun. 1995. The SWI-SNF complex: A chromatin remodeling machine? *Trends Biochem. Sci.* **20:** 143–146.

Pirrotta, V., C.S. Chan, D. McCabe, and S. Qian. 1995. Distinct parasegmental and imaginal enhancers and the establishment of the expression pattern of the *Ubx* gene. *Genetics* **141:** 1439–1450.

Platero, J.S., T. Hartnett, and J.C. Eissenberg. 1995. Functional analysis of the chromo domain of HP1. *EMBO J.* **14:** 3977–3986.

Powers, J.A. and J.C. Eissenberg. 1993. Overlapping domains of the heterochromatin-associated protein HP1 mediate nuclear localization and heterochromatin binding. *J. Cell Biol.* **120:** 291–299.

Raff, J.W., R. Kellum, and B. Alberts. 1994. The *Drosophila* GAGA transcription factor is associated with specific regions of heterochromatin throughout the cell cycle. *EMBO J.* **13:** 5977–5983.

Rastelli, L., C.S. Chan, and V. Pirrotta. 1993. Related chromosome binding sites for zeste, suppressors of zeste and Polycomb group proteins in *Drosophila* and their dependence on *Enhancer of zeste* function. *EMBO J.* **12:** 1513–1522.

Reijnen, M.J., K.M. Hamer, J.L. den Blaauwen, C. Lambrechts, I. Schoneveld, R. van Driel, and A.P. Otte. 1995. Polycomb and bmi-1 homologs are expressed in overlapping patterns in *Xenopus* embryos and are able to interact with each other. *Mech. Dev.* **53:** 35–46.

Sathe, S.S. and P.J. Harte. 1995a. The *extra sex combs* protein contains WD motifs essential for its function as a repressor of the homeotic genes. *Mech. Dev.* **52:** 77–87.

———. 1995b. The *extra sex combs* protein is highly conserved between *Drosophila virilis* and *Drosophila melanogaster*. *Mech. Dev.* **52:** 225–232.

Sedkov, Y., S. Tillib, L. Mizrokhi, and A. Mazo. 1994. The bithorax complex is regulated by *trithorax* earlier during *Drosophila* embryogenesis than is the Antennapedia complex, correlating with a bithorax-like expression pattern of distinct early trithorax transcripts. *Development* **120:** 1907–1917.

Shearn, A. 1989. The *ash-1*, *ash-2* and *trithorax* genes of *Drosophila melanogaster* are functionally related. *Genetics* **121:** 517–525.

Simon, J. 1995. Locking in stable states of gene expression:transcriptional control during *Drosophila* development. *Curr. Opin. Cell Biol.* **7:** 376–385.

Simon, J., D. Bornemann, K. Lunde, and C. Schwartz. 1995. The extra sex combs product contains WD40 repeats and its time of action implies a role distinct from other Polycomb group products. *Mech. Dev.* **53:** 197–208.

Simon, J., M. Peifer, W. Bender, and M. O'Connor. 1990. Regulatory elements of the bithorax complex that control expression along the anterior-posterior axis. *EMBO J.* **9:** 3945–3956.

Soeller, W.C., C.E. Oh, and T.B. Kornberg. 1993. Isolation of cDNAs encoding the *Drosophila* GAGA transcription factor. *Mol. Cell. Biol.* **13:** 7961–7970.

Stassen, M.J., D. Bailey, S. Nelson, V. Chinwalla, and P.J. Harte. 1995. The *Drosophila* trithorax proteins contain a novel variant of the nuclear receptor type DNA binding domain and an ancient conserved motif found in other chromosomal proteins. *Mech. Dev.* **52:** 209–224.

Struhl, G. and D. Brower. 1982. Early role of the *esc*+ gene product in the determination of segments in *Drosophila*. *Cell* **31:** 285–292.

Tamkun, J.W. 1995. The role of brahma and related proteins in transcription and development. *Curr. Opin. Genet. Dev.* **5:** 473–477.

Tamkun, J.W., R. Deuring, M.P. Scott, M. Kissinger, A.M. Pattatucci, T.C. Kaufman, and J.A. Kennison. 1992. *brahma*: A regulator of *Drosophila* homeotic genes structurally related to the yeast transcriptional activator SNF2/SWI2. *Cell* **68:** 561–572.

Tkachuk, D.C., S. Kohler, and M.L. Cleary. 1992. Involvement of a homolog of *Drosophila* trithorax by 11q23 chromosomal translocations in acute leukemias. *Cell* **71:** 691–670.

Travers, A.A. 1992. The reprogramming of transcriptional competence. *Cell* **69:**

573–575.

Tschiersch, B., A. Hofmann, V. Krauss, R. Dorn, G. Korge, and G. Reuter. 1994. The protein encoded by the *Drosophila* position-effect variegation suppressor gene *Su(var)3-9* combines domains of antagonistic regulators of homeotic gene complexes. *EMBO J.* **13:** 3822–3831.

Tsukiyama, T. and C. Wu. 1995. Purification and properties of an ATP-dependent nucleosome remodeling factor. *Cell* **83:** 1011–1020.

Tsukiyama, T., P.B. Becker, and C. Wu. 1994. ATP-dependent nucleosome disruption at a heat-shock promoter mediated by binding of GAGA transcription factor. *Nature* **367:** 525–532.

Tsukiyama, T., C. Daniel, J. Tamkun, and C. Wu. 1995. ISWI, a member of the SWI2/SNF2 ATPase family, encodes the 140 kDa subunit of the nucleosome remodeling factor. *Cell* **83:** 1021–1026.

van der Lugt, N.M.T., J. Domen, K. Linders, M. van Roon, E. Robanus-Maandag, H. te Riele, M. van der Valk, J. Deschamps, M. Sofroniew, M. van Lohuizen, and A. Berns. 1994. Posterior transformation, neurological abnormalities, and severe hematopoietic defects in mice with a targeted deletion of the *bmi-1* proto-oncogene. *Genes Dev.* **8:** 757–769.

van Lohuizen, M., M. Frasch, E. Wientjens, and A. Berns. 1991. Sequence similarity between the mammalian *bmi-1* proto-oncogene and the *Drosophila* regulatory genes *Psc* and *Su(z)2*. *Nature* **353:** 353–355.

Weeks, J.R., S.E. Hardin, J. Shen, J.M. Lee, and A.L. Greenleaf. 1993. Locus-specific variation in phosphorylation state of RNA polymerase II in vivo: Correlations with gene activity and transcript processing. *Genes Dev.* **7:** 2329–2344.

Wilson, C.J., D.M. Chao, A.N. Imbalzano, G.R. Schnitzler, R.E. Kingston, and R.A. Young. 1996. RNA polymerase II holoenzyme contains SWI/SNF regulators involved in chromatin remodeling. *Cell* **84:** 235–244.

Winston, F. and M. Carlson. 1992. Yeast SNF/SWI transcriptional activators and the SPT/SIN chromatin connection. *Trends Genet.* **8:** 387–391.

Yu, B.D., J.L. Hess, S.E. Horning, G.A. Brown, and S.J. Korsmeyer. 1995. Altered Hox expression and segmental identity in *Mll*-mutant mice. *Nature* **378:** 505–508.

Zink, D. and R. Paro. 1995. *Drosophila* Polycomb-group regulated chromatin inhibits the accessibility of a *trans*-activator to its target DNA. *EMBO J.* **14:** 5660–5671.

Zollman, S., D. Godt, G.G. Prive, J.L. Couderc, and F.A. Laski. 1994. The BTB domain, found primarily in zinc finger proteins, defines an evolutionarily conserved family that includes several developmentally regulated genes in *Drosophila*. *Proc. Natl. Acad. Sci.* **91:** 10717–10721.

Histones, Histone Modifications, and the Inheritance of Chromatin Structure

Stefan U. Kass and Alan P. Wolffe
Laboratory of Molecular Embryology
National Institute of Child Health and Human Development
National Institutes of Health
Bethesda, Maryland 20892-2710

The replication of the genome during S phase represents a crucial period for the establishment and maintenance of programs of differential gene activity. Duplication of the chromosome requires the reassembly of the precise nucleoprotein complexes directing gene transcription or repression. The maintenance of chromosomal structures through replication reflects the commitment of a defined cell type or line to a particular state of determination. This influence of preexisting protein/DNA interactions in a parental chromosome on the structure and function of daughter chromosomes therefore reflects an epigenetic imprint. In this chapter, we consider the events occurring at the eukaryotic replication fork, their consequences for preexisting chromosomal structures, and how an epigenetic imprint might be maintained.

ACTIVE AND REPRESSED STATES OF EUKARYOTIC GENES

The primary requirement for transcription is the association of the basal transcriptional machinery (including the TATA-binding protein, TBP) with the promoter. This machinery directs the recruitment of RNA polymerase (I, II, or III) dependent on the exact TBP-associated factors (TAFs). A wide variety of sequence-specific transcription factors influence the recruitment of TBP, TAFs, the general transcription factors (GTFs; e.g., TFIIB), or RNA polymerase to the promoter and their subsequent activity once they have been recruited. The recognition sites for sequence-specific transcription factors can be found many thousands of base pairs of DNA away from the promoter. They generally occur in clusters of binding sites spread over several hundred base pairs, known as enhancers and locus control regions (LCRs). The compaction of DNA mediated by various aspects of chromatin structure may facilitate communication between these dispersed regulatory elements and potentiate the transcription process.

Epigenetic Mechanisms of Gene Regulation
 0-87969-490-4/96 $5 + .00

The histone proteins have the major structural role in compacting chromosomal DNA through the assembly of nucleosomes and the chromatin fiber. Each nucleosome contains more than 160 bp of DNA wrapped in two superhelical turns around an octamer containing two molecules of each of the four core histones (H2A, H2B, H3, H4). A fifth histone, H1, binds to the outside of this structure and interacts with the linker DNA between nucleosomes. The association of histone H1 stabilizes the folding of the nucleosomal array into the chromatin fiber (Fig. 1). This structure is a loosely packed solenoid with approximately six to seven nucleosomes per turn. Each turn includes at least 1000 bp of DNA. Beyond the nucleosome, the precise architecture of chromosomes is much less well defined. The chromatin fiber is organized into looped domains of 20–120 kb. Chromatin domains are most probably defined by the intrinsic topological constraints of coiling the chromatin fiber and by attachments to the nuclear matrix at their edges or boundaries. Individual chromosomal domains are potentially insulated from each other in terms of the action of regulatory elements (Chung et al. 1993) and are further defined structurally through the targeted modification of their constituent structural proteins (Hebbes et al. 1994; see below). Insulator elements

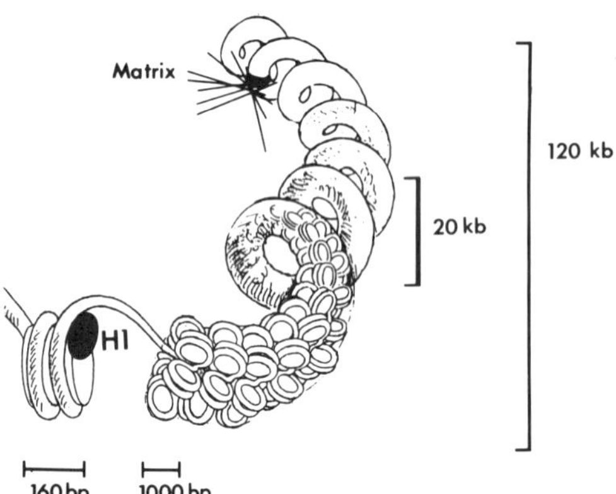

Figure 1 Chromatin structure. Ascending hierarchies of DNA compaction within chromatin are shown. Each histone octamer organizes 160 bp of DNA. Histone H1 binds to the linker DNA between nucleosomes. Each turn of the chromatin fiber contains at least 1000 bp, and chromatin domains or loops contain from 20 kb to 120 kb of DNA. Chromatin domains are probably defined by the intrinsic topological constraints of coiling the chromatin fiber and by attachments to the nuclear matrix. Insulator elements represent the boundaries of chromosomal domains; they punctuate chromatin structure most probably by interactions with the nuclear matrix.

represent important DNA regulatory elements that define the boundaries of chromosomal domains. They prevent the state of transcriptional activity in one domain from influencing that in an adjacent domain. They may represent attachment sites to the nuclear matrix (Chung et al. 1993; Hebbes et al. 1994).

Chromosomal DNA packaged into nucleosomes, the chromatin fiber, and chromatin domains provides the natural template for the transcription process. Much of the regulation of gene activity is dependent on the reversible unpackaging of genes or the action of proteins that stabilize repressive chromatin structures. The assembly of the large nucleoprotein complex regulating transcription may be facilitated by the histones (Schild et al. 1993) or by other proteins that bend or distort DNA, such as HMGs 1, 2, and I/Y (Giese et al. 1992; Tjian and Maniatis 1994; Wolffe 1994a). Probably more than a hundred different proteins and several thousand base pairs of DNA contribute to the regulation of the final transcription activation process. Multiple features may also contribute to maintaining a gene in a transcriptionally inactive or repressed state. A common mechanism appears to be a deficiency in an essential component required for the assembly of the active complex. If this component is a DNA-binding protein, the cognate DNA sequence may instead become associated with the histone proteins. If a nucleosome includes the binding site for the basal transcriptional machinery, the gene is repressed. This is because the basal transcriptional machinery itself cannot disrupt a nucleosome (Workman and Roeder 1987; Clark and Wolffe 1991). Nevertheless, certain sequence-specific transcription factors can associate with nucleosomal templates (Perlmann and Wrange 1988). These proteins initiate a chain of events that leads to nucleosomal disruption over the promoter elements recognized by the basal transcriptional machinery (Workman et al. 1991; Wolffe 1994b). Under physiological conditions, the association of transcription factors with repressed chromatin templates appears to depend on prior nucleosome positioning, which maintains the necessary recognition sites in an accessible configuration (Richard-Foy and Hager 1987; Straka and Horz 1991). Transcription factors and intrinsic DNA structure contribute to the exact placement of histone/DNA contacts (Wolffe and Drew 1989; Simpson 1991). This is a "local position effect" in which gene expression depends on the exact association of the histones with particular DNA sequences and on the three-dimensional folding of DNA directed by the histones.

The role of chromosomal domain organization in the regulation of transcription has been increasingly recognized through experiments in *Drosophila* that make use of the phenomenon of position-effect variegation (Schaffer et al. 1993). These experiments employ a powerful combination of genetic analysis and cytological observation of the large polytene chromosomes. The introduction of a normally active gene into a chromosome at a position adjacent to a transcriptionally inactive, condensed heterochromatin domain will lead to significant repression of the transcription process. This is believed to occur through a spreading of heterochromatin structure into the normally active gene. Thus, the expression of a gene depends on its chromosomal position, hence the term "position effect." In *Drosophila* and yeast, many of the genes influencing position effect have been characterized and have been found either to encode structural components of chromatin itself or to be capable of modifying the organization of chromatin through enzymatic mechanisms. For example, position-effect variegation depends on the presence of normal levels of the histone proteins and posttranslational modifications of the amino-terminal tails of the core histones such as acetylation (Mottus et al. 1980). Acetylation of amino-terminal tails alters the interaction of the histones with DNA in the nucleosome and the chromatin fiber (Bauer et al. 1994; Garcia-Ramirez et al. 1995). Position-effect variegation also depends on proteins that appear to recognize nucleosomal arrays, but not naked DNA, such as chromodomain (*chr*omatin *m*odification *o*rganizer) proteins: HP1 and Polycomb (see Schaffer et al. 1993). HP1 is associated only with inactive chromosomal regions, but Polycomb interacts with at least 50 different sites within *Drosophila* chromosomes, including developmentally important loci encoding homeodomain proteins (Orlando and Paro 1993). However, mutations in a second trithorax gene family, some members of which interact with the histone proteins, allow genes associated with Polycomb to remain active (Tamkun et al. 1992). Thus, Polycomb is likely to exert its effects on gene expression via chromatin structures dependent on the histone proteins (see Paro and Harte; Pirotta; both this volume).

Position-effect variegation is recognized as a universal phenomenon in eukaryotic chromosomes. Genes integrated into yeast chromosomes near the silent mating loci or close to the telomeres are repressed in a way that reflects their proximity to these sites (Gottschling et al. 1990). In mammalian chromosomes, highly variable levels of transcriptional activity follow from the random integration of reporter genes into the genome (Grosveld et al. 1987; Steif et al. 1989). For certain genes, such as the tissue-specific globin locus in vertebrates, these effects can be relieved by the introduction of LCRs. These are clusters of sequence-specific transcription-factor-binding sites that exert a dominant transcriptional activation function over an entire domain. The mechanism of this activation function remains to be determined; however, communication of the LCR with enhancers and promoters either directly and/or through modification of chromatin structural components are favored hypotheses (Fig. 2) (Felsenfeld 1992; Wolffe and Dimitrov 1993).

The definition of chromosomal domains as sites of targeted histone modification has followed from the use of antibodies against acetylated core histones. In *Saccharomyces cerevisiae*, most of the genome is transcriptionally active and contains hyperacetylated core histones (Clark et al. 1993). However, the transcriptionally inactive domains including the mating-type cassettes and telomeres contain histone H4 that is hypoacetylated (Braunstein et al. 1993). Similarly, hyperacetylated histone H4 also is particularly prevalent over the active chromatin domain containing the chicken β-globin gene family (Hebbes et al. 1994). These results are indicative of a very specific targeting of both histone

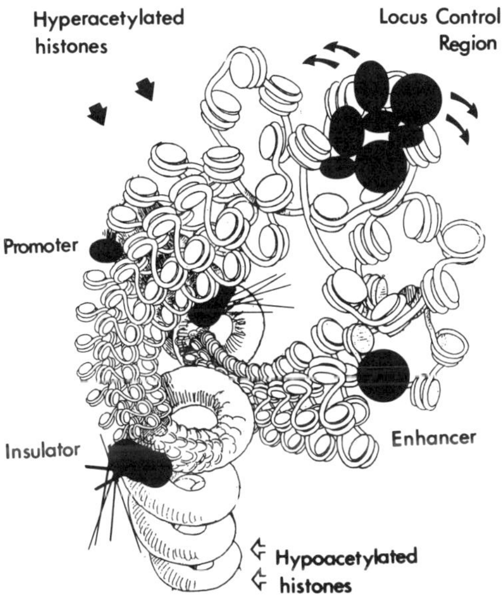

Figure 2 A model for a functional chromatin domain. The LCR disrupts local chromatin structure, as do nucleoprotein complexes associated with promoters and enhancers. The LCR has the capacity to facilitate directly or indirectly chromatin structural changes such as histone acetylation. In certain instances, the entire chromatin domain accumulates hyperacetylated histones, as indicated by a looser structure. Outside of the domain, histones are hypoacetylated, and the chromatin fiber is more condensed. The LCR can also make physical contact with promoter and enhancer complexes. The boundaries of LCR function are defined by insulator elements that may make contact with the nuclear matrix.

deacetylase and histone acetyltransferase activities to particular chromatin domains. Immunolabeling of polytene chromosomes in *Drosophila* also reveals a nonrandom distribution of histone H4 acetylation correlating with transcriptional activity (Turner et al. 1992; Bone et al. 1994). Within female mammals, the transcriptionally inactive X chromosome is distinguished by a lack of histone H4 acetylation (Jeppesen and Turner 1993). Therefore, several independent experimental approaches have shown that actively transcribed and potentially active chromatin domains are selectively enriched in hyperacetylated histones, whereas transcriptionally inactive chromatin contains hypoacetylated histones (Fig. 2).

An aspect of chromosomal organization that is very pertinent to epigenetic effects in mammalian chromosomes is DNA methylation. DNA that is enriched in methylated CpG sequences is typically transcriptionally inactive and is associated with chromatin structures containing hypoacetylated core histones and histone H1. In contrast, the promoters of many eukaryotic genes contain islands of CpG sequences that are unmethylated and that are assembled into chromatin structures enriched in hyperacetylated histones and deficient in histone H1 (Tazi and Bird 1990). The selective repression of promoters containing methylated CpG has been attributed either to the association of methylated DNA-binding proteins such as MeCP1 with the promoter (Boyes and Bird 1991, 1992), or to the generation of a repressive chromatin structure that may or may not contain such proteins (Kass et al. 1993). Histone H1 may also interact preferentially with chromatin containing methylated DNA (Levine et al. 1993; but see Nightingale and Wolffe 1995).

FATE OF NUCLEOPROTEIN COMPLEXES CONFERRING ACTIVE AND REPRESSED STATES DURING REPLICATION

It has long been thought that once established, the nucleoprotein complexes determining gene activity or repression may be capable of reproducing these structures during replication (Tsanev and Sendov 1971; Brown 1984; Weintraub 1985). The capacity of these complexes to duplicate clearly depends on the fate of a particular complex during the replication process.

The basal transcriptional machinery appears to be removed from promoter elements by the passage of a replication fork (Wolffe and Brown 1986). Replication supercedes the transcription process, and a direct consequence of replication fork progression through an active 5S rRNA gene is the displacement of transcription factors. Several correlations from in vivo work support the generality of this observation. There is a clear antagonism between transcription and replication on efficiently replicating SV40 DNA molecules (Lebkowski et al. 1985; Lewis and Manley 1985). Replication forks invade the transcriptionally active ribosomal RNA genes in yeast (Saffer and Miller 1986; Lucchini and Sogo 1994). Nucleosomes are also disrupted by replication (Sogo et al. 1986; Gruss et al. 1993). Furthermore, histones are at least transiently displaced from DNA, and nucleosomes appear to reassemble de novo (see below; Gruss et al. 1990; Jackson 1990). Thus, replication transiently erases both functional transcription complexes and repressive nucleoprotein histone/DNA interactions. The component protein molecules that assemble activating or repressive structures may, however, rapidly reassociate with DNA.

Despite the local disruption of nucleosomes during replication fork transit, the histones originally associated with a region of replicating DNA (1–2 kb) tend to remain associated with that same region of DNA (Randall and Kelly 1992). This means that histones possessing a particular marked state of modification remain in the vicinity of a particular gene and certainly within a preexisting chromosomal domain (Perry et al. 1993). Moreover, nucleosomes within arrays segregate randomly to either of the two daughter DNA molecules, but do so in small groups (3–4 nucleosomes) (Sogo et al. 1986). Thus, a regional state of histone modification, such as acetylation extending over more than 1000 bp, is likely to be discontinuously maintained after replication. This could provide a platform for the retention of chromatin-associated methylated DNA-binding proteins, chromodomain proteins such as Polycomb, histone acetyltransferases, or deacetylases. Nevertheless, preexisting chromatin proteins can only account for the assembly of 50% of the daughter DNA into chromatin. Thus, preexisting chromatin can only serve to "seed" the reassembly of similar structures within a domain. If enzymes that modify chromatin, such as histone acetyltransferases, are retained, their ensuing activity will lead to modification of the entire domain. Alternatively, cooperative interaction between proteins such as Polycomb might occur to reestablish heterochromatin structure. These latter events would provide an attractive mechanism to maintain an epigenetic imprint (see below).

NUCLEOSOME AND TRANSCRIPTION COMPLEX ASSEMBLY ON REPLICATING DNA

In vivo, the process of chromatin assembly is coupled to the replication of DNA. Chromatin assembly on nascent DNA is studied by radiolabeling nascent DNA using [^{3}H]thymidine and then developing methods to fractionate this radiolabeled DNA together with associated proteins. This type of fractionation often employs sucrose gradients or chemicals such as formaldehyde that link DNA to protein covalently. Newly synthesized DNA is immediately packaged by histone proteins through the action of molecular chaperones associated with the replication fork (Smith and Stillman 1991). This replication-coupled chromatin assembly facilitates the establishment of a transcriptionally repressed state (Almouzni and Wolffe 1993). Nevertheless, nascent chromatin is assembled into a chromatin structure that is more sensitive to nucleases and potentially more accessible to other DNA-binding proteins than mature chromatin. The maturation of this nascent chromatin from a nuclease-sensitive conformation takes approximately 10–20 minutes in a mammalian cell (Kempnauer et al. 1980; Cusick et al. 1983). Fractionation of nascent chromatin at various times following replication demonstrated that histones H3 and H4 are sequestered onto DNA before histones H2A and H2B. Histone H1 is the last histone to be stably incorporated into chromatin (Worcel et al. 1978; Cremisi and Yaniv 1980). This observation is related to the structure of the nucleosome, since histones H3 and H4 form the core of the structure, whereas histones H2A and H2B bind at the periphery of the nucleosome, and histone H1 can only associate in its proper place after two superhelical turns of DNA are wrapped around the core histones (Hayes et al. 1991, 1994). Other proteins, such as the chromodomain proteins or those that recognize methylated DNA within chromatin, may only associate once the core histones and histone H1 have bound. These last proteins to bind may establish a more stable state of repression (Wolffe 1989; Chipev and Wolffe 1992; Bouvet et al. 1994; Ura et al. 1995).

Nascent chromatin structures may be more accessible to certain transcription factors than mature chromatin (Wolffe 1991). The histone H4 incorporated into nascent chromatin is posttranslationally modified by acetylation before nucleosome assembly (Ruiz-Carillo et al. 1975). Acetylation of the core histones and deficiency in histones H2A/H2B and histone H1 facilitate transcription factor access to DNA (Lee et al. 1993; Ura et al. 1995). Proteins exist that organize nucleosomes into particular positions (Chasman et al. 1990; Roth et al. 1992). The histone tetramer $(H3/H4)_2$ is also capable of recognizing DNA sequences that make major contributions to the eventual positioning of the entire nucleosome. All of these features facilitate the reassociation of sequence-specific DNA-binding proteins with daughter DNA molecules. The competition between these potentially activating and repressive influences determines the eventual association and utilization of the basal transcriptional machinery at the regulatory elements of a gene.

HYPOTHESES TO EXPLAIN THE INHERITANCE OF CHROMATIN STATES

As discussed above, the available evidence suggests that the transient disruption of preexisting active and repressive chromatin structures occurs during the replication process. Nevertheless, experimental work demonstrates that proteins displaced by the elongating replication fork also have a window of opportunity to reassociate with the daughter DNA molecule.

Several scenarios can be considered. (1) If a superabundance of transcription factors were available throughout the cell cycle including S phase, then constitutive gene activity might result, uninterrupted by DNA replication. These transcription factors could continuously recruit the basal transcription machinery to the gene and prevent nucleosome assembly. (2) If transcription factor complexes were limiting, but bound to regulatory sequences in multimeric complexes, these complexes could be divided between daughter DNA strands, thereby nucleating subsequent complex reassembly in the face of ongoing chromatin assembly (Fig. 3, step 1) (Brown 1984). In either case, failure to maintain the transcription factor complex would only result in heritable loss of transcription if the complex could not displace histones from the DNA de novo.

The maintenance of specific transcription factor/DNA interactions may be facilitated by considering the promoter, the enhancer, and LCRs not as separate entities, but as contributory components to a single structure (Fig. 4) (Wolffe 1990). This could be achieved through protein-protein interactions between the distinct nucleoprotein complexes assembled at each regulatory element. One reason for the separation of these regulatory elements over extensive distances may be that any one structure may be independently disrupted by DNA replication, while the other remains intact. If protein binding to one sequence element influences the binding of proteins to the other, then the intact nucleoprotein complex may facilitate the re-formation of the disrupted one.

Stably propagated covalent modifications of DNA, such as cytosine methylation, may influence the association of transcription factors directly (Ben-Hattar et al. 1989), maintain the association of repressive proteins such as MeCP1 and MeCP2 (Fig. 3, step 2) (Boyes and Bird 1991, 1992), or influence nucleosome positioning (Englander et al. 1993). If a positioned nucleosome includes a key DNA regulatory element so that a particular transcription factor cannot bind to it, the gene

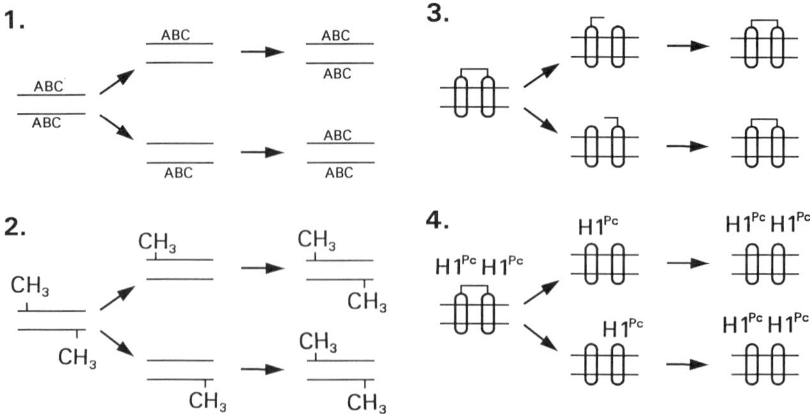

Figure 3 Hypothetical mechanisms for imprinting. (1) Maintenance of specific *trans*-acting factor (ABC) interactions through replication (*first arrows*) followed by cooperative sequestration of additional transcription factors (*second arrows*). (2) Maintenance of methylation. Methylated cytosine in a CpG sequence and methylated cytosine binding proteins (MeCP) are segregated to daughter DNA duplexes following replication (*first arrows*). Methyltransferase preferentially methylates CpG at a hemimethylated site (*second arrows*). MeCP rebinds to methylated DNA. (3) Maintenance of nucleosomal structure. Following replication (*first arrows*), positioned nucleosomes segregate to daughter DNA duplexes, but direct the phasing of adjacent nucleosomes (*second arrows*). (4) Maintenance of long-range chromatin structure. Following replication (*first arrows*), repressive proteins such as histone H1 or Polycomb (Pc) are reduced in abundance. Alternatively, activating features of chromatin such as acetylated histones are dispersed to both daughter chromatids. Direct or indirect cooperativity between these proteins, or structure-dependent features, may direct the sequestration of additional H1 or Pc or acetylated histones (*second arrows*).

will be repressed. Histones themselves may bind to nucleosome positioning sequences or interact with transcription factors such that nucleosome positioning occurs (Fig. 3, step 3) (Simpson 1991).

We have seen how histone molecules in small nucleosome arrays can be distributed to daughter chromosomes locally during replication, thereby preserving their association with particular DNA sequences. If the proteins responsible for higher-order structure (acetylases, chromodomain proteins, and so on) were partitioned in the same way, larger domains might be conserved and propagated also (Fig. 3, step 4).

Chromatin organization outside of the origin recognition complex may also have significance for the initiation of replication and the timing of this initiation in S phase. If replication disrupts both active and repressed chromatin structures, then the entire nucleus has to be remodeled after each replication event. We have suggested a means of accomplishing this remodeling; however, the re-formation of nuclear structures

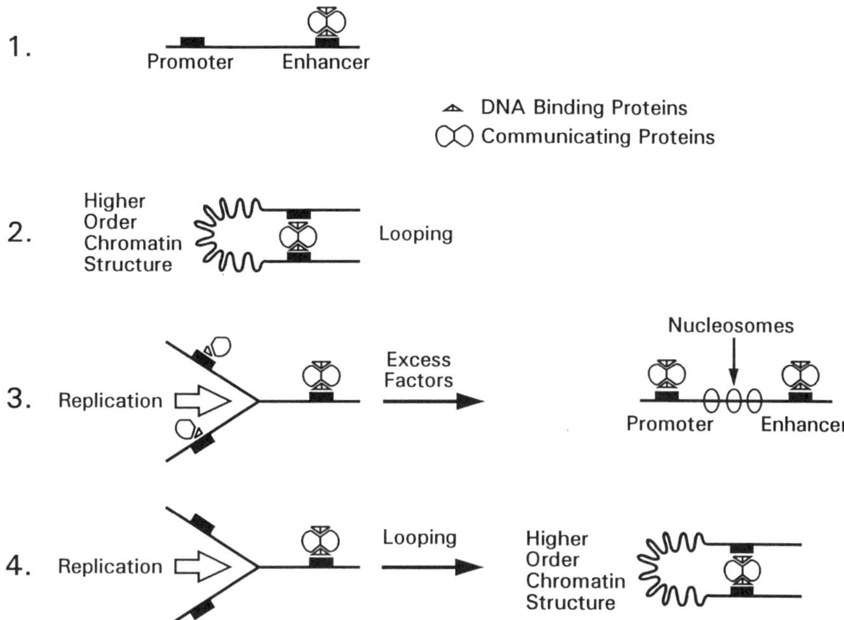

Figure 4 Models for the maintenance of transcription complexes through replication. The regulatory regions of a gene are shown as a promoter and an enhancer (*bars*). Open boxes are DNA-binding or non-DNA-binding components of a transcription complex. (1) In this case, similar factors are shared between enhancers and promoters. (2) The sequestration of transcription factors onto the promoter (and its activation) is facilitated by looping of the intervening DNA between enhancer and promoter. Common DNA-binding proteins associate with both the enhancer and promoter. The bound proteins interact with a common non-DNA-binding protein. (3) DNA replication disrupts the transcription complex on the promoter, splitting it in half. The remaining transcription factors can sequester free factors from solution, generating a complete transcription complex on each daughter chromatid. In this case, the enhancer is required for establishment, but not maintenance, of the transcription complex. (4) Alternatively, DNA replication again disrupts the complex on the promoter, displacing transcription factors; however, because of the distance between enhancer and promoter, the enhancer complex remains intact. DNA looping can establish a new transcription complex on the promoter. Here, the enhancer is required for both establishment and maintenance of the transcription complex through cell division.

has other implications. If there are limiting transcription factors available in a cell, then a gene that is replicated early in S phase has more opportunity for the assembly of an active transcription complex than a gene that replicates late. This is simply because the gene that replicates early is available for transcription factors to bind before all of the early-replicating portion of the genome has sequestered these factors. A late-replicating gene therefore experiences a relative deficiency in factor availability (Gottesfeld and Bloomer 1982; Wormington et al. 1982). Conversely, it is also possible that the type of chromatin assembled early in S phase is more accessible to transcription factors than chromatin assembled late in S phase. For example, early-replicating chromatin may sequester histones that are more highly acetylated and, consequently, favor continued transcription activity. Transcriptionally active genes replicate early in S phase (Goldman et al. 1984; Gilbert 1986; Guinta and Korn 1986; Wolffe 1993). The reason for this early replication is unknown, but possibilities include the local disruption of chromatin structure by transcription complexes, such that the DNA within those chromatin domains becomes more accessible to the replication machinery (Wolffe and Brown 1988). Many transcription factors are also replication factors (DePamphilis 1988); consequently, local concentrations of transcription factors may favor the assembly of replication initiation complexes.

TESTING THE HYPOTHESES WITH EXPERIMENT: THE EXAMPLES OF THE X CHROMOSOME, IGF-2, IGF-2R, H19, AND SNRPN GENES

Within each somatic cell nucleus of the female mammal lie two X chromosomes; one is transcriptionally active and the other is not. Active and inactive states are maintained through replication. These two different X chromosomes provide an excellent case of both epigenetic effects and the inheritance of chromatin states. Do any of the hypotheses we have discussed explain the selective transcriptional inactivity of one of the pair of identical X chromosomes, and the maintenance of this inactivation? In addition, imprinting within the paternal and maternal genomes in the mouse has been studied in detail for Igf2, Igf2r, H19, and Snrpn genes. Substantial recent process has been made in examining the mechanism of this imprinting (Surani 1993), and useful similarities and distinctions emerge from comparison with the X chromosome.

Many genes in the inactive X chromosome are heavily methylated, in contrast to the active X chromosome (Grant and Chapman 1988). However, inactivation of the X-linked genes occurs before methylation during the differentiation of embryonic female somatic cells (Lock et al. 1987). Thus, other mechanisms than methylation may exist to repress transcription within the inactive X chromosome. In contrast, if methylation is impaired in the developing mouse embryo, major changes occur in the expression pattern of imprinted genes (Li et al. 1993). Reduction in DNA methylation leads to activation of the normally silent paternal allele of H19, the repression of the normally active paternal allele of the Igf2 gene, and the repression of the normally active maternal allele of the Igf2r gene. These results dramatically demonstrate a key role for DNA methylation in imprinting.

The active X chromosome normally replicates early in S phase, whereas the inactive X chromosome replicates late (Takagi 1974). However, female lymphoma cell lines have been isolated in which the converse occurs (Yoshida et al. 1993). Thus, the inactive X chromosome does not have to replicate late in S phase in order to be transcriptionally quiescent. Likewise, although the replication timing patterns of maternal and paternal alleles of the imprinted genes Igf2, Igf2r, H19, and Snrpn differ, with the paternal allele always being early replicating, there is no simple correlation with gene activity (Kitsberg et al. 1993). Nevertheless, the differential replication timing phenomenon is useful in demonstrating that the imprinted genes are embedded in large chromatin domains that have distinct replication patterns. Formation of such domains may provide a structural imprint on gene activity (Kitsberg et al. 1993).

Promoters in the inactive X chromosome appear to be incorporated into positioned nucleosomes, whereas promoters in the active X chromosome are free of such structures and have transcription factors bound to them (Riggs and Pfeifer 1992). Thus, specific chromatin structures clearly appear to have a role in regulating differential gene activity between the two X chromosomes. Moreover, nucleosomes on the active X chromosome contain predominantly acetylated histones, whereas those on the inactive X chromosome are not acetylated (Jeppesen and Turner 1993). Thus, the establishment and maintenance of specific chromatin structures containing modified histones comprise an excellent candidate mechanism for establishing and maintaining differential expression of genes between the two X chromosomes. Differential methylation and replication timing may serve to stabilize these different states of gene activity.

CONCLUSIONS

The packaging of regulatory DNA within the eukaryotic chromosome has considerable potential, not only for modulating the transcriptional activity of genes, but also for propagating states that are permissive or restrictive for transcription. Sequence-specific transcription factors, histones and their modifications, chromodomain proteins and enzymes that modify histones, DNA methylation and proteins that recognize methylated DNA within chromatin may all play independent or interrelated roles in regulating gene activity. They also have the potential for propagating their interactions with nascent DNA following replication. This is clearly seen for the imprinted Igf2, Igf2r, and H19 genes in the mouse. In contrast, observations on the phenomenon of X-chromosome inactivation suggest that the formation and stability of specific histone/DNA interactions through replication may be central to the inheritance of chromatin states, and that other molecular mechanisms have supporting roles. The future offers the exciting prospect of reconstructing the propagation of stable active or repressed chromatin states in vitro and, consequently, understanding the events occurring at the replication fork in molecular detail.

REFERENCES

Almouzni, G. and A.P. Wolffe. 1993. Replication coupled chromatin assembly is required for the repression of basal transcription *in vivo*. *Genes Dev.* **7:** 2033–2047.

Bauer, W.R., J.J. Hayes, J.H. White, and A.P. Wolffe. 1994. Nucleosome structural changes due to acetylation. *J. Mol. Biol.* **236:** 685–690.

Ben-Hattar, J., P. Beard, and J. Jirieny. 1989. Cytosine methylation in CTF and SP1 recognition sites of an HSV tk promoter: Effects on transcription *in vivo* and on factor binding *in vitro*. *Nucleic Acids Res.* **17:** 10179–10190.

Bone, J.R., J. Lavender, R. Richman, M.J. Palmer, M.B. Turner, and M.L. Kuroda. 1994. Acetylated histone H4 on the male X chromosome is associated with dosage compensation in *Drosophila*. *Genes Dev.* **8:** 96–104.

Bouvet, P., S. Dimitrov, and A.P. Wolffe. 1994. Specific regulation of chromosomal 5S rRNA gene transcription *in vivo* by histone H1. *Genes Dev.* **8:** 1147–1159.

Boyes, J. and A. Bird. 1991. DNA methylation inhibits transcription indirectly via a methyl-CpG binding protein. *Cell* **64:** 1123–1134.

———. 1992. Repression of genes by DNA methylation depends on CpG density and promoter strength: Evidence for involvement of a methyl-CpG binding protein. *EMBO J.* **11:** 327–333.

Braunstein, M., A.B. Rose, S.G. Holmes, C.D. Allis, and J.R. Broach. 1993. Transcriptional silencing in yeast is associated with reduced histone acetylation. *Genes Dev.* **7:** 592–604.

Brown, D.D. 1984. The role of stable complexes that repress and activate eukaryotic genes. *Cell* **37:** 359–365.

Chasman, D.I., N.F. Lue, A.R. Buchman, J.W. LaPointe, Y. Lorch, and R.D. Kornberg. 1990. A yeast protein that influences the chromatin structure of UAS_G and functions as a powerful auxiliary gene activator. *Genes Dev.* **4:** 503–514.

Chipev, C.C. and A.P. Wolffe. 1992. Chromosomal organization of *Xenopus laevis* oocyte and somatic 5S rRNA genes *in vivo. Mol. Cell. Biol.* **12:** 45–55.

Chung, J.H., M. Whiteley, and G. Felsenfeld. 1993. A 5′ element of the chicken β-globin domain serves as an insulator in human erythroid cells and protects against position effect in *Drosophila. Cell* **74:** 505–514.

Clark, D.J. and A.P. Wolffe. 1991. Superhelical stress and nucleosome mediated repression of 5S RNA gene transcription *in vitro. EMBO J.* **10:** 3419–3428.

Clark, L.J., L.P. O'Neill, and B.M. Turner. 1993. Selective use of H4 acetylation sites in the yeast *Saccharomyces cerevisiae. Biochem. J.* **294:** 557–561.

Cremisi, C. and M. Yaniv. 1980. Sequential assembly of newly synthesized histones on replicating SV40 DNA. *Biochem. Biophys. Res. Commun.* **92:** 1117–1123.

Cusick, M.E., K.S. Lee, M.L. DePamphilis, and P.M. Wasserman. 1983. Structure of chromatin at deoxyribonucleic acid replication forks: Nuclease hypersensitivity results from both prenucleosomal deoxyribonucleic acid and an immature chromatin structure. *Biochemistry* **22:** 3873–3884.

De Pamphilis, M.L. 1988. Transcriptional elements as components of eukaryotic origins of DNA replication. *Cell* **52:** 635–638.

Englander, E.W., A.P. Wolffe, and B.H. Howard. 1993. Nucleosome interactions with a human Alu element transcriptional repression and effects of template methylation. *J. Biol. Chem.* **268:** 19565–19573.

Felsenfeld, G. 1992. Chromatin: An essential part of the transcriptional apparatus. *Nature* **355:** 219–224.

Garcia-Ramirez, M., C. Rocchini, and J. Ausio. 1995. Modulation of chromatin folding by histone acetylation. *J. Biol. Chem.* **270:** 17923–17928.

Giese, K., J. Cox, and R. Grosschedl. 1992. The HMG domain of lymphoid enhancer factor 1 bends DNA and facilitates assembly of functional nucleoprotein structures. *Cell* **69:** 185–195.

Gilbert, D.M. 1986. Temporal order of replication of *Xenopus laevis* 5S ribosomal RNA genes in somatic cells. *Proc. Natl. Acad. Sci.* **83:** 2924–2928.

Goldman, M.A., G.P. Holmquist, M.C. Gray, L.A. Caston, and A. Nag. 1984. Replication timing of genes and middle repetitive sequences. *Science* **224:** 686–692.

Gottesfeld, J. and L.S. Bloomer. 1982. Assembly of transcriptionally active 5S RNA gene chromatin *in vitro. Cell* **28:** 781–791.

Gottschling, D.E., O.M. Aparicio, B.L. Billington, and V.A. Zakian. 1990. Position effect at *S. cerevisiae* telomeres: Reversible repression of pol II transcription. *Cell* **63:** 751–762.

Grant, S.G. and V.M. Chapman. 1988. Mechanisms of X-chromosome regulation. *Annu. Rev. Genet.* **22:** 199–233.

Grosveld, F., G. Blom van Assendelft, D. Greaves, and G. Kollias. 1987. Position-independent, high level expression of the human β-globin gene in transgenic mice. *Cell* **51:** 975–985.

Gruss, C., J. Wu, T. Koller, and J.M. Sogo. 1993. Disruption of nucleosomes at replication forks. *EMBO J.* **12:** 4533–4545.

Gruss, C., C. Gutierrez, W.C. Burhans, M.L. De Pamphilis, T. Koller, and J.M. Sogo. 1990. Nucleosome assembly in mammalian cell extracts before and after DNA replication. *EMBO J.* **9:** 2911–2922.

Guinta, D.R. and L.J. Korn. 1986. Differential order of replication of *Xenopus laevis* 5S RNA genes. *Mol. Cell. Biol.* **6:** 2536–2542.

Hayes, J.J., D.J. Clark, and A.P. Wolffe. 1991. Histone contributions to the structure of

DNA in the nucleosome. *Proc. Natl. Acad. Sci.* **88:** 6829–6833.

Hayes, J.J., D. Pruss, and A.P. Wolffe. 1994. Histone domains required to assemble a chromatosome including the *Xenopus borealis* somatic 5S rRNA gene. *Proc. Natl. Acad. Sci.* **91:** 7817–7821.

Hebbes, T.R., A.L. Clayton, A.W. Thorne, and C. Crane-Robinson. 1994. Core histone hyperacetylation co-maps with generalized DNase I sensitivity in the chicken β-globin chromosomal domain. *EMBO J.* **13:** 1823–1830.

Jackson, V. 1990. *In vivo* studies on the dynamics of histone-DNA interaction: Evidence for nucleosome dissolution during replication and transcription and a low level of dissolution independent of both. *Biochemistry* **29:** 719–731.

Jeppesen, P. and B.M. Turner. 1993. The inactive X chromosome in female mammals is distinguished by a lack of histone H4 acetylation, a cytogenetic marker for gene expression. *Cell* **74:** 281–291.

Kass, K.U., J.P. Goddard, and R.L.P. Adams. 1993. Inactive chromatin spreads from a focus of methylation. *Mol. Cell. Biol.* **13:** 7372–7379.

Kempnauer, K.H., E. Fanning, B. Oho, and R. Knippers. 1980. Maturation of newly replicated chromatin of simian virus 40 and its host cell. *J. Mol. Biol.* **136:** 359–374.

Kitsberg, D., S. Selig, M. Brandeis, I. Simon, I. Keshet, D.J. Driscoll, R.D. Nicholls, and H. Cedar. 1993. Allele-specific replication timing of imprinted gene regions. *Nature* **364:** 459–463.

Lebkowski, J.S., S. Clancy, and M.P. Calos. 1985. Simian virus 40 replication in adenovirus-transformed human cells antagonizes gene expression. *Nature* **317:** 169–171.

Lee, D.Y., J.J. Hayes, D. Pruss, and A.P. Wolffe. 1993. A positive role for histone acetylation in transcription factor binding to nucleosomal DNA. *Cell* **72:** 73–84.

Levine, A., A. Yeivin, E. Ben-Asher, Y. Aloni, and A. Razin. 1993. Histone H1-mediated inhibition of transcription initiation of methylated templates *in vitro*. *J. Biol. Chem.* **268:** 21754–21759.

Lewis, E.D. and J.L. Manley. 1985. Repression of simian virus 40 early transcription by viral DNA replication in human 293 cells. *Nature* **317:** 172–175.

Li, E., C. Beard, and R. Jaenisch. 1993. Role of DNA methylation in genomic imprinting. *Nature* **366:** 362–365.

Lock, L.F., N. Takagi, and G.R. Martin. 1987. Methylation of the hprt gene on the inactive X occurs after chromosome inactivation. *Cell* **48:** 36–46.

Lucchini, R. and J.M. Sogo. 1994. Chromatin structure and transcriptional activity around the replication forks arrested at the 3′end of the yeast rRNA genes. *Mol. Cell. Biol.* **14:** 318–326.

Mottus, R., R. Reeves, and T.A. Grigliatti. 1980. Butyrate suppression of position effect variegation in *Drosophila melanogaster*. *Mol. Gen. Genet.* **178:** 465–469.

Nightingale, K. and A.P. Wolffe. 1995. Methylation at CpG sequences does not influence histone H1 binding to a nucleosome including a *Xenopus borealis* 5S RNA gene. *J. Biol. Chem.* **270:** 4197–4200.

Orlando, V. and R. Paro. 1993. Mapping Polycomb-repressed domains in the bithorax complex using in vivo formaldehyde cross-linked chromatin. *Cell* **75:** 1187–1198.

Perlmann, T. and O. Wrange. 1988. Specific glucocorticoid receptor binding to DNA reconstituted in a nucleosome. *EMBO J.* **7:** 3073–3083.

Perry, C.A., C.D. Allis, and A.T. Annunziato. 1993. Parental nucleosomes segregated to newly replicated chromatin are underacetylated relative to those assembled *de novo*.

Biochemistry **32:** 13615–13623.
Randall, S.K. and T.J. Kelly. 1992. The fate of parental nucleosomes during SV40 DNA replication. *J. Biol. Chem.* **267:** 14259–14265.
Richard-Foy, H. and G.L. Hager. 1987. Sequence-specific positioning of nucleosomes over the steroid-inducible MMTV promoter. *EMBO J.* **6:** 2321–2328.
Riggs, A.P. and G.P. Pfeifer. 1992. X-chromosome inactivation and cell memory. *Trends Genet.* **8:** 169–174.
Roth, S.Y., M. Shimizu, L. Johnson, M. Grunstein, and R.T. Simpson. 1992. Stable nucleosome positioning and complete repression by the yeast α2 repressor are disrupted by amino-terminal mutation in histone H4. *Genes Dev.* **6:** 411–425.
Ruiz-Carrillo, A., L.J. Wangh, and V.G. Allfrey. 1975. Processing of newly synthesized histone molecules. *Science* **190:** 117–128.
Saffer, L.D. and O.L. Miller, Jr. 1986. Electron microscopic study of *Saccharomyces cerevisiae* rDNA chromatin replication. *Mol. Cell. Biol.* **6:** 1147–1157.
Schaffer, C.D., L.L. Wallrath, and S.C. Elgin. 1993. Regulating genes by packaging domains: Bits of heterochromatin in euchromatin? *Trends Genet.* **9:** 35–37.
Schild, C., F.-X. Claret, W. Wahli, and A.P. Wolffe. 1993. A nucleosome-dependent static loop potentiates estrogen-regulated transcription from the *Xenopus* vitellogenin B1 promoter *in vitro. EMBO J.* **12:** 423–433.
Simpson, R.T. 1991. Nucleosome positioning: Occurrence, mechanisms and functional consequences. *Prog. Nucleic Acid Res. Mol. Biol.* **40:** 143–184.
Smith, S. and B. Stillman. 1991. Stepwise assembly of chromatin during DNA replication *in vitro. EMBO J.* **10:** 971–980.
Sogo, J.M., H. Stahl, T. Koller, and R. Knippers. 1986. Structure of the replicating SV40 minichromosomes: The replication fork, core histone segregation and terminal structures. *J. Mol. Biol.* **189:** 189–204.
Steif, A., D.M. Winter, W.H. Stratling, and A.E. Sippel. 1989. A nuclear DNA attachment element mediates elevated and position-independent gene activity. *Nature* **341:** 343–345.
Straka, C. and W. Horz. 1991. A functional role for nucleosomes in the repression of a yeast promoter. *EMBO J.* **10:** 361–368.
Surani, M.A. 1993. Silence of the genes. *Nature* **366:** 302–303.
Takagi, N. 1974. Differentiation of X chromosomes in early female mouse embryos. *Exp. Cell. Res.* **86:** 127–135.
Tamkun, J.W., R. Deuring, M.P. Scott, M. Kissinger, A.M. Patlatucci, T.C. Kaufman, and J.A. Kennison. 1992. Brahma: A regulator of *Drosophila* homeotic genes structurally related to the yeast transcriptional activator SNF2/SWI2. *Cell* **68:** 561–572.
Tazi, J. and A. Bird. 1990. Alternative chromatin structure at CpG islands. *Cell* **60:** 902–920.
Tjian, R. and T. Maniatis. 1994. Transcriptional activation: A complex puzzle with few easy pieces. *Cell* **77:** 5–8.
Tsanev, R. and B. Sendov. 1971. Possible molecular mechanism for cell differentiation in multicellular organisms. *J. Theor. Biol.* **30:** 337–393.
Turner, B.M., A.J. Birley, and J. Lavender. 1992. Histone H4 isoforms acetylated at specific lysine residues define individual chromosomes and chromatin domains in *Drosophila* polytene nuclei. *Cell* **69:** 375–384.
Ura, K., J.J. Hayes, and A.P. Wolffe. 1995. A positive role for nucleosome mobility in the transcriptional activity of chromatin templates: Restriction by linker histones.

EMBO **J. 14:** 3752–3765.

Weintraub, H. 1985. Assembly and propagation of repressed and derepressed chromosomal states. *Cell* **42:** 705–711.

Wolffe, A.P. 1989. Dominant and specific repression of *Xenopus* oocyte 5S RNA genes and satellite I DNA by histone H1. *EMBO J.* **8:** 527–537.

———. 1990. Transcription complexes. *Prog. Clin. Biol. Res.* **322:** 171–186.

———. 1991. Implications of DNA replication for eukaryotic gene expression. *J. Cell Sci.* **99:** 201–206.

———. 1993. Replication timing and *Xenopus* 5S RNA gene transcription *in vitro*. *Dev. Biol.* **157:** 224–231.

———. 1994a. Architectural transcription factors. *Science* **264:** 1100–1101.

———. 1994b. Transcription: In tune with the histones. *Cell* **77:** 13–16.

Wolffe, A.P. and D.D. Brown. 1986. DNA replication *in vitro* erases a *Xenopus* 5S RNA gene transcription complex. *Cell* **47:** 217–227.

———. 1988. Developmental regulation of two 5S ribosomal RNA genes. *Science* **241:** 1626–1632.

Wolffe, A.P. and S. Dimitrov. 1993. Histone modulated gene activity: Developmental implications CRC. *Crit. Rev. Eukaryotic Gene Expr.* **3:** 167–191.

Wolffe, A.P. and H.R. Drew. 1989. Initiation of transcription on nucleosomal templates. *Proc. Natl. Acad. Sci.* **86:** 9817–9821.

Worcel, A., S. Han, and M.L. Wong. 1978. Assembly of newly replicated chromatin. *Cell* **15:** 969–977.

Workman, J.L. and R.G. Roeder. 1987. Binding of transcription factor TFIID to the major late promoter during *in vitro* nucleosome assembly potentiates subsequent initiation by RNA polymerase II. *Cell* **51:** 613–622.

Workman, J.L., I.C. Taylor, and R.E. Kingston. 1991. Activation domains of stably bound GAL4 derivatives alleviate repression of promoters by nucleosomes. *Cell* **64:** 533–544.

Wormington, W.M., M. Schlissel, and D.D. Brown. 1982. Developmental regulation of *Xenopus* 5S RNA genes. *Cold Spring Harbor Symp. Quant. Biol.* **47:** 879–884.

Yoshida, I., N. Kashio, and N. Takagi. 1993. Cell fusion-induced quick change in replication time of the inactive mouse X chromosome: An implication for the maintenance mechanism of late replication. *EMBO J.* **12:** 4397–4405.

The Silencing of Human Immunodeficiency Virus

Daniel P. Bednarik
Human Genome Sciences, Inc.
Department of Cellular Biology
Rockville, Maryland 20850

A PERSPECTIVE OF RETROVIRAL GENOME METHYLATION

Ten years ago, when I first began to study the methylation of retrovirus genomes, we had quite literally stumbled upon the observation that certain cell lines had the ability to silence viral promoters that were artificially introduced via the introduction of foreign DNA or "transfection." Most of these experiments were performed using a heterologous promoter fused to the chloramphenicol acetyl transferase (CAT) reporter gene. In the case of the HIV-1 long terminal repeat (LTR) linked to the CAT gene, we found that African green monkey cells (VERO) stably transfected with this construct lost the ability to express the CAT mRNA (Bednarik et al. 1987). Rather than discard our stable cell lines that lacked CAT expression, we attempted to stimulate expression by addition of the 5-methylcytosine analog, 5-azacytidine. The significance of this drug is its ability to cause the demethylation or "hypomethylation" of genomic DNA during replication. To our amazement, CAT activity was quickly restored, and even more interesting was the observation that 5-azacytidine worked in a very short period of time, as soon as 8 hours after addition of the drug. This observation launched the next plethora of experiments, which involved testing a spectrum of reagents and physiochemical stimuli, all targeted at LTR reactivation. In this chapter, I present a brief perspective of viral genome methylation and then focus on our most recent observations regarding the methylation of the HIV LTR and how this process might be involved in the establishment of transcriptionally silenced retrovirus species.

Retroviruses, which have been the main focus of our work, are positive-strand RNA viruses in the retroviridae family, which includes the oncoviruses (oncovirinae), lentiviruses (lentivirinae), and spumaviruses (spumavirinae), all of which are essentially movable genetic elements or retrotransposons (Varmus 1983; Varmus and Brown 1990). Retroviruses integrate randomly into the host cellular genome and become susceptible to modification by DNA methylation depending on the immediate DNA environment. The relationship of DNA methylation to retrovirus pathophysiology became apparent when Harbers et al. (1981) demonstrated that the murine leukemia virus (MLV) LTRs could become hypermethylated after integration into chromosomal DNA, with subsequent inactivation of viral transcription and expression. Molecular cloning of the integrated provirus resulted in the propagation of infectious virions when the molecular clone was transfected. The integrated provirus could be maintained in a latent, unexpressed state and was influenced by DNA methylation in a manner that bears consideration with respect to the clinical relevance of this process and human disease.

SILENCING OF VIRAL DNA; POSSIBLE RELATIONSHIP TO DISEASE STATES

The most obvious role for DNA methylation in virus-mediated disease is as a part of the mechanism for latency. Viruses that are transciptionally inactive after integration and subsequent methylation include retroviruses, hepatitis-B virus (HBV), and the adenovirus genome (Doerfler 1983, 1984; Bednarik et al. 1987, 1990, 1991; Miyoshi et al. 1992; Gutekunst et al. 1993; Doerfler et al. 1995). During the course of our studies, we had encountered clinical situations in which patients infected with HIV were seronegative, but positive for proviral DNA as determined by polymerase chain reaction (PCR) (H. Farzedagan, pers. comm.). One possible explanation for this phenomenon is that the integrated provirus became defective via mutation or recombination. The other possibility is that the provirus became cryptically silent after being hypermethylated. The continuing work of Doerfler et al. implicates the process of DNA methylation as a very old mechanism by which cells inactivated foreign DNA internalized via the gastrointestinal tract (Doerfler et al. 1995). These studies involved the administration of a target DNA molecule to cells to measure the rate of incorporation and subsequent methylation. For this purpose, the bacterial M13 λ phage DNA was administered to mice, which later could be shown to have incorporated some of the exogenous DNA into lymphocytes in a hypermethylated form. These observations are intriguing, since they imply that this mechanism can sometimes identify foreign gene sequences and silence them. Exactly how the cell can accomplish this is unknown, and whether cells contain a system for "scanning" such foreign DNA remains to be seen.

METHYLATION OF VIRAL PROMOTER ELEMENTS AND TRANSCRIPTIONAL INACTIVATION: SUSCEPTIBILITY OF RETROVIRUSES TO DNA METHYLATION

Previous studies have shown that the in vitro methylation of two CpG sites within the HIV LTR could disrupt the ability of the LTR to direct the expression of viral genes (Bednarik et al. 1990). This effect could be overcome *in trans* by the viral *trans*-activator TAT. Other studies have demonstrated that the inactivation of CpG-island-associated genes by DNA methylation is mediated by a methyl-CpG-binding protein, MeCP-1 (Boyes and Bird 1991, 1992; Lewis et al. 1992). This protein binds to DNA containing multiple, symmetrically methylated CpGs and is believed to cause transcriptional repression via the exclusion of cellular transcription factors (Meehan et al. 1989; Boyes and Bird 1991). More recently, similar studies have shown that the density of DNA methylation plays a critical role in the ability of MeCP-1 to repress transcription (Boyes and Bird 1992; Lewis et al. 1992). When DNA is only sparsely or infrequently methylated, MeCP-1 forms a weak complex with the DNA that can easily be dissociated, thus overcoming the transcriptional block. However, densely methylated DNA forms strong complexes with MeCP-1 and therefore remains transcriptionally inactive.

HIV Genome Structure

Figure 1 illustrates the overall genomic structure of HIV-1. Upon integration of the virus into the host DNA genome, transcription is driven by the 5′LTR and employs several transcription proteins not discussed in detail here. For the purposes of the following experiments, the HIV LTR was fused to the CAT gene, which serves as a reporter gene to measure the transcriptional activity of the LTR under various conditions. The activity of the LTR is greatly enhanced by the *trans*-acting protein TAT. This protein lends itself to the rapid processivity of the viral mRNA. The most noted aspect of mRNA processivity is the interaction of TAT with the *trans*-acting responsive region, or TAR (Rosen et al. 1985). This region is usually identified as a bulge and stem-loop structure formed by the homologous pairing of bases into the form of a hairpin (Fig. 1). Studies that have provided valuable insight into the function of this structure were performed by creating mutants that contained altered base composition within the TAR structure or by deleting part or all of TAR (Rosen et al. 1985; Muesing et al. 1987). For our studies, the entire HIV LTR was simply cloned upstream of the CAT gene to establish a report construct. Two important concepts in HIV expression are re-

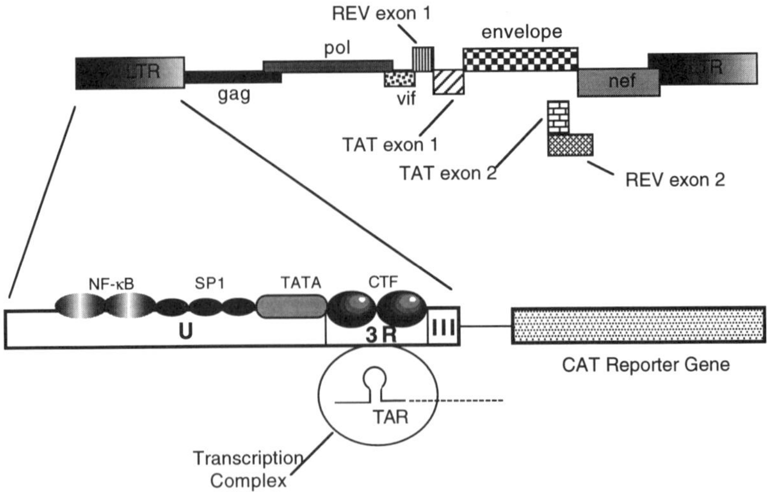

Figure 1 HIV-1 genome structure and transcriptional complex. Specific regions of the genome that encode viral proteins are illustrated. An enlargement of the LTR is shown, along with the binding sites of specific transcription proteins relative to one another. The TAR element is also shown as a stem-loop structure embedded in the transcription complex.

quired when performing transcription studies. These are the terms "*cis*" and "*trans*." "*cis*" refers to the transcription from within the LTR itself and usually is indicative of a basal level of transcription. "*trans*" refers to transcription in the presence of another factor or protein which somehow guides, or influences, the rate of mRNA synthesis.

Earlier work by Doerfler et al. (1995) strongly suggested that the insertion of viral elements into the eukaryotic genome could result in the spreading of methylation from one site outward. Studies were designed to examine transcription from the HIV-1 LTR when CpG dinucleotides were either sparsely or densely methylated and to assay the ability of TAT to *trans*-activate these sequences both in vivo and in vitro. Our findings confirmed earlier observations in which we described that the enzymatic methylation of two CpG dinucleotides localized in *Hpa*II/*Msp*I sites in the LTR inhibited transcription, but that this effect could be overcome *in trans* by TAT in vivo (Bednarik et al. 1990). This phenomenon was time-dependent, and TAT was not able to overcome the methylation-induced transcriptional block within the limits of detection sensitivity.

The rationale behind the following experiments was to test the efficacy of DNA methylation on the suppression of HIV expression both in vivo and in vitro. In either case, the HIV LTR was subjected to artificial, enzymatic methylation with either a site-specific CpG methylase, such as *Hpa*II methylase, or an enzyme that methylates all CpG sites, such as CpG methylase. Both enzymes were commercially obtained from New England BioLabs, Inc. For in vivo transfection experiments, the LTR contained in the plasmid pU3RIIICAT was enzymatically methylated and subsequently introduced into target cells. For in vitro transcription experiments, the same procedure was followed, except that an in vitro transcription reaction was employed.

Sensitivity of *cis* Elements in the HIV LTR to CpG Methylation

The HIV LTR contains CpG dinucleotides at 17 positions spanning 700 nucleotides. Two of these are localized in *Hpa*II/*Msp*I sites upstream of the transcriptional start site, whereas the others are scattered between positions 180 and 700. Previous work by Boyes and Bird (1991) demonstrated the density-dependent relationship of transcriptional activity and CpG methylation. To assess the relationship of methyl-CpG density upon transcription from the HIV LTR, we employed the enzymes *Hpa*II and *Sss*I methylase, respectively. *Hpa*II methylase places a methyl group on the second cytosine in the sequence CCGG, and *Sss*I methylase recognizes all CpG dinucleotides. A plasmid DNA reporter construct, containing the HIV LTR fused to the CAT gene (pU3RIIICAT), was enzymatically methylated in vitro by using either *Hpa*II or *Sss*I methylase, the latter of which more closely approximates the activity of the human de novo methyltransferase.

Methylated DNA was electroporated either alone or in combination with a plasmid DNA containing the gene that encodes the *trans*-activator TAT (pIIIexTATIII) into the human CEM T-cell line A3.01 (Bednarik et al. 1990). When CAT expression was analyzed at various times thereafter, we found that expression was increased as much as 20-fold in the presence of TAT (Fig. 2). TAT was only able to *trans*-activate CAT expression when the TAT gene was transfected within 24 hours after introduction of the CAT reporter plasmid. This implied that the LTR was no longer amenable to TAT-mediated *trans*-activation at later times. *trans*-Activation of the HIV LTR by TAT when methylated at two CpG sites by *Hpa*II methylase was significantly less than that of *trans*-activated controls at 24 hours posttransfection. After 24 hours, *trans*-activation by TAT was no longer apparent. In contrast to these observations, when all CpG dinucleotides were methylated in vitro by employing *Sss*I methylase, *trans*-activation of the HIV LTR by TAT was below the level of de-

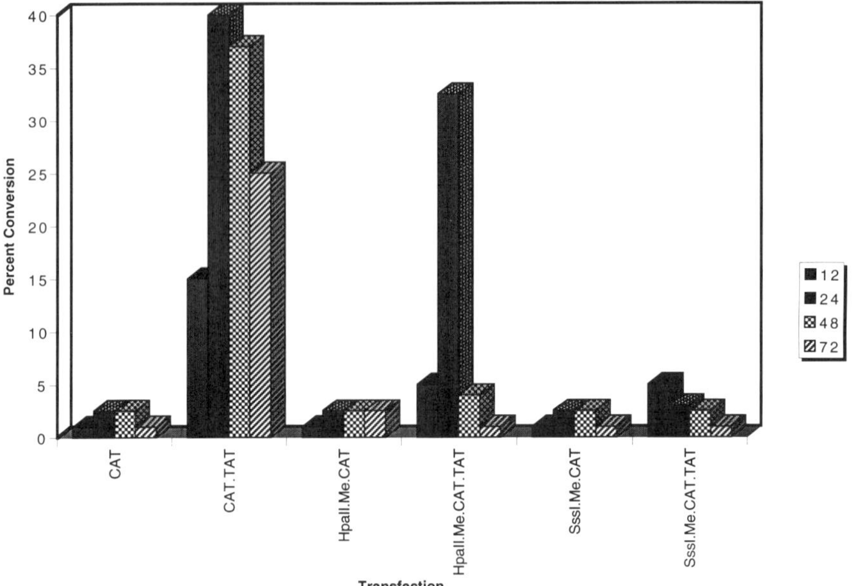

Figure 2 In vivo methylation-mediated inhibition of transcription. The CAT reporter plasmid DNA, pU3RIIICAT, was transfected in the methylated or unmethylated form with or without TAT. Data are expressed as a function of hours after transfection with either pU3RIIICAT (CAT), pU3RIIICAT+pIIIexTATIII (CAT.TAT), *Hpa*II-methylated pU3RIIICAT (*Hpa*II.Me.CAT), *Hpa*II-methylated pU3RIIICAT+pIIIexTATIII (*Hpa*II.Me.CAT.TAT), *Sss*I-methylated pU3RIIICAT (*Sss*I.Me.CAT), or *Sss*I-methylated pU3RIIICAT+pIIIexTATIII (*Sss*I.Me.CAT.TAT).

tection sensitivity. This experiment demonstrated the similarity of our model system to the experiments performed by Boyes and Bird (1991), in that a methyl-CpG density phenomenon is also applicable to the transcription of viral DNA. We have previously demonstrated that CpG methylation localized within the core enhancer region facilitated the inactivation of the HIV LTR *in cis*, with this transcriptional block being rapidly obviated *in trans* by TAT (Bednarik et al. 1990).

Methylation Modulates LTR *trans*-Activation

The apparent time-dependent loss of observable *trans*-activation by TAT when the LTR was methylated by *Hpa*II methylase at two sites in the LTR suggests a secondary event that does not permit detectable levels of *trans*-activation. These may include the interaction of DNA-binding proteins that are inhibitory to transcription, such as MeCP-1 (Boyes and Bird 1992), and/or the more progressive hypermethylation of the LTR secondary to the initial methylation of two CpG sites, which might attract the binding of proteins such as MeCP-1 and -2. The methylation of the LTR at two CpG sites eventually causes the LTR to behave as though it were methylated at all CpG sites, as in the case of the *Sss*I-methylated pU3RIIICAT, thereby suggesting either the occurrence of further de novo methylation or simply a slower rate of association of transcriptional inhibitory protein(s).

To further determine whether TAT activity could be directly inhibited by DNA methylation, we analyzed the effect of CpG methylation of the HIV LTR on TAT expression by enzymatically methylating the plasmid pIIIexTATIII, followed by transfection into A3.01 CEM T-cells that contained a single, stably integrated copy of the pU3RIIICAT reporter construct (Bednarik et al. 1991). When the LTR directing the expression of TAT was methylated, either by *Hpa*II methylase or by *Sss*I methylase, TAT was not expressed, as evidenced by the inability of the plasmid to *trans*-activate expression of the stably integrated CAT gene (Fig. 3). In this case, there was essentially no transient increase in CAT expression, since the expression of TAT was effectively silenced by methylation of two CpG sites via *Hpa*II methylase or by methylation of all CpG sites via *Sss*I methylase. In this experiment, we demonstrate that CpG methylation silences LTR-directed transcription of pIIIexTATIII, thereby lowering the level of intracellular TAT protein required for HIV LTRCAT *trans*-activation to below the limits of detection.

LTR Methylation Regulates Transcription in a Cell-free System

The factors that govern the transcription and *trans*-activation of the HIV LTR have been recently dissected by analyzing transcription in a cell-free system (Marciniak et al. 1990; Marciniak and Sharp 1991; Remenick et al. 1991). In the in vitro assay, TAT specifically stimulates HIV transcription in a dose-dependent manner. The mechanism by which TAT *trans*-activates the HIV LTR involves the increase in processivity and elongation of nascent HIV transcripts (Marciniak et al. 1990; Marciniak and Sharp 1991). We employed in vitro transcription experiments to determine whether our in vivo observations would hold true for a cell-free system (Fig. 4) (Gutekunst et al. 1993). The introduction of methyl-CpG dinucleotide sequences reduced the level of basal transcription such that *trans*-activation of the HIV LTR by TAT in vitro was not apparent, regardless of whether only two CpG sites or all CpG sites were methyl-

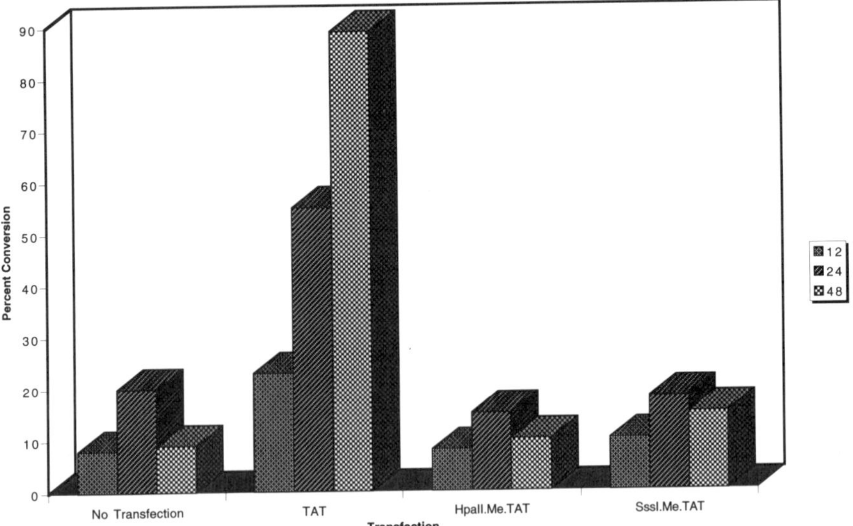

Figure 3 Effect of DNA CpG methylation on TAT expression. Bar graph represents percent conversion of [^{14}C]chloramphenicol versus time for each of four transfections. Cells (A3.01) containing a single copy of stably integrated pU3RIIICAT were electroporated with 5 μg of pIIIexTATIII, which was either unmethylated (TAT), *Hpa*II-methylated (*Hpa*II.Me.TAT), or *Sss*I-methylated (*Sss*I.Me.TAT). Each transfection was subsequently assayed for CAT activity at 12, 24, and 48 hr. (Reprinted, with permission, from Gutekunst et al. 1993.)

ated. These data are in contrast to results obtained in vivo, where it was found that TAT could initially overcome the transcriptional block introduced when only two CpG dinucleotides were methylated. Procedures involved in the preparation of whole-cell extracts, or perhaps the in vitro reaction mixture, may permit more rapid transcriptional inactivation of the *Hpa*II-methylated plasmid DNA. An event secondary to the initial methylation of the two *Hpa*II sites might explain the reduction in CAT expression *in trans*. Experiments testing the stability of methylated plasmid DNA have ruled out any possibility of degradation (Bednarik et al. 1990). These results are intriguing, since methylation of the HIV LTR at two or all CpG sites effectively incapacitates LTR-directed expression both *in cis* and *in trans* and may implicate the involvement of other cellular regulatory proteins such as MeCP-1 and -2 (Antequera et al. 1990; Boyes and Bird 1991, 1992; Lewis et al. 1992). Transcription of the HIV LTR *in cis* results in the formation of nascent transcripts that contain a three-nucleotide bulge in the stem-loop structure of TAR RNA (Dingwall et al. 1990; Roy et al. 1990a,b). Previous experiments in which this

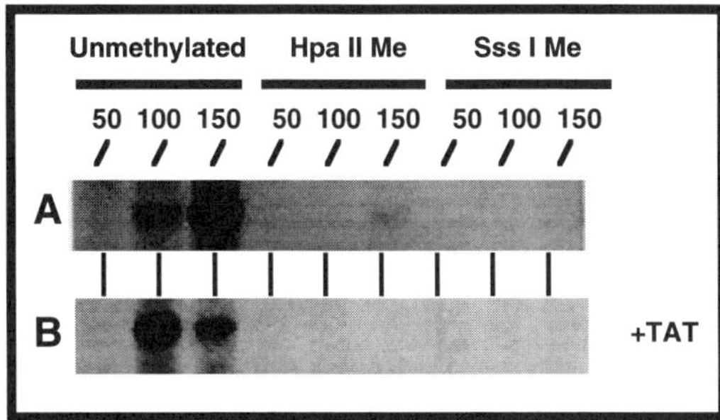

Figure 4 Effect of DNA CpG LTR on in vitro transcriptional run-off analysis of pU3RIIICAT in whole-cell extracts. (*A*) Basal, *cis*-level transcription activity of template DNA using either 50, 100, or 150 ng of linear plasmid DNA, which was either unmethylated, *Hpa*II-methylated, or *Sss*I-methylated. (*B*) Effect of the addition of purified TAT protein to the transcription reactions. Template DNA (50 ng) was analyzed for in vitro transcription activity in the presence of either 0, 100, or 200 ng of purified TAT protein. (Reprinted, with permission, from Gutekunst et al. 1993.)

bulge structure was eliminated or mutated resulted in a complete loss of TAT function in vitro (Berkhout et al. 1989; Selby et al. 1989; Dingwall et al. 1990; Roy et al. 1990a,b; Berkhout and Jeang 1991). The methylation of the LTR may lower the *cis*-level accumulation of nascent TAR RNA structures in vitro, thereby depriving TAT of an RNA substrate. Alternatively, it is possible that methylation of the HIV template interferes with the ability of TAT to assemble the initiation complex.

During the course of these studies, we have examined HIV-LTR-directed transcription when LTR sequences were either sparsely or densely methylated at CpG dinucleotides both in vivo and in vitro. These data confirmed our previous results in which methylation of two CpG sites within the HIV LTR inhibited basal transcription (Bednarik et al. 1990). Interestingly, the obviating effect of TAT was lost when CAT levels were determined at times later than 24 hours. In contrast, hypermethylation of the LTR at all CpG sites using the enzyme *Sss*I methylase, which more closely approximates the activity of the human de novo DNA methyltransferase, resulted in a complete loss of detectable transcriptional activity *in trans*, regardless of the time at which CAT levels were assayed. Previous experiments that employed the rescue of methylated or unmethylated plasmid DNA from nuclei demonstrated that methylation does not destabilize plasmid DNA or increase its susceptibility to degradation (Bednarik et al. 1990). One possible explanation for these results is illustrated in Figure 5. Methylation of the HIV LTR at two CpG sites might result in a transcriptional block, presumably by the formation of a complex between the DNA and some methylation-dependent binding protein such as MeCP. The net effect would be the inhibition of transcriptional activity *in cis* to a level where the processive effect of TAT can no longer be detected. This phenomenon may be most pronounced when the frequency of CpG methylation is very high (Doerfler 1983; Razin and Szyf 1984; Besser et al. 1990). One could speculate, alternatively, that the initial methylation at one or two sites may act as a signal for the de novo methylation of the DNA, thereby resulting in comparable levels of hypermethylation analogous to that obtained with the enzyme *Sss*I. Once the DNA is hypermethylated, the complexes formed between the methylation-dependent DNA-binding proteins and the DNA

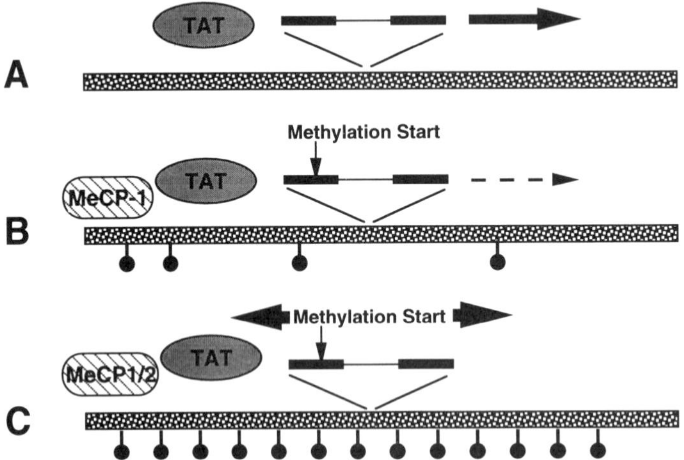

Figure 5 Methyl-CpG density-dependent inhibition of HIV LTR transcription model. Schematic representation of a possible mechanism to explain the loss of transcriptional activity *in trans* of integrated HIV proviral DNA which was either unmethylated (*A*), sparsely methylated (*B*), or densely methylated (*C*). The filled circles represent 5-methylcytosine. *trans*-Activation of the provirus is readily apparent in the absence of CpG methylation (*A*); methylation at a low density reduces the *cis*-level expression of the HIV LTR such that only a transient increase in *trans*-activation is observed (*B*). This may be explained by a slower and/or weaker association of a cellular methylcytosine-specific inhibitory protein (MeCP-1 or -2). Hypermethylation of all CpG sites (*C*) results in the rapid association of MeCP, which strongly inhibits *cis*-level expression of the LTR and therefore reduces the observable *trans* effect below the level of detection. The start of DNA methylation may occur outward from a single site as has been demonstrated previously (Doerfler et al. 1995).

cannot be disrupted, thereby maintaining a very low level of, or completely inhibiting, *cis*-level transcription. This model is consistent with the hypothesis proposed by Boyes and Bird (1992), and Lewis and Bird (1992), in that the DNA-binding protein MeCP-1 has a lower affinity for DNA that is sparsely methylated than for DNA that is densely methylated.

In a cell-free system, the presence of methylated CpG sites also inhibits RNA-polymerase-II-dependent transcription. In addition, TAT was unable to detectably *trans*-activate these methylated sequences in vitro. This does not necessarily mean that TAT is directly inhibited; it could be that *trans*-activation occurs at a very low level, below the limit of detection sensitivity in these assays. Further work will be required to completely elucidate the mechanism(s) involved in this phenomenon. Previous work has shown that TAT can bind to nascent transcript RNA sequences that contain three essential nucleotides in a putative "bulge" structure localized within the TAR region of HIV RNA (Muesing et al. 1987; Feng and Holland 1988; Berkhout et al. 1989). The interaction of TAT with TAR on the nascent RNA results in increased transcriptional processivity and elongation from the site of transcriptional initiation (Marciniak et al. 1990; Marciniak and Sharp 1991). When the DNA is densely hypermethylated, the kinetics of transcription may be so dramatically reduced that the levels of nascent RNA structures containing the TAR region are limited, and therefore activation by TAT is not detectable.

CONCLUSIONS

Over the last several years, great strides have been made in understanding the mechanism of DNA methylation, and it is clear that DNA methylation can modulate virus expression and infectivity. Human diseases, such as AIDS, result from very complicated multifactorial processes. It is reasonable to assume that these viral diseases can be influenced by the methylation of the viral genome. It is very difficult to demonstrate a direct relationship between virus genome methylation and the latency or absence of disease in vivo, even when elements of the virus are known to be present clinically or biochemically. Recent technological advances in the PCR determination of methylcytosine may shed light on any clinical relationships of viral genome methylation and disease. Exactly how cells identify foreign DNA and then silence these sequences and, in contrast, why cells do not always silence foreign or viral genomes, will continue to be an intriguing focus of study.

ACKNOWLEDGMENTS

I acknowledge the dedicated, hard work of Dr. Karen Gutekunst, who performed many of these experiments and who always stood by me during difficult times. Many thanks also to Cindi Thir, who helped in the typing of the monograph.

REFERENCES

Antequera, F., D. Macleod, and A.P. Bird. 1990. Specific protection of methylated CpGs in mammalian nuclei. *Cell* **58:** 509–517.

Bednarik, D.P., J.A. Cook, and P.M. Pitha. 1990. Inactivation of the HIV LTR by DNA CpG methylation: Evidence for a role in latency. *EMBO J.* **9:** 1157–1164.

Bednarik, D.P., J.D. Mosca, and N.B.K. Raj. 1987. Methylation as a modulator of expression of the human immunodeficiency virus. *J. Virol.* **61:** 1253–1257.

Bednarik, D.P., C. Duckett, S.U. Kim, V.L. Perez, K. Griffis, P.C. Guenthner, and T.M. Folks. 1991. DNA CpG methylation inhibits binding of NF-κB proteins to the HIV-1 long terminal repeat cognate DNA motifs. *New Biol.* **3:** 969–976.

Berkhout, B. and K.T. Jeang. 1991. Detailed mutational analysis of TAR RNA: Critical spacing between the bulge and loop recognition domains. *Nucleic Acids Res.* **19:** 6169–6176.

Berkhout, B., R.H. Silverman, and K.T. Jeang. 1989. TAT *trans*-activates the human immunodeficiency virus through a nascent RNA target. *Cell* **59:** 273–282.

Besser, D., F. Gotz, K. Schulze-Forster, H. Wagner, H. Kroger, and D. Simon. 1990. DNA methylation inhibits transcription by RNA polymerase III of a tRNA gene, but not of a 5S rRNA gene. *FEBS Lett.* **269:** 358–362.

Boyes, J. and A.P. Bird. 1991. DNA methylation inhibits transcription indirectly via a methyl-CpG binding protein. *Cell* **64:** 1123–1134.

———. 1992. Repression of genes by DNA methylation depends on CpG density and promoter strength: Evidence for involvement of a methyl-CpG binding protein. *EMBO J.* **11:** 327–333.

Dingwall, C., I. Ernberg, M.J. Gait, S.M. Green, S. Heaph, J. Karn, A.D. Lowe, M. Singh, and M.A. Skinner. 1990. HIV-1 TAT protein stimulates transcription by binding to a U-rich bulge in the stem of the TAR RNA structure. *EMBO J.* **9:** 4145–4153.

Doerfler, W. 1983. DNA methylation and gene activity. *Annu. Rev. Biochem.* **52:** 93–124.

———. 1984. DNA methylation and its functional significance: Studies on the adenovirus system. *Curr. Top. Microbiol. Immunol.* **108:** 79–98.

Doerfler, W., G. Orend, R. Schubbert, K. Fechteler, H. Heller, P. Wilgenbus, and J. Schroer. 1995. On the insertion of foreign DNA into mammalian genomes: Mechanisms and consequences. *Gene* **157:** 241–245.

Feng, S. and E.C. Holland. 1988. HIV-1 TAT *trans*-activation requires the loop sequence within tar. *Nature* **334:** 165–167.

Gutekunst, K.A., F. Kashanchi, J.N. Brady, and D. Bednarik. 1993. Transcription of the HIV-1 LTR is regulated by the density of DNA CpG methylation. *J. Acquired Immune Defic. Syndr.* **6:** 541–549.

Harbers, K., N. Schieke, H. Stuhlmann, and D. Jahner. 1981. DNA methylation and gene expression: Endogenous retroviral genome becomes infectious after molecular cloning.

Proc. Natl. Acad. Sci. **78:** 7609–7613.

Lewis, J.D., R.R. Meehan, W.J. Henzel, I. Maurer-Fogy, P. Jeppesen, F. Klein, and A. Bird. 1992. Purification, sequence, and cellular localization of a novel chromosomal protein that binds to methylated DNA. *Cell* **69:** 905–914.

Marciniak, R.A. and P.A. Sharp. 1991. HIV-1 TAT protein promotes formation of more processive elongation complexes. *EMBO J* **10:** 4189–4196.

Marciniak, R.A., B.J. Calnan, A.D. Frankel, and P.A. Sharp. 1990. HIV-1 TAT protein *trans*-activates transcription in vitro. *Cell* **63:** 791–802.

Meehan, R.R., J.D. Lewis, S. McKay, E.L. Kleiner, and A.P. Bird. 1989. Identification of a mammalian protein that binds specifically to DNA containing methylated CpGs. *Cell* **58:** 499–507.

Miyoshi, E., J. Fujii, N. Hayashi, K. Ueda, T. Towata, H. Fusamoto, T. Kamada, and N. Tanaguchi. 1992. Enhancement of hepatitis-B surface antigen expression by 5-azacytidine in a hepatitis-B-virus-transfected cell line. *Int. J. Cancer* **52:** 137–140.

Muesing, M.A., D.H. Smith, and D.J. Capon. 1987. Regulation of mRNA accumulation by a human immunodeficiency virus trans-activator protein. *Cell* **48:** 691–701.

Razin, A. and M. Szyf. 1984. DNA methylation patterns. Formation and function. *Biochim. Biophys. Acta* **782:** 313–342.

Remenick, J., M.R. Radonovich, and J.N. Brady. 1991. Human immunodeficiency virus TAT *trans*activation: Induction of a tissue-specific enhancer in a nonpermissive cell line. *J. Virol.* **65:** 5641–5646.

Rosen, C.A., J.G. Sodroski, and W.A. Haseltine. 1985. The location of *cis*-acting regulatory sequences in the human T-cell lymphotropic virus type III long terminal repeat. *Cell* **41:** 813–823.

Roy, S., U. Delling, C.H. Chen, C.A. Rosen, and N. Sonenberg. 1990a. A bulge structure in HIV-1 TAR RNA is required for TAT binding and TAT-mediated *trans*-activation. *Genes Dev.* **4:** 1365–1373.

Roy, S., N.T. Parkin, C. Rosen, J. Itovitch, and N. Sonenberg. 1990b. Structural requirements for *trans* activation of human immunodeficiency virus type 1 long terminal repeat-directed gene expression by TAT: Importance of base pairing, loop sequence, and bulges in the TAT-responsive sequence. *J. Virol.* **64:** 1402–1406.

Selby, M.J., E.S. Bain, P.A. Luciw, and B.M. Peterlin. 1987. Structure, sequence, and position of the stem-loop in tar determine transcriptional elongation by TAT through the HIV-1 long terminal repeat. *Genes Dev.* **3:** 547–558.

Varmus, H.E. 1983. Retroviruses. In *Mobile genetic elements* (ed. J. Shapiro), pp. 411–503. Academic Press, New York.

Varmus, H.E. and P. Brown. 1990. Retroviruses. In *Mobile DNA* (ed. D.E. Berg and M.M. Howe), pp. 53–108. American Society for Microbiology, Washington, D.C.

CpG Suppression in HIV-1 Versus HIV-2: Correlation with Pathogenicity and Possible Implications for the Design of Antiretroviral Vaccines

Jonathan W. Nyce
Department of Molecular Pharmacology and Therapeutics
EpiGenesis Pharmaceuticals, Inc.
Greenville, North Carolina 27834

The remarkable process by which a single cell, a fertilized egg, develops into an adult human being with more than 200 different cell types relies on the ability of cells to control gene expression. In every human cell there are within the chromosomes approximately 100,000 genes. Yet, each specialized cell of the human body usually expresses only a small subset of these 100,000 genes. Unless something goes wrong, most genes are switched off. DNA methylation represents a major mechanism by which cells undergo such selective gene silencing (Sager and Kitchin 1975). In animal cells, virtually all of this methylation occurs in CpG dinucleotides, and methylation of specific CpG dinucleotides is associated with transcriptional inactivation in many genes.

The use of DNA methylation in the control of gene expression in vertebrate cells must provide strong advantages, for it is not without its consequences. One such evolutionary consequence of DNA methylation in vertebrate cells results from the "spontaneous" deamination of methylated cytosines, creating thymidine. This results, over long evolutionary periods, in CpG suppression, the replacement of methylated CpG dinucleotides in vertebrate DNA with their deamination product TpG. CpG dinucleotides in the DNA of higher vertebrates thus occur at a frequency of about 1% of all dinucleotides, compared to an expected frequency of 4% based on G + C content (Cedar 1984).

This propensity for 5-methylcytosine to undergo deamination has severe consequences for human health. A large percentage of mutations in cancer and other diseases can be traced to replacement of 5-methylcytosine with its deamination product, thymidine (Maddalena et al. 1988; Abadie et al. 1989; Cooper and Krawczak 1990; Koeberl et al. 1989; Hirschhorn et al. 1990; Perutz 1990; Rideout et al. 1990). Either significant advantages must accrue in organisms utilizing 5-methyl-

Epigenetic Mechanisms of Gene Regulation

cytosine to control gene expression, or complex, multi-tissue body plans and developmental phases may be impossible without it.

METHYLATION ALSO TRANSCRIPTIONALLY SILENCES VIRAL GENOMES

There is an extensive literature demonstrating that host DNA methylation plays a critical role in the expression of viral genomes having a DNA phase in their life cycle (Fynan et al. 1993; Nonkwelo and Long 1993; Cassens et al. 1994; Farza et al. 1994; Graessman et al. 1994; List et al. 1994; Minarovits et al. 1994; Challita et al. 1995). Methylation of long terminal repeat (LTR) regions in retroviruses maintains latency of integrated proviral genomes (Bednarik et al. 1991; Singh and Pauza 1992; Gutekunst et al. 1993), and exposure to demethylating agents causes latent proviral genomes to rapidly become expressed (Rascati 1988; Saggioro et al. 1990; O'Brien et al. 1995). The density of CpG methylation directly correlates with expression of viral genes in Epstein-Barr virus (Araki et al. 1990; Ernberg 1990), herpes simplex virus (Ben-Hattar and Jiricny 1988; Stephanopoulos et al. 1988), human T-lymphotrophic virus I (HTLV-1; Kitamura et al. 1985; Cassens et al. 1994), adenovirus (Doerfler 1993), human papillomavirus (Rosl et al. 1993), hepatitis B virus (Jansson et al. 1992), Marek's disease virus (Fynan et al. 1993), simian foamy virus (Schwiezer et al. 1993), Moloney murine leukemia virus (Hoeben et al. 1991), a variety of plant viruses (Hershkovitz et al. 1990; Brough et al. 1992; Ermak et al. 1993), polyomavirus (Liboi and Basilico 1984), and most importantly for this discussion, HIV-1 (Bednarik et al. 1987, 1991; Gutekunst et al. 1993). The correlation between host methylation and the expression of viral DNA has thus been repeatedly demonstrated by many laboratories in a large array of dissimilar viruses.

Viruses subject to CpG-methylation-mediated transcriptional regulation also show CpG suppression, typically in a manner and to a degree similar to their hosts (Doerfler et al. 1984). We became interested in studying CpG suppression in HIV-1 for two reasons:

1. Several laboratories had demonstrated that expression of HIV-1 provirus is regulated by the density of its CpG methylation (Bednarik et al. 1987, 1991; Gutekunst et al. 1993).
2. It was shown to be possible to manipulate the latency period following HIV-1 infection by manipulating the density of CpG methylation of the integrated provirus using demethylating agents (Bednarik et al. 1987, 1991; Gutekunst et al. 1993).

CPG SUPPRESSION IN HUMAN IMMUNODEFICIENCY VIRUSES: CORRELATION OF CpG CONTENT WITH VIRULENCE

There are two forms of HIV, HIV-1 and HIV-2. We became interested in the different levels of pathogenicity associated with these two viruses and the possible role that host DNA methylation might play in this difference. HIV-1 infection has almost invariably proven to be lethal. In contrast, it has been convincingly established that HIV-2 is substantially less virulent. For example, in a recent study, 33% of Senegalese women infected with HIV-1 had developed AIDS within 5 years, whereas not one of more than 100 women infected with HIV-2 developed AIDS over the same time period (Marlink et al. 1994). We asked, Could such differences in virulence be related to the "methylatability" of the proviral genomes integrated into host DNA? That is, could differences in lethality of the two viruses be related to the ability of the host to methylate the proviral genome and inactivate the transcription of viral proteins?

HIV-1 IS EVOLVING TO REMOVE HOST-METHYLATABLE CpGs FROM ITS GENOME

To explore this hypothesis, we first examined the number of CpGs in many strains of HIV-1 and compared them to the number of CpGs in strains of HIV-2. We were intrigued to find that HIV-1 demonstrated a concerted evolutionary mechanism for selectively removing methylatable CpGs from its genome. HIV-1 isolates (eight examined) all have far fewer CpGs than all HIV-2 isolates (four examined; see Fig. 1). In terms of the variation from expected number of CpGs, calculated as CpG expected/observed ($CpG_{exp./obs.}$), HIV-1 isolates showed an average $CpG_{exp./obs.}$ ratio of 4.92 ± 0.10, whereas the $CpG_{exp./obs.}$ for HIV-2 isolates was 2.98 ± 0.22 ($CpG_{obs.}$ = 0.88 for HIV-1, 1.68 for HIV-2; CpG_{exp} for HIV 1 was 1.33% and for HIV-2 was 5.0%). HIV-1 appears to be evolving in such a way as to remove methylatable CpGs from its genome, thereby escaping the host's capacity to inactivate viral DNA by methylating it. The degree to which HIV-1 has depleted methylatable CpGs from its genome appears to surpass that encountered elsewhere in the animal kingdom and its many viruses.

A PROKARYOTIC PRECEDENT

A similar strategy to evade host defense mechanisms against foreign DNA is employed by phage that infect certain bacteria (Kruger et al. 1988; Shpaer and Mullins 1990). The restriction/modification (RM) sys-

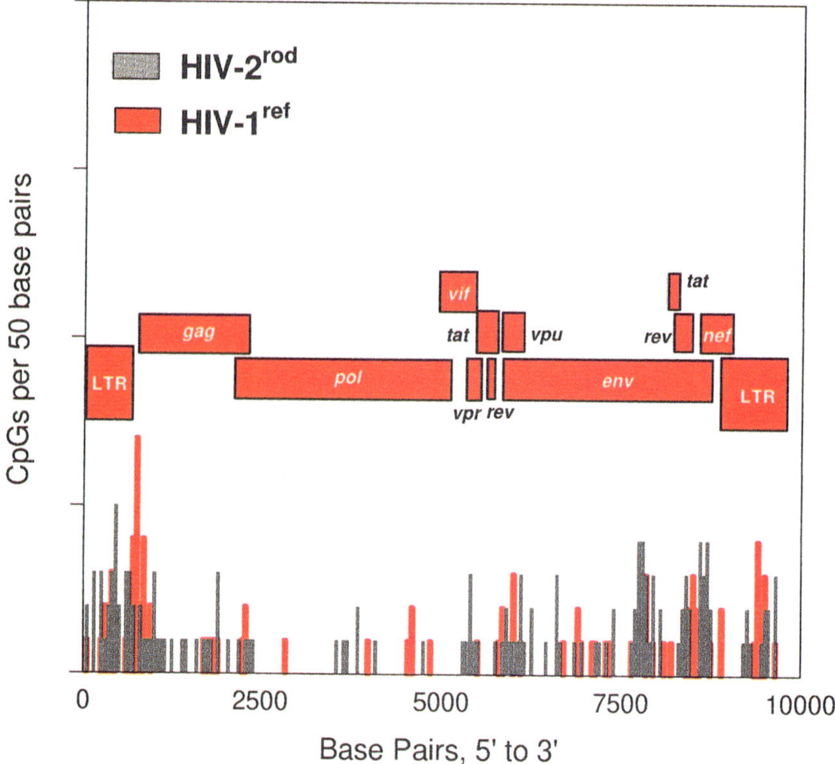

Figure 1 HIV-2^{rod} contains 147 methylatable CpGs, HIV-1^{ref} only 95, typical, respectively, of HIV-2 and HIV-1 genomes. The relative position of genes in HIV-1 is given above the CpG histograms. The CpG histograms quantitate CpG content in HIV-2^{rod} and HIV-1^{ref} in 50-bp increments from the 5′ LTR to the 3′ LTR.

tems of bacteria, which include a DNA methylase coupled to a restriction endonuclease with identical or overlapping sequence specificity, have evolved at least in part to identify and neutralize invading phage DNA, thereby decreasing the effectiveness of phage infection (Kruger and Bickle 1983). One mechanism that certain phage have developed to counteract the RM systems of their intended bacterial hosts is counter-selection of the RM target sequence from within their genomes (Sharp 1986). Bacteriophages T3 and T7, for example, have strongly counter-selected a variety of recognition sites for the type II RM systems of their host bacteria (Sharp et al. 1985). Similarly, selection against dam methylation sites in the genomes of DNA enterobacteriophages has also been reported (McClelland 1985). This strategy of counterselecting against the target sequences of host DNA methylase enzymes appears to be widespread among bacterial viruses. Other strategies employed by bacteriophage include the expression of proteins that directly interfere with the function of the host DNA methylase (Kruger et al. 1978, 1985; Schroeder et al. 1984) or that hydrolyze *S*-adenosylmethionine (SAM) and thus eliminate the methyl donor of DNA methylation reactions (Hughes et al. 1987). Many bacteriophage employ multiple systems to counteract bacterial RM systems. Thus, the *ocr* gene of bacteriophage T7 encodes a protein that inhibits a specific group of host cell DNA methylases, and the recognition sequences of those host DNA methylases that are not subject to inhibition by *ocr* are significantly suppressed in the T7 genome (Kruger et al. 1989). It should not be too surprising, therefore, if similar strategies to avoid host DNA methylase "surveillance" of viral genomes also exist in animal cells.

"HEAVY BREATHING" AND DEAMINATION OF METHYLATED CYTOSINES

The deamination of methylated cytosines in vitro occurs preferentially within single-stranded DNA (Adams and Eason 1984), and evidence has been presented in support of the concept that the melting of AT-rich double-stranded DNA to short single-stranded stretches may account for the enhanced CpG suppression observed in such regions (Adams and Eason 1984). This "heavy breathing" concept of CpG suppression thus suggests that AT-rich DNA is characterized by enhanced potential for the introduction of single-strand "bubbles," and that methylated cytosines within such bubbles may show particular sensitivity toward spontaneous chemical deamination. HIV-1 DNA, particularly over genomic regions relatively devoid of CpG, is AT-rich as compared to bulk human DNA, and integrated HIV-1 proviral DNA would be expected to undergo more rigorous breathing as compared to bulk human DNA, and hence, potentially, more facile deamination of 5-methylcytosine and depletion of methylatable CpG dinucleotides. In support of this concept, regions within the HIV-1 genome that are devoid, or nearly devoid, of CpGs contain counterbalancing islands of the deamination product, TpG.

The concept of preferred deamination of methylated cytosines occurring within regions of proviral DNA undergoing heavy breathing, rather than as a consequence of events occurring at the level of viral RNA, is most apparent in the HIV-1 genome. HIV-1 evolution is clearly proceeding along the lines of depleting host methylatable CpGs from its genome. Neither HIV-2, SIV, nor related retroviruses show the enhanced level of

CpG suppression seen in HIV-1. Although other viral genomes do appear to be sensitive to host-mediated transcriptional inactivation of their genomes, they do not appear to be evolving to remove methylatable CpGs from their genomes—at least not at a rate similar to HIV-1.

Analysis shows that members of the family *Flaviviridae* are mildly CpG-suppressed, even though a DNA stage is not part of their life cycle. Since dinucleotide suppression appears to be specific for CpG in these genomes as well, it is possible that CpG dinucleotides play a general role in the structure of both DNA and RNA genomes. Independent of their role as the preferred substrate for DNA methylase, they may mediate binding of host regulatory factors to both DNA and RNA genomes. This concept suggests selection against CpG dinucleotides by exclusively RNA viral genomes in order to avoid accessibility to such host regulatory factors. Alternatively, CpG dinucleotide suppression specific for CpG in exclusively RNA genomes suggests that host DNA methylase enzymes (or an RNA methylase) conduct molecular surveillance for the presence of diverse foreign nucleic acids. A cytosine deaminase that functions in RNA editing has been described recently, but its reported recognition sequence does not contain CpG, at least for apoB mRNA (Navaratnam et al. 1995). A further possibility to explain CpG suppression in RNA genomes lacking a DNA phase in their life cycle is that such genomes evolved from CpG-suppressed vertebrate genes, and such CpG suppression was simply maintained in the virus. However, the rapid evolutionary rates of RNA viruses argue against this possibility. In any event, it is important to note that such exclusively RNA viral genomes do not show CpG suppression to the level observed in HIV-1.

DE-EVOLVING HIV-1

Our hypothesis that the extreme CpG suppression in HIV-1 genomes represents a viral strategy for escaping host methylation-mediated transcriptional inactivation of the proviral genome led us to postulate that we could reverse this process and modulate HIV-1 transcription by adding CpGs back to its genome. This CpG insertion could be done in such a way that no changes would occur at the amino acid level; i.e., the DNA would contain many more CpGs, but the amino acid code would remain unchanged ("synonomous mutations"). The synthetic virion produced by the insertion of CpGs could thus be transcriptionally impaired with respect to the non-CpG-inserted parental virion. If so, its protein structure would remain unchanged, but its genome would be hypersensitive to methylation-mediated transcriptional regulation via the newly inserted CpG dinucleotides.

POTENTIAL APPLICATION OF THIS CONCEPT TO RETROVIRAL VACCINE DESIGN

Clinical trials of potential HIV-1 vaccines have produced unsatisfactory results (Belshe et al. 1994; Cohen 1994; Hoth et al. 1994). More than a dozen large projects, utilizing either peptide vaccines or killed, denatured virus or viral fragments, have failed to produce reliable immunization. In contrast, success has been reported using a live attenuated virus to protect against simian immunodeficiency virus (SIV) infection in macaques. In these studies the *nef* gene was removed from the virus, producing an intact virus capable of replicating and infecting cells, but which lacked the *nef* protein. It was found that rhesus monkeys (*Macaca mulatta*) inoculated with this live attenuated virus vaccine were protected from subsequent infection with extremely large doses of SIV (Daniel et al. 1992). Recently, similar success has been reported using a live attenuated HIV-2 virus to prevent simian AIDS (Putkonen et al. 1995). No other HIV vaccine model has produced the positive result of these live attenuated retroviral vaccines (Cranage et al. 1992; Desrosiers 1992).

Interestingly, a natural experiment appears to have verified that removal of the *nef* gene may produce an effective live attenuated vaccine in man as well. Thus, small numbers of individuals were identified who had been infected with HIV-1 for more than a decade, but in whom such infection showed no signs of progressing to AIDS (Huang et al. 1995). These people appeared to have been naturally immunized by strains of HIV-1 containing deletions in the *nef* gene.

The results of the experiments in monkeys and the discovery of the natural experiment in man demonstrated the possibility of immunization by infection with HIV-1 lacking the *nef* protein and have prompted considerable debate over whether or not a live attenuated vaccine for HIV-1 should be considered (Bolognesi 1994; Desrosiers 1994). One question regarding the safety of *nef* deletions to attenuate HIV-1 is whether or not the attenuation is durable over the lifetime of the vaccinee (Bolognesi 1994). Another question relates to the recent observation that infant macaques appear to be susceptible to *nef*-gene-deleted SIV (Baba et al. 1995; Ruprecht et al. 1995).

POTENTIAL DEVELOPMENT OF A LIVE, DOUBLY ATTENUATED HIV-1 VACCINE WITH IMPROVED SAFETY FEATURES

As we have discussed, HIV-1 appears to have evolved to remove methylatable CpGs from its genome, an evolutionary strategy that may prevent transcriptional regulation of its proviral DNA by host DNA methylases. Therefore, one strategy that may enhance transcriptional regulation of HIV-1 proviral DNA by host-mediated DNA methylation is to "de-evolve" it by adding synonymous CpGs back into its genome.

The virulence of HIV-1 has recently been linked to its high replication rate via transcription of proviral DNA (Ho et al. 1995; Wei et al. 1995). If the number of methylatable CpGs in HIV-2 as compared to HIV-1 relates, as we suggest, to the different levels of virulence of these two viruses, then the general virulence of HIV-1 might be attenuated by increasing the number of its CpGs to that found in HIV-2. Theoretically, one could increase the number of silent CpGs to the point at which proviral DNA becomes transcriptionally incompetent, and hence avirulent. In terms of vaccine development, however, this might not be advantageous, since the power of live attenuated viral vaccines comes in large measure from their ability to provide continuous antigenic stimulation and hence long-lasting immunity. Therefore, moderate CpG insertion sufficient to permit low-level transcription of proviral DNA may be more appropriate. If it is true that HIV-2's reduced virulence is due to the additional CpGs it possesses as compared to HIV-1, then the virulence of HIV-1 might very well be reduced by addition of appropriately placed CpG dinucleotides, such placement modeled after that observed in HIV-2. CpG insertion would thus likely have both qualitative and quantitative aspects.

This strategy has the following underlying tenets:

- HIV-2 is substantially less pathogenic than HIV-1.
- HIV-2 has nearly twofold more methylatable CpGs than does HIV-1.
- The increased numbers of host methylatable CpGs in HIV-2 as compared to HIV-1 may account for the reduced pathogenicity of the former, and the increased pathogenicity of the latter.
- Insertion of synonomous CpGs into the HIV-1 genome may modulate transcription of the HIV-1 genome and hence the level of infection/pathogenicity.
- Correct modulation of the level of infection may prevent viral production from overwhelming host defenses.
- Insertion of silent CpGs sufficient to prevent viral production from overwhelming host defenses, and yet sufficient to provide continuous antigenic stimulation, may provide long-lasting immunity.
- CpG insertion may constitute a plausible form of "secondary attenuation," added to a primary form of attenuation such as *nef* gene deletion.

SUMMARY

Substantial circumstantial evidence suggests that DNA methylation may play a role as a natural epigenetic defense mechanism that can limit the virulence of certain pathogenic viruses—a sort of molecular surveillance at the nucleic acid level to parallel immune surveillance at the cellular level. If the increased number of methylatable CpGs in HIV-2 genomes underlies its reduced virulence as compared to HIV-1, then it may be possible to regulate the pathogenicity of HIV-1 by increasing its CpG content. Since viral load following HIV-1 infection is primarily a function of transcription of proviral DNA, and since HIV-1 appears to be evolving such that methylatable CpGs are being selected against at an enhanced rate, attenuation of proviral transcription in "CpG-reconstituted" HIV-1 may provide a plausible strategy for reducing the pathogenicity of this virus. Such a strategy could be used as a secondary level of attenuation, e.g., in addition to *nef* gene deletion, in order to increase the safety of live attenuated HIV vaccines. This strategy presupposes that total inactivation of proviral transcription would be inappropriate. Rather, HIV-1 proviral DNA transcription should be reduced (sufficient to permit continuous low-level immune stimulation, but insufficient to permit the high virus loads required to produce disease), but not eliminated. Finally, whereas HIV-1 appears to be the prototype (thus far) of viruses that are evolving more rapidly than the norm with respect to CpG suppression, this strategy of inserting synonymous CpGs to enhance host transcriptional regulation capacity should be applicable to any virus in which the presence of CpG plays a role in genomic template function.

ACKNOWLEDGMENTS

This work was supported in part by National Institutes of Health grant RO1-CA-47217 to J.W.N.

REFERENCES

Abadie, V., S. Lyonnet, N. Maurin, M. Berthelon, C. Caillaud, F. Giraud, J.F. Mattei, J. Rey, F. Rey, and A. Munnich. 1989. CpG dinucleotides are mutation hotspots in phenylketonuria. *Genomics* **5:** 936–939.

Adams, R.L. and R. Eason. 1984. Increased G+C content of DNA stablizes methyl CpG dinucleotides. *Nucleic Acids Res.* **12:** 5869–5877.

Araki, K., K. Akagi, J. Miyazaki, K. Matsubara, and K. Yamamura. 1990. Correlation of tissue-specific methylation with gene inactivity in hepatitis B virus transgenic mice. *Jpn. J. Cancer Res.* **81:** 1265–1271.

Baba, T.W., Y.S. Jeong, D. Pennick, R. Bronson, M.F. Greene, and R.M. Ruprecht. 1995. Pathogenicity of live, attenuated SIV after mucosal infection of neonatal macaques. *Science* **267:** 1820–1825.

Bednarik, D.P., J.D. Mosca, and N.B. Raj. 1987. Methylation as a modulator of expression of human immunodeficiency virus. *J. Virol.* **61:** 1253–1257.

Bednarik, D.P., C. Duckett, S.U. Kim, V.L. Perez, K. Griffis, P.C. Guenthner, and T.M. Folks. 1991. DNA CpG methylation inhibits binding of NF-kappa B proteins to the HIV-1 long terminal repeat cognate DNA motifs. *New Biol.* **3:** 969–976.

Belshe, R.B., D.P. Bolognesi, M.L. Clements, L. Corey, R. Dolin, J. Mestecky, M. Mulligan, D. Stablein, and P. Wright. 1994. HIV infection in vaccinated volunteers. *J. Am. Med. Assoc.* **272:** 431.

Ben-Hattar, J. and J. Jiricny. 1988. Methylation of single CpG dinucleotides within a promoter element of the herpes simplex virus tk gene reduces its transcription in vivo. *Gene* **65:** 219–227.

Bolognesi, D.A. 1994. A live attenuated vaccine for HIV-1? *J. NIH Res.* **8:** 410–420.

Brough, C.L., W.E. Gardiner, N.M. Inamdar, X.Y. Zhang, M. Ehrlich, D.M. Bisaro. 1992. DNA methylation inhibits propagation of tomato golden mosaic virus DNA in transfected protoplasts. *Plant Mol. Biol.* **18:** 703–712.

Cassens, S., U. Ulrich, P. Beimling, and D. Simon. 1994. Inhibition of human T cell leukemia virus I long terminal repeat expression by DNA methylation: Implications for latency. *J. Gen. Virol.* **75:** 3255–3259.

Cedar, H. 1994. DNA methylation and gene expression. In*: DNA methylation: Biochemistry and biological significance* (ed. A. Razin et al.), pp. 147–164. Springer-Verlag, New York.

Challita, P.M., D. Skelton, A. el-Khoueiry, X.J. Yu, K. Weinberg, and D.B. Kohn. 1995. Multiple modifications in *cis* elements of the long terminal repeat of retroviral vectors lead to increased expression and decreased DNA methylation in embryonic carcinoma cells. *J. Virol.* **69:** 748–755.

Cohen, J. 1994. AIDS vaccines: Are researchers racing towards success, or crawling? *Science* **265:** 1373–1375.

Cooper, D.N. and M. Krawczak. 1990. The mutational spectrum of single base-pair substitutions causing human genetic disease: Patterns and predictions. *Hum. Genet.* **85:** 55–74.

Cranage, M.P., L.A. Ashworth, P.J. Greenaway, M. Murphey-Corb, and R.C. Desrosiers. 1992. AIDS vaccine developments. *Nature* **355:** 685–686.

Daniel, D., F. Kirchhoff, S.C. Czajak, P.K. Sehgal, and R.C. Desrosiers. 1992. Protective effects of a live attenuated SIV vaccine with a deletion in the *nef* gene. *Science* **258:** 1938–1941.

Desrosiers, R.C. 1992. HIV with multiple gene deletions as a live attenuated vaccine for AIDS. *AIDS Res. Hum. Retroviruses* **8:** 411–421.

———. 1994. Safety issues facing development of a live attenuated, multiply deleted HIV-1 vaccine. *AIDS Res. Hum. Retroviruses* **10:** 331–332.

Doerfler, W. 1993. Patterns of de novo DNA methylation and promoter inhibition: Studies on the adenovirus and the human genome. *Exper. Suppl.* **64:** 262–299.

Doerfler, W., K.D. Langner, I. Kruczek, L. Vardimon, and D. Renz. 1984. In *DNA Methylation: Biochemistry and biological significance.* (ed. A. Razin et al.), pp. 222–247. Springer-Verlag, New York.

Ermak, G., U. Paszkowski, M. Wohlmuth, O.M. Scheid, and J. Paszkowski. 1993.

Cytosine methylation inhibits replication of African cassava mosaic virus by two distinct mechanisms. *Nucleic Acids Res.* **21:** 3445–3450.

Ernberg, I. 1990. Epstein-Barr virus latency and activation in vivo. *Curr. Top. Microbiol. Immunol.* **166:** 337–346.

Farza, H., T.A. Dragani, T. Metzler, G. Manenti, P. Tiollais, G. Della Porta, and C. Pourcel. 1994. Inhibition of hepatitis B virus surface antigen gene expression in carcinogen-induced liver tumors from transgenic mice. *Mol. Carcinog.* **9:** 185–192.

Fynan, E.F., D.L. Ewert, and T.M. Block. 1993. Latency and reactivation of Marek's disease virus in B lymphocytes transformed by avian leukosis virus. *J. Gen. Virol.* **74:** 2163–2170.

Graessman, A., G. Sandberg, E. Guhl, and M. Graessmann. 1994. Methylation of single sites within the herpes simplex virus tk coding region and the simian virus 40 T-antigen intron causes gene inactivation. *Mol. Cell. Biol.* **14:** 2004–2010.

Gutekunst, K.A., F. Kashanchi, J.N. Brady, and D.P. Bednarik. 1993. Transcription of the HIV-1 LTR is regulated by the density of DNA CpG methylation. *J. Acquired Immune Defic. Syndr.* **6:** 541-549.

Hershkovitz, M., Y. Gruenbaum, P. Renbaum, A. Razin, and A. Loyter. 1990. Effect of CpG methylation on gene expression in transfected plant protoplasts. *Gene* **94:** 189–193.

Hirschhorn, R., S. Tzall, and A. Ellenbogen. 1990. Hot spot mutations in adenosine deaminase deficiency. *Proc. Natl. Acad. Sci.* **87:** 6171–6175.

Ho, D.D., A.U. Neumann, A.S. Perelson, W. Chen, J.M. Leonard, and M. Markowitz. 1995. *Nature* **373:** 123–126.

Hoeben, R.C., A.A. Migchielsen, R.C. van der Jagt, H. van Ormondt, and A.J. van der Eb. 1991. Inactivation of the Moloney murine leukemia virus long terminal repeat in murine fibroblast cell lines is associated with methylation and dependent on its chromosomal position. *J. Virol.* **65:** 904–912.

Hoth, D.F., D.P. Bolognesi, L. Corey, and S.H. Vermund. 1994. NIH Conference. HIV vaccine development: A progress report. *Ann. Intern. Med.* **121:** 603–611.

Huang, Y., L. Zhang, and D.D. Ho. 1995. Characterization of *nef* sequences in long-term survivors of human immunodeficiency virus type 1 infection. *J. Virol.* **69:** 93–100.

Hughes, J.A., L.R. Brown, and A.J. Ferro. 1987. Expression ot the cloned coliphage T3 S–adenosylmethionine hydrolase gene inhibits DNA methylation and polyamine biosynthesis in *Escherichia coli. J. Bacteriol.* **169:** 3625–3632.

Jansson, A., M. Masucci, and L. Rymo. 1992. Methylation of discrete sites within the enhancer region regulates the activity of the Epstein-Barr virus BamH1 W promoter in Burkitt lymphoma lines. *J. Virol.* **66:** 62–69.

Kitamura, T., M. Takano, H. Hoshino, K. Shimotohno, M. Miwa, F. Takaku, and T. Sugimura. 1985. Methylation pattern of human T-cell leukemia virus in vivo and in vitro: px and LTR regions are hypomethylated in vivo. *Int. J. Cancer* **35:** 629–635.

Koeberl, D.D., C.D. Bottema, J.M. Buerstedde, and S.S. Sommer. 1989. Functionally important regions of the factor IX gene have a low rate of polymorphism and a high rate of mutation in the dinucleotide CpG. *Am. J. Hum. Genet.* **45:** 448–457.

Kruger, D.H. and T.A. Bickle. 1983. Bacteriophage survival: Multiple mechanisms for avoiding the DNA restriction systems of their hosts. *Microbiol. Rev.* **47:** 345–360.

Kruger, D.H., G.J. Barcak, and H.O. Smith. 1988. Abolition of DNA recognition site resistance to the restriction endonuclease EcoRII. *Biomed. Biochim. Acta* **47:** K1–5.

Kruger, D.H., G. Gola, I. Weisshuhn, and S.J. Hansen. 1978. The ocr gene function of bacterial viruses T3 and T7 prevents host-controlled modification. *J. Gen. Virol.* **41:** 189–192.

Kruger, D.H., C. Schroeder, M. Santibanez-Koref, and M. Reuter. 1989. Avoidance of DNA methylation. A virus-encoded methylase inhibitor and evidence for counterselection of methylase recognition sites in viral genomes. *Cell Biophys.* **15:** 87–95.

Kruger, D.H., C. Schroeder, M. Reuter, I.G. Bogdarina, Y.I. Buryanov, and T.A. Bickle. 1985. DNA methylation of bacterial viruses T3 and T7 by different DNA methylases in *Escherichia coli* K12 cells. *Eur. J. Biochem.* **150:** 323–330.

Liboi, E. and C. Basilico. 1984. Inhibition of polyoma gene expression in transformed mouse cells by hypermethylation. *Virology* **135:** 440–451.

List, H.J., V. Patzel, U. Zeidler, A. Schopen, G. Ruhl, J. Stollwerk, and G. Klock. 1994. Methylation sensitivity of the enhancer from the human papillomavirus type 16. *J. Biol. Chem.* **269:** 11902–11911.

Maddalena, A., J.E. Specnce, W.E. O'Brien, and R.L. Nussbaum. 1988. Characterization of point mutations in the same arginine codon in three unrelated patients with ornithine transcarbamylase deficiency. *J. Clin. Invest.* **82:** 1353–1358.

Marlink, R., P. Kanki, I. Thior, K. Travers, G. Eisen, T. Siby, I. Traore, C.-C. Hsieh, M.C. Dia, E.-H. Gueye, J. Hellinger, A. Guèye-Ndiaye, J.-L. Sankalé, I. Ndoye, S. Mboup, and M. Essex. 1994. Reduced rate of disease development after HIV-2 infection as compared to HIV-1. *Science* **265:** 1587–1590.

McClelland, M. 1985. Selection against dam methylation sites in the genomes of DNA of enterobacteriophages. *J. Mol. Evol.* **21:** 317–322.

Minarovits, J., L.F. Hu, S. Minarovits-Kormuta, G. Klein, and I. Ernberg. 1994. Sequence-specific methylation inhibits the activity of the Epstein-Barr virus LMP 1 and BCR2 enhancer-promoter regions. *Virology* **200:** 661–667.

Navaratnam, N., S. Bhattacharya, T. Fujino, D. Patel, A.L. Jarmuz, and J. Scott. 1995. Evolutionary origins of apoB mRNA editing: Catalysis by a cytidine deaminase that has acquired a novel RNA-binding motif at its active site. *Cell* **81:** 187–195.

Nonkwelo, C.B. and W.K. Long. 1993. Regulation of Epstein-Barr virus BamH1-H divergent promoter by DNA methylation. *Virology* **197:** 205–215.

O'Brien, M.C., T. Ueno, N. Jahan, M. Zajac-Kaye, and H. Mitsuya. 1995. HIV-1 expression induced by anti-cancer agents in latently HIV-1-infected ACH2 cells. *Biochem. Biophys. Res. Commun.* **207:** 903-909.

Perutz, M.F. 1990. Frequency of abnormal human haemoglobins caused by C→T transitions in CpG dinucleotides. *J. Mol. Biol.* **213:** 203–206.

Putkonen, P., L. Walther, Y.J. Zhang, S.L. Li, C. Nilsson, J. Albert, P. Biberfeld, R. Thorstensson, and G. Biberfeld. 1995. Long-term protection against SIV-induced disease in macaques vaccinated with a live attenuated HIV-2 vaccine. *Nat. Med.* **1:** 914–918.

Rascati, R.J. 1988. Effects of cytidine analogs on methylation of DNA and retrovirus induction. *Mutat. Res.* **208:** 21–25.

Rideout, W.M., G.A. Coetzee, A.F. Olumi, and P.A. Jones. 1990. 5-Methylcytosine as an endogenous mutagen in the human LDL receptor and p53 genes. *Science* **249:** 1288–1290.

Rosl, F:, A. Arab, B. Klevenz, and H. zur Hausen. 1993. The effect of DNA methylation on gene regulation of human papillomaviruses. *J. Gen. Virol.* **74:** 791–801.

Ruprecht, R.M., T.W. Baba, and M.F. Greene. 1995. Attenuated vaccines for AIDS? *Lancet* **346:** 177–178.

Sager, R. and R. Kitchin. 1975. Selective silencing of eukaryotic DNA. *Science*. **189:** 426-433.

Saggioro, D., M. Panozzo, and L. Chieco-Bianchi. 1990. Human T-lymphotropic virus type I transcriptional regulation by methylation. *Cancer Res.* **50:** 4968–4973.

Schroeder, C., M. Reuter, and D.H. Kruger. 1984. DNA methylation of T3 virus ocr+ strains in *Escherichia coli* cells harbouring the EcoK DNA host specificity system. *Biomed. Biochim. Acta* **43:** K1–5.

Schwiezer, M., U. Fleps, A. Jackle, R. Renne, R. Turek, and D. Neumann-Haefelin. 1993. Simian foamy virus type 3 in latently infected vero cells: Reactivation by demethylation of proviral DNA. *Virology* **192:** 663–666.

Sharp, P.M. 1986. Molecular evolution of bacteriophages: Evidence of selection against the recognition sites of host restriction enzymes. *Mol. Biol. Evol.* **3:** 75–83.

Sharp, P.M., M.S. Rogers, and D.J. McConnel. 1985. Selection pressures on codon usage in the complete genome of bacteriophage T7. *J. Mol. Evol.* **21:** 150–160.

Shpaer, E.G. and J.I. Mullins. 1990. Selection against CpG dinucleotides in lentiviral genes: A possible role of methylation in regulation of viral expression. *Nucleic Acids Res.* **18:** 5793–5797.

Singh, M.K. and C.D. Pauza. 1992. Extrachromosomal human immunodeficiency virus type 1 sequences are methylated in latently infected U937 cells. *Virology* **188:** 451–458.

Stephanopoulos, D.E., J.C. Kappes, and D.I. Bernstein. 1988. Enhanced in vitro reactivation of herpes simplex virus type 2 from latently infected guinea-pig neural tissues by 5-azacytidine. *J. Gen. Virol.* **69:** 1079–1083.

Wei, X., S.K. Ghosh, M.E. Taylor, V.A. Johnson, E.A. Emini, P. Deutsch, J.D. Lifson, S. Bonhoeffer, M.A. Nowak, and B.H. Hahn. 1995. Viral dynamics in human immunodeficiency virus type 1 infection. *Nature* **373:** 117–122.

Epigenetic Regulation of the Maize *Spm* Transposable Element

Nina Fedoroff
Biotechnology Institute
The Pennsylvania State University
University Park, Pennsylvania 16802

Maize transposons, the mobile genes discovered by Barbara McClintock half a century ago, were the first plant genes known to be epigenetically silenced. Early in the study of the *Suppressor-mutator* (*Spm*) transposable element, McClintock recognized that the transposable element could be silenced heritably but not irreversibly, an odd and important observation (McClintock 1957, 1958). A silent *Spm* transposon showed none of its usual genetic activities. It could neither itself move or cause another transposable element to move, nor could it change the expression of genes with *Spm* insertions, as could an active transposable element (McClintock 1956). However, a silent transposable element was different from a mutant transposable element because it could return to a genetically active form, reacquiring all of its activities simultaneously. McClintock explored the ability of the *Spm* element to alternate between active and silent phases in considerable genetic detail over the subsequent decade, noting that an unrelated transposon, the *Activator* (*Ac*) element, could be similarly silenced (McClintock 1959, 1961, 1962, 1964, 1965a,b, 1971). McClintock's most striking deduction was that an active *Spm* could reactivate an inactive one (McClintock 1971).

We now know that both transcription and transposition of the *Spm* element are subject to negative epigenetic control by a molecular mechanism that involves methylation of transposable element regulatory sequences (Fedoroff 1994). The transposable element itself encodes an autoregulatory protein that promotes the reactivation of an epigenetically silenced transposable element. I review the early genetic clues in McClintock's work pointing to the existence of both epigenetic and autoregulatory control mechanisms, then summarize our present understanding of the underlying molecular mechanisms. The *Spm* element's epigenetic regulatory system is a paradigm uniquely accessible to both genetic and molecular analysis because it controls genes that are not essential for survival, but it can easily be anticipated from what is known about plant development and physiological adaptations to changing en-

Epigenetic Mechanisms of Gene Regulation

vironmental conditions that similar mechanisms will surface in the integral regulatory circuitry of plants.

THE GENETIC DISCOVERY OF EPIGENETIC SILENCING

Spm is the transposable element that McClintock studied in the greatest genetic detail, perhaps because of the fascinating complexity of the genetic "crosstalk" between the transposable elements of this family. The results of her meticulous genetic analyses provided many clues to the *Spm* element's functions. McClintock clearly recognized the regulatory implications of her observations, although her interpretation that the transposons themselves are displaced regulatory elements appears less likely than that her mutants revealed the inner workings of the transposable element's own regulatory system.

Decades ago, McClintock observed that an *Spm* element could change or "mutate" to one of a number of different forms in which it either was silent or showed a pattern of alternation between active and inactive phases during development (McClintock 1957, 1958, 1959, 1961, 1962). Intrigued by the observation that any particular developmental pattern of transposable element expression was heritable, McClintock devoted considerable effort to understanding the genetic basis of the patterning phenomenon. In the course of her studies, McClintock recognized different developmental patterns of transposable element expression and understood that a given pattern is both heritable and yet somewhat labile, capable of changing to still different heritable expression patterns. Moreover, she observed that transposable elements could remain silent for several plant generations, only occasionally returning to an active form.

Even in the results of some of McClintock's earliest studies leading to the discovery of transposition there is a hint that transposons are normally in genomes, but not active (McClintock 1946). McClintock later successfully tested the hypothesis that previously undetectable transposable elements could be converted to an active form under circumstances of extensive chromosome breakage (McClintock 1950a,b, 1951). All of these early observations pointed to the existence of a mechanism by which the transposable elements could be heritably maintained in a silent, yet genetically intact form, precisely the definition of an epigenetic mechanism. Because the differential expression patterns are heritable, but at the same time change at a relatively high frequency to different types of heritable patterns, these too appear to have an epigenetic basis. The multiplicity of developmental patterns of transposon expression that McClintock's studies revealed suggested either that there were many regulatory systems acting on the transposable element or that it was controlled by a single system of remarkable versatility.

GENETIC EVIDENCE FOR AN EPIGENETIC REGULATORY CIRCUIT

Initial Evidence of Regulatory Interactions between Transposable Elements

Although seemingly unrelated to transposable element regulation at the time, McClintock's extensive studies of what she referred to as the "states" of a locus (which we now call a gene) provide insight into other facets of *Spm* regulation (McClintock 1965b). In retrospect, these studies constitute some of the earliest genetic data on interactions between a regulatory protein and its target DNA sequence (McClintock 1955). Like other transposable element families, the *Spm* family comprises fully functional autonomous transposable elements and a host of moderately to severely disabled relatives. The nonautonomous family members, designated *defective Spm* (*dSpm*) elements, are dependent on the autonomous transposable elements for mobility. McClintock found that *Spm* could both promote excision of a *dSpm* element from a mutant gene and change its phenotypic expression. Her studies focused on *dSpm* insertions into the maize *a* and *a2* loci, both of which encode enzymes in the biosynthetic pathway leading to the production of the plant's purple anthocyanin pigments. McClintock's *dSpm* insertion alleles fell into two different categories. In one type of allele, later designated *Spm-suppressible*, the transposon-disrupted gene continued to be expressed, but only in the absence of the autonomous *Spm* element. When *Spm* was present in the same plant, this type of transposon-disrupted gene was not expressed. In the other type of allele, later designated *Spm-dependent*, the gene with the transposon insertion was generally expressed only in the presence of a fully functional *Spm* (McClintock 1965b; Masson et al. 1987).

Although each allele exhibits its own unique response pattern to the *trans*-activating *Spm*, the overall implication is simple and clear: *Spm* codes for *trans*-acting proteins that bind to the transposon sequence inserted into or near the gene. Binding of these proteins either increases or decreases expression of the transposon-disrupted gene. In some cells, these proteins, probably acting together with other cellular proteins, excise the transposon from the gene and move it to a different location.

Some of the *dSpm* insertion alleles behave like contemporary promoter-reporter gene fusions, showing how the transposon's own regulatory circuit works. Some of the *Spm-dependent* alleles of the *a* gene had an odd and quite confusing property which McClintock called "presetting." The *a* gene in these cases is expressed when *Spm* is present, so they belong in the *Spm-dependent* category. However, this type of *dSpm*-disrupted gene is usually not expressed when a complete element is not present in the same plant. Yet McClintock noticed that some of these alleles can be "preset" to be expressed immediately after the *trans*-activating *Spm* element is removed by transposition or genetic segregation (McClintock 1963, 1964, 1965a, 1967). The preset gene is expressed in kernels receiving the *dSpm*-disrupted gene, but not the autonomous *Spm* that preset it. The preset pattern is almost never heritable: When a kernel showing the preset expression pattern is grown, the progeny kernels the plant produces do not show the preset pattern (rare exceptional kernels exhibit the preset pattern in subsequent generations), but presetting is quite reproducible. If the *dSpm*-disrupted gene is brought together with a fully functional *Spm* element by a genetic cross, once again the *dSpm*-disrupted gene will be preset to be expressed in the progeny kernels of this plant which receive the *dSpm*-disrupted gene, but not the *Spm* that preset it.

Genetic Evidence for an Epigenetic Regulator

The ability of a gene with an *Spm* insertion to be preset eventually provided the key to understanding regulation of *Spm*, although it was not at all clear at the time. The first and most obvious implication of presetting is that the mutant gene is not inactivated by the insertion itself, but by some type of heritable, yet reversible, hence epigenetic, modification of the insertion. Following this line of reasoning, the autonomous *Spm* must encode a regulatory macromolecule that interacts with the inserted *dSpm* element to activate gene expression while it is present. However, it must alter the epigenetic modification of the transposable element's regulatory sequence at the gene, albeit transiently, to permit expression of the gene after the source of the regulatory gene product is gone. Recalling the observations reviewed earlier that *Spm* elements themselves are subject to epigenetic inactivation, this interpretation leads to the prediction that epigenetically silenced transposable elements might respond to active transposable elements as do *Spm-dependent* alleles caused by insertion of *dSpm* elements. Indeed, McClintock's sophisticated genetic monitoring of the interactions between active and inactive transposable elements provides the confirming evidence. An active transposable element transiently reactivates an inactive one, so long as both are together in the same plant (McClintock 1971).

Is there also evidence that transposable elements preset each other? McClintock reported that the active and inactive transposable elements subsequently segregate from each other unchanged (McClintock 1971), but a careful reexamination of the question revealed that an active *Spm* exerts both a transient *and* a heritable effect on a silent *Spm* (Fedoroff 1989b). Even a deeply silent *Spm*, which we have called a *cryptic Spm*, can eventually be fully reactivated by an active *Spm*. The cryptic form is extremely stable in the genetic sense. That is, it is inherited in the silent form generation after generation. A *cryptic Spm* has never been reported to reactivate spontaneously, but there are physical and chemical agents, such as irradiation, as well as genetic conditions that promote element reactivation. Among these is exposure to an active *Spm*, suggesting that the *Spm* codes for a protein or other kind of gene product that promotes reactivation of the *cryptic Spm*. The transition from the cryptic form to a heritably active form is a curiously slow one. Even in the presence of an active *Spm* and with selection for active progeny, the transition takes several plant generations (Fedoroff 1989b). Prior to full activation, the silent *Spm* element exists in an intermediate form in which it is expressed only in the presence of another, fully active *Spm*. Such an *Spm*, although still silent on its own, can be preset by the active element. It then remains active in the immediate progeny kernels that received it, but when the kernel showing the preset pattern is grown, the *Spm* is once again inherited in a silent form (Fedoroff 1989b). The types of heritable developmental patterns of *Spm* expression described by McClintock emerge somewhat later in the reactivation process (McClintock 1957).

Taken together, these genetic observations suggested that the *Spm* element is regulated by a novel epigenetic mechanism. This epigenetic mechanism can maintain the *cryptic Spm* in a silent form generation after generation, but it also appears to be capable of establishing and maintaining different heritable patterns in which the *Spm* element shows a regular "program" of expression during development (Fedoroff and Banks 1988). The ability of an active *Spm* to influence the "epigenetic" state of an inactive or cryptic *Spm* has the interesting implication that *Spm* codes for a regulatory protein which can activate an epigenetically inactive *Spm* both transiently and heritably. Molecular genetic tests of the hypothesis became possible after the transposable element was cloned and as its genetic components were first dissected and then reassembled in model transgenic systems.

MOLECULAR BIOLOGY OF THE *Spm* TRANSPOSON

Structure of the *Spm*

As noted earlier, *Spm* elements comprise a "family" of related sequences. These include fully functional *Spm* elements and transposition-defective elements (*dSpm*). Most of the *dSpm* elements that have been analyzed are *Spm* elements with internal deletions. The maize genome contains many copies of sequences with some homology to *Spm*, few of which have been analyzed. It also contains a few copies of full-length elements, some of which are genetically silent, as described above, and a few of which are active. Fully functional active transposable elements of the *Spm* family are interchangeable genetically. It was reported by Peterson many years ago that McClintock's *Spm* element was genetically identical to the *Enhancer* (*En*) element that he had studied (Peterson 1965). We now know that the DNA sequences of *Spm* and *En* are also almost identical (Pereira et al. 1986; Masson et al. 1987). Each is about 8.4 kb in length and superficially simple in organization (Fig. 1a). Only a single

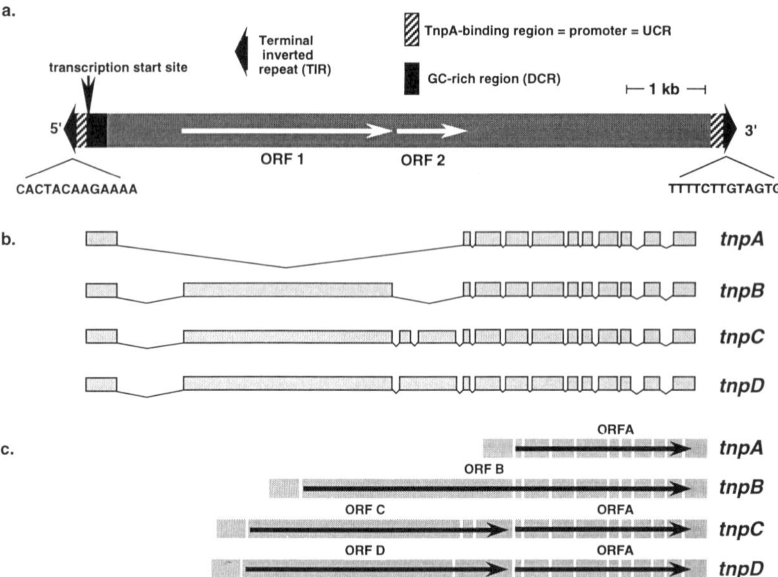

Figure 1 Diagrammatic representation of the *Spm* element's structure (*a*), transcripts (*b*), and ORFs (*c*). Boxes represent exons. Arrows show the extent of each ORF. ORF C and ORF D contain sequences derived from ORF 1 and ORF 2, located in the first intron on the *tnpA* transcript. (TIR) Terminal inverted repeat; (UCR) upstream control region; (DCR) downstream control region; (ORF) open reading frame.

transcription unit has been identified so far, and it occupies most of the transposable element's length. The few hundred base pairs outside the transcription unit consist of sequences essential for both transcription and transposition of the transposable element. The very ends comprise inverted repeats of a short sequence, CACTACAAGAAAA. The presence of short terminal inverted repeats (or TIRs) is characteristic of one category of transposable element, and the TIRs demarcate the segment of DNA that transposes. Between the *Spm*'s TIRs and its transcription unit at both transposable element ends, there are multiple similar or identical copies of a short sequence that is a binding site for one of the transposable element-encoded proteins (Gierl et al. 1988). This is designated the TnpA-binding region in Figure 1. At the left end of the element is the transcription start site, and this is referred to as the 5′ end of the *Spm* element. The TnpA-binding region at the 5′ end of *Spm* contains the promoter and has also been called the upstream control region or UCR (Fig. 2).

Expression of *Spm*

Spm's seemingly simple sequence organization belies considerable genetic complexity. The *Spm*'s primary transcript is alternatively spliced, yielding mRNA sequences with different open reading frames (ORFs), some with extensive overlaps (Fig. 1b). Four different transcripts are produced by different patterns of splicing from the primary transcript (Fig.

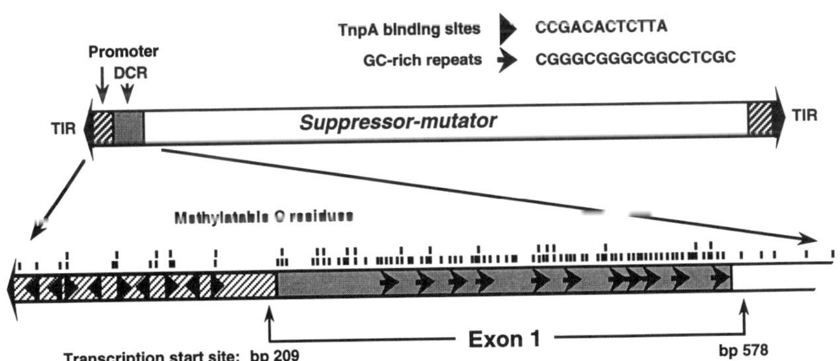

Figure 2 The structure of the *Spm* element's 5′ end. The vertical marks above the lower diagram represent the C residues in the methylatable CG dinucleotides and CNG trinucleotides on one strand. The marks are represented in two tiers in cases where methylatable C residues are located sufficiently close together to preclude resolution at this scale.

1c). These have been designated *tnpA*, *tnpB*, *tnpC*, and *tnpD* in order of increasing size (Masson et al. 1989). The *tnpA* transcript is the shortest and the first one identified (Pereira et al. 1986). It is by far the transposable element's most abundant transcript, being 50–100 times more abundant than any of the other transcripts (Masson et al. 1989). The spliced transcripts and their ORFs are shown in parts b and c of Figure 1. Two of the transcripts contain a single ORF, and the others contain two ORFs. Each transcript has one ORF that is unique to it, and it is this coding sequence that has been given the transcript's name. Thus, the potential proteins encoded by these transcripts are called TnpA, TnpB, TnpC, and TnpD.

The *Spm* Promoter

Studies to identify the *Spm* element's promoter sequence were carried out using an assay in which sequences from the *Spm*'s 5′ end were fused to a firefly luciferase gene and introduced into cultured tobacco cells by bombarding them with tungsten particles coated with DNA (Raina et al. 1993). The *Spm*'s promoter is the 0.2-kb 5′-terminal sequence (initially designated the UCR) which is also the TnpA-binding region (Fig. 2). Promoter activity is lost gradually as increasing numbers of TnpA-binding sites are deleted, but some promoter activity remains until the last TnpA-binding site is removed. The promoter does not have a conventional TATA box. Addition of the GC-rich sequence encoding the transposable element's first untranslated exon (called the downstream control region or DCR) makes no difference to the ability of the promoter to be expressed in the short-term transient assay (Fig. 2). However, when the DCR is included, the promoter is insensitive to an enhancer sequence placed upstream of it (Raina et al. 1993). This suggests that the DCR acts as a special kind of insulator sequence, protecting the transposon's weak promoter from external enhancers into or near which it might insert.

Transposable Element-encoded Proteins

The TnpA and TnpD proteins are both required to mobilize a *dSpm* element in transgenic tobacco plants (Masson and Fedoroff 1989; Frey et al. 1990; Masson et al. 1991). To date, the other proteins (TnpB and TnpC) potentially encoded by the transposable element have been found neither to help nor to hinder transposition (M. Schläppi and N.V. Fedoroff, unpubl.). Because TnpA does not bind to the transposable element's TIRs, but to its subterminal binding sites, it has been postulated that TnpD is the transposable element-encoded protein responsible for correct cleavage of the transposon ends (Masson et al. 1991). The results of recent experiments reveal that TnpD does not bind *Spm* directly, but rather interacts with TnpA bound to DNA containing TnpA-binding sites (R. Raina et al., unpubl.). It has also been postulated that TnpA serves to bring the ends of the transposable element together for transposition (Gierl et al. 1988; Masson et al. 1991), a role for which there is some direct evidence (R. Raina and N. Fedoroff, unpubl.). Much remains to be learned about the manner in which TnpA and TnpD interact with each other and with the *Spm* element to promote transposition.

TnpA Is a DNA-binding Protein

TnpA binds to the subterminal repetitive regions of *Spm*, but not to its TIRs (Gierl et al. 1988). The target binding sequence has been identified as the 12-bp consensus subterminal repetitive sequence. TnpA was first reported to bind as a dimer, optimally to a tail-to-tail dimer of the 12-bp binding site (Trentmann et al. 1993). Regions of the protein required for DNA binding and protein dimerization were defined in these studies through the use of deletions. More recently, it has been found that TnpA binds equally well to monomeric and multimeric binding sites, but that the ability of the protein to mediate the formation of higher-order DNA/protein complexes through protein-protein interactions increases with the number of binding sites in the target DNA (R. Raina and N. Fedoroff, unpubl.). In assays using synthetic oligonucleotides, binding is reduced if the cytosine residues of the CG dinucleotides and CNG trinucleotides within the binding site are methylated (Gierl et al. 1988).

MOLECULAR BASIS OF EPIGENETIC SILENCING OF *Spm*

Spm Methylation in Maize

Studies on the molecular differences between genetically active and inactive *Spm* elements began with the realization that the transposable element contains an extremely GC-rich sequence. The sequence just downstream from the transcription start site of the transposable element is 80% G+C (Fig. 1) and contains many sequences in which the C residues can be modified by methylation (Pereira et al. 1986; Masson et al. 1987). These include CG dinucleotides and the symmetrical CNG trinucleotides. DNA from plants with either an active or an inactive transposable element was cut with restriction enzymes whose ability to cut DNA is inhibited by methylation of C residues in their cleavage sites. It was found that the *Spm* element was methylated throughout most of its length, whether or not it was in the active form (Banks et al. 1988). However, when the *Spm* was active, a short region of its 5′ end was demethylated. This region included the 0.2-kb promoter and the 0.35-kb DCR just downstream from the promoter (Fig. 2). When the *Spm* was inactivated genetically, the promoter was methylated first. As the *Spm* element became more stably (heritably) inactive, methylation of the DCR increased progressively. Conversely, the conversion of a *cryptic Spm* to an active one was associated with decreasing methylation of the promoter and DCR. This implies either that methylation is causal to transposable element inactivation or that it is a direct reflection of it (Banks et al. 1988; Banks and Fedoroff 1989).

Spm Is Inactivated and Methylated in Transgenic Tobacco

Achieving a deeper understanding of *Spm* inactivation and reactivation necessitated the ability to reproduce the associated genetic phenomena in a transgenic system lacking the many background *Spm* elements present in maize. We chose tobacco plants for this purpose because they do not contain sequences resembling *Spm*. We found that the *Spm* element is inactivated reversibly in tobacco plants as it is in maize (Schläppi et al. 1994). Again as in maize, inactivation is accompanied by methylation of the *Spm*'s 5′ end (Schläppi et al. 1994). Once epigenetic silencing (and the associated methylation) had been observed in transgenic plants lacking homologous transposable elements, it was possible to ask whether a transposable element-encoded protein could reactivate a silent, methylated *Spm* element. This was done by introducing cDNAs for each of the four identified *Spm* ORFs, expressed from a strong promoter, into transgenic plants with an inactive transposable element. The results of such experiments revealed that TnpA, the most abundant transposable element-encoded protein, could not only reactivate the silent *Spm* element, but that its presence also led to demethylation of its 5′-terminal UCR and DCR sequences (Schläppi et al. 1993). None of the other proteins had any effect on the genetic activity of the transposable element. Thus, TnpA was identified as the transposable element's positive autoregulatory protein.

Presetting Revisited

It was observed during analysis of the progeny of TnpA-containing transgenic tobacco plants that the reactivated *Spm* element commonly segregates in its original inactive form, just as McClintock had observed in maize (McClintock 1971; Schläppi et al. 1993). Yet in some plants, the *Spm* element had been preset for continued expression following its exposure to TnpA protein, and this was reflected in reduced methylation of its 5′ end. Similar presetting of reactivating *Spm* elements by fully active ones had been observed in earlier maize experiments (Fedoroff 1989b). These observations resemble McClintock's presetting phenomenon, observed with *dSpm* insertions (McClintock 1963, 1964, 1965a, 1967). Thus, TnpA may interfere with methylation of the *Spm* element's promoter while it is present. The demethylated state may persist through meiosis in some cases, giving rise to progeny that still have an active transposable element. Remethylation of the element's promoter occurs during development of the subsequent generation, returning the element to an inactive form.

The *Spm*'s Promoter and DCR Sequence Suffice for De Novo Methylation

To identify the minimal sequence required for de novo methylation and silencing of the adjacent transcription unit, sequences from the transposable element's 5′ end were fused to a firefly luciferase gene and stably transformed into tobacco cells (Schläppi et al. 1994). The transposable element's promoter is rapidly inactivated and methylated in transformed plant cells, but only if the DCR is present. If the transposable element's promoter lacks the DCR, the luciferase gene continues to be expressed. In earlier studies, it had been found that although the DCR sequence has no effect on the adjacent *Spm* promoter, the presence of the DCR makes the promoter insensitive to an upstream enhancer (Raina et al. 1993). It had also been observed that the DCR can effectively silence a heterologous promoter (Raina et al. 1993). Thus, the DCR sequence has emerged as a central component of the *Spm* element's regulatory mechanism. Not only does the DCR insulate the transposable element from external promoters and enhancers, but it also appears to be required for silencing and methylation of the transposable element's promoter.

REGULATORY ACTIVITIES OF TnpA

TnpA Represses an Unmethylated *Spm* Promoter

Early genetic data reviewed above imply that *Spm* encodes both positive and negative regulatory gene products (McClintock 1965b). The effect of

TnpA on an unmethylated promoter was investigated using *Spm* promoter/luciferase reporter gene fusions both in short-term transient assays and in stably transformed tobacco cells. TnpA represses the active, unmethylated *Spm* promoter by about 80% in both types of assays (Schläppi et al. 1994). It had previously been reported that when a dimeric TnpA-binding site was inserted between a promoter and a reporter gene, TnpA inhibited expression of the reporter gene (Grant et al. 1990). The latter observation suggests that the binding of TnpA to the *Spm* promoter sequences is sufficiently strong to interfere with transcription from the promoter of a gene into which a *dSpm* element has inserted, providing a possible explanation for McClintock's initial observation that certain *dSpm* insertion mutations have a null phenotype when an autonomous *Spm* element is present in the same genome (McClintock 1953, 1954). Deletion of TnpA coding sequences other than those required for DNA binding and dimerization has no effect on its repressor activity, suggesting that TnpA represses transcription by competing with transcription factors for binding sites on the *Spm* promoter (Trentmann et al. 1993; Schläppi et al. 1994).

TnpA Activates a Methylated *Spm* Promoter

In sharp contrast to its effect on an unmethylated *Spm* promoter, TnpA activates a methylated, inactive *Spm* promoter (Trentmann et al. 1993; Schläppi et al. 1994). Moreover, promoter reactivation is invariably accompanied by a decrease in promoter methylation. This apparent paradox might be attributable to differences in the binding of TnpA to methylated and unmethylated DNAs. That is, TnpA might activate transcription if it is bound weakly, as it appears to bind to methylated binding-site-containing oligonucleotides or *Spm* promoter fragments (Gierl et al. 1988; R. Raina and N. Fedoroff, unpubl.), but inhibit transcription when it is bound more tightly to the many binding sites in the unmethylated promoter. This hypothesis was tested by substituting a different DNA-binding domain, that of the yeast Gal4 protein, for that of TnpA and investigating the ability of the hybrid protein to activate a luciferase reporter gene whose promoter contains the cognate Gal4-binding sites (Schläppi and Fedoroff 1995). The results of these and other domain-swapping experiments have shown that TnpA exhibits only repressor activity. This conclusion is underscored by the observation that TnpA can be converted to a strong transcriptional activator of the *Spm* promoter by addition of a heterologous viral activation domain to its DNA-binding and dimerization domains. Its ability to activate the methylated *Spm* promoter therefore involves additional cellular proteins or affects the methylation of the *Spm* promoter by a mechanism other than direct binding and transcription activation.

EPIGENETIC REGULATION: SYNTHESIS OF THE GENETIC AND MOLECULAR HYPOTHESES

Epigenetic Silencing

A multistep negative epigenetic mechanism was postulated from genetic data to explain the range of inactive *Spm* forms, as well as those that exhibit differential patterns of transposable element expression during development (Masson et al. 1987; Fedoroff and Banks 1988). The results of subsequent molecular studies suggest that negative epigenetic regulation is attributable to methylation of the *Spm* element's promoter and GC-rich DCR sequence at multiple sites. Promoter methylation does not require transposable-element-encoded gene products and can occur in tobacco cells (Schläppi et al. 1993, 1994). Rapid methylation requires only the promoter and the adjacent DCR sequence. There is a close correlation between methylation and the heritability of the inactive form of the transposable element, as well as a similar correlation between the genetic activity of the transposable element during development and promoter methylation (Banks et al. 1988; Banks and Fedoroff 1989). The existence of more than 100 methylatable C residues on each strand within the first 0.6 kb of the transposable element's 5′ end suggests that differences in both the heritability of the inactive state and the diverse developmental patterns of transposable element expression may find their molecular explanation in the extent or pattern of methylation (Fig. 2).

TnpA, the Transposable Element-encoded Autoregulator

It was initially postulated from genetic data that *Spm* is regulated by an epigenetic mechanism that can either inactivate it very stably and heritably or "program" it to be expressed in a regular pattern during plant development. Moreover, the first evidence that *Spm* encodes a regulatory gene product that can activate a silent *Spm* element was genetic (Fedoroff 1983; Masson et al. 1987). The results of subsequent molecular studies revealed that methylation of the *Spm*'s promoter and downstream GC-rich DCR is the most likely molecular candidate to explain epigenetic silencing and programming. Moreover, TnpA has been identified as the epigenetic regulator. Although its mechanism of action is not yet understood, it is clearly able both to activate a methylated, inactive

Spm promoter and to promote its demethylation. Molecular experiments have further established that TnpA is also a *repressor* of the *Spm*'s promoter when it is not methylated. Although the mechanisms that regulate *Spm* transcription and transposition are not yet fully understood, it is becoming increasingly evident that they are intimately interrelated (Fig. 3). The transposable element's promoter sequence is coextensive with the 5′-terminal TnpA-binding region (Fig. 2). Thus, it is not surprising that binding of TnpA to the unmethylated promoter represses it (Raina et al. 1993). There are many binding sites for TnpA at both ends of the transposable element, and deletion of even a few at either end diminishes transposition frequency (Fedoroff 1989a). This genetic observation, together with the results of TnpA-binding studies, suggests that TnpA binding first represses transcription and then serves to bring the transposable element's termini together for transposition (Fig. 3).

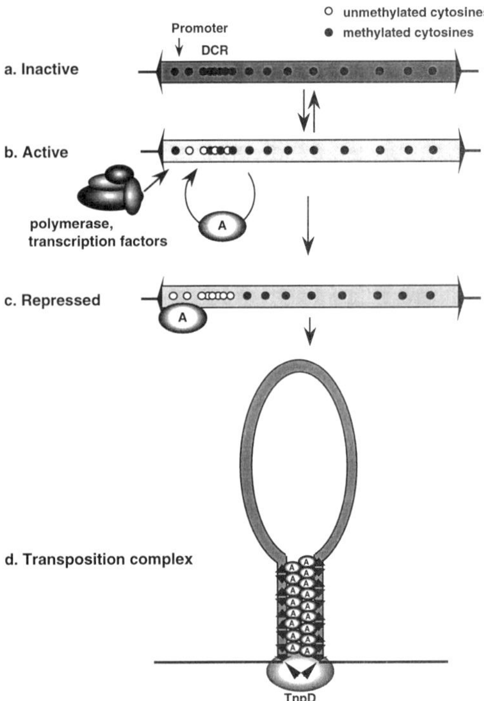

Figure 3 Transposition and regulation of *Spm*: an hypothesis. Filled and open circles represent methylated and unmethylated C residues in the element's sequence. Arrowheads within the oval representing the TnpD protein indicate the sites at which the transposon ends are cleaved during transposition.

An Hypothesis

The foregoing observations are consistent with the simple hypothesis that transcription from the *Spm* promoter leads to the accumulation of TnpA, with progressive occupation of TnpA-binding sites eventually inhibiting further transcription from the promoter and resulting in formation of a transposition complex containing both transposable element ends held together by DNA-bound TnpA dimers. Supporting this hypothesis is the recent observation that higher-order DNA/protein complexes form at lower TnpA concentrations as the number of TnpA-binding sites per target DNA molecule increases (R. Raina and N. Fedoroff, unpubl.). Consistent with this observation, the negative effect on transposition in vivo of both DCR and TnpA-binding-site deletions might be attributable to changes in the concentration of TnpA required to form a transposition complex (Fig. 3).

Methylation of the *Spm* promoter and DCR inhibits transposable element transcription and decreases the ability of the transposable element to transpose even when supplied with TnpA and TnpD *in trans* (Banks et al. 1988). An intriguing and poorly understood aspect of *Spm*'s epigenetic regulatory mechanism is the role of the DCR in promoting transposable element methylation and inactivation. The DCR facilitates methylation in the promoter region upstream of it. Conversely, TnpA facilitates demethylation of both the promoter and the downstream DCR sequence. These observations suggest that the DCR is an important element in the epigenetic inactivation and reactivation of the *Spm* element. Its influence may be mediated by its high GC content, its effect on DNA conformation, or its ability to interact with proteins.

Although we know that TnpA can reactivate an inactive, methylated *Spm* promoter and facilitate its demethylation, we know little about the underlying molecular mechanism. The observations that TnpA lacks a conventional transcriptional activation domain and promotes demethylation of sequences at considerable distance from its binding sites rule out the simplest type of explanation in which demethylation and activation are a consequence of competition between a transcription factor, TnpA, and a methylase for binding to the same or overlapping sites. The mechanism by which TnpA activates the methylated *Spm* promoter is of interest both because of its novelty and because of the possibility that similar mechanisms are used in plant development and differentiation.

SUMMARY

Expression and transposition of the *Spm* transposon of maize are controlled by interacting epigenetic and autoregulatory mechanisms. Methylation of critical transposable element sequences prevents both transcription and transposition in maize, heritably inactivating the *Spm* element. The promoter, comprising the terminal 0.2 kb of the transposable element, and an 0.35-kb highly GC-rich downstream sequence are the methylation target sequences. The transposable element encodes two proteins necessary for transposition, TnpA and TnpD. There are multiple TnpA-binding sites both in the 5′-terminal promoter region and at the transposable element's 3′ end. In addition to its role in transposition, TnpA is both a positive and a negative regulator of transcription. TnpA represses the transposable element's promoter when it is not methylated. When the transposable element is inactive and its promoter is methylated, TnpA activates the methylated promoter and facilitates both its transient and heritable demethylation.

ACKNOWLEDGMENT

This work was supported by National Institutes of Health grant GM-34296 to N.F.

REFERENCES

Banks, J.A. and N. Fedoroff. 1989. Patterns of developmental and heritable change in methylation of the *Suppressor-mutator* transposable element. *Dev. Genet.* **10:** 425–437.

Banks, J.A., P. Masson, and N. Fedoroff. 1988. Molecular mechanisms in the developmental regulation of the maize *Suppressor-mutator* transposable element. *Genes Dev.* **2:** 1364–1380.

Fedoroff, N.V. 1983. Controlling elements in maize. In *Mobile genetic elements* (ed. J. Shapiro), pp. 1–63. Academic Press, New York.

———. 1989a. Maize transposable elements. In *Mobile DNA* (ed. M. Howe and D. Berg), pp. 375–411. American Society for Microbiology, Washington, D.C.

———. 1989b. The heritable activation of *cryptic Suppressor-mutator* elements by an active element. *Genetics* **121:** 591–608.

———. 1994. DNA methylation and activity of the maize *Spm* transposable element. In *Gene silencing in higher plants and related phenomena in other eukaryotes* (ed. P. Meyer), pp. 143–164. Springer-Verlag, New York.

Fedoroff, N.V. and J.A. Banks. 1988. Is the *Suppressor-mutator* element controlled by a basic developmental regulatory mechanism? *Genetics* **120:** 559–577.

Frey, M., J. Reinecke, S. Grant, H. Saedler, and A. Gierl. 1990. Excision of the *En/Spm* transposable element of *Zea mays* requires two element-encoded proteins. *EMBO J.* **9:** 4037–4044.

Gierl, A., S. Lutticke, and H. Saedler. 1988. TnpA product encoded by the transposable element *En-1* of *Zea mays* is a DNA binding protein. *EMBO J.* **7:** 4045–4053.

Grant, S.R., A. Gierl, and H. Saedler. 1990. *En/Spm* encoded tnpA protein requires a specific target sequence for suppression. *EMBO J.* **9:** 2029–2035.

Masson, P. and N.V. Fedoroff. 1989. Mobility of the maize *Suppressor-mutator* element in transgenic tobacco cells. *Proc. Natl. Acad. Sci.* **86:** 2219–2223.

Masson, P., M. Strem, and N. Fedoroff. 1991. The tnpA and tnpD gene products of the *Spm* element are required for transposition in tobacco. *Plant Cell* **3:** 73–85.

Masson, P., G. Rutherford, J.A. Banks, and N. Fedoroff. 1989. Essential large transcripts of the maize *Spm* element are generated by alternative splicing. *Cell* **58:** 755–765.

Masson, P., R. Surosky, J. Kingsbury, and N.V. Fedoroff. 1987. Genetic and molecular analysis of the *Spm-dependent a-m2* alleles of the maize *a* locus. *Genetics* **117:** 117–137.

McClintock, B. 1946. Maize genetics. *Carnegie Inst. Wash. Year Book* **45:** 176–186.

———. 1950a. Mutable loci in maize. *Carnegie Inst. Wash. Year Book* **49:** 157–167.

———. 1950b. The origin and behavior of mutable loci in maize. *Proc. Natl. Acad. Sci.* **36:** 344–355.

———. 1951. Mutable loci in maize. *Carnegie Inst. Wash. Year Book* **50:** 174–181.

———. 1953. Mutation in maize. *Carnegie Inst. Wash. Year Book* **52:** 227–237.

———. 1954. Mutations in maize and chromosomal aberrations in *Neurospora*. *Carnegie Inst. Wash. Year Book* **53:** 254–260.

———. 1955. Controlled mutation in maize. *Carnegie Inst. Wash. Year Book* **54:** 245–255.

———. 1956. Controlling elements and the gene. *Cold Spring Harbor Symp. Quant. Biol.* **21:** 197–216.

———. 1957. Genetic and cytological studies of maize. *Carnegie Inst. Wash. Year Book* **56:** 393–401.

———. 1958. The suppressor-mutator system of control of gene action in maize. *Carnegie Inst. Wash. Year Book* **57:** 415–429.

———. 1959. Genetic and cytological studies of maize. *Carnegie Inst. Wash. Year Book* **58:** 452–456.

———. 1961. Further studies of the suppressor-mutator system of control of gene action in maize. *Carnegie Inst. Wash. Year Book* **60:** 469–476.

———. 1962. Topographical relations between elements of control systems in maize. *Carnegie Inst. Wash. Year Book* **61:** 448–461.

———. 1963. Further studies of gene-control systems in maize. *Carnegie Inst. Wash. Year Book* **62:** 486–493.

———. 1964. Aspects of gene regulation in maize. *Carnegie Inst. Wash. Year Book* **63:** 592–602.

———. 1965a. Components of action of the regulators *Spm* and *Ac*. *Carnegie Inst. Wash. Year Book* **64:** 527–536.

———. 1965b. The control of gene action in maize. *Brookhaven Symp. Biol.* **18:** 162–184.

———. 1967. Regulation of pattern of gene expression by controlling elements in maize. *Carnegie Inst. Wash. Year Book* **65:** 568–578.

———. 1971. The contribution of one component of a control system to versatility of gene expression. *Carnegie Inst. Wash. Year Book* **70:** 5–17.

Pereira, A., H. Cuypers, A. Gierl, Z. Schwarz-Sommer, and H. Saedler. 1986. Molecular analysis of the *En/Spm* transposable element system of *Zea mays*. *EMBO J.* **5:** 835–841.

Peterson, P.A. 1965. A relationship between the *Spm* and *En* control systems in maize. *Am. Nat.* **99:** 391–398.

Raina, R., D. Cook, and N. Fedoroff. 1993. Maize *Spm* transposable element has an enhancer-insensitive promoter. *Proc. Natl. Acad. Sci.* **90:** 6355–6359.

Schläppi, M. and N. Fedoroff. 1995. Analysis of the *Spm*-encoded TnpA and TnpD proteins using single- and double-hybrid protein fusions in a plant transient assay. *Maize Genet. Coop. Newsl.* **69:** 15–17.

Schläppi, M., R. Raina, and N. Fedoroff. 1994. Epigenetic regulation of the maize Spm transposable element: Novel activation of a methylated promoter by TnpA. *Cell* **77:** 427–437.

Schläppi, M., D. Smith, and N. Fedoroff. 1993. TnpA *trans*-activates methylated maize *Suppressor-mutator* transposable elements in transgenic tobacco. *Genetics* **133:** 1009–1021.

Trentmann, S.M., H. Saedler, and A. Gierl. 1993. The transposable element *En/Spm*-encoded TNPA protein contains a DNA binding and a dimerization domain. *Mol. Gen. Genet.* **238:** 201–208.

Epigenetic Silencing of *Mu* Transposable Elements in Maize

Robert A. Martienssen
Cold Spring Harbor Laboratory
Cold Spring Harbor, New York 11724

Transposable genetic elements interact with specific DNA-binding proteins to facilitate their excision from the chromosome and their integration elsewhere. Autonomous transposons carry genes that encode some of these proteins, and so provide the *trans*-acting functions, such as transposase, required for their own transposition and that of related nonautonomous elements. Transposons are frequently subject to epigenetic regulation and share many of the properties associated with chromosomal genes that are epigenetically regulated: They are found in multiple copies; they are integrated at diverse sites and are often subject to position effects; they sometimes encode multivalent DNA-binding proteins that can alter chromatin structure; and they are often subject to DNA methylation. It is possible that some epigenetic mechanisms of gene regulation evolved as a means of controlling transposons and viruses that were otherwise capable of widespread integration throughout the genome (for review, see Martienssen and Richards 1995).

Epigenetic regulation of transposons can be monitored in vivo because the expression of chromosomal genes disrupted by transposon insertion is influenced by transposase in two different ways. (1) Transposon excision depends on transposase and often restores wild-type chromosomal gene expression. (2) Transposons inserted in regulatory regions of some genes interact with transposase, resulting in alterations in chromosomal gene expression without transposon excision (McClintock 1965b; Peterson 1966; Rutledge et al. 1988; Boeke 1989; Robertson and Engels 1989; Gierl 1990; Corces and Geyer 1991). These regulatory interactions are particularly convenient assays for transposon activity, because they affect gene expression in a uniform fashion without relying on occasional transposon excision.

In maize, one of the best-studied transposable element gene control systems is the *Suppressor-mutator* (*Spm*)/*Enhancer* (*En*) system (see Fedoroff, this volume). *Spm* elements are 8-kb transposons that encode at least two proteins, TnpA and TnpD. Both proteins are required for efficient transposition of full-length *Spm* and deletion-derivative *dSpm* elements. Mutations caused by *Spm* insertions can be used to monitor autonomous *Spm/En* activity (McClintock 1971; Peterson 1995), and the *cis*- and *trans*-acting components involved are well understood genetically and biochemically (Frey et al. 1990; Grant et al. 1990; Schlappi et al. 1993).

Robertson's *Mutator* (*Mu*) transposable elements in maize can also influence gene expression via interactions between *Mu* transposons and factors required for transposition (Martienssen et al. 1990; Chandler and Hardeman 1992; Bennetzen et al. 1993). Because of the widespread use of this transposable element system in gene tagging, many different chromosomal genes have been brought under the control of *Mu* transposons in maize, and this has revealed a diversity of epigenetic influences on *Mu* activity.

There are at least seven classes of *Mu* elements that all share the same 200-bp terminal inverted repeats (TIRs). One of these classes (*MuDR*) encodes the functions required for transposition of the other *Mu* elements (Chomet et al. 1991; Hershberger et al. 1991; Qin et al. 1991). Loss of *MuDR* is correlated with extensive cytosine methylation of *Mu1* elements and loss of transposon activity (Chandler and Walbot 1986; Bennetzen 1987). When *MuDR* is reintroduced by crosses, this methylation is abolished and transposon activity is usually restored (Martienssen et al. 1990; Brown and Sundaresan 1992; Lisch et al. 1995). However, permanent losses of *Mu* activity have been reported (Robertson 1986; Walbot 1986; Bennetzen 1987).

MuDR elements can be lost genetically by segregation or by spontaneous internal deletion (Chomet et al. 1991; Lisch et al. 1995), but loss of *MuDR* activity can also occur epigenetically during plant development (Bennetzen et al. 1988; Martienssen et al. 1990; Martienssen and Baron 1994). These changes in *Mu* activity can be monitored using mutations in genes that depend on *Mu* activity for their phenotypic effects. Examples of such mutations, the mechanisms by which they come under control of *Mu* activity, and the patterns of epigenetic variation that they reveal are reviewed below.

SUPPRESSIBLE MUTATIONS CAUSED BY ROBERTSON'S MUTATOR (*Mu*) ELEMENTS

hcf106-mum1 and Other Recessive Suppressible Mutations

The *hcf106* gene (*high chlorophyll fluorescence* 106) is required for chloroplast development in maize (Martienssen et al. 1989). In the *hcf106-mum1* allele, a nonautonomous *Mu1* element has inserted in the promoter region of the *hcf106* gene, at the third of three closely spaced transcript initiation sites (Barkan and Martienssen 1991). This results in failure to accumulate transcripts from the gene, and a pale-green seedling-lethal phenotype in homozygous *hcf106-mum1* plants. However, the lethal phenotype is only observed when the autonomous element *MuDR* is present elsewhere in the genome (in "*Mu*-on" plants). When *MuDR* is lost (in "*Mu*-off" plants), *hcf106* gene expression is restored, resulting in phenotypic suppression of the *hcf106-mum1* mutation (Martienssen et al. 1990). Transcripts in phenotypically suppressed plants initiate over a 40-bp region that spans the junction between the *Hcf106* gene and the *Mu1* element, suggesting that an outward reading promoter in the terminus of *Mu1* has been activated (Barkan and Martienssen 1991). Phenotypic suppression depends on the presence of the *Mu1* element in the promoter of the gene, as derivative insertion and deletion alleles of *hcf106* are not suppressed in this way (Das and Martienssen 1995).

When *MuDR* activity is lost, *Mu1* elements become extensively methylated (Chandler and Walbot 1986; Bennetzen 1987), and the same is true of the *Mu1* element in the *hcf106* gene (Martienssen et al. 1990). Cytosine methylation at the *hcf106* locus is not restricted to the *Mu1* transposon but extends into the promoter region up to 100 bp upstream of the *Mu1* element (Martienssen et al. 1990). A more detailed analysis of methylation-sensitive restriction sites is shown in Figure 1. *Pst*I (CTGCAG) and *Pvu*II (CAGCTG) sites located between and upstream of two putative TATA box sequences in the *hcf106* promoter are fully methylated in *hcf106-mum1* plants lacking *MuDR* activity, but are unmethylated in mutant plants that have *MuDR* activity. The same sites are unmethylated at the wild-type locus that has no *Mu1* element. A *Pvu*I site (CGATCG) 50 bp downstream from the *Mu1* element shows a similar pattern of *MuDR*-dependent loss of methylation, but in this case methylation is only partial in the absence of *MuDR* activity. When active *MuDR* is reintroduced, methylation at each of these sites is lost once more (not shown).

The presence of *Mu1* thus brings flanking DNA sequences under the control of *Mu*-dependent methylation. This suggests a possible mechanism for phenotypic suppression. Perhaps, in the presence of *MuDR*, the *hcf106* promoter is unmethylated and active, resulting in transcriptional interference between the gene's promoter and the promoter contained within the *Mu1* element. This prevents transcription at the locus. In the absence of *MuDR*, methylation of the wild-type promoter results in nega-

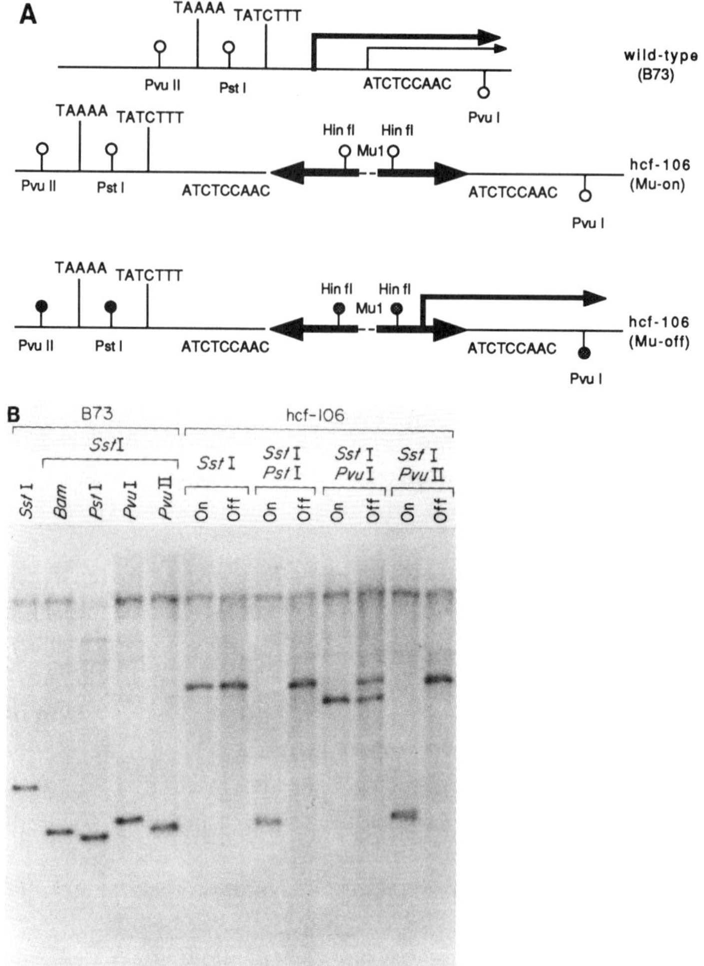

Figure 1 The promoter region of the *hcf106* gene from mutant, wild-type, and suppressed alleles. The mutant allele, *hcf106-mum1*, is shown along with the wild-type allele from maize inbred line B73. Transcription start sites are shown as arrows, and putative TATA boxes (TAAA and TATCTTT) are found upstream of each site. 9 bp of the *hcf106* gene are duplicated on either side of the *Mu1* transposon insertion (ATCTCCAAC). Modified restriction enzyme sites are filled circles, unmodified are open circles, and partially modified are mottled. (*B*) The Southern blot shows DNA from wild-type (B73), homozygous mutant *hcf106* (*Mu*-on), and homozygous suppressed *hcf106* (*Mu*-off) seedlings digested with restriction enzymes and hybridized to a probe from the upstream flanking region of the *hcf106* gene. The faintly hybridizing upper band is a reduced-stringency homolog unlinked to *hcf106*.

tive regulation, relieving interference from the promoter contained within the *Mu1* element. One argument against this model is that transcripts that accumulate from the suppressed allele are still regulated by light and tissue-specific factors, in the same way as the wild-type gene is regulated (Barkan and Martienssen 1991). This indicates that at least some *cis*-acting regulatory sequences must still be functional when the *hcf106* locus is extensively methylated, even if the promoter is not.

Since the discovery that *hcf106-mum1* was suppressed in the absence of *Mu* activity, numerous other *Mu*-suppressible mutants of maize have been identified, affecting both leaf and kernel phenotypes. The *A1* gene, for example, encodes an enzyme required for anthocyanin pigment biosynthesis. A 1.4-kb *Mu1* element is inserted at the *A1* locus in the *a1-mum2* allele (O'Reilly et al. 1985).

In the presence of *MuDR*, *a1-mum2* kernels and plants are colorless with sectors of pigmented tissue resulting from transposon excision. In the absence of *MuDR*, vegetative tissues (leaves, stems, and flowers) are fully pigmented: i.e., the mutation is suppressed (Chomet et al. 1991). In the kernel, however, the aleurone layer of the endosperm remains colorless in the absence of *MuDR*. The start site of transcription has been determined for the wild-type gene (Schwarz-Sommer et al. 1987), and it is about 100 bp downstream from the *Mu1* insertion site. This suggests that, as in the case of *hcf106*, the *Mu1* element has inserted into upstream regulatory sequences.

The *VP1* gene encodes a transcription factor required for the expression of pigment genes in the aleurone layer of developing kernels and for the expression of dormancy genes in the embryo. In the *vp1-mum3* allele, a 1.4-kb *Mu* element has inserted within 50 bp of the transcript start site (McCarty et al. 1989). *vp1-mum3* kernels have a tightly regulated pattern of spots in the aleurone resulting from excision of the *Mu* element in single aleurone cells (McCarty et al. 1989). However, kernels that have lost *Mu* activity are fully colored (Fig. 2B). This phenotype is stably inherited in subsequent generations but can revert to an unstable (spotted) phenotype when *Mu* activity is restored (R. Martienssen, unpubl.; C. Carson and D. McCarty, pers. comm.). These results show that the *vp1-mum3* allele is suppressible in the same way as *hcf106*. Interestingly, in the presence of *Mu* activity, a proportion of *vp1-mum3* kernels are variegated with a distinctive pattern (Fig. 2B). This variegation is discussed below. Recessive suppressible mutations comprise the largest class of mutants that depend on *Mu* activity and may account for as many as one in three mutations isolated from *Mu* lines (Martienssen and Baron 1994).

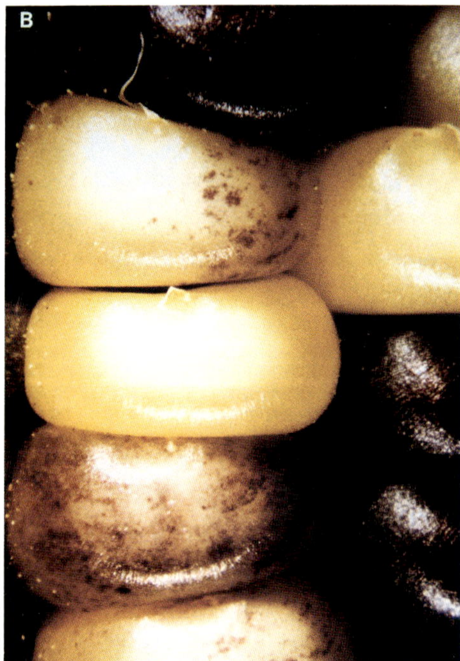

Figure 2 (See facing page for legend)

Dominant Suppressible Mutations

The dominant mutants of maize, *Lesion mimic 27* and *Lesion mimic 28*, depend on *Mu* activity for their phenotypic effects (Martienssen and Baron 1994; W. Kammerer et al., pers. comm.). These mutations mimic the hypersensitive response to many maize pathogens: temperature- and light-dependent chlorotic lesions form on the leaves of heterozygous and homozygous plants, even in the absence of detectable pathogen infestation (Neuffer and Calvert 1975). The lesion-mimic phenotype is only observed in plants that carry the autonomous *MuDR* element elsewhere in the genome. When this element is lost, *Les27* and *Les28* plants revert to wild type. The molecular basis for this genetic interaction is unknown.

Mutator insertions in the third intron of the maize homeobox gene *knotted1* result in dominant suppressible alterations in leaf cell fate (Greene et al. 1994). In at least two out of ten dominant alleles caused by insertion of *Mu1* or *Mu8* elements, the dominant knotted phenotype depends on *Mu* activity. Poor penetrance may complicate this phenomenon for some of the other alleles. Because these and other insertions all fall within a small region of the intron, Greene et al. (1994) proposed that these insertions interrupt a negative "silencer" element that represses gene activity in the leaves of wild-type plants. Perhaps this repression is disrupted by transposon insertion, resulting in ectopic derepression in the leaf and a dominant phenotype. Presumably, derepression depends on interaction with *MuDR*-encoded proteins. An alternative model is that *MuDR* actively promotes ectopic *knotted* expression in some way (Greene et al. 1994).

Figure 2 (*A*) Lesion-mimic and Knotted phenotypes are coordinately regulated in sectored double mutants. Two double-mutant plants carrying the dominant *Mu*-dependent *Lesion-mimic 28* and *Knotted1-mum2* mutations are shown. *Les28* causes the formation of spots of necrotic tissue (lesions) on the leaf, and *Kn1* causes the formation of ectopic pigmented structures (knots) on the leaf blade. The plant on the left is entirely mutant. The plant on the right is variegated for *Mu* activity and has two double-mutant sectors on an otherwise wild-type leaf. The phenotypes are coordinately regulated in these sectors. (*B*) The suppressible mutant *viviparous1-mum3*. Kernels segregating the suppressible mutant *vp1-mum3* are shown. *vp1* controls anthocyanin pigmentation in the aleurone layer. Solid-color kernels have lost *Mutator* activity (*Mu*-off) and spotted kernels have retained it (*Mu*-on). Variegated kernels have nonclonal patterns of pigmentation reflecting loss of *Mu* activity (4th kernel from the top). Two kernels are shown in which half the kernel is variegated and half is spotted (2nd and 5th kernels from the top).

Suppressible Revertants of Dominant Mutations

Ectopic expression of the related homeodomain proteins Knotted1, Roughsheath, Liguless3, and Liguless4 results in changes in the specification and timing of leaf cell fate (Freeling 1992; Hake 1992). Spontaneous dominant alleles have been isolated at each of these loci, and these behave as neomorphs because of this ectopic expression (Freeling 1992). Transposons have been used to revert these dominant mutations by insertion into regulatory regions required for ectopic expression (Lowe et al. 1992; Schneeburger et al. 1995; Fowler et al. 1996). In the case of *knotted1* and *liguless3*, many of these revertant alleles turned out to be suppressible. That is, the presence of *MuDR* elsewhere in the genome was required for the revertant, wild-type phenotype. In the absence of *MuDR*, the effects of the transposon insertion were suppressed, so that the allele conditioned a dominant mutant phenotype once again. Given that revertants were insertions into regulatory regions, the high frequency of suppressible mutations might be anticipated if the mechanism of suppression depends on regulatory interactions with *MuDR* transposase (see above).

THE GENETICS OF VARIEGATION

Somatic loss of *Mu* activity can be visualized by variegation for the *hcf106* pale green mutant phenotype (Martienssen et al. 1990). Variegated plants have a distinctive pattern of loss in which groups of cells that have lost *Mu* activity arise progressively during development of the plant. These cells have a normal dark green phenotype, contrasting with the pale green mutant phenotype of adjacent tissue (Martienssen et al. 1990). Leaves that emerge late during development have larger and more numerous sectors of normal cells, until the upper leaves (the last to emerge) are completely normally pigmented. The sectors of normal cells resemble clonal sectors in their size and position in successive leaves (Martienssen et al. 1990).

By analogy with *Spm*, two types of phenotypic variegation associated with *MuDR* elements might be responsible for the pattern of variegation observed in plants homozygous for *hcf106-mum1*: "cycling" of autonomous elements between active and inactive phases (McClintock 1958; Peterson 1966) and "presetting" of gene expression patterns influenced by transposons at the locus (McClintock 1964, 1965b). Cycling gives rise to variegated patterns because some groups of cells inherit active autonomous elements while other groups inherit inactive elements. This results in variegated gene expression patterns from suppressible mutations in pigment biosynthetic genes. In contrast, preset patterns are an inherent property of a subset of transposon-induced mutations. They only require exposure to an autonomous element in the previous generation, and the patterns of variegation are variable and nonclonal in nature. These two possibilities can be distinguished in plants that carry two mutations whose phenotypic effects depend on transposon activity. Cycling of autonomous transposon activity would effect all suppressible mutants coordinately, whereas presetting is specific to the locus in question.

The dominant *Les28* mutation conditions a Lesion-mimic phenotype in the presence of *Mu* activity (see above). In *Les28* heterozygotes, sectors of necrotic lesions on the lower leaves, but not the upper leaves, are frequently observed. This variegated pattern closely resembles that found in plants homozygous for *hcf106-mum1*. In families segregating both *hcf106-mum1* and *Les28*, the ratio of single to double mutants showed that the two mutations were coregulated by *MuDR in trans* (Martienssen and Baron 1994). Furthermore, doubly mutant plants lost *Mu* activity and the double-mutant phenotype coordinately during development, resulting in variegated plants that had mutant sectors with both phenotypes interspersed among wild-type sectors with neither phenotype. DNA from mutant tissue was found to have unmethylated *Mu1* and *MuDR* elements, whereas DNA from the phenotypically wild-type sectors had methylated elements. These methylated elements were not otherwise rearranged. Plants that lost activity somatically also lost activity germinally, resulting in differential transmission of *Mu* activity from male and female inflorescences, as these are located at different positions with respect to the main axis of the plant (Martienssen et al. 1990; Martienssen and Baron 1994). Progeny from chimeric plants that lost *Mu* activity were clustered on the ear so that germinal as well as somatic loss appeared to be clonally inherited from single cells early in development.

Variegation thus affects two unlinked mutations coordinately, and so resembles cycling rather than presetting (McClintock 1968). If gene expression from the *hcf106* locus were preset by *MuDR* in the same way as *Spm* presets expression at the *a1* and *c2* loci, then the pattern of *hcf106* gene expression would segregate independently of the *Les28* pattern of expression, and coordinate sectors would not be observed. The pattern of variegation reflects the epigenetic loss of *MuDR* activity during development, rather than an inherent property of the *hcf106* and *Les28* genes themselves.

One or a few dominant factors appear to be responsible for the progressive loss of *Mu* activity in these plants (Martienssen and Baron

1994), and deletion derivatives of *MuDR* elements are good candidates for these factors (Lisch et al. 1995). Deletion derivatives of autonomous transposons in maize (Cuypers et al. 1988) and *Drosophila* (Robertson and Engels 1989) are known to dominantly antagonize full-length elements.

MOSAIC ANALYSIS

The coordinate nature of the sectoring we observe can be used as a tool for mosaic analysis (Martienssen and Baron 1994; Fowler et al. 1996). Double mutants with two suppressible mutations will have coordinate sectors of mutant and wild-type tissue if the cell layer in which each gene product is required is the same. If the layer (or lineage) is different, sectors of one phenotype will overlap or be distinct from sectors of the other phenotype, as most clonal sectors are restricted to one or another layer for at least a part of their length. If the cell layer is known in which one suppressible mutation acts, then the site of action of another suppressible mutant can be deduced from the pattern of sectors in the double mutant. For example, the *hcf106* mutant affects chlorophyll pigmentation, and the gene product is found in the photosynthetic cells of the leaf, which are primarily subepidermal (A. Yonetani and R. Martienssen, unpubl.). In *hcf106-mum1 Les28* mutants, all mutant sectors are doubly mutant for both phenotypes (coordinate), so that the *Lesion-mimic 28* gene product must be required in the same cell layer as *Hcf106* (Martienssen and Baron 1994). This suggests that the *Les28* gene product is required in the subepidermal rather than the epidermal layer. This method has advantages over classic methods that make use of radiation-induced terminal deficiencies to uncover linked markers (Hake and Freeling 1986). First, linkage to the marker is not required; second, several unlinked markers can be used at the same time; and third, sectors arise throughout development so that late and early events can be studied.

Fowler et al. (1996) have constructed double mutants between *a1-mum2*, a suppressible marker mutation that effects anthocyanin pigmentation, and suppressible revertant alleles of *Lg3*, a dominant mutant that results in changes in leaf cell fate. In these plants, sectors arise that have lost *MuDR* activity, and cells in these sectors are pigmented (because of suppression of *a1-mum2*). These sectors have morphological abnormalities because of suppression of the *Mu*-induced revertant of *Lg3*. Sectors in internal layers of the leaf have phenotypic consequences for epidermal cell fate, and effects on neighboring cells lead the authors to conclude that *Lg3* is non-cell-autonomous.

The dominant *Kn1-mum2* mutant of maize is caused by the insertion of a *Mu8* element into the third intron of the *Knotted* homeobox gene (see above; Greene et al. 1994). The mutant phenotype in this case depends on *Mu* activity, as for *Les28*. Families have been constructed that segregate both *Les28* and *Kn-mum2* (R. Martienssen and S. Hake, unpubl.). Knotted and Lesion-mimic phenotypes were found to be coregulated in families segregating both mutants, as was found for *Les28* and *hcf106* (Martienssen and Baron 1994). In addition, many double-mutant plants were sectored, having large sectors that had lost the Lesion-mimic phenotype. These sectors were almost always free of knots, pigmented veins, ectopic ligule, and other phenotypes associated with the *Kn1-mum2* mutation (Greene et al. 1994). Examples are shown in Figure 2A. Many *lesion-mimic* sectors had no visible knots, however, and two small sectors lost the Lesion-mimic phenotype but still had one or two knots. Non-concordance in these sectors may reflect a requirement for *knotted* in the innermost layers of the leaf (Sinha and Hake 1990) and a requirement for *Les28* in subepidermal mesophyll cells (Martienssen and Baron 1994).

THE ORIGIN OF PATTERNS OF VARIEGATION

Lisch et al. (1995) have demonstrated that plants can lose *MuDR* activity by spontaneous deletion of the autonomous *MuDR* element. When monitored with a suppressible mutation such as *a1-mum2*, this type of loss is stochastic, with large and small sectors of cells arising more-or-less randomly during development (Chomet et al. 1991). In contrast, the pattern of variegation in plants that are losing *MuDR* activity epigenetically indicates that somatic loss of *Mu* activity occurs nonrandomly. Sectors become larger and more frequent in successive leaves, until the entire upper portion of the plant has lost *Mu* activity.

Maize leaves emerge as a growing ring around the shoot apical meristem, so that leaf cells are recruited from the entire circumference of the shoot (Sharman 1942). Thus, the pattern of sectors observed in leaves serves as a fate map for cells in the shoot meristem that have lost *Mu* activity (Martienssen et al. 1990). Sectors pass from leaf to leaf, indicating that they arose from cells in the meristem, rather than in individual leaves. The pattern of sectors indicates that epigenetic loss of *Mu* activity occurs progressively in cells that occupy the central zone of the vegetative meristem, until all the cells in the meristem are affected. These cells divide slowly, in a stem-cell-like manner, giving rise to daughter cells that contribute to rapidly growing leaf primordia. In contrast, sectors restricted to a single leaf are much rarer, indicating that those cells that do retain *Mu* activity in the meristem give rise to clones that retain *Mu* activity throughout leaf development, resulting in a sector.

Cells in the central zone of the vegetative meristem have a longer cell cycle than other meristematic cells, which might allow heritable epigenetic modifications to occur uninterrupted by cell division. De novo DNA methylation, for example, might be strand-specific and easily lost by strand segregation (Martienssen and Baron 1994). Only in cells undergoing relatively slow cell divisions could epigenetic modifications affect both strands at all loci, resulting in permanent change. Alternatively, the central cells might be connected via specific cell-cell interactions and have a tendency to adopt a given epigenetic state by a "community effect." A single cell that stochastically loses *Mu* activity might thus influence its neighbors in the central zone to do the same. This interaction would have to be restricted to the meristem, as clonal descendants in the leaf do not influence the phenotype of neighboring cells. Recent observations that cellular proteins and RNA molecules can pass between cells might provide a plausible mechanism for this effect (Lucas et al. 1995).

Epigenetic variegation is also observed in endosperm tissues pigmented under control of the suppressible allele *vp1-mum3* (Fig. 2B). This allele of the *VP1* transcription factor gene conditions pigmented aleurone cell layers in kernels without *MuDR* and colorless, mutable aleurones in kernels with *MuDR* (see above). Variegated kernels are observed in some families and can occur in small clusters on otherwise fully pigmented ears (not shown). Occasionally, large regions of variegated aleurone are found adjacent to regions of colorless tissue within a single kernel (Fig. 2B). These observations indicate that variegation itself may be controlled by an unstable factor. However, the pattern of variegation differs from that observed in *hcf106* plants. Variegated *vp1-mum3* kernels have groups of cells that accumulate variable levels of anthocyanin pigment in nonclonal patterns.

These patterns resemble "mottled" patterns resulting from the inheritance of certain alleles of the *r* (red color) gene from the pollen parent (see Kermicle, this volume), as well as "preset" patterns of variegation conditioned by *Spm* insertions in genes involved in pigment biosynthesis (McClintock 1968). These observations and others have led to the suggestion that transposons may be involved in paramutation and gene silencing in plants (McClintock 1965b; Martienssen et al. 1990).

TRANSPOSONS AND GENE SILENCING

Transposable elements have been found in the promoters of genes that are subject to paramutation in maize and in snapdragon (see Kermicle, this volume). In principle, epigenetic interactions between autonomous elements (integrated in paramutagenic alleles) and their inactive derivatives (in paramutable alleles) could account for some aspects of paramutation (Martienssen 1996). DNA-binding proteins encoded by the autonomous transposon could bind the nonautonomous transposon, thereby influencing adjacent gene expression in the paramutable allele.

Transposons and retrotransposons have been found to exert epigenetic influences on gene regulation when integrated near promoters in animals and fungi, as well as in plants (for review, see Martienssen and Richards 1995). Furthermore, transposons and retrotransposons are widespread components of plant and animal genomes in general, and of heterochromatin in particular. Integration of transgenes into the plant genome via transformation is likely to result in the juxtaposition of transposable element and transgene sequences much of the time. This could lead to epigenetic effects similar to those observed when transposons integrate near genes. Transposon-mediated gene silencing could thus be more prevalent than commonly supposed.

SUMMARY

Epigenetic variegation in patterns of gene expression is a widespread but poorly understood phenomenon in plants, animals, and microbes. Mitotically inherited, but unstable, levels of gene activity can lead to clonal and to nonclonal patterns of variegation. Unlike many animal cells, plant cells undergo little migration relative to one another so that clonal lineages resulting from epigenetic changes can be conveniently mapped during somatic development. Furthermore, the germ line differentiates so late in development that many somatic lineages give rise to gametes. This allows the heritability of epigenetic changes to be readily determined, making plants uniquely suited to studying the genetic and developmental basis underlying patterns of epigenetic variegation.

Epigenetic variegation due to silencing of Robertson's *Mutator* elements results in complex patterns of gene expression during plant development. Because of the large number of mutations caused by *Mu* insertion, these patterns can be monitored using a large variety of developmental markers, including genes affecting pigmentation and morphogenesis. These mutants provide a useful tool for mosaic analysis, and clonal and nonclonal patterns of variegation have been observed. At least some of these patterns reflect cycling of multiple *MuDR* elements between active and inactive phases. Direct molecular analysis of sectors from variegated plants has been used to show that cytosine methylation differences accompany phenotypic variegation.

In the shoot meristem, the length of the cell division cycle and cell-cell interactions might influence the imposition and maintenance of epigenetic imprints. The mechanisms underlying epigenetic variegation are key to understanding how patterns of cell division and differentiation influence gene expression in normal plants and animals. Transposable elements may have a widespread role in mediating as well as revealing some of these effects.

ACKNOWLEDGMENTS

I thank Alice Barkan, Mike Freeling, Sarah Hake, and Don McCarty, and past and present members of their respective laboratories, for sharing their observations and seed stocks, and for communicating data prior to publication. Work in my laboratory was supported by a grant from the Department of Energy (DE-FG02-91ER20047).

REFERENCES

Barkan, A. and R.A. Martienssen. 1991. Inactivation of maize transposon *Mu* suppresses a mutant phenotype by activating an outward-reading promoter near the end of *Mu1*. *Proc. Natl. Acad. Sci.* **88:** 3502–3506.

Bennetzen, J.L. 1987. Covalent DNA modification and the regulation of *Mutator* element transposition in maize. Mol. Gen. Genet. **208:** 45–51.

Bennetzen, J.L., W.E. Brown, and P.S. Springer. 1988. The state of DNA modification within and flanking maize transposable elements. In *Plant transposable elements* (ed. O.E. Nelson), pp. 237–251. Plenum Press, New York.

Bennetzen, J.L., P.S. Springer, A.D. Cresse, and M. Hendrickx. 1993. Specificity and regulation of the *Mutator* transposable element system in maize. *Crit. Rev. Plant Sci.* **12:** 57–95.

Boeke, J. 1989. Transposable elements in *Saccharomyces cerevisiae*. In *Mobile DNA* (ed. D.E. Berg and M.M. Howe), pp. 335–374. American Society for Microbiology, Washington, D.C.

Brown, J. and V. Sundaresan. 1992. Genetic study of the loss and restoration of *Mutator* transposon activity in maize: Evidence against dominant-negative regulator associated with loss of activity. *Genetics* **130:** 889–898.

Chandler, V.L. and K. Hardeman. 1992. The *Mu* elements of *Zea mays*. *Adv. Genet.* **30:** 77–122.

Chandler, V.L. and V. Walbot. 1986. DNA modification of a maize transposable element correlates with loss of activity. *Proc. Natl. Acad. Sci.* **83:** 1767–1771.

Chomet, P.S., D. Lisch, K. Hardeman, V.L. Chandler, and M. Freeling. 1991. Identification of a regulatory transposon that controls the *Mutator* transposable element system in maize. *Genetics* **129:** 261–270.

Corces, V. and P.K. Geyer. 1991. Interactions of retrotransposons with the host genome: The case of the *gypsy* element of *Drosophila*. *Trends Genet.* **7:** 86–90.

Cuypers, H., S. Dash, P.A. Peterson, H. Saedler, and A. Gierl. 1988. The defective En-

I102 element encodes a product reducing the mutability of the En/Spm transposable element system of *Zea mays*. *EMBO J.* **7:** 2953–2960.

Das, L. and R.A. Martienssen. 1995. Site-selected transposon mutagenesis at the *hcf106* locus in maize. *Plant Cell* **7:** 287–294.

Fowler, J.E., G.J. Muehlbauer, and M. Freeling. 1996. Mosaic analysis of the Liguless3 mutant phenotype in maize by coordinate suppression of *Mutator*-insertion alleles. *Genetics* **143:** 489–503.

Freeling, M. 1992. A conceptual framework for maize leaf development. *Dev. Biol.* **153:** 44–58.

Frey, M, J. Reinecke, S. Grant, H. Saedler, and A. Gierl. 1990. Excision of the En/Spm transposable element of *Zea mays* requires two element-encoded proteins. *EMBO J. 9:* 4037–4044.

Gierl, A. 1990. How maize transposable elements escape negative selection. *Trends Genet.* **6:** 155–158.

Grant, S.R., A. Gierl, and H. Saedler. 1990. En/Spm encoded tnpA protein requires a specific target sequence for suppression. *EMBO J.* **9:** 2029–2035.

Greene, B., R. Walko, and S.Hake. 1994. *Mutator* insertions in an intron of the maize *knotted1* gene result in dominant suppressible mutations. *Genetics* **138:** 1275–1285.

Hake, S. 1992. Unravelling the knots in plant development. *Trends Genet.* **8:** 109–114.

Hake, S. and M. Freeling. 1986. Analysis of genetic mosaics shows that the extra epidermal cell divisions in *Knotted* mutant maize plants are induced by adjacent mesophyll cells. *Nature* **320:** 21–23.

Hershberger, R.J., C.A. Warren, and V. Walbot. 1991. *Mutator* activity in maize correlates with the presence and expression of the *Mu* transpsosable element *Mu9*. *Proc. Natl. Acad. Sci.* **88:** 10198–10202.

Lisch, D., P. Chomet, and M. Freeling. 1995. Genetic characterization of the *Mutator* system in maize: Behaviour and regulation of *Mu* transposons in a minimal line. *Genetics* **139:** 1777–1796.

Lowe, B, J. Mathern, and S. Hake. 1992. Active *Mutator* elements suppress the Knotted phenotype and increase recombination at the *Kn1-0* tandem duplication. *Genetics* **132:** 813–822.

Lucas, W.J., S. Bouche-Pillon, D.J. Jackson, L. Nguyen, L. Baker, B. Ding, and S. Hake. 1995. Selective trafficking of *KNOTTED1* homeodomain protein and its mRNA through plasmodesmata. *Science* **270:** 1980–1983.

Martienssen, R.A. 1996 Paramutation and gene silencing in plants. *Curr. Biol.* **6:** 810–813.

Martienssen, R.A. and A. Baron. 1994. Coordinate suppression of mutations caused by Robertson's *Mutator* transposons in maize. *Genetics* **136:** 1157–1170.

Martienssen, R.A. and E.J. Richards. 1995. DNA methylation in eukaryotes. *Curr. Opin. Genet. Dev.* **5:** 234–242.

Martienssen, R.A., A. Barkan, M. Freeling, and W.C. Taylor. 1989. Molecular cloning of a nuclear gene in maize involved in thylakoid organization, and its regulation by Robertson's *mutator*. *EMBO J.* **8:** 1633–1640.

Martienssen, R.A., A. Barkan, W.C. Taylor, and M. Freeling. 1990. Somatically heritable switches in DNA modification of *Mu1* transposable elements in maize revealed by a suppressible mutant. *Genes Dev.* **4:** 331–343.

McCarty, D.R., C.B. Carson, M. Lazar, and S.C. Simonds. 1989. Transposable element-induced mutations of the *viviparous-1* gene in maize. *Dev. Genet.* **10:** 473.

McClintock, B. 1958. The Suppressor-mutator system of control of gene action in maize. *Carnegie Inst. Wash. Year Book* **57:** 415–429.

———. 1964. Aspects of gene regulation in maize. *Carnegie Inst. Wash. Year Book* **63:** 592–602.

———. 1965a. Components of action of the regulators *Spm* and *Ac*. *Carnegie Inst. Wash. Year Book* **64:** 527–536.

———. 1965b. The control of gene action in maize. *Brookhaven Symp. Biol.* **18:** 162–184.

———. 1968. The states of a gene locus in maize. *Carnegie Inst. Wash. Year Book* **66:** 20–28.

———. 1971. The contribution of one component of a control system to versatility of gene expression. *Carnegie Inst. Wash. Year Book* **70:** 5–17.

Neuffer, M.G. and O.H. Calvert. 1975. Dominant disease lesion mimics in maize. *J. Hered.* **66:** 265–270.

O'Reilly, C., N.S. Shepherd, A. Pereira, Z. Schwarz-Sommer, I.Bertram, D.S. Robertson, P.A. Peterson, and H. Saedler. 1985. Molecular cloning of the *a1* locus of *Zea mays* using the transposable elements *En* and *Mu*. *EMBO J.* **4:** 591–597.

Peterson, P.A. 1966. Phase variation of regulatory elements in maize. *Genetics* **54:** 249–266.

———. 1995. Genetic analysis of the functions of the transposable element *En* in *Zea mays*: Limited transposase elicits a differential response on reporter alleles. *Genetics* **141:** 1135–1145.

Qin, M., D.S. Robertson, and A. Ellingboe. 1991. Cloning of the *Mutator* transposable element *MuA2*, a putative regulator of somatic mutability of the *a1-mum2* allele in maize. *Genetics* **129:** 845–854.

Robertson, D.S. 1986. Genetic studies on the loss of *Mu* mutator activity in maize. *Genetics* **113:** 765–773.

Robertson, H.M. and W.R. Engels. 1989. Modified *P* elements that mimic the P cytotype in *Drosophila melanogaster*. *Genetics* **123:** 815–824.

Rutledge, B.J., M.A. Mortin, E. Schwarz, D. Thierry-Mieg, and M. Meselson. 1988. Genetic interactions of modifier genes and modifiable alleles in *Drosophila melanogaster*. *Genetics* **119:** 391–397.

Schlappi, M., D. Smith, and N.V. Fedoroff. 1993. TnpA *trans*-activates methylated maize *Suppressor-mutator* elements in transgenic tobacco. *Genetics* **133:** 1009–1021.

Schneeburger, R.G., P. Becraft, S. Hake, and M. Freeling. 1995. Ectopic expression of the *knox* homeobox gene *roughsheath1* alters cell fate in the maize leaf. *Genes Dev.* **9:** 2292–2304.

Schwartz-Sommer, Z., N. Shepherd, E. Tacke, A. Gierl, W. Rohde, L. Leclercq, M. Mattes, R. Berndtgen, P. Peterson, and H. Saedler. 1987. Influence of transposable elements on the structure and function of the *A1* gene of *Zea mays*. *EMBO J.* **6:** 287–294.

Sharman, B.C. 1942. The developmental anatomy of the shoot of *Zea mays*. *Ann. Bot.* **6:** 245–284.

Sinha, N. and S. Hake. 1990. Mutant characters of *Knotted* maize leaves are determined in the innermost tissue layers. *Dev. Biol.* **141:** 203–210.

Walbot, V. 1986. Inheritance of Mutator activity in *Zea mays* as assayed by somatic instability of the *bz2-mu1* allele. *Genetics* **114:** 1293–1312.

Appendices

Essentials of Plant Development

Robert A. Martienssen
Cold Spring Harbor Laboratory
Cold Spring Harbor, New York 11724

In this appendix, features of plant development are reviewed that are important to various epigenetic phenomena. The flowering plants described in this book fall into two major classes, namely monocotyledonous and dicotyledonous, which diverged at least 100 MYA. Petunia, tobacco, and *Arabidopsis* are dicots, whereas maize is a monocot and belongs to the grass family. In the following summary, maize and *Arabidopsis* are used as examples of each class. An excellent review of the principles of plant development can be found in *Patterns in Plant Development* (Steeves and Sussex 1989), and detailed descriptions of maize and *Arabidopsis* development can be found in *Arabidopsis* (Meyerowitz and Somerville 1994) and *The Mutants of Maize* (Neuffer et al. 1996).

There are a number of features of plant development that make plants uniquely suited to epigenetics research. First, because of their indeterminate growth habit, epigenetic patterns of gene regulation in plants can be established throughout the life cycle. Once established, these patterns can be inherited by somatic cell lineages in much the same way as in other organisms.

Second, unlike many animal cells, plant cells undergo little migration relative to one another so that clonal lineages can be conveniently mapped during somatic development. Because of these well-characterized lineages, it is possible to discern non-clonal as well as clonal epigenetic patterns in plants, and examples of each are described in several chapters in this book. Third, the germ line in plants differentiates so late in plant development that many somatic lineages give rise to gametes. This allows the heritability of epigenetic changes to be readily determined.

There are two distinct stages in the life cycle of higher plants: the diploid sporophytic stage and the haploid gametophytic stage. During the sporophytic stage, the plant body (the sporophyte) is elaborated from the zygote via embryogenesis and meristematic growth (see below). Reproductive growth completes this stage, and meiosis occurs in male and female reproductive organs in the flower. However, in contrast to animal cells, the haploid products of meiosis undergo mitotic divisions, giving rise to a haploid, gametophytic phase of the plant life cycle. These aspects of plant development are summarized below.

GAMETOPHYTES AND GERM CELLS

The haploid products of female meiosis are called megaspores, and the products of male meiosis are called microspores. In the haploid gametophytic stage of the plant life cycle, these spores undergo 2–3 rounds of cell division to form the female gametophyte (the embryo sac, which is completely enclosed in the flower) and the male gametophyte (the pollen grain), respectively. These gametophytes contain the germ cells, which are among the terminal products of the haploid gametophytic mitoses. The microspore nucleus divides once to give rise to vegetative and generative daughter nuclei. The generative nucleus divides once more, leading to two sperm cells in the mature pollen grain. The megaspore undergoes three rounds of division, resulting in eight nuclei. Cellularization results in one large, binucleate central cell and six smaller cells, including the egg cell. In plants, germ cells are set aside extremely late in development, quite unlike the situation in animals. For example, the very small number of cells that constitute the haploid gametophyte are the targets of parental imprinting in plants (see Kermicle, this volume).

FERTILIZATION AND SEED DEVELOPMENT

The two sperm cells migrate down the pollen tube as it grows and are released into the embryo sac during fertilization. One sperm fuses with the egg cell, and the other fuses with the central cell. This second fertilization results in the formation of the triploid endosperm, an extraembryonic storage tissue that terminally differentiates after a series of rapid nuclear divisions, cellularization, and endoreduplication. In maize and other grasses, the endosperm accumulates starch and storage proteins and undergoes programmed maturation and dessication, leading to the characteristic starchy endosperm in the dry seed (Fig.1). The outer layer of the endosperm remains viable, however, and is known as the aleurone layer. This layer accumulates anthocyanin pigments, and the readily visible pigmentation of individual kernels on the ear has made the genetics of pigment formation a paradigm for genetic and epigenetic studies in plants. A high proportion of the regulatory and structural genes that control pigment formation have been cloned molecularly in maize, petunia, snapdragon, and other plants, allowing detailed molecular analysis of epigenetic processes in these plants.

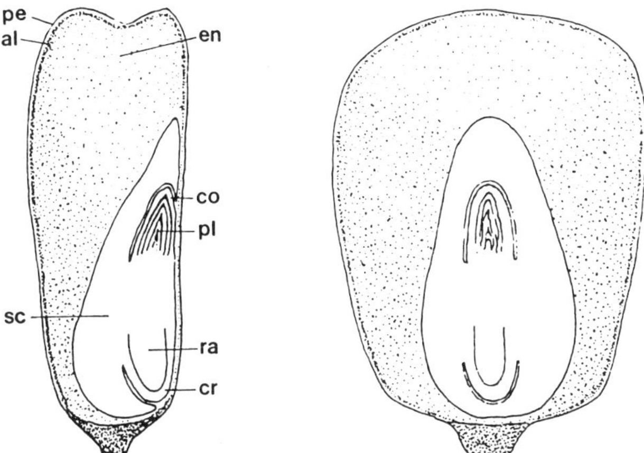

Figure 1 Longitudinal sections of a mature seed of maize. (pe) Pericarp; (en) endosperm; (al) aleurone; (sc) scutellum; (co) coleoptile; (pl) plumule; (ra) radicle; (er) coleorhiza. The plumule comprises 5–6 immature leaves in the dry seed before germination. (Reprinted, with permission, from Poethig 1982 [redrawn from Kesselbach 1949].)

In most higher plants, the zygote divides asymmetrically, giving rise to apical and basal cells that will contribute to the embryo proper and the suspensor, respectively. The apical cell undergoes a series of divisions resulting in a globular proembryo connected to the embryo sac wall by the suspensor, a connecting stalk that ultimately degenerates. The proembryo rapidly reorganizes to establish root and shoot meristems, and one or two cotyledons ("seed leaves") in monocots and dicots, respectively. In *Arabidopsis*, petunia, and tobacco, these cotyledons resemble leaves, with many of the same tissues and cell types. However, true leaves arise from the shoot meristem during late embryogenesis, resulting in two (*Arabidopsis*) to six (maize) leaf primordia in the dry seed (Fig.1). The embryo and endosperm can remain dormant for many years before germination is induced and postembryonic growth is initiated.

MERISTEMS AND POSTEMBRYONIC GROWTH

Pattern formation in plants occurs largely postembryonically, so that all the major cell and tissue types must be established during rapid growth. This is accomplished via the action of shoot and root meristems: Meristems are small groups of dividing cells (typically a few hundred) that retain stem-cell-like properties. That is, cells in the meristem divide to give rise to daughter cells, some of which remain in the meristem while others go on to contribute to determinate structures such as leaves and flowers. In the shoot apex, these "stem cells" are confined to the central zone of the meristem, a group of slowly dividing cells at the summit. Cells in this zone that undergo genetic or epigenetic mutation give rise to clonally related sectors in several consecutive leaves derived from daughter cells. These sectors exhibit characteristic patterns that allow clonal and non-clonal sectors to be distinguished in many cases. Lineage analysis using these mosaic plants has demonstrated that stem cells can persist for many cell generations, but that their identity is dependent on their position in the apex rather than on their lineage. This is because many sectors eventually terminate, meaning that the stem cell that gave rise to it was replaced by neighboring cells during the extended period of meristematic growth.

Organ initiation occurs in the surrounding "peripheral zone," where rapidly dividing cells are partitioned into leaf and floral primordia that undergo subsequent determinate growth. Fate mapping has established that cells destined to form these primordia are not set aside until just before organ initiation, resulting in considerable flexibility during plant growth and allowing rapid reorganization of the plant body following environmental, pathogenic, predatory, or other external stimuli.

Meristems comprise two or more outer layers of cells that divide within the plane of the layer (anticlinally) and so are maintained as separate lineages. This organization typically breaks down during organ formation, so that inner tissues near the tips and margins of the leaves include contributions from upper layers of the meristem, following out-of-plane (periclinal) divisions. Outer layers whose division patterns are anticlinal in the meristem are known as tunica layers, while the underlying cells are known as the corpus. Maize has only one tunica layer (the LI) whereas *Arabidopsis* has two (the LI and LII). When there are multiple tunica layers, the LI typically contributes to epidermal but not subepidermal tissues in the leaf. However, exceptional divisions occur (and can be induced) without disrupting tissue organization, reinforcing the notion that plant cells adopt their ultimate fates very late in development, according to their position in the organ primordia, rather than their clonal history in the meristem.

Axillary, or subsidiary, meristems arise as buds in the axils of leaves (i.e., at the junction between the leaf and the stem). These buds have similar properties to the shoot meristem, but many higher plants (including maize and *Arabidopsis*) exhibit apical dominance, so that inflorescences arise from the primary shoot earlier than from the axillary branches. If the primary meristem is damaged or removed, axillary meristems develop much earlier.

FLOWER DEVELOPMENT

The shoot apical meristem undergoes a number of phase changes during shoot growth, which results in changes in the types of lateral organs that are initiated on its flanks. Vegetative phases include a juvenile and an adult stage, which have characteristic (and sometimes dramatic) differences in leaf morphology. The transition to flowering (the reproductive phase) is under complex genetic and environmental control. It begins with the transformation of the shoot apical meristem into an inflorescence meristem, which gives rise to bracts (modified leaves), inflorescence branches, and floral primordia. Floral primordia give rise to perianth organs (sepals, petals, and soft glumes) and reproductive organs (stamens and carpels).

In maize, the shoot apical meristem is transformed into an inflorescence meristem about 3–4 weeks after germination. This event is irreversible and, in most cultivated varieties, is relatively independent of day length and other environmental factors. The main inflorescence meristem develops into the many-branched tassel found at the top of the mature plant. A complex series of sub-branches bud from the main meristem. Spikelet pairs bud from these branches, and primary and secondary florets develop from each spikelet. This complex pattern of branching results in the dense and highly organized maize male inflorescence. Early floral primordia have both male and female organs, but female organs abort early in tassel development. In each flower, three stamens develop and eventually dehisce, releasing many thousands of yellow pollen grains. In contrast, the female inflorescence develops from a side shoot of the primary shoot meristem. After initiation of a number of specialized leaves (the husks) and shortened internodes, the maize ear develops as an unbranched structure adorned with spikelet pairs. Each spikelet develops normally, but one floret in each spikelet aborts, resulting in an ordered array of double rows of flowers along the ear. The male organs degenerate in these flowers, resulting in female flowers in which the three central carpels are fused to form an ovary containing a single ovule. A megaspore mother cell is specified within each ovule, which undergoes meiosis, resulting in a linear tetrad of four meiotic products. The most distal three meiotic products degenerate, leaving a single megaspore that goes on to form the female gametophyte (see above). When pollinated, each flower gives rise to a kernel, so that each kernel represents the product of a separate meiosis and an individual progeny of the parent plant. The ordered arrangement of these individual progeny on an ear of corn allows somatically arising epigenetic changes to be mapped onto clonally related progeny by looking for sectors on the ear.

Once initiated, floral primordia undergo pattern formation via the action of a series of homeotic and regulatory genes to establish the primary floral organs. This process has been intensively studied at the molecular level in *Arabidopsis* and in snapdragon (*Antirrhinum*), in which highly conserved regulatory genes lead to very different floral morphologies. Floral organs are arranged in these species in four concentric whorls, comprising sepals, petals, stamens, and carpels, respectively. The identity of organs within these whorls is determined by the combinatorial action of a limited set of homeotic genes. Homeotic genes contribute to three overlapping functions: A, B, and C. Function A specifies the perianth organs (petals and sepals), function B specifies petals and stamens, and function C specifies carpels and stamens. In addition, various regulatory genes interact to control these homeotic functions: Most of the homeotic genes that have been molecularly identified encode transcription factors of various classes that are found in animals and yeast, as well as some that may be unique to plants.

Within the carpels, ovule primordia arise from the carpel wall, and a megaspore mother cell differentiates within each ovule, as described above. In the anthers, pollen mother cells arise in the microsporangium (pollen sac) within each anther. These undergo meiosis, and all four products from each tetrad go on to develop into pollen grains (see above).

PLANT CELL ARCHITECTURE

Plant cells differ from animal cells in several important respects: They have rigid cell walls that can control the orientation of cell expansion and cell division to some extent. They also contain large vacuoles that can mediate rapid changes in physiological levels of ions and other small molecules. Finally, and perhaps most importantly, plant cells are connected via cytoplasmic channels known as plasmodesmata. There are at least two classes of plasmodesmata that arise during and following cell division, respectively. They allow cytoplasmic and endomembrane connections between plant cells, connections that can be regulated by certain viral, and perhaps cellular, proteins. These connections result in "symplastic domains" or groups of cells that share pools of molecules that can pass between them. These pools include certain transcription factors and perhaps other regulatory molecules, such as RNA. Many of these differences reflect the sessile nature of most plant cells: They are cemented in place relative to each other, effectively prohibiting cell migration.

CONCLUSIONS

The differences between plant and animal cellular structure and development suggest that different developmental mechanisms might mediate pattern formation in plants. However, genetic and cellular parallels between the two kingdoms suggest that many ancient mechanisms are conserved. The comparison of epigenetic processes in plants and animals might be expected to reveal some of these mechanisms, given the common roles of DNA methylation, chromatin structure, and transposable elements in epigenetic phenomena in animals, plants, and microbes. It is possible that a combination of these common components underlies the basis for some epigenetic mechanisms of gene regulation across these widely divergent organisms.

REFERENCES

Kiesselbach, T.A. 1949. The structure and reproduction of corn. *Nebr. Agric. Exp. Stn. Res. Bull.* **161**.

Meyerowitz, E.M. and C.R. Somerville, eds. 1994. Arabidopsis. Cold Spring Harbor Laboratory Press, Cold Spring Harbor, New York.

Neuffer, M.G., E.H. Coe, and S.R. Wessler, eds. 1996. *Mutants of maize*. Cold Spring Harbor Laboratory Press, Cold Spring Harbor, New York. (In press.)

Poethig, R.S. 1982. Maize—The plant and its parts. In *Maize for biological research* (ed. W.F. Sheridan), p. 9–18. Plant Molecular Biology Association, Charlottesville, Virginia.

Steeves, T.A. and I.M. Sussex, eds. 1989. *Patterns in plant development*, 2nd edition. Cambridge University Press, United Kingdom.

Essentials of Mouse Development

Arthur D. Riggs
Biology Department
Beckman Research Institute of the City of Hope
Duarte, California 91010

Many scientists interested in epigenetics, especially those coming from a background of microbial or plant genetics, may not be sufficiently familiar with the features and nomenclature of mouse development to easily understand the chapters in this book devoted to mammalian epigenetic phenomena. The purpose of this chapter is to provide in a few pages the essentials of mouse development, with emphasis on the features that seem to be of importance for epigenetics. As a review and practical introduction to mouse embryology, a book entitled *Manipulating the Mouse Embryo* (Hogan et al. 1994) is highly recommended. Most epigenetic phenomena in mammals, including cell determination, X-chromosome inactivation, and imprinting, are established in the early embryo and then maintained in somatic cells of the adult. Cell determination and differentiation during mammalian development is also highly dependent on cell memory. In general, the response of a mammalian cell to a signal depends on developmental history. Events that occurred many cell generations earlier are remembered, presumably by somatically heritable epigenetic mechanisms.

FERTILIZATION AND PREIMPLANTATION DEVELOPMENT

Fertilization takes place in the oviduct approximately 1 hour post coitus (p.c.) (see Fig. 1). Development of the resulting zygote is slow until implantation, which does not take place until about 4.5 days p.c. Full maturation of the egg, including completion of the second meiotic division and extrusion of the second polar body, normally takes place after fertilization and is not complete until 2–5 hours p.c. During egg maturation and for some time thereafter, the male and female nuclei remain as separate entities within the single-cell zygote. Nuclear membranes eventually assemble around each chromosomal set, with the female pronucleus forming at 4–7 hours p.c. and the male pronucleus forming at 6–9 hours p.c. Over a period of many hours, the haploid pronuclei then move toward the center of the egg, with DNA replication taking place during the migration. After completion of DNA synthesis, 17–20 hours

Epigenetic Mechanisms of Gene Regulation

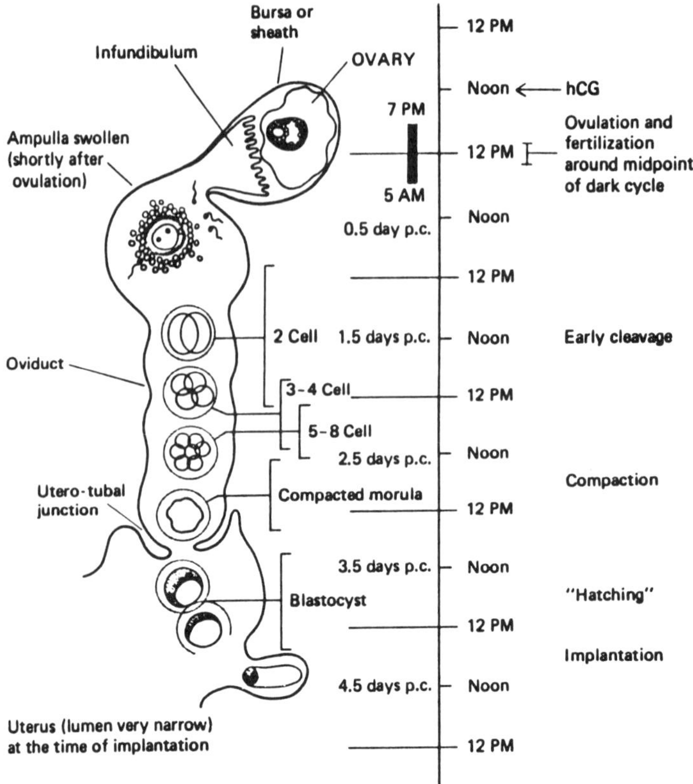

Figure 1 Preimplantation development of the mouse. (Reprinted, with permission, from Hogan et al. 1986.)

p.c., the nuclear membranes of the pronuclei break down, the chromosomes assemble on the spindle, and first cleavage takes place. Twelve hours later, 2 days p.c., the first "normal" mitosis is completed, forming the 4-cell embryo, or morula. At the 8- to 16-cell stage (3 days p.c.) the morula compacts, and then the first cell-lineage determinations are made. Cells on the outside of the compacted morula differentiate to form the trophectoderm and the extraembryonic lineage, whereas cells on the inside remain undifferentiated. By 3.5–4 days p.c., the embryo has developed to the blastocyst stage, and at about 4.5 days, the blastocyst implants into the uterine wall.

POSTIMPLANTATION DEVELOPMENT

At 4.5 days and implantation, the blastocyst has developed a cavity, the blastocoel, and formed several identifiable cell types—polar and mural

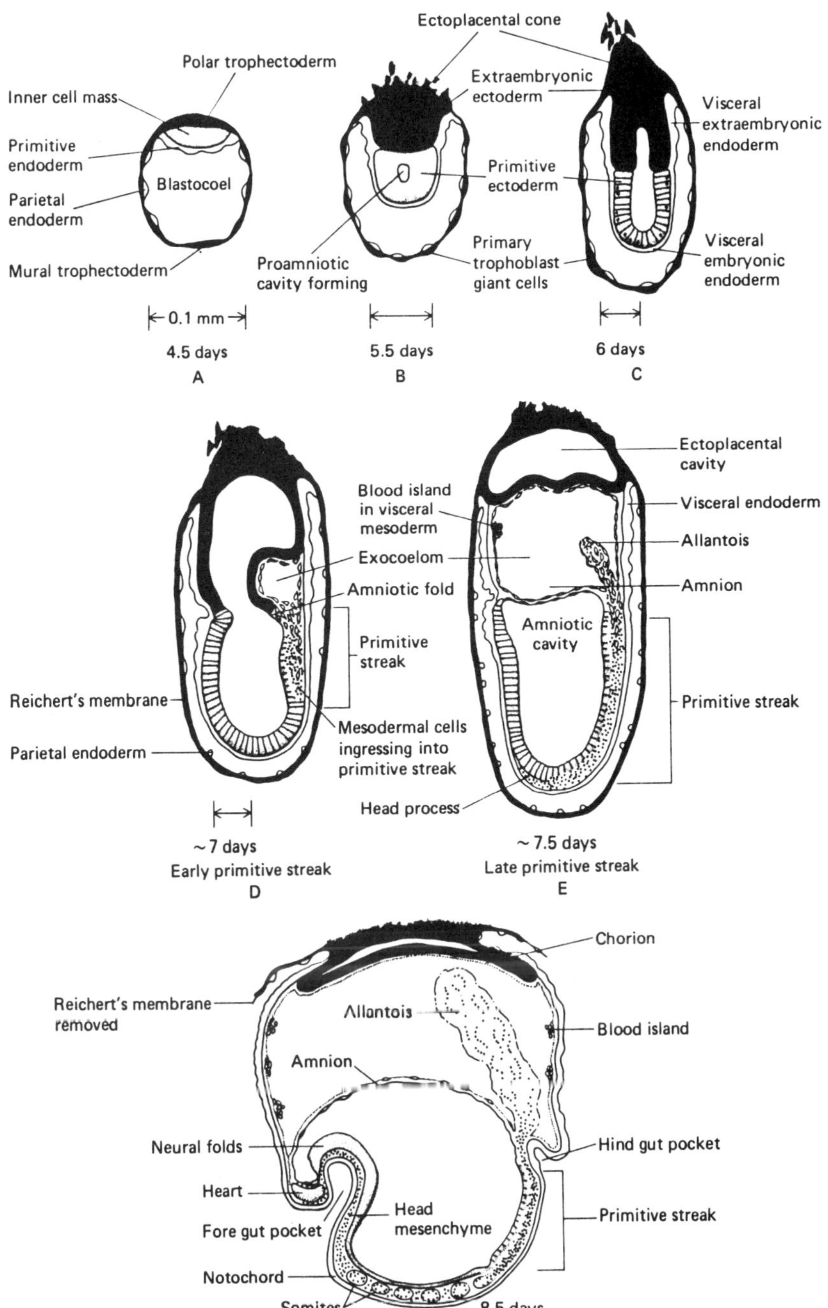

Figure 2 Postimplantation development of the mouse. (Reprinted, with permission, from Hogan et al. 1986.)

trophectoderm, parietal and primitive endoderm, and the inner cell mass (see Fig. 2). Only the 20–40 cells in the inner cell mass (ICM) form the cells of the primitive ectoderm (often called the epiblast), which is the precursor for all tissues of the adult mouse; the other cells of the blastocyst only form extraembryonic membranes and placental tissues which eventually are discarded. Implantation begins a stage of rapid growth of the embryo, with cell cycle times as short as 4–5 hours. As illustrated in Figure 2, various tissues form rapidly, with the tissues of the embryo being derived solely from descendants of ICM cells, and the extraembryonic tissues being derived from both the ICM and the trophectoderm lineages (see Fig. 3). At about day 7 the primitive streak has formed and the three major cell lineages that later form the adult mouse—endoderm, mesoderm, and ectoderm—are being established. By 8.5 days p.c., neural folds, notochord, somites, and primordial organs can be discerned. By day 10, the basic body plan is in place. The mesoderm has become segmented into paired somites that will later form vertebrae and muscle; the ectoderm has formed the neural tube; and the heart, circulatory system, and limb buds have been formed. Birth is at 19 days p.c. After birth, there is continued maturation of many tissues and organs such as the gonads. Importantly, cell division never stops, as several adult organs such as the skin, gut, and hematopoietic system are maintained by the continued division of stem cells.

THE GERM LINE AND GAMETOGENESIS

Primordial germ cells arise from the primitive ectoderm and are first detectable at day 7. By day 8.5, they have become a cluster of about 75 cells located in the base of the allantois (see Fig. 2F). By the time the germ cells can be discerned, they have begun a migration in which they will eventually move to the genital ridges located on the pronephros, the primitive kidney. The germ cells are mitotically active during the migration, dividing about every 16 hours and increasing in number to about 25,000 cells by the time they populate the genital ridge at day 13.5.

The fate of male or female germ cells differs after their arrival at the genital ridges. Female germ cells enter meiosis, and by 5 days after birth all have become immature oocytes arrested at the diplotene stage of meiosis and surrounded by follicle cells. After sexual maturity, hormonal signals periodically cause some oocytes to complete the first meiotic division, to extrude the first polar body, and to be released from the ovary into the oviduct.

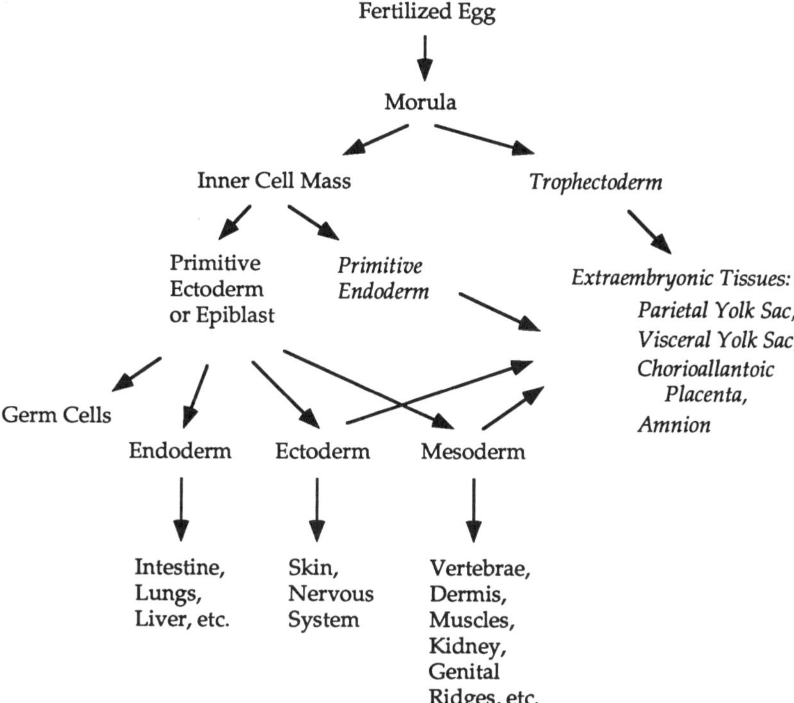

Figure 3 Cell lineages in mouse development. Extraembryonic lineages are shown in italics.

Male germ cells undergo mitotic arrest in the G_1 phase of the cell cycle upon arrival at the genital ridges and then remain in arrest until puberty, at which time spermatogenesis begins. Spermatogenesis is a complex process. Type-A spermatogonia function as stem cells, undergoing an asymmetric mitotic division to form type-B spermatogonia, which subsequently become spermatocytes and enter meiosis. After meiosis, there is a haploid stage, lasting for 12 days, during which the postmeiotic spermatid cells extrude cytoplasm and undergo transformation to mature sperm.

CONCLUSIONS AND COMPARISON WITH OTHER ORGANISMS

In mammals, as in most organisms, male and female gametes are structurally and biochemically very different; in addition, they are created by two very different processes that are related mainly in that they both include meiosis. The DNA methylation system is active during gametogenesis and the maturation of oocytes and sperm, so it is not surprising that methylation patterns are different between mature male and female gametes (see Shemer and Razin, this volume). Thus, methylation patterns are strong candidates for the primary imprint mark that enables the male or female parental source of a gamete to be "remembered" postfertilization. It may be less well appreciated that the male and female haploid genomes remain separate and subject to different treatment and environments for 17–20 hours after fertilization. Thus, there is ample opportunity for imprinting marks, either methylation or other heritable, *cis*-acting chromosome structures, to be established or modified post-fertilization.

Continued division of differentiated cells is a prominent feature of mammalian development and adult function. For example, it has been estimated that an intestinal crypt cell, a stem cell, undergoes more than 5000 divisions in a human lifetime (Greaves 1996), and by age 40, a man's spermatogonial stem cells have divided about 1000 times (Farrer et al. 1991). It is clear that good cell memory of differentiated states is of paramount importance for mammals. Higher plants share with mammals and other vertebrates a dependence on high-fidelity cell memory, as cells in the apical meristem, for example, continue dividing for the life of a plant and can even contribute to the germ line after countless divisions over hundreds of years. This need for extremely stable heritable epigenetic switches is likely to be much less, however, for an organism such as *Drosophila* that has a relatively small genome (20-fold less than mammalian) and undergoes a very limited number of cell divisions. On average, only 3 cell divisions occur in the developmental process that creates a *Drosophila* larva. After fertilization of a *Drosophila* egg, the diploid nucleus undergoes a series of very rapid nuclear divisions, but separate cells are not formed until after 13 nuclear divisions (Edgar and O'Farrell 1989). The nuclei then migrate to the egg periphery, where cells are then formed, creating the cellular blastoderm stage. On average, only 3 more cell divisions take place, and then, again, DNA replication takes place without cell division, forming the polytene chromosomes typical of *Drosophila*. The point of this discussion is that major differences in the developmental processes between organisms should be kept in mind when trying to evaluate results and to generalize with regard to epigenetic mechanisms.

REFERENCES

Edgar, B.A. and P.H. O'Farrell. 1989. Genetic control of cell division patterns in the *Drosophila* embryo. *Cell* **57:** 177–187.

Farrer, L.A., L.A. Cupples, L. Connor, P.A. Wolf, and J.H. Growdon. 1991. Association of decreased paternal age and late-onset Alzheimer's disease. *Arch. Neurol.* **48:** 599–604.

Greaves, M. 1996. Is telomerase activity in cancer due to selection of stem cells and differentiation arrest? *Trends Genet.* **12:** 127–128.

Hogan, B., F. Constantini, and E. Lacy. 1986. *Manipulating the mouse embryo: A laboratory manual.* Cold Spring Harbor Laboratory, Cold Spring Harbor, New York.

Hogan, B., R. Beddington, F. Constantini, and E. Lacy. 1994. *Manipulating the mouse embryo: A laboratory manual,* 2nd edition. Cold Spring Harbor Laboratory Press, Cold Spring Harbor, New York.

Reprints

Paramutation: Directed Genetic Change

Paramutation occurs in somatic cells and heritably alters the functional state of a locus.

R. Alexander Brink, E. Derek Styles, John D. Axtell

It was observed about 10 years ago that an R^r gene in maize, which conditions anthocyanin formation in seed and plant, invariably has lowered anthocyanin-forming potential following passage through a heterozygote with the stippled (R^{st}) allele. The changed form of R^r was gametically transmissible. It reverted toward the standard type when made homozygous, but only partially. This unusual kind of heritable change was termed paramutation. The R^r allele in question was said to be paramutable, and the stippled factor was described as paramutagenic (*1*).

The initial studies revealed that paramutation involved chromosome components that do not conform to the ordinary rules of gene stability and integrity in heterozygotes. Furthermore, the induced heritable changes were directed. R paramutation appeared to be comparable in these respects to certain phenomena reported much earlier in *Pisum* by Bateson and Pellew (*2*), in *Malva* by Lilienfeld (*3*), and in *Oenothera* by Renner (*4*) as anomalous exceptions to Mendelian inheritance. The R case in maize offered advantages for experimental analysis which these other systems lacked, and it has been intensively investigated. The present article is limited to discussion of the R studies, for the most part, but the results appear to be meaningful for paramutation in general.

The biological significance of paramutation rests on the fact that the phenomenon involves constraint on gene expression during development of the individual which is exerted by factors located within the chromosome itself. The illuminating advances made in recent decades in characterizing the genetic substance biochemically have disclosed genomic components in bacteria that function *in situ* to regulate the action of genes specifying protein structure (*5*). The chromosomes of higher organisms are vastly larger and more complex than their counterparts in bacteria, and one may expect to find that they too embody devices whereby gene action is regulated locally. Paramutation is of general interest for the clues it may give to the nature of gene-controlling mechanisms.

The authors are, respectively, professor of genetics, University of Wisconsin, Madison; assistant professor of biology, University of Victoria, Victoria, Canada; and assistant professor of agronomy, Purdue University, Lafayette, Indiana.

Induction of Paramutation

The general problem which paramutation presents is illustrated by the results of a simple experiment. Pollen of three kinds of maize plants, with a common highly inbred background, is applied to the silks of $r^g r^g$ individuals as follows:

1) $r^g r^g$ ♀ × $R^r R^r$ ♂
2) $r^g r^g$ ♀ × $R^{r'} R^{r'}$ ♂
3) $r^g r^g$ ♀ × $F_1 R^r R^{r'}$ ♂

R alleles condition anthocyanin formation in the plant and in the aleurone—the outer cell layer of the endosperm. The recessive r^g allele conditions development of colorless aleurone and colorless plant (lacking in anthocyanin). R^r, termed "standard R^r," represents a class of factors, of wide geographic distribution, that conditions pigment formation in both aleurone and plant. The endosperm is the product of a fertilization separate from that which gives rise to the accompanying embryo. It receives a double complement of genes from the female parent, and one set from the male parent, and so is a triploid structure. Standard R^r in single dose ($R^r\ r\ r$) gives darkly mottled aleurone, and in either two doses ($R^r R^r r$) or three doses ($R^r R^r R^r$), gives self-colored aleurone. R^{st} (stippled) gives a spotting pattern in the aleurone, directly proportional to dosage.

The endosperm genotypes which result from the above three testcrosses are

1) All $R^r r^g r^g$
2) All $R^{r'} r^g r^g$
3) 50 percent $R^r r^g r^g$ plus 50 percent $R^{r'} r^g r^g$

Half of the seeds ($R^r r^g r^g$) resulting from mating 3 should conform genetically to those from mating 1, and the other half should conform to the seed resulting from mating 2. The latter expectation is fulfilled: the stippled kernels resulting from testcrosses 2 and 3 are indistinguishable from each other. The correspondence, however, does not extend to the $R^r r^g r^g$ seeds from matings 1 and 3. All the $R^r r^g r^g$ kernels from testcross 3 are more weakly pigmented than those from testcross 1. The difference is illustrated in Fig. 1. The basis of the phenotypic difference is a heritable alteration in pigmenting potential of R^r by virtue of passage through the $R^r R^{st}$ heterozygote; R^r has been changed to a more weakly pigmenting form symbolized by $R^{r\prime}$. The change from R^r to $R^{r\prime}$ is attributed to paramutation.

The $R^r r^g r^g$ kernels from both mating 1 and mating 3 are mottled, and the degree of mottling varies somewhat within each group. A scoring procedure that has proved adequate to deal with the continuous variation in amount of anthocyanin pigmentation in kernels of these classes involves matching of each kernel in a 50-kernel random sample of the $R^r r^g r^g$ seed from a testcross ear against a set of six standard seeds ranging progressively from near-colorless to near-self-color, and defining seven classes. The mean aleurone-color score for the sample is then used as the measure of the pigment-producing potential of the R^r male gametes formed by the staminate plant under test.

The mean score for the $R^r r^g r^g$ kernels from a type 1 testcross in a representative experiment was 5.9. The value for the corresponding class of seeds ($R^{r\prime} r^g r^g$) from the type 3 mating was 2.4. Thus passage of R^r through the heterozygote with the stippled factor reduced pigment-producing potential of the allele by more than three units on the seven-grade aleurone color scale. The fall in pigmenting potential characterized all, or nearly all, the R^r gametes from the $R^r R^{st}$ parent. The change is thus unique in two respects: it is unidirectional and also invariable in occurrence.

As mentioned above, the stippled kernels resulting from matings 2 and 3 are alike in appearance. Supplementary tests prove that the R^{st} allele also is unaltered in strength of paramutagenic action by passage through an $R^r R^{st}$ heterozygote. Paramutation thus affects only the R^r factor in $R^r R^{st}$ plants.

A marked difference in pigment-producing action of R^r and $R^{r\prime}$ is again expressed if plants are grown from the $R^r r^g$ and $R^{r\prime} r^g$ seeds resulting from the matings referred to in the preceding section. The strength of action of $R^{r\prime}$, however, is found to have shifted partway back to that of R^r in standard form. The $R^{r\prime}$ genes from $R^{r\prime} R^{r\prime}$ homozygotes derived by self-pollination of $R^r R^{st}$ plants likewise have significantly weaker aleurone-pigmenting action than standard R^r genes from homozygous stock cultures. The difference is illustrated in Fig. 2. Thus $R^{r\prime}$ is gametically transmissible through the embryo, but not necessarily in the highly repressed form characteristic of $R^{r\prime}$ resulting directly from $R^r R^{st}$ heterozygotes.

Kermicle (6) observed that, in two $R^{r\prime} R^{r\prime}$ sublines derived by selfing an $R^r R^{st}$ plant, the partial reversion in strength of $R^{r\prime}$ action mainly occurred in the first $R^{r\prime} R^{r\prime}$ generation. Thereafter the $R^{r\prime}$ aleurone-pigmentation potential remained stable at a level about halfway between that of standard R^r and that of the primary $R^{r\prime}$ paramutant.

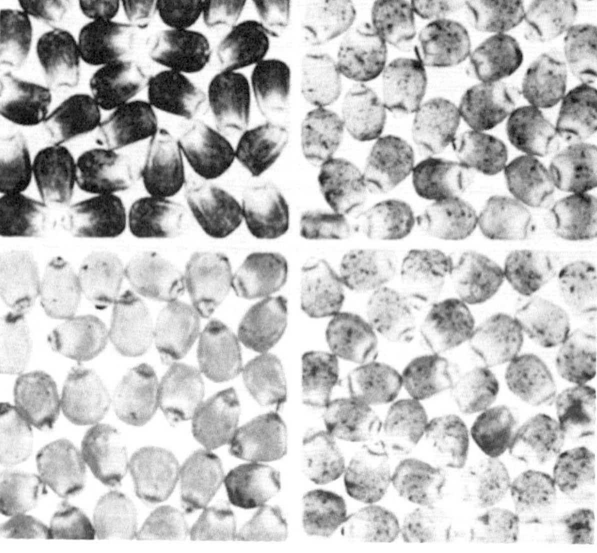

Fig. 1. Effect on seed color of passing the standard R^r factor through a heterozygote with the R^{st} allele. The two groups of kernels shown in the top row resulted from matings on a colorless $r^g r^g$ strain, with (top left) pollen from a standard $R^r R^r$ stock culture and (top right) pollen from an $R^{st} R^{st}$ stock culture. Note that the $R^r r^g r^g$ testcross kernels are darkly mottled. The two groups of kernels shown in the bottom row resulted from the application of F_1 $R^r R^{st}$ pollen to the same colorless female strain. The seed phenotype associated with the R^{st} allele is unchanged, as is evident from a comparison of the spotting pattern of the $R^{st} r^g r^g$ testcross kernels (bottom right) with patterns of the $R^{st} r^g r^g$ controls (top right). The $R^{r\prime} r^g r^g$ testcross kernels (bottom left), on the other hand, are lightly mottled as compared with the standard $R^r r^g r^g$ testcross kernels (top left). The change in phenotype in this case reflects the markedly reduced aleurone-pigmenting potential of R^r that regularly follows passage of R^r through a stippled heterozygote.

Characteristics of the R Paramutation System

An analysis of the earlier work on paramutation at the R locus in maize was published in 1964 by Brink (7). Several of the questions discussed at that time are dealt with here only briefly, except as new evidence on them has been published. Many of the data on which the new conclusions in these cases rest may be found in the general review mentioned above (7) and in the primary articles cited therein. Particular attention in this article is accorded the results of recent studies of special significance for the interpretation of paramutation.

Association of the R Locus with Paramutation

The genetic elements involved in *R* paramutability are intimately connected with determinants situated at the *R* locus in the long arm of chromosome 10. The *R* locus is complex, and there is now extensive evidence supporting Stadler's (*8*) view that two, among several, components are (S) and (P), conditioning seed and plant color, respectively. Tests show that paramutability at the locus is closely associated, but probably not coincident, with (S). There is evidence also that paramutability is unaffected by loss from the locus of (P). These conclusions rest in part on the results of a comparative study of the response to R^{st} action of two types of mutants from standard R^r : R^g (colored aleurone, green plant) and r^r (colorless aleurone, red plant). Stadler and Nuffer (*9*) postulated that one of the ways in which R^g and r^r mutants could arise is through crossing over, following oblique synapsis of (S) and (P) in $R^r R^r$ plants; such crossing over would result in chromosomes deficient in one or the other of these elements. Emmerling (*10*) showed that a high proportion of R^g mutants from R^r were, in fact, deficient in (P). Brown (*11*) found that nine R^g mutants from standard R^r were indistinguishable from each other and from the parent R^r allele in sensitivity to R^{st} action and in the ability to acquire "secondary paramutagenic potential" following passage through heterozygotes with the stippled factor. [Brown and Brink (*12*) observed that passage of standard R^r through an $R^r R^{st}$ heterozygote not only reduced the pigment-producing potential of R^r but also rendered the allele paramutagenic, although weakly so as compared with R^{st}. They termed this phenomenon secondary paramutation.] Application by Bray and Brink (*13*) of Emmerling's (*10*) test proved that some, but not all, of these R^g mutants were deficient in (P). It is apparent, therefore, that the potential for paramutation at the *R* locus is fully conserved in spite of changes in, and even loss of, the (P) component.

The response of the locus to action of the stippled factor, on the other hand, regularly is lost when R^r mutates to r^r (colorless seed, red plant). Brown (*11*) showed that all members of a random sample of 20 r^r mutants from standard R^r gave a negative result when tested for acquisition of secondary paramutagenic activity in $R^{st} r^r$ plants, whereas R^r and mutant R^g controls all gave positive results. Eleven of the r^r mutants were tested to determine whether, in any instance, the stippled factor reduced the plant-pigmenting potential of r^r as it does that of R^r. Again the results were negative throughout. Evidently loss of the (S) function, presumably either by mutation of (S) to (s) or by (S) deficiency, abolishes sensitivity of the locus to action of the stippled factor.

Ashman (*14*) found that 13 colorless-aleurone, red-plant mutants isolated among the offspring of $R^r R^{st}$ plants, most of which were crossovers between markers bracketing the *R* locus, were nonparamutagenic throughout. All the evidence at present available, there-

Fig. 2 (above). Testcross ears from (left) $r^g r^g$ ♀ × $R^r R^r$ ♂ and (right) $r^g r^g$ ♀ × $R^{r'} R^{r'}$ ♂ matings. The lower degree of kernel pigmentation of the kernels on the ear at right illustrates the genetically persistent reduction in aleurone-pigmenting potential of R^r that regularly follows passage of the factor through a heterozygote with the stippled allele. Fig. 3 (right). Representative ears from testcrosses on $r^g r^g$ females of plants in two $R^{r'} r^g$ families, with pollen collected from separate branches in the tassel of a particular $R^r R^{st}$ individual. Half the kernels on each ear are colorless by virtue of homozygosity for the r^g allele. The darkly colored kernels ($R^{r'} r^g r^g$) on the ear at right show that little change in R^r pigmenting potential occurred in the corresponding tassel branch of the $R^r R^{st}$ plant. The numerous colorless and near-colorless $R^{r'} r^g r^g$ kernels on the ear at left show that marked reduction in R^r aleurone-pigmenting potential occurred in another tassel branch on the same $R^r R^{st}$ plant. [From Sastry, Cooper, and Brink (*16*)]

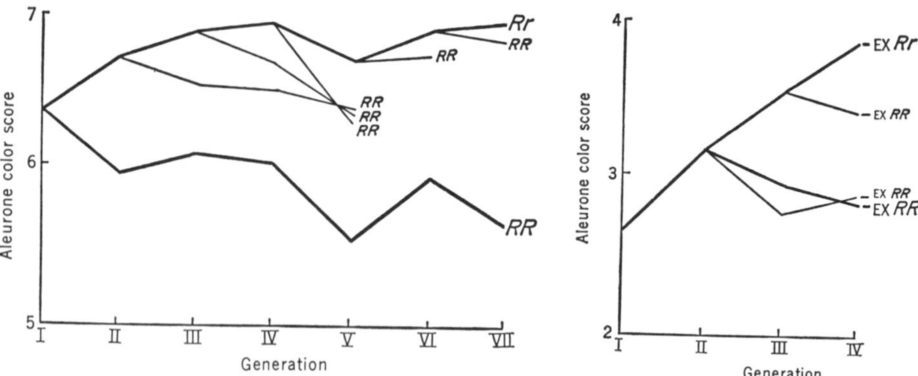

Fig. 4 (left). Progressive divergence in R^r aleurone color expression through six generations of matings, without selection, of $R^r r^g$ and $R^r R^r$ males with $r^g r^g$ females. The R^r allele in question was derived initially from a single source, and the background inheritance throughout was that of the W22 inbred strain. The $R^r r^g$ and $R^r R^r$ lineages were separately propagated by self-pollination of the two kinds of plants within the respective $R^r r^g$ and $R^r R^r$ successions illustrated in the diagram. [From Styles and Brink (*21*)] Fig. 5 (right). Mean aleurone scores of $R^{r'}$ derivatives from standard R^r stocks maintained homozygous and heterozygous with r^g through four generations, showing changes in sensitivity of standard R^r to paramutation when subsequently tested in $R^r R^{st}$ heterozygotes. [From Styles and Brink (*21*)]

fore, supports the conclusion that the mutation of colored to colorless seed leads to a null condition at the locus for paramutation.

The property of overt paramutagenicity likewise resides at the *R* locus. McWhirter and Brink (*15*) found that when the stippled factor (R^{st}) mutates to self-colored aleurone (R^{sc})—an event that occurs with a frequency of about two occurrences per 1000 gametes—paramutagenicity may be unaltered, lost entirely, or changed in varying degrees. The mutation of R^{st} to R^{sc} in some instances without alteration in paramutagenicity demonstrates that the genetic bases of the stippled phenotype and paramutagenicity are distinct from each other. The fact that, in other cases, both functions change together suggests that the two kinds of elements in question are close enough to each other in the chromosome to be involved frequently in a single mutation.

Additional evidence for the compound nature and divisibility of the genetic material involved in *R* paramutagenicity has been adduced by Ashman (*14*) from a study of intralocus recombination at meiosis. He characterized three classes of seed-color mutants among the offspring of

$$+ R^r + / g R^{st} M^{st} \times g r^g +$$

crosses, one of which, comprising 12 near-colorless-aleurone, red-plant individuals, is of particular interest in this context. (M^{st} is a modifier of R^{st} aleurone spotting, about six units distal to the *R* locus.) All 12 mutants were recombinants between the outside markers. It may be inferred from the phenotype that they carried the basic aleurone-color component of R^{st} and also the red-plant-color element of R^r. If these inferences are valid, then we may conclude that the mutants were products of recombination within the *R* locus itself. Ashman showed that all 12 members of the group were paramutagenic, but less so than R^{st}, and, furthermore, that they differed significantly from each other in degree of paramutagenicity. Ashman interpreted this evidence as indicating that the physical basis of paramutagenicity was a chromosomal region that was partitionable by crossing over.

Bray and Brink (*13*) explored the relation between mutation and paramutation at the *R* locus and found it to be complex. Certain regularities, however, were brought to light. The normal form of the standard R^r allele, for example, was observed to mutate to r^r (colorless aleurone) at the rate of 12.7×10^{-4}, whereas the rate for paramutant $R^{r'}$ was much lower—namely, 7.4×10^{-4}. Similarly, the paramutant forms of four out of five R^g mutants from standard R^r had lower mutation rates than their normal counterparts. For the fifth, the mutation rates for R^g and $R^{g'}$ were equal. Paramutant $R^{r'}$ tends to revert toward the pigmenting potential characteristic of standard R^r following extraction from $R^r R^{st}$ heterozygotes, as Kermicle (*6*) reported. Bray and Brink observed that, as the pigmenting ability of $R^{r'}$ rises in such cases, the mutability to r^r also increases.

Paramutation as a Somatic Phenomenon

Sastry, Cooper, and Brink (*16*) found, on sampling separately the pollen in different tassel branches, that many individuals heterozygous for a paramutable and a paramutagenic *R* allele were clearly mosaic with respect to aleurone-pigmenting potential of the paramutable member of the pair (Fig. 3). The occurrence of somatic mosaicism also was established in $R^{r'} r^g$ plants resulting from $r^g r^g$ ♀ × $R^r R^{st}$ ♂ matings, in which partial reversion of the $R^{r'}$ factor toward the pigmenting potential of standard R^r is known to occur. Sectoring for paramutation in the tassel would be expected if paramutation occurs during vegetative growth of the plant and also takes place independently in different cell lineages. The results obtained by Sastry, Cooper, and Brink show that easily recognizable differences in strength of *R* pigmenting potential between tassel sectors within a plant in which paramutation is occurring are sufficiently frequent to give credence to the conclusion that all such plants are paragenetically mosaic in one degree or another.

McWhirter and Brink (*15*) found that a given paramutable *R* allele may be brought to different average levels of aleurone-pigmenting potential by being passed through a series of heterozygotes carrying R^{sc} alleles of varying paramutagenicities. They observed also that paramutation was progressive through successive plant generations. Since the only somatic structures that may be sampled directly in maize are terminal products of growth—namely, tassel and ear shoot—the course of paramutation as plant development proceeds cannot be ascertained directly in this species. The indirect evidence strongly suggests, however, that paramutation is progressive during ontogeny. The occurrence of mosaicism shows that the amount of change varies in different cell lineages, and it may be supposed that the progression is irregular.

Paramutation-like phenomena that were described long before the term *paramutation* was used afford the most direct evidence for the occurrence of paramutation during development of the plant. Hybrids that were morphologically inconstant were found in *Pisum* by Bateson and Pellew (*2*), in *Malva* by Lilienfeld (*3*), and in *Oenothera* by Renner (*4*). Progeny tests, in which seed from phenotypically different branches within a single mosaic plant were used, showed, in each of these cases, that the distribution of offspring varied in accordance with the phenotype of the parent somatic tissue. More recently Hagemann (*17*) has found that in *Lycopersicon* also the heritable changes induced by paramutagenic alleles may appear initially as sectors during vegetative growth.

If paramutation at the *R* locus in maize is comparable to the phenomena described in *Pisum, Malva, Oenothera*, and *Lycopersicon*, pollen collected from an entire $R^r R^{st}$ tassel will embody an array of paramutant $R^{r'}$ alleles representing the terminal states in all the cell lineages that extend into the sporogenous tissue, and in which successive changes in R^r potential during plant development have been accumulated. If the reduction in R^r potential is far advanced in all cell lineages within a plant, all the resulting $R^{r'}$ testcross kernels will have low aleurone-pigmentation scores. Testcrosses involving heterozygotes between standard R^r (an allele highly sensitive to paramutation) and R^{st} (a strongly paramutagenic factor) often yield $R^{r'}$ kernel populations which appear to have uniformly low aleurone-color scores. By means of progeny tests, Kermicle (*6*) has shown, however, that even such seemingly uniform kernel samples may be heterogeneous for level of *R* pigmenting potential. Evidence of this sort suggests that somatic mosaicism for *R* paramutation is the rule. Much variability of *R* pigmenting potential between different cell lineages that arises early in plant development may be extinguished at later stages. The tests of pollen from different tassel branches within the plant made by Sastry, Cooper, and Brink demonstrate, however, a large residuum of mosaicism at the flowering stage.

Recently Coe (*18*) has postulated that paramutation at the *B* locus on chromosome 2 in maize does not occur during vegetative growth of the plant but occurs late in ontogeny, probably at meiosis. Coe reports that F_1 $B\,B'$ hybrids between normal $B\,B$ and homozygous paramutant $B'\,B'$ plants are paramutant in phenotype, and produce only B' gametes. He suggests, however, that the basis of the paramutant phenotype, in this instance, is not change of the $B\,B'$ to the $B'\,B'$ genotype during plant growth but dominance of B' over B. Experiments designed to test this suggestion did not provide clear evidence on the question. Coe (*18*) reports, however, that the comparatively infrequent primary mutation of normal *B* to the paramutagenic form, B', has a somatic basis. He found that such mutants appear as visible sectors during plant development. It may be concluded, therefore, that there is no established exception to the general rule exemplified in *Pisum, Malva, Oenothera*, and *Lycopersicon*, and also at the *R* locus on chromosome 10 in maize, that paramutation takes place in vegetative cells during plant growth.

Effect of Abnormal Chromosome Structure on R Paramutation

The *R* locus lies in the distal half of the long arm of chromosome 10 (10L). Reciprocal translocations involving a break in 10L, and a chromosome carrying a large, distinctive, terminal knob, known as K10, both influence the function and paramutability of *R*.

Insertion of standard R^r into a translocated chromosome (*T*) involving a break in 10L either proximal or distal to the *R* locus, and not necessarily close to the latter, leads to two changes in *R* action: (i) R^r aleurone-pigment-producing potential is enhanced and (ii) the allele is rendered relatively insensitive to paramutation in TR^r/R^{st} heterozygotes. These changes are irregularly expressed, and they persist for at least one generation following return of R^r to a structurally normal chromosome by crossing over. The latter fact proves that the changes in sensitivity are not position effects in the conventional sense of the term (*19*).

Brink and Weyers (*20*) tested a single crossover of standard R^r into a K10 chromosome and observed that sensitivity of the allele to paramutation was significantly reduced as compared with sensitivity of the same factor in a structurally normal chromosome 10. The possibility that the observed change in R^r sensitivity was due, not to the knob, but to some linked factor incidentally present in this particular K10 chromosome has since been explored by testing 12 additional crossovers of paramutable *R* alleles into K10 chromosomes. Three mutant R^g alleles from standard R^r, termed $R^g{}_1$, $R^g{}_2$, and $R^g{}_3$, which coincided with standard R^r in aleurone-pigmenting action and in sensitivity to R^{st} action, were used. The outcome of the experiment was decisive; the R^g alleles that had been incorporated in the knob-carrying chromosome were significantly less sensitive to paramutation in R^{st} heterozygotes than their R^g controls in structurally normal chromosomes in all 12 of the cases tested.

Additional tests have shown that R^r is not made relatively insensitive to paramutation by passage through a plant in which R^r and K10 are present in repulsion—that is, are borne by homologous chromosomes. A number of $R^r\,R^r$ and r^rK10/r^r plants were mated with each other, and the two resulting classes of offspring, R^r/r^rK10 and R^r/r^r, were then crossed with R^{st} R^{st}. Testcrosses showed that $R^{r'}$ alleles derived from the two groups of R^r/R^{st} plants were indistinguishable from each other in aleurone-pigmenting potential.

The experimental results for chromosome K10 suggest, but do not prove, that the terminal knob exerts an indirect effect on sensitivity of R^r to action of the stippled factor by genetically altering the component of the *R* locus immediately involved in paramutation. Since the *R* locus and K10 are widely separated from each other in 10L, and a condition of the change in question is that the two factors be present in the nucleus in coupling, it is inferred that the stimulus exerted by the K10 knob is propagated along the chromosome to the R locus, where the paramutable component then responds by changing quantitatively.

Inherent Metastability of Paramutable R Alleles

Recent investigations have revealed that the functional state of a paramutable R allele can change heritably without association of the allele in a heterozygote with a paramutagenic factor such as R^{st}. The capacity for heritable change in action is a built-in characteristic. The R^{st} allele in a standard $R^r R^{st}$ heterozygote does not "contaminate," or confer any new property on, standard R^r; rather it directs and amplifies a kind of genetic change that standard R^r is capable of undergoing quite independently of R^{st}, or any other partner.

The chain of evidence which leads to this conclusion began with the confirmation by Styles and Brink (*21*) of an earlier observation by Brink and Blackwood (*22*) that R^r genes derived from otherwise comparable $R\,r$ and $R\,R$ plants differed regularly, and the demonstration that the difference resulted from an inherited change in R action. Standard R^r, initially derived from stock $R^r R^r$ cultures, becomes progressively enhanced in aleurone-pigmenting potential when the allele is maintained through successive generations in $R^r r^g$ heterozygotes (Fig. 4). The further fact was established that the sensitivity of standard-R^r action to repression in $R^r R^{st}$ heterozygotes becomes progressively reduced as the aleurone-pigmenting potential of the allele becomes progressively enhanced (Fig. 5). The evidence suggests, therefore, that both phenomena reflect a single kind of change in standard R^r.

Styles and Brink (*21*) noted that the heritable enhancement in pigmenting potential of standard R^r in $R^r r^g$ plants is a paramutational process in that it is a regular and directed change. Standard R^r is thus subject to heritable alterations in both directions in terms of aleurone-pigmenting potential, and the direction of change is controllable. The possibility exists that r is influential in directing the enhancement of standard R^r action, just as R^{st} is influential in directing its repression. Proof that r is paragenetically amorphic in $R\,r$ plants, however, has now been obtained by Styles and Brink (*23*) in a comparative study of the aleurone-pigmenting potentials of $R^g r^r$ heterozygotes and $R^g/ r\text{–}x_1$ hemizygotes. (The symbol $r\text{–}x_1$ represents a small deficiency spanning the R locus; the deficiency is female-, but not male-transmissible.)

In that study, $R^g{}_2$, a mutant from standard R^r which conditions the development of colored aleurone and green plant and is indistinguishable from R^r in sensitivity to paramutation, was maintained for three successive generations in a lineage of $R^g{}_2\, r^r$ heterozygotes and also in an otherwise closely comparable lineage of plants hemizygous for the R locus. The two lineages were maintained by recurrent pollinations of $r^r/r\text{–}x_1$ individuals by $R^g{}_2\, r^r$ and $R^g{}_2/r-x_1$ plants. The effects on $R^g{}_2$ aleurone-pigmenting potential were determined in matings on $r^g r^g$ females. Mean testcross scores for control $R^g{}_2\, R^g{}_2$ plants, of stock-culture origin, ranged between 5.63 and 5.85. Mean testcross scores from $R^g{}_2\, r^r$ lineages in generations 1, 2, and 3 were 5.72, 6.07, and 6.48, respectively. The corresponding mean values from the parallel $R^g{}_2/r\text{–}x_1$ testcrosses were 5.80, 6.03, and 6.42. Thus the aleurone-pigmenting action of $R^g{}_2$ was progressively enhanced in both experimental lineages and, on the average, by almost identical amounts (Fig. 6). It is evident, therefore, that enhancement in the pigmenting potential of a paramutable R allele can occur autonomously, and that paramutational change in the upward direction is not necessarily directed by a partner allele.

Tests with R alleles in paramutant (partially repressed), rather than normal, form provide further evidence of the capacity of paramutable R factors to undergo heritable changes spontaneously. Kermicle (*6*) showed that paramutant R' alleles represented by first-generation progeny of $R\,R^{st}$ plants undergo partial reversion in an R' hemizygote. The amount of reversion is equivalent to that occurring in $R'\,r$ individuals. This conclusion has been amply confirmed by Styles and Brink (*23*) in tests extended over three generations.

The more recent experiment was made in triplicate; the respective $R^{g\prime}\, r^r$ and $R^{g\prime}/r\text{—}x_1$ offspring from three $r^r/r\text{—}x_1$ ♀ × $R^g R^{st}$ ♂ matings were used. The mean $R^{g\prime}$ testcross score for the three initial testcrosses was 2.29 on the commonly used seven-grade aleurone color scale. Matings were made in series, and $r^r/r\text{—}x_1$ plants were used recurrently as the female parents throughout, continuous $R^{g\prime}\, r^r$ and $R^{g\prime}/r\text{—}x_1$ lineages being thereby produced in each of the three replicates. The mean testcross scores in generation 3 for the corresponding $R^{g\prime}\, r^r$ and $R^{g\prime}/r\text{—}x_1$ lineages were 5.04 and 4.62. Thus, pronounced reversion in level of $R^{g\prime}$ action occurred in both lineages. The difference between the two values was no greater than that between values from several parallel mixed lineages, a fact which suggests that the disparity was not meaningful for the comparison in question.

The important deduction from the experimental evidence summarized in this section is that a paramutable R allele, in standard as well as paramutant form, may change spontaneously in directed fashion. The basic condition underlying paramutation is the metastability inherent in sensitive alleles.

Interactions between Paramutable R Alleles

Recent studies by Styles (*24*) have shown that paramutable R alleles of unlike geographic origin and initially unlike in strength of aleurone-pigmenting action are not necessarily permanently unlike. Furthermore, such R alleles were found to interact with each other in heterozygotes in a characteristic way.

Styles chose for detailed study four R alleles from different geographic sources that, in single dose, give complete, or nearly complete, self-color and four that give distinctly mottled aleurone, in plants of the W22 inbred strain. Some of the alleles carried the red-seedling marker; the others gave green seedlings. All eight alleles were shown to be inherently metastable, like standard R^r. The significant fact newly established was that the initial differences between the eight factors were impermanent. Each allele could be altered, by simple mating procedures, until it was essentially indistinguishable from any other. Furthermore, heritable, but not necessarily stable, differences in aleurone-color expression could be produced in different sublines carrying the same allele. The R^r or R^g factors that initially, in single dose, conditioned development of very dark, or self-colored, aleurone gave progressively more mottling when bred as homozygotes, but retained their potential for conditioning self-color when kept heterozygous for r. Alleles that initially conditioned aleurone mottling, in single dose gave progressively divergent degrees of mottling when propagated as homozygotes or heterozygotes. The stocks bred as $R\,r$ heterozygotes usually became more darkly mottled, whereas corresponding stocks bred as homozygotes either became more lightly mottled or did not change in R expression.

Styles (*25*) also observed that paramutable R alleles which differed significantly from each other in aleurone-pigmenting potential when propagated routinely in stock cultures of the W22 inbred strain changed genetically in a characteristic way when maintained as heterozygotes with each other. If R_1 represents a paramutable R allele of relatively high aleurone-pigmenting potential in stock culture and R_2 represents one of lower potential, then, in successive generations of self-fertilized $R_1\,R_2$ plants, the level of pigmenting potential of R_1 tends to fall to that of R_2. Eventually expression of the two alleles becomes indistinguishable.

The parallel between the effect of R_2 on R_1 in $R_1\,R_2$ heterozygotes and the phenomenon which Brown and Brink (*12*) called secondary paramutation is apparent.

It thus becomes clear that the metastability inherent in paramutable R alleles may be manifested in a variety of ways. Apparently uniformity of R expression in these cases, if achieved, is a contingent condition, and does not reflect permanent stabilization of the factor.

Model of a Paramutable R Factor

Brink (*7*) and Sastry, Cooper, and Brink (*16*) have attempted to account for the experimental evidence on R paramutation in terms of the following model. The R locus is assumed to comprise two main components: (i) a structural gene (more probably a gene complex) specifying a particular enzyme involved in anthocyanin formation, and (ii) a heterochromatic segment closely associated with the structural gene, which affects expression of the latter during seed and plant development. The heterochromatic segment, termed the repressor segment, is assumed to consist of varying numbers of a common repeating unit, called a metamere, the effect of which is to repress R action. The degree of repression is assumed to be proportional to the number of metameres making up the repressor segment. An R gene accompanied by a large number of metameres is only weakly expressed, whereas R action is strong if only a few metameres are present. All paramutable R alleles are assumed to possess at least one metamere.

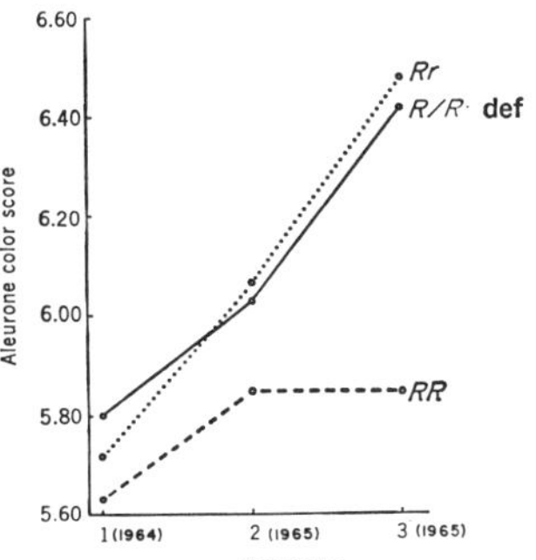

Fig. 6. Aleurone testcross scores following pollination of $r^g\,r^g$ plants with generations 1, 2, and 3 of Rr and R/R-deficient individuals, and scores for $R\,R$ controls. It is apparent that the R-pigmenting potential rises, as the generations advance, at the same rate in R/R-deficient hemizygotes and Rr heterozygotes.

Paramutation is assumed to be a change in the number of metameres making up the repressor segment. It is postulated that such change results from misreplication within the segment during somatic mitosis. The copying of a given repressor segment more than once in a mitotic cycle would lead to an increase, and failure to copy some part of the segment would result in a decrease, in the total number of metameres. The potential strength of R expression would be altered accordingly.

The directed nature of paramutation is assumed to reflect sensitivity of the metamere replication process to regulation in definite ways. It may be supposed, for example, that the stippled allele in an $R^r\,R^{st}$ plant establishes a metabolic condition in the nucleus that favors overcopying of metameres at the R locus in the homologous chromosome. The immediate effect of R^{st} might be to prolong the period in mitosis during which the metameres in the repressor segment serve as templates for the formation of daughter metameres; or first copies of the parent metameres might not be sequestered at once from the replication process but might serve in turn as templates for second copies within the same mitotic cycle. Hemizygosity for R^r, or heterozygosity of R^r for the paragenetically amorphic allele r^r, on the other hand, is assumed to shift the balance of conditions affecting replication in a direction that favors undercopying of metameres.

The postulated effects on metamere replication, however, must be assumed to be permissive rather than determinative. The presence of R^{st} in the nucleus of an $R^r\,R^{st}$ plant, for example, merely increases the probability that, at a given mitosis, the number of metameres associated with R^r will be increased. It does not ensure that such change will always occur, nor does it regulate the amount of change at each mitosis. The mosaicism, in $R^r\,R^{st}$ plants, for level of R^r pigmenting potential is explainable on the assumption that changes in metamere number occur at varying rates in different cell lineages within the individual.

Experimental Tests of the Model

An early observation by Stadler (*26*), since verified by others, suggested a method of testing this model of a paramutable R allele. Stadler found that in maize, contrary to the results with various other plants and animals, high-energy radiations do not ordinarily induce point mutations that give viable homozygotes. Such induced mutations appear to be small deficiencies that behave like recessive lethal genes. If the level of pigmenting potential of a paramutable R factor is a function of the number of metameres in a repressor segment adjacent to R, and if ionizing radiations always cause deficiencies, then effective treatment with x-rays of individuals carrying a partially repressed R allele should always shift the R phenotype in the plus direction. That is, such an R allele, if altered at all, should tend to be derepressed by virtue of the loss of some metameres. It is assumed that such loss of metameres would be only partial in most instances, hence not lethal to the plant.

Fig. 7 (left). Testcross ears from matings on strain-W22 $r^r r^r$ females with (left) pollen from a plant grown from untreated strain-W22 $R_c' R_c'$ seed and (right) pollen from a plant reared from $R_c' R_c'$ sib seed treated with diethyl sulfate. [From Axtell and Brink (*27*)] Fig. 8 (right). Testcross ears from matings on strain-W22 $r^g r^g$ females with (left) pollen from a plant grown from untreated $R^r R^{st}$ seed and (right) pollen from a plant grown from $R^r R^{st}$ sib seed treated with diethyl sulfate. The very lightly spotted kernels, on both ears, are the stippled segregates. It is evident that only the $R^{r'} r^g r^g$ class of testcross kernels has been affected by the diethyl sulfate treatment. [From Axtell and Brink (*27*)]

Experiments were initiated in our laboratory to determine whether in fact high-energy radiations cause unidirectional change in the strength of the action of paramutated *R* genes. Parallel tests were undertaken with chemical mutagens. The two series of investigations are currently in progress.

Chemical Paramutagenesis

Experiments with two chemical mutagens, diethyl sulfate and ethyl methanesulfonate, have advanced to a stage at which a meaningful assessment of the results can be made. The following summary account is based upon recent tests by Axtell and Brink (*27*).

1) The strength of the pigmenting action of paramutated *R* alleles is altered markedly and with high frequency by treatment of seed with either of these chemicals in fresh aqueous solution. The treatment of pollen with saturated diethyl sulfate vapor for periods such that subsequent seed set is reduced even to about 20 percent of normal does not affect the strength of paramutant *R* action.

2) More than 90 percent of the Ml plants reared from diethyl-sulfate-treated seed homozygous for a particular highly repressed paramutant *R* allele, designated R_c', yielded kernels with higher levels of aleurone pigmentation (Fig. 7). Anderson *et al.* (*28*) observed that the average size of tassel sector showing partial fertility following exposure of maize seed to high-energy radiation was approximately two tassel branches. If this observation is generally valid, then pollen collected from a single tassel branch usually would represent a cell lineage in the mature plant based upon a single cell in the embryo of the treated seed. Pollen from individual lateral branches of three R_c' R_c' plants reared from diethyl-sulfate-treated seed was used in testcrosses on colorless individuals. Nine of the ten branches yielded kernels more heavily pigmented than the control seeds. Thus, considered on a cellular basis, the incidence of change in response to the chemical treatment is very high. Obviously the frequency of change is far higher than that characteristic of ordinary mutation, which diethyl sulfate is known to promote also.

3) The altered forms of the R_c' allele recovered from plants reared from seeds treated with diethyl sulfate were heritable, and the deviant phenotypes were found to be dependent on the *R* locus in each example tested.

4) Repressed *R* alleles whose aleurone-pigmenting potential had been raised by treatment with diethyl sulfate were again repressed by passage through R^{st} heterozygotes. This result shows that the effect of the alkylating agent is on the *R*-locus component concerned with paramutation.

5) Paramutation of R^r to $R^{r'}$ in R^r R^{st} plants grown from seed treated with diethyl sulfate is regularly and markedly reduced (see Table 1 and Fig. 8). Significant amounts of change in the strength of *R* action were observed following testcrosses involving all the treated Ml plants. Present data do not serve to distinguish between three possible interpretations of this result: (i) the sensitivity of the R^r allele to R^{st} action may have been reduced; (ii) paramutation of R^r to $R^{r'}$ may have occurred in response to R^{st} action in the heterozygote and then have been reversed by the treatment with diethyl sulfate; or (iii) the paramutagenic action of R^{st} may have been impaired by the chemical.

6) Two lines of evidence suggest that the changes induced in R' action by an alkylating agent, such as diethyl sulfate, are directed. First, the mean pigment-producing potential of R' alleles of initially intermediate potential is regularly raised by treatment of the seed, but not to the original standard-*R* value. Although the average change in phenotype is upward on the pigmenting scale, the possibility that R' may sometimes be changed by the treatment to a more weakly pigmenting form is not excluded. The second body of data supporting the conclusion that the changes are directed is derived from the experiment in which $R^r R^{st}$ seed were treated with diethyl sulfate (Table 1). Scores for the entire population of $R^{r'}$ r^g r^g testcross kernels

derived from each of the ten treated individuals were shifted upward by nearly two grades on the seven-grade aleurone-pigmentation scale. There was no indication in any instances of a reduction in level of $R^{r'}$ pigmenting potential in the progeny of the treated $R^r R^{st}$ plants below that of the untreated $R^r R^{st}$ controls.

It is significant, of course, that the effect of diethyl-sulfate treatment of seed on the pigmenting potential of the R^r allele in $R^r R^{st}$ heterozygotes is opposite to the effect on the potential by the R^{st} allele. The R^{st} allele invariably causes repression of R^r action, whereas the alkylating agent gives a response in the other direction relative to the average $R^{r'}$, or paramutant, value. The changes are directed in both instances, upward on the aleurone-pigment scale in one case, downward in the other.

Table 1. Frequency distributions and mean aleurone color scores for the $R^r r^g r^g$ kernels from testcrosses on $r^g r^g$ females of $R^r R^r$ (untreated), $R^r R^{st}$ (untreated), and $R^r R^{st}$ (diethyl-sulfate-treated) plants. The amount of reduction in level of R^r action ordinarily associated with passage of the allele through a stippled heterozygote is greatly reduced if the F_1 $R^r R^{st}$ seed from which the test plants are grown is treated with diethyl sulfate. [From Axtell and Brink (*27*)]

Staminate plant No.	Aleurone color class							Mean score*
	1	2	3	4	5	6	7	
			R^r R^r, *untreated*					
38-3						6	34	6.85
38-4						8	32	6.80
38-6						20	20	6.50
38-8						14	26	6.65
39-1						12	28	6.70
39-3						17	23	6.58
39-7						10	30	6.75
39-10						16	24	6.60
			R^r R^{st}, *untreated*					
103-1		7	15	11	7			3.45
103-2		3	11	14	12			3.88
103-3		3	6	11	19	1		4.23
103-4	2	2	12	16	8			3.65
103-5	1	2	3	10	17	7		4.53
103-6		1	9	10	20			4.23
103-7	1	4	20	10	5			3.35
			R^r R^{st}, *treated*†					
98-1				2	17	9	12	5.78
98-2				2	5	14	19	6.25
98-3		1	5	8	15	8	3	4.83
98-4						13	27	6.68
98-5			1	6	8	17	8	5.63
98-6			1	7	21	7	4	5.15
98-7			1		11	16	12	5.95
98-8				1	9	16	14	6.08
98-9					9	19	12	6.08
98-10				1	13	18	8	5.83

* Overall mean scores: families 38 and 39, 6.55; family 103, 3.90; family 98, 5.83. † Treatment: 2 ml of diethyl sulfate per 1000 ml of H_2O per 400 minutes ($1.5 \times 10^{-2}M$).

Discussion and Conclusions

A conclusion of major interest to which the recent studies on the *R* locus in maize leads is that paramutability is an intrinsic property of certain alleles. Paramutagenic factors do not confer the means for this kind of genetic change on their partners in heterozygotes; they alter the expression of the paramutability the basis for which the partner already embodies. A paramutagenic factor may exert an influence which leads to a major change in level of pigmenting potential of its partner in a heterozygote, but it is not self-sufficient in this respect. The paramutagenic factor does not instigate paramutation; it acts as an adjuvant to the process, which can occur independently of it. The primary consideration in paramutation is thus the metastability inherent in sensitive alleles.

This point of view has been implicit in earlier reports from our laboratory (*7*). It is now supported by critical evidence. The aleurone-pigmenting potential of a paramutable *R* allele from an *R R* stock-culture origin rises significantly, and at the same rate, in successive generations of otherwise identical *R r* heterozygotes and *R*/*R*-deficient hemizygotes. The capacity for paramutation, as a directed form of heritable change, is inherent in the paramutable allele itself. Paramutation is not dependent, in heterozygotes, on paramutagenic factors. The latter modify the process, and the modifications result in changes in level of pigmenting potential—changes which vary over a wide range.

The evidence for change in the functional state of *R* in plants hemizygous for the locus rules out, of course, any form of interallelic particle transfer as the basis of paramutation. Conversion, as visualized in 1930 by Winkler (*29*) or as shown by Mitchell (*30*) to result from intragenic recombination, does not apply to *R* paramutation. Excluded also is the hypothetical form of conversion recently suggested by Coe (*18*) in seeking an interpretation of paramutation at the *B* locus in maize.

Obviously conversion has no relevance for a hemizygous locus, and if paramutation can occur at such a locus, as has now been proved in the *R* case in maize, the essential basis of the phenomenon cannot be any form of copying from a model in a homologous chromosome or particle transfer between alleles. The term *conversion* as applied by some other investigators to paramutation is misleading in its basic implication, and its use in this context should be abandoned.

It is not yet possible to formulate a general rule applicable to the spectrum of interallelic relations exhibited by the *R* paramutation system. Variation in level of pigmenting potential of paramutable *R* alleles appears to become minimal in lineages of *R R* plants maintained as homozygotes by continued self-fertilization. The two representatives of a given *R* allele in diploid individuals seem to attain a condition of metastable equilibrium in relation to each other under this mating system. The observed relative uniformity of *R* expression in such cases, however, clearly is contingent, and does not reflect permanent stabilization of the allele. The metastability inherent in all such factors regularly comes to expression in the descendants, following intercrosses between different paramutable *R R* stocks and after outcrosses, for example, to $R^{st} R^{st}$, $r^r r^r$, or $r^r/r\text{-}x_1$ plants.

The general significance of the extraordinary sensitivity of the *R* paramutation system to alkylating agents remains to be determined. The available data do not provide a decisive test of the validity of the model described above, wherein level of pigmenting potential of a paramutable *R* factor is assumed to be a function of the number of metameres making up a repressor segment associated with the *R* gene. The seemingly regular increases in pigmenting potential of paramutable *R′* alleles resulting from treatment of seeds with diethyl sulfate could be due to loss of metameres either by direct deletion or though an effect on the chromosome replication process that leads indirectly to metamere losses. The direct deletion of metameres is considered unlikely because of the high frequency of induced changes appearing in the offspring following treatment of the seed from which the parent plant was reared, and because of the insensitivity of *R′* alleles in the sperm present in pollen grains to the action of alkylating agents. Perhaps the striking effects of alkylation result from impairment of the repressor segment as a template in somatic cells—impairment such that undercopying of metameres, and hence partial derepression of *R′*, during chromosome replication results.

References and Notes

1. R. A. Brink, *Cold Spring Harbor Symp. Quant. Biol.* **23**, 379 (1958).
2. W. Bateson and C. Pellew, *Proc. Roy. Soc. London* **B89**, 174 (1958).
3. F. A. Lilienfeld, *Bibliotheca Genet.* **13**, 1 (1929).
4. O. Renner, *Z. Induktive Abstammungs-Vererbungslehre* **74**, 91 (1937).
5. G. Buttin, F. Jacob, J. Monod, in *Heritage from Mendel*, R. A. Brink. Ed. (Univ. of Wisconsin Press, Madison, 1967), pp. 155–177.
6. J. Kermicle, thesis, University of Wisconsin, 1963.
7. R. A. Brink, in *Role of the Chromosomes in Development*, M. Locke, Ed. (Academic Press, New York, 1964), pp. 183–230.
8. L. J. Stadler, *Cold Spring Harbor Symp. Quant. Biol.* **16**, 49 (1951).
9. L. J. Stadler and M. G. Nuffer, *Science* **117**, 471 (1953).
10. M. Emmerling, *Cold Spring Harbor Symp. Quant. Biol.* **23**, 393 (1958).
11. D. F. Brown, *Genetics* **54**, 899 (1966).
12. ——— and R. A. Brink, *ibid.* **45**, 1313 (1960).
13. R. A. Bray and R. A. Brink, *ibid.* **54**, 137 (1966).
14. R. B. Ashman, *ibid.* **51**, 305 (1965); *ibid.* **52**, 835 (1965).
15. K. S. McWhirter and R. A. Brink, *ibid.* **47**, 1053 (1962).
16. G. R. K. Sastry, H. B. Cooper, Jr., R. A. Brink, *ibid.* **52**, 407 (1965).
17. R. Hagemann, *Z. Vererbungslehre* **89**, 587 (1958).
18. E. H. Coe, Jr., *Genetics* **53**, 1035 (1966).
19. R. A. Brink, *ibid.* **46**, 1207 (1961); ——— and M. Blackwood, *ibid.*, p. 1185; R. A. Brink and N. K. Notani, *ibid.*, p. 1223; R. A. Brink and J. Venkateswarlu, *ibid.* **51**, 585 (1965); D. E. Kester and R. A. Brink, *ibid.* **54**, 1401 (1966).
20. R. A. Brink and W. J. Weyers, *ibid.* **45**, 1445 (1960).
21. E. D. Styles and R. A. Brink, *ibid.* **54**, 433 (1966).
22. R. A. Brink and M. Blackwood, *ibid.* **46**, 1185 (1961).
23. E. D. Styles and R. A. Brink, unpublished work.
24. E. D. Styles, *Genetics* **55**, 339 (1967).
25. ———, *ibid.*, p. 411.
26. L. J. Stadler, *Cold Spring Harbor Symp. Quant. Biol.* **9**, 168 (1941).
27. J. D. Axtell and R. A. Brink, *Proc. Nat. Acad. Sci. U.S.* **58**, 181 (1967).
28. E. G. Anderson, A. E. Longley, C. H. Li, K. L. Retherford, *Genetics* **34**, 639 (1949).
29. H. Winkler, *Die Konversion der Gene* (Fischer, Jena, 1930).
30. M. Mitchell, *Proc. Nat. Acad. Sci. U.S.* **41**, 935 (1955).
31. This article is paper No. 1136 from the Laboratory of Genetics, University of Wisconsin. The studies reported were aided by grants from the research committee of the Graduate School of funds supplied by the Wisconsin Alumni Research Foundation; by grants from the National Science Foundation and the Atomic Energy Commission [contract AT (11-1)-1300]; and by a Public Health Service Fellowship (1-F3-GM-39, 065-1) from the National Institute of General Medical Science to J. D. Axtell. The initial version of this article was drafted by R. A. Brink while a guest in the Genetics Section, Division of Plant Industry, Commonwealth Scientific and Industrial Research Organization, Canberra, Australia, under a National Science Foundation Senior Post-Doctoral Fellowship and on study leave from the University of Wisconsin. The hospitality and support are gratefully acknowledged. The preliminary manuscript served as the basis of a lecture by R. A. Brink before the Australian Genetics Society, Melbourne, 13 January 1967.

DNA Modification Mechanisms and Gene Activity during Development

Developmental clocks may depend on the enzymic modification of specific bases in repeated DNA sequences.

R. Holliday and J. E. Pugh

It is generally accepted that the differentiated state of a given type of cell is associated with the activity of a particular set of genes, together with the total inactivity of those sets associated with the differentiation of other cell types. It is also clear that the differentiated state of dividing or nondividing cells is often extremely stable. In this article we suggest mechanisms that may account for this stability and that also attempt to explain the ordered switching on or off of genes during development.

The phenotype of the organism depends on the genotype, and the genetic contribution from both parents is in almost all cases equal. Since the ultimate control of development resides in the genetic material, the actual program must be written in base sequences in the DNA. It is also clear that cytoplasmic components can have a powerful or overriding influence on genomic activity in particular cells, yet these cytoplasmic components are, of course, usually derived from the activity of genes at some earlier stage in development. A continual interaction between cytoplasmic enzymes and DNA sequences is an essential part of the model to be presented.

Modification Enzymes

In bacteria, enzymes exist which modify DNA by methylating adenine in the 6-position (*1*). These enzymes are extremely specific in their action; they modify bases at particular positions in short defined sequences of DNA, which, at least in some instances, form a palindrome. (A palindrome in DNA is an inverted duplication, with twofold rotational symmetry. The $3' \rightarrow 5'$ base sequence is the same on each strand.) These modifications prevent the DNA being degraded by restriction enzymes, which are equally specific in their action. In higher organisms, bases are also modified: 5-methylcytosine is a common component of DNA (*2*), and 6-methyladenine has been identified in simple eukaryotes (*3*). It is not yet known whether these modifications occur at specific sites. In the case of transfer RNA (tRNA) of both bacteria and higher organisms, a number of bases are modified at specific sites (*4*).

The methylation of adenine in DNA is not heritable in the usual sense, but a bacterium with a modification enzyme could, in principle, have a very different phenotype to one without if the presence or absence of methylation affected transcription. Hawthorne (*5*) and Scarano (*6*) have suggested that certain other base modifications could lead to heritable changes in base sequences and that these could control the activity of adjacent structural genes. We explore these possibilities further and suggest that such changes could operate developmental clocks which turn genes on or off after a specific number of cell divisions. In addition, we propose that the same ordered control of the transcription of genes could be achieved by the methylation of bases, without changes in sequence. The modification mechanisms are as follows.

1) To explain the instability of the mating type loci in certain strains of yeast, Hawthorne (*5*) has suggested that an operator region could exist in two alternative states. One state has A · T (adenine · thymine) base pairs at particular sites in the controlling region, and the other has G · C (guanine · cytosine) base pairs at the same sites. The transition from A · T to G · C or G · C to A · T requires cell division, and it occurs as follows. The modification of adenine at particular sites could occur by the removal of an amino group at the 6-position. This gives rise to inosine (I), and it is known that inosine base pairs with cytosine rather than thymine (*7*). Therefore one round of replication after modification will produce A · T and I · C, and at the following replication a G · C base pair is formed. The reverse transition occurs by the action of a second modifying enzyme, which removes an amino group from the 6-position of cytosine to give uracil (U). After two rounds of replication, the original A · T base pair is restored. These transitions are illustrated in Fig. 1a. The change in base sequence that occurs is irreversible if only one of the two modifying enzymes is present. Since there is now genetic evidence that mismatched bases in DNA are repaired to give normal base pairing (*8*), Hawthorne suggests that the modification occurs in the short stretches of single-stranded DNA in the replication fork. There would therefore be no opportunity for the repair of mismatched bases such as I · T or G · U.

2) Another possibility, which has been discussed by Scarano (*6*), in connection with the problem of differentiation, depends on the methylation of cytosine at the 5-position, followed by deamination at the 6-position to give thymine. In this way a G · C pair would be changed to an A · T pair after replication (Fig. 1b). The amination of thymine to 5-methylcytosine, which pairs with G, will give the reverse change.

3) To maintain methylated bases in DNA, a modification enzyme must always be present. To segregate methylated from nonmethylated cells, two enzymes are necessary. One model is illustrated in Fig. 2. The first enzyme, E1, methylates one strand within a stretch of palindromic DNA and the other strand just outside this stretch. This does not provide a substrate for the second enzyme, E2, until replication occurs, and then only one of the daughter molecules is methylated by E2. This enzyme resembles bacterial modification enzymes in adding a methyl group to the other half of the palindrome, but it differs in not acting on nonmethylated DNA. [The same discrimination between half-methylated and nonmethylated DNA is shown by restriction enzymes of bacteria (*1*).] Therefore, once both strands of the se-

R. Holliday is head of the Genetics Division, National Institute for Medical Research, Mill Hill, London NW7 1AA, England, and J. E. Pugh is a Medical Research Council Scholar in the same division.

quence are methylated, all subsequent progeny are modified, provided that E2 remains present. The other product of the first division segregates further methylated cells as long as E1 is present, but in its absence only nonmethylated progeny will be formed. The methylated state could be extremely stable, as the methyl groups would be diluted out only if the modification enzyme is lost through mutation in its structural gene. However, an essential part of our model is the switching on or off of the genes for modification enzymes during development or differentiation.

In bacteria and their viruses, specific mechanisms for the control of gene activity at the level of transcription are well known, and it has been shown that operator regions have palindromic features (*9*). It is generally believed that similar control mechanisms must exist in higher organisms. There are several simple ways in which changes in base sequence or methylation could determine whether or not a particular gene is transcribed, some of which have already been discussed by Venner and Reinert (*10*). One possibility is that the sequence where modification occurs is also an operator sequence to which a repressor binds. In one state of the DNA the repressor binds to the operator and the contiguous structural gene is inactive. In the other state the repressor does not bind to the operator and transcription occurs. Alternatively, modification could occur in the promoter sequence, that is, the short region of DNA to which the transcribing RNA polymerase first binds; in one modified state the gene would be transcribed and in the other it would not. [It is known in several instances that promoter regions contain short palindromes, since they can be attacked by restriction enzymes (*11*).] Binding sites for RNA polymerase will be common to many or all structural genes, yet the modification enzyme is specifically inactivating or activating particular genes. We must therefore postulate that the specificity of binding is provided by a defined sequence adjacent to the promoter, but that modification actually occurs in the promoter region. A third possibility, even simpler, is that base changes would simply introduce (or remove) a polypeptide chain terminating sequence within a structural gene.

In the subsequent discussion we often refer to enzymes which modify DNA as controlling enzymes and to their substrates as controlling sequences.

Somatic Segregation of Gene Activities

The modifications outlined in Figs. 1 and 2 can result in the formation of an unaltered cell, together with one in which a particular gene is activated or inactivated. This situation is like that of a stem line cell which continually divides to form cells with new functions. The stem line cell is unstable, but the daughter cells which are modified are quite stable. However, although the switching on of a single gene may commit the cell to differentiation, it is unlikely to be sufficient to bring about all the changes required for differentiation. One obvious possibility is that the first activated gene codes for another modifying enzyme that is active at several other sites in the genome, which have the same controlling sequence. This may, for instance, shut off genes whose activity is necessary for cell division and turn on others which synthesize those proteins that give the cell its particular properties. It is easy to see how somatic segregation could lead to the triggering of sequential regulatory events, or the type of cascade regulation discussed by Britten and Davidson (*12*) and Pontecorvo (*13*). A further possibility is where a gene is switched on transiently. A controlling enzyme may modify the controlling sequence adjacent to its own structural gene. The enzyme is first switched on by the action of some other controlling enzyme, but as soon as it is synthesized it overrides the action of the first enzyme by reversing the modification. In this way a controlling enzyme would only be present for one or two cell divisions, but during this time it could, of course, affect the activity of other genes.

Certain complications could arise when one considers the possible segre-

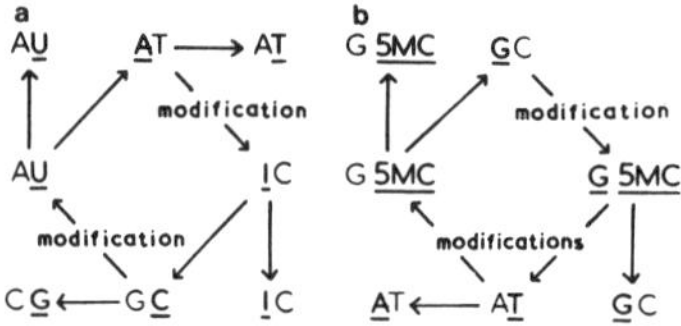

Fig. 1. (a) The transition from an A · T to a G · C base pair and the reverse transition, based on the scheme of Hawthorne (*5*). The modification of adenine occurs by the removal of an amino group at the 6-position to give inosine (I), which pairs with cytosine (C). The modification of cytosine occurs by the removal of an amino group at the 6-position to give uracil (U), which pairs with adenine. The modifications occur at the time of replication. Bases in conserved strands are underlined. (b) The transition from a G · C to an A · T base pair and the reverse transition, based on the scheme of Scarano (*6*). Cytosine is methylated at the 5-position (5 MC) and then deaminated at the 6-position to give thymine. Only one replication is necessary for the change from G · C to A · T. The reverse change involves the addition of an amino group to thymine at the 6-position to give 5-methylcytosine, which pairs with G at the subsequent replication.

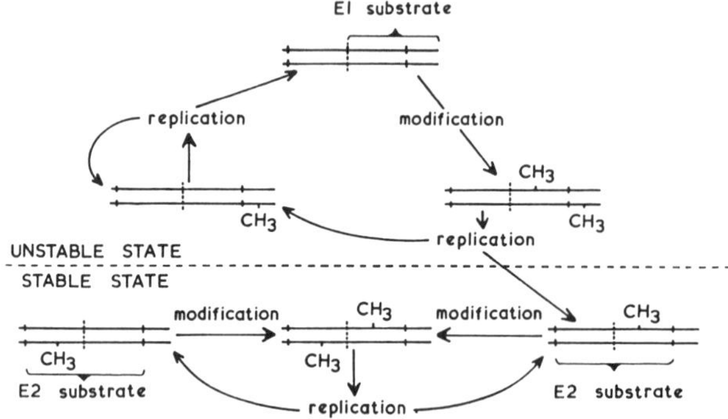

Fig. 2. The segregation of methylated DNA from an unmethylated precursor. The first modification enzyme, E1, methylates one-half of a palindromic sequence and an adjacent sequence in the complementary strand. Replication provides one substrate for the second enzyme, E2, which methylates the other half of the palindrome and all subsequent progeny molecules. In the presence of E1 and E2, unmodified or partially modified cells continually give rise to stable modified ones. If E1 is inactivated or disappears, stable modified and unmodified cells are formed.

gation events that can occur in the division of a diploid cell. Each controlling sequence occurs on each homolog, and therefore modification at both sites will result in the production of daughters with one, two, or no modifications in the ratio 1 : 2 : 1. In many instances this may not matter, as a single activated gene may set in train all the required changes. Another possibility which we favor is that controlling sequences are modified and activate genes on only one of the two homologous chromosomes. In differentiated antibody forming cells, only one of the two structural genes in a heterozygote is transcribed (*14*), and in female mammals only one of the two X chromosomes is active (*15*). The mechanisms discussed suggest how modification could occur in only one of two homologous chromosomes in a cell. Suppose the controlling sequence is a substrate for both a repressor and a modification enzyme, E1, but the affinity of the repressor is very much greater. There is therefore a low probability of modification, and a very much lower probability that both controlling sequences on both chromosomes will be modified in one cell generation. Once modification has occurred, it prevents repressor binding and allows transcription of the adjacent structural genes. The products of these inactivate E1 and switch off its synthesis. Provided that the initial modification is in both arms of a palindromic sequence, a maintenance enzyme, E2, keeps one chromosome methylated in all subsequent generations, as in the lower half of Fig. 2, whereas its homolog is unmethylated and remains so. We do not propose that this simple model will alone account for whole X chromosome inactivation.

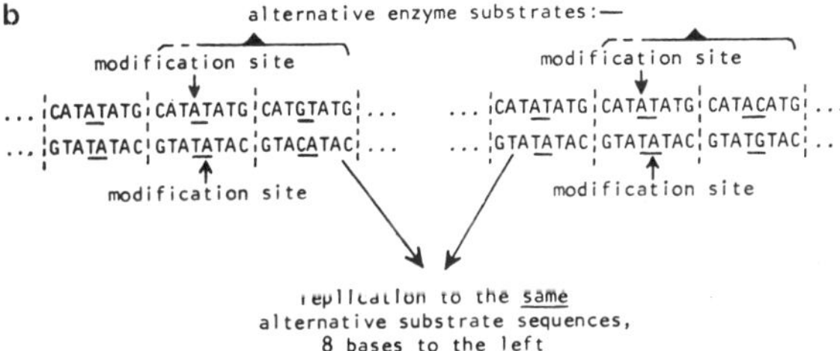

Fig. 3. (a) A mechanism for counting cell divisions based on the A · T → G · C transition in Fig. 1a. The modification enzyme recognizes the first sequence of eight bases, because it contains G at the 5-position, together with the whole or some part of the sequence to its left. The A at the 5-position of this second sequence is changed to G to give a new recognition sequence eight bases to the left. When all the sequences have been successively modified, a structural gene at the extreme left of the repeated controlling sequences (not shown here) is activated. (b) The modification of both strands of a controlling sequence. The controlling enzyme recognizes a sequence which has a G · C pair at either the 4- or 5-position. (If each controlling sequence forms a short palindrome, the recognition sequences can be structurally identical, as shown here.) In both cases it modifies A at the 4- and 5-position of the sequence to the left. Both these modified sequences then become recognition sequences after replication. The modifications therefore move progressively from right to left and count cell divisions as in (a).

Segregation of gene activities is not the only important event, as we have to consider also the mechanisms whereby all the progeny from a particular cell are altered or differentiated in the same way at a particular time in development. The application of the model to this situation is developed below.

Developmental Clocks

It can be readily seen how in principle the modification mechanism could enable a cell to count the number of divisions it has gone through during a particular stage in development. Consider the hypothetical repeating sequences shown in Fig. 3a. At the right end there is a sequence to which the modification enzyme binds. This sequence is first modified by an A → I → G transition. When this has occurred, the site of action for the enzyme has now moved eight bases to the left. This process will be repeated as many times as there are repeats of the sequence. At the end of the precisely determined number of divisions, the operator or promoter site is altered in the way that has been mentioned and the developmental switch comes into operation.

Such a developmental clock will not operate precisely if the bases modified are on one strand. In this case modified and unmodified strands segregate, and the subsequent progeny of a single cell will have modified a varying number of control sequences. This difficulty is avoided if both strands are modified (Fig. 3b). There is a binding sequence for the controlling enzyme which can exist in two forms, differing in at least two base pairs. It is adjacent to a very similar sequence which will be modified in both strands by the enzyme. These sequences when modified become the same as the binding sequences. Therefore the modifications move progressively from one end of the region of repetitive DNA to the other, and the divisions are counted.

Methylating enzymes may also count cell divisions. Of several possible mechanisms we describe one (Fig. 4). The clock is started by E1, which acts on a specific substrate at one end of the repeated sequences. It methylates one strand, and this is an essential signal for the second enzyme, E2, which inserts further methyl groups on both strands within the next sequence. This enzyme cannot act on DNA methylated on both strands in one sequence, but it does so after replication again provides substrates with only one methylated strand. In this way an additional segment of the clock is modified at every division until the end is reached, whether or not all the sequences behind the growing points are methylated.

In both these types of clock, all the offspring from a progenitor cell will reach the same stage of development after they have gone through some specified number of divisions. The clock may, of course, trigger one or several segregation events that lead to specific differences in cell types within the clone. Separate clocks could run sequentially at the same time, overlapping each other in time within one cell lineage.

Britten and Davidson (*12*) have pointed out that the existence of multiple repeats of DNA sequences in the genome suggests that common regulatory sequences may be adjacent to many different structural genes. The developmental clocks that we have described would suggest an additional function for repetitive DNA which is not transcribed. These sequences would be tandem repeats of palindromes. Evidence for the existence of many such sequences in the DNA of higher organisms has been obtained (*16*).

Development of the Chick Limb Bud

The recent experiments on the early development of the chick wing (*17*) provide a convincing example of a developmental clock. The tip of the limb bud, which is called the progress zone, contains dividing cells, and the products of division form in strict sequence the various structures of the limb from its base to the extremity. If the progress zone from a limb in which the basic structures are nearly fully formed is transplanted to a very young limb from which the progress zone has been removed, then none of the structures are produced. On the other hand, if a young progress zone replaces one on the end of a wing which has already laid down all essential structures, then another wing is formed at the end of the first. In this case, the sequence of bone rudiments would be humerus, radius or ulna, hand, humerus, radius or ulna, hand. These results show that there is a temporal order in the laying down of successive structures, and this order might very well be related to the number of cell divisions that have elapsed in the cells of the progress zone.

A Clock for Aging?

The life-span of an organism is under genetic control, and it has frequently been asserted that there must be a developmental program for aging. More specifically, it has been suggested that the aging program might be related to division potential of cells, because diploid cells in culture have a clearly defined life-span which is dependent on the number of population doublings rather than chronological time (*18*). Current interest in mechanisms of aging has centered around error theories, for which some evidence has been published. If, instead, the life-span of these cells is programed, we think that a clock of the type outlined in Figs. 3 and 4 might provide the necessary specificity in doubling potential before senescence and cell death occurred. When the clock runs out, there are many possible deleterious or lethal events that might be triggered. For instance, the enzymes for chromosome replication or any other essential cellular function may be switched off; alternatively, there may be a general reduction in the accuracy of information transfer between macromolecules.

There is no doubt that programed death of certain tissues or groups of cells is a normal component of embryogenesis and development (*19*). This program could be based on the clock mechanisms we suggest; furthermore if restriction enzymes (specific deoxyribonucleases) (*1*) occur in higher organisms, substrates for these might be created by the loss or gain of modification enzymes in particular cells, and this would be followed by the degradation of the DNA and death of these cells.

The Developmental Program

The combination of developmental clocks and precise segregation mechanisms which together determine which genes will be activated provides the essential requirement for an ordered genetic program for development. One can describe the determined changes as being part of a developmental tree, where, at precise times during development, cells branch out into different

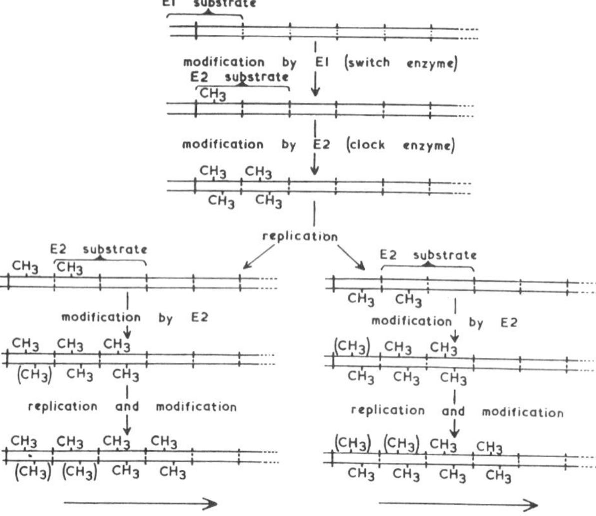

Fig. 4. A mechanism for counting cell divisions based on the methylation of palindromic controlling sequences. The first enzyme, E1, switches on the clock by recognizing a starter sequence, at the extreme left, which is adjacent to the first of the repeated sequences of the clock. One strand of this sequence is methylated by E1, and this provides a substrate for E2, which inserts three more methyl groups in the first two controlling sequences. E2 does not act further once both strands are modified. However, after replication new substrates of E2 are formed, allowing the next sequence to be methylated. (All the sequences behind the "growing point" may become modified, but this does not affect the clock mechanism.)

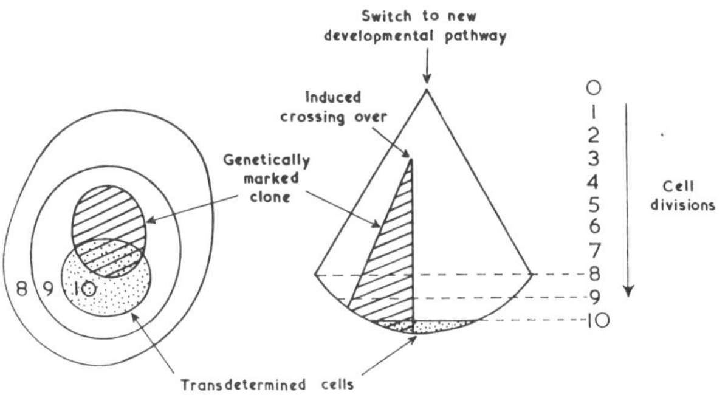

Fig. 5. An explanation for transdetermination based on an event in a single cell. The diagram on the right represents a clone derived from a cell in which a developmental clock has been triggered. It takes ten divisions for the clock to induce transdetermination in a group of cells. The induction of mitotic crossing-over by irradiation occurs after the clock has started and it produces a genetically marked clone (or patch of cells, as shown on the left) which can overlap the transdetermined region (*33*).

sublines that later themselves become subdivided into more diverse classes. At each stage the cells become more and more tied to a specific pathway of development and more and more distinct from cells derived from branches of the tree. Moreover, once a differentiated state has been reached, the model explains why it is so stable. Base changes are obviously stable, and methylated bases are maintained provided that the necessary enzyme is present. They would be lost only as a result of mutation in the structural gene for the modification enzyme.

Although the model suggests that development is clonal, it should be noticed that specific events can occur in groups of cells rather than individual ones. For instance, after fertilization a developmental clock or clocks may be set so that after n divisions one or more segregation events are triggered. At this time the 2^n cells that have been formed may segregate into two or more types of cell. Embryonic cells with specific cell surfaces can recognize each other (*20*); these cell types may therefore aggregate together into groups. This is possibly what happens when different embryos are fused to form mosaic allophenic mice (*21*).

The model described is more likely to provide an explanation for the ordered development of embryos of the mosaic type than it is for those of the regulative type. In the former, exemplified by *Drosophila* (*22*), the parts of the embryo are rigidly determined to develop into particular larval or adult structures. Removal of groups of cells of the determined embryo results in loss of specific differentiated structures. In the latter, removal of parts of the developing organism, or the inhibition of cell division, may simply result in the formation of a smaller complete organism at some later stage in development (*23*).

We do not wish to underestimate the importance of a cell's environment in the determination of its subsequent fate during development. It is, for instance, widely believed that the pattern of development is determined by fields or gradients set up by organizing or signaling cells or groups of cells (*24*). We would simply point out that the origin of such a situation must initially depend on programed differences between cells in the developing organism, and some of these differences could come about in the way suggested. Moreover, some of the switches or clocks we have discussed could be triggered by hormonal or other influences, or alternatively they may determine how a cell will respond to such stimuli. It may be significant that some cells can accept positional information only if they are first appropriately conditioned by cell division (*25*).

Totipotent Nuclei

In extensive experiments by Gurdon and his associates (*26*), nuclei from differentiated cells were injected into anucleate eggs. For instance, nuclei from tadpole intestinal epithelial cells supported embryonic development to a stage where many types of differentiated cell were present, and in some instances adult animals were formed. In other experiments, nuclei derived from adult skin tissue were successfully transplanted into anucleate eggs. These results show that nuclei in differentiated cells can be reprogramed by egg cytoplasm: they are totipotent because they can subsequently give rise to all other types of cell. There are, however, types of differentiated cell such as neurons, which contain nuclei that do not support normal development after transplantation, and in these cases the changes in genetic activity that occur during differentiation may be irreversible.

The modification mechanisms described are all reversible; therefore it is possible that a battery of specific enzymes exists in the egg cytoplasm which recognize controlling sequences and reverse many—although not necessarily all—modifications. The specificity of these enzymes may be somewhat less than those which originally introduced the modifications during development. For instance, all the modified bases in the repeated sequences of a developmental clock could be erased at a single step. It is not at all unlikely that there is a special mechanism for reprograming in the egg cytoplasm, since apart from the transplantation experiments just mentioned, the oocyte and the spermatozoan are highly specialized products of meiotic division, the nuclei of which themselves have to be reprogramed. Nevertheless, we find it hard to believe that reprograming could actually involve base changes in the DNA, and for this reason we tend to prefer the version of our model that depends on methylation of particular bases. A general demethylating enzyme is a possibility (provided that it was removed before the embryo started development), but we prefer the following alternative.

If the egg cytoplasm contains no maintenance enzyme, then methyl groups will simply be diluted out during the early cleavage divisions. At each nuclear division the number of chromosomes containing methyl groups will be reduced by half. After x divisions the probability of any one chromosome remaining modified is $2n/2^x$, where n is the haploid chromosome number. Where n is between 10 and 30, it would need between 11 and 13 divisions to reduce the number of cells containing at least one modified chromosome to 1 percent. We suggest that an initial clock, immune from the diluting out process, would set in train the whole process of development after approximately this number of nuclear divisions has occurred.

Effect of Bromodeoxyuridine on Differentiation

Perhaps the strongest evidence that DNA is directly implicated in differentiation comes from numerous studies with the thymidine analog bromodeoxyuridine (BrdU). It has frequently been shown that low concentrations of BrdU which are nontoxic to cells specifically inhibit differentiation or development. There is no effect if excess thymidine is added at the same time as the analog, but in several instances once the BrdU is incorporated into DNA, the block in differentiation cannot be reversed by adding excess thymidine. Only a few of the many examples of the specific action of BrdU can be mentioned [for a full review, see (*27*)]. Myogenic cells can be cultivated in vitro for several days. After this time, DNA synthesis ceases, the cells fuse to form multinucleate tubules and synthesize the contractile proteins actin and myosin. Bromodeoxyuridine does not prevent the myoblasts from proliferating, but its presence, even for one cell division, completely inhibits their differentiation (*28*). Mesoderm of the chick limb bud differentiates into cartilage in cell culture, and this differentiation is irreversibly blocked by the substitution of approximately 2 percent of the thymine in DNA with the bromo analog. After treatment with BrdU is terminated, the analog rapidly disappears from the dividing cells, but even so differentiation does not then occur (*29*). In other instances, the analog is diluted out by replication and differentiation follows. Finally, it has been shown that BrdU blocks the development of embryos if applied at an early cleavage stage (*30*).

We propose that these effects are brought about by the substitution of a bromine atom for the methyl group on the 5-position of thymine, and this prevents the normal modification of controlling sequences during development. This could occur either by preventing the loss or gain of a methyl group of a particular pyrimidine base (for instance, in the change from cytosine to thymine previously mentioned) or, more generally, by altering the action of modification enzymes on controlling sequences containing BrdU-substituted DNA.

Determination and Transdetermination in Drosophila

The stability of differentiated cells has already been mentioned, but we now turn to the remarkable studies of Hadorn and his associates (*22, 31*), who have demonstrated that the determined state of undifferentiated larval cells can be very stable. The adult structures of *Drosophila* are formed during metamorphosis from imaginal discs in the larva. Imaginal disc tissue can be grown in the abdomen of adult flies and continuously propagated by transfer of pieces of tissue to fresh adults. Disc tissue reimplanted in larvae differentiates during metamorphosis to produce a particular adult structure, such as part of a wing, leg, or antenna. This is triggered by the hormone ecdysone, which activates the developmental program and allows the further events required for differentiation to proceed. A particular line of disc cells is determined to produce a specific adult structure. This is inherited from cell to cell, as disc fragments have been subcultured for more than 70 transfer generations over a period of several years without any change in determined state. We suggest that this stability is due to the inheritance of appropriate modifications in their DNA. Sometimes disc tissue that is determined to develop in one direction spontaneously changes to another determined state. This transdetermination never occurs in the absence of proliferation; indeed, its frequency is related to the number of cell divisions which have occurred. If during growth the modifications are occasionally lost, then these cells may move into a determined state distinct from the first one. It is a characteristic of transdetermination that specific changes occur more frequently than others and that successive changes follow particular pathways.

Kauffman (*32*) has presented a detailed model for determination based on the setting of a number of bistable states, or developmental switches. The various pathways for transdetermination are explicable if the setting of the switches alters with given frequencies, one state changing to the other more frequently than the reverse change. His analysis is quite consistent with our model if the switches are modified or unmodified states of particular controlling sequences and one change, for instance the failure to methylate DNA, is more frequent than the reverse.

By use of mitotic crossing-over to mark particular groups of cells, it is possible to show that a patch of tissue in which transdetermination has occurred can occasionally overlap one which has arisen as a result of mitotic crossing-over in a single cell. It is therefore impossible for each patch to be an individual clone, and it has consequently been argued that transdetermination occurs in groups of cells (*33*). However, this argument no longer holds if transdetermination depends on two events: first, the reversion in one cell to a predetermined state, then a given amount of proliferation to a new determined state. If only a proportion of the cells in the clone have undergone sufficient divisions to reach the new determined state (and such cells are known to aggregate together), then it is quite possible for the patch from mitotic crossing-over to be included within this larger clone, only part of which has undergone transdetermination (Fig. 5).

Homoeotic mutants are those that produce developmental defects analogous to transdetermination. For instance, the mutation *aristapedia* in *Drosophila* results in the development of a leg structure in place of part of an antenna (*34*). Such mutants may have a defect either in a controlling enzyme, which fails to recognize a particular controlling sequence, or alternatively they might have an altered controlling sequence which is not recognized by the appropriate controlling enzyme. As a result, cells are channeled into an alternative developmental pathway. It has been shown that a homoeotic mutant can mimic transdetermination in that the developmental abnormality originates in a group of cells rather than in one. But in this case the cells are part of a larger clone, the whole organism, with a particular genetic defect. In a similar way a patch of transdetermined cells could originate from a larger clone derived from a cell with altered DNA.

Conclusions

We are aware that no direct evidence exists for specific modification enzymes in eukaryotes, let alone that such enzymes might exercise control of gene activities. Nevertheless, in view of our almost complete ignorance of the mechanism for the unfolding of the genetic program during development, it seems justifiable to suggest speculative hypotheses that may lead to meaningful experimental approaches, particularly when these hypotheses are based on some of the known features of modification systems in bacteria. It is significant that Sager (*35*) has argued, from a quite different viewpoint, that restriction and modification mechanisms may exist in higher organisms.

A direct search for specific modification enzymes and modified bases in specific sequences will be difficult, as the number of controlling sequences of any one type in the genome may be only one or a few. Methylases have been identified in sea urchin embryos (*36*), and there is evidence that the distribution of methyl groups in DNA is not random. It may be significant that the doublet CpG is the most highly methylated (*6, 36*), but occurs much less frequently than expected from the overall base composition of eukaryotic DNA (*37*). A search for the transition of cytosine to thymine by methylation and deamination has not so far been successful (*38*).

Although further study of methylases and the pattern of methylation of certain families of reiterated DNA in different tissues or at different stages of development might well be profitable, we feel that it is unlikely that biochemical studies alone will provide direct evidence for our model. The use of developmental mutants is probably essential, since by comparison with wild-type organisms it may be possible to identify the nature of their biochemical defects. We would predict two general classes of mutant: those with altered controlling sequences, which may be dominant (as in the case of operator constitutive mutations); and those with altered controlling enzymes, which would usually be recessive and obtainable in temperature sensitive form. Analysis of developmental pathways can be assisted by the use of homoeotic mutants, and in this connection we agree with McClintock (*39*), who has emphasized that, if the ordered processes of development are deranged, then genes which usually become active at very specific times may instead be activated spasmodically or in random fashion during development. Her studies with maize [for a review, see (*40*)] have led to the discovery of unstable states and controlling elements. The latter not only control the stability and level of expression of nearby genes, but also transpose from one chromosomal location to another. The possibility of transposition of genetic elements has also been discussed in connection with the problem of immunoglobulin synthesis from genes coding for constant and variable regions (*41*). Many of the properties of such systems as McClintock's could, we believe, be explained on the basis of repeated sequences of controlling DNA, which could dissociate from and reassociate with several chromosomal sites by means of genetic recombination. What may now be needed is an examination of these genetic elements in a higher organism in which both biochemical and genetic studies can be undertaken.

References and Notes

1. W. Arber, in *The Bacteriophage Lambda*, A. D. Hershey, Ed. (Cold Spring Harbor Laboratory, Cold Spring Harbor, N.Y., 1971), p. 83; M. Meselson, R. Yuan, J. Heywood, *Annu. Rev. Biochem.* **41**, 447 (1972); J. D. Smith, W. Arber, U. Kühnlein, *J. Mol. Biol.* **63**, 1 (1972).
2. J. Doskocil and F. Sorm, *Biochim. Biophys. Acta* **55**, 953 (1962); G. R. Wyatt, *Biochem. J.* **48**, 584 (1951).
3. M. A. Gorovsky, S. Hattman, G. L. Pleger, *J. Cell Biol.* **56**, 697 (1973).
4. D. Soll, *Science* **173**, 293 (1971).
5. D. C. Hawthorne, personal communication and in preparation.
6. E. Scarano, *Adv. Cytopharmacol.* **1**, 13 (1971).
7. D. R. Davies and A. Rich, *J. Am. Chem. Soc.* **80**, 1003 (1958).
8. R. Holliday, *Genetics*, in press.
9. W. Gilbert, N. Maizels, A. Maxam, *Cold Spring Harbor Symp. Quant. Biol.* **38**, 845 (1973); T. Maniatis, B. G. Barrell, M. Ptashne, J. Donelson, *Nature (Lond.)* **250**, 394 (1974); B. Lewin, *Cell* **2**, 1 (1974).
10. H. Venner and H. Reinert, *Z. Allg. Mikrobiol.* **13**, 613 (1973).
11. B. Allet, R. J. Roberts, R. F. Gesteland, R. Solem, *Nature (Lond.)* **249**, 217 (1974); R. Maurer, T. Maniatis, M. Ptashne, *ibid.*, p. 221.
12. R. J. Britten and E. H. Davidson, *Science* **165**, 349 (1969); E. H. Davidson and R. J. Britten, *Q. Rev. Biol.* **48**, 565 (1973).
13. G. Pontecorvo, *Proc. R. Soc. Lond. Ser. B* **158**, 1 (1963).
14. B. Pernis, *Cold Spring Harbor Symp. Quant. Biol.* **33**, 333 (1967).
15. M. F. Lyon, *Biol. Rev. (Camb.)* **47**, 1 (1972).
16. D. A. Wilson and C. A. Thomas, *J. Mol. Biol.* **84**, 115 (1974); T. R. Cech, A. Rosenfeld, J. E. Hearst, *ibid.* **81**, 299 (1973).
17. D. Summerbell, J. Lewis, L. Wolpert, *Nature (Lond.)* **244**, 492 (1973).
18. L. Hayflick, *Exp. Cell Res.* **37**, 614 (1965); J. Ponten, *Inst. Natl. Santé Rech. Med.* **27**, 53 (1973).
19. A. Glucksmann, *Biol. Rev. (Camb.)* **26**, 59 (1951); J. W. Saunders, Jr., *Science* **154**, 604 (1966).
20. S. Roth, *Dev. Biol.* **18**, 602 (1968); R. Nöthiger, *Wilhelm Roux' Arch. Entwicklungsmech. Org.* **155**, 269 (1964).
21. B. Mintz, *Symp. Soc. Exp. Biol.* **25**, 345 (1971); M. N. Nesbitt and S. M. Gartler, *Annu. Rev. Genet.* **5**, 143 (1971).
22. H. Ursprung and R. Nöthiger, *Biology of Imaginal Discs* (Springer Verlag, Berlin, 1972).
23. J. S. Huxley and G. R. de Beer, *The Elements of Experimental Embryology* (Cambridge Univ. Press, Cambridge, 1934); J. Cooke, *Nature (Lond.)* **243**, 55 (1973).
24. L. Wolpert, *J. Theor. Biol.* **25**, 1 (1969); F. H. C. Crick, *Symp. Soc. Exp. Biol.* **25**, 429 (1971).
25. P. A. Lawrence, F. H. C. Crick, H. Munro, *J. Cell Sci.* **11**, 815 (1972).
26. J. B. Gurdon, *J. Embryol. Exp. Morphol.* **10**, 622 (1962); ——— and R. A. Laskey, *ibid.* **24**, 227 (1970); J. B. Gurdon and V. Vehlinger, *Nature (Lond.)* **210**, 1240 (1966).
27. H. Holtzer, H. Weintraub, R. Mayne, B. Mochan, *Curr. Top. Dev. Biol.* **7**, 229 (1972).
28. R. Bischoff and H. Holtzer, *J. Cell Biol.* **48**, 523 (1971).
29. D. Levitt and A. Dorfman, *Proc. Natl. Acad. Sci. U.S.A.* **69**, 1253 (1972).
30. M. Gontcharoff and D. Mazia, *Exp. Cell Res.* **46**, 315 (1967); R. Tenser and J. Brachet, *Differentiation* **1**, 51 (1973).
31. E. Hadorn, *Brookhaven Symp. Biol.* **18**, 148 (1965); H. Wildermuth, *Sci. Prog.* **58**, 329 (1970); W. Gehring, *J. Embryol. Exp. Morphol.* **79**, 731 (1973).
32. S. A. Kauffman, *Science* **181**, 310 (1973).
33. W. Gehring, *Dev. Biol.* **16**, 438 (1967).
34. E. Balkaschina, *Wilhelm Roux' Arch. Entwicklungsmech. Org.* **115**, 448 (1929).
35. R. Sager and R. Kitchin, personal communication and in preparation; R. Sager and Z. Ramanis, *Theor. Appl. Genet.* **43**, 101 (1973).
36. P. Grippo, M. Iaccarino, E. Parisi, E. Scarano, *J. Mol. Biol.* **36**, 195 (1968).
37. J. H. Subak-Sharpe, *Br. Med. Bull.* **23**, 161 (1967); J. M. Morrison, H. M. Keir, H. Subak-Sharpe, L. V. Crawford, *J. Gen. Virol.* **1**, 101 (1967).
38. T. W. Sneider, *J. Mol. Biol.* **79**, 731 (1973).
39. B. McClintock, *Cold Spring Harbor Symp. Quant. Biol.* **16**, 13 (1951); *ibid.* **21**, 197 (1956); *Brookhaven Symp. Biol.* **8**, 58 (1955); *ibid.* **18**, 164 (1965).
40. J. R. S. Fincham and G. R. K. Sastry, *Annu. Rev. Genet.*, in press.
41. W. J. Dreyer and J. C. Bennett, *Proc. Natl. Acad. Sci. U.S.A.* **54**, 864 (1965); E. S. Lennox and M. Cohn, *Annu. Rev. Biochem.* **36**, 365 (1967); J. A. Gally and G. M. Edelman, *Annu. Rev. Genet.* **6**, 1 (1972).

42. We thank Drs. B. Alberts, J. Cooke, F. H. C. Crick, R. M. Gaze, J. B. Gurdon, B. Lewin, A. McLaren, Zh. A. Medvedev, and L. E. Orgel for their helpful comments.

X inactivation, differentiation, and DNA methylation

A.D. RIGGS

City of Hope National Medical Center, Duarte, Calif.

Abstract

A model based on DNA methylation is proposed to explain the initiation and maintenance of mammalian X inactivation and certain aspects of other permanent events in eukaryotic cell differentiation. A key feature of the model is the proposal of sequence-specific DNA methylases that methylate unmethylated sites with great difficulty but easily methylate half-methylated sites. Although such enzymes have not yet been detected in eukaryotes, they are known in bacteria. An argument is presented, based on recent data on DNA-binding proteins, that DNA methylation should affect the binding of regulatory proteins. In support of the model, short reviews are included covering both mammalian X inactivation and bacterial restriction and modification enzymes.

In this paper I attempt to explain mammalian X inactivation on the bases of DNA methylation and the properties of bacterial DNA methylases. Since most readers familiar with bacterial methylases may not be familiar with the X inactivation phenomenon, and vice versa, it was thought necessary to review briefly both areas before presenting new models.

The X inactivation phenomenon

Female mammals have two X chromosomes, whereas males have only one, creating a potential gene-dosage differential. In fact, however, this differential does not exist because a mechanism has evolved to maintain gene dosage effectively constant in both sexes. This mechanism, called X inactiva-

Supported by grants from the National Institutes of Health (HD-04420) and the National Science Foundation (GB-26517).

Request reprints from: Dr. ARTHUR D. RIGGS, City of Hope National Medical Center, 1500 East Duarte Road, Duarte, CA 91010 (USA).

tion, results in one, and only one, active X chromosome per diploid autosomal set of chromosomes. Any additional X chromosomes are genetically inactive, heteropyknotic, and late replicating. The condensed, inactive X chromosome forms the sex chromatin, or Barr body, usually seen in interphase and prophase and located near the nuclear membrane. The inactivation phenomenon (OHNO et al., 1959; LYON, 1961; BEUTLER et al., 1962) is well documented at the cytological and genetic level and is the subject of several recent reviews (OHNO, 1969, 1974; EICHER, 1970; COOPER, 1971; LYON, 1971, 1972; BROWN and CHANDRA, 1973).

For this paper, I will make the following assumptions about X inactivation in eutherian mammals: (1) *Activation.* The key step is activation of one chromosome. Initially, both chromosomes of the zygote are inactive, but by the time of implantation (50 to 400 cells), one and only one X chromosome has been activated. The condensation and late-replicating pattern of one X chromosome can be considered as a secondary consequence of the lack of activation. (2) *Randomness.* On the average, half of the cells of an individual have the maternal X chromosome active and half have the paternal X chromosome active. (3) *Permanence.* Except in the germ line, the initial differentiation event is irreversible. The inactive X chromosome in a particular cell remains inactive in all progeny cells. Inactivation is maintained throughout all successive cell divisions and concomitant DNA replication events. (4) *Totality.* The entire X chromosome is activated or inactivated as a unit. In mice, there is good evidence for a single primary X inactivation center in the X chromosome (RUSSELL, 1963; RUSSELL and MONTGOMERY, 1965; CATTANACH et al., 1970; DREWS et al., 1974). Translocated X chromosome fragments lacking the inactivation center remain active (CACHEIRO et al., 1973). Thus, inactivation apparently spreads from the inactivation center to cover all (or almost all) genes on the X chromosome. Although there may not be unanimous agreement on all these points, I believe the above statements are in keeping with the current consensus. For a more detailed discussion of these points, the reader is referred to the recent reviews listed above.

X inactivation occurs early in development, when the fate of embryonic cells is being determined, and thus can be considered an event in embryonic differentiation. The phenomenon of X inactivation evolved fairly recently, at about the time of emergence of the common precursor of mammals and marsupials (OHNO, 1969). A mechanism for the inactivation of entire chromosomes has also evolved in some insects, e.g., mealy bugs (BROWN and NUR, 1964; CHANDRA, 1971). OHNO (1969, 1973) argues that since evolution is a very conservative process, seldom using truly new mechanisms, X inactivation probably results from minor variations of the basic mechanisms of gene regulation and differentiation. Thus, there is reason to think that an explanation of X inactivation may throw some light on the mechanisms of differentiation.

Review of present theories

Several attempts have been made to explain X inactivation at the molecular level. However, rather than attempt a comprehensive review, I will select only two types of models that I think are particularly successful in explaining the basic facts of X inactivation. Then I will try to make some generalizations about all models. Finally, I will point out some weaknesses in these models.

The earliest model proposed, the episomal activator model of GRUMBACH et al. (1963), remains one of the most attractive. As illustrated in fig. 1, they proposed that an episomal factor becomes incorporated at a specific receptor

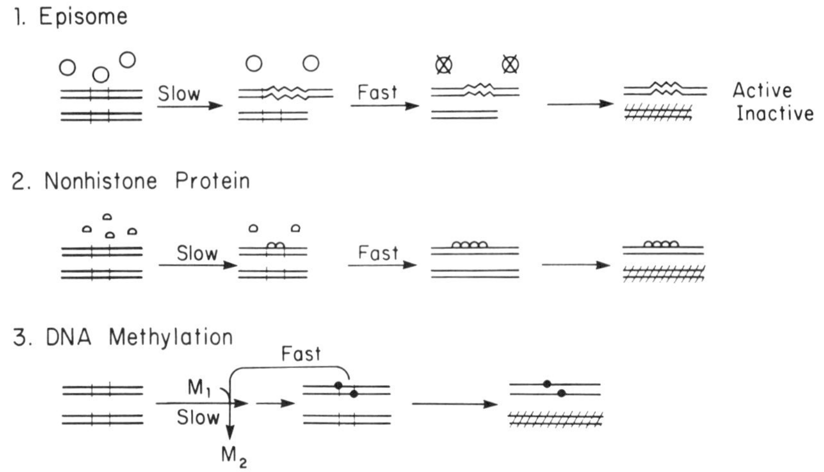

Fig. 1. Molecular models for X inactivation. Model 1 proposes the integration of an episome (circle) into the inactivation center of one X chromosome. The X chromosome with the integrated episome remains active, whereas the other X chromosome becomes inactive. Model 2 proposes that nonhistone proteins keep one X chromosome active. Model 3 proposes that methylation of the inactivation center of one X chromosome causes it to remain active. See text for additional details on all three models.

locus on the X. This leads to both the activation of the X with the integrated episome and also the inactivation of any other unincorporated episomal factors in the cell. It was also necessary to postulate that the episome was removed from the active X during oogenesis or spermatogenesis so that the inactivation would again be random in the next generation.

On the other hand, OHNO (1969, 1973) and LYON (1971, 1972) have proposed models based on sequence-specific DNA-binding proteins (nonhistone proteins) that bind at a certain site (the inactivation center), thereby leading to activation of the chromosome. OHNO's model (DREWS et al., 1974; OHNO, 1974) is also illustrated in fig. 1. He proposes that the autosomes produce a limited number of protein molecules designed to bind cooperatively at the inactivation center. The first molecule binds randomly, but because of the high cooperativity, all others bind adjacent to the first molecule on the same chromosome. Later, a generalized X chromosome repressor substance is produced to inactivate any chromosome not having the inactivation center covered by the cooperatively binding activator protein.

What do most of the present models have in common? First, to explain the randomness of inactivation (activation), a single receptor site on the X chromosome is postulated at which some entity binds specifically. Second, most models propose either (a) a *single* entity (protein, episome, or membrane attachment site) or (b) a *slow* initial activation process followed by a fast secondary process. The initial activation step must be slow because if this were not the case, the activation of both X chromosomes often would occur. In the episome models, the slow process is the integration of the first episome; the fast process is the destruction of any extra episomes. In the cooperative-binding protein model, the slow process is the binding of the initial protein molecules; the fast process is the cooperative binding of all other activator proteins adjacent to the first.

What are some of the problems with the present models? The episome activator model of GRUMBACH et al. (1963) can explain most aspects of X inactivation, especially randomness and permanence. It offers no good explanation for the totality of inactivation, but then, none of the present models (including the model to be presented here) offers a convincing explanation of the spread of inactivation to cover the entire X chromosome. Perhaps the major difficulty with episome models is that they have not stimulated biochemical experimentation. These models also have the defect of being rather unappealing as general models for differentation; to have the expression of most genes controlled by the integration of episomes seems unattractive.

X activation by the binding of sequence-specific nonhistone proteins has the advantage of being rather easily generalized to the regulation of most genes during differentiation. However, I argue that these models have difficulty explaining the permanence of X chromosome differentiation. DNA replication should create a crisis, often leading to loss of differentiation. During replication, sequence-specific proteins are almost certainly "swept off" the DNA. It is reasonable to think that at least 100 base pairs are disturbed (strands separated; DNA:RNA hybrids; covered by polymerase, etc.) at each replication fork. DNA replication is a relatively slow process, proceeding at 30 to 50 nucleotides per second in mammalian cells (HUBERMAN and RIGGS, 1968; GAUTSCHI et al., 1973; HAND and TAMM, 1973). During 2 seconds, diffusion would cause a protein to become randomly distributed in the nucleus. A cluster of cooperative-binding proteins adjacent to one another along the DNA also would be destroyed as a replication fork moved through. Another difficulty is that new proteins would need to be made for each round of replication.

Occasionally a binding event should occur on the previously inactive X chromosome, leading to its activation. Of course, one can invoke late replication of the inactive X chromosome as essential for the maintenance of the differentiated state. However, in the mouse there is recent evidence that a change in the staining properties of one X may precede late replication by several days and replication events (TAKAGI, 1973). Furthermore, in some species, such as the shrew, the inactive X chromosome is not late replicating (RAO et al., 1970). Even in the adult mouse and rat, late replication may not be an invariant property of the inactive X (TIEPOLO et al., 1967).

There is another feature of the X inactivation phenomenon that I find particularly intriguing, and this is the evidence for "imprinting," a term introduced by CROUSE (1960) to describe changes in the state of a chromosome that occurs during one generation that allows it to be recognized as different in the next generation. In marsupials, X inactivation is not random, but rather the paternal X chromosome is always inactivated (COOPER et al., 1971; SHARMAN, 1971). It receives an "imprint" during passage through the male that allows it to be distinguished from the maternal X chromosome. The DNA of the paternal chromosome is not permanently altered, for the chromosome is active in the next generation. Integrated episomes (or the lack of integrated episomes) provide one reasonable model to explain imprinting. But are there alternate ways to change DNA permanently but reversibly?

I suggest that there is another reasonable model, involving DNA methylation, that can explain most aspects of X inactivation, including imprinting, and has the additional advantage of providing an interesting model for gene regulation during differentiation. It will be seen that the known properties of bacterial DNA methylation enzymes are almost ideally suited for the proposed model. Thus, the evolution of the X inactivation system would be simple, requiring no new mechanisms. Because my model depends heavily on the remarkable properties of bacterial DNA modification (methylation) enzymes, a brief review of their properties will be presented first, followed by an argument that DNA methylation should have regulatory functions.

Bacterial restriction and modification enzymes

There have been several recent reviews of bacterial restriction and modification enzymes (Arber, 1971; Boyer, 1971; Meselson et al., 1972). Many elegant genetic and biochemical experiments have established the following: When foreign DNA (usually viral) enters *E. coli*, there begins a race between two events. The vast majority of the time a highly sequence-specific endonuclease (restriction nuclease) binds to a site on the DNA and cleaves the DNA, leading to the destruction of the foreign DNA molecule. Very rarely, the DNA is not cleaved, but instead the site in the DNA is methylated by a modification methylase. The sequence recognized by one restriction and modification system (*E. coli* RI) is shown in fig. 2. Many restriction and modification enzymes are now known, each specifically recognizing a different sequence (see Smith and Nathans, 1973, for a summary and suggested nomenclature). The product of the methylation event can be either N^6-methyladenine or 5-methylcytosine, depending on the methylase (fig. 3). For example, the *E. coli* RI and *E. coli* K modification methylases form N^6-methyladenine, whereas the *E. coli* RII enzyme forms 5-methylcytosine. As shown in fig. 2, the RI modification methylase is known to methylate two adenines symmetrically situated, one on each strand, inside the binding (or recognition) sequence (Dugaiczyk et al., 1974). *S*-adenosyl-L-methionine serves as the methyl donor for all known DNA methylases. Methylation of the recognition site completely protects the DNA from cleavage (Meselson and Yuan, 1968). For the *E. coli* K host-controlled restriction and modification system, the endonuclease and the methylase are part of the same multimeric protein (Haberman et al., 1972). Both enzyme activities share a protein subunit thought to determine binding specificity.

↓ *
5'... A / Tp Gp Ap Ap Tp Tp Cp T / A... 3'
3'... T / Ap Cp Tp Tp Ap Ap Gp A / T... 5'
* ↑

Fig. 2. The sequence of the recognition site for the *E. coli* RI DNA methylase and endonuclease, as determined by DUGAICZYK et al. (1974). The asterisks show which bases are methylated; arrows show where endonuclease cleavage occurs if the site is unmethylated.

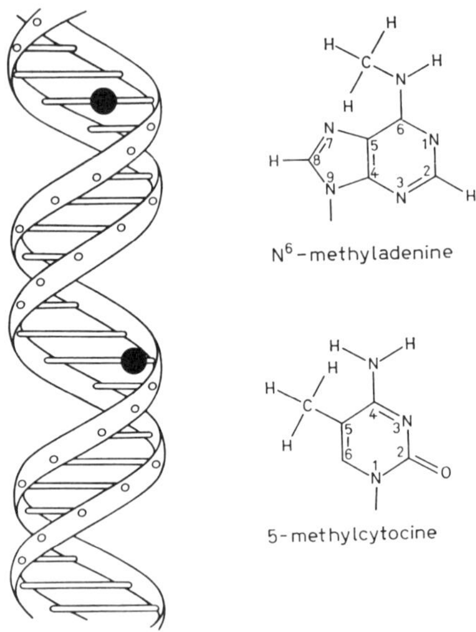

Fig. 3. Methylated bases in DNA. The location of the methyl group in the major groove of DNA is shown to the left of the chemical structure of the methylated bases.

The methylation of an unmethylated site is apparently a difficult task for the enzyme. The turnover time for the methylase activity on an unmethylated site has been estimated to be about 10 hours for the K enzyme (MESELSON and YUAN, 1972) or 7 hours for the B enzyme (LAUTENBERGER and LINN, 1972). Thus, methylation can be a very slow process. The genome of *E. coli* K contains many sequences (perhaps 1000) recognized by the K restriction endonuclease. Why doesn't the cell commit suicide every time its DNA is replicated? Suicide does not occur because a half-methylated site (one strand methylated) is fully protected from endonuclease activity (MESELSON and YUAN, 1968; VOVIS et al., 1973), and the methylation of a half-methylated site occurs very rapidly, probably within 60 seconds, as estimated from in vivo work (BILLEN, 1968; LARK, 1968). The appropriate in vitro work has not yet been reported.

To summarize, there is evidence that the host-controlled DNA methylation enzymes of *E. coli* have the following properties: (1) highly specific recognition of base sequence in DNA; (2) very slow methylation of unmethylated sites; (3) fast methylation of half-methylated sites. Some of these and additional points will be considered in more detail later; in particular, I will consider how these properties can be used to establish and maintain a differentiated state, such as that seen for X inactivation.

Methylation should have a regulatory function

The substrate for the *E. coli* host-controlled modification methylase is double-stranded DNA having exactly the correct base sequence. The enzyme does not act on denatured DNA (see MESELSON et al., 1972). In these respects, the properties of restriction and modification methylases are very similar to those of the lac repressor, which recognizes base sequences in duplex DNA (GILBERT and MÜLLER-HILL, 1967; RIGGS et al., 1968, 1970a, b).

The products of methylation (5-methylcytosine or N^6-methyladenine) are capable of participating in normal Watson-Crick type base pairing and would not be expected to disrupt the DNA duplex, since the methyl groups are located in the major groove (see fig. 3).

I have argued that the lac repressor binds to the outside of duplex DNA and reads the edges of the bases exposed in the grooves (RIGGS et al., 1970a). Recent data on repressor binding to synthetic DNAs (RIGGS et al., 1972) and to isolated operator fragments (GILBERT, 1972; GILBERT and MAXAM, 1973) support this possibility; and ADLER et al. (1972) have proposed that the N-terminal region of the repressor fits into the major groove. Also, STEITZ and coworkers have obtained crystallographic data suggesting a model where the lac repressor binds to the outside of duplex DNA (STEITZ et al., 1974). A peptide fitting into the major groove might form hydrogen bonds using the N-6 hydrogen and the N-7 nitrogen of adenine. Methylation of adenine to form N^6-methyladenine causes three major changes: (1) a relatively large hydrophobic group is introduced in the middle of the major groove, (2) the N-6 hydrogen not needed for Watson-Crick base pairing is eliminated, and (3) there is steric hindrance between the methyl group and atom N-7, probably eliminating the H-bonding capability of N-7. Thus, the formation of N^6-methyladenine *radically* alters the major groove and ought to affect regulatory protein binding. 5-Methylcytosine also should effect regulatory protein binding because it is well established that the lac repressor is very sensitive to minor changes in the major groove, such as those caused by the substitution of bromodeoxyuridine for thymidine (LIN and RIGGS, 1970, 1972). Therefore, it is reasonable to think that methylation will change the affinity of DNA-binding proteins for their binding sites. The argument is even stronger because it is well established that the *E. coli* K restriction and modification enzyme has a strong, easily detectable affinity for an unmethylated site but no detectable affinity for a methylated site (YUAN and MESELSON, 1970). Therefore, *methylation is known to affect the binding of sequence-specific proteins to DNA*. It is interesting to note that most methylation need not be regulatory, but those methylations that occur in operator sites are very likely to affect regulatory protein binding.

A DNA methylation model for X inactivation

First, I will outline the model as it applies to eutherian mammals and then consider each of the essential points in more detail. As illustrated in figs. 1 and 4, I propose that in the zygote, before the differentiation of the X chromosome, the primary inactivation center is not methylated. The methylation of an unmethylated inactivation center is proposed to be a slow process, requiring, on the average, many hours. However, eventually one of the primary inactivation centers is methylated. This results in the activation of that chromosome (perhaps the methylase no longer binds and acts as a repressor). As a direct or indirect result of this activation event, two proteins are synthesized (or one protein with two functions). One protein functions to change the methylase so that the methylation of an unmethylated inactivation center becomes impossible (it was very slow to begin with). The altered methylase still can methylate quickly a half-methylated inactivation center that has one strand methylated. A second protein functions to condense and inactivate any X chromosome with an unmethylated inactivation center. As shown in fig. 4, given a methylase capable of methylating a half-methylated inactivation center, but unable to methylate an unmethylated inactivation center, the differentiation of the X chromosomes will be maintained easily through successive cell divisions and DNA replication events. Now I will consider some of the essential postulates in more detail.

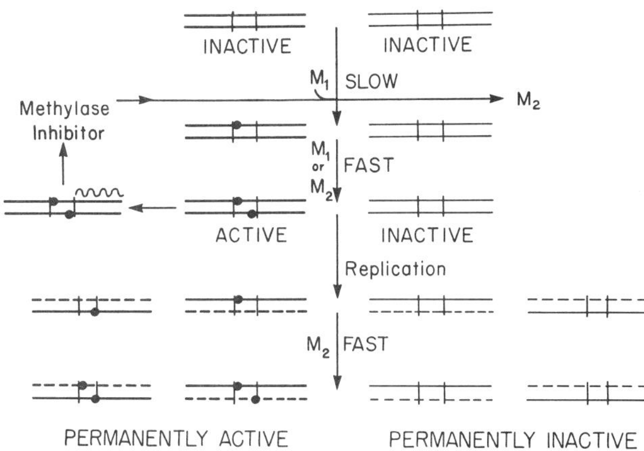

Fig. 4. A DNA methylation model for X inactivation. Given a sequence-specific DNA methylase that acts efficiently only on half-methylated sites, the differentiated state of the X inactivation center can easily be maintained through DNA replication. See text for additional details.

1. One primary X inactivation center. This is in keeping with most data and other models, and seems essential to explain the randomness of inactivation and the O^{hv} mutation (see below). It is convenient to have some symmetry in the base sequence of this site. That is, a sequence of bases on one strand in the left half of the inactivation center is repeated in reverse order on the other strand in the right half of the binding site. Such symmetry insures that the two half-methylated sites produced by replication of a fully methylated site are identical with regard to recognition by the methylase. Such symmetry in the recognition sites for bacterial modification methylases is well established (see fig. 2).

2. Sequence-specific DNA methylase. The most important postulate is that an autosomally located DNA methylase exists that (a) is highly sequence specific, (b) methylates the inactivation center symmetrically on both strands, and (c) methylates an unmethylated inactivation center with great difficulty but readily methylates a hybrid center (one strand methylated). As already pointed out, these are properties in keeping with present data on the *E. coli* host-controlled DNA methylases.

3. Methylase inhibitor. In order to explain that only one X chromosome is activated *per chromosomal set*, it is convenient to postulate an X-linked function. I propose that the X chromosome codes for a protein that interacts with the methylase (M_1, see figs. 1 and 4) and slightly inhibits its activity so that methylation of an unmethylated inactivation center no longer occurs to any significant extent. However, it is proposed that the altered methylase (M_2) still can quickly methylate a hybrid, half-methylated site. There is no need to postulate a small number of methylases; it is only necessary to have the methylase inhibitor produced at the same or slightly higher level than the methylase. If the methylase and methylase inhibitor bind stoichiometrically, this simple feedback loop results in one active X per diploid autosomal set.

The model proposed here seems to account for the following aspects of X inactivation: *Permanence.* Once established, the differentiated state of the X inactivation center would be maintained. *Random inactivation in eutherian mammals.* Randomness follows easily if the critical methylation event takes place after fertilization and if the X inactivation center is unmethylated in the sperm. *Paternal inactivation in marsupials.* Preferential maternal activation would result if methylation of the inactivation center occurred before fertilization. Note that, in general, "imprinting" can be explained by the state of methylation of DNA in the egg and sperm. *The* O^{hv} *mutation.* Recently a mutation (O^{hv}) was found, mapping in or near the inactivation center of the mouse X chromosome, that results in preferential activation of the X carrying the mutation (DREWS et al., 1974). This mutation may increase the affinity of the inactivation center for the methylase or increase the probability of methylation. *Aneuploids and polyploids.* These are adequately accounted for by the simple feedback loop described above. The level of X-linked methylase inhibitor is automatically adjusted to match the number of sets of autosomal methylase genes. For example, in tetraploids, methylation would continue until two X chromosomes were activated. *Reversibility for the next generation.* The differentiated state resulting from methylation during developmental history is erased whenever DNA methylation is prevented during DNA replication. This may occur during oogenesis or spermatogenesis or during the rapid DNA replication cycles following fertilization.

The model presented above does *not* explain the totality of inactivation, i.e., the mechanism whereby inactivation and condensation spreads from the inactivation center to encompass the entire X chromosome. However, the mechanism for the spread of inactivation may well be entirely different from that for the initiation and maintenance of the differentiated state of the inactivation center. Thus, the mechanism of the spreading effect is not a major consideration of this paper.

Methylation and differentiation

The general idea that DNA methylation might have a regulatory role is not new. Many authors have suggested such a role (SCARANO et al., 1967; SNEIDER and POTTER, 1969; ADAMS, 1970; MAMELAK and BOYER, 1970; VANYUSHIN et al., 1970; HARRISSON, 1971; COMINGS, 1972). However, to my knowledge, eukaryotic methylation has not been discussed in light of the recent evidence accumulating about the *E. coli* lac repressor and DNA methylases, nor has emphasis been placed on the advantages of methylation for permanent changes in regulation occurring during differentiation. An important exception is that SCARANO has proposed a model for differentiation based on DNA methylation (SCARANO et al., 1967; SCARANO, 1971; TOSI et al., 1972). He suggests that 5-methylcytosine produced by DNA methylases might be deaminated at the DNA level to produce thymine. One round of replication would lead to a transition from a G-C base pair to an A-T base pair. Thus, methylation would result in directed mutation, an event that, of course, would have permanent regulatory consequences when occurring in operators. This is an interesting model, but at present there is no firm evidence for the deamination of 5-methylcytosine to thymine at the DNA level (see SNEIDER, 1973).

The major point to be made here is that the properties of restriction and modification methylases are almost ideally suited to the establishment of stable differentiated states, without the necessity of proposing mutational events. Since the methylation of an unmethylated site is quite difficult, it is reasonable to think that very stringent conditions must be met for the initial methylation event to occur. However, once a site is methylated, it remains methylated because the methylation of a half-methylated site is easy and probably not sensitive to minor environmental fluctuation. Thus, "irreversible" differentiation is easily programmed; stable circuits are easily constructed. I suggest that the reader see COOK (1974) and substitute the words DNA methylation for "superstructure."

Present and future experiments in eukaryotes

Many of the points emphasized in this paper are, at least in principle, experimentally testable. A search of the literature showed that a number of interesting experiments already have been done. The results of some of these experiments will be summarized here along with suggestions for future work.

Methylation should be regulatory and lead to the activation of genes. DNA isolated from the genetically active macronucleus of *Tetrahymena* contains N^6-methyladenine, whereas DNA from the inactive micronucleus contains at least 10-fold lower levels of this methylated base (GOROVSKY et al., 1973). Additional experiments comparing active and inactive DNA are of obvious interest. It would be particularly interesting to look at the content of methylated bases in the inactive set of chromosomes in the mealybug (see CHANDRA, 1971). PARSA et al. (1973) have reported recently that embryonic rudiments of rat pancreas fail to differentiate in a methionine-deficient medium, even though growth is not affected. Their studies also suggest that it is not a deficiency of methionine per se that blocks differentiation but rather a deficiency of *S*-adenosyl-L-methionine, which is the methyl donor for DNA methylase.

Sequence-specific DNA methylases should exist. A cytosine methylase has been studied in mammalian cells (KALOUSEK and MORRIS, 1969; DRAHOVSKY and MORRIS, 1971a, b), but this enzyme methylates single-stranded DNA and does not seem to have properties desirable for the establishment and maintenance of differentiated states. 5-Methylcytosine is the only methylated base yet found in mammalian DNA (VANYUSHIN et al., 1970; LAWLEY et al., 1972), and its level is fairly high, 2–4% of total DNA cytosine. Because of the high level of methylation, it seems likely that most DNA methylation is not involved in specific gene regulation. However, the content of 5-methylcytosine may be tissue specific in the organs of the developing chick embryo (KAPPLER, 1971). The high background of 5-methylcytosine and the presence of the cytosine methylase studied by DRAHOVSKY and MORRIS (1971a, b) will make it difficult to search for additional DNA methylases producing 5-methylcytosine. Fortunately, the level of N^6-methyladenine in mammalian cells is very low (VANYUSHIN, 1970; LAWLEY et al., 1972), and this may aid in future searches for methylases methylating only certain operators, promoters, chromosomes, or the X inactivation center.

The DNA of egg, sperm, and the inactive X chromosome should be undermethylated. It was interesting to find that sperm DNA is, indeed, 50% undermethylated when compared to other tissues of the same species (VANYUSHIN, et al., 1970). My model requires that at least the inactivation center of the inactive X chromosome be unmethylated. Of course, it will be very difficult to measure the methylation of just this site. However, there is already evidence in the literature suggesting that the entire late-replicating X chromosome may be undermethylated. The inactive X chromosome is late replicating and probably comprises a sizable percentage of the late-replicating DNA in most species. In cells of the Chinese hamster and mouse, it has been reported that late replicating DNA is 20–50% undermethylated (ADAMS, 1971; COMINGS, 1972).

Acknowledgements

In particular, I thank Dr. SUSUMU OHNO for many interesting, provocative discussions and for providing an environment where ideas are encouraged rather than suppressed. I also thank JUDITH SINGER for her interest and intelligent questioning.

References

ADAMS, R.L.P.: The relationship between synthesis and methylation of DNA in mouse fibroblasts. Biochim. biophys. Acta *254:* 205–212 (1971).

ADAMS, R.L.P.: Delayed methylation of DNA in developing sea urchin embryos. Nature New Biol. *244:* 27–29 (1973).

ADLER, K.; BEYREUTHER, K.; FANNING, E.; GEISLER, N.; GRONENBORN, B.; KLEMM, A.; MÜLLER-HILL, B.; PFAHL, M. and SCHMITZ, A.: How lac repressor binds to DNA. Nature, Lond. *237:* 322–327 (1972).

ARBER, W.: Host-controlled variation, pp. 83–96. *In* A.D. HERSHEY, ed.: The bacteriophage lambda (Cold Spring Harbor Laboratory, New York 1971).

BEUTLER, E.; YEH, M. and FAIRBANKS, V.F.: The normal human female as a mosaic of X chromosome activity: studies using the gene for G-6-PD deficiency as a marker. Proc. nat. Acad. Sci., Wash. *48:* 9–16 (1962).

BILLEN, D.: Methylation of the bacterial chromosome: an event at the "replication point"? J. molec. Biol. *31:* 477–486 (1968).

BOYER, H.W.: DNA restriction and modification mechanisms in bacteria. Annu. Rev. Microbiol. *25:* 153–176 (1971).

BROWN, S.W. and CHANDRA, H.S.: Inactivation system of the mammalian X chromosome. Proc. nat. Acad. Sci., Wash. *70:* 195–199 (1973).

BROWN, S.W. and NUR, U.: Heterochromatic chromosomes in the coccids. Science *145:* 130–137 (1964).

CACHEIRO, N.L.A.; SWARTOUT, M.S. and RUSSELL, L.B.: Replication of DNA in X autosome translocation in the mouse. Genetics *74*(Suppl. 2): s36 (1973).

CATTANACH, B.M.; PEREZ, J.N. and POLLARD, C.D.: Controlling elements in the mouse X chromosome. II. Location in the linkage map. Genet. Res. *15:* 183–195 (1970).

CHANDRA, H.S.: Inactivation of whole chromosomes in mammals and coccids: some comparisons. Genet. Res. *18:* 265–276 (1971).

COMINGS, D.E.: Methylation of euchromatic and heterochromatic DNA. Exp. Cell Res. *74:* 383–390 (1972).

COOK, P.R.: On the inheritance of differentiated traits. Biol. Rev. *49:* 51–84 (1974).

COOPER, D.W.: Directed genetic change model for X chromosome inactivation in eutherian mammals. Nature, Lond. *230:* 292–294 (1971).

COOPER, D.W.; VANDEBERG, J.L.; SHARMAN, G.B. and POOLE, W.E.: Phosphoglycerate kinase polymorphism in kangaroos provides further evidence for paternal X inactivation. Nature New Biol. *230:* 155–157 (1971).

CROUSE, H.V.: The controlling element in sex chromosome behavior in *Sciara.* Genetics *45:* 1429–1443 (1960).

DRAHOVSKY, D. and MORRIS, N.R.: Mechanism of action of rat liver DNA methylase. I. Interaction with double-stranded methyl-acceptor DNA. J. molec. Biol. *57:* 475–489 (1971a).

DRAHOVSKY, D. and MORRIS, N.R.: Mechanism of action of rat liver DNA methylase. II. Interaction with single-stranded methyl-acceptor DNA. J. molec. Biol. *61:* 343–356 (1971b).

DREWS, U.; BLECHER, S.R.; OWEN, D.A. and OHNO, S.: Genetically directed preferential X activation seen in mice. Cell *1:* 3–8 (1974).

DUGAICZYK, A.; HEDGPETH, J.; BOYER, H.W. and GOODMAN, H.M.: Physical identity of the SV40 deoxyribonucleic acid sequence recognized by the *Eco* restriction endonuclease and modification methylase. Biochemistry *13:* 502–512 (1974).

EICHER, E.M.: X autosome translocations in the mouse: total inactivation versus partial inactivation of the X chromosome. Advanc. Genet. *15:* 175–259 (1970).

GAUTSCHI, J.R.; KERN, R.M. and PAINTER, R.B.: Modification of replicon operation in HeLa cells by 2,4-dinitrophenol. J. molec. Biol. *80:* 393–402 (1973).

GILBERT, W.: The lac repressor and the lac operator. Ciba Symp. *7:* 245–256 (1972).

GILBERT, W. and MAXAM, A.: The nucleotide sequence of the lac operator. Proc. nat. Acad. Sci., Wash. *70:* 3581–3584 (1973).

GILBERT, W. and MÜLLER-HILL, B.: The lac operator is DNA. Proc nat. Acad. Sci., Wash. *58:* 2415–2421 (1967).

GOROVSKY, M.A.; HATTMAN, S. and PLEGER, G.L.: [^{6}N]methyl adenine in the nuclear DNA of a eucaryote, *Tetrahymena pyriformis.* J. Cell Biol. *56:* 697–701 (1973).

GRUMBACH, M.M.; MORISHIMA, A. and TAYLOR, J.H.: Human sex chromosome abnormalities in relation to DNA replication and heterochromatinization. Proc. nat. Acad. Sci., Wash. *49:* 581–589 (1963).

HABERMAN, A.; HEYWOOD, J. and MESELSON, M.: DNA modification methylase activity of *Escherichia coli* restriction endonucleases K and P. Proc. nat. Acad. Sci., Wash. *69:* 3138–3141 (1972).

HAND, R. and TAMM, I.: DNA replication: direction and rate of chain growth in mammalian cells. J. Cell Biol. *58:* 410–418 (1973).

HARRISSON, C.M.H.: The arrangement of chromatin in the interphase nucleus with reference to cell differentiation and repression in higher organisms. Tissue & Cell *3:* 523–550 (1971).

HUBERMAN, J. and RIGGS, A.D.: On the mechanism of DNA replication in mammalian chromosomes. J. molec. Biol. *32:* 327–341 (1968).

ISSA, M.; BLANK, C.E. and ATHERTON, G.W.: The temporal appearance of sex chromatin and of the late-replicating X chromosome in blastocysts of the domestic rabbit. Cytogenetics *8:* 219–237 (1969).

KALOUSEK, F. and MORRIS, N.R.: The purification and properties of deoxyribonucleic acid methylase from rat spleen. J. biol. Chem. *244:* 1157–1163 (1969).

KAPPLER, J.W.: The 5-methylcytosine content of DNA: tissue specificity, J. Cell Physiol. *78:* 33–38 (1971).

LARK, C.: Studies on the *in vivo* methylation of DNA in *Escherichia coli* 15T⁻. J. molec. Biol. *31:* 389–399 (1968).

LAUTENBERGER, J.A. and LINN, S.: The deoxyribonucleic acid modification and restriction enzymes of *Escherichia coli.* J. biol. Chem. *247:* 6176–6182 (1972).

LAWLEY, P.D.; CARTHORN, A.R.; SHAH, S.A. and SMITH, B.: Biomethylation of deoxyribonucleic acid in cultured human tumour cells (HeLa): methylated bases other than 5-methylcytosine not detected. Biochem. J. *128:* 133–138 (1972).

LIN, S. and RIGGS, A.: Lac repressor binding to DNA not containing the lac operator and to synthetic poly dAT. Nature, Lond. *228:* 1184–1186 (1970).

LIN, S. and RIGGS, A.D.: Lac operator analogues: bromodeoxyuridine substitution in the lac operator affects the rate of dissociation of the lac repressor. Proc. nat. Acad. Sci., Wash. *69:* 2574–2576 (1972).

LYON, M.F.: Gene action in the X chromosome of the mouse (*Mus musculus* L). Nature, Lond. *190:* 372–373 (1961).

LYON, M.F.: Possible mechanisms of X chromosome inactivation. Nature New Biol. *232:* 229–232 (1971).

LYON, M.F.: X chromosome inactivation and developmental patterns in mammals. Biol. Rev. *47:* 1–35 (1972).

MAMELAK, L. and BOYER, H.W.: Genetic control of the secondary modification of deoxyribonucleic acid in *Escherichia coli.* J. Bacteriol. *104:* 57–62 (1970).

MESELSON, M. and YUAN, R.: DNA restriction enzyme from *E. coli.* Nature, Lond. *217:* 1110–1114 (1968).

MESELSON, M.; YUAN, R. and HEYWOOD, J.: Restriction and modification of DNA. Annu. Rev. Biochem. *41:* 447–466 (1972).

OHNO, S.: Evolution of sex chromosomes in mammals. Annu. Rev. Genet. *3:* 495–524 (1969).

OHNO, S.: Ancient linkage groups and frozen accidents. Nature, Lond. *244:* 259–262 (1973).

OHNO, S.: Conservation of ancient linkage groups and some insight into the genetic regulatory mechanisms of X inactivation. Cold Spr. Harb. Symp. quant. Biol. *38:* 155–164 (1974).

OHNO, S.; KAPLAN, W.D. and KINOSITA, R.: Formation of the sex chromatin by a single X chromosome in liver cells of *Rattus norvegicus.* Exp. Cell Res. *18:* 415–418 (1959).

PARSA, I.; MARSH, W.H. and FITZGERALD, P.J.: Pancreas acinar cell differentiation. VIII. Effect of methionine on DNA synthesis of pancreas anlage in organ culture. Exp. Cell Res. *82:* 466–468 (1973).

RAO, S.R.V.; SHARMA, V.K. and SHAH, V.C.: DNA synthesis in duplicate-type sex chromosomes of the Indian house shrew, *Suncus murinus* (Insectivora). Cytogenetics *9:* 384–395 (1970).

RIGGS, A.D.; BOURGEOIS, S. and COHN, M.: The lac repressor-operator interaction. III. Kinetic studies. J. molec. Biol. *53:* 401–417 (1970a).

RIGGS, A.D.; BOURGEOIS, S.; NEWBY, R.F. and COHN, M.: DNA binding of the lac repressor. J. molec. Biol. *34:* 365–368 (1968).

RIGGS, A.D.; LIN, S. and WELLS, R.D.: Lac repressor binding to synthetic DNAs of defined nucleotide sequence. Proc. nat. Acad. Sci., Wash. *69:* 761–764 (1972).

RIGGS, A.D.; SUZUKI, H. and BOURGEOIS, S.: Lac repressor-operator interaction. I. Equilibrium studies. J. molec. Biol. *48:* 67–83 (1970b).

RUSSELL, L.B.: Mammalian X chromosome action: inactivation limited in spread and in region of origin. Science *140:* 976–978 (1963).

RUSSELL, L.B. and MONTGOMERY, C.S.: The use of an X-autosome translocation in locating the X chromosome inactivation center. Genetics *52:* 470–471 (1965).

SCARANO, E.: The control of gene function in cell differentiation and in embryogenesis. Advanc. Cytopharmacol. *1:* 13–24 (1971).

SCARANO, E.; IACCARINO, M.; GRIPPO, P. and PARISI, E.: The heterogeneity of thymine methyl group origin in DNA pyrimidine isostichs of developing sea urchin embryos. Proc. nat. Acad. Sci., Wash. *57:* 1394–1400 (1967).

SHARMAN, G.B.: Late DNA replication in the paternally derived X chromosome of female kangaroos. Nature, Lond. *230:* 231–232 (1971).

SMITH, H.O. and NATHANS, D.: A suggested nomenclature for bacterial host modification and restriction systems and their enzymes. J. molec. Biol. *81:* 419–423 (1973).

SNEIDER, T.W.: On the source of "minor thymine" in DNA from a Novikoff rat hepatoma cell line. J. molec. Biol. *79:* 731–734 (1973).

SNEIDER, T.W. and POTTER, V.R.: Methylation of mammalian DNA: studies on Novikoff hepatoma cells in tissue culture. J. molec. Biol. *42:* 271–284 (1969).

STEITZ, T.A.; RICHMOND, T.J.; WISE, E. and ENGELMAN, D.: The lac repressor protein: molecular shape, subunit structure, and proposed model for operator interaction based on structural studies of microcrystals. Proc. nat. Acad. Sci., Wash. *71:* 593-597 (1974).

TAKAGI, N.: Differentiation of X chromosomes in the female mouse. Genetics *74*(Suppl.2): s269 (1973).

TIEPOLO, L.; FRACCARO, M.; HULTEN, M.; LINDSTEN, J.; MANNINI, A. and MING, P.L.: Timing of sex chromosome replication in somatic and germ line cells of the mouse and rat. Cytogenetics *6:* 51–66 (1967).

TOSI, L.; GRANIERI, A. and SCARANO, E.: Enzymatic DNA modifications in isolated nuclei from developing sea urchin embryos. Exp. Cell Res. *72:* 257–264 (1972).

VANYUSHIN, B.F.; TKACHEVA, S.G. and BELOZERSKY, A.N.: Rare bases in animal DNA. Nature, Lond. *225:* 948–949 (1970).

VOVIS, G.F.; HORIUCHI, K.; HARTMAN, N. and ZINDER, N.D.: Restriction endonuclease B and fl heteroduplex DNA. Nature New Biol. *246:* 13–16 (1973).

YUAN, R. and MESELSON, M.: A specific complex between a restriction endonuclease and its DNA substrate. Proc. nat. Acad. Sci., Wash. *65:* 357–362 (1970).

Manuscript received 5 June 1974; accepted for publication 20 August 1974.

Glossary

Words set in **bold type** are defined elsewhere in this glossary.

ACS: ARS consensus sequence; an 11-bp element essential for the function of origins of DNA replication in *S. cerevisiae*. ARS stands for autonomous replication sequence, an element that is essential for replication of yeast plasmids.

Adenoviruses: DNA viruses belonging to the family Adenoviridae and causing respiratory infections in humans and animals.

AdoHcy: *S*-Adenosylhomocysteine; formed as a product of the reaction catalyzed by **DNA methyltransferase**, for which **AdoMet** is a cofactor.

AdoMet: *S*-Adenosylmethionine; a cofactor for many methyltransferase enzymes and the source of the methyl group added to DNA cytosine by **DNA methyltransferase**.

Aleurone: One or more layers of cells found on the outside of the **endosperm** in cereal seeds such as maize kernels. In maize, the aleurone layer can accumulate anthocyanin pigments whose biosynthesis is controlled by a small number of genes. This makes the aleurone layer an ideal tissue for genetic studies.

Alu element: A repetitive DNA element approximately 300 bp long that is abundantly dispersed throughout the genome of primates. The name is derived from the *Alu*I restriction enzyme cleavage site that is within most Alu elements.

Androgenetic embryo (androgenote): An embryo that contains two paternally derived sets of chromosomes and no maternally derived chromosomes. Androgenetic mouse embryos can be made by nuclear transplantation.

Angelman syndrome: An inherited genetic disorder in humans that is characterized by severe mental retardation, small brain, seizures, movement disorder, inappropriate laughter, and characteristic facial appearance.

Anther: *See* Stamen.

Apical dominance: In many plant species, including *Arabidopsis*, secondary **meristems** in the leaf axils do not develop into shoots until flowering on the primary (central) **inflorescence** shoot is well advanced. Loss of apical dominance refers to parallel, rather than sequential, development of primary and secondary inflorescence meristems.

Apoptosis: A specific cell-suicide mechanism in which the nucleus condenses and the chromatin is fragmented, ultimately into nucleosomes. Often referred to as programmed cell death in certain developmental contexts.

Autocrine effects: The effects of a substance secreted by a cell on that cell itself.

Autosomal gene: A gene located on an autosome, a chromosome which is not a sex-determining chromosome. In many organisms, including mammals and *Drosophila*, this means a gene not on the X or Y chromosomes.

Auxin: A class of plant growth hormones which have many physiological effects. In combinations with **cytokinin**, auxin is required for the sustained proliferation of many cultured plant tissues.

5-Azacytidine: A cytidine analog in which the 5 carbon of the cytosine ring has been replaced with nitrogen. 5-Azacytidine is a potent inhibitor of mammalian **DNA methyltransferase**.

Bifilarly methylated DNA: Double-stranded DNA that is symmetrically methylated on both strands.

Bisulfite genomic sequencing: A procedure in which bisulfite is used to deaminate cytosine to uracil in genomic DNA. Conditions are chosen so that 5-methylcytosine is not changed. PCR amplification and subsequent DNA sequencing reveals which cytosines are methylated in genomic DNA.

Bromodomain: A 77-amino-acid motif conserved in many eukaryotic regulatory proteins, mostly transcriptional activators or adapters. Its function remains to be elucidated.

Cadastral gene: A plant gene that controls the expression of floral **homeotic genes**.

Carpel: Floral organs that make up the **gynoecium**, or female reproductive structure in plants. Typically, one or more fused carpels contribute to the ovary, which contains the ovules. The ovule supports the female gametophyte and, hence, the egg cell.

Cauline leaf: A leaf formed on the **inflorescence** stem of a rosette plant such as *Arabidopsis*.

CG island: *See* CpG island.

CG ocean: *See* CpG ocean.

CG suppression: The underrepresentation of CG dinucleotides, such as observed in the mammalian genome and believed to be due to deamination-caused conversion over evolutionary time of 5′ m^5CG/3′ Gm^5C dinucleotide pairs in duplex DNA to 5′TG/3′AC or 5′ CA/3′ GT dinucleotide pairs.

Chlorosis: The loss of chlorophyll and associated pigments from small lesions or from whole leaves.

Chorion: The outer membrane that encloses the embryo in reptiles, birds, and mammals. In placental mammals it contributes to the development of the placenta.

Chorionic villi: Fingerlike projections of the **chorion** that contribute to the formation of the placenta in mammals.

Choroid plexus: Specialized structure in the mammalian brain having vascular bundles that project into the ventricular cavity. Cerebrospinal fluid is produced by the ultrafiltration of blood through the capillaries.

Chromatin: A general term for the complex of nucleic acids and proteins that make up chromosomes.

Chromatin domain: A region of **chromatin**, the exact character and size of which depends on experimental context. A chromatin domain can be a single nucleosome or can extend to an array of more than 100 nucleosomes.

Chromatin insulator: A DNA element that protects a gene from **position effects**.

Chromodomain: The prefix chromo is an acronym for *chro*matin-*m*odifier-*o*rganizer. The chromodomain is an amino acid motif consisting of 36 amino acids that are conserved within a family of proteins believed to bind to chromatin, modify the structure of chromatin, and be involved in heterochromatin formation and transcriptional repression. These proteins include Polycomb (Pc) and heterochromatin protein 1 (HP1).

Chromosomal domain: A region of a chromosome, the exact nature and size of which depend on experimental context. A domain can be a region of chromosomal packaging, such as a loop extending from two adjacent attachments to a chromosomal axis, and can vary in size from an array of less than 100 nucleosomes (~30 kb) to potentially more than 500 nucleosomes. A chromosomal domain might also represent a functional unit of chromosomal structure defined by boundary elements or insulators at the edges of the domain.

Clonal expansion: The population of cells produced from a single cell. It is synonymous with clone, but is used in particular contexts. For example, a mammalian cell with a particular chromosomal abnormality can, by clonal expansion, produce a population of the same type of cell within the organism.

Congenic strain: A variant mouse strain that is obtained by backcrossing a donor mouse strain to an inbred parental strain for at least ten generations while maintaining by appropriate selection the presence of a small genetic region derived from the donor strain.

Cosuppression: *See* Suppression.

CpG island: A region of G+C-rich DNA that covers several hundred base pairs and has a CpG dinucleotide content close to the level expected from the base composition (i.e., similar to the GpC dinucleotide content). These regions are not subject to **CG suppression**, as is most of the mammalian genome. Typically, CpG islands are not methylated and often are found at the 5′ ends of genes.

CpG ocean: A region of the genome in which CpG dinucleotides are underrepresented because of **CG suppression**.

Crown gall: A tumor disease of plants caused by virulent strains of the soil bacterium, *Agrobacterium tumefaciens*.

Cultivar: An established variety of a cultivated plant species with uniform characteristics.

CWG island: A region of the genome containing close to the expected number of CWG trinucleotides, where W stands for either A or T.

CWG ocean: A region of the genome with much less than the expected frequency of CWG trinucleotides, where W stands for either A or T. This may reflect a similar mechanism to **CG suppression** in organisms, such as plants, that methylate CWG as well as CG.

Cytokinin: A class of plant growth hormones with many physiological effects which, in combination with **auxin**, is required for the sustained proliferation of many cultured plant tissues.

Dam methylation: Methylation at the N6 position of adenine residues in double-stranded 5′-GATC sequences by an enzyme called Dam methyltransferase. Dam methylation is present in wild-type *E. coli*.

Dcm methylation: Methylation at the N5 position of the inner cytosine residues in double-stranded sequences 5′-CCAGG-3′ and 5′-CCTGG-3′ by an enzyme called Dcm methyltransferase. Dcm methylation is present in wild-type *E. coli*.

DCR: Downstream control region; a G+C-rich sequence contained within the first exon of the maize **Spm** element.

Disomy: *See* Monosomy/disomy/trisomy.

DMR: Differentially methylated region; a region in the mouse Igf2 gene exhibiting methylation differences between the paternal and maternal chromosomes.

DNA methyltransferase: An enzyme, DNA (cytosine-5)-methyltransferase, that transfers methyl groups from **AdoMet** to the 5 position of cytosine in DNA.

DSB: Double strand break; a break in both strands of the DNA double helix that is required to initiate mating-type switching in yeast.

dSpm: A transposition-defective **Spm** element, generally an internally deleted Spm.

Ectopic chromatin: Cytologically observable, drawn-out fibers that connect specific sites within and between *Drosophila* polytene chromosome arms.

Ectopic expression: The abnormal expression of a gene in an inappropriate context. Often applied to a **luxury protein** that might be synthesized in the wrong type of cell; e.g., hemoglobin in a neuron. It can also be applied to tumor cells, which often express fetal antigens on their surface or have other protein abnormalities.

Egg cylinder stage: An early, postimplantation stage of mouse embryogenesis (day 5 post coitum), corresponding to a human embryo of 7–12 days of age.

Electroporation: The permeabilization of the cell membrane by a brief electric pulse. A technique commonly used to transfect DNA into bacterial, plant, and mammalian cells.

Embryo proper: That part of a mouse or other mammalian embryo that develops into, or contributes cells to, the adult organism.

Embryonic stem cells: *See* ES cells.

En: *See* Spm.

Endogene: An endogenous gene, i.e., a gene normally present in the genome; not a **transgene**.

Endosperm: Terminally differentiated extraembryonic triploid storage tissue found in the seeds of higher plants. The endosperm originates by fertilization of the diploid central cell of the female gametophyte (the embryo sac) by one of two sperm cells found in the pollen. The other sperm cell fertilizes the egg.

Enhancer: A DNA element that regulates the rate of transcription initiation by interacting with promoter elements. Enhancers can be located considerable distances, up to several kilobases, upstream or downstream from promoters.

Epialleles: Alternative states of a gene that have an identical DNA sequence but differ in methylation or **chromatin** structure and, hence, level of expression.

Epigenetic regulation: Regulation of gene activity mediated by a reversible change in DNA modification or **chromatin** structure.

Epigenetics: The study of mitotically or meiotically heritable variations in gene function that cannot be explained by changes in DNA sequence.

Epimutation: A heritable change in gene activity that is not due to a change in base sequence. Generally applied to an abnormal change in gene activity; for example, in a tumor or a cell in culture. In development, normal changes in gene activity occur and these are often somatically inherited. Such changes are not epimutations. The pattern of epimutations in an individual or in a cell is referred to as the epigenotype.

Erythroleukemia cells: Abnormal precursors of red blood cells. They can be induced with certain chemicals to produce hemoglobin.

ES cells: Embryonic stem cells; cultured pluripotential cells derived from the inner cell mass of **blastocyst**-stage mouse embryos. These cells are often used for gene targeting by homologous recombination in tissue culture. After gene targeting, ES cells can be reintroduced into blastocysts to make transgenic mouse lines.

Estradiol: A steroid hormone produced by the ovary. In mammals, the major effects of estradiol are in the differentiation of the uterus and mammary gland.

Euchromatin: Originally a cytological term for chromatin that is not **heterochromatin**. Heterochromatin is more condensed than euchromatin during interphase. Most genes are in euchromatin.

Explant: Excised tissue used to establish cultured cell and tissue lines.

Expression variegation: A type of variation in gene expression during development that causes streaks or patches of cells to have different phenotypes. The coat-color patches of a calico cat are an example of variegated expression.

Extraembryonic endoderm: An extraembryonic tissue that is derived from cells of the **primitive endoderm** around days 4.5–5.5 post coitum and contributes to the yolk sac.

Extraembryonic tissues: Tissues, such as the yolk sac, amnion, and placenta, that are derived from the early embryo and form support tissues for the embryo, but do not remain as part of the adult animal.

Familial hypercholesteremia: A common human autosomal dominant disorder characterized by a markedly elevated serum cholesterol level. The disease is caused by a germ-line mutation of the gene for the *l*ow-*d*ensity *l*ipoprotein (LDL) receptor.

FISH: Fluorescent in situ hybridization; a technique for visual detection in the microscope of specific DNA sequences on cytologically fixed chromosomes, after hybridization with DNA probes labeled with a fluorochrome. FISH can be done on both interphase and metaphase mammalian cells.

Floral meristem: A **meristem** that gives rise to a flower.

Functional hemizygosity: An autosomal gene is functionally hemizygous if only one of the two copies in a diploid cell is expressed. Almost all genes on the mammalian X chromosome are also functionally hemizygous, either because only one X is present in the male, or because of X-chromosome inactivation in the female.

GAGA factor: A *Drosophila* transcriptional activator protein involved in remodeling chromatin structure. The factor binds to $(GA)_n$-rich sites. It is encoded by the Trithorax-like gene.

Gametophyte: The haploid, sexual phase of the plant life cycle. In higher plants, gametophytes arise from the haploid products of meiosis via a few rounds of mitotic division. The male gametophyte (the pollen grain) produces the male gametes (sperm cells), and the female gametophyte (the embryo sac) produces the female gametes (egg cells).

Gene conversion: A nonreciprocal recombination process occurring during meiosis by which the DNA sequence of one allele is changed (converted) to that of the other allele.

Gene silencing: Inactivation of gene expression, usually with the implication that inactivation is not due to mutation, but rather to epigenetic modification such as methylation or change in chromatin structure. The inactive state can be heritable under most conditions yet reversible under special conditions, such as treatment with inhibitors of DNA methylation.

Genital ridge: A site in the early mouse embryo to which **primordial germ cells** migrate prior to gonad formation.

Genomic imprinting: The phenomenon whereby genes function differently depending on whether they are inherited from the maternal or paternal parent. This is thought to be caused by information superimposed on DNA sequences which is different in male and female gametes. Such information is transmitted, or inherited, in somatic cells but usually erased and reset in the germ line.

Germ band extension stage: During the development of the *Drosophila* embryo, the germ band (the part of the embryo containing the developing segments) extends posteriorly, curling up along the dorsal side of the embryo between 3 and 7 hours of development. The germ band then retracts, bringing the segments back on the ventral side of the embryo between 7 and 10 hours of development.

Gynoecium: The female reproductive structure in the flower. It is made up of one or more carpels.

Gynogenetic embryo (gynogenote): An embryo in which both sets of chromosomes are of maternal origin.

Habituation: A type of cell-heritable modification leading to a loss in the requirement for a specific growth hormone.

Hairy-root syndrome: The collection of symptoms (dwarfing, highly wrinkled leaves, excessive number of roots) that are shown by plants regenerated from cells transformed with the Ri-plasmid of *Agrobacterium rhizogenes*.

Haplotype: A particular set of alleles at linked loci that are found associated on an individual chromosomal homolog.

Hemimethylation: The state in which a DNA duplex is methylated in one strand but unmethylated in the other.

Hepatitis virus: A DNA virus belonging to the family Flaviviridae (hepatitis C), or Hepadnaviridae (hepatitis A/B). These viruses infect hepatocytes causing liver disease and sometimes cancer.

Heterochromatin: Cytologically defined as **chromatin** that remains condensed and differentially stainable during interphase of the cell cycle; other characteristics include replication late in S phase and transcriptional silence.

Heterothallic: Describes yeast strains that have a fixed mating type (P or M in *S. pombe*) and as a result can only mate when mixed with a strain of the opposite mating type.

Heuristic model: A formal model proposed for purposes of teaching or provoking discussion.

HLH protein: A protein containing a basic helix-loop-helix motif shared among many known transcription factors from a wide range of organisms.

HM loci: The silent mating-type loci in *S. cerevisiae*, **HML** (left) and **HMR** (right).

HML: A locus on the left end of *S. cerevisiae* chromosome III containing transcriptionally silent versions of genes specifying cell mating type.

HMR: A locus on the right end of *S. cerevisiae* chromosome III containing transcriptionally silent versions of genes specifying cell mating type.

Homeobox: A characteristic DNA sequence of 180 bp that codes for the 60-amino-acid DNA-binding domain (homeodomain) of some developmentally important regulatory genes. Mutations in the homeobox can have homeotic effects.

Homeodomain: *See* Homeobox.

Homeostatic mechanism: A physiological process that contributes to the maintenance of a relatively stable internal environment in a multicellular organism.

Homeotic genes: A class of genes that determine the identity of an organ, segment, or other structural unit during development. In animals, clustered sets of homeotic genes (hox gene clusters) encode proteins that determine the identity of structures that develop from repeating embryonic elements or segments. In plants, homeotic genes control the identity of, e.g., floral organs.

Homeotic mutation: A mutation that alters the identity of a particular organ or other structural unit, causing it to adopt the identity of another such organ or unit. For example, antennapedia mutants in *Drosophila* have legs in place of antennae. In *Arabidopsis*, apetala3 mutants have **sepals** in place of petals and **carpels** in place of **stamens**.

Homothallic strains: Yeast strains which are capable of switching mating type and thus of mating with themselves, forming zygotes, asci, and spores under appropriate conditions.

Hotspot for mutation: A DNA sequence in which frequent mutations are observed.

Housekeeping enzymes: Enzymes present in all cells capable of normal metabolism. They are essential for the synthesis or breakdown of proteins, nucleic acids, and lipids; for glycolysis and respiration; and for many other standard metabolic pathways.

Hox gene clusters: Clustered sets of certain **homeobox** genes. *Drosophila* has only one set, whereas vertebrates have multiple sets.

Hydrolytic deamination: Methylcytosine and cytosine undergo deamination in vitro in the presence of hot acid and alkali. Deamination of cytosine and methylcytosine leads to the production of uracil and thymine, respectively. In vivo, the rate of spontaneous deamination of methylcytosine is twice that of cytosine.

Hypersensitive site: A site in chromatin that is preferentially sensitive to deoxyribonucleases. Hypersensitive sites are often found in active promoters or enhancers.

Idioblast: A plant cell committed to develop into a cell type that differs from the surrounding tissue.

Imaginal disc: A structure found in insects such as *Drosophila* consisting of undifferentiated epithelial cells set aside in each of the embryonic segments from which the adult structures differentiate during metamorphosis. Thoracic segments contain both a dorsal disc (e.g., wing) and a ventral disc (leg).

Imprinted gene: A gene that functions differently depending on whether it is of paternal or maternal origin. Imprinted genes in diploid mammalian cells are generally expressed from only one allele.

Imprinting domain: A chromosomal region containing **imprinted genes**, which may be controlled by a common regulatory element(s).

Inflorescence: The flowering part of a plant; a cluster of flowers.

Inflorescence meristem: The relatively undifferentiated, dividing plant tissue that gives rise to the **inflorescence.**

Isodicentric chromosome: A structurally abnormal chromosome containing a duplication of part of the chromosome including the centromere. The resulting structure contains two centromeres and a point of symmetry that depends on the position of the breakpoint. Chromosome (Xp-) isodicentrics, for instance, may have breakpoints anywhere on the short arm of the human X chromosome between the telomeric extremity (Xp22) and the juxtacentromeric region (Xp111)

Isoparental embryos: Embryos that contain two sets of chromosomes from one parental gender. *See* **Gynogenetic embryo; Androgenetic embryo.**

Isoschizomer restriction enzymes: Restriction enzymes from different bacteria which recognize the same target sequence in DNA. Often these enzymes respond differently to a modified base, or bases, within the sequence. Thus, *Msp*I cuts both CCGG and Cm^5CGG, whereas *Hpa*II cuts only the unmethylated sequence.

LCR: Locus control region; a key regulatory element that controls a **chromosomal domain** which includes several genes. A LCR may be required for a linked transgene to be expressed independent of copy number and chromosomal position.

Lentivirus: A virus belonging to the family Retroviridae. Members of this family of retroviruses cause immunodeficiency diseases in man and animals. Examples are human immunodeficiency virus (man), visna (sheep), and equine infectious anemia virus (horse).

Leptomeninges: A tissue that comprises the arachnoid and pia mater layers of the meninges, the connective tissues surrounding the mammalian nervous system.

Li-Fraumeni syndrome: A rare, dominant familial cancer disorder that has been linked to a germ-line mutation of the p53 tumor suppressor gene. Affected families are susceptible to a spectrum of childhood and adult neoplasms, which include breast cancer, soft-tissue sarcomas, brain tumors, osteosarcomas, and leukemias.

Ligule: A fringe of epidermal tissue found at the boundary between the sheath and the blade of a maize leaf.

LTR: Long terminal repeat; a region of a retrovirus that contains sequences necessary for transcription and DNA integration. This region is present at both the 5′ and 3′ ends of the viral genome.

Luxury enzymes or proteins: Proteins uniquely produced by specialized, differentiated cells. Thus, hemoglobin is a luxury protein in red blood cells and is not present in other specialized cells, such as neurons or epithelial cells.

M2 population: The progeny derived from selfing M1 plants, which themselves are progeny that arise by selfing plants grown from mutagenized seed. Recessive mutations, resulting from the seed mutagenesis, are detected in M2 plants, which are homozygous for the mutation.

MAT: A locus near the middle of *S. cerevisiae* chromosome III containing expressed versions of genes specifying cell mating type.

MDBP-2-H1: Methylated DNA-binding protein 2; a protein that has sequence homology with histone H1.

MeCP-1 and MeCP-2: Methyl CpG-binding protein; a class of nuclear proteins that recognize densely methylated DNA (MeCP-1) or sparsely methylated DNA (MeCP-2). These proteins are believed to bind to methylated DNA and inhibit transcription.

Meristem: A group of undifferentiated plant cells that gives rise to differentiated structures such as flowers or leaves. Meristem cells are stem-cell-like, in that some daughter cells contribute to the differentiated structures while others remain meristematic. Examples include the **shoot meristem** (which gives rise to the shoot), the root meristem (which gives rise to the root), and the **inflorescence meristem** (which gives rise to flowers). Meristems can change identity during development; e.g., a vegetative shoot meristem turns into an inflorescence meristem on floral induction.

N6-Methyladenine: An adenine derivative in which one of the hydrogens on the exocyclic amino group has been replaced by a methyl group.

5-Methylcytosine (m^5C): A cytosine derivative with the hydrogen at carbon 5 replaced with a methyl group.

5-Methylcytosine deaminase: An enzyme that deaminates 5-methylcytosine at position 4 to give free thymine. This reaction occurs not at the DNA level but at the free base level.

5-Methylcytosine-DNA glycosylase: An enzyme that removes 5-methylcytosine from double-stranded DNA.

Mismatch repair: Any of the several cellular mechanisms for correction of mispaired nucleotides in double-stranded DNA.

Mismatch-specific thymine-DNA glycosylase: An enzyme that specifically removes the thymine from a T:G mismatch in DNA.

Monosomy/disomy/trisomy: Terms that describe the number, either 1, 2, or 3, respectively, of a certain chromosome in a cell. An abnormal chromosome number can arise during somatic cell divisions by mitotic nondisjunction: i.e., one daughter receives three copies of a chromosome (trisomy) and the other only one (monosomy). During further divisions, the normal number of two chromosomes (disomy) can be reestablished from either a monosomic or trisomic condition, but these two chromosomes can now be identical to each other. Hence, a cell with two different alleles, a and b, of a gene (heterozygous) can eventually by this process produce daughters that are either aa or bb (homozygous).

Murine leukemia viruses: Viruses that belong to the family Retroviridae and cause leukemia in mice.

Myelodysplasia: An abnormal condition that involves cells of the bone-marrow-derived myeloid lineage and often progresses to malignancy.

Myoblasts: Muscle precursor cells. In cell culture they differentiate under certain conditions to myotubes and myofibers.

Myotome: The muscle primordium contained in each **somite** of a vertebrate embryo.

Ncd: Noncomplementing diploids.

Nearest-neighbor analysis: A procedure that reveals the frequencies of two adjacent bases in DNA. Thus, the depletion of the doublet CpG in mammalian DNA was discovered when the frequency of the three other C-containing doublets (CpC, CpA, and CpT) was also measured.

Neomorph: A dominant mutation with a novel effect that cannot be attenuated by altering the dose of the wild-type gene in mutant individuals. For example, a neomorphic mutation might cause **ectopic expression** in a cell type that does not express the wild-type gene.

Neoplastic character: The collective properties of tumor cells responsible for their cancer-causing phenotype.

Neoplastic transformation: The change from a normal cell to a cancer cell. The term transformation is usually applied to a set of changes that can include abnormal karyotype and altered morphology, surface characteristics, and response to growth factors. (Transformation used in this context must be distinguished from transformation by DNA, in which a genetically marked DNA fragment is taken up by a cell and specifically alters its phenotype, and, if integrated into a chromosome, also its genotype.)

Nuclear periphery: Location within the nucleus at or near the nuclear membrane.

Nucleoid body: The nucleoprotein area of a bacterial cell, which contains the bacterial chromosome in a highly compacted state.

Oncovirus: A virus that is capable of transforming mammalian cells to an immortalized and often cancerous phenotype. Oncoviruses are usually of the Retroviridae family, e.g., human T-cell lymphotropic virus (HTLV).

Palindromic DNA sequence: An inverted repeat in which the nucleotide sequence of one strand reads in reverse order to that of the complementary strand. For example, the complementary strand of 5′ AATGCGCGCATT (which is 3′ TTACGCGCGTAA) will also read 5′ to 3′ as AATGCGCGCATT. Such sequences can form side arm loops (hairpin or cruciform structures) by intrastrand foldback and complementary hydrogen bonding within each strand.

Paracrine: An adjective describing a substance produced by cells and acting at short range on neighboring cells.

Paramutable allele: An unstable allele where the phenotypic consequences are enhanced by the presence of a **paramutagenic allele** in a heterozygote.

Paramutagenic allele: An allele possessing the ability to cause **paramutation**.

Paramutation: A mitotically and meiotically heritable change in the phenotypic expression of one allele of a gene that can occur spontaneously or be induced by a **paramutagenic allele** of the same gene in a heterozygote. The paramutation does not change the base sequence of the gene.

Parasegment: A repeating structural unit in the body plan of insects such as *Drosophila*, where a parasegment consists of the posterior half of a segment and the anterior half of the following segment. The parasegment is the organizational unit for **homeotic gene** regulation.

Parthenogenetic embryos (parthenogenotes): Embryos derived from unfertilized eggs. These embryos usually have two maternal sets of chromosomes, one set from what would have been an extruded polar body.

PcG: *See* Polycomb group genes.

Penetrance: The penetrance of a mutation is the proportion of genotypically mutant individuals that have the associated mutant phenotype. When penetrance is less than 100%, some genotypically mutant individuals are phenotypically normal. Penetrance is influenced by genetic factors (modifier genes) as well as environmental factors.

Pith parenchyma: The ground tissue of the pith; i.e., the innermost region of the plant stem.

Plasmodesmata: Cellular junctions that allow cytoplasmic connections between plant cells. Plasmodesmata facilitate **symplastic coupling.**

Polycomb group genes (PcG): A group of genes identified as repressors of **homeotic genes**. Mutant phenotypes caused by loss of PcG show homeotic transformations due to **ectopic expression** of homeotic genes. PcG proteins interact in multimeric complexes that are thought to form transcriptionally silent chromatin structures.

Polytene chromosomes: Multistranded chromosomes, often with a thousand or more DNA strands, that result from successive rounds of replication without intervening mitosis or cell divisions. The resulting giant chromosomes show banded structures indicative of precise pairing of chromatids and homologs.

Position effect: A phenomenon by which a gene's activity is influenced by its location on a chromosome.

Position-effect variegation (PEV): A type of clonally heritable variation in gene expression that leads to streaks or patches of cells with altered expression and which is caused by the position of the gene in the chromosome. Often seen in *Drosophila* when a gene not normally located near **heterochromatin** is translocated to the vicinity of heterochromatin; abnormal inactivation then occurs randomly in cells of the early embryo and gives rise to clones of cells with altered phenotype.

POZ/BTB domain: A 120-amino-acid motif named after the proteins in which similar sequences were first identified: Poxvirus zinc finger, Bric-a-brac, Tramtrack, and Broad complex. The domain mediates homo- and heterodimerization of certain transcriptional activators.

Prader-Willi syndrome: An inherited genetic disorder in humans characterized by growth retardation, characteristic facial appearance, small hands and feet, and hypothalamic dysfunction causing obesity.

PRE: **PcG** response element; a DNA element to which PcG protein complexes specifically bind. PREs are thought to act as transcriptional silencers by nucleating the formation of transcriptionally inactive **chromatin** structures.

Primary diploid fibroblasts: Small tissue biopsies placed in culture medium show outgrowth of normal diploid cells, often fibroblasts. These newly established cultures containing cells that have undergone only a few divisions in vitro are referred to as primary cultures; these cultures continue to grow for some time in an unchanged state but eventually become senescent and either die out or are taken over by **transformed cells**.

Primitive endoderm: A tissue in the early mouse embryo that contributes to the formation of the yolk sac, amnion, and placenta.

Primordial germ cells: Cells of the early mouse embryo that are the progenitors of spermatogonia or oogonia and, thus, the gametes produced by the adult gonad.

Protoplast: A cell from which the cell wall has been removed.

Puff: A cytologically prominent, decondensed site (band) in a **polytene chromosome**; now known to be associated with highly active transcription.

Redundant gene: A gene that is present in many functional copies, so that one copy can complement the loss of another copy.

Regenerant: An entire plant grown (i.e., regenerated) from a single cell.

Replicon: A DNA region, the replication of which is controlled as a unit; usually thought of as a region replicated from a single replication origin. Eukaryotic replicons range from 20 to 300 kb.

Retrovirus: A positive-strand RNA virus belonging to the family Retroviridae. This family of viruses uses an enzyme called reverse transcriptase to make a DNA copy that integrates into the host cellular genome.

RIP: Repeat-induced point mutation; a cellular process operating in the period between fertilization and karyogamy in the sexual cycle of *Neurospora crassa*. DNA sequence duplications are detected and mutated in a specific way (C:G to T:A changes), and frequently the cytosines in the mutated sequences are methylated.

Root meristem: *See* Meristem.

Sclerotome: A loose mesenchymal cell in the mammalian embryo that migrates from the inner and ventral edges of the **myotomes** and contributes to the formation of the axial skeleton.

Sepal: The green leaf-like organ that encloses the flower bud prior to opening of the flower.

SET domain: A conserved 150-amino-acid sequence motif named after the first three *Drosophila* proteins in which it was found: Su(var)3-9, E(z), and trx. It is present in more than 20 other proteins as well. Its function remains to be elucidated.

Shoot meristem: A meristem located at the apex of shoots, and from which the aerial parts of a plant are derived. *See* Meristem.

SIR: Silent information repressor; a group of four genes in *S. cerevisiae* responsible for mediating transcriptional repression at telomeres and at the silent mating-type loci.

Somaclonal variation: Phenotypic variability observed among plants regenerated from cell or tissue culture.

Somite: A segmental mass of mesoderm in the vertebrate embryo, occurring in pairs along the notochord and developing into muscles and vertebrae.

SPB: Spindle pole body; a yeast structure equivalent to the mammalian centrosome. It nucleates the formation of astral and spindle microtubule fibers.

Spm: The Suppressor-mutator transposable element of maize. Also known as En (Enhancer).

Spongiotrophoblast: The outer layer of the mouse placental trophoblast in which most of the cells are diploid. Other trophoblast cells are giant and polyploid.

Sporophyte: The diploid, spore-producing phase of the plant life cycle; the plant body.

Spumaviruses: Viruses belonging to the family Retroviridae, more commonly known as "foamy viruses."

Stamen: The male reproductive organ of a plant; it consists of a filament supporting the anther, which produces pollen, the male gamete.

Suppression: A mutation is suppressed in individuals that have the mutant genotype but are phenotypically normal because of genetic or epigenetic modifiers. When applied to dominant transgenes in plants, the term cosuppression usually refers to reduction in gene expression of both the transgene and a homologous partner when both are present in the same genome. The term has become widely used for many types of transgene interactions in plants.

Supracellular phenomenon: A phenomenon that occurs at levels of organization higher than the cell level; e.g., interactions between cells, tissues, and organs.

Symplastic coupling: A special type of cell-cell interaction in which two or more plant cells, connected by plasmodesmata, can exchange macromolecules such as introduced dyes, proteins, or RNA. Cells interconnected in this way can form a network of cells that behave as a supracellular domain.

Syngamy: Fusion of the paternal and maternal pronuclei in the fertilized egg.

T1 generation: A term used in plant genetics; refers to the progeny resulting from self-pollination of the primary transformant regenerated from tissue culture.

T2 generation: Progeny resulting from self-pollination of T1 plants through succeeding generations.

TAS: Telomere-associated sequences; DNA sequences lying internal to the terminal repeats at most *S. pombe* telomeres.

Teratocarcinoma cells: Cancerous, embryo germ-line-derived cells which, during their growth in vitro or in vivo, can differentiate to give rise to many cell types and recognizable, but disorganized, tissues. Teratomas (teratocarcinomas) are tumors that arise in vivo from teratocarcinoma cells.

Teratoma: *See* Teratocarcinoma cells.

TPE: Telomere position effect; a phenomenon of transcriptional silencing adjacent to telomeres.

trans-Activation: A process whereby proteins or nucleic acids encoded by one gene can act in *trans* to influence expression of a different gene. For example, the HIV TAT gene product "*trans*-activates" the HIV LTR.

Transcriptional activation domain: The protein domain of a transcription factor that mediates its ability to activate transcription. Some activator domains function by protein-protein interactions with the RNA polymerase holoenzyme complex.

Transfection: The experimental methodology that permits the introduction of foreign DNA or RNA into a cell.

Transformed cells: A transformed mammalian cell (*see* Neoplastic transformation) has several altered characteristics, including the ability to grow indefinitely, in contrast to primary diploid cells, which have only finite growth in culture.

Transgene: A gene introduced into a host genome by transfection or other similar means (such as *Agrobacterium*-mediated transformation of plant cells).

Transient expression: Expression of a transgene seen shortly after its introduction into a recipient cell but without stable integration into the genome.

Transposable element: A DNA segment with the special property of occasionally being able to move from one chromosomal location to another.

Transposon: A **transposable element.**

Trisomy: *See* Monosomy/disomy/ trisomy.

Trithorax group genes (trxG): A group of 20 known genes needed for the normal, continuous activity of **homeotic genes** and which exhibit antagonistic genetic effects with the **PcG genes**. Some molecularly characterized members have been found to encode proteins affecting chromatin structure.

Trophectoderm: One of the first tissue lineages generated by the early mammalian embryo. Along with the **primitive endoderm**, the trophectoderm develops to form a major part of the placenta and the extraembryonic yolk sac.

Vernalization: Promotion of flowering in response to prolonged periods of low temperatures.

VERO cell line: A SV40 virus-transformed African Green Monkey kidney cell line.

Vitellogenin genes (I and II): In oviparous vertebrates these genes code for a major egg yolk protein. The genes are only expressed in the liver, where they are under the control of **estradiol.**

WD repeats: A structural amino acid motif found in a variety of proteins, including both nuclear and cytoplasmic proteins. It consists of a unit of about 40 amino acids rich in tryptophan (W) and aspartate (D). WD repeat proteins usually contain 5–8 nonidentical WD repeats, each of which is thought to mediate interactions with other proteins.

Whorl: Leaves or flower parts arranged in a circle around the stem.

YAC: Yeast artificial chromosome; a cloning vector derived from yeast plasmids and used for cloning 50- to 2000-kb fragments of foreign DNA in *S. cerevisiae*.

Z-DNA: A left-handed, alternate form of double-stranded DNA in which the backbone phosphates zigzag.

Zinc finger: A protein motif involved in the recognition of DNA. The structure contains a complexed zinc ion and consists of an antiparallel protein-chain hairpin and loop followed by a helix.

Index

In the mouse the names of the genes are written with the first letter capital, in human all the letters of the genes are capitals; e.g., Igf2 for mouse and IGF2 for human.